Introductory and Intermediate Algebra

Margaret L. Lial
American River College

John Hornsby
University of New Orleans

Charles D. Miller

▲ **ADDISON-WESLEY**

An imprint of Addison Wesley Longman, Inc.

Reading, Massachusetts • Menlo Park, California • New York • Harlow, England
Don Mills, Ontario • Sydney • Mexico City • Madrid • Amsterdam

Publisher: Jason Jordan

Acquisition Editor: Rita Ferrandino

Assistant Editor: K. B. Mello

Developmental Editor: Sandi Goldstein

Managing Editor: Ron Hampton

Production Services: Elm Street Publishing Services, Inc.

Art Development: Meredith Nightingale

Marketing Manager: Liz O'Neil

Prepress Services Buyer: Caroline Fell

Senior Manufacturing Manager: Roy Logan

Manufacturing Manager: Ralph Mattivello

Text and Cover Designer: Susan Carsten

Cover Photograph: © M. Rutherford/SuperStock

Introductory and Intermediate Algebra

Library of Congress Cataloging-in-Publication Data

Lial, Margaret L.
    Introductory and intermediate algebra / Margaret L. Lial, John Hornsby,
Charles D. Miller.
        p.   cm.
    Includes index.
    ISBN 0-321-01926-1
    1. Algebra.   I. Hornsby, John.   II. Miller, Charles David.
III. Title.
QA152.2.L545   1998
512.9—dc21
                                                                97-16486
                                                                    CIP

123456789-DOW-01 00 99 98 97

Contents

Preface   **vii**

Feature Walk-Through   **xii**

An Introduction to the Scientific Calculator   **xv**

To the Student: Success in Algebra   **xix**

CHAPTER 1   **The Real Number System**   **1**

1.1   Exponents, Order of Operations, and Inequality   1
1.2   Variables, Expressions, and Equations   9
1.3   Real Numbers and the Number Line   17
1.4   Addition of Real Numbers   25
1.5   Subtraction of Real Numbers   33
1.6   Multiplication of Real Numbers   41
1.7   Division of Real Numbers   49
    Numbers in the Real World: The Magic Number in Sports   **54**
1.8   Properties of Addition and Multiplication   59
    Numbers in the Real World: Calendar Mathematics: A Perpetual Calendar Algorithm   **66**
1.9   Simplifying Expressions   71
Chapter 1 Summary   **77**
Chapter 1 Review Exercises   **81**
Chapter 1 Test   **87**

CHAPTER 2   **Linear Equations and Applications**   **89**

2.1   The Addition Property of Equality   89
2.2   The Multiplication Property of Equality   95
2.3   More on Solving Linear Equations   101
2.4   An Introduction to Applied Problems   109
    Numbers in the Real World: Just How Is Windchill Factor Determined?   **114**

2.5   Formulas and Geometry Applications   119
2.6   Ratio and Proportion; Percent   129
    Numbers in the Real World: The Changing Face of the United States   **134**
Chapter 2 Summary   **139**
    Numbers in the Real World: Calendar Mathematics: Number Patterns   **142**
Chapter 2 Review Exercises   **143**
Chapter 2 Test   **147**
Cumulative Review Exercises Chapters 1–2   **149**

CHAPTER 3   **Linear Inequalities and Absolute Value**   **151**

3.1   Linear Inequalities in One Variable   151
3.2   Set Operations and Compound Inequalities   161
3.3   Absolute Value Equations and Inequalities   169
    Summary Exercises on Solving Linear and Absolute Value Equations and Inequalities   181
Chapter 3 Summary   **183**
Chapter 3 Review Exercises   **185**
Chapter 3 Test   **189**
Cumulative Review Exercises Chapters 1–3   **191**

CHAPTER 4   **Linear Equations and Inequalities in Two Variables; Functions**   **193**

4.1   The Rectangular Coordinate System   193
4.2   The Slope of a Line   203
4.3   Linear Equations in Two Variables   215
4.4   Linear Inequalities in Two Variables   229
    Numbers in the Real World: A Graphing Calculator Minicourse Lesson 1: Solution of Linear Equations   **234**

4.5 An Introduction to Functions   239
Numbers in the Real World:
A Graphing Calculator Minicourse
Lesson 2: Solution of Linear Inequalities   246
4.6 An Application of Functions: Variation   253
Numbers in the Real World:
A Graphing Calculator Minicourse
Lesson 3: Tables   258
Chapter 4 Summary   263
Chapter 4 Review Exercises   267
Chapter 4 Test   271
Cumulative Review Exercises Chapters 1–4   273

CHAPTER 5   Systems of
Linear Equations   275
5.1 Solving Systems of Linear Equations by
Graphing   275
5.2 Solving Systems of Linear Equations by
Addition   283
5.3 Solving Systems of Linear Equations by
Substitution   293
5.4 Linear Systems of Equations in Three
Variables   301
Numbers in the Real World:
A Graphing Calculator Minicourse
Lesson 4: Solving a Linear System
of Equations   306
5.5 Applications of Linear Systems   311
5.6 Determinants   323
5.7 Solution of Linear Systems of Equations
by Determinants—Cramer's Rule   331
Chapter 5 Summary   337
Chapter 5 Review Exercises   341
Chapter 5 Test   347
Cumulative Review Exercises Chapters 1–5   349

CHAPTER 6   Exponents and
Polynomials   351
6.1 The Product Rule and Power Rules for
Exponents   351
Numbers in the Real World: Do You Have
the Knack for Percent?   356

6.2 Integer Exponents and the Quotient Rule   359
6.3 An Application of Exponents: Scientific
Notation   367
6.4 Addition and Subtraction of Polynomials   371
6.5 Multiplication of Polynomials   381
6.6 Products of Binomials   387
6.7 Division of a Polynomial by a Monomial   393
6.8 The Quotient of Two Polynomials   397
Chapter 6 Summary   403
Chapter 6 Review Exercises   405
Chapter 6 Test   409
Cumulative Review Exercises Chapters 1–6   411

CHAPTER 7   Factoring   413
7.1 Factors; The Greatest Common Factor   413
Numbers in the Real World: Number
Patterns in Our World   418
7.2 Factoring Trinomials   421
Numbers in the Real World: Using
Number Properties in Identification   424
7.3 More on Factoring Trinomials   427
Numbers in the Real World: The Influence
of Spanish Coinage on Stock Prices   432
7.4 Special Factorizations   435
Summary Exercises on Factoring   441
7.5 Solving Quadratic Equations by Factoring   443
Numbers in the Real World: Using
Computers to Break Codes   448
7.6 Applications of Quadratic Equations   451
Chapter 7 Summary   459
Chapter 7 Review Exercises   461
Chapter 7 Test   467
Cumulative Review Exercises Chapters 1–7   469

CHAPTER 8   Rational Expressions   471
8.1 The Fundamental Property of Rational
Expressions   471
Numbers in the Real World: The Dangers
of Ultraviolet Rays   476
8.2 Multiplication and Division of Rational
Expressions   479

8.3 Least Common Denominators   485

Numbers in the Real World: Statistics
Require Careful Interpretation   **488**

8.4 Addition and Subtraction of Rational
Expressions   491

Numbers in the Real World: Number Magic   **496**

8.5 Complex Fractions   499

8.6 Equations Involving Rational Expressions   505

Numbers in the Real World: Geometry
in Our World   **510**

Summary Exercises on Rational Expressions   513

8.7 Applications of Rational Expressions   515

Numbers in the Real World: Big Numbers
in Our World (and Others)... And Just
How Big Is a Googol?   **522**

Chapter 8 Summary   **527**

Chapter 8 Review Exercises   **531**

Chapter 8 Test   **535**

Cumulative Review Exercises Chapters 1–8   **537**

CHAPTER 9   Roots and Radicals   541

9.1 Finding Roots   541

9.2 Multiplication and Division of Radicals   551

9.3 Addition and Subtraction of Radicals   559

9.4 Rationalizing the Denominator   563

9.5 Simplifying Radical Expressions   569

9.6 Equations with Radicals   577

Numbers in the Real World: Are You a
Good "Guesstimator"?   **582**

9.7 Complex Numbers   587

Chapter 9 Summary   **597**

Chapter 9 Review Exercises   **601**

Chapter 9 Test   **607**

Cumulative Review Exercises Chapters 1–9   **609**

CHAPTER 10   Quadratic Equations
and Inequalities   611

10.1 Solving Quadratic Equations by the
Square Root Property   611

Numbers in the Real World: A Mathematical
Model of the Spread of AIDS   **614**

10.2 Solving Quadratic Equations by
Completing the Square   617

10.3 Solving Quadratic Equations by the
Quadratic Formula   625

Summary Exercises on Quadratic
Equations   635

10.4 Equations Quadratic in Form   637

10.5 Formulas and Applications Involving
Quadratic Equations   647

10.6 Nonlinear and Fractional Inequalities   655

Chapter 10 Summary   **665**

Chapter 10 Review Exercises   **669**

Chapter 10 Test   **675**

Cumulative Review Exercises Chapters 1–10   **677**

CHAPTER 11   Graphs of Nonlinear
Functions and Conic
Sections   681

11.1 Graphs of Quadratic Functions; Vertical
Parabolas   681

11.2 More about Quadratic Functions;
Horizontal Parabolas   691

11.3 Graphs of Elementary Functions and
Circles   703

11.4 Ellipses and Hyperbolas   713

Numbers in the Real World:
A Graphing Calculator Minicourse
Lesson 5: Matrices   **720**

11.5 Nonlinear Systems of Equations   725

11.6 Second-Degree Inequalities; Systems
of Inequalities   733

Numbers in the Real World:
A Graphing Calculator Minicourse
Lesson 6: Solution of Quadratic Equations
and Inequalities   **736**

Chapter 11 Summary   **743**

Chapter 11 Review Exercises   **747**

Chapter 11 Test   **753**

Cumulative Review Exercises Chapters 1–11   **757**

CHAPTER 12   Exponential and
Logarithmic Functions   761

12.1 Inverse Functions   761

12.2 Exponential Functions   771

Numbers in the Real World:
A Graphing Calculator Minicourse
Lesson 7: Solving a Nonlinear System
of Equations   **776**

12.3 Logarithmic Functions   779

12.4   Properties of Logarithms   787

12.5   Evaluating Logarithms   793

12.6   Exponential and Logarithmic Equations
and Their Applications   801
Numbers in the Real World:
A Graphing Calculator Minicourse
Lesson 8: Inverse Functions   806

Chapter 12 Summary   811
Numbers in the Real World:
A Graphing Calculator Minicourse
Lesson 9: Solution of Exponential and
Logarithmic Equations   814

Chapter 12 Review Exercises   815

Chapter 12 Test   819

Cumulative Review Exercises Chapters 1–12   823

Answers to Selected Exercises   A-1
Index   I-1
Video Correlation Guide   I-11

Preface

This first edition of *Introductory and Intermediate Algebra* is designed for college students who require review of the basic concepts of algebra before taking additional courses in mathematics, science, business, nursing, or computer science. The objectives of this text include introducing students to the basic concepts of algebra, with emphasis on solving equations and inequalities. Polynomials and factoring, operations with rational expressions, and operations with roots and radicals are also covered. We provide an early introduction to the important concept of functions and their graphs. Linear, polynomial, exponential, and logarithmic equations and both linear and nonlinear systems are studied so students can solve related applications. Conic sections are also covered, and emphasis on graphing is found throughout the text.

This text is designed for a two-semester sequence, and it comprises selected material from the current editions of our popular text/workbook series (the sixth editions of *Introductory Algebra* and *Intermediate Algebra*). It was published in response to the many requests for such a volume, so that students could use the same text for two courses. The widespread acceptance of our first edition of a similar title in hardback form further indicated that there is a need for such a text/workbook.

This text retains the successful features of previous editions of *Introductory Algebra* and *Intermediate Algebra*. The hallmark of any mathematics text is the quality of its exercise sets. We have written exercise sets that provide ample opportunity for drill and at the same time test conceptual understanding. In preparing this edition, we have also addressed the concerns of the **National Council of Teachers of Mathematics** and the **American Mathematical Association of Two-Year Colleges** by including many new exercises focusing on concepts, writing, and analysis of data obtained from sources in the world around us.

The following pages describe some of the key features of this text.

New and Updated Features

▶ *Numbers in the Real World: Collaborative Investigations* and *A Graphing Calculator Minicourse* pages show students how mathematics is used in everyday life. Students read and interpret articles and graphs, and are introduced to the most relevant features of graphing calculators. These pages may be done by individual students but are well-suited as collaborative assignments for pairs or small groups of students, for open-ended discussion by an entire class, and for general instruction on graphing calculator use.

▶ *Many new exercises* use data from real-life sources. They are designed to show students how algebra is used to describe and understand data in everyday life.

▶ *Mathematical Connections exercise groups* appear in many sections. These exercises are designed to be worked in sequential order, and they tie together various concepts so that students can see the connections among various topics in mathematics. For example, they may show how algebra and geometry are connected, or how a graph of a linear equation in two variables is related to the solution of the corresponding linear equation in one variable.

▶ While graphing calculators are not required for this text, it is likely that students will go on to courses that use them. For this reason, we have included a feature entitled Interpreting Technology in selected exercise sets that illustrates the power of graphing calculators and allows the student to interpret typical results seen on graphing calculator screens.

▶ Many students are seeing algebra for the first time or for the first time in many years. Furthermore, today's student population more than ever before includes individuals whose native language is not English. To assist these students, we have provided phonetic spellings of many of the important terms that appear in the book. A separate *Spanish Glossary* is also available.

▶ Instructors and students will have access to a World Wide Web site (**www.mathnotes.com**), where additional support material is available.

Continuing Features

We have retained popular features of the previous editions of the series. Some of these features are as follows:

▶ *Ample and varied exercise sets* Students of basic algebra require a large number and variety of exercises. This text meets that need with approximately 5800 exercises.

▶ *Learning objectives* Each section begins with clearly stated, numbered objectives, and material in the section is keyed to these objectives. In this way, students know exactly what is being covered in each section.

▶ *Margin problems* Margin problems, with answers immediately available, are found in every section. These problems allow the student to practice the material covered in the section in preparation for the exercise set that follows.

▶ *Cautions and Notes* We often give students warnings of common errors and emphasize important ideas in Cautions and Notes that appear throughout the exposition.

▶ *Ample opportunity for review* Each section contains a *chapter summary, chapter review exercises* keyed to individual sections as well as mixed review exercises, and a *chapter test*. Furthermore, following every chapter after Chapter 1, there is a set of cumulative review exercises that covers material going back to the first chapter. Students always have an opportunity to review material that appears earlier in the text, and this provides an excellent way to prepare for the final examination in the course.

▶ *Answers and solutions* Answers to all margin problems are provided at the bottom of the page on which the problem appears. Furthermore, answers to odd-numbered exercises in numbered sections, answers to every exercise in Mathematical Connections groups, and answers to all chapter test exercises, review exercises, and cumulative review exercises are provided at the back of the text. In this edition, we have also included sample answers to writing exercises.

All-New Supplements Package

Our extensive new supplements package includes testing materials, solutions, software, and videotapes.

FOR THE INSTRUCTOR

Instructor's Resource Guide

The *Instructor's Resource Guide* includes short-answer and multiple-choice versions of a pretest; eight forms of chapter tests for each chapter, including six open-response and two multiple-choice forms; short-answer and multiple-choice forms of a final examination; and an extensive set of additional exercises, providing 10 to 20 exercises for each textbook objective, which instructors can use as an additional source of questions for tests, quizzes, or student review of difficult topics. In addition, a section containing teaching tips is included for the instructor's convenience.

Instructor's Solutions Manual

This book includes solutions to all of the even-numbered section exercises (except the Connections exercises, which are located in the *Student's Solutions Manual*). The two solutions manuals provide detailed, worked-out solutions to each exercise and margin problem in the book.

Answer Book

This manual includes answers to all exercises.

TestGen EQ with QuizMaster EQ

This test generation software is available in Windows and Macintosh versions and is fully networkable. TestGen EQ's friendly, graphical interface enables instructors to easily view, edit, and add questions; transfer questions to tests; and print tests in a variety of fonts and forms. Search and sort features let instructors quickly locate questions and arrange them in a preferred order. Six question formats are available, including short-answer, true-false, multiple-choice, essay, matching, and bimodal formats. A built-in question editor gives users power to create graphs, import graphics, insert mathematical symbols and templates, and insert variable numbers or text. Computerized testbanks include algorithmically defined problems organized according to each textbook.

QuizMaster EQ enables instructors to create and save tests using TestGen EQ so students can take them for practice or a grade on a computer network. Instructors can set preferences for how and when tests are administered. QuizMaster EQ automatically grades the exams, stores results on disk, and allows instructors to view or print a variety of reports for individual students, classes, or courses.

InterAct Mathematics Plus—Management System

InterAct Math Plus combines course management and online testing with the features of the basic InterAct Math tutorial software to create an invaluable teaching resource. Consult your Addison Wesley Longman representative for details.

FOR THE STUDENT

Student's Solutions Manual

This book contains solutions to every other odd-numbered section exercise (those not included at the back of the textbook) as well as solutions to all margin problems, Connections exercises, chapter review exercises, chapter tests, and cumulative review exercises. (ISBN 0-321-01927-X)

InterAct Mathematics Tutorial Software

InterAct Math tutorial software has been developed and designed by professional software engineers working closely with a team of experienced developmental math educators.

InterAct Math tutorial software includes exercises that are linked with every objective in the textbook and require the same computational and problem-solving skills as their companion exercises in the text. Each exercise has an example and an interactive guided solution that are designed to involve students in the solution process and to help them identify precisely where they are having trouble. In addition, the software recognizes common student errors and provides students with appropriate customized feedback.

With its sophisticated answer-recognition capabilities, InterAct Math tutorial software recognizes appropriate forms of the same answer for any kind of input. It also tracks student activity and scores for each section, which can then be printed out. Available for Windows and Macintosh computers, the software is free to qualifying adopters or can be bundled with books for sale to students. (Macintosh: ISBN 0-321-02933-X, Windows: ISBN 0-321-02932-1)

Videotapes

A videotape series entitled Real to Reel has been developed to accompany *Introductory and Intermediate Algebra*. In a separate lesson for each section in the book, the series covers all objectives, topics, and problem-solving techniques discussed within the text. (ISBN 0-321-01994-6)

Spanish Glossary

A separate *Spanish Glossary* is now being offered as part of the supplements package for this textbook series. This book contains the key terms from each of the four texts in the series and their Spanish translations. (ISBN 0-321-01647-5)

Acknowledgments

For a textbook series to succeed through six editions, it is necessary for the authors to rely on comments and suggestions of users, non-users, instructors, and students. We are grateful for the many responses that we have received over the years. We wish to thank the following individuals who reviewed the material that comprises the text:

Nancy Ballard, *Mineral Area College*

Solveig Bender, *William Rainey Harper College*

Jean Bolyard, *Fairmont State College*

Richard DeCesare, *Southern Connecticut University*

Susan Eagleton, *Lee College*

Mimi Elwell, *Lake Michigan College*

Phyllis Faw, *Oklahoma City Community College*

Joseph Gutel, *Oklahoma State University—Oklahoma City*

Joseph Hargray, *St. Petersburg Junior College*

Mary Lou Hart, *Brevard Community College*

Michael Karelius, *American River College*

Linda Laningham, *Rend Lake College*

Keith Lathrop, *Valencia Community College—West Campus*

Robert Mooney, *Salem State College*

Elizabeth Morrison, *Valencia Community College*

Janice Rech, *University of Nebraska at Omaha*

Don Reichman, *Mercer County College*

Nancy Ressler, *Oakton Community College*

P. T. Sanjivamurthy, *Cuyahoga Community College—Metro Campus*

Linda Stein, *Hudson Valley Community College*

Steven Tarry, *Ricks College*

Lucy C. Thrower, *Francis Marion University*

Nina Verheyden, *Kilgore College*

Clif Ware, *Hillsborough Community College*

No author can write a text of this magnitude without the help of many other individuals. Our sincere thanks go to Rita Ferrandino of Addison Wesley Longman who coordinated the package of texts of which this book is a part. We are appreciative of the support given us by Greg Tobin, also of Addison Wesley Longman, who helped make the transition from our former publisher go as smoothly as one could ever imagine. We wish to thank Sandi Goldstein for her efforts in coordinating the reviews and working with us in the early stages of preparation for production. Theresa McGinnis and Abby Tanenbaum helped in checking for the accuracy of the answers. Cathy Wacaser of Elm Street Publishing Services provided her usual excellent production work. She is indeed one of the best in the business. As usual, Paul Van Erden created an accurate, useful index. We are also grateful to Tommy Thompson who made suggestions for the feature "To the Student: Success in Algebra."

Paul Eldersveld of the College of DuPage has coordinated our print supplements for many years. The importance of his job cannot be overestimated, and we want to thank him for his work over the years. He is a wonderful friend and colleague, and we are happy to have him as part of our publishing family.

Margaret L. Lial
John Hornsby

Feature Walk-Through

A new design updates and refreshes the overall flow of the text. New, full-color situational art throughout the text enhances student comprehension of the material.

Section Objectives: Each section begins with stated objectives that guide student learning.

Student Resources: Found at the opening of each section, this feature cross-references relevant material in the student supplements package, providing each student with a rich variety of extra help and resources.

World Wide Web: Instructors and students will have access to a World Wide Web site (www.mathnotes.com), where additional support material is available.

Margin Exercises: Exercises appear in the margin for immediate practice and reinforcement. Answers to the margin exercises are available at the bottom of the page.

Cautions and Notes: Common student errors are anticipated and highlighted with the heading "Caution" or "Note." This helps clarify concepts for students and assists them in identifying common errors.

6.2 Integer Exponents and the Quotient Rule

OBJECTIVE 1 In the previous section we studied the product rule for exponents. In all of our work, exponents were positive integers. To develop meanings for exponents other than positive integers (such as 0 and negative integers), we want to define them in such a way that rules for exponents are the same, regardless of the kind of number used for the exponents.

Suppose we want to find a meaning for an expression such as

$$6^0,$$

where 0 is used as an exponent. If we were to multiply this by 6^2, for example, we would want the product rule to still be valid. Therefore, we would have

$$6^0 \cdot 6^2 = 6^{0+2} = 6^2.$$

So multiplying 6^2 by 6^0 should give 6^2. Because 6^0 is acting as if it were 1 here, we should define 6^0 to equal 1. This is the definition for 0 used as an exponent with any nonzero base.

Definition of Zero Exponent

For any nonzero real number a, $a^0 = 1$.
Example: $17^0 = 1$

EXAMPLE 1 Using Zero Exponents

Evaluate each exponential expression.
(a) $60^0 = 1$
(b) $(-60)^0 = 1$
(c) $-60^0 = -(1) = -1$
(d) $y^0 = 1$, if $y \neq 0$
(e) $-r^0 = -1$, if $r \neq 0$

Caution
Notice the difference between parts (b) and (c) of Example 1. In Example 1(b) the base is -60 and the exponent is 0. Any nonzero base raised to a zero exponent is 1. But in Example 1(c), the base is 60. Then $60^0 = 1$, and $-60^0 = -1$.

WORK PROBLEM 1 AT THE SIDE. ▶▶

OBJECTIVE 2 Now let us consider how we can define negative integers as exponents. Suppose that we want to give a meaning to

$$6^{-2}$$

so that the product rule is still valid. If we multiply 6^{-2} by 6^2, we get

$$6^{-2} \cdot 6^2 = 6^{-2+2} = 6^0 = 1.$$

The expression 6^{-2} is acting as if it were the reciprocal of 6^2, because their product is 1. The reciprocal of 6^2 may be written $\frac{1}{6^2}$, leading us to define 6^{-2} as $\frac{1}{6^2}$. This is a particular case of the definition of negative exponents.

OBJECTIVES
1. Use zero as an exponent.
2. Use negative numbers as exponents.
3. Use the quotient rule for exponents.
4. Use combinations of rules.

FOR EXTRA HELP

Tutorial   Tape 11   SSM, Sec. 6.2

1. Evaluate.
   (a) 28^0

   (b) $(-16)^0$

   (c) -7^0

   (d) m^0, $m \neq 0$

   (e) $-p^0$, $p \neq 0$

ANSWERS
1. (a) 1 (b) 1 (c) −1 (d) 1 (e) −1

359

NUMBERS IN THE
Real World   *collaborative investigations*

The Changing Face of the United States

While the United States has long been considered a "melting pot" of cultures, it has never been so obvious as it is today. In particular, immigration is leading to a population boom in the West and South. According to a computer analysis of U.S. Census Bureau data by Paul Overberg here are the projected population–percentage changes of racial groups and Hispanics in 1995–2000 in the most populous states.

State	White	Black	Asian	Native American	Hispanic
California	1.1%	.5%	18.3%	−2.3%	15.7%
Texas	6.5%	11.0%	25.5%	13.1%	13.6%
New York	−1.8%	3.4%	18.6%	5.8%	10.4%
Florida	6.5%	11.9%	22.5%	13.3%	22.3%
Pennsylvania	.3%	4.8%	22.5%	12.5%	19.7%
Illinois	1.0%	2.9%	18.2%	4.0%	16.2%
Ohio	.7%	5.6%	21.7%	0.0%	13.0%
Michigan	.5%	4.1%	23.5%	3.4%	12.0%
New Jersey	.6%	7.6%	27.3%	0.0%	16.5%
Georgia	7.5%	12.9%	26.8%	6.3%	26.0%

FOR GROUP DISCUSSION

1. It is expected that California, the nation's most populous state with 31.6 million people, will have 17.7 million more people by 2025, which represents a 56% growth. How was this percent obtained using these figures?

2. Based on the chart above, what is the fastest-growing racial group overall?

3. It is expected that people of Hispanic origin will account for 44% of the 72 million additional people in the United States in the year 2025. How many people is this?

4. As a class, discuss the problems that minorities face in mathematics courses.

134

Numbers in the Real World: Many chapters are highlighted with "Numbers in the Real World," a collaborative-learning feature that illustrates how mathematics is encountered and applied in real-world settings and that addresses the concerns of NCTM and AMATYC. This feature not only provides examples that incorporate real data, but also includes issues that students are likely to encounter in their everyday lives.

To address the increasing use of graphing calculators in the classroom, this feature includes nine lessons that provide general guidelines on how they can be used effectively in algebra. We provide explanations on how solution of equations, inequalities, and systems can be supported, how matrices and determinants can be analyzed and evaluated, and how a table of values of a function can be generated at the touch of a key.

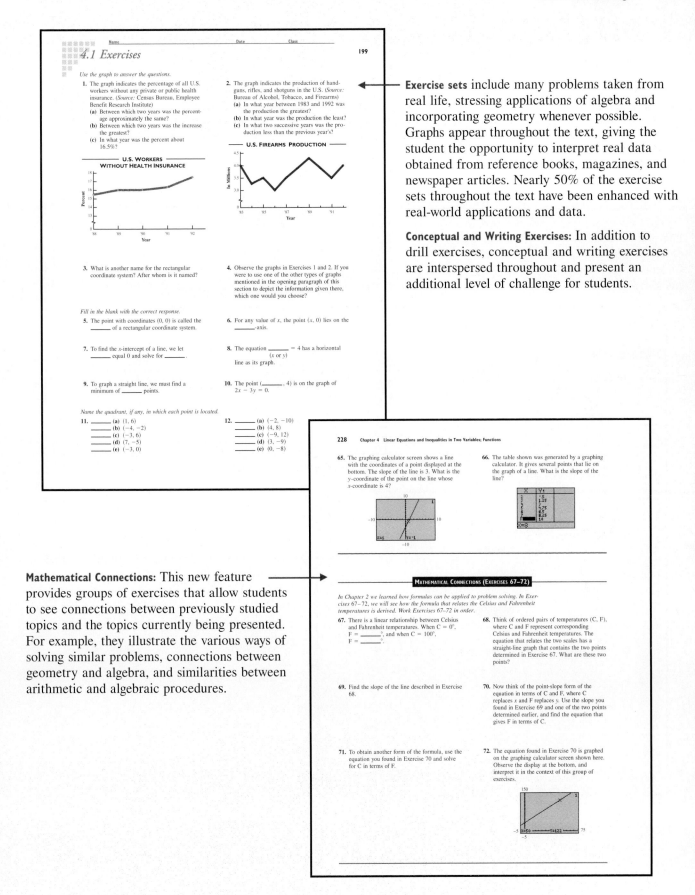

Exercise sets include many problems taken from real life, stressing applications of algebra and incorporating geometry whenever possible. Graphs appear throughout the text, giving the student the opportunity to interpret real data obtained from reference books, magazines, and newspaper articles. Nearly 50% of the exercise sets throughout the text have been enhanced with real-world applications and data.

Conceptual and Writing Exercises: In addition to drill exercises, conceptual and writing exercises are interspersed throughout and present an additional level of challenge for students.

Mathematical Connections: This new feature provides groups of exercises that allow students to see connections between previously studied topics and the topics currently being presented. For example, they illustrate the various ways of solving similar problems, connections between geometry and algebra, and similarities between arithmetic and algebraic procedures.

Art Program: The art program has been refreshed with a new, appealing, pedagogical layout. Extensive additions to the art program keep students interested in the mathematical material.

Interpreting Technology: A new feature, Interpreting Technology demonstrates the power of graphing calculators and allows students to view typical graphing calculator-generated screens and interpret results.

End-of-Chapter Material: In order to reinforce concepts learned and assist students with their retention, the end-of-chapter material includes a summary of Key Terms and New Symbols, a Quick Review of concepts learned, Review Exercises, a Chapter Test, and a Cumulative Review.

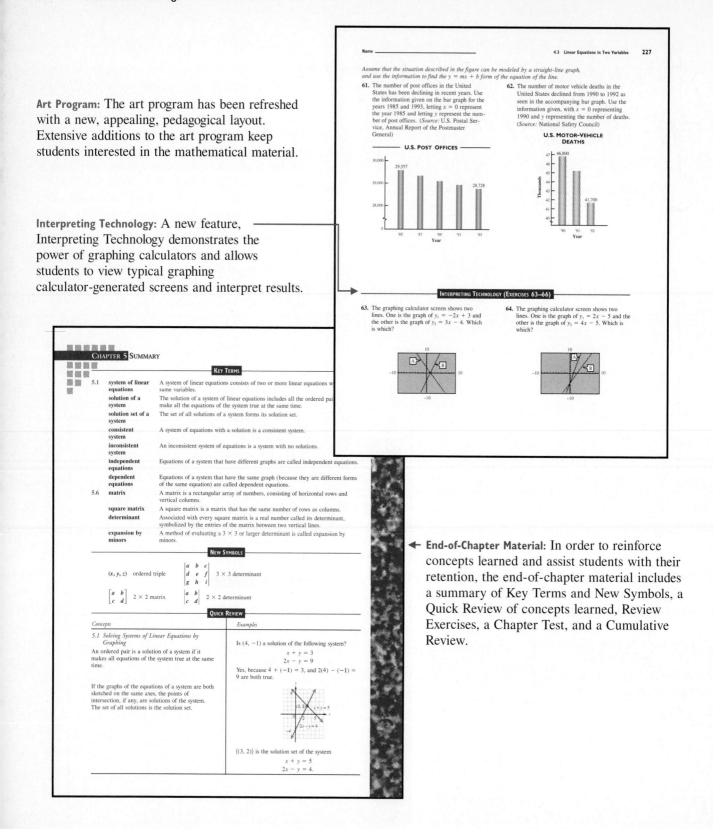

An Introduction to Scientific Calculators

The emphasis placed on paper-and-pencil computation, which has long been a part of the school mathematics curriculum, is not as important as it once was. The search for easier ways to calculate and compute has culminated in the development of hand-held calculators and computers. Professional organizations devoted to the teaching and learning of mathematics recommend that this technology be incorporated throughout the mathematics curriculum. In light of this recommendation, this text assumes that all students have access to calculators, and the authors recommend that students become computer literate to the extent their academic resources and individual finances allow.

In this text we view calculators as a means of allowing students to spend more time on the conceptual nature of mathematics and less time on the drudgery of computation with paper and pencil. Calculators come in a large array of different types, sizes, and prices, making it difficult to decide which machine best fits your needs. For the general population, a calculator that performs the operations of arithmetic and a few other functions is sufficient. These are known as **four-function calculators.** Students who take higher mathematics courses (engineers, for example) usually need the added power of **scientific calculators. Programmable calculators,** which allow short programs to be written and performed, and **graphing calculators,** which actually plot graphs on small screens, are also available. Keep in mind that calculators differ from one manufacturer to the other. For this reason, remember the following.

> Always refer to your owner's manual if you need assistance in performing an operation with your calculator.

Because of their relatively inexpensive cost and features that far exceed those of four-function calculators, scientific calculators probably provide students the best value for their money. For this reason, we will give a short synopsis of the major functions of scientific calculators.

Most scientific calculators use *algebraic logic*. (Models sold by Texas Instruments, Sharp, Casio, and Radio Shack, for example, use algebraic logic.) A notable exception is Hewlett Packard, a company whose calculators use *Reverse Polish Notation* (RPN). In this introduction, we explain how to use calculators with algebraic logic.

Arithmetic Operations

To perform an operation of arithmetic, simply enter the first number, touch the operation key ([+], [−], [×], or [÷]), enter the second number, and then touch the [=] key. For example, to add 4 and 3, use the following keystrokes.

(The final answer is displayed in color.)

Change Sign Key

The key marked [±] allows you to change the sign of a display. This is particularly useful when you wish to enter a negative number. For example, to enter −3, use the following keystrokes.

Parentheses Keys

These keys, [(] and [)], are designed to allow you to group numbers in arithmetic operations as you desire. For example, if you wish to evaluate 4 × (6 + 3), use the following keystrokes.

Memory Key

Scientific calculators can hold a number in memory for later use. The label of the memory key varies among models; two versions are [M] and [STO]. [M+] and [M−] allow you to add to or subtract from the value currently in memory. The memory recall key, labeled [MR], [RM], or [RCL], allows you to retrieve the value stored in memory.

Suppose that you wish to store the number 5 in memory. Enter 5, then touch the key for memory. You can then perform other calculations. When you need to retrieve the 5, touch the key for memory recall.

If a calculator has a constant memory feature, the value in memory will be retained even after the power is turned off. Some advanced calculators have more than one memory. It is best to read the owner's manual for your model to see exactly how memory is activated.

Clearing/Clear Entry Keys

These keys allow you to clear the display or clear the last entry entered into the display. They are usually marked [C] and [CE]. In some models, touching the [C] key once will clear the last entry, while touching it twice will clear the entire operation in progress.

Second Function Key

This key is used in conjunction with another key to activate a function that is printed *above* an operation key (and not on the key itself). It is usually marked [2nd]. For example, suppose you wish to find the square of a number, and the squaring function (explained in more detail later) is printed above another key. You would need to touch [2nd] before the desired squaring function can be activated.

Some newer models of scientific calculators (the TI-35X by Texas Instruments, for example) even provide a third function key, marked [3rd], which is used in a manner similar to the one described for the second function key.

Square Root and Cube Root Keys

Touching the square root key, $\boxed{\sqrt{x}}$, will give the square root (or an approximation of the square root) of the number in the display. For example, to find the square root of 36, use the following keystrokes.

The square root of 2 is an example of an irrational number (see Section 9.1). The calculator will give an approximation of its value, since the decimal for $\sqrt{2}$ never terminates and never repeats. The number of digits shown will vary among models. To find an approximation of $\sqrt{2}$, use the following keystrokes.

The cube root key, $\boxed{\sqrt[3]{x}}$, is used in the same manner as the square root key. To find the cube roots of 64 and 93, use the following keystrokes.

Note: Calculators differ in the number of digits provided in the display. For example, one calculator gives the approximation of the cube root of 93 as 4.530654896, showing more digits than shown above.

Squaring and Cubing Keys

The squaring key, $\boxed{x^2}$, allows you to square the entry in the display. For example, to square 35.7, use the following keystrokes.

The squaring key and the square root key are often found on the same key, with one of them being a second function (that is, activated by the second function key, described above).

The cubing key, $\boxed{x^3}$, allows you to cube the entry in the display. To cube 3.5, follow these keystrokes.

Reciprocal Key

The key marked $\boxed{1/x}$ (or $\boxed{x^{-1}}$) is the reciprocal key. (When two numbers have a product of 1, they are called *reciprocals*.) Suppose you wish to find the reciprocal of 5. Use the following keystrokes.

Inverse Key

Some calculators have an inverse key, marked $\boxed{\text{INV}}$. Inverse operations are operations that "undo" each other. For example, the operations of squaring and taking the square root are inverse operations. The use of the $\boxed{\text{INV}}$ key varies among different models of calculators, so read your owner's manual carefully.

Exponential Key

This key, marked $\boxed{x^y}$ or $\boxed{y^x}$, allows you to raise a number to a power. For example, if you wish to raise 4 to the fifth power (that is, find 4^5), use the following keystrokes.

Root Key

Some calculators have this key specifically marked $\boxed{\sqrt[x]{x}}$ or $\boxed{\sqrt[x]{y}}$; with others, the operation of taking roots is accomplished by using the inverse key in conjunction with the exponential key. Suppose, for example, your calculator is of the latter type, and you wish to find the fifth root of 1024. Use the following keystrokes.

Notice how this "undoes" the operation explained in the exponential key discussion earlier.

Pi Key

The number π is important in mathematics. It occurs, for example, in the area and circumference formulas for a circle. By touching the $\boxed{\pi}$ key, you can get the display of the first few digits of π. (Because π is irrational, the display shows only an approximation.) One popular model gives the following display when the $\boxed{\pi}$ key is activated: $\boxed{3.1415927}$. As mentioned before, calculators will vary in the number of digits they give for π in their displays.

Other Considerations

When decimal approximations are shown on scientific calculators, they are either *truncated* or *rounded*. To see which of these a particular model is programmed to do, evaluate 1/18 as an example. If the display shows .0555555 (last digit 5), it truncates the display. If it shows .0555556 (last digit 6), it rounds off the display.

When very large or very small numbers are obtained as answers, scientific calculators often express these numbers in scientific notation. For example, if you multiply 6,265,804 by 8,980,591, the display might look like this:

$$\boxed{5.6270623 \quad 13}$$

The "13" at the far right means that the number on the left is multiplied by 10^{13}. This means that the decimal point must be moved 13 places to the right if the answer is to be expressed in its usual form. Even then, the value obtained will only be an approximation: 56,270,623,000,000.

Two features of advanced scientific calculators are programmability and graphing capability. A programmable calculator has the capability of running small programs, much like a mini-computer. A graphing calculator can be used to plot graphs of functions on a small screen. One of the issues in mathematics education today deals with how graphing calculators should be incorporated into the curriculum. Their availability in the 1990s parallels the availability of scientific calculators in the 1980s, and they are providing a major influence in mathematics education as we move into the twenty-first century.

To the Student: Success in Algebra

The main reason students have difficulty with mathematics is that they don't know how to study it. Studying mathematics *is* different from studying subjects like English or history. The key to success is regular practice.

This should not be surprising. After all, can you learn to play the piano or to ski well without a lot of regular practice? The same thing is true for learning mathematics. Working problems nearly every day is the key to becoming successful. Here is a list of things that will help you succeed in studying algebra.

1. *Attend class regularly*. Pay attention to what your teacher says and does in class, and take careful notes. In particular, note the problems the teacher works on the board and copy the complete solutions. Keep these notes separate from your homework to avoid confusion when you read them over later.

2. Don't hesitate to ask questions in class. It is not a sign of weakness, but of strength. There are always other students with the same question who are too shy to ask.

3. *Read your text carefully*. Many students read only enough to get by, usually only the examples. Reading the complete section will help you solve the homework problems. Most exercises are keyed to specific examples or objectives that will explain the procedures for working them.

4. Before you start on your homework assignment, rework the problems the teacher worked in class. This will reinforce what you have learned. Many students say, "I understand it perfectly when you do it, but I get stuck when I try to work the problem myself."

5. Do your homework assignment only *after* reading the text and reviewing your notes from class. Check your work against the answers in the back of the book. If you get a problem wrong and are unable to understand why, mark that problem and ask your instructor about it. Then practice working additional problems of the same type to reinforce what you have learned.

6. Work as neatly as you can. Write your symbols clearly, and make sure the problems are clearly separated from each other. Working neatly will help you to think clearly and also make it easier to review the homework before a test.

7. After you complete a homework assignment, look over the text again. Try to identify the main ideas that are in the lesson. Often they are clearly highlighted or boxed in the text.

8. Use the chapter test at the end of each chapter as a practice test. Work through the problems under test conditions, without referring to the text or the answers until you are finished. You may want to time yourself to see how long it takes you. When you finish, check your answers against those in the back of the book, and study the problems you missed. Answers are keyed to the appropriate sections of the text.

9. Keep any quizzes and tests that are returned to you, and use them when you study for future tests and the final exam. These quizzes and tests indicate what your instructor considers most important. Be sure to correct any problems on these tests that you missed, so you will have the corrected work to study.

10. Don't worry if you do not understand a new topic right away. As you read more about it and work through the problems, you will gain understanding. Each time you review a topic you will understand it a little better. No one understands each topic completely right from the start.

The Real Number System

1

1.1 Exponents, Order of Operations, and Inequality

OBJECTIVE 1 In the study of elementary arithmetic when we find prime factorizations, we often find products that have the same factors appearing many times. In algebra, we use a raised dot to indicate multiplication. For example, when 81 is written in prime factored form as

$$81 = 3 \cdot 3 \cdot 3 \cdot 3,$$

the factor 3 appears four times. Repeated factors can be written in a short form by using an *exponent*. For example, the factorization of 81 is written with an exponent as

$$\underbrace{3 \cdot 3 \cdot 3 \cdot 3}_{\text{4 factors of 3}} = 3^4 \quad \begin{array}{l} \text{Exponent} \\ \text{Base} \end{array}$$

The number 4 is the **exponent** (EX-poh-nunt) and 3 is the **base** in the **exponential** (EX-poh-NEN-shul) **expression** 3^4. Exponents are also called **powers.** We read 3^4 as "3 to the fourth power" or simply "3 to the fourth."

E X A M P L E 1   **Finding the Value of an Exponential Expression**

Find the values of the following.

(a) 5^2

$$\underbrace{5 \cdot 5}_{} = 25$$

5 is used as a factor 2 times.

Read 5^2 as "5 squared."

(b) 6^3

$$\underbrace{6 \cdot 6 \cdot 6}_{} = 216$$

6 is used as a factor 3 times.

Read 6^3 as "6 cubed."

(c) 2^5

$$2 \cdot 2 \cdot 2 \cdot 2 \cdot 2 = 32 \qquad \text{2 is used as a factor 5 times.}$$

Read 2^5 as "2 to the fifth power."

CONTINUED ON NEXT PAGE

1. Find the value of each exponential expression.

(a) 6^2

(b) 3^5

(c) $\left(\dfrac{3}{4}\right)^2$

(d) $\left(\dfrac{1}{2}\right)^4$

(e) $(.4)^3$

2. Find the value of each expression.

(a) $3 \cdot 8 + 7$

(b) $9 + 12 \cdot 6$

(d) 7^4

$$7 \cdot 7 \cdot 7 \cdot 7 = 2401 \qquad \text{7 is used as a factor 4 times.}$$

Read 7^4 as "7 to the fourth power."

(e) $\left(\dfrac{2}{3}\right)^3$

$$\frac{2}{3} \cdot \frac{2}{3} \cdot \frac{2}{3} = \frac{8}{27} \qquad \tfrac{2}{3} \text{ is used as a factor 3 times.}$$

◀◀ **WORK PROBLEM I AT THE SIDE.**

OBJECTIVE 2 ▶ Many problems involve more than one operation. For example, in finding the value of

$$5 + 2 \cdot 3,$$

which should be done first—multiplication or addition? The following **order of operations** has been agreed on as the most reasonable. (This is the order used by most calculators and computers.)

Order of Operations

Step 1 If possible, simplify within parentheses and above and below fraction bars, using the following steps.

Step 2 Apply all exponents.

Step 3 Do any multiplications or divisions in the order in which they occur, working from left to right.

Step 4 Do any additions or subtractions in the order in which they occur, working from left to right.

EXAMPLE 2   Using the Order of Operations

Find the value of $5 + 2 \cdot 3$.

Using the order of operations given above, first multiply 2 and 3, and then add 5.

$$5 + 2 \cdot 3 = 5 + 6 \qquad \text{Multiply.}$$
$$= 11 \qquad \text{Add.}$$

◀◀ **WORK PROBLEM 2 AT THE SIDE.**

A dot has been used to show multiplication; another way to show multiplication is with parentheses. For example, $3(7)$ means $3 \cdot 7$ or 21. Also $3(4 + 5)$ means 3 times the sum of 4 and 5. By the order of operations, the sum in parentheses must be found first, then the product. The next example shows the use of parentheses for multiplication and parentheses and fraction bars for grouping.

EXAMPLE 3   Using the Order of Operations

Find the value of each of the following.

(a) $9(6 + 11)$

Work first inside the parentheses.

$$9(6 + 11) = 9(17) \qquad \text{Add inside parentheses.}$$
$$= 153 \qquad \text{Multiply.}$$

CONTINUED ON NEXT PAGE

(b) $2(5 + 6) + 7 \cdot 3 = 2(11) + 7 \cdot 3$ Add inside parentheses.

$\qquad\qquad\qquad = 22 + 21$ Multiply.

$\qquad\qquad\qquad = 43$ Add.

(c) $\dfrac{4(5 + 3) + 3}{2(3) - 1}$

Simplify the numerator and denominator separately.

$\dfrac{4(5 + 3) + 3}{2(3) - 1} = \dfrac{4(8) + 3}{2(3) - 1}$ Add inside parentheses.

$\qquad\qquad\quad = \dfrac{32 + 3}{6 - 1}$ Multiply.

$\qquad\qquad\quad = \dfrac{35}{5}$ Add and subtract.

$\qquad\qquad\quad = 7$ Divide.

(d) $9 + 2^3 - 5$

Following the order of operations, we calculate 2^3 first.

$9 + 2^3 - 5 = 9 + 8 - 5$ Use the exponent.

$\qquad\qquad\; = 12$ Add, then subtract.

WORK PROBLEM 3 AT THE SIDE. ▶▶

Note

Parentheses and fraction bars are used as grouping symbols to indicate an expression that represents a single number. That is why we must first simplify within parentheses and above and below fraction bars.

Calculators follow the order of operations given in this section. You may want to try some of the examples to see that your calculator gives the same answers. Be sure to use the parentheses keys to insert parentheses where they are needed. To work Example 3(c) with a calculator, you must put parentheses around the numerator and the denominator.

OBJECTIVE 3 ▶ An expression with double parentheses, such as the expression $2(8 + 3(6 + 5))$, can be confusing. We can avoid confusion by using square brackets, $[\;\;]$, in place of one pair of parentheses.

E X A M P L E 4   Using Brackets

Simplify $2[8 + 3(6 + 5)]$.

Begin inside the parentheses. Then follow the order of operations.

$2[8 + 3(6 + 5)] = 2[8 + 3(11)]$ Add.

$\qquad\qquad\qquad = 2[8 + 33]$ Multiply.

$\qquad\qquad\qquad = 2[41]$ Add.

$\qquad\qquad\qquad = 82$ Multiply.

WORK PROBLEM 4 AT THE SIDE. ▶▶

3. Find the value of each expression.

(a) $2 \cdot 9 + 7 \cdot 3$

(b) $7 \cdot 6 - 3(8 + 1)$

(c) $\dfrac{2(7 + 8) + 2}{3 \cdot 5 + 1}$

(d) $2 + 3^2 - 5$

4. Find the value of each expression.

(a) $4[7 + 3(6 + 1)]$

(b) $9[(4 + 8) - 3]$

ANSWERS

3. (a) 39 **(b)** 15 **(c)** 2 **(d)** 6
4. (a) 112 **(b)** 81

5. Write each statement in words, then decide whether it is true or false.

(a) $7 < 5$

(b) $12 > 6$

(c) $4 \neq 10$

(d) $28 \neq 4 \cdot 7$

6. Tell whether each statement is true or false.

(a) $30 \leq 40$

(b) $25 \geq 10$

(c) $40 \leq 10$

(d) $21 \leq 21$

(e) $3 \geq 3$

OBJECTIVE 4 ▶ So far we have used only the symbols of arithmetic, such as $+$, $-$, $\times$ (or $\cdot$), and $\div$. Another common symbol is the one for equality, $=$, which says that two numbers are equal. This symbol with a slash through it, $\neq$, means "is *not* equal to." For example,

$$7 \neq 8$$

indicates that 7 is not equal to 8.

If two numbers are not equal, then one of the numbers must be less than the other. The symbol $<$ represents "is less than," so "7 is less than 8" is written

$$7 < 8.$$

Also, we write "6 is less than 9" as $6 < 9$.

The symbol $>$ means "is greater than." We write "8 is greater than 2" as

$$8 > 2.$$

The statement "17 is greater than 11" becomes $17 > 11$.

Keep the meanings of the symbols $<$ and $>$ clear by remembering that the symbol always points to the smaller number.

$$\text{Smaller number} \rightarrow 8 < 15$$

$$15 > 8 \leftarrow \text{Smaller number}$$

◀◀ **WORK PROBLEM 5 AT THE SIDE.**

Two other symbols, $\leq$ and $\geq$, also represent the idea of inequality. The symbol $\leq$ means "is less than or equal to," so

$$5 \leq 9$$

means "5 is less than or equal to 9." If either the $<$ part or the $=$ part is true, then the inequality $\leq$ is true. The statement $5 \leq 9$ is true because $5 < 9$ is true.

The symbol $\geq$ means "is greater than or equal to";

$$9 \geq 5$$

is true because $9 > 5$ is true. Also, $8 \leq 8$ is true because $8 = 8$ is true. But $13 \leq 9$ is not true because neither $13 < 9$ nor $13 = 9$ is true.

E X A M P L E   5   Using the Symbols $\leq$ and $\geq$

Tell whether each statement is true or false.

(a) $15 \leq 20$ The statement $15 \leq 20$ is true because $15 < 20$.

(b) $25 \geq 30$ Both $25 > 30$ and $25 = 30$ are false; therefore, $25 \geq 30$ is false.

(c) $12 \geq 12$ Since $12 = 12$, this statement is true.

◀◀ **WORK PROBLEM 6 AT THE SIDE.**

ANSWERS

5. **(a)** Seven is less than five. False
   (b) Twelve is greater than six. True
   (c) Four is not equal to ten. True
   (d) Twenty-eight is not equal to four times seven. False

6. **(a)** true **(b)** true **(c)** false
   (d) true **(e)** true

OBJECTIVE 5 Word phrases or statements often must be converted to symbols in algebra. The next example shows how to do this.

E X A M P L E 6   Converting Words to Symbols

Write each word statement in symbols.

(a) Twelve **equals** ten **plus** two. $12 = 10 + 2$

(b) Nine **is less than** ten. $9 < 10$
Compare this with 9 less than 10, which is written $10 - 9$.

(c) Fifteen **is not equal to** eighteen. $15 \neq 18$

(d) Seven **is greater than** four. $7 > 4$

(e) Thirteen **is less than or equal to** forty. $13 \leq 40$

(f) Six **is greater than or equal to** six. $6 \geq 6$

WORK PROBLEM 7 AT THE SIDE. ▶▶

OBJECTIVE 6 Any statement with $<$ can be converted to one with $>$, and any statement with $>$ can be converted to one with $<$. We do this by reversing both the order of the numbers and the direction of the symbol. For example, the statement $6 < 10$ can be written as $10 > 6$.

Exchange numbers.

$6 < 10$ becomes $10 > 6$

Reverse symbol.

E X A M P L E 7   Converting between $<$ and $>$

The following list shows the same statements written in two equally correct ways.

(a) $9 < 16$ $16 > 9$

(b) $5 > 2$ $2 < 5$

(c) $3 \leq 8$ $8 \geq 3$

(d) $12 \geq 5$ $5 \leq 12$

WORK PROBLEM 8 AT THE SIDE. ▶▶

7. Write in symbols.

(a) Nine equals eleven minus two.

(b) Seventeen is less than thirty.

(c) Eight is not equal to ten.

(d) Fourteen is greater than twelve.

(e) Thirty is less than or equal to fifty.

(f) Two is greater than or equal to two.

8. Write each statement with the inequality symbol reversed.

(a) $8 < 10$

(b) $3 > 1$

(c) $9 \leq 15$

(d) $6 \geq 2$

ANSWERS

7. (a) $9 = 11 - 2$ (b) $17 < 30$
(c) $8 \neq 10$ (d) $14 > 12$
(e) $30 \leq 50$ (f) $2 \geq 2$
8. (a) $10 > 8$ (b) $1 < 3$ (c) $15 \geq 9$
(d) $2 \leq 6$

Here is a summary of the symbols of equality and inequality.

SYMBOLS OF EQUALITY AND INEQUALITY

Symbol	Meaning	Symbol	Meaning
$=$	is equal to	$>$	is greater than
$\neq$	is not equal to	$\leq$	is less than or equal to
$<$	is less than	$\geq$	is greater than or equal to

Caution
The symbols of equality and inequality are used to write mathematical *sentences*. They differ from the symbols for operations ($+$, $-$, $\cdot$, and $\div$), discussed earlier, which are used to write mathematical *expressions* that represent a number. For example, compare the sentence $4 < 10$, which gives the relationship between 4 and 10, with the expression $4 + 10$, which tells how to operate on 4 and 10 to get the number 14.

1.1 Exercises

Decide whether each of the following is true or false. If it is false, explain why.

1. The exponential expression 3^5 means $5 \cdot 5 \cdot 5$.

2. $(4 + 6) \cdot 8$ and $4 + 6 \cdot 8$ both equal 80.

3. In an inequality using $<$ or $>$, the inequality symbol should point toward the smaller number for the inequality to be true.

4. $6 + 8 = 14$ is a mathematical sentence, while $6 + 8 - 14$ is a mathematical expression.

Find the value of each exponential expression. See Example 1.

5. 7^2

6. 4^2

7. 12^2

8. 14^2

9. 4^3

10. 5^3

11. 10^3

12. 11^3

13. 3^4

14. 6^4

15. 4^5

16. 3^5

17. $\left(\dfrac{2}{3}\right)^4$

18. $\left(\dfrac{3}{4}\right)^3$

19. $(.04)^3$

20. $(.05)^4$

21. Explain in your own words how to evaluate a power of a number, such as 6^3.

22. Which of the following are not grouping symbols? parentheses, brackets, fraction bars, exponents

Find the value of each of the following expressions. See Examples 2–4.

23. $9 \cdot 5 - 13$

24. $7 \cdot 6 - 11$

25. $\dfrac{1}{4} \cdot \dfrac{2}{3} + \dfrac{2}{5} \cdot \dfrac{11}{3}$

26. $\dfrac{9}{4} \cdot \dfrac{2}{3} + \dfrac{4}{5} \cdot \dfrac{5}{3}$

27. $9 \cdot 4 - 8 \cdot 3$

28. $11 \cdot 4 + 10 \cdot 3$

29. $(4.3)(1.2) + (2.1)(8.5)$

30. $(2.5)(1.9) + (4.3)(7.3)$

31. $5[3 + 4(2^2)]$

32. $6[2 + 8(3^3)]$

33. $3^2[(11 + 3) - 4]$

34. $4^2[(13 + 4) - 8]$

35. $\dfrac{8 + 6(3^2 - 1)}{3 \cdot 2 - 2}$

36. $\dfrac{8 + 2(8^2 - 4)}{4 \cdot 3 - 10}$

37. $\dfrac{4(6 + 2) + 8(8 - 3)}{6(4 - 2) - 2^2}$

38. $\dfrac{6(5 + 1) - 9(1 + 1)}{5(8 - 6) - 2^3}$

39. Explain why, in the expression $3 + 4 \cdot 6$, the product $4 \cdot 6$ should be found *before* the addition is performed.

40. When evaluating $(4^2 + 3^3)^4$, what is the *last* exponent that would be applied?

Tell whether each statement is true or false. In Exercises 43–52, first simplify each expression involving an operation. See Example 5.

41. $8 \geq 17$

42. $10 \geq 41$

43. $17 \leq 18 - 1$

44. $12 \geq 10 + 2$

45. $6 \cdot 8 + 6 \cdot 6 \geq 0$

46. $4 \cdot 20 - 16 \cdot 5 \geq 0$

47. $6[5 + 3(4 + 2)] \leq 70$

48. $6[2 + 3(2 + 5)] \leq 135$

49. $\dfrac{9(7 - 1) - 8 \cdot 2}{4(6 - 1)} > 3$

50. $\dfrac{2(5 + 3) + 2 \cdot 2}{2(4 - 1)} > 1$

51. $8 \leq 4^2 - 2^2$

52. $10^2 - 8^2 > 6^2$

Write each word statement in symbols. See Example 6.

53. Fifteen is equal to five plus ten.

54. Twelve is equal to twenty minus eight.

55. Nine is greater than five minus four.

56. Ten is greater than six plus one.

57. Sixteen is not equal to nineteen.

58. Three is not equal to four.

59. Two is less than or equal to three.

60. Five is less than or equal to nine.

Write each statement in words and decide whether it is true or false.

61. $7 < 19$

62. $9 < 10$

63. $3 \neq 6$

64. $9 \neq 13$

65. $8 \geq 11$

66. $4 \leq 2$

67. Construct a true statement that involves an addition on the left side, the symbol $\geq$, and a multiplication on the right side.

68. Construct a false statement that involves subtraction on the left side, the symbol $\leq$, and a division on the right side. Then tell why the statement is false and how it could be changed to become true.

Write each statement with the inequality symbol reversed. See Example 7.

69. $5 < 30$

70. $8 > 4$

71. $12 \geq 3$

72. $25 \leq 41$

The table shows the five top grossing feature films in 1994, in millions of dollars.

Rank	Title	Gross Receipts
1	*The Lion King*	298.9
2	*Forrest Gump*	298.1
3	*True Lies*	146.3
4	*The Santa Clause*	130.5
5	*The Flintstones*	121.7

Source: *Universal Almanac 1996,* John W. Wright, General Editor, Andrews and McMeel, Kansas City, p. 234.

73. Which films grossed greater than 298 million dollars?

74. Which films grossed less than 130.5 million dollars?

75. For which films were the gross receipts less than or equal to 130.5 million dollars?

1.2 Variables, Expressions, and Equations

A **variable** is a symbol, usually a letter, such as x, y, or z, used to represent any unknown number. An **algebraic** (al-juh-BRAY-ik) **expression** is a collection of numbers, variables, symbols for operations, and symbols for grouping, such as parentheses, square brackets, or division bars. For example,

$$x + 5, \quad 2m - 9, \quad \text{and} \quad 8p^2 + 6(p - 2)$$

are all algebraic expressions. In the algebraic expression $2m - 9$, the expression $2m$ means $2 \cdot m$, the product of 2 and m, and $8p^2$ shows the product of 8 and p^2. Also, $6(p - 2)$ means the product of 6 and $p - 2$.

OBJECTIVE 1 An algebraic expression has different numerical values for different values of the variables.

E X A M P L E   1   **Finding the Value of an Expression Given a Value of the Variable**

Find the values of the following algebraic expressions if $m = 5$ and if $m = 9$.

(a) $8m$

Replace m with 5, to get

$$8m = 8 \cdot 5 \qquad \text{Let } m = 5.$$
$$= 40. \qquad \text{Multiply.}$$

If $m = 9$,

$$8m = 8 \cdot 9 \qquad \text{Let } m = 9.$$
$$= 72. \qquad \text{Multiply.}$$

(b) $3m^2$

If $m = 5$,

$$3m^2 = 3 \cdot 5^2 \qquad \text{Let } m = 5.$$
$$= 3 \cdot 25 \qquad \text{Square.}$$
$$= 75. \qquad \text{Multiply.}$$

If $m = 9$,

$$3m^2 = 3 \cdot 9^2$$
$$= 3 \cdot 81 = 243.$$

Caution
In Example 1(b), it is important to notice that $3m^2$ means $3 \cdot m^2$; it *does not* mean $3m \cdot 3m$. Unless parentheses are used, the exponent refers only to the variable or number just before it.

WORK PROBLEM 1 AT THE SIDE. ▶▶

E X A M P L E   2   **Finding the Value of an Expression with More Than One Variable**

Find the value of each expression if $x = 5$ and $y = 3$.

(a) $2x + 5y$

$$2x + 5y = 2 \cdot 5 + 5 \cdot 3 \qquad \text{Replace } x \text{ with 5 and } y \text{ with 3.}$$
$$= 10 + 15 \qquad \text{Multiply.}$$
$$= 25 \qquad \text{Add.}$$

— **CONTINUED ON NEXT PAGE**

OBJECTIVES

1 ▶ Find the value of algebraic expressions, given values for the variables.

2 ▶ Convert phrases from words to algebraic expressions.

3 ▶ Identify solutions of equations.

4 ▶ Translate word statements to equations.

5 ▶ Distinguish between expressions and equations.

FOR EXTRA HELP

Tutorial   Tape 1   SSM, Sec. 1.2

1. Find the value of each expression if $p = 3$.

(a) $6p$

(b) $p + 12$

(c) $5p^2$

ANSWERS

1. (a) 18 (b) 15 (c) 45

2. Find the value of each expression if $x = 6$ and $y = 9$.

(a) $4x + 7y$

(b) $\dfrac{4x - 2y}{x + 1}$

(c) $2x^2 + y^2$

(b) $\dfrac{9x - 8y}{2x - y}$

$$\dfrac{9x - 8y}{2x - y} = \dfrac{9 \cdot 5 - 8 \cdot 3}{2 \cdot 5 - 3} \qquad \text{Replace } x \text{ with 5 and } y \text{ with 3.}$$

$$= \dfrac{45 - 24}{10 - 3} \qquad \text{Multiply.}$$

$$= \dfrac{21}{7} \qquad \text{Subtract.}$$

$$= 3 \qquad \text{Divide.}$$

(c) $x^2 - 2y^2 = 5^2 - 2 \cdot 3^2 \qquad$ Replace x with 5 and y with 3.

$\qquad = 25 - 2 \cdot 9 \qquad$ Use the exponents.

$\qquad = 25 - 18 \qquad$ Multiply.

$\qquad = 7 \qquad$ Subtract.

◀◀ **WORK PROBLEM 2 AT THE SIDE.**

An Introduction to Scientific Calculators in the front of this book explains how to evaluate exponentials with a calculator.

OBJECTIVE 2 In the previous section, we wrote word phrases in symbols. The next example shows how variables are used to translate words to symbols. This process will be very important later for solving applied problems.

E X A M P L E   3   **Using Variables to Change Word Phrases into Algebraic Expressions**

Change the following word phrases to algebraic expressions. Use x as the variable to represent the number.

(a) The sum of a number and 9

"Sum" is the answer to an addition problem. This phrase translates as

$$x + 9 \quad \text{or} \quad 9 + x.$$

(b) 7 minus a number

"Minus" indicates subtraction, so the answer is

$$7 - x.$$

Note that $x - 7$ would *not* be correct because we cannot do a subtraction in either order and get the same results.

(c) A number subtracted from 12

Since a number is subtracted *from* 12, write this as

$$12 - x.$$

Compare this result with "12 is subtracted from 25," which is $25 - 12$.

(d) The product of 11 and a number

$$11 \cdot x \quad \text{or} \quad 11x$$

CONTINUED ON NEXT PAGE

ANSWERS

2. (a) 87 **(b)** $\dfrac{6}{7}$ **(c)** 153

(e) 5 divided by a number

$$\frac{5}{x}$$

(f) The product of 2 and the difference between a number and 8

$$2(x - 8)$$

Caution

Notice that in translating the words "the difference between a number and 8" the order is kept the same: $x - 8$. "The difference between 8 and a number" would be written $8 - x$.

> **WORK PROBLEM 3 AT THE SIDE.** ▶▶

OBJECTIVE 3 ▶ An **equation** states that two expressions are equal. Examples of equations are

$$x + 4 = 11, \qquad 2y = 16, \qquad \text{and} \qquad 4p + 1 = 25 - p.$$

To **solve** an equation, we must find all values of the variable that make the equation true. The values of the variable that make the equation true are called the **solutions** of the equation.

EXAMPLE 4   Deciding Whether a Number Is a Solution of an Equation

Decide whether the given number is a solution of the equation.

(a) $5p + 1 = 36$; 7

$$
\begin{aligned}
5p + 1 &= 36 \\
5 \cdot 7 + 1 &= 36 && \text{Replace } p \text{ with 7.} \\
35 + 1 &= 36 && \text{Multiply.} \\
36 &= 36 && \text{True}
\end{aligned}
$$

The number 7 is a solution of the equation.

(b) $9m - 6 = 32$; 4

$$
\begin{aligned}
9m - 6 &= 32 \\
9 \cdot 4 - 6 &= 32 && \text{Replace } m \text{ with 4.} \\
36 - 6 &= 32 && \text{Multiply.} \\
30 &= 32 && \text{False}
\end{aligned}
$$

The number 4 is not a solution of the equation.

> **WORK PROBLEM 4 AT THE SIDE.** ▶▶

OBJECTIVE 4 ▶ We have seen how to translate phrases from words to expressions. Sentences given in words are translated as equations.

---

3. Write as an algebraic expression. Use x as the variable.

(a) The sum of 5 and a number

(b) A number minus 4

(c) A number subtracted from 48

(d) The product of 6 and a number

(e) 9 multiplied by the sum of a number and 5

4. Decide whether the given number is a solution of the equation.

(a) $p - 1 = 3$; 2

(b) $2k + 3 = 15$; 7

(c) $8p - 11 = 5$; 2

ANSWERS

3. (a) $5 + x$ (b) $x - 4$
(c) $48 - x$ (d) $6x$ (e) $9(x + 5)$
4. (a) no (b) no (c) yes

5. Change each sentence to an equation. Use x as the variable.

(a) Three times the sum of a number and 13 is 19.

(b) Five times a number is subtracted from 21, giving 15.

6. Decide whether each of the following is an equation or an expression.

(a) $2x + 5y - 7$

(b) $\dfrac{3x - 1}{5}$

(c) $2x + 5 = 7$

(d) $\dfrac{x}{y - 3} = 4x$

E X A M P L E 5 Translating a Word Sentence to an Equation

Change each word sentence to an equation. Use x as the variable.

(a) Twice the sum of a number and four is six.

The word "is" suggests equals. With x as the variable, translate as follows.

$$
\begin{array}{ccccc}
\text{Twice} & \text{a number and four} & \text{is} & \text{six.} \\
\downarrow & \downarrow & \downarrow & \downarrow \\
2\cdot & (x + 4) & = & 6
\end{array}
$$

the sum of

$$2(x + 4) = 6$$

(b) Nine more than five times a number is 49.

"Nine more than" means "nine is added to." Use x to represent the unknown number.

$$
\begin{array}{ccccc}
\text{Nine} & \text{more than} & \text{five times a number} & \text{is} & 49. \\
\downarrow & \downarrow & \downarrow & \downarrow & \downarrow \\
9 & + & 5x & = & 49
\end{array}
$$

$$9 + 5x = 49$$

 **WORK PROBLEM 5 AT THE SIDE.**

OBJECTIVE 5 Students often have trouble distinguishing between equations and expressions. Remember that an equation is a sentence; an expression is a phrase.

$$
\begin{array}{cc}
4x + 5 = 9 & 4x + 5 \\
\uparrow & \uparrow \\
\text{Equation} & \text{Expression}
\end{array}
$$

E X A M P L E 6 Distinguishing between Equations and Expressions

Decide whether each of the following is an equation or an expression.

(a) $2x - 5y$

There is no equals sign, so this is an expression.

(b) $2x = 5y$

Because of the equals sign, this is an equation.

WORK PROBLEM 6 AT THE SIDE.

ANSWERS

5. (a) $3(x + 13) = 19$
  (b) $21 - 5x = 15$
6. (a) expression **(b)** expression
  (c) equation **(d)** equation

1.2 Exercises

Identify each of the following as an expression or an equation. See Example 6.

1. $3x + 2(x - 4)$　　　　　　**2.** $6(m + 3) + m$　　　　　　**3.** $5p + 2(p - 4) = 7$

4. $5(2r - 3) + r = 12$　　　　**5.** $x + y = 3$　　　　　　**6.** $x + y - 3$

7. Why are "5 less than a number" and "5 is less than a number" translated differently?

8. Why is $2x^3$ not the same as $2x \cdot 2x \cdot 2x$?

*Find the numerical values of the following if (**a**) $x = 4$ and (**b**) $x = 6$. See Example 1.*

9. $4x^2$
　(a)　　　　　　　**(b)**

10. $5x^2$
　(a)　　　　　　　**(b)**

11. $\dfrac{3x - 5}{2x}$
　(a)　　　　　　　**(b)**

12. $\dfrac{4x - 1}{3x}$
　(a)　　　　　　　**(b)**

13. $\dfrac{6.459x}{2.7}$ (to the nearest thousandth)
　(a)　　　　　　　**(b)**

14. $\dfrac{.74x^2}{.85}$ (to the nearest thousandth)
　(a)　　　　　　　**(b)**

15. $3x^2 + x$
　(a)　　　　　　　**(b)**

16. $2x + x^2$
　(a)　　　　　　　**(b)**

*Find the numerical values of the following if (**a**) $x = 2$ and $y = 1$ and (**b**) $x = 1$ and $y = 5$. See Example 2.*

17. $3(x + 2y)$
　(a)　　　　　　　**(b)**

18. $2(2x + y)$
　(a)　　　　　　　**(b)**

19. $x + \dfrac{4}{y}$
　(a)　　　　　　　**(b)**

20. $y + \dfrac{8}{x}$
　(a)　　　　　　　**(b)**

21. $\dfrac{x}{2} + \dfrac{y}{3}$

    (a)   **(b)**

22. $\dfrac{x}{5} + \dfrac{y}{4}$

    (a)   **(b)**

23. $\dfrac{2x + 4y - 6}{5y + 2}$

    (a)   **(b)**

24. $\dfrac{4x + 3y - 1}{2x + y}$

    (a)   **(b)**

25. $2y^2 + 5x$

    (a)   **(b)**

26. $6x^2 + 4y$

    (a)   **(b)**

27. $\dfrac{3x + y^2}{2x + 3y}$

    (a)   **(b)**

28. $\dfrac{x^2 + 1}{4x + 5y}$

    (a)   **(b)**

29. $.841x^2 + .32y^2$

    (a)   **(b)**

30. $.941x^2 + .2y^2$

    (a)   **(b)**

Change the word phrases to algebraic expressions. Use x as the variable to represent the number. See Example 3.

31. Twelve times a number

32. Thirteen added to a number

33. Two subtracted from a number

34. Eight subtracted from a number

35. Four times a number subtracted from seven

36. Three times a number subtracted from fourteen

37. The difference between twice a number and 6

38. The difference between 6 and half a number

39. 12 divided by the sum of a number and 3

40. The difference between a number and 5, divided by 12

41. The product of 6 and four less than a number

42. The product of 9 and five more than a number

43. In the phrase "four more than the product of a number and 6," does the word *and* signify the operation of addition? Explain.

44. What value of x would cause the expression $2x + 3$ to equal 9?

45. There are many pairs of values of x and y for which $2x + y$ will equal 6. Name two such pairs.

46. Suppose that the directions on a test read "Solve the following expressions." How would you politely correct the person who wrote these directions?

Decide whether the given number is a solution of the equation. See Example 4.

47. $p - 5 = 12$; 7

48. $x + 6 = 15$; 10

49. $5m + 2 = 7$; 1

50. $3r + 5 = 8$; 1

51. $6p + 4p + 9 = 11$; $\dfrac{1}{5}$

52. $2x + 3x + 8 = 20$; $\dfrac{12}{5}$

53. $2y + 3(y - 2) = 14; 3$ **54.** $6a + 2(a + 3) = 14; 2$ **55.** $\dfrac{z + 4}{2 - z} = \dfrac{13}{5}; \dfrac{1}{3}$

56. $\dfrac{x + 6}{x - 2} = \dfrac{37}{5}; \dfrac{13}{4}$ **57.** $3r^2 - 2 = 53.47; 4.3$ **58.** $2x^2 + 1 = 28.38; 3.7$

Change the word sentences to equations. Use x as the variable. See Example 5.

59. The sum of a number and 8 is 18.   **60.** A number minus three equals 1.

61. Five more than twice a number is 5.   **62.** The product of a number and 3 is 6.

63. Sixteen minus three-fourths of a number is 13.   **64.** The sum of six-fifths of a number and 2 is 14.

65. Three times a number is equal to 8 more than twice the number.   **66.** Twelve divided by a number equals $\frac{1}{3}$ times that number.

MATHEMATICAL CONNECTIONS (EXERCISES 67–70)

A mathematical model is an equation that can be used to determine the relationship between two quantities. It is used to approximate unknown values of one quantity from known values of the other quantity. For example, based on data from McDonald's, the number of McDonald's locations is given by $y = .32x - 2.5$, where y is in thousands and $x = 0$ corresponds to the year 1955, $x = 10$ corresponds to 1965, and so on.* Use this model to determine the approximate number of McDonald's locations in the given year.

67. 1965   **68.** 1975   **69.** 1985   **70.** 1995

---

*Chicago Tribune, Sunday, January 21, 1996, Section 5, p. 1.

1.3 *Real Numbers and the Number Line*

OBJECTIVES

1 ▶ Use integers to express numbers in applications.

2 ▶ Graph rational numbers on the number line.

3 ▶ Tell which of two real numbers is smaller.

4 ▶ Find the opposite of a number.

5 ▶ Find the absolute values of real numbers.

FOR EXTRA HELP

Tutorial   Tape I   SSM, Sec. 1.3

We now introduce the set of whole numbers.

Whole Numbers

$$\{0, 1, 2, 3, 4, 5, \dots\}$$

The numbers used for counting are called the **natural numbers.**

Natural Numbers

$$\{1, 2, 3, 4, 5, \dots\}$$

These numbers, along with many others, can be represented on **number lines** like the one in Figure 1. We draw a number line by choosing any point on the line and labeling it 0. Choose any point to the right of 0 and label it 1. The distance between 0 and 1 gives a unit of measure used to locate other points, as shown in Figure 1. The points labeled in Figure 1 correspond to the first few whole numbers.

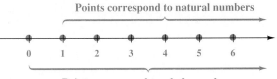

FIGURE I

OBJECTIVE 1 ▶ The natural numbers are located to the right of 0 on the number line. But numbers may also be placed to the left of 0. For each natural number we can place a corresponding number to the left of 0. These numbers, written $-1, -2, -3, -4$, and so on, are shown in Figure 2. Each is the **opposite** or **negative** of a natural number. The natural numbers, their opposites, and zero form a new set of numbers, called the **integers** (IN-te-jurs).

Integers

$$\{ \dots -3, -2, -1, 0, 1, 2, 3 \dots \}$$

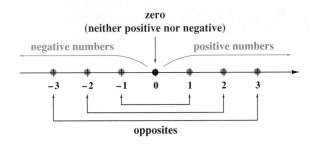

FIGURE 2 *Points that correspond to integers*

There are many practical applications of negative numbers. For example, a Fahrenheit temperature on a cold January day might be $-10°$, and a business that spends more than it takes in has a negative "profit."

1. Use an integer to express the number(s) in each application.

(a) Erin discovers that she has spent $53 more than she has in her checking account.

(b) The record Fahrenheit high temperature in the U.S. was 134° in Death Valley, California, July 10, 1913.

(c) A football team gained 5 yards, then lost 10 yards on the next play.

2. Graph the following numbers on the number line.

$$-3, -2.75, -\frac{3}{4}, 1\frac{1}{2}, \frac{17}{8}$$

+—+—+—+—+—+—+—+
−3 −2 −1 0 1 2 3

EXAMPLE 1 Using Negative Numbers in Applications

Use integers to express the numbers in the following applications.

(a) The lowest Fahrenheit temperature ever recorded in meteorological records was 128.6° below zero at Vostok, Antarctica, on July 22, 1983.
Use −128.6 because "below zero" indicates a negative number.

(b) The shore surrounding the Dead Sea is 1312 feet below sea level. Again, "below sea level" indicates a negative number, −1312.

◀◀ WORK PROBLEM 1 AT THE SIDE.

OBJECTIVE 2 Not all numbers are integers. For example, $\frac{1}{2}$ is not; it is a number halfway between the integers 0 and 1. Also, $3\frac{1}{4}$ is not an integer. These numbers and others that are quotients of integers are *rational numbers*. (The name comes from the word *ratio*, which indicates a quotient.)

> **Rational Numbers**
>
> {numbers that can be written as quotients of integers, with denominator not 0}

Since any integer can be written as the quotient of itself and 1, all integers are rational numbers. A decimal number that comes to an end (terminates), such as .23, is a rational number: $.23 = \frac{23}{100}$. Decimal numbers that repeat in a fixed block of digits, such as .3333 . . . = $.\overline{3}$ and .454545 . . . = $.\overline{45}$, are also rational numbers. For example, $.\overline{3} = \frac{1}{3}$.

To **graph** a number we place a dot on the number line at the point that corresponds to the number.

EXAMPLE 2 Graphing Rational Numbers

Graph the following numbers on the number line.

$$-\frac{3}{2}, -\frac{2}{3}, \frac{1}{2}, 1\frac{1}{3}, \frac{23}{8}, 3\frac{1}{4}$$

To locate the improper fractions on the number line, write them in the form of mixed numbers or decimals. The graph is shown in Figure 3.

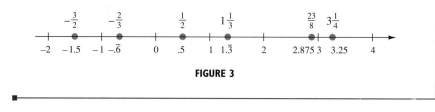

FIGURE 3

◀◀ WORK PROBLEM 2 AT THE SIDE.

Although a great many numbers are rational, there are also numbers that are not. For example, a floor tile 1 foot on a side has a diagonal whose length is the square root of 2 (written $\sqrt{2}$). See Figure 4. It can be shown

ANSWERS

1. (a) −53 **(b)** 134 **(c)** 5, −10

2.

−2.75 $-\frac{3}{4}$ $1\frac{1}{2}$ $\frac{17}{8}$
◆—•—+—+—•—+—•—+—+→
−3 −2 −1 0 1 2 3

that $\sqrt{2}$ cannot be written as a quotient of integers, so it is an example of a number that is not rational. It is **irrational** (ear-RASH-un-ul).

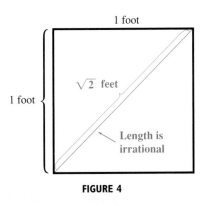

1 foot

1 foot

$\sqrt{2}$ feet

Length is irrational

FIGURE 4

Irrational Numbers

{nonrational numbers represented by points on the number line}

The decimal form of an irrational number never terminates and never repeats. Some examples of irrational numbers include $\sqrt{3}$, $\sqrt{7}$, .10110111011110 . . . , $-\sqrt{10}$, and π, which is the ratio of the distance around a circle to the distance across it. These numbers lie between the rational numbers on the number line. Irrational numbers are discussed in Chapter 9.

Finally, *all* numbers that can be represented by points on the number line are called **real numbers.**

Real Numbers

{all numbers that can be represented by points on the number line} or {all rational and irrational numbers}.

All the numbers mentioned above are real numbers. The relationships between the various types of numbers are shown in two different ways in Figure 5. Notice that any real number is either a rational number or an irrational number.

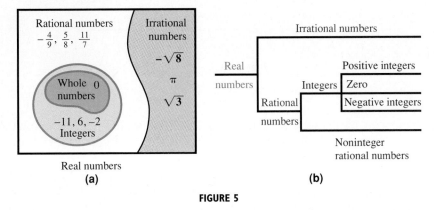

FIGURE 5

OBJECTIVE 3 Given any two whole numbers, we can tell which number is smaller. But what about two negative numbers, as in the set of integers? Moving from zero to the right along a number line, the positive numbers corresponding to the points on the number line *increase.* For example, $8 < 12$, and 8 is to the left of 12 on a number line. This ordering is extended to all real numbers by definition.

3. Tell whether each statement is true or false.

(a) $-2 < 4$

(b) $6 > -3$

(c) $-9 < -12$

(d) $-4 \geq -1$

(e) $-6 \leq 0$

The Ordering of the Real Numbers

For any two real numbers a and b, **a is less than b** if a is to the left of b on the number line.

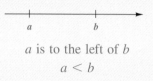

a is to the left of b

$a < b$

This means that any negative number is smaller than 0, and any negative number is smaller than any positive number. Also, 0 is smaller than any positive number.

EXAMPLE 3   Determining the Order of Real Numbers

Is it true that $-3 < -1$?

To find out, locate -3 and -1 on a number line, as shown in Figure 6. Because -3 is to the left of -1 on the number line, -3 is smaller than -1. The statement $-3 < -1$ is true.

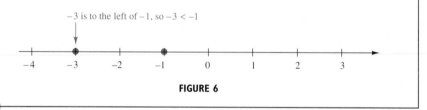

FIGURE 6

◀◀ WORK PROBLEM 3 AT THE SIDE.

OBJECTIVE 4 Earlier, we saw that every positive integer has a negative integer that is its opposite or negative. This is true for every real number except 0, which is its own opposite. A characteristic of pairs of opposites is that they are the same distance from 0 on the number line. See Figure 7.

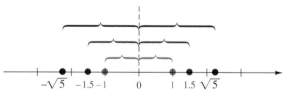

FIGURE 7   *Pairs of opposites*

We indicate the opposite of a number by writing the symbol $-$ in front of the number. For example, the opposite of 7 is -7 (read "negative 7"). We could write the opposite of -4 as $-(-4)$, but we know that 4 is the opposite of -4. Since a number can have only one opposite, 4 and $-(-4)$ must represent the same number, so that

$$-(-4) = 4.$$

A generalization of this idea is given below.

Double Negative Rule

For any real number a,

$$-(-a) = a.$$

┌─ **E X A M P L E 4   Finding the Opposite of a Number**

The following chart shows several numbers and their opposites.

Number	Opposite
-4	$-(-4)$, or 4
-3	3
0	0
5	-5
19	-19

Example 4 suggests the following rule.

> The opposite of a number is found by changing the sign of the number.

WORK PROBLEM 4 AT THE SIDE. ▶▶

OBJECTIVE 5 As mentioned above, opposites are numbers the same distance from 0 on the number line but on opposite sides of 0. Another way to say this is to say that opposites have the same absolute value. The **absolute value** of a number is the undirected distance between 0 and the number on the number line. The symbol for the absolute value of the number a is $|a|$, read "the absolute value of a." For example, the distance between 2 and 0 on the number line is 2 units, so

$$|2| = 2.$$

Also, the distance between -2 and 0 on the number line is 2, so

$$|-2| = 2.$$

Since distance is a physical measurement, which is never negative, we can make the following statement.

> The absolute value of a number can never be negative.

For example,

$$|12| = 12 \quad \text{and} \quad |-12| = 12$$

because both 12 and -12 lie at a distance of 12 units from 0 on the number line. Since 0 is a distance 0 units from 0, we have

$$|0| = 0.$$

┌─ **E X A M P L E 5   Evaluating Absolute Value**

Simplify by removing absolute value symbols.

(a) $|5| = 5$

(b) $|-5| = 5$

(c) $-|-5| = -(5) = -5$ Replace $|-5|$ with 5.

(d) $-|-13| = -(13) = -13$

└─ **CONTINUED ON NEXT PAGE**

4. Find the opposite of each number.

(a) 6

(b) 15

(c) -9

(d) -12

(e) 0

ANSWERS

4. (a) -6 **(b)** -15 **(c)** 9 **(d)** 12 **(e)** 0

5. Simplify by removing absolute value symbols.

(a) $|-6|$

(b) $|9|$

(c) $-|15|$

(d) $-|-9|$

(e) $|9 - 4|$

(f) $-|32 - 2|$

(e) $|8 - 5|$

Simplify within the absolute value bars first.

$$|8 - 5| = |3| = 3$$

(f) $-|8 - 5| = -|3| = -3$

(g) $-|12 - 3| = -|9| = -9$

Parts (e)–(g) in Example 5 suggest that absolute value bars also act as grouping symbols.

◀◀ **WORK PROBLEM 5 AT THE SIDE.**

1.3 Exercises

Decide whether each statement is true or false.

1. Every whole number is an integer.

2. Every rational number is a real number.

3. No number can be both rational and irrational.

4. No natural number is negative.

5. Every whole number is positive.

6. Some real numbers are not rational.

7. Every natural number is a whole number.

8. The number 0 is irrational.

9. Some whole numbers are not integers.

10. Not every rational number is positive.

Use an integer to express each number in the following applications of numbers. See Example 1.

11. Between 1970 and 1980, the population of the state of New York decreased by 683,226.

12. Death Valley lies 282 feet below sea level.

13. The city of New Orleans lies 8 feet below sea level.

14. Alex Fedotov, in 1977, flew a jet airplane at an altitude of 123,524 feet.

Graph each group of numbers on a number line. See Example 2.

15. $0, 3, -5, -6$

16. $2, 6, -2, -1$

17. $-2, -6, -4, 3, 4$

18. $-5, -3, -2, 0, 4$

19. $\dfrac{1}{4}, 2\dfrac{1}{2}, -3\dfrac{4}{5}, -4, -1\dfrac{5}{8}$

20. $5\dfrac{1}{4}, 4\dfrac{5}{9}, -2\dfrac{1}{3}, 0, -3\dfrac{2}{5}$

Select the smaller number in each pair. See Example 3.

21. $-11, -4$ **22.** $-9, -16$ **23.** $-21, 1$ **24.** $-57, 3$

25. $0, -100$ **26.** $-215, 0$ **27.** $-\dfrac{2}{3}, -\dfrac{1}{4}$ **28.** $-\dfrac{3}{8}, -\dfrac{9}{16}$

Decide whether each statement is true or false. See Example 3.

29. $8 < -16$ **30.** $12 < -24$ **31.** $-3 < -2$ **32.** $-10 < -9$

*For each of the following, **(a)** find the opposite of the number and **(b)** find the absolute value of the number. See Examples 4 and 5.*

33. -2 **34.** -8 **35.** 6

36. 11 **37.** $-\dfrac{3}{4}$ **38.** $-\dfrac{1}{3}$

Simplify by removing absolute value symbols. See Example 5.

39. $|-7|$ **40.** $|-3|$ **41.** $|4|$ **42.** $|10|$

43. $-|12|$ **44.** $-|23|$ **45.** $-|-14|$ **46.** $-|-19|$

47. $|13 - 4|$ **48.** $|8 - 7|$

49. Students often say "The absolute value of a number is always positive." Is this true? If not, explain.

50. If the absolute value of a number is equal to the number itself, what must be true about the number?

The table shows several categories of necessary consumer expenses and the change in the producer price index from the previous year for two recent years.

Category	Change 1992	Change 1993
Food	−2.7	2.5
Fuels	−1.8	−.4
Textiles and apparel	2.9	.2
Metals and metal products	3.8	0
Furniture and household durables	3.1	1.5
Transportation and equipment	8.9	3.3

Source: Bureau of Labor Statistics, U.S. Dept. of Labor.

51. What category of which year represents the greatest decrease?

52. Which represents a smaller decrease: the change for fuels in 1992 or 1993?

53. True or false? The absolute value of the change for food products in 1993 is less than the absolute value of the change for food products in 1992.

54. True or false? The absolute value of the change in 1993 fuels is less than the absolute value of the change in 1993 textiles and apparel.

1.4 Addition of Real Numbers

OBJECTIVE 1 We can use the number line to explain the addition of real numbers. Later, we will give the rules for addition.

E X A M P L E 1   Adding with the Number Line

Use the number line to find the sum $2 + 3$.

Add the positive numbers 2 and 3 by starting at 0 and drawing an arrow two units to the *right*, as shown in Figure 8. This arrow represents the number 2 in the sum $2 + 3$. Next, from the right end of this arrow draw another three units to the right. The number below the end of this second arrow is 5, so $2 + 3 = 5$.

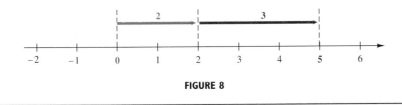

FIGURE 8

E X A M P L E 2   Adding with the Number Line

Use the number line to find the sum $-2 + (-4)$. (Parentheses are placed around the -4 to avoid the confusing use of $+$ and $-$ next to each other.)

To add the negative numbers -2 and -4 on the number line, we start at 0 and draw an arrow two units to the *left*, as shown in Figure 9. We draw the arrow to the left to represent the addition of a *negative* number. From the left end of this first arrow, we draw a second arrow four units to the left. The number below the end of this second arrow is -6, so $-2 + (-4) = -6$.

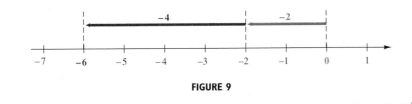

FIGURE 9

WORK PROBLEM I AT THE SIDE. ▶▶

In Example 2, we found that the sum of the two negative numbers -2 and -4 is a negative number whose distance from 0 is the sum of the distance of -2 from 0 and the distance of -4 from 0. That is, *the sum of two negative numbers is the negative of the sum of their absolute values.*

$$-2 + (-4) = -(|-2| + |-4|) = -(2 + 4) = -6$$

> To add two numbers having the same signs, add the absolute values of the numbers. Give the result the same sign as the numbers being added. Example: $-4 + (-3) = -7$.

E X A M P L E 3   Adding Two Negative Numbers

Find the sums.

(a) $-2 + (-9) = -11$ The sum of two negative numbers is negative.

— **CONTINUED ON NEXT PAGE**

— **CONTINUED ON NEXT PAGE**

OBJECTIVES

1 ▶ Add two numbers with the same sign on a number line.

2 ▶ Add positive and negative numbers.

3 ▶ Add mentally.

4 ▶ Use the order of operations with real numbers.

5 ▶ Interpret words and phrases that indicate addition.

FOR EXTRA HELP

Tutorial   Tape 2   SSM, Sec. 1.4

1. Use the number lines to find the sums.

(a) $1 + 4$

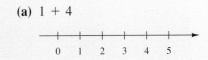

(b) $-2 + (-5)$

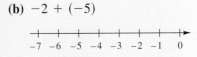

ANSWERS

1. (a) $1 + 4 = 5$

(b) $-2 + (-5) = -7$

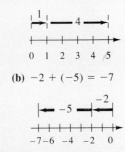

25

2. Find the sums.

(a) $-7 + (-3)$

(b) $-12 + (-18)$

(c) $-15 + (-4)$

3. Use the number lines to find the sums.

(a) $6 + (-3)$

```
+---+---+---+---+---+---+--->
0   1   2   3   4   5   6
```

(b) $-5 + 1$

```
+---+---+---+---+---+---+--->
-5  -4  -3  -2  -1   0
```

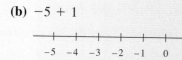

(b) $-8 + (-12) = -20$
(c) $-15 + (-3) = -18$

◀◀ **WORK PROBLEM 2 AT THE SIDE.**

OBJECTIVE 2 ▶ We use the number line again to illustrate the sum of a positive number and a negative number.

E X A M P L E 4   Adding Numbers with Different Signs

Use the number line to find the sum $-2 + 5$.

We find the sum $-2 + 5$ on the number line by starting at 0 and drawing an arrow two units to the left. From the left end of this arrow, we draw a second arrow five units to the right, as shown in Figure 10. The number below the end of this second arrow is 3, so $-2 + 5 = 3$.

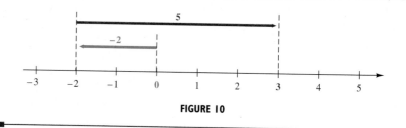

FIGURE 10

◀◀ **WORK PROBLEM 3 AT THE SIDE.**

Addition of numbers with different signs also can be defined using absolute value.

> To add numbers with different signs, first find the difference between the absolute values of the numbers. Give the answer the same sign as the number with the larger absolute value. Example: $-12 + 6 = -6$.

For example, to add -12 and 5, we find their absolute values: $|-12| = 12$ and $|5| = 5$; then we find the difference between these absolute values: $12 - 5 = 7$. Since $|-12| > |5|$, the sum will be negative, so the final answer is $-12 + 5 = -7$.

The $\boxed{-}$ or $\boxed{+/-}$ key is used to input a negative number in the calculator. Try using your calculator to add negative numbers.

OBJECTIVE 3 ▶ While a number line is useful in showing the rules for addition, it is important to be able to find sums mentally.

E X A M P L E 5   Adding a Positive Number and a Negative Number

Check each answer, trying to work the addition mentally. If you get stuck, use a number line.

(a) $7 + (-4) = 3$

(b) $-8 + 12 = 4$

(c) $-\dfrac{1}{2} + \dfrac{1}{8} = -\dfrac{4}{8} + \dfrac{1}{8} = -\dfrac{3}{8}$ Remember to find a common denominator first.

CONTINUED ON NEXT PAGE

(d) $\dfrac{5}{6} + \left(-\dfrac{4}{3}\right) = \dfrac{5}{6} + \left(-\dfrac{8}{6}\right) = -\dfrac{3}{6} = -\dfrac{1}{2}$

(e) $-4.6 + 8.1 = 3.5$

WORK PROBLEM 4 AT THE SIDE. ▶▶

The rules for adding signed numbers are summarized below.

Adding Signed Numbers

Like signs Add the absolute values of the numbers. Give the sum the same sign as the numbers being added.
Unlike signs Find the difference between the larger absolute value and the smaller. Give the answer the sign of the number having the larger absolute value.

OBJECTIVE 4 Sometimes a problem involves square brackets, [ ]. As we mentioned earlier, brackets are treated just like parentheses. We do the calculations inside the brackets until a single number is obtained. Remember to use the order of operations given in Section 1.1 for adding more than two numbers.

EXAMPLE 6 Adding with Brackets

Find the sums.

(a) $-3 + [4 + (-8)]$

First work inside the brackets. Follow the rules for the order of operations given in Section 1.1.

$$-3 + [4 + (-8)] = -3 + (-4) = -7$$

(b) $8 + [(-2 + 6) + (-3)] = 8 + [4 + (-3)] = 8 + 1 = 9$

WORK PROBLEM 5 AT THE SIDE. ▶▶

OBJECTIVE 5 Let's now look at the interpretation of words and phrases that involve addition. Problem solving often requires translating words and phrases into symbols. We began this process with translating simple phrases in Section 1.1.

The word *sum* indicates addition. There are other key words and phrases that also indicate addition. Some of these are given in the chart below.

Word or Phrase	Example	Numerical Expression and Simplification
Sum	The *sum of* −3 and 4	$-3 + 4 = 1$
Added to	5 *added to* −8	$-8 + 5 = -3$
More than	12 *more than* −5	$(-5) + 12 = 7$
Increased by	−6 *increased by* 13	$-6 + 13 = 7$

4. Check each answer, trying to work the addition in your head. If you get stuck, use a number line.

(a) $-8 + 2 = -6$

(b) $-15 + 4 = -11$

(c) $17 + (-10) = 7$

(d) $\dfrac{3}{4} + \left(-\dfrac{11}{8}\right) = -\dfrac{5}{8}$

(e) $-9.5 + 3.8 = -5.7$

5. Find the sums.

(a) $2 + [7 + (-3)]$

(b) $6 + [(-2 + 5) + 7]$

(c) $-9 + [-4 + (-8 + 6)]$

ANSWERS

4. All are correct.

5. (a) 6 (b) 16 (c) −15

6. Write a numerical expression for each phrase, and simplify the expression.

(a) 4 more than −12

(b) The sum of 6 and −7

(c) −12 added to −31

(d) 7 increased by the sum of 8 and −3

E X A M P L E 7 Interpreting Words and Phrases

Write a numerical expression for each phrase, and simplify the expression.

(a) The **sum of** −8 and 4 and 6

$$-8 + 4 + 6 = [-8 + 4] + 6 = -4 + 6 = 2$$

Notice that brackets were placed around −8 + 4, and this addition was done first, using the order of operations given earlier. In fact, the same result would be obtained if the brackets were placed around 4 + 6. This idea will be discussed further in Section 1.8.

$$-8 + 4 + 6 = -8 + [4 + 6] = -8 + 10 = 2$$

(b) 3 **more than** −5, **increased by** 12

$$-5 + 3 + 12 = [-5 + 3] + 12 = -2 + 12 = 10$$

◀◀ **WORK PROBLEM 6 AT THE SIDE.**

Gains (or increases) and losses (or decreases) sometimes appear in applied problems. When they do, the gains may be interpreted as positive numbers and the losses as negative numbers.

E X A M P L E 8 Interpreting Gains and Losses

A football team gained 3 yards on the first play from scrimmage, lost 12 yards on the second play, and then gained 13 yards on the third play. How many yards did the team gain or lose altogether?

The gains are represented by positive numbers and the loss by a negative number.

$$3 + (-12) + 13$$

Add from left to right.

$$3 + (-12) + 13 = [3 + (-12)] + 13 = (-9) + 13 = 4$$

The team gained 4 yards altogether.

◀◀ **WORK PROBLEM 7 AT THE SIDE.**

7. A football team lost 8 yards on the first play from scrimmage, lost 5 yards on the second play, and then gained 7 yards on the third play. How many yards did the team gain or lose altogether?

ANSWERS

6. (a) −12 + 4; −8 **(b)** 6 + (−7); −1
 (c) −31 + (−12); −43
 (d) 7 + [8 + (−3)]; 12
7. The team lost 6 yards.

1.4 Exercises

Fill in the blank with the correct response.

1. The sum of two negative numbers will always be a _____ number.
   (positive/negative)

2. The sum of a number and its opposite will always be _____ .

3. If I am adding a positive number and a negative number, and the negative number has the larger absolute value, the sum will be a _____ number.
   (positive/negative)

4. By the rules for order of operations, to simplify $4[3(-2 + 5) - 1]$, the first step is to add _____ and _____ .

Find the following sums. See Examples 1–6.

5. $6 + (-4)$

6. $8 + (-5)$

7. $12 + (-15)$

8. $4 + (-8)$

9. $-7 + (-3)$

10. $-11 + (-4)$

11. $-10 + (-3)$

12. $-16 + (-7)$

13. $-12.4 + (-3.5)$

14. $-21.3 + (-2.5)$

15. $10 + [-3 + (-2)]$

16. $13 + [-4 + (-5)]$

17. $5 + [14 + (-6)]$

18. $7 + [3 + (-14)]$

19. $-3 + [5 + (-2)]$

20. $-7 + [10 + (-3)]$

21. $-8 + [3 + (-1) + (-2)]$

22. $-7 + [5 + (-8) + 3]$

23. $\dfrac{9}{10} + \left(-\dfrac{3}{5}\right)$

24. $\dfrac{5}{8} + \left(-\dfrac{17}{12}\right)$

25. $-\dfrac{1}{6} + \dfrac{2}{3}$

26. $-\dfrac{6}{25} + \dfrac{19}{20}$

27. $2\dfrac{1}{2} + \left(-3\dfrac{1}{4}\right)$

28. $-4\dfrac{3}{8} + 6\dfrac{1}{2}$

29. $7.8 + (-9.4)$

30. $14.7 + (-10.1)$

31. $-7.1 + [3.3 + (-4.9)]$

32. $-9.5 + [-6.8 + (-1.3)]$

33. $[-8 + (-3)] + [-7 + (-7)]$

34. $[-5 + (-4)] + [9 + (-2)]$

35. Is it possible to add a negative number to another negative number and get a positive number? If so, give an example.

36. Under what conditions will the sum of a positive number and a negative number be a number which is neither negative nor positive?

Perform each operation, and then determine whether the statement is true or false. Try to do all work in your head. See Examples 5 and 6.

37. $-11 + 13 = 13 + (-11)$

38. $16 + (-9) = -9 + 16$

39. $-10 + 6 + 7 = -3$

40. $-12 + 8 + 5 = -1$

41. $18 + (-6) + (-12) = 0$

42. $-5 + 21 + (-16) = 0$

43. $|-8 + 10| = -8 + (-10)$

44. $|-4 + 6| = -4 + (-6)$

45. $\dfrac{11}{5} + \left(-\dfrac{6}{11}\right) = -\dfrac{6}{11} + \dfrac{11}{5}$

46. $-\dfrac{3}{2} + \dfrac{5}{8} = \dfrac{5}{8} + \left(-\dfrac{3}{2}\right)$

47. $-7 + [-5 + (-3)] = [(-7) + (-5)] + 3$

48. $6 + [-2 + (-5)] = [(-4) + (-2)] + 5$

MATHEMATICAL CONNECTIONS (EXERCISES 49–52)

Recall the rules for adding signed numbers introduced in this section, and answer Exercises 49–52 in order.

49. Suppose that the sum of two numbers is negative and you know that one of the numbers is positive. What can you conclude about the other number?

50. If you are asked to solve the equation $x + 5 = -7$ from a set of numbers, why could you immediately eliminate any positive numbers as possible solutions? (Remember how you answered Exercise 49.)

51. Suppose that the sum of two numbers is positive, and you know that one of the numbers is negative. What can you conclude about the other number?

52. If you are asked to solve the equation $x + (-8) = 2$ from a set of numbers, why could you immediately eliminate any negative numbers as possible solutions? (Remember how you answered Exercise 51.)

Write a numerical expression for each phrase, and simplify the expression. See Example 7.

53. The sum of −5 and 12 and 6

54. The sum of −3 and 5 and −12

55. 14 added to the sum of −19 and −4

56. −2 added to the sum of −18 and 11

57. The sum of −4 and −10, increased by 12

58. The sum of −7 and −13, increased by 14

59. 4 more than the sum of 8 and −18

60. 10 more than the sum of −4 and −6

Solve each problem. See Example 8.

61. Kramer owed Jerry $10 for snacks raided from the refrigerator. Kramer later borrowed $70 from George to finance his latest get-rich scheme. What positive or negative number represents Kramer's financial status?

62. Shalita's checking account balance is $54.00. She then takes a gamble by writing a check for $89.00. What is her new balance? (Write the balance as a signed number.)

63. The surface, or rim, of a canyon is at altitude 0. On a hike down into the canyon, a party of hikers stops for a rest at 130 meters below the surface. They then descend another 54 meters. What is their new altitude? (Write the altitude as a signed number.)

64. A pilot announces to the passengers that the current altitude of their plane is 34,000 feet. Because of some unexpected turbulence, the pilot is forced to descend 2100 feet. What is the new altitude of the plane? (Write the altitude as a signed number.)

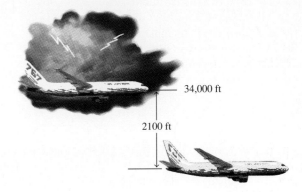

65. The lowest temperature ever recorded in Little Rock, Arkansas, was −5°F. The highest temperature ever recorded there was 117°F more than the lowest. What was this highest temperature?

66. On a series of three consecutive running plays, Steve Young gained 4 yards, lost 3 yards, and lost 2 yards. What positive or negative number represents his total net yardage for the series of plays?

67. On three consecutive passes, Troy Aikman of the Dallas Cowboys passed for a gain of 6 yards, was sacked for a loss of 12 yards, and passed for a gain of 43 yards. What positive or negative number represents the total net yardage for the plays?

68. On January 23, 1943, the temperature rose 49°F in two minutes in Spearfish, South Dakota. If the starting temperature was −4°F, what was the temperature two minutes later?

69. Jennifer owes $153 to a credit card company. She makes a $14 purchase with the card, and then pays $60 on the account. What amount does she still owe?

70. A female polar bear weighed 660 pounds when she entered her winter den. She lost 45 pounds during each of the first two months of hibernation, and another 205 pounds before leaving the den with her two cubs in March. How much did she weigh when she left the den?

71. Kim Falgout owes $870.00 on her Master Card account. She returns two items costing $35.90 and $150.00 and receives credits for these on the account. Next, she makes a purchase of $82.50, and then two more purchases of $10.00 each. She finally makes a payment of $500.00. How much does she still owe?

72. A welder working with stainless steel must use precise measurements. Suppose a welder attaches two pieces of steel that are each 3.60 inches in length, and then attaches an additional three pieces that are each 9.10 inches long. She finally cuts off a piece that is 7.60 inches long. Find the length of the welded piece of steel.

73. Use the ideas presented in this section to guess the solution of the equation $8 + t = 5$.

74. What number makes the equation $p + 4 = 1$ a true statement?

75. Make up your own equation similar to the ones in Exercises 73–74 that has −2 as a solution.

1.5 Subtraction of Real Numbers

OBJECTIVES

1. Find a difference on the number line.
2. Use the definition of subtraction.
3. Work subtraction problems that involve brackets.
4. Interpret words and phrases that indicate subtraction.

FOR EXTRA HELP

Tutorial   Tape 2   SSM, Sec. 1.5

OBJECTIVE 1 As we mentioned earlier, the answer to a subtraction problem is called a **difference.** Differences between signed numbers can be found by using a number line. Addition and subtraction are opposite operations. Thus, because *addition* of a positive number on the number line is shown by drawing an arrow to the *right, subtraction* of a positive number is shown by drawing an arrow to the *left*.

E X A M P L E   1   Subtracting with the Number Line

Use the number line to find the difference $7 - 4$.

To find the difference $7 - 4$ on the number line, begin at 0 and draw an arrow 7 units to the right. From the right end of this arrow, draw an arrow 4 units to the left, as shown in Figure 11. The number at the end of the second arrow shows that $7 - 4 = 3$.

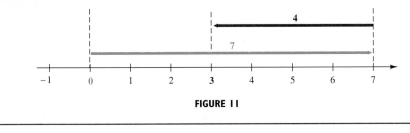

FIGURE 11

WORK PROBLEM 1 AT THE SIDE. ▶▶

OBJECTIVE 2 The procedure used in Example 1 to find $7 - 4$ is exactly the same procedure that would be used to find $7 + (-4)$, so that

$$7 - 4 = 7 + (-4).$$

This shows that *subtraction* of a positive number from a larger positive number is the same as *adding* the opposite of the smaller number to the larger. This result is extended as the definition of subtraction for all real numbers.

Definition of Subtraction

For any real numbers a and b,

$$a - b = a + (-b).$$

Example: $4 - 9 = 4 + (-9) = -5$.

That is, to **subtract** b from a, *add the* opposite (or *negative*) of b to a. This definition leads to the following procedure for subtracting signed numbers.

Subtracting Signed Numbers

Step 1 Change the subtraction symbol to addition.

Step 2 Change the sign of the number being subtracted.

Step 3 Add, as in the previous section.

---

1. Use the number line to find the differences.

(a) $5 - 1$

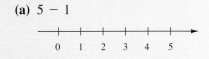

(b) $6 - 2$

ANSWERS

1. (a) $5 - 1 = 4$

(b) $6 - 2 = 4$

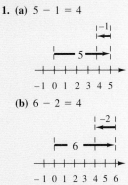

33

2. Subtract.

(a) 6 − 10

(b) −2 − 4

(c) 3 − (−5)

(d) −8 − (−12)

3. Work each problem.

(a) 2 − [(−3) − (4 + 6)]

(b) [(5 − 7) + 3] − 8

(c) 6 − [(−1 − 4) − 2]

E X A M P L E   2   **Using the Definition of Subtraction**

Subtract.

Change − to +.
No change ⌐   ⌐ Opposite of 3

(a) 12 − 3 = 12 + (**−3**) = 9

(b) 5 − 7 = 5 + (−7) = −2

(c) 8 − 15 = 8 + (−15) = −7

Change − to +.
No change ⌐   ⌐ Opposite of −5

(d) −3 − (−5) = −3 + (**5**) = 2

(e) −6 − (−9) = −6 + (9) = 3

(f) 8 − (−5) = 8 + (5) = 13

◀◀ **WORK PROBLEM 2 AT THE SIDE.**

Subtraction can be used to reverse the result of an addition problem. For example, if 4 is added to a number and then subtracted from the sum, the original number is the result.

$$12 + 4 = 16 \quad \text{and} \quad 16 − 4 = 12$$

The symbol − has now been used for three purposes:

1. to represent subtraction, as in 9 − 5 = 4;

2. to represent negative numbers, such as −10, −2, and −3;

3. to represent the opposite (or negative) of a number, as in "the opposite (or negative) of 8 is −8."

We may see more than one use in the same problem, such as −6 − (−9), where −9 is subtracted from −6. The meaning of the symbol depends on its position in the algebraic expression.

OBJECTIVE 3 As before, with problems that have both parentheses and brackets, first do any operations inside the parentheses and brackets. Work from the inside out. Because subtraction is defined in terms of addition, the order of operations rules from Section 1.1 can still be used.

E X A M P L E   3   **Subtracting with Grouping Symbols**

Work each problem.

(a) −6 − [2 − (**8 + 3**)] = −6 − [2 − **11**]

= −6 − [2 + (−11)]   Change − to +.

= −6 − (−9)

= −6 + (9) = 3

(b) 5 − [(−3 − 2) − (4 − 1)] = 5 − [(−3 + (−2)) − 3]

= 5 − [(−5) − 3]

= 5 − [(−5) + (−3)]

= 5 − (−8)

= 5 + 8 = 13

◀◀ **WORK PROBLEM 3 AT THE SIDE.**

OBJECTIVE 4 Now we translate words and phrases that involve subtraction of real numbers. *Difference* is one of them. Some others are given in the chart below.

Word or Phrase	Example	Numerical Expression and Simplification
Difference	The *difference* between −3 and −8	$-3 - (-8) = -3 + 8 = 5$
Subtracted from	12 *subtracted from* 18	$18 - 12 = 6$
Less than	6 *less than* 5	$5 - 6 = 5 + (-6) = -1$
Decreased by	9 *decreased by* −4	$9 - (-4) = 9 + 4 = 13$

Caution
When you are subtracting two numbers, it is important that you write them in the correct order, because, in general, $a - b \neq b - a$. For example, $5 - 3 \neq 3 - 5$. For this reason, it is important to *think carefully before interpreting an expression involving subtraction!* (This problem did not arise for addition.)

E X A M P L E 4 Interpreting Words and Phrases

Write a numerical expression for each phrase, and simplify the expression.

(a) The **difference between** −8 and 5

It is conventional to write the numbers in the order they are given when "difference between" is used.

$$-8 - 5 = -8 + (-5) = -13$$

(b) 4 **subtracted from** the sum of 8 and −3

Here the operation of addition is also used, as indicated by the word *sum*. First, add 8 and −3. Next, subtract 4 from this sum.

$$[8 + (-3)] - 4 = 5 - 4 = 1$$

(c) 4 **less than** −6

Be careful with order here. 4 must be taken *from* −6.

$$-6 - 4 = -6 + (-4) = -10$$

Notice that "4 less than −6" differs from "4 *is less than* −6." "4 is less than −6" is symbolized as $4 < -6$ (which is a false statement).

(d) 8, **decreased by** 5 **less than** 12

First, write "5 less than 12" as $12 - 5$. Next, subtract $12 - 5$ from 8.

$$8 - (12 - 5) = 8 - 7 = 1$$

> **WORK PROBLEM 4 AT THE SIDE.** ▶▶

We have seen a few applications of signed numbers in earlier sections. The next example involves subtraction of signed numbers.

4. Write a numerical expression for each phrase, and simplify the expression.

(a) The difference between −5 and −12

(b) −2 subtracted from the sum of 4 and −4

(c) 7 less than −2

(d) 9, decreased by 10 less than 7

ANSWERS
4. **(a)** $-5 - (-12)$; 7
   (b) $[4 + (-4)] - (-2)$; 2
   (c) $-2 - 7$; −9
   (d) $9 - (7 - 10)$; 12

5. The highest elevation in Argentina is Mt. Aconcagua, which is 6960 meters above sea level. The lowest point in Argentina is the Valdes Peninsula, 40 meters below sea level. Find the difference between the highest and lowest elevations.

E X A M P L E 5   Solving a Problem Involving Subtraction

The record high temperature of 134° Fahrenheit in the United States was recorded at Death Valley, California, in 1913. The record low was −80°F, at Prospect Creek, Alaska, in 1971. See Figure 12. What is the difference between these highest and lowest temperatures?

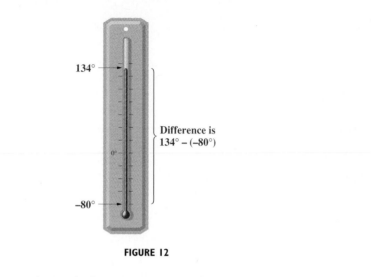

FIGURE 12

We must subtract the lowest temperature from the highest temperature.

$$134 - (-80) = 134 + 80 \qquad \text{Use the definition of subtraction.}$$
$$= 214 \qquad \text{Add.}$$

The difference between the two temperatures is 214°F.

◀◀ WORK PROBLEM 5 AT THE SIDE.

1.5 Exercises

Fill in the blank with the correct response.

1. By the definition of subtraction, in order to perform the subtraction problem $-6 - (-8)$, we must add the opposite of _____ to _____ .

2. By the rules for order of operations, to simplify $8 - [3 - (-4 - 5)]$, the first step is to subtract _____ from _____ .

3. "The difference between 7 and 12" translates as _____, while "the difference between 12 and 7" translates as _____ .

4. $-9 - (-3) = -9 +$ _____   **5.** $-8 - 4 = -8 +$ _____   **6.** $-19 - 22 = -19 +$ _____

Find the differences. See Examples 1–3.

7. $-7 - 3$ **8.** $-12 - 5$ **9.** $-10 - 6$ **10.** $-13 - 16$

11. $7 - (-4)$ **12.** $9 - (-6)$ **13.** $6 - (-13)$ **14.** $13 - (-3)$

15. $-7 - (-3)$ **16.** $-8 - (-6)$ **17.** $3 - (4 - 6)$ **18.** $6 - (7 - 14)$

19. $-3 - (6 - 9)$ **20.** $-4 - (5 - 12)$ **21.** $\dfrac{1}{2} - \left(-\dfrac{1}{4}\right)$

22. $\dfrac{1}{3} - \left(-\dfrac{4}{3}\right)$ **23.** $-\dfrac{3}{4} - \dfrac{5}{8}$ **24.** $-\dfrac{5}{6} - \dfrac{1}{2}$

25. $\dfrac{5}{8} - \left(-\dfrac{1}{2} - \dfrac{3}{4}\right)$ **26.** $\dfrac{9}{10} - \left(\dfrac{1}{8} - \dfrac{3}{10}\right)$ **27.** $4.4 - (-9.2)$

28. $6.7 - (-12.6)$ **29.** $-7.4 - 4.5$ **30.** $-5.4 - 9.6$

31. $-5.2 - (8.4 - 10.8)$ **32.** $-9.6 - (3.5 - 12.6)$

33. $[(-3.1) - 4.5] - (.8 - 2.1)$ **34.** $[(-7.8) - 9.3] - (.6 - 3.5)$

35. Explain in your own words how to subtract signed numbers.

36. We know that in general, $a - b \neq b - a$. Can you give values for a and b so that $a - b$ is *equal to* $b - a$? If so, give two such pairs.

Work each problem. See Example 3.

37. $(-3 - 8) - (7 - 4)$ **38.** $(-5 - 9) - (7 - 2)$ **39.** $-10 - [(5 - 4) - (-5 - 8)]$

40. $-12 - [(9 - 2) - (-6 - 3)]$ **41.** $-4 + [(-6 - 9) - (-7 + 4)]$ **42.** $-8 + [(-3 - 10) - (-4 + 1)]$

43. $\left(-\dfrac{3}{4} - \dfrac{5}{2}\right) - \left(-\dfrac{1}{8} - 1\right)$ **44.** $\left(-\dfrac{3}{8} - \dfrac{2}{3}\right) - \left(-\dfrac{9}{8} - 3\right)$

45. $[-34.99 + (6.59 - 12.25)] - 8.33$ **46.** $[-12.25 - (8.34 + 3.57)] - 17.88$

47. Make up a subtraction problem so that the difference between two negative numbers is a negative number.

48. Make up a subtraction problem so that the difference between two negative numbers is a positive number.

Write a numerical expression for each phrase and simplify. See Example 4.

49. The difference between 4 and −8

50. The difference between 7 and −14

51. 8 less than −2

52. 9 less than −13

53. The sum of 9 and −4, decreased by 7

54. The sum of 12 and −7, decreased by 14

55. 12 less than the difference between 8 and −5

56. 19 less than the difference between 9 and −2

Solve each problem. See Example 5.

57. The coldest temperature recorded in Chicago, Illinois, was −27° in 1985. The record low in Huron, South Dakota, was set in 1994 and was 14° lower than −27°. What was the record low in Huron?

58. No one knows just why humpback whales love to heave their 45-ton bodies out of the water, but leap they do. Mark and Debbie, two researchers based on the island of Maui, noticed that one of their favorite whales, "Pineapple," leaped 15 feet above the surface of the ocean while her mate cruised 12 feet below the surface. What is the difference between these two distances?

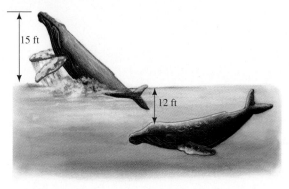

15 ft

12 ft

59. The top of Mount Whitney, visible from Death Valley, has an altitude of 14,494 feet above sea level. The bottom of Death Valley is 282 feet below sea level. Using zero as sea level, find the difference between these two elevations.

60. A chemist is running an experiment under precise conditions. At first, she runs it at $-174.6°F$. She then lowers it by $2.3°F$. What is the new temperature for the experiment?

61. Chris owed his brother $10. He later borrowed $70. What positive or negative number represents his present financial status?

62. Francesca has $15 in her purse, and Emilio has a debt of $12. Find the difference between these amounts.

63. For the year of 1993, one health club showed a profit of $76,000, while another showed a loss of $29,000. Find the difference between these.

64. At 1:00 A.M., a plant worker found that a dial reading was 7.904. At 2:00 A.M., she found the reading to be -3.291. By how much had the reading declined?

The bar graph gives a representation of the number of stolen bases by baseball player Rickey Henderson. Use a signed number to represent the change in the number of stolen bases from one year to the next. For example, from 1990 to 1991, the change was $58 - 65 = -7$.

65. From 1991 to 1992

66. From 1992 to 1993

67. From 1989 to 1990

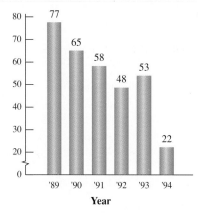

STEALS BY RICKEY HENDERSON

Year

68. From 1993 to 1994

In Exercises 69–72, suppose that a represents a positive number and b represents a negative number. Determine whether the given expression must represent a positive number or a negative number.

69. $a - b$ **70.** $b - a$ **71.** $a + |b|$ **72.** $b - |a|$

1.6 Multiplication of Real Numbers

OBJECTIVES

1 ▸ Find the product of a positive number and a negative number.

2 ▸ Find the product of two negative numbers.

3 ▸ Use the order of operations.

4 ▸ Evaluate expressions that use variables.

5 ▸ Interpret words and phrases that indicate multiplication.

FOR EXTRA HELP

Tutorial   Tape 2   SSM, Sec. 1.6

Multiplication of real numbers is defined to preserve the properties of multiplication of positive numbers that we are familiar with. The result of multiplication is called the **product.** In this section we learn how to multiply with positive and negative numbers. We already know the rule for multiplying two positive numbers. The product of two positive numbers is positive. We also know that the product of 0 and any positive number is 0, so we extend that property to all real numbers.

For any number a,

$$a \cdot 0 = 0.$$

OBJECTIVE 1 In order to define the product of a positive and a negative number so that the result is consistent with the multiplication of two positive numbers, look at the following pattern.

$$3 \cdot 5 = 15$$ The
$$3 \cdot 4 = 12$$ numbers
$$3 \cdot 3 = 9$$ decrease
$$3 \cdot 2 = 6$$ by
$$3 \cdot 1 = 3$$ 3.
$$3 \cdot 0 = 0$$
$$3 \cdot (-1) = ?$$

What should $3(-1)$ equal? The product $3(-1)$ represents the sum

$$-1 + (-1) + (-1) = -3$$

so the product should be -3. Also,

$$3(-2) = -2 + (-2) + (-2) = -6.$$

WORK PROBLEM 1 AT THE SIDE. ▶▶

The results from Problem 1 maintain the pattern in the list above, which suggests the following rule.

> The product of a positive number and a negative number is negative. Example: $6 \cdot -3 = -18$.

E X A M P L E 1   Multiplying a Positive Number and a Negative Number

Find the following products using the multiplication rule given above.

(a) $8(-5) = -(8 \cdot 5) = -40$

(b) $(-7)(2) = -(7 \cdot 2) = -14$

(c) $(-9)\left(\dfrac{1}{3}\right) = -3$

(d) $(-6.2)(4.1) = -25.42$

WORK PROBLEM 2 AT THE SIDE. ▶▶

1. Find each product by finding the sum of three numbers.

(a) $3(-3)$

(b) $3(-4)$

(c) $3(-5)$

2. Find the products.

(a) $2(-6)$

(b) $7(-8)$

(c) $(-9)(2)$

(d) $(-16)\left(\dfrac{5}{32}\right)$

(e) $(4.56)(-10)$

ANSWERS

1. (a) -9 (b) -12 (c) -15
2. (a) -12 (b) -56 (c) -18
   (d) $-\dfrac{5}{2}$ (e) -45.6

3. Find the products.

(a) $(-5)(-6)$

(b) $(-7)(-3)$

(c) $(-8)(-5)$

(d) $(-11)(-2)$

(e) $(-17)(3)(-7)$

(f) $(-41)(2)(-13)$

OBJECTIVE 2 The product of two positive numbers is positive, and the product of a positive number and a negative number is negative. What about the product of two negative numbers? Let's look at another pattern.

$$(-5)(4) = -20 \quad \text{The}$$
$$(-5)(3) = -15 \quad \text{numbers}$$
$$(-5)(2) = -10 \quad \text{increase}$$
$$(-5)(1) = -5 \quad \text{by}$$
$$(-5)(0) = 0 \quad 5.$$
$$(-5)(-1) = ?$$

The numbers on the left of the equals sign (in color) decrease by 1 for each step down the list. The products on the right increase by 5 for each step down the list. To maintain this pattern, $(-5)(-1)$ should be 5 more than $(-5)(0)$, or 5 more than 0, so

$$(-5)(-1) = 5.$$

The pattern continues with

$$(-5)(-2) = 10$$
$$(-5)(-3) = 15$$
$$(-5)(-4) = 20$$
$$(-5)(-5) = 25,$$

and so on. This pattern suggests that we should multiply two negative numbers as follows.

The product of two negative numbers is positive.
Example: $-5 \cdot -4 = 20$.

EXAMPLE 2 Multiplying Two Negative Numbers

Find the products using the multiplication rule given above.

(a) $(-9)(-2) = 18$

(b) $(-6)(-12) = 72$

(c) $(-2)(4)(-1) = (-8)(-1) = 8$

(d) $(3)(-5)(-2) = (-15)(-2) = 30$

◄◄ WORK PROBLEM 3 AT THE SIDE.

Here is a summary of the results for multiplying signed numbers.

Multiplying Signed Numbers

The product of two numbers having the *same* signs is *positive*, and the product of two numbers having *different* signs is *negative*.

OBJECTIVE 3 In the next example, we use the order of operations (discussed in Section 1.1) with positive and negative numbers.

ANSWERS

3. (a) 30 **(b)** 21 **(c)** 40 **(d)** 22 **(e)** 357 **(f)** 1066

EXAMPLE 3 Using the Order of Operations

Simplify.

(a) $(-9)(2) - (-3)(2)$

First find all products, working from left to right.

$$(-9)(2) - (-3)(2) = -18 - (-6)$$

Now perform the subtraction. $-18 - (-6) = -18 + 6 = -12$

(b) $-5(-2 - 3) = -5(-5) = 25$

WORK PROBLEM 4 AT THE SIDE. ▶▶

OBJECTIVE 4 The next examples show numbers substituted for variables where the rules for multiplying signed numbers must be used.

EXAMPLE 4 Evaluating an Expression for Numerical Values

Evaluate $(3x + 4y)(-2m)$ if $x = -1$, $y = -2$, and $m = -3$.

First substitute the given values for the variables. Then find the value of the expression. Put parentheses around the number for each variable.

$(3x + 4y)(-2m)$

$= [3(-1) + 4(-2)][-2(-3)]$ Use parentheses around given values.

$= [-3 + (-8)][6]$ Find the products.

$= (-11)(6)$ Use order of operations.

$= -66$ Multiply.

EXAMPLE 5 Evaluating an Expression for Numerical Values

Evaluate $2x^2 - 3y^2$ if $x = -3$ and $y = -4$.

Use parentheses as shown.

$2(-3)^2 - 3(-4)^2 = 2(9) - 3(16)$ Square -3 and -4.

$= 18 - 48$ Multiply.

$= -30$ Subtract.

WORK PROBLEM 5 AT THE SIDE. ▶▶

OBJECTIVE 5 The word *product* refers to multiplication. In the following chart, we give some other key words and phrases that indicate multiplication.

Word or Phrase	Example	Numerical Expression and Simplification
Product	The *product of* -5 and -2	$(-5)(-2) = 10$
Times	13 *times* -4	$13(-4) = -52$
Twice (meaning "2 times")	*Twice* 6	$2(6) = 12$
Of (used with fractions)	$\frac{1}{2}$ *of* 10	$\frac{1}{2}(10) = 5$
Percent of	12% *of* -16	$.12(-16) = -1.92$

4. Perform the indicated operations.

(a) $(-3)(4) - (2)(6)$

(b) $-7(-2 - 5)$

(c) $-8[-1 - (-4)(-5)]$

5. Evaluate the following expressions.

(a) $2x - 7(y + 1)$
if $x = -4$ and $y = 3$

(b) $(-3x)(4x - 2y)$
if $x = 2$ and $y = -1$

(c) $2x^2 - 4y^2$
if $x = -2$ and $y = -3$

ANSWERS
4. (a) -24 **(b)** 49 **(c)** 168
5. (a) -36 **(b)** -60 **(c)** -28

6. Write a numerical expression for each phrase and simplify.

(a) The product of 6 and the sum of −5 and −4

(b) Twice the difference between 8 and −4

(c) Three-fifths of the sum of 2 and −7

(d) 20% of the sum of 9 and −4

E X A M P L E 6 Interpreting Words and Phrases

Write a numerical expression for each phrase and simplify. Use the order of operations.

(a) The product of 12 and the sum of 3 and −6
Here 12 is multiplied by "the sum of 3 and −6."

$$12[3 + (-6)] = 12(-3) = -36$$

(b) Three times the difference between 4 and −11

$$3(4 - (-11)) = 3(4 + 11) = 3(15) = 45$$

(c) Two-thirds of the sum of −5 and −3

$$\frac{2}{3}[-5 + (-3)] = \frac{2}{3}[-8] = -\frac{16}{3}$$

(d) 15% of the difference between 14 and −2
Remember that 15% = .15.

$$.15[14 - (-2)] = .15(14 + 2) = .15(16) = 2.4$$

◀◀ **WORK PROBLEM 6 AT THE SIDE.**

ANSWERS
6. (a) $6[(-5) + (-4)]$; −54
(b) $2[8 - (-4)]$; 24
(c) $\frac{3}{5}[2 + (-7)]$; −3
(d) $.20[9 + (-4)]$; 1

1.6 Exercises

Fill in the blanks with the correct responses.

1. The product of two negative numbers is a _____ number.

2. The product of a _____ number and a _____ number is a negative number.

3. When the sum of two negative numbers is multiplied by a positive number, the product is a _____ number.

4. When the sum of two positive numbers is multiplied by a negative number, the product is a _____ number.

5. When a negative number is squared, the result is a _____ number.

6. When a negative number is raised to the fifth power, the result is a _____ number.

Find the products. See Examples 1 and 2.

7. $(-7)(4)$ **8.** $(-8)(5)$ **9.** $5(-6)$ **10.** $(-4)(-20)$

11. $(-4)(2)(6)$ **12.** $(-6)(2)(-2)$ **13.** $\left(-\dfrac{3}{8}\right)\left(-\dfrac{20}{9}\right)$ **14.** $\left(-\dfrac{5}{4}\right)\left(-\dfrac{6}{25}\right)$

15. $(-6.8)(.35)$　　　　**16.** $(-4.6)(.24)$　　　　**17.** $(-6)\left(-\dfrac{1}{4}\right)$　　　　**18.** $(-8)\left(-\dfrac{1}{2}\right)$

Perform the indicated operations. See Example 3.

19. $7 - 3 \cdot 6$　　　　　　**20.** $8 - 2 \cdot 5$　　　　　　**21.** $-10 - (-4)(2)$

22. $-11 - (-3)(6)$　　　**23.** $15(8 - 12)$　　　　**24.** $3(4 - 16)$

25. $-7(3 - 8)$　　　　　**26.** $-5(4 - 7)$　　　　**27.** $(12 - 14)(1 - 4)$

28. $(8 - 9)(4 - 12)$　　　**29.** $(7 - 10)(10 - 4)$　　　**30.** $(5 - 12)(19 - 4)$

31. $(-2 - 8)(-6) + 7$　　　**32.** $(-9 - 4)(-2) + 10$　　　**33.** $3(-5) - (-7)$

34. $4(-8) - (-11)$　　　**35.** $(-9 - 3)(-5) - (-4)$　　　**36.** $(-7 - 4)(-9) - (-2)$

37. Explain the method you would use to evaluate $3x + 2y$ if $x = -3$ and $y = 4$.

38. If x and y are both replaced by negative numbers, is the value of $4x + 8y$ positive or negative?

Evaluate the following expressions if $x = 6$, $y = -4$, and $a = 3$. See Examples 4 and 5.

39. $5x - 2y + 3a$

40. $6x - 5y + 4a$

41. $(2x + y)(3a)$

42. $(5x - 2y)(-2a)$

43. $\left(\dfrac{1}{3}x - \dfrac{4}{5}y\right)\left(-\dfrac{1}{5}a\right)$

44. $\left(\dfrac{5}{6}x + \dfrac{3}{2}y\right)\left(-\dfrac{1}{3}a\right)$

45. $(-5 + x)(-3 + y)(3 - a)$

46. $(6 - x)(5 + y)(3 + a)$

47. $-2y^2 + 3a$

48. $5x - 4a^2$

49. $3a^2 - x^2$

50. $4y^2 - 2x^2$

Write a numerical expression for each phrase and simplify. See Example 6.

51. The product of -9 and 2, added to 9

52. The product of 4 and -7, added to -12

53. Twice the product of -1 and 6, subtracted from -4

54. Twice the product of -8 and 2, subtracted from -1

55. Nine subtracted from the product of 7 and -12

56. Three subtracted from the product of -2 and 5

57. The product of 12 and the difference between 9 and -8

58. The product of -3 and the difference between 3 and -7

59. Four-fifths of the sum of -8 and -2

60. Three-tenths of the sum of -2 and -28

Find the solution for each equation by using your knowledge of multiplication of signed numbers.

61. $-2x = -6$ **62.** $-3x = -9$ **63.** $-5x = 10$ **64.** $-2x = 2$

MATHEMATICAL CONNECTIONS (EXERCISES 65–68)

In this section we used a pattern to justify that the product of a positive number and a negative number is a negative number. Work Exercises 65–68 in order to see another way to justify this property.

65. Multiplication of two positive integers can be interpreted as repeated addition. For example, 3×5 can be thought of as using 3 as a term five times: $3 + 3 + 3 + 3 + 3$. If we add five positive numbers what must be the sign of the sum?

66. Multiplication of a negative integer and a positive integer can also be interpreted as repeated addition, with the negative factor being used as the repeated term. For example, -3×5 can be thought of as using -3 as a term five times: $-3 + (-3) + (-3) + (-3) + (-3)$. If we add five negative numbers what must be the sign of the sum?

67. Interpreting -3×5 as repeated addition as shown in Exercise 66, what is the sum when -3 is used as the added term five times?

68. Because there is only one answer to the multiplication problem -3×5, what must this product be, based on your answer to Exercise 67? In general, what can we say about the product of a negative number and a positive number?

1.7 Division of Real Numbers

OBJECTIVE 1 The difference between two numbers is found by adding the opposite of the second number to the first. Division is related to multiplication in a similar way. The *quotient* of two numbers is found by *multiplying* by the *reciprocal*. By definition, since

$$8 \cdot \frac{1}{8} = \frac{8}{8} = 1 \quad \text{and} \quad \frac{5}{4} \cdot \frac{4}{5} = \frac{20}{20} = 1,$$

the reciprocal of 8 is $\frac{1}{8}$, and that of $\frac{5}{4}$ is $\frac{4}{5}$.

> Pairs of numbers whose product is 1 are called **reciprocals** of each other.

E X A M P L E I Finding the Reciprocal

The following chart shows several numbers and the reciprocal (if it exists) of each number.

Number	Reciprocal
4	$\frac{1}{4}$
-5	$\frac{1}{-5}$ or $-\frac{1}{5}$
$\frac{3}{4}$	$\frac{4}{3}$
$-\frac{5}{8}$	$-\frac{8}{5}$
0	None

Why is there no reciprocal for the number 0? Suppose that k is to be the reciprocal of 0. Then $k \cdot 0$ should equal 1. But $k \cdot 0 = 0$ for any number k. Because there is no value of k that is a solution of the equation $k \cdot 0 = 1$, the following statement can be made.

> 0 has no reciprocal.

WORK PROBLEM I AT THE SIDE. ▶▶

OBJECTIVE 2 By definition, the *quotient* of a and b is the product of a and the reciprocal of b.

> **Definition of Division**
>
> For any real numbers a and b, with $b \neq 0$,
>
> $$\frac{a}{b} = a \cdot \frac{1}{b}.$$

The definition above indicates that b, the number to divide by, cannot be 0. The reason is that 0 has no reciprocal, so $\frac{1}{0}$ is not a number. Thus any quotient with a zero denominator is *undefined*—that is, it does not satisfy the definition.

> Division by 0 is undefined and is never permitted.

OBJECTIVES

1 Find the reciprocal of a number.
2 Divide with signed numbers.
3 Simplify numerical expressions.
4 Interpret words and phrases that indicate division.
5 Translate simple sentences into equations.

FOR EXTRA HELP

Tutorial   Tape 3   SSM, Sec. 1.7

1. Complete the chart.

Number	Reciprocal
(a) 6	
(b) -2	
(c) $\frac{2}{3}$	
(d) $-\frac{1}{4}$	
(e) 0	

ANSWERS

1. (a) $\frac{1}{6}$ (b) $-\frac{1}{2}$ or $\frac{1}{-2}$
(c) $\frac{3}{2}$ (d) -4 (e) none

49

2. Find the quotients.

(a) $\dfrac{42}{7}$

(b) $\dfrac{-36}{(-2)(-3)}$

(c) $\dfrac{-12.56}{-.4}$

(d) $\dfrac{10}{7} \div -\dfrac{24}{5}$

(e) $\dfrac{-3}{0}$

> **Note**
> If a division problem turns out to involve division by 0, we indicate by writing "undefined."

Because division is defined in terms of multiplication, all the rules for multiplication of signed numbers also apply to division.

E X A M P L E 2   Using the Definition of Division

Write each quotient as a product and evaluate.

(a) $\dfrac{12}{3} = 12 \cdot \dfrac{1}{3} = 4$

(b) $\dfrac{(5)(-2)}{2} = \dfrac{-10}{2} = -10 \cdot \dfrac{1}{2} = -5$

(c) $\dfrac{(-1)(-8)}{-4} = \dfrac{8}{-4} = 8 \cdot \left(\dfrac{1}{-4}\right) = -2$

(d) $\dfrac{-1.47}{-7} = -1.47\left(-\dfrac{1}{7}\right) = .21$

(e) $-\dfrac{2}{3} \div -\dfrac{5}{4} = -\dfrac{2}{3} \cdot -\dfrac{4}{5} = \dfrac{8}{15}$

(f) $\dfrac{-10}{0}$ Undefined

◀◀ **WORK PROBLEM 2 AT THE SIDE.**

Multiplying by the reciprocal to divide integers is awkward. It is easier to divide in the usual way. On the other hand, to divide by a fraction, it is easier to multiply by the reciprocal. The following rule for division can be used instead of multiplying by the reciprocal.

> **Dividing Signed Numbers**
> The quotient of two numbers having the *same* sign is *positive;* the quotient of two numbers having *different* signs is *negative.*
> Examples: $\dfrac{-15}{-5} = 3$ and $\dfrac{-15}{5} = -3$.

E X A M P L E 3   Dividing Signed Numbers

Find the quotients.

(a) $\dfrac{8}{-2} = -4$

(b) $\dfrac{-4.5}{-.09} = 50$

(c) $-\dfrac{1}{8} \div \left(-\dfrac{3}{4}\right) = -\dfrac{1}{8} \cdot \left(-\dfrac{4}{3}\right) = \dfrac{1}{6}$

ANSWERS

2. (a) 6 (b) −6 (c) 31.4
(d) $-\dfrac{25}{84}$ (e) undefined

WORK PROBLEM 3 AT THE SIDE. ▶▶

From the definitions of multiplication and division of real numbers,

$$\frac{-40}{8} = -40 \cdot \frac{1}{8} = -5$$

and

$$\frac{40}{-8} = 40\left(\frac{1}{-8}\right) = -5$$

so

$$\frac{-40}{8} = \frac{40}{-8}.$$

Based on this example, the quotient of a positive and a negative number can be written in any of the following three forms.

For any positive real numbers a and b,

$$\frac{-a}{b} = \frac{a}{-b} = -\frac{a}{b}.$$

The form $\frac{a}{-b}$ is seldom used.

Similarly, the quotient of two negative numbers can be expressed as the quotient of two positive numbers.

For any positive real numbers a and b,

$$\frac{-a}{-b} = \frac{a}{b}.$$

The rules for operations with signed numbers are summarized here.

Operations with Signed Numbers

Addition

Like signs   Add the absolute values of the numbers. The result is given the same sign as the numbers.
Unlike signs   Subtract the smaller absolute value from the larger absolute value. Give the result the sign of the number having the larger absolute value.

Subtraction

Add the opposite of the second number to the first number, using the addition rules.

Multiplication and Division

Like signs   The product or quotient of two numbers with like signs is positive.
Unlike signs   The product or quotient of two numbers with unlike signs is negative.
Division by 0 is undefined.

3. Find the quotients.

(a) $\dfrac{-8}{-2}$

(b) $\dfrac{-16.4}{2.05}$

(c) $\dfrac{1}{4} \div \left(-\dfrac{2}{3}\right)$

ANSWERS

3. (a) 4 **(b)** -8 **(c)** $-\dfrac{3}{8}$

4. Perform the indicated operations.

(a) $\dfrac{6(-4) - 2(5)}{3(2 - 7)}$

OBJECTIVE 3 In the next example, we simplify numerical expressions involving quotients.

E X A M P L E   4   Simplifying Expressions Involving Division

Simplify each expression.

(a) $\dfrac{5(-2) - (3)(4)}{2(1 - 6)}$

Follow the order of operations. Simplify the numerator and denominator separately. Then divide or write in lowest terms.

$$\dfrac{5(-2) - (3)(4)}{2(1 - 6)} = \dfrac{-10 - 12}{2(-5)} \qquad \begin{array}{l}\text{Multiply in numerator.}\\ \text{Subtract in denominator.}\end{array}$$

$$= \dfrac{-22}{-10} \qquad \begin{array}{l}\text{Subtract in numerator.}\\ \text{Multiply in denominator.}\end{array}$$

$$= \dfrac{11}{5} \qquad \text{Express in lowest terms.}$$

(b) $\dfrac{-6(-8) + (-3)9}{(-2)[4 - (-3)]}$

(b) $\dfrac{-3^2 - 6^2}{5(-3 + 2)}$

$$\dfrac{-3^2 - 6^2}{5(-3 + 2)} = \dfrac{-9 - 36}{5(-1)} \qquad \begin{array}{l}\text{Square 3 and 6.}\\ \text{Add } -3 \text{ and 2.}\end{array}$$

$$= \dfrac{-45}{-5} \qquad \begin{array}{l}\text{Subtract in numerator.}\\ \text{Multiply in denominator.}\end{array}$$

$$= 9 \qquad \text{Divide.}$$

(c) $\dfrac{5^2 + 3^2}{3(-4) - 5}$

◀◀ WORK PROBLEM 4 AT THE SIDE.

OBJECTIVE 4 The word *quotient* refers to the result obtained in a division problem. In algebra, quotients are usually represented with a fraction bar. The symbol ÷ is seldom used.

The following chart gives some key words and phrases associated with division.

Word or Phrase	Example	Numerical Expression and Simplification
Quotient	The *quotient of* −24 and 3	$\frac{-24}{3} = -8$
Divided by	−16 *divided by* −4	$\frac{-16}{-4} = 4$

It is customary to write the first number named as the numerator and the second as the denominator when interpreting a phrase involving division.

ANSWERS

4. (a) $\dfrac{34}{15}$ (b) $-\dfrac{3}{2}$ (c) -2

┌─
E X A M P L E 5   Interpreting Words and Phrases

Write a numerical expression for each phrase, and simplify the expression.

(a) The **quotient** of 14 and the sum of -9 and 2

"Quotient" indicates division. The number 14 is the numerator and "the sum of -9 and 2" is the denominator.

$$\frac{14}{-9 + 2} = \frac{14}{-7} = -2$$

(b) The product of 5 and -6, **divided by** the difference between -7 and 8

The numerator of the fraction representing the division is obtained by multiplying 5 and -6. The denominator is found by subtracting -7 and 8.

$$\frac{5(-6)}{-7 - 8} = \frac{-30}{-15} = 2$$
└─

WORK PROBLEM 5 AT THE SIDE. ▶▶

OBJECTIVE 5 In this section and the preceding three sections, important words and phrases involving the four operations of arithmetic have been introduced. We can use these words and phrases to interpret sentences that translate into equations. The ability to do this will help us to solve the types of applied problems found in Section 2.4.

┌─
E X A M P L E 6   Translating a Sentence into an Equation

Write the following in symbols, using x as the variable.

(a) Three **times** a number **is** -18.

The word *times* indicates multiplication, and the word *is* translates as the equals sign ($=$).

$$3x = -18$$

(b) The **sum** of a number and 9 **is** 12.

$$x + 9 = 12$$

(c) The **difference between** a number and 5 **is** 0.

$$x - 5 = 0$$

(d) The **quotient of** 24 and a number **is** -2.

$$\frac{24}{x} = -2$$
└─

WORK PROBLEM 6 AT THE SIDE. ▶▶

Caution

It is important to recognize the distinction between the types of problems found in Example 5 and Example 6. In Example 5, the phrases translate as *expressions,* while in Example 6, the sentences translate as *equations.* Remember that an equation is a sentence, while an expression is a phrase. For example,

$$\frac{5(-6)}{-7 - 8} \text{ is an expression,}$$

$$3x = -18 \text{ is an equation.}$$

---

5. Write a numerical expression for each phrase, and simplify the expression.

(a) The quotient of 20 and the sum of 8 and -3

(b) The product of -9 and 2, divided by the difference between 5 and -1

6. Write the following in symbols, using x as the variable.

(a) Twice a number is -6.

(b) The difference between -8 and a number is -11.

(c) The sum of 5 and a number is 8.

(d) The quotient of a number and -2 is 6.

ANSWERS

5. (a) $\dfrac{20}{8 + (-3)}$; 4

   (b) $\dfrac{(-9)(2)}{5 - (-1)}$; -3

6. (a) $2x = -6$
   (b) $-8 - x = -11$
   (c) $5 + x = 8$
   (d) $\dfrac{x}{-2} = 6$

NUMBERS IN THE
Real World collaborative investigations

National League

East Division

	W	L	Pct.	GB
Atlanta	76	44	.633	—
Phila.	61	60	.504	14½
Montreal	58	62	.483	18
Florida	54	64	.458	21
New York	52	67	.437	23½

Central Division

	W	L	Pct.	GB
Cincinnati	74	45	.622	—
Chicago	61	59	.508	13½
Houston	61	59	.508	13½
Pittsburgh	51	69	.425	23½
St. Louis	50	71	.413	25

West Division

	W	L	Pct.	GB
Los Angeles	63	58	.521	—
Colorado	62	58	.517	½
San Diego	59	60	.496	2½
San Fran.	57	63	.475	5

American League

East Division

	W	L	Pct.	GB
Boston	74	45	.622	—
New York	60	60	.500	14½
Baltimore	56	65	.463	19
Toronto	50	70	.417	24½
Detroit	50	70	.417	24½

Central Division

	W	L	Pct.	GB
Cleveland	82	37	.689	—
Kansas City	61	58	.513	21
Milwaukee	59	61	.492	23½
Chicago	54	64	.458	27½
Minnesota	45	74	.378	37

West Division

	W	L	Pct.	GB
California	68	54	.557	—
Seattle	61	59	.508	6
Texas	60	60	.500	7
Oakland	59	62	.488	8½

The Magic Number in Sports

Near the end of a major league baseball season, fans are often interested in the current first-place team's **magic number.** The magic number is the sum of the required number of wins of the first-place team and the number of losses of the second-place team for the remaining games necessary to clinch the pennant. (In a regulation major league season, each team plays 162 games.)

To calculate the magic number M for a first-place team prior to the end of a season, we can use the formula

$$M = W_2 + N_2 - W_1 + 1, \text{ where}$$

W_2 = the current number of wins of the second-place team;
N_2 = the number of remaining games of the second-place team; and
W_1 = the current number of wins of the first-place team.

On Wednesday, September 6, 1995, baseball fans woke up to the standings seen at the left.

In 1995, due to the strike that shortened the season, each team played 144 games rather than the usual 162 games. To calculate Atlanta's magic number, we note that $W_2 = 61$ (the number of wins for Philadelphia), $N_2 = 144 - (61 + 60) = 23$ (the number of games Philadelphia had remaining), and $W_1 = 76$ (the number of wins Atlanta had). Therefore, the magic number for Atlanta was

$$M = 61 + 23 - 76 + 1 = 9.$$

Later in the season, when the total of Atlanta wins and Philadelphia losses became 9, Atlanta clinched the pennant.

FOR GROUP DISCUSSION

Divide the class into groups, and have each group choose one of the following teams: Boston, Cleveland, California, or Cincinnati. Determine the magic number for the team chosen. (*Note:* California "folded" and Seattle won the West Division of the American League!)

1.7 Exercises

Decide whether each of the following statements is true or false.

1. The quotient of two numbers with the same sign is positive.

2. The quotient of two numbers with unlike signs is negative.

3. If two negative numbers are multiplied and their product is then divided by a negative number, the result is greater than zero.

4. If a positive number is multiplied by a negative number and their product is then divided by a positive number, the result is greater than zero.

5. The reciprocal of a positive number is negative.

6. The reciprocal of a negative number is negative.

Find the reciprocal, if one exists, for each number. See Example 1.

7. 11

8. 12

9. -5

10. -3

11. $\dfrac{5}{6}$

12. $\dfrac{9}{10}$

13. $3 - 3$

14. $5 + (-5)$

15. $-\dfrac{8}{7}$

16. $-\dfrac{9}{7}$

17. $.4$

18. $.50$

19. Which one of the following expressions is undefined?

(a) $\dfrac{5 - 5}{5 + 5}$ (b) $\dfrac{5 + 5}{5 + 5}$ (c) $\dfrac{5 - 5}{5 - 5}$ (d) $\dfrac{5 - 5}{5}$

20. Explain why 0 has no reciprocal.

Find the quotients. See Examples 2 and 3.

21. $\dfrac{-15}{5}$

22. $\dfrac{-18}{6}$

23. $\dfrac{20}{-10}$

24. $\dfrac{28}{-4}$

25. $\dfrac{-160}{-10}$

26. $\dfrac{-260}{-20}$

27. $\dfrac{0}{-3}$

28. $\dfrac{0}{-6}$

29. $\dfrac{-10.252}{-.4}$

30. $\dfrac{-29.584}{-.8}$

31. $\left(-\dfrac{3}{4}\right) \div \left(-\dfrac{1}{2}\right)$

32. $\left(-\dfrac{3}{16}\right) \div \left(-\dfrac{5}{8}\right)$

33. $(-6.8) \div (-2)$

34. $(-23.5) \div (-5)$

35. $\dfrac{18}{3-9}$

36. $\dfrac{21}{4-7}$

37. $\dfrac{-50}{7-(-3)}$

38. $\dfrac{-36}{4-(-5)}$

39. $\dfrac{-12-36}{-12}$

40. $\dfrac{-18-2}{-5}$

Simplify the numerators and denominators separately. Then find the quotients. See Example 4.

41. $\dfrac{-5(-6)}{9-(-1)}$

42. $\dfrac{-12(-5)}{7-(-5)}$

43. $\dfrac{-21(3)}{-3-6}$

44. $\dfrac{-40(3)}{-2-3}$

45. $\dfrac{-10(2)+6(2)}{-3-(-1)}$

46. $\dfrac{8(-1)+6(-2)}{-6-(-1)}$

47. $\dfrac{-27(-2)-(-12)(-2)}{-2(3)-2(2)}$

48. $\dfrac{-13(-4)-(-8)(-2)}{(-10)(2)-4(-2)}$

49. $\dfrac{1^2+4^2}{3^2+5^2}$

50. $\dfrac{3^2+4^2}{6^2+8^2}$

51. $\dfrac{2^2-8^2}{6(-4+3)}$

52. $\dfrac{3^2-4^2}{7(-8+9)}$

Write a numerical expression for each phrase, and simplify the expression. See Example 5.

53. The quotient of -36 and -9

54. The quotient of -48 and -6

55. The quotient of -12 and the sum of -5 and -1

56. The quotient of -20 and the sum of -8 and -2

57. The sum of 15 and -3, divided by the product of 4 and -3

58. The sum of -18 and -6, divided by the product of 2 and -4

59. The product of -34 and 7, divided by -14

60. The product of -25 and 4, divided by -10

Write the following sentences in symbols, using x as the variable. See Example 6.

61. Six times a number is -42.

62. Four times a number is -36.

63. The quotient of a number and 3 is -3.

64. The quotient of a number and 4 is -1.

65. 6 is 2 less than a number.

66. 7 is 5 less than a number.

67. When 15 is divided by a number, the result is -5.

68. When 6 is divided by a number, the result is -3.

The operation of division plays an important role in divisibility tests. By using a divisibility test, we can determine whether a given natural number is divisible (without remainder) by another natural number. For example, an integer is divisible by 2 if its last digit is divisible by 2, and not otherwise. Exercises 69–73 introduce some of the simpler divisibility tests.

69. Tell why   **(a)** 3,473,986 is divisible by 2 and   **(b)** 4,336,879 is not divisible by 2.

70. An integer is divisible by 3 if the sum of its digits is divisible by 3, and not otherwise. Show that   **(a)** 4,799,232 is divisible by 3 and   **(b)** 2,443,871 is not divisible by 3.

71. An integer is divisible by 4 if its last two digits form a number divisible by 4, and not otherwise. Show that   **(a)** 6,221,464 is divisible by 4 and   **(b)** 2,876,335 is not divisible by 4.

72. An integer is divisible by 5 if its last digit is divisible by 5, and not otherwise. Show that **(a)** 3,774,595 is divisible by 5 and   **(b)** 9,332,123 is not divisible by 5.

73. An integer is divisible by 6 if it is divisible by both 2 and 3, and not otherwise. Show that **(a)** 1,524,822 is divisible by 6 and   **(b)** 2,873,590 is not divisible by 6.

1.8 Properties of Addition and Multiplication

If you are asked to find the sum

$$3 + 89 + 97$$

you might mentally add $3 + 97$ to get 100, and then add $100 + 89$ to get 189. While the rule for order of operations says to add from left to right, it is a fact that we may change the order of the terms and group them in any way we choose without affecting the sum. These are examples of shortcuts that we use in everyday mathematics. These shortcuts are justified by the basic properties of addition and multiplication, which are discussed in this section. In the following statements, a, b, and c represent real numbers.

OBJECTIVE 1 **Commutative** (cuh-MEW-tuh-tiv) **properties** The word *commute* means to go back and forth. Many people commute to work or to school. If you travel from home to work and follow the same route from work to home, you travel the same distance each time. The commutative properties say that if two numbers are added or multiplied in any order, they give the same result.

$$a + b = b + a$$
$$ab = ba$$

E X A M P L E 1 Using the Commutative Properties

Use a commutative property to complete each statement.

(a) $-8 + 5 = 5 +$ _____

By the commutative property for addition, the missing number is -8 because $-8 + 5 = 5 + (-8)$.

(b) $(-2)(7) =$ _____ (-2)

By the commutative property for multiplication, the missing number is 7, since $(-2)(7) = (7)(-2)$.

WORK PROBLEM 1 AT THE SIDE. ▶▶

OBJECTIVE 2 **Associative** (uh-SOH-shi-a-tiv) **properties** When we *associate* one object with another, we tend to think of those objects as being grouped together. The associative properties say that when we add or multiply three numbers, we can group them in any manner and get the same answer.

$$(a + b) + c = a + (b + c)$$
$$(ab)c = a(bc)$$

E X A M P L E 2 Using the Associative Properties

Use an associative property to complete each statement.

(a) $8 + (-1 + 4) = (8 +$ _____ $) + 4$

The missing number is -1.

(b) $[2 \cdot (-7)] \cdot 6 = 2 \cdot$ _____

The completed expression on the right should be $2 \cdot [(-7) \cdot 6]$.

WORK PROBLEM 2 AT THE SIDE. ▶▶

OBJECTIVES

1 ▶ Use the commutative properties.
2 ▶ Use the associative properties.
3 ▶ Use the identity properties.
4 ▶ Use the inverse properties.
5 ▶ Use the distributive property.

FOR EXTRA HELP

Tutorial   Tape 3   SSM, Sec. 1.8

1. Complete each statement. Use a commutative property.

   (a) $x + 9 = 9 +$ _____

   (b) $(-12)(4) =$ _____ (-12)

   (c) $5x = x \cdot$ _____

2. Complete each statement. Use an associative property.

   (a) $(9 + 10) + (-3)$
   $= 9 + [$_____$ + (-3)]$

   (b) $-5 + (2 + 8)$
   $= ($_____$) + 8$

   (c) $10 \cdot [(-8) \cdot (-3)]$
   $=$ _____

ANSWERS

1. **(a)** x **(b)** 4 **(c)** 5
2. **(a)** 10 **(b)** $-5 + 2$
   (c) $[10 \cdot (-8)] \cdot (-3)$

59

3. Decide whether each statement is an example of a commutative property, an associative property, or both.

(a) $2(4 \cdot 6) = (2 \cdot 4)6$

(b) $(2 \cdot 4)6 = (4 \cdot 2)6$

(c) $(2 + 4) + 6 = 4 + (2 + 6)$

By the associative property of addition, the sum of three numbers will be the same no matter which way the numbers are "associated" in groups. For this reason, parentheses can be left out in many addition problems. For example, both

$$(-1 + 2) + 3 \quad \text{and} \quad -1 + (2 + 3)$$

can be written as

$$-1 + 2 + 3.$$

In the same way, parentheses also can be left out of many multiplication problems.

E X A M P L E 3   **Distinguishing between the Associative and Commutative Properties**

(a) Is $(2 + 4) + 5 = 2 + (4 + 5)$ an example of the associative property?

The order of three numbers is the same on both sides of the equals sign. The only change is in the grouping, or association, of the numbers. Therefore, this is an example of the associative property.

(b) Is $6(3 \cdot 10) = 6(10 \cdot 3)$ an example of the associative property or the commutative property?

The same numbers, 3 and 10, are grouped on each side. On the left, however, the 3 appears first in $(3 \cdot 10)$. On the right, the 10 appears first. Since the only change involves the order of the numbers, this statement is an example of the commutative property.

(c) Is $(8 + 1) + 7 = 8 + (7 + 1)$ an example of the associative property or the commutative property?

In the statement, both the order and the grouping are changed. On the left the order of the three numbers is 8, 1, and 7. On the right it is 8, 7, and 1. On the left the 8 and 1 are grouped, and on the right the 7 and 1 are grouped. Therefore, both the associative and the commutative properties are used.

◀◀ **WORK PROBLEM 3 AT THE SIDE.**

We can use the commutative and associative properties to simplify expressions.

E X A M P L E 4   **Using the Commutative and Associative Properties to Simplify Expressions**

Simplify $16 + 2x + 5$.

By the order of operations, we should add from left to right, but we can't combine 16 and $2x$ into a single term. The commutative and associative properties are used to group 16 and 5, so we can add them.

$(16 + 2x) + 5$	Order of operations
$(2x + 16) + 5$	Commutative property
$2x + (16 + 5)$	Associative property
$2x + 21$	Add.

We don't usually show all these steps, we just add. But this example shows how the properties make it possible.

WORK PROBLEM 4 AT THE SIDE. ▶▶

4. Simplify $8 + 4y + 10$.

OBJECTIVE **3** **Identity properties** If a child wears a costume to masquerade on Halloween, the child's appearance is changed, but his or her *identity* is unchanged. The identity of a real number is left unchanged when identity properties are applied. The identity properties say that the sum of 0 and any number equals that number, and the product of 1 and any number equals that number.

$$a + 0 = a \quad \text{and} \quad 0 + a = a$$
$$a \cdot 1 = a \quad \text{and} \quad 1 \cdot a = a$$

The number 0 leaves the identity, or value, of any real number unchanged by addition. For this reason, 0 is called the **identity element for addition.** Since multiplication by 1 leaves any real number unchanged, 1 is the **identity element for multiplication.**

5. Use an identity property to complete each statement.

(a) $9 + 0 =$ _____

E X A M P L E 5   Using the Identity Properties

These statements are examples of the identity properties.

(a) $-3 + 0 = -3$ **(b)** $1 \cdot 25 = 25$

WORK PROBLEM 5 AT THE SIDE. ▶▶

(b) _____ $+ (-7) = -7$

We use the identity property for multiplication to write fractions in lowest terms and to get common denominators.

E X A M P L E 6   Using the Identity Property of 1 to Simplify Expressions

Simplify the following expressions.

(a) $\dfrac{49}{35}$

$$\dfrac{49}{35} = \dfrac{7 \cdot 7}{5 \cdot 7} \qquad \text{Factor.}$$

$$= \dfrac{7}{5} \cdot \dfrac{7}{7} \qquad \text{Write as a product.}$$

$$= \dfrac{7}{5} \cdot 1 \qquad \text{Property of 1}$$

$$= \dfrac{7}{5} \qquad \text{Identity property}$$

(c) $\dfrac{1}{4} \cdot$ _____ $= \dfrac{3}{12}$

(b) $\dfrac{3}{4} + \dfrac{5}{24}$

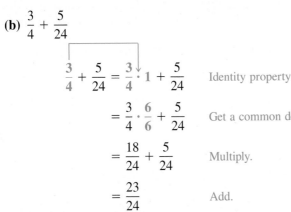

$$\dfrac{3}{4} + \dfrac{5}{24} = \dfrac{3}{4} \cdot 1 + \dfrac{5}{24} \qquad \text{Identity property}$$

$$= \dfrac{3}{4} \cdot \dfrac{6}{6} + \dfrac{5}{24} \qquad \text{Get a common denominator.}$$

$$= \dfrac{18}{24} + \dfrac{5}{24} \qquad \text{Multiply.}$$

$$= \dfrac{23}{24} \qquad \text{Add.}$$

(d) _____ $\cdot 1 = 5$

6. Use the identity property to simplify each expression.

(a) $\dfrac{85}{105}$

(b) $\dfrac{9}{10} - \dfrac{53}{50}$

7. Complete the statements so that they are examples of either an identity property or an inverse property. Tell which property is used.

(a) $-6 + $ _____ $= 0$

(b) $\dfrac{4}{3} \cdot$ _____ $= 1$

(c) $-\dfrac{1}{9} \cdot$ _____ $= 1$

(d) $275 + $ _____ $= 275$

8. Simplify.

(a) $5m - 3 - 5m$

(b) $\left(\dfrac{4}{3}\right)(-7)\left(\dfrac{3}{4}\right)$

◀◀ WORK PROBLEM 6 AT THE SIDE.

OBJECTIVE 4 Inverse properties Each day before you go to work or school, you probably put on your shoes before you leave. Before you go to sleep at night, you probably take them off, and this leads to the same situation that existed before you put them on. These operations from everyday life are examples of inverse operations. The inverse properties of addition and multiplication lead to the additive and multiplicative identities, respectively. The opposite of a, $-a$, is the **additive inverse** of a and the reciprocal of a, $\frac{1}{a}$, is the **multiplicative inverse** of the nonzero number a. The sum of the numbers a and $-a$ is 0, and the product of the nonzero numbers a and $\frac{1}{a}$ is 1.

$$a + (-a) = 0 \quad \text{and} \quad -a + a = 0$$

$$a \cdot \frac{1}{a} = 1 \quad \text{and} \quad \frac{1}{a} \cdot a = 1 \quad (a \neq 0)$$

E X A M P L E 7   Using the Inverse Properties

The following statements are examples of the inverse properties.

(a) $\dfrac{2}{3} \cdot \dfrac{3}{2} = 1$

(b) $(-5)\left(-\dfrac{1}{5}\right) = 1$

(c) $-\dfrac{1}{2} + \dfrac{1}{2} = 0$

(d) $4 + (-4) = 0$

◀◀ WORK PROBLEM 7 AT THE SIDE.

E X A M P L E 8   Using the Additive Inverse Property to Simplify an Expression

Simplify $-2x + 10 + 2x$.

$$
\begin{aligned}
-2x + 10 + 2x &= (-2x + 10) + 2x & \text{Order of operations} \\
&= [10 + (-2x)] + 2x & \text{Commutative property} \\
&= 10 + [(-2x) + 2x] & \text{Associative property} \\
&= 10 + 0 & \text{Inverse property} \\
&= 10 & \text{Identity property}
\end{aligned}
$$

Again, although this example shows how we can use the properties to change the order of operations, in practice we would not show all the steps.

◀◀ WORK PROBLEM 8 AT THE SIDE.

OBJECTIVE 5 Distributive property The everyday meaning of the word *distribute* is "to give out from one to several." An important property of real number operations involves this idea.

ANSWERS

6. (a)  $\dfrac{17}{21}$ (b) $-\dfrac{4}{25}$

7. (a) 6; inverse (b) $\dfrac{3}{4}$; inverse

(c) -9; inverse (d) 0; identity

8. (a) -3 (b) -7

Look at the following statements.

$$2(5 + 8) = 2(13) = 26$$
$$2(5) + 2(8) = 10 + 16 = 26$$

Since both expressions equal 26,

$$2(5 + 8) = 2(5) + 2(8).$$

This result is an example of the *distributive property,* the only property involving *both* addition and multiplication. With this property, a product can be changed to a sum or difference. This idea is illustrated by the divided rectangle in Figure 13.

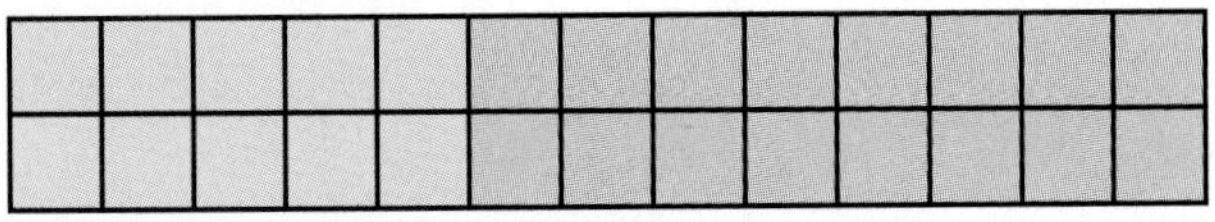

Area of the left part is 2(5) = 10
Area of the right part is 2(8) = 16
Area of total is 2(13) = 2(5 + 8) = 26
2(5 + 8) = 2(5) + 2(8)

FIGURE 13

The distributive property says that multiplying a number a by a sum of numbers $b + c$ gives the same result as multiplying a by b and a by c and then adding the two products.

$$a(b + c) = ab + ac \quad \text{and} \quad (b + c)a = ba + ca$$

As the arrows show, the a outside the parentheses is "distributed" over the b and c inside. Another form of the distributive property is valid for subtraction.

$$a(b - c) = ab - ac \quad \text{and} \quad (b - c)a = ba - ca$$

The distributive property also can be extended to more than two numbers.

$$a(b + c + d) = ab + ac + ad$$

E X A M P L E   9   Using the Distributive Property

Use the distributive property to rewrite each expression.

(a) $5(9 + 6) = 5 \cdot 9 + 5 \cdot 6$ Multiply both terms by 5.

$\qquad\qquad = 45 + 30$ Multiply.

$\qquad\qquad = 75$ Add.

(b) $4(x + 5 + y) = 4x + 4 \cdot 5 + 4y$ Distributive property

$\qquad\qquad\qquad = 4x + 20 + 4y$ Multiply.

(c) $-2(x + 3) = -2x + (-2)(3)$ Distributive property

$\qquad\qquad = -2x - 6$ Multiply.

(d) $3(k - 9) = 3k - 3 \cdot 9$ Distributive property

$\qquad\qquad = 3k - 27$ Multiply.

CONTINUED ON NEXT PAGE

9. Use the distributive property to rewrite each expression.

(a) $2(p + 5)$

(b) $-4(y + 7)$

(c) $5(m - 4)$

(d) $9 \cdot k + 9 \cdot 5$

(e) $3a - 3b$

(f) $7(2y + 7k - 9m)$

10. Write without parentheses.

(a) $-(3k - 5)$

(b) $-(2 - r)$

(c) $-(-5y + 8)$

(d) $-(-z + 4)$

ANSWERS
9. **(a)** $2p + 10$
   (b) $-4y - 28$ **(c)** $5m - 20$
   (d) $9(k + 5)$ **(e)** $3(a - b)$
   (f) $14y + 49k - 63m$
10. **(a)** $-3k + 5$ **(b)** $-2 + r$
    (c) $5y - 8$ **(d)** $z - 4$

(e) $6 \cdot 8 + 6 \cdot 2$

We can use the distributive property in reverse here. Since each term has a factor of 6,

$$6 \cdot 8 + 6 \cdot 2 = 6(8 + 2) \qquad \text{Distributive property}$$
$$= 6(10) = 60. \qquad \text{Add, then multiply.}$$

(f) $4x - 4m = 4(x - m)$ Distributive property

(g) $8(3r + 11t + 5z) = 8(3r) + 8(11t) + 8(5z)$ Distributive property
$$= (8 \cdot 3)r + (8 \cdot 11)t + (8 \cdot 5)z \qquad \text{Associative property}$$
$$= 24r + 88t + 40z$$

When the distributive property is used in reverse as in Examples 9(e) and (f), the process is called **factoring.**

◄◄ **WORK PROBLEM 9 AT THE SIDE.**

The distributive property is used to remove the parentheses from expressions such as $-(2y + 3)$. We do this by first writing $-(2y + 3)$ as $-1 \cdot (2y + 3)$.

$$-(2y + 3) = -1 \cdot (2y + 3) \qquad \text{Identity property}$$
$$= -1 \cdot (2y) + (-1) \cdot (3) \qquad \text{Distributive property}$$
$$= -2y - 3 \qquad \text{Multiply.}$$

E X A M P L E 10   **Using the Distributive Property to Remove Parentheses**

Write without parentheses.

(a) $-(7r - 8) = -1(7r) + (-1)(-8)$ Distributive property
$$= -7r + 8 \qquad \text{Multiply.}$$

(b) $-(-9w + 2) = -1(-9w + 2) = 9w - 2$

◄◄ **WORK PROBLEM 10 AT THE SIDE.**

The identity property is often used with the distributive property to simplify expressions.

E X A M P L E 11   **Using the Identity Property with the Distributive Property**

Simplify each expression.

(a) $12p - p$

$$12p - p = 12p - 1p \qquad \text{Identity property}$$
$$= (12 - 1)p \qquad \text{Distributive property}$$
$$= 11p \qquad \text{Subtract.}$$

(b) $y + y$

$$y + y = 1y + 1y \qquad \text{Identity property}$$
$$= (1 + 1)y \qquad \text{Distributive property}$$
$$= 2y \qquad \text{Add.}$$

35. $-6(x + 7)$; distributive

36. $-5(y + 2)$; distributive

37. $(w + 5) + (-3)$; associative

38. $(b + 8) + (-10)$; associative

Use the properties of this section to simplify the following expressions. See Examples 4 and 8.

39. $9 + 3x + 7$

40. $-8 + 5t + 10$

41. $6t + 8 - 6t + 3$

42. $9r + 12 - 9r + 1$

43. $-3w + 7 + 3w$

44. $-5t + 3 + 5t$

45. $\frac{2}{3}x - 11 + 11 - \frac{2}{3}x$

46. $\frac{1}{5}y + 4 - 4 - \frac{1}{5}y$

47. $\left(\frac{9}{7}\right)(-.38)\left(\frac{7}{9}\right)$

48. $\left(\frac{4}{5}\right)(-.73)\left(\frac{5}{4}\right)$

49. $t + (-t) + \frac{1}{2}(2)$

50. $w + (-w) + \frac{1}{4}(4)$

Use the distributive property to rewrite each expression. Simplify if possible. See Examples 9 and 11.

51. $5x + x$

52. $6q + q$

53. $4(t + 3)$

54. $5(w + 4)$

55. $-8(r + 3)$

56. $-11(x + 4)$

57. $-5(y - 4)$

58. $-9(g - 4)$

59. $-\frac{4}{3}(12y + 15z)$

60. $-\frac{2}{5}(10b + 20a)$

61. $8 \cdot z + 8 \cdot w$

62. $4 \cdot s + 4 \cdot r$

63. $7(2v) + 7(5r)$ **64.** $13(5w) + 13(4p)$ **65.** $8(3r + 4s - 5y)$

66. $2(5u - 3v + 7w)$ **67.** $q + q + q$ **68.** $m + m + m + m$

69. $-5x + x$ **70.** $-9p + p$

Use the distributive property to write each of the following without parentheses. See Example 10.

71. $-(4t + 5m)$ **72.** $-(9x + 12y)$ **73.** $-(-5c - 4d)$

74. $-(-13x - 15y)$ **75.** $-(-3q + 5r - 8s)$ **76.** $-(-4z + 5w - 9y)$

77. The operations of "getting out of bed" and "taking a shower" are not commutative. Give an example of another pair of everyday operations that are not commutative.

78. Are the operations "going upstairs" and "going downstairs" commutative?

79. True or false: "preparing a meal" and "eating a meal" are commutative operations.

80. The phrase "dog biting man" has two different meanings, depending on how the words are associated:

$$\text{(dog biting) man} \qquad \text{dog (biting man)}$$

Give another example of a three-word phrase that has different meanings depending on how the words are associated.

81. Use parentheses to show how the associative property can be used to give two different meanings to "foreign sales clerk."

82. Use parentheses to show two different meanings for "new cook book."

1.9 Simplifying Expressions

OBJECTIVE 1 As we saw in the previous section, we can use the properties of addition and multiplication to simplify algebraic expressions.

E X A M P L E   I   Simplifying Expressions

Simplify the following expressions.

(a) $4x + 8 + 9$

Since $8 + 9 = 17$, $4x + 8 + 9 = 4x + 17$.

(b) $4(3m - 2n)$

Use the distributive property.

$$4(3m - 2n) = 4(3m) - 4(2n)$$
$$= 12m - 8n$$

(c) $6 + 3(4k + 5) = 6 + 3(4k) + 3(5)$ Distributive property
$$= 6 + 12k + 15 \qquad \text{Multiply.}$$
$$= 21 + 12k \qquad \text{Add.}$$

(d) $5 - (2y - 8) = 5 - 1(2y - 8)$ Identity property
$$= 5 - 2y + 8 \qquad \text{Distributive property}$$
$$= 13 - 2y \qquad \text{Add.}$$

> **Note**
> Although the steps were not shown, in Examples 1(c) and 1(d) we mentally used the commutative and associative properties to add in the last step. In practice, these steps are usually left out, but we should realize that they are used whenever the ordering in a sum is rearranged.

WORK PROBLEM I AT THE SIDE. ▶▶

OBJECTIVE 2 A **term** is a single number, or a product of a number and one or more variables raised to powers. Examples of terms include

$$-9x^2, \quad 15y, \quad -3, \quad 8m^2n, \quad \text{and} \quad k.$$

The **numerical coefficient** (koh-uh-FISH-ent) of the term $9m$ is 9; the numerical coefficient of $-15x^3y^2$ is -15; the numerical coefficient of x is 1; and the numerical coefficient of 8 is 8. A coefficient is a multiplier.

> **Caution**
> It is important to be able to distinguish between *terms* and *factors*. For example, in the expression $8x^3 + 12x^2$, there are two terms. They are $8x^3$ and $12x^2$. Terms are separated by a $+$ or $-$ sign. On the other hand, in the term $(8x^3)(12x^2)$, $8x^3$ and $12x^2$ are *factors*. Factors are multiplied.

1. Simplify each expression.

(a) $9k + 12 - 5$

(b) $7(3p + 2q)$

(c) $2 + 5(3z - 1)$

(d) $-3 - (2 + 5y)$

71

2. Give the numerical coefficient of each term.

(a) $15q$

(b) $-2m^3$

(c) $-18m^7q^4$

(d) $-r$

(e) $\dfrac{5x}{4}$

3. Identify each pair of terms as *like* or *unlike*.

(a) $9x, 4x$

(b) $-8y^3, 12y^2$

(c) $7x^2y^4, -7x^2y^4$

(d) $13kt, 4tk$

EXAMPLE 2   Identifying the Numerical Coefficient of a Term

Give the numerical coefficient of each term.

Term	Numerical Coefficient
$-7y$	-7
$34r^3$	34
$-26x^5yz^4$	-26
$-k$	-1
r	1
$\dfrac{3x}{8} = \dfrac{3}{8}x$	$\dfrac{3}{8}$

 WORK PROBLEM 2 AT THE SIDE.

OBJECTIVE 3 Terms with exactly the same variables (including the same exponents) are called **like terms.** For example, $9m$ and $4m$ have the same variables and are like terms. Also, $6x^3$ and $-5x^3$ are like terms. The terms $-4y^3$ and $4y^2$ have different exponents and are **unlike terms.**

Here are some examples of like terms.

$$5x \quad \text{and} \quad -12x \qquad 3x^2y \quad \text{and} \quad 5x^2y$$

Here are some examples of unlike terms.

$$4xy^2 \quad \text{and} \quad 5xy \qquad -7w^3z^3 \quad \text{and} \quad 2xz^3$$

WORK PROBLEM 3 AT THE SIDE.

OBJECTIVE 4 We add or subtract like terms by using the distributive property. For example,

$$3x + 5x = (3 + 5)x = 8x.$$

This process is called **combining terms.**

> **Caution**
> Remember that *only like terms* may be combined.

EXAMPLE 3   Combining Like Terms

Combine terms in the following expressions.

(a) $6r + 3r + 2r$

Use the distributive property to combine like terms.

$$6r + 3r + 2r = (6 + 3 + 2)r = 11r$$

(b) $4x + x = 4x + 1x = 5x$ Identity property

(c) $3a - 2b + 7 + 5a + 4b - 9 = (3 + 5)a + (-2 + 4)b + (7 - 9)$
$$= 8a + 2b - 2$$

Look for like terms and combine them.

(d) $32y + 10y^2$ cannot be simplified because $32y$ and $10y^2$ are unlike terms. The exponents on y are different.

ANSWERS

2. **(a)** 15 **(b)** -2 **(c)** -18 **(d)** -1 **(e)** $\dfrac{5}{4}$

3. **(a)** like **(b)** unlike **(c)** like **(d)** like

WORK PROBLEM 4 AT THE SIDE. ▶▶

4. Combine terms.

(a) $4k + 7k$

E X A M P L E   4   Simplifying Expessions Involving Like Terms

Simplify the following expressions.

(a) $14y + 2(6 + 3y) = 14y + 2(6) + 2(3y)$ Distributive property

$\qquad = 14y + 12 + 6y$ Multiply.

(b) $4r - r$

$\qquad = 20y + 12$ Combine like terms.

(b) $9k - 6 - 3(2 - 5k) = 9k - 6 - 3(2) - 3(-5k)$ Distributive property

$\qquad = 9k - 6 - 6 + 15k$ Multiply.

(c) $5z + 9z - 4z$

$\qquad = 24k - 12$ Combine like terms.

(c) $-(2 - r) + 10r = -1(2 - r) + 10r$ $-(2 - r) = -1(2 - r)$

$\qquad = -1(2) - 1(-r) + 10r$ Distributive property

$\qquad = -2 + r + 10r$ Multiply.

(d) $8p + 8p^2$

$\qquad = -2 + 11r$ Combine like terms.

(d) $5(2a^2 - 6a) - 3(4a^2 - 9) = 10a^2 - 30a - 12a^2 + 27$ Distributive property

$\qquad = -2a^2 - 30a + 27$ Combine like terms.

(e) $5x - 3y + 2x - 5y - 3$

WORK PROBLEM 5 AT THE SIDE. ▶▶

OBJECTIVE 5 ▶ In the next example, we translate a word phrase to a mathematical expression and then simplify it. We will need to do this in the next chapter when we solve applied problems.

5. Simplify.

(a) $10p + 3(5 + 2p)$

E X A M P L E   5   Converting Words to a Mathematical Expression

Write the following phrase as a mathematical expression and simplify: four times a number, subtracted from the sum of twice the number and 4.

Let x represent the number.

(b) $7z - 2 - (1 + z)$

The sum of twice   Four times
the number and 4   the number
$\qquad\downarrow\qquad\qquad\downarrow$
$(2x + 4) - 4x$ Write with symbols.

$\qquad = -2x + 4$ Combine terms.

(c) $-(3k^2 + 5k) + 7(k^2 - 4k)$

6. Write the following phrases as mathematical expressions, and simplify by combining terms.

◀◀ WORK PROBLEM 6 AT THE SIDE.

(a) Three times a number, subtracted from the sum of the number and 8

(b) Twice a number added to the sum of 6 and the number

1.9 Exercises

Choose the letter of the correct response in each of the following.

1. Which one of the following is correct?
  (a) $6 + 2x = 8x$
  (b) $6 - 2x = 4x$
  (c) $6x - 2x = 4x$
  (d) $3 + 8(4t - 6) = 11(4t - 6)$

2. Which one of the following is an example of a pair of like terms?
  (a) $6t, 6w$
  (b) $-8x^2y, 9xy^2$
  (c) $5ry, 6yr$
  (d) $-5x^2, 2x^3$

3. Which one of the following is an example of a term with numerical coefficient 5?
  (a) $5x^3y^7$
  (b) x^5
  (c) $\dfrac{x}{5}$
  (d) 5^2xy^3

4. Which one of the following is a correct translation for "six times a number, subtracted from the product of eleven and the number" (if x represents the number)?
  (a) $6x - 11x$
  (b) $11x - 6x$
  (c) $(11 + x) - 6x$
  (d) $6x - (11 + x)$

Simplify each expression. See Example 1.

5. $4r + 19 - 8$

6. $7t + 18 - 4$

7. $8(4q - 3t)$

8. $12(9m - 7n)$

9. $5 + 2(x - 3y)$

10. $8 + 3(s - 6t)$

11. $-2 - (5 - 3p)$

12. $-10 - (7 - 14r)$

Give the numerical coefficient of each of the following terms. See Example 2.

13. $14x$

14. $9x$

15. $-12k$

16. $-23y$

17. $5m^2$

18. $-3n^6$

19. xw

20. pq

21. $-x$

22. $-t$

23. 74

24. 98

25. Give an example of a pair of like terms with the variable x, such that one of them has a negative numerical coefficient, one has a positive numerical coefficient, and their sum has a positive numerical coefficient.

26. Give an example of a pair of unlike terms such that each term has x as the only variable.

Identify each group of terms as like *or* unlike.

27. $8r, -13r$

28. $-7a, 12a$

29. $5z^4, 9z^3$

30. $8x^5, -10x^3$

31. $4, 9, -24$

32. $7, 17, -83$

33. x, y

34. t, s

35. There is an old saying "You can't add apples and oranges." Explain how this saying can be applied to the goal of Objective 3 in this section.

36. Explain how the distributive property is used in combining $6t + 5t$ to get $11t$.

Simplify each expression, and combine like terms. See Examples 3 and 4.

37. $4k + 3 - 2k + 8 + 7k - 16$

38. $9x + 7 - 13x + 12 + 8x - 15$

39. $-\dfrac{4}{3} + 2t + \dfrac{1}{3}t - 8 - \dfrac{8}{3}t$

40. $-\dfrac{5}{6} + 8x + \dfrac{1}{6}x - 7 - \dfrac{7}{6}$

41. $-5.3r + 4.9 - 2r + .7 + 3.2r$

42. $2.7b + 5.8 - 3b + .5 - 4.4b$

43. $2y^2 - 7y^3 - 4y^2 + 10y^3$

44. $9x^4 - 7x^6 + 12x^4 + 14x^6$

45. $13p + 4(4 - 8p)$

46. $5x + 3(7 - 2x)$

47. $-4(y - 7) - 6$

48. $-5(t - 13) - 4$

49. $-5(5y - 9) + 3(3y + 6)$

50. $-3(2t + 4) + 8(2t - 4)$

Convert the following phrases into mathematical expressions. Use x as the variable. Combine like terms when possible. See Example 5.

51. Five times a number, added to the sum of the number and three

52. Six times a number, added to the sum of the number and six

53. A number multiplied by -7, subtracted from the sum of 13 and six times the number

54. A number multiplied by 5, subtracted from the sum of 14 and eight times the number

55. Six times a number added to -4, subtracted from twice the sum of three times the number and 4

56. Nine times a number added to 6, subtracted from triple the sum of 12 and 8 times the number

57. Write the expression $9x - (x + 2)$ using words, as in Exercises 51–56.

58. Write the expression $2(3x + 5) - 2(x + 4)$ using words, as in Exercises 51–56.

MATHEMATICAL CONNECTIONS (EXERCISES 59–62)

A manufacturer has fixed costs of $1000 to produce widgets. Each widget costs $5 to make. The fixed cost to produce gadgets is $750 and each gadget costs $3 to make. Work Exercises 59–62 in order.

59. Write an expression for the cost to make x widgets. (*Hint:* The cost will be the sum of the fixed cost and the cost per item times the number of items.)

60. Write an expression for the cost to make x gadgets.

61. Write an expression for the total cost to make x widgets and x gadgets.

62. Simplify the expression you wrote in Exercise 61.

1.1	**exponent**	An exponent, or power, is a number that indicates how many times a factor is repeated.
	base	The base is the number that is a repeated factor when written with an exponent.
	exponential expression	A number written with an exponent is an exponential expression.
1.2	**variable**	A variable is a symbol, usually a letter, used to represent an unknown number.
	algebraic expression	An algebraic expression is a collection of numbers, variables, symbols for operations, and symbols for grouping.
	equation	An equation is a statement that says two expressions are equal.
	solution	A solution of an equation is any replacement for the variable that makes the equation true.
1.3	**whole numbers**	The set of whole numbers is $\{0, 1, 2, 3, 4, 5, \ldots\}$.
	natural numbers	The set of natural numbers includes the numbers used for counting: $\{1, 2, 3, 4, \ldots\}$.
	negative number	A negative number is located to the *left* of 0 on the number line.
	positive number	A positive number is located to the *right* of 0 on the number line.
	signed numbers	Signed numbers are either positive or negative.
	integers	The set of integers is $\{\ldots, -3, -2, -1, 0, 1, 2, 3, \ldots\}$.
	rational numbers	Rational numbers can be written as quotients of two integers, with denominator not 0.
	irrational numbers	Irrational numbers are nonrational numbers represented by points on the number line.
	real numbers	Real numbers include all numbers that can be represented by points on the number line, or all rational and irrational numbers.
	opposite	The opposite of a number a is the number that is the same distance from 0 on the number line as a, but on the opposite side of 0. This number is also called the **negative** of a or the **additive inverse** of a.
	absolute value	The absolute value of a number is the distance between 0 and the number on the number line.
1.4	**sum**	The answer to an addition problem is called the sum.
1.5	**difference**	The answer to a subtraction problem is called the difference.
1.6	**product**	The result of multiplication is called the product.
1.7	**reciprocal**	Pairs of numbers whose product is 1 are called reciprocals or **multiplicative inverses** of each other.
	quotient	The answer to a division problem is called the quotient.
1.8	**identity element for addition**	When the identity element for addition, which is 0, is added to a number, the number is unchanged.
	identity element for multiplication	When a number is multiplied by the identity element for multiplication, which is 1, the number is unchanged.
	additive inverse	The opposite of a number is also called the additive inverse of the number.
	multiplicative inverse	The reciprocal of a number is also called the multiplicative inverse of the number.
	factoring	When the distributive property is used to write a sum or a difference as a product, the process is called factoring.

1.9	**term**	A term is a single number, or a product of a number and one or more variables raised to powers.
	numerical coefficient	The numerical factor in a term is its numerical coefficient.
	like terms	Terms with exactly the same variables (including the same exponents) are called like terms.

NEW SYMBOLS

a^n	n factors of a	$a(b), (a)(b), a \cdot b,$ or ab	a times b
$=$	is equal to	$\dfrac{a}{b}, a/b$ or $a \div b$	a divided by b
$\neq$	is not equal to		
$<$	is less than	$\{\ \}$	set braces
$\leq$	is less than or equal to	$\lvert x \rvert$	absolute value of x
$>$	is greater than		
$\geq$	is greater than or equal to		

QUICK REVIEW

Concepts	Examples
1.1 Exponents, Order of Operations, and Inequality	
Order of Operations	
Step 1 If possible, simplify within parentheses and above and below fraction bars, using the following steps.	
Step 2 Apply all exponents.	$36 - 4(2^2 + 3) = 36 - 4(4 + 3)$
Step 3 Do any multiplications or divisions in the order in which they occur, working from left to right.	$= 36 - 4(7)$ $= 36 - 28$
Step 4 Do any additions or subtractions in the order in which they occur, working from left to right.	$= 8$
1.2 Variables, Expressions, and Equations	
Evaluate an expression with a variable by substituting a given number for the variable.	Evaluate $2x + y^2$ if $x = 3$ and $y = -4$. $2x + y^2 = 2(3) + (-4)^2$ $= 6 + 16$ $= 22$
Values of a variable that make an equation true are solutions of the equation.	Is 2 a solution of $5x + 3 = 18$? $5(2) + 3 = 18$ $13 = 18$ False 2 is not a solution.

Concepts	Examples
1.3 Real Numbers and the Number Line **Ordering Real Numbers** *a* is less than *b* if *a* is to the left of *b* on the number line.	 $-2 < 3$ $3 > 0$ $0 < 3$
The opposite or additive inverse of *a* is $-a$.	$-(5) = -5$ $-(-7) = 7$ $-0 = 0$
The absolute value of *a*, $\lvert a \rvert$, is the distance between *a* and 0 on the number line.	$\lvert 13 \rvert = 13$ $\lvert 0 \rvert = 0$ $\lvert -5 \rvert = 5$
1.4 Addition of Real Numbers To add two numbers with the same sign, add their absolute values. The sum has that same sign.	$9 + 4 = 13$ $-8 + (-5) = -13$
To add two numbers with different signs, subtract their absolute values. The sum has the sign of the number with larger absolute value.	$7 + (-12) = -5$ $-5 + 13 = 8$
1.5 Subtraction of Real Numbers **Definition of Subtraction** $a - b = a + (-b)$ Subtracting signed numbers: **1.** Change the subtraction symbol to addition. **2.** Change the sign of the number being subtracted. **3.** Add, as in the previous section.	$5 - (-2) = 5 + 2 = 7$ $-3 - 4 = -3 + (-4) = -7$ $-2 - (-6) = -2 + 6 = 4$ $13 - (-8) = 13 + 8 = 21$
1.6, 1.7 Multiplication of Real Numbers **and** *Division of Real Numbers* **Multiplying and Dividing Signed Numbers** The product (or quotient) of two numbers having the *same sign* is *positive;* the product (or quotient) of two numbers having *different signs* is *negative.*	$6 \cdot 5 = 30$ $(-7)(-8) = 56$ $\dfrac{10}{2} = 5$ $\dfrac{-24}{-6} = 4$ $(-6)(5) = -30$ $(6)(-5) = -30$ $\dfrac{-18}{9} = -2$ $\dfrac{49}{-7} = -7$
Division *by* 0 is undefined.	$\dfrac{0}{5} = 0$ $\dfrac{5}{0}$ is undefined.

Concepts	Examples
1.8 Properties of Addition and Multiplication	
Commutative	
$a + b = b + a$	$7 + (-1) = -1 + 7$
$ab = ba$	$5(-3) = (-3)5$
Associative	
$(a + b) + c = a + (b + c)$	$(3 + 4) + 8 = 3 + (4 + 8)$
$(ab)c = a(bc)$	$[(-2)(6)](4) = (-2)[(6)(4)]$
Identity	
$a + 0 = a \qquad 0 + a = a$	$-7 + 0 = -7 \qquad 0 + (-7) = -7$
$a \cdot 1 = a \qquad 1 \cdot a = a$	$9 \cdot 1 = 9 \qquad 1 \cdot 9 = 9$
Inverse	
$a + (-a) = 0 \qquad -a + a = 0$	$7 + (-7) = 0 \qquad -7 + 7 = 0$
$a \cdot \dfrac{1}{a} = 1 \qquad \dfrac{1}{a} \cdot a = 1 \; (a \neq 0)$	$-2\left(-\dfrac{1}{2}\right) = 1 \qquad -\dfrac{1}{2}(-2) = 1$
Distributive	
$a(b + c) = ab + ac$	$5(4 + 2) = 5(4) + 5(2)$
$(b + c)a = ba + ca$	$(4 + 2) \cdot 5 = 4(5) + 2(5)$
1.9 Simplifying Expressions	
Only like terms may be combined.	$-3y^2 + 6y^2 + 14y^2 = 17y^2$
	$4(3 + 2x) - 6(5 - x) = 12 + 8x - 30 + 6x$
	$= 14x - 18$

CHAPTER 1 REVIEW EXERCISES

[1.1] *Find the value of each exponential expression.**

1. 5^4

2. $(.03)^4$

3. $.21^3$

4. $\left(\dfrac{5}{2}\right)^3$

Find the value of each of the following expressions.

5. $8 \cdot 5 - 13$

6. $5[4^2 + 3(2^3)]$

7. $\dfrac{7(3^2 - 5)}{2 \cdot 6 - 16}$

8. $\dfrac{3(9 - 4) + 5(8 - 3)}{2^3 - (5 - 3)}$

Write each word statement in symbols.

9. Thirteen is less than seventeen.

10. Five plus two is not equal to ten.

[1.2] *Find the numerical values of the following if* $x = 6$ *and* $y = 3$.

11. $2x + 6y$

12. $4(3x - y)$

13. $\dfrac{x}{3} + 4y$

14. $\dfrac{x^2 + 3}{3y - x}$

Change the word phrases to algebraic expressions. Use x as the variable to represent the number.

15. Six added to a number

16. A number subtracted from eight

17. Nine subtracted from six times a number

18. Three-fifths of a number added to 12

Decide whether the given number is a solution of the equation.

19. $5x + 3(x + 2) = 22$; 2

20. $\dfrac{t + 5}{3t} = 1$; 6

Change the word sentences to equations. Use x as the variable.

21. Six less than twice a number is 10.

22. The product of a number and 4 is 8.

[1.3] *Graph each group of numbers on a number line.*

23. $-4, -\dfrac{1}{2}, 0, 2.5, 5$

24. $-2, -3, |-3|, |-1|$

25. $-3\dfrac{1}{4}, 2\dfrac{4}{5}, -1\dfrac{1}{8}, \dfrac{5}{6}$

26. $|-4|, -|-3|, -|-5|, -6$

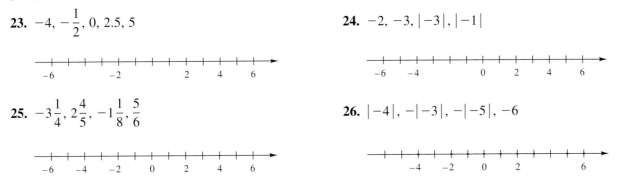

---
* For help with any of these exercises, refer to the section given in brackets.

Select the smaller number in each pair.

27. $-10, 5$

28. $-8, -9$

29. $-\dfrac{2}{3}, -\dfrac{3}{4}$

30. $0, -|23|$

Decide whether each statement is true or false.

31. $12 > -13$

32. $0 > -5$

33. $-9 < -7$

34. $-13 > -13$

For each of the following, (a) find the opposite of the number and (b) find the absolute value of the number.

35. -9

36. 0

37. 6

38. $-\dfrac{5}{7}$

Simplify by removing absolute value symbols.

39. $|-12|$

40. $-|3|$

41. $-|-19|$

42. $-|9 - 2|$

[1.4] *Find the following sums.*

43. $-10 + 4$

44. $14 + (-18)$

45. $-8 + (-9)$

46. $\dfrac{4}{9} + \left(-\dfrac{5}{4}\right)$

47. $[-6 + (-8) + 8] + [9 + (-13)]$

48. $(-4 + 7) + (-11 + 3) + (-15 + 1)$

Write a numerical expression for each phrase, and simplify the expression.

49. 19 added to the sum of -31 and 12

50. 13 more than the sum of -4 and -8

Solve each problem.

51. Tri Nguyen has \$18 in his checking account. He then writes a check for \$26. What negative number represents his balance?

52. The temperature at noon on an August day in Houston was 93°F. After a thunderstorm, it dropped 6°. What was the new temperature?

[1.5] *Find the following differences.*

53. $-7 - 4$

54. $-12 - (-11)$

55. $5 - (-2)$

56. $-\dfrac{3}{7} - \dfrac{4}{5}$

57. $2.56 - (-7.75)$

58. $(-10 - 4) - (-2)$

59. $(-3 + 4) - (-1)$

60. $-(-5 + 6) - 2$

Write a numerical expression for each phrase, and simplify the expression.

61. The difference between −4 and −6

62. Five less than the sum of 4 and −8

Solve each problem.

63. Eric owed his brother $28. He repaid $13 but then borrowed another $14. What positive or negative amount represents his present financial status?

64. If the temperature drops 7° below its previous level of −3°, what is the new temperature?

65. Explain in your own words how the subtraction problem −8 − (−6) is performed.

66. Can the difference of two negative numbers be positive? Explain with an example.

The line graph gives a representation of the number of stolen bases by baseball player Vince Coleman each year. Use a signed number to represent the change in the number of stolen bases from one year to the next. For example, from 1989 to 1990 the change was
77 − 66 = 11.

67. from 1990 to 1991

68. from 1991 to 1992

69. from 1992 to 1993

70. from 1993 to 1994

────── **STEALS BY** ──────
VINCE COLEMAN

[1.6] *Perform the indicated operations.*

71. (−12)(−3)

72. 15(−7)

73. $\left(-\dfrac{4}{3}\right)\left(-\dfrac{3}{8}\right)$

74. (−4.8)(−2.1)

75. 5(8 − 12)

76. (5 − 7)(8 − 3)

77. 2(−6) − (−4)(−3)

78. 3(−10) − 5

Evaluate the following expressions if x = −5, y = 4, and z = −3.

79. $6x - 4z$ **80.** $5x + y - z$ **81.** $5x^2$ **82.** $z^2(3x - 8y)$

Write a numerical expression for each phrase, and simplify the expression.

83. Nine less than the product of −4 and 5

84. Five-sixths of the sum of 12 and −6

[1.7] *Find the following quotients.*

85. $\dfrac{-36}{-9}$ **86.** $\dfrac{220}{-11}$ **87.** $-\dfrac{1}{2} \div \dfrac{2}{3}$ **88.** $-33.9 \div (-3)$

89. $\dfrac{-5(3) - 1}{8 - 4(-2)}$ **90.** $\dfrac{5(-2) - 3(4)}{-2[3 - (-2)] - 1}$ **91.** $\dfrac{10^2 - 5^2}{8^2 + 3^2 - (-2)}$ **92.** $\dfrac{(.6)^2 + (.8)^2}{(-1.2)^2 - (-.56)}$

Write a numerical expression for each phrase, and simplify the expression.

93. The quotient of 12 and the sum of 8 and −4

94. The product of −20 and 12, divided by the difference between 15 and −15

Write the following in symbols, using x as the variable.

95. 8 times a number is −24.

96. The quotient of a number and 3 is −2.

97. 3 less than a number is −7.

98. The sum of a number and 5 is −6.

[1.8] *Decide whether each statement is an example of the commutative, associative, identity, inverse, or distributive property.*

99. $6 + 0 = 6$ **100.** $5 \cdot 1 = 5$ **101.** $-\dfrac{2}{3}\left(-\dfrac{3}{2}\right) = 1$

102. $17 + (-17) = 0$ **103.** $5 + (-9 + 2) = [5 + (-9)] + 2$ **104.** $w(xy) = (wx)y$

105. $3x + 3y = 3(x + y)$

106. $(1 + 2) + 3 = 3 + (1 + 2)$

Use the distributive property to rewrite each expression. Simplify if possible.

107. $7y + y$

108. $-12(4 - t)$

109. $3(2s) + 3(4y)$

110. $-(-4r + 5s)$

111. Evaluate $25 - (5 - 2)$ and $(25 - 5) - 2$. Use this example to explain why subtraction is not associative.

112. Evaluate $180 \div (15 \div 5)$ and $(180 \div 15) \div 5$. Use this example to explain why division is not associative.

[1.9] *Use the distributive property as necessary and combine like terms.*

113. $16p^2 - 8p^2 + 9p^2$

114. $4r^2 - 3r + 10r + 12r^2$

115. $-8(5k - 6) + 3(7k + 2)$

116. $2s - (-3s + 6)$

117. $-7(2t - 4) - 4(3t + 8) - 19(t + 1)$

118. $3.6t^2 + 9t - 8.1(6t^2 + 4t)$

Convert the following phrases into mathematical expressions. Use x as the variable, and combine like terms when possible.

119. Seven times a number subtracted from the product of -2 and three times the number

120. The quotient of 9 more than a number and 6 less than the number

121. In Exercise 119, does the word *and* signify addition? Explain.

122. Write the expression $3(4x - 6)$ using words, as in Exercises 119–120.

MIXED REVIEW EXERCISES*

Perform the indicated operations.

123. $[(-2) + 7 - (-5)] + [-4 - (-10)]$

124. $\left(-\dfrac{5}{6}\right)^2$

125. $-|(-7)(-4)| - (-2)$

126. $\dfrac{6(-4) + 2(-12)}{5(-3) + (-3)}$

127. $\dfrac{3}{8} - \dfrac{5}{12}$

128. $\dfrac{12^2 + 2^2 - 8}{10^2 - (-4)(-15)}$

129. $\dfrac{8^2 + 6^2}{7^2 + 1^2}$

130. $-16(-3.5) - 7.2(-3)$

131. $2\dfrac{5}{6} - 4\dfrac{1}{3}$

132. $-8 + [(-4 + 17) - (-3 - 3)]$

133. $-\dfrac{12}{5} \div \dfrac{9}{7}$

134. $(-8 - 3) - 5(2 - 9)$

135. $[-7 + (-2) - (-3)] + [8 + (-13)]$

136. $\dfrac{15}{2} \cdot \left(-\dfrac{4}{5}\right)$

Write a numerical expression for each, and simplify it if possible. Use x as the variable if one is needed.

137. In 1997, a company spent $13,600 on advertising. In 1998, the amount spent on advertising was reduced by $1400. How much was spent on advertising in 1998?

138. The quotient of a number and 14 less than three times the number

---

* The order of exercises in this final group does not correspond to the order in which topics occur in the chapter. This random ordering should help you prepare for the chapter test in yet another way.

CHAPTER 1 TEST

Decide whether the statement is true or false.

1. $4[-20 + 7(-2)] \leq 135$

2. $(-3)^2 + 2^2 = 5^2$

3. Graph the group of numbers $-1, -3, |-4|, |-1|$ on the number line.

Select the smaller number from each pair.

4. $6, -|-8|$

5. $-.742, -1.277$

6. Write in symbols: The quotient of -6 and the sum of 2 and -8. Simplify the expression.

7. If a and b are both negative, is $\dfrac{a + b}{a \cdot b}$ positive or negative?

Perform the indicated operations whenever possible.

8. $-2 - (5 - 17) + (-6)$

9. $-5\dfrac{1}{2} + 2\dfrac{2}{3}$

10. $-6 - [-7 + (2 - 3)]$

11. $4^2 + (-8) - (2^3 - 6)$

12. $(-5)(-12) + 4(-4) + (-8)^2$

13. $\dfrac{-7 - (-6 + 2)}{-5 - (-4)}$

1. _____

2. _____

3.

4. _____

5. _____

6. _____

7. _____

8. _____

9. _____

10. _____

11. _____

12. _____

13. _____

14. _____

$$14. \ \frac{30(-1-2)}{-9[3-(-2)]-12(-2)}$$

Evaluate the following expressions, given $x = -2$ and $y = 4$.

15. _____

15. $3x - 4y^2$

16. _____

$$16. \ \frac{5x + 7y}{3(x + y)}$$

Solve the following problem.

17. _____

17. The highest Fahrenheit temperature ever recorded in Idaho was 118°, while the lowest was −60°. What is the difference between these highest and lowest temperatures?

Match the property in Column I with the example of it in Column II.

I	II

18. _____

18. Commutative   A. $3x + 0 = 3x$

19. _____

19. Associative   B. $(5 + 2) + 8 = 8 + (5 + 2)$

20. _____

20. Inverse   C. $-3(x + y) = -3x + (-3y)$

21. _____

21. Identity   D. $-5 + (3 + 2) = (-5 + 3) + 2$

22. _____

22. Distributive   E. $-\dfrac{5}{3}\left(-\dfrac{3}{5}\right) = 1$

23. _____

23. Simplify by using the distributive property and combining like terms:

$$-2(3x^2 + 4) - 3(x^2 + 2x)$$

24. _____

24. What property is used to show that $3(x + 1) = 3x + 3$?

25. (a) _____

   (b) _____

25. Consider the expression $-6[5 + (-2)]$.
   (a) Evaluate it by first working within the brackets.
   (b) Evaluate it by using the distributive property.
   (c) Why must the answers in items (a) and (b) be the same?

   (c) _____

Linear Equations and Applications

2.1 The Addition Property of Equality

To solve applied problems, we must be able to solve equations. The simplest type of equation is a *linear equation.* Methods for solving linear equations will be introduced in this section. We will be using the definitions and properties of real numbers that we learned in Chapter 1.

OBJECTIVE 1 Before we can solve a linear equation, we must be able to recognize one.

> A **linear equation** can be written in the form
> $$Ax + B = C$$
> for real numbers A, B, and C, with $A \neq 0$.

As we saw in Chapter 1, a solution of an equation is a number that makes the equation a true statement when it replaces the variable. Equations that have exactly the same solutions are **equivalent equations.** Linear equations are solved by using a series of steps to produce a simpler equivalent equation of the form

$$x = \text{a number.}$$

OBJECTIVE 2 In the equation $x - 5 = 2$, both $x - 5$ and 2 represent the same number because this is the meaning of the equal sign. To solve the equation, we change the left side from $x - 5$ to just x. This is done by adding 5 to $x - 5$. To keep the two sides equal, we must also add 5 to the right side.

$x - 5 = 2$	Given equation
$x - 5 + 5 = 2 + 5$	Add 5 to each side.
$x + 0 = 7$	Additive inverse property
$x = 7$	Identity property

The solution of the given equation is 7. Check by replacing x with 7 in the given equation.

$x - 5 = 2$		Given equation
$7 - 5 = 2$	?	Let $x = 7$.
$2 = 2$		True

Since the final statement is true, 7 checks as the solution.

www.mathnotes.com

OBJECTIVES

1. Identify linear equations.
2. Use the addition property of equality.
3. Simplify equations, and then use the addition property of equality.
4. Solve equations that have no solution or infinitely many solutions.

FOR EXTRA HELP

Tutorial   Tape 4   SSM, Sec. 2.1

1. Solve.

(a) $m - 2.9 = -6.4$

(b) $y - 4.1 = 6.3$

To solve the equation above, we added the same number to each side. The **addition property of equality** justifies this step.

> **Addition Property of Equality**
>
> If A, B, and C are real numbers, then the equations
>
> $$A = B \quad \text{and} \quad A + C = B + C$$
>
> are equivalent equations. That is, we can add the same number to each side of an equation without changing the solution.

In the addition property, C represents a real number. This means that any quantity that represents a real number can be added to both sides of an equation to change it to an equivalent equation.

The set of all solutions of an equation is called its **solution set**. We write the solution in *braces*. For example, the solution set of $x - 5 = 2$ is $\{7\}$.

EXAMPLE 1 Using the Addition Property of Equality

Solve $x - 16.2 = 7.5$.

If the left side of this equation were just x, the solution would be known. Get x alone by using the addition property of equality, adding 16.2 to each side.

$$x - 16.2 = 7.5$$
$$x - 16.2 + \mathbf{16.2} = 7.5 + \mathbf{16.2} \qquad \text{Add 16.2 on each side.}$$
$$x = 23.7 \qquad \text{Combine terms.}$$

Here we combined the steps that change $x - 16.2 + 16.2$ to $x + 0$ and $x + 0$ to x. We will combine these steps from now on. Check by substituting 23.7 for x in the original equation.

$$x - 16.2 = 7.5 \qquad \text{Given equation}$$
$$\mathbf{23.7} - 16.2 = 7.5 \qquad ? \qquad \text{Let } x = 2.37.$$
$$7.5 = 7.5 \qquad \text{True}$$

Since the check results in a true statement, $\{23.7\}$ is the solution set.

2. Solve.

(a) $a + 2 = -3$

◀◀ **WORK PROBLEM 1 AT THE SIDE.**

The addition property of equality says that the same number may be *added* to each side of an equation. In Chapter 1, subtraction was defined as addition of the opposite. Thus, we can also use the following rule when solving an equation.

> The same number may be subtracted from each side of an equation without changing the solution.

(b) $r + 16 = 22$

◀◀ **WORK PROBLEM 2 AT THE SIDE.**

EXAMPLE 2 Subtracting a Variable Expression

Solve $\frac{3}{5}k + 17 = \frac{8}{5}k$.

Get all terms with variables on the same side of the equation. One way to do this is to subtract $\frac{3}{5}k$ from each side.

ANSWERS

1. (a) $\{-3.5\}$ **(b)** $\{10.4\}$
2. (a) $\{-5\}$ **(b)** $\{6\}$

CONTINUED ON NEXT PAGE

$$\frac{3}{5}k + 17 = \frac{8}{5}k$$

$$\frac{3}{5}k + 17 - \frac{3}{5}k = \frac{8}{5}k - \frac{3}{5}k \qquad \text{Subtract } \tfrac{3}{5}k.$$

$$17 = 1k \qquad \text{Combine terms.}$$

$$17 = k \qquad \text{Identity property}$$

The solution set is {17}. From now on we will skip the step that changes $1k$ to k. Check the solution by replacing k with 17 in the original equation.

Another way to solve the equation in Example 2 is to first subtract $\frac{8}{5}k$ from each side.

$$\frac{3}{5}k + 17 = \frac{8}{5}k$$

$$\frac{3}{5}k + 17 - \frac{8}{5}k = \frac{8}{5}k - \frac{8}{5}k \qquad \text{Subtract } \tfrac{8}{5}k.$$

$$17 - k = 0 \qquad \text{Combine terms.}$$
$$\qquad\qquad\qquad\qquad\quad \text{Additive identity}$$

$$17 - k - 17 = 0 - 17 \qquad \text{Subtract 17.}$$

$$-k = -17 \qquad \text{Combine terms.}$$

This result gives the value of $-k$, but not of k itself. However, this result does say that the additive inverse of k is -17, which means that k must be 17, the same result we obtained in Example 2.

$$-k = -17$$

$$k = 17$$

This situation may be generalized as shown below.

If a is a number and $-x = a$, then $x = -a$.

> **WORK PROBLEM 3 AT THE SIDE.** ▶▶

OBJECTIVE 3 Sometimes an equation must be simplified as a first step in its solution.

E X A M P L E   3   Using the Distributive Property to Simplify an Equation

Solve $3(2 + 5x) - (1 + 14x) = 6$.

$$3(2 + 5x) - 1(1 + 14x) = 6 \qquad -(1 + 14x) = -1(1 + 14x)$$

$$3(2) + 3(5x) - 1(1) - 1(14x) = 6 \qquad \text{Distributive property}$$

$$6 + 15x - 1 - 14x = 6 \qquad \text{Multiply.}$$

$$x + 5 = 6 \qquad \text{Combine terms.}$$

$$x = 1 \qquad \text{Subtract 5 from each side.}$$

Check by substituting 1 for x in the original equation. The solution set is {1}.

> **WORK PROBLEM 4 AT THE SIDE.** ▶▶

---

3. Solve $\dfrac{7}{2}m + 1 = \dfrac{9}{2}m$.

4. Solve.

    (a) $-(5 - 3r) + 4(-r + 1) = 1$

    (b) $-3(m - 4) + 2(5 + 2m) = 29$

ANSWERS

3. {1}

4. (a) {−2} **(b)** {7}

5. Solve each equation.

(a) $2(x - 6) = 2x - 12$

OBJECTIVE ▶4 The equations solved so far each have had exactly one solution. Sometimes this is not the case, as shown in the next examples.

E X A M P L E   4   Solving an Equation That Has Infinitely Many Solutions

Solve $5x - 15 = 5(x - 3)$.

$$5x - 15 = 5(x - 3)$$

$$5x - 15 = 5x - 15 \qquad \text{Distributive property}$$

$$5x - 15 + \mathbf{15} = 5x - 15 + \mathbf{15} \qquad \text{Add 15 to each side.}$$

$$5x = 5x \qquad \text{Combine terms.}$$

$$5x - \mathbf{5x} = 5x - \mathbf{5x} \qquad \text{Subtract } 5x \text{ from each side.}$$

$$0 = 0$$

The final step leads us to an equation that contains no variables ($0 = 0$ in this case). Whenever such a statement is true, as it is in this example, *any* real number is a solution. (Try several replacements for x in the given equation to see that they all satisfy the equation.) An equation with both sides exactly the same, like $0 = 0$, is called an **identity.** An identity is true for all replacements of the variables. We indicate the solution set as {all real numbers}.

Caution

When you are solving an equation like the one in Example 4, do not write "0" as the solution. While 0 is a solution, there are infinitely many other solutions.

(b) $3x + 6(x + 1) = 9x - 4$

E X A M P L E   5   Solving an Equation That Has No Solution

Solve $2x + 3(x + 1) = 5x + 4$.

$$2x + 3(x + 1) = 5x + 4$$

$$2x + 3x + 3 = 5x + 4 \qquad \text{Distributive property}$$

$$5x + 3 = 5x + 4 \qquad \text{Combine terms.}$$

$$5x + 3 - \mathbf{5x} = 5x + 4 - \mathbf{5x} \qquad \text{Subtract } 5x \text{ from each side.}$$

$$3 = 4 \qquad \text{Combine terms.}$$

Again, the variable has disappeared, but this time a *false* statement ($3 = 4$) results. Whenever this happens in solving an equation, it is a signal that the equation has no solution, and we write the symbol for the *empty* or *null set*, $\emptyset$.

◀◀ **WORK PROBLEM 5 AT THE SIDE.**

ANSWERS

5. (a) {all real numbers}
  (b) $\emptyset$

2.1 Exercises

Decide whether or not the equation is a linear equation. You may need to simplify the equation first.

1. $3x + 5 = 8$

2. $-x - 5 = 12$

3. $4x + 3 = x - 1$

4. $16 - 2x = 5x$

5. $x^2 + 4 = 16$

6. $\dfrac{9}{x} = 3$

7. Refer to the definition of *linear equation* given in this section. Why is the restriction $A \neq 0$ necessary?

8. Which of the pairs of equations are equivalent equations?
  (a) $x + 2 = 6$ and $x = 4$ **(b)** $10 - x = 5$ and $x = -5$ **(c)** $x + 3 = 9$ and $x = 6$
  (d) $4 + x = 8$ and $x = -4$

Solve each equation by using the addition property of equality. Check each solution. See Examples 1, 2, 4, and 5.

9. $x - 4 = 8$

10. $x - 8 = 9$

11. $7 + r = -3$

12. $8 + k = -4$

13. $\dfrac{9}{7}r - 3 = \dfrac{2}{7}r$

14. $\dfrac{8}{5}w - 6 = \dfrac{3}{5}w$

15. $5.6x + 2 = 4.6x$

16. $9.1x - 5 = 8.1x$

17. $3p + 6 = 10 + 2p$

18. $8b - 4 = -6 + 7b$

19. $1.2y - 4 = .2y - 4$

20. $7.7r + 6 = 6.7r + 6$

21. $3x + 9 = 3x + 8$

22. $-2x + 5 = -2x$

23. $8x + 1 = 1 + 8x$

24. $4w - 5 = -5 + 4w$

Solve the following equations. First simplify each side of the equation as much as possible. Check each solution. See Examples 3, 4, and 5.

25. $10x + 5x + 7 - 8 = 12x + 3 + 2x$

26. $7p + 4p + 13 - 7 = 7p + 9 + 3p$

27. $6x + 5 - 7x + 3 = 5x - 6x - 4$

28. $4x - 3 - 8x + 1 = 5x - 9x + 7$

29. $5.2q - 4.6 - 7.1q = -2.1 - 1.9q - 2.5$

30. $-4.0x + 2.7 - 1.6x = 1.3 - 5.6x + 1.4$

31. $\dfrac{5}{7}x + \dfrac{1}{3} = \dfrac{2}{5} - \dfrac{2}{7}x + \dfrac{2}{5}$

32. $\dfrac{6}{7}s - \dfrac{3}{4} = \dfrac{4}{5} - \dfrac{1}{7}s + \dfrac{1}{6}$

33. $(5y + 6) - (3 + 4y) = 10$

34. $(8r - 3) - (7r + 1) = -6$

35. $2(p + 5) - (9 + p) = -3$

36. $4(k - 6) - (3k + 2) = -5$

37. $-6(2b + 1) + (13b - 7) = 0$

38. $-5(3w - 3) + (1 + 16w) = 0$

39. $10(-2x + 1) = -14(x + 2) + 38 - 6x$

40. $2(2 - 3r) = 5(1 - r) - r - 1$

41. $-2(8p + 2) - 3(2 - 7p) = 2(4 + 2p)$

42. $-5(1 - 2z) + 4(3 - z) = 7(3 + z)$

43. $4(7x - 1) + 3(2 - 5x) = 4(3x + 5) - 6$

44. $9(2m - 3) - 4(5 + 3m) = 5(4 + m) - 3$

45. In your own words, state how you would find the solution of a linear equation if your next-to-last step reads "$-x = 5$."

46. If the final step in solving a linear equation leads to the statement $0 = 0$, why is it incorrect to say that 0 is the solution of the equation? What are the solutions of the equation?

47. Write an equation where 6 must be added to both sides to solve the equation, and the solution set consists of a single negative number.

48. Write an equation where $\dfrac{1}{2}$ must be subtracted on both sides, and the solution set consists of a single positive number.

2.2 The Multiplication Property of Equality

The addition property of equality by itself is not enough to solve some equations, such as $3x + 2 = 17$.

$$3x + 2 = 17$$
$$3x + 2 - 2 = 17 - 2 \qquad \text{Subtract 2 from each side.}$$
$$3x = 15 \qquad \text{Simplify.}$$

Instead of just x on the left side, the equation has $3x$. Another property is needed to change $3x = 15$ to $x = $ a number.

OBJECTIVE 1 ▶ If $3x = 15$, then $3x$ and 15 both represent the same number. Multiplying both $3x$ and 15 by the same number will also result in an equality. The **multiplication property of equality** states that we can multiply each side of an equation by the same number without changing the solution.

Multiplication Property of Equality

If A, B, and C ($C \neq 0$) represent real numbers, the equations

$$A = B \quad \text{and} \quad AC = BC$$

have exactly the same solution.
  In other words, we can multiply each side of an equation by the same nonzero number without changing the solution.

This property can be used to solve $3x = 15$. The $3x$ on the left must be changed to $1x$, or x, instead of $3x$. To obtain x alone, multiply each side of the equation by $\frac{1}{3}$. We use $\frac{1}{3}$ because $\frac{1}{3} \cdot 3 = \frac{3}{3} = 1$, since $\frac{1}{3}$ is the reciprocal of 3.

$$3x = 15$$
$$\frac{1}{3}(3x) = \frac{1}{3} \cdot 15 \qquad \text{Multiply each side by } \tfrac{1}{3}.$$
$$\left(\frac{1}{3} \cdot 3\right)x = \frac{1}{3} \cdot 15 \qquad \text{Associative property}$$
$$1x = 5 \qquad \text{Multiplicative inverse property}$$
$$x = 5 \qquad \text{Multiplicative identity property}$$

The solution set of the equation is $\{5\}$. We can check this result in the original equation. As in the previous section, we shall combine the last two steps shown in the example above.

WORK PROBLEM 1 AT THE SIDE. ▶▶

Just as the addition property of equality permits *subtracting* the same number from each side of an equation, the multiplication property of equality permits *dividing* each side of an equation by the same nonzero number. For example, the equation $3x = 15$, which we first solved by multiplication, also could be solved by dividing each side by 3, as follows.

$$3x = 15$$
$$\frac{3x}{3} = \frac{15}{3} \qquad \text{Divide by 3.}$$
$$x = 5$$

OBJECTIVES

1 ▶ Use the multiplication property of equality.

2 ▶ Use the multiplication property of equality to solve equations with decimals.

3 ▶ Simplify equations, and then use the multiplication property of equality.

4 ▶ Solve equations such as $-r = 4$.

FOR EXTRA HELP

Tutorial   Tape 4   SSM, Sec. 2.2

1. Check that $\{5\}$ is the solution set of $3x = 15$.

ANSWERS

1. Since $3(5) = 15$, the solution set of $3x = 15$ is $\{5\}$.

2. Solve.

   (a) $-6p = -14$

We can divide each side of an equation by the same nonzero number without changing the solution.

Note

In practice, it is usually easier to multiply on each side if the coefficient of the variable is a fraction, and divide on each side if the coefficient is an integer. For example, to solve

$$-\frac{3}{4}x = 12$$

it is easier to multiply by $-\frac{4}{3}$ than to divide by $-\frac{3}{4}$. On the other hand, to solve

$$-5x = -20$$

it is easier to divide by -5 than to multiply by $-\frac{1}{5}$.

(b) $3r = -12$

EXAMPLE 1   Dividing Each Side of an Equation by a Nonzero Number

Solve $25p = 30$.

Get p (instead of $25p$) on the left by using the multiplication property of equality. Divide each side of the equation by 25, the coefficient of p.

$$25p = 30$$

$$\frac{25p}{25} = \frac{30}{25} \qquad \text{Divide by 25.}$$

$$p = \frac{30}{25} = \frac{6}{5} \qquad \text{Reduce to lowest terms.}$$

To check, substitute $\frac{6}{5}$ for p in the given equation.

$$25p = 30$$

$$\frac{25}{1}\left(\frac{6}{5}\right) = 30 \qquad ? \qquad \text{Let } p = \frac{6}{5}.$$

$$30 = 30 \qquad\qquad \text{True}$$

(c) $-2m = 16$

The solution set is $\left\{\frac{6}{5}\right\}$.

◀◀ **WORK PROBLEM 2 AT THE SIDE.**

In the next two examples, multiplication produces the solution more quickly than division would.

EXAMPLE 2   Using the Multiplication Property of Equality

Solve $\frac{a}{4} = 3$.

Replace $\frac{a}{4}$ by $\frac{1}{4}a$, since division by 4 is the same as multiplication by $\frac{1}{4}$. To get a alone on the left, multiply each side by 4, the reciprocal of the coefficient of a.

CONTINUED ON NEXT PAGE

$$\frac{a}{4} = 3$$

$$\frac{1}{4}a = 3 \qquad \text{Change } \tfrac{a}{4} \text{ to } \tfrac{1}{4}a.$$

$$4 \cdot \frac{1}{4}a = 4 \cdot 3 \qquad \text{Multiply by 4.}$$

$$1a = 12 \qquad \text{Multiplicative inverse property}$$

$$a = 12 \qquad \text{Multiplicative identity property}$$

Check the answer.

$$\frac{a}{4} = 3 \qquad \text{Given equation}$$

$$\frac{12}{4} = 3 \quad ? \quad \text{Let } a = 12.$$

$$3 = 3 \qquad \text{True}$$

The solution set is {12}.

WORK PROBLEM 3 AT THE SIDE. ▶▶

E X A M P L E 3   Using the Multiplication Property of Equality

Solve $\frac{3}{4}h = 6$.

Get h alone on the left by multiplying each side of the equation by $\frac{4}{3}$. Use $\frac{4}{3}$ because $\frac{4}{3} \cdot \frac{3}{4}h = 1 \cdot h = h$.

$$\frac{3}{4}h = 6$$

$$\frac{4}{3}\left(\frac{3}{4}h\right) = \frac{4}{3} \cdot 6 \qquad \text{Multiply by } \tfrac{4}{3}.$$

$$1 \cdot h = \frac{4}{3} \cdot \frac{6}{1} \qquad \text{Multiplicative inverse property}$$

$$h = 8 \qquad \text{Multiplicative identity property}$$

The solution set is {8}. Check the answer by substitution in the given equation.

WORK PROBLEM 4 AT THE SIDE. ▶▶

OBJECTIVE 2 ▶ In the next example, we solve an equation with decimals.

E X A M P L E 4   Solving an Equation with Decimals

Solve $2.1x = 6.09$.

Divide both sides by 2.1.

$$\frac{2.1x}{2.1} = \frac{6.09}{2.1}$$

You may use a calculator to simplify the work at this point.

$$1x = 2.9 \qquad \text{Divide.}$$

$$x = 2.9 \qquad \text{Multiplicative identity property}$$

Check that the solution set is {2.9}.

WORK PROBLEM 5 AT THE SIDE. ▶▶

3. Solve.

(a) $\dfrac{y}{5} = 5$

(b) $\dfrac{p}{4} = -6$

4. Solve.

(a) $-\dfrac{5}{6}t = -15$

(b) $\dfrac{3}{4}k = -21$

5. Solve.

(a) $-.7m = -5.04$

(b) $12.5k = -63.75$

ANSWERS

3. (a) {25} **(b)** {−24}
4. (a) {18} **(b)** {−28}
5. (a) {7.2} **(b)** {−5.1}

6. Solve.

(a) $4r - 9r = 20$

(b) $7m - 5m = -12$

◀◀ WORK PROBLEM 6 AT THE SIDE.

7. Solve.

(a) $-m = 2$

(b) $-p = -7$

OBJECTIVE **3▶** In the next example, it is necessary to simplify the equation before using the multiplication property of equality.

E X A M P L E 5   Simplifying Terms in an Equation

Solve $5m + 6m = 33$.

$$5m + 6m = 33$$

$$11m = 33 \qquad \text{Combine terms.}$$

$$\frac{11m}{11} = \frac{33}{11} \qquad \text{Divide by 11.}$$

$$1m = 3 \qquad \text{Divide.}$$

$$m = 3 \qquad \text{Multiplicative identity property}$$

The solution set is {3}. Check this solution.

OBJECTIVE **4▶** The following example shows how to use the multiplication property of equality to solve equations such as $-r = 4$.

E X A M P L E 6   Using −1 with the Multiplication Property of Equality

Solve $-r = 4$.

On the left side, change $-r$ to r by first writing $-r$ as $-1 \cdot r$.

$$-r = 4$$

$$-1 \cdot r = 4 \qquad -r = -1 \cdot r$$

$$-1(-1 \cdot r) = -1 \cdot 4 \qquad \text{Multiply by } -1 \text{, since } -1 \cdot -1 = 1.$$

$$[(-1)(-1)] \cdot r = -4 \qquad \text{Associative property}$$

$$1 \cdot r = -4 \qquad \text{Multiplicative inverse property}$$

$$r = -4 \qquad \text{Multiplicative identity property}$$

Check this solution.

$$-r = 4 \qquad \text{Given equation}$$

$$-(-4) = 4 \qquad ? \qquad \text{Let } r = -4.$$

$$4 = 4 \qquad \text{True}$$

The solution, -4, checks, and the solution set is {-4}.

◀◀ WORK PROBLEM 7 AT THE SIDE.

ANSWERS
6. (a) {-4} **(b)** {-6}
7. (a) {-2} **(b)** {7}

2.2 Exercises

Decide whether the given item is an expression or an equation.

1. $3x + 7 - 2x + 6$

2. $-8y + 13 + 9y - 4$

3. $3x + 7 - 2x = 6$

4. $-8y + 13 + 9y = 4$

By what number is it necessary to multiply both sides of the equation in order to obtain just x on the left side? Do not actually solve these equations.

5. $\dfrac{2}{3}x = 9$

6. $\dfrac{4}{5}x = 7$

7. $.1x = 2$

8. $.01x = 7$

9. $-\dfrac{9}{2}x = -3$

10. $-\dfrac{8}{3}x = -10$

11. $-x = .45$

12. $-x = .23$

By what number is it necessary to divide both sides of the equation in order to obtain just x on the left side? Do not actually solve these equations.

13. $6x = 4$

14. $7x = 9$

15. $-4x = 10$

16. $-13x = 5$

17. $.12x = 36$

18. $.21x = 42$

19. $-x = 32$

20. $-x = 94$

21. In the statement of the multiplication property of equality in this section, there is a restriction that $C \neq 0$. What would happen if you should multiply both sides of an equation by 0?

22. Which one of the equations that follow does not require the use of the multiplication property of equality?

    (a) $3x - 5x = 6$ **(b)** $-\dfrac{1}{4}x = 12$ **(c)** $5x - 4x = 7$ **(d)** $\dfrac{x}{3} = -2$

Solve each equation and check your solution. See Examples 1–6.

23. $2m = 15$

24. $3m = 10$

25. $3a = -15$

26. $5k = -70$

27. $10t = -36$

28. $4s = -34$

29. $-6x = -72$

30. $-8x = -64$

31. $2r = 0$

32. $5x = 0$

33. $-y = 12$

34. $-t = 14$

35. $.2t = 8$ **36.** $.9x = 18$ **37.** $\frac{1}{4}y = -12$ **38.** $\frac{1}{5}p = -3$

39. $\frac{x}{7} = -5$ **40.** $\frac{k}{8} = -3$ **41.** $-\frac{7}{9}c = \frac{3}{5}$ **42.** $-\frac{5}{6}d = \frac{4}{9}$

43. $4x + 3x = 21$ **44.** $9x + 2x = 121$ **45.** $3r - 5r = 10$ **46.** $9p - 13p = 24$

47. $5m + 6m - 2m = 63$ **48.** $11r - 5r + 6r = 168$

49. $5x + 2 = 8x + 8$ **50.** $2y - 4 = 7y + 1$

51. $9w - 5w + 3 = -w$ **52.** $7 + 6k - 2k = 8$

53. Write an equation that requires the use of the multiplication property of equality, where both sides must be multiplied by $\frac{2}{3}$, and the solution set consists of a single negative number.

54. Write an equation that requires the use of the multiplication property of equality, where both sides must be divided by 100, and the solution set does not contain an integer.

Write an equation using the information given in the problem. Use x as the variable. Then solve the equation.

55. Three times a number is 18 more than five times the number. Find the number.

56. If four times a number is added to three times the number, the result is the sum of five times the number and 10. Find the number.

2.3 *More on Solving Linear Equations*

OBJECTIVES

1 ► Learn the four steps for solving a linear equation and how to use them.

2 ► Solve equations by clearing fractions and decimals.

FOR EXTRA HELP

Tutorial   Tape 4   SSM, Sec. 2.3

OBJECTIVE 1 ► In this section we use the addition and multiplication properties together to solve more complicated equations. We will use the following four-step method.

> **Solving Linear Equations**
>
> *Step 1* Clear parentheses using the distributive property, if needed; combine terms.
>
> *Step 2* Use the addition property to simplify further if necessary, so that the variable term is on one side of the equation and a number is on the other.
>
> *Step 3* Use the multiplication property if necessary to get the equation in the form $x = $ a number.
>
> *Step 4* Check your answer by substituting into the *original* equation.

The check is used only to catch errors in carrying out the steps.

E X A M P L E   1   Using the Four Steps to Solve an Equation

Solve the equation $3r + 4 - 2r - 7 = 4r + 3$.

We use the four steps described above.

$$\text{Step 1} \quad 3r + 4 - 2r - 7 = 4r + 3$$
$$r - 3 = 4r + 3 \qquad \text{Combine like terms.}$$

$$\text{Step 2} \quad r - 3 + 3 = 4r + 3 + 3 \qquad \text{Add 3.}$$
$$r = 4r + 6$$
$$r - 4r = 4r + 6 - 4r \qquad \text{Subtract } 4r.$$
$$-3r = 6 \qquad \text{Combine terms.}$$

$$\text{Step 3} \quad \frac{-3r}{-3} = \frac{6}{-3} \qquad \text{Divide by } -3.$$
$$r = -2 \qquad \text{Reduce.}$$

Step 4 Substitute -2 for r in the original equation.

$$3r + 4 - 2r - 7 = 4r + 3$$
$$3(-2) + 4 - 2(-2) - 7 = 4(-2) + 3 \qquad ? \quad \text{Let } r = -2.$$
$$-6 + 4 + 4 - 7 = -8 + 3 \qquad ? \quad \text{Multiply.}$$
$$-5 = -5 \qquad \text{True}$$

The solution set is $\{-2\}$.

In Step 2 of Example 1, we added and subtracted the terms in such a way that the variable term ended up on the left side of the equation. Choosing differently would have put the variable term on the right side of the equation. Usually there is no real advantage either way.

WORK PROBLEM 1 AT THE SIDE. ▶▶

E X A M P L E   2   Using the Four Steps to Solve an Equation

Solve the equation $4(k - 3) - k = k - 6$.

$$\text{Step 1} \quad 4(k - 3) - k = k - 6$$
$$4k - 12 - k = k - 6 \qquad \text{Distributive property}$$
$$3k - 12 = k - 6 \qquad \text{Combine terms.}$$

── **CONTINUED ON NEXT PAGE**

1. Solve.

(a) $5y - 7y + 6y - 9$
$= 3 + 2y$

(b) $-3k - 5k - 6 + 11$
$= 2k - 5$

2. Solve.

(a) $7(p - 2) + p = 2p + 4$

Step 2

$$3k - 12 + 12 = k - 6 + 12 \qquad \text{Add 12.}$$
$$3k = k + 6 \qquad \text{Combine terms.}$$
$$3k - k = k + 6 - k \qquad \text{Subtract } k.$$
$$2k = 6 \qquad \text{Combine terms.}$$

Step 3

$$\frac{2k}{2} = \frac{6}{2} \qquad \text{Divide by 2.}$$
$$k = 3 \qquad \text{Reduce.}$$

Step 4 Check this answer by substituting 3 for k in the given equation. Remember to do all the work inside the parentheses first.

$$4(k - 3) - k = k - 6$$
$$4(3 - 3) - 3 = 3 - 6 \quad ? \quad \text{Let } k = 3.$$
$$4(0) - 3 = 3 - 6 \quad ? \quad 3 - 3 = 0$$
$$0 - 3 = 3 - 6 \quad ? \quad 4(0) = 0$$
$$-3 = -3 \qquad \text{True}$$

The solution set is $\{3\}$.

◀◀ **WORK PROBLEM 2 AT THE SIDE.**

E X A M P L E   3   Using the Four Steps to Solve an Equation

Solve the equation $8a - (3 + 2a) = 3a + 1$.

Step 1 Simplify.

$$8a - (3 + 2a) = 3a + 1$$
$$8a - 1 \cdot (3 + 2a) = 3a + 1 \qquad \text{Multiplicative identity property}$$
$$8a - 3 - 2a = 3a + 1 \qquad \text{Distributive property}$$
$$6a - 3 = 3a + 1 \qquad \text{Combine terms.}$$

(b) $3(m + 5) - 1 + 2m$
$\quad = 5(m + 2)$

Step 2 First, add 3 to each side; then subtract $3a$.

$$6a - 3 + 3 = 3a + 1 + 3 \qquad \text{Add 3.}$$
$$6a = 3a + 4 \qquad \text{Combine terms.}$$
$$6a - 3a = 3a + 4 - 3a \qquad \text{Subtract } 3a.$$
$$3a = 4 \qquad \text{Combine terms.}$$

Step 3

$$\frac{3a}{3} = \frac{4}{3} \qquad \text{Divide by 3.}$$
$$a = \frac{4}{3} \qquad \tfrac{3}{3} = 1;\ 1a = a$$

Step 4 Check that the solution set is $\left\{\frac{4}{3}\right\}$.

Caution

Be very careful with signs when solving an equation like the one in Example 3. When clearing parentheses in the expression

$$8a - (3 + 2a),$$

remember that the $-$ sign acts like a factor of -1, changing the sign of *every* term in the parentheses. Thus,

$$8 - (3 + 2a) = 8 - 3 - 2a.$$
$$\qquad\qquad\quad \uparrow \quad \uparrow$$
$$\text{Change to } - \text{ in both terms.}$$

ANSWERS

2. (a) $\{3\}$ **(b)** $\emptyset$

WORK PROBLEM 3 AT THE SIDE. ▶▶

3. Solve.

(a) $7m - (2m - 9) = 39$

E X A M P L E 4   Using the Four Steps to Solve an Equation

Solve the equation $4(8 - 3t) = 32 - 8(t + 2)$.

Step 1 $\quad 4(8 - 3t) = 32 - 8(t + 2)$

$\qquad\quad 32 - 12t = 32 - 8t - 16$ Distributive property

$\qquad\quad 32 - 12t = 16 - 8t$ Combine terms.

Step 2 $32 - 12t - \mathbf{32} = 16 - 8t - \mathbf{32}$ Subtract 32.

$\qquad\qquad\quad -12t = -16 - 8t$ Combine terms.

$\qquad\quad -12t + \mathbf{8t} = -16 - 8t + \mathbf{8t}$ Add 8t.

$\qquad\qquad\quad -4t = -16$ Combine terms.

Step 3 $\qquad \dfrac{-4t}{-4} = \dfrac{-16}{-4}$ Divide by -4.

$\qquad\qquad\quad t = 4$ Reduce.

(b) $4x + 2(3 - 2x) = 6$

Step 4 Check this solution in the given equation.

$\qquad\quad 4(8 - 3t) = 32 - 8(t + 2)$

$\qquad 4(8 - 3 \cdot \mathbf{4}) = 32 - 8(\mathbf{4} + 2)$?   Let $t = 4$.

$\qquad 4(8 - 12) = 32 - 8(6)$?   Combine terms.

$\qquad\quad 4(-4) = 32 - 48$?   Combine terms.

$\qquad\quad\;\; -16 = -16$ True

The solution, 4, checks, and the solution set is $\{4\}$.

4. Solve.

(a) $2(4 + 3r)$
$\quad = 3(r + 1) + 11$

WORK PROBLEM 4 AT THE SIDE. ▶▶

OBJECTIVE 2▶ We can clear an equation of fractions by multiplying both sides by the LCD of all the fractions in the equation. It is a good idea to do this before starting the four-step method to avoid working with fractions.

E X A M P L E 5   Clearing an Equation of Fractions

Solve $\frac{2}{3}x - \frac{1}{2}x = -\frac{1}{6}x - 2$.

The least common denominator of all the fractions in the equation is 6. Start by multiplying both sides of the equation by 6.

$$\frac{2}{3}x - \frac{1}{2}x = -\frac{1}{6}x - 2$$

$$6\left(\frac{2}{3}x - \frac{1}{2}x\right) = 6\left(-\frac{1}{6}x - 2\right) \qquad \text{Multiply by 6.}$$

$$6\left(\frac{2}{3}x\right) + 6\left(-\frac{1}{2}x\right) = 6\left(-\frac{1}{6}x\right) + 6(-2) \qquad \text{Distributive property}$$

$$4x - 3x = -x - 12$$

(b) $2 - 3(2 + 6z)$
$\quad = 4(z + 1) + 18$

CONTINUED ON NEXT PAGE

5. Solve $\frac{1}{4}x - 4 = \frac{3}{2}x + \frac{3}{4}x$.

Now use the four steps to solve this equivalent equation.

Step 1	$x = -x - 12$	Combine like terms.
Step 2	$x + x = x - x - 12$	Add x.
	$2x = -12$	
Step 3	$\dfrac{2x}{2} = \dfrac{-12}{2}$	Divide by 2.
	$x = -6$	

Step 4 Check by substituting -6 for x in the original equation.

$$\frac{2}{3}(-6) - \frac{1}{2}(-6) = -\frac{1}{6}(-6) - 2 \quad ? \quad \text{Let } x = -6.$$

$$-4 + 3 = 1 - 2 \qquad ?$$

$$-1 = -1 \qquad \text{True}$$

The solution set is $\{-6\}$.

> **Caution**
> When clearing equations of fractions, be sure to multiply *every* term on both sides of the equation by the LCD.

◀◀ **WORK PROBLEM 5 AT THE SIDE.**

The multiplication property can also be used to clear an equation of decimals.

6. Solve $.06(100 - y) + .04y = .05(92)$.

E X A M P L E 6   Clearing an Equation of Decimals

Solve $.10t + .05(20 - t) = .09(20)$.

Since the decimals are all hundredths, start the solution by multiplying both sides of the equation by 100. A number can be multiplied by 100 by moving the decimal point two places to the right.

$$.10t + .05(20 - t) = .09(20)$$

$$10t + 5(20 - t) = 9(20)$$

Now use the four steps.

Step 1	$10t + 5(20) + 5(-t) = 180$	Distributive property
	$10t + 100 - 5t = 180$	
	$5t + 100 = 180$	Combine terms.
Step 2	$5t + 100 - 100 = 180 - 100$	Subtract 100.
	$5t = 80$	Combine terms.
Step 3	$\dfrac{5t}{5} = \dfrac{80}{5}$	Divide by 5.
	$t = 16$	

Step 4 Check to see that $\{16\}$ is the solution set by substituting 16 into the original equation.

◀◀ **WORK PROBLEM 6 AT THE SIDE.**

ANSWERS

5. $\{-2\}$

6. $\{70\}$

2.3 Exercises

Solve each equation, and check your solution. See Examples 1–4.

1. $5m + 8 = 7 + 4m$

2. $4r + 2 = 3r - 6$

3. $10p + 6 = 12p - 4$

4. $-5x + 8 = -3x + 10$

5. $7r - 5r + 2 = 5r - r$

6. $9p - 4p + 6 = 7p - 3p$

7. $x + 3 = -(2x + 2)$

8. $2x + 1 = -(x + 3)$

9. $4(2x - 1) = -6(x + 3)$

10. $6(3w + 5) = 2(10w + 10)$

11. $6(4x - 1) = 12(2x + 3)$

12. $6(2x + 8) = 4(3x - 6)$

13. $3(2x - 4) = 6(x - 2)$

14. $3(6 - 4x) = 2(-6x + 9)$

15. After working correctly through several steps of the solution of a linear equation, a student obtains the equation $7x = 3x$. Then the student divides both sides by x to get $7 = 3$ and gives $\emptyset$ as the solution set. Is this correct? If not, explain why.

16. Which one of the following linear equations does *not* have {all real numbers} as its solution set?

    (a) $5x = 4x + x$ **(b)** $2(x + 6) = 2x + 12$ **(c)** $\frac{1}{2}x = .5x$ **(d)** $3x = 2x$

17. Explain in your own words the major steps used in solving a linear equation that does not contain fractions or decimals as coefficients.

18. Explain in your own words the major steps used in solving a linear equation that contains fractions or decimals as coefficients.

Solve each equation by first clearing it of fractions or decimals. See Examples 5 and 6.

19. $\dfrac{3}{5}t - \dfrac{1}{10}t = t - \dfrac{5}{2}$

20. $-\dfrac{2}{7}r + 2r = \dfrac{1}{2}r + \dfrac{17}{2}$

21. $-\dfrac{1}{4}(x - 12) + \dfrac{1}{2}(x + 2) = x + 4$

22. $\dfrac{1}{9}(y + 18) + \dfrac{1}{3}(2y + 3) = y + 3$

23. $\dfrac{2}{3}k - \left(k + \dfrac{1}{4}\right) = \dfrac{1}{12}(k + 4)$

24. $-\dfrac{5}{6}q - \left(q - \dfrac{1}{2}\right) = \dfrac{1}{4}(q + 1)$

25. $.20(60) + .05x = .10(60 + x)$

26. $.30(30) + .15x = .20(30 + x)$

27. $1.00x + .05(12 - x) = .10(63)$

28. $.92x + .98(12 - x) = .96(12)$

29. $.06(10,000) + .08x = .072(10,000 + x)$

30. $.02(5000) + .03x = .025(5000 + x)$

MATHEMATICAL CONNECTIONS (EXERCISES 31–36)

Work Exercises 31–36 in order.

31. Evaluate the term $100ab$ for $a = 2$ and $b = 4$.

32. Will you get the same answer as in Exercise 31 if you evaluate $(100a)b$ for $a = 2$ and $b = 4$? Why or why not?

33. Is the term $(100a)(100b)$ equivalent to $100ab$? Why or why not?

34. If your answer to Exercise 33 is *no,* explain why the distributive property is not involved.

35. The simplest way to solve the equation $.05(x + 2) + .10x = 2.00$ is to begin by multiplying both sides by 100. If we do this, the first term on the left becomes $100 \cdot .05(x + 2)$. Is this expression equivalent to $[100 \cdot .05](x + 2)$? (*Hint:* Compare to Exercises 31 and 32 with $a = .05$ and $b = x + 2$.)

36. Students often want to "distribute" the 100 to both $.05$ and $(x + 2)$ in the expression $100 \cdot .05(x + 2)$. Is this correct? (*Hint:* See Exercises 34 and 35.)

Solve each equation, and check your solution. See Examples 1–6.

37. $-2(2s - 4) - 8 = -3(4s + 4) - 1$ 　　　　　**38.** $-3(5z + 24) + 2 = 2(3 - 2z) - 4$

39. $-(4y + 2) - (-3y - 5) = 3$ 　　　　　**40.** $-(6k - 5) - (-5k + 8) = -3$

41. $\dfrac{1}{2}(x + 2) + \dfrac{3}{4}(x + 4) = x + 5$

42. $\dfrac{1}{3}(x + 3) + \dfrac{1}{6}(x - 6) = x + 3$

43. $.10(x + 80) + .20x = 14$

44. $.30(x + 15) + .40(x + 25) = 25$

45. $4(x + 8) = 2(2x + 6) + 20$

46. $4(x + 3) = 2(2x + 8) - 4$

47. $9(v + 1) - 3v = 2(3v + 1) - 8$

48. $8(t - 3) + 4t = 6(2t + 1) - 10$

Write the answer to the problem as an algebraic expression.

49. Two numbers have a sum of 11. One number is q. Find the other number.

50. The product of two numbers is 9. One number is k. What is the other number?

51. A bank teller has t dollars in five-dollar bills. How many five-dollar bills does the teller have?

52. A plane ticket costs b dollars for an adult and d dollars for a child. Find the total cost of 3 adult and 2 child tickets.

2.4 An Introduction to Applied Problems

Earlier, we practiced translating words, phrases, and sentences into mathematical expressions and equations. Now we will begin to use these translations in solving applied problems using algebra. Some of the problems will seem contrived, and to some extent they are. But the skills you develop in solving simple problems will help you in solving more realistic problems in chemistry, biology, business, and other fields. We suggest the following general procedure for solving applied problems.

OBJECTIVES

1 ▶ Translate sentences of an applied problem into an equation, and then solve the problem.

2 ▶ Learn the definitions of complementary and supplementary angles and use them to write equations to solve applied problems.

FOR EXTRA HELP

Tutorial   Tape 5   SSM, Sec. 2.4

Solving Applied Problems

Step 1 Read the problem carefully, and choose a variable to represent the numerical value that you are asked to find—the unknown number. *Write down* what the variable represents.

Step 2 *Write down* a mathematical expression using the variable for any other unknown quantities. Draw figures or diagrams if they apply.

Step 3 Translate the problem into an equation.

Step 4 Solve the equation.

Step 5 Answer the question asked in the problem.

Step 6 Check your solution by using the words of the original problem. Be sure that the answer is appropriate and makes sense.

OBJECTIVE 1 ▶ The third step in solving an applied problem is often the hardest. Begin to translate the problem into an equation by writing the given phrases as mathematical expressions. In transforming an applied problem into an algebraic equation, replace any words that mean *equal* or *same* with an $=$ sign. Other forms of the verb "to be," such as *is, are, was,* and *were,* also translate this way. The $=$ sign leads to an equation to be solved.

┌ **E X A M P L E   1   Finding an Unknown Number**

If three times the sum of a number and 4 is decreased by twice the number, the result is -6. Find the number.

Step 1   Read the problem carefully. We are asked to find a number, so we write

Let x represent the number.

Step 2 $3(x + 4)$ = three times the sum of the number and 4

$2x$ = twice the number

Step 3   Translate the information given in the problem.

Three times
the sum of a   decreased   twice the
number and 4   by   number   is -6.

$\downarrow$ $\downarrow$ $\downarrow$ $\downarrow$ $\downarrow$

$3(x + 4)$ $-$ $2x$ $=$ -6

Step 4   Solve the equation.

$$3(x + 4) - 2x = -6$$
$$3x + 12 - 2x = -6 \quad \text{Distributive property}$$
$$x + 12 = -6 \quad \text{Combine terms.}$$
$$x = -18 \quad \text{Subtract 12.}$$

Solution set: $\{-18\}$

└ **CONTINUED ON NEXT PAGE**

1. Write an equation, find the solution set, and then solve the problem. Use x as the variable. If 5 is added to the product of 9 and a number, the result is 19 less than the number. Find the number.

Step 5 The number is -18.

Step 6 Check that -18 is the correct answer by substituting this result into the words of the original problem. Three times the sum of -18 and 4 is $3(-18 + 4) = 3(-14) = -42$. Twice -18 is -36; subtract -36 from -42 to get $-42 - (-36) = -6$, as required.

◀◀ **WORK PROBLEM 1 AT THE SIDE.**

See if you can tell how the six steps are used in the rest of the examples.

E X A M P L E 2   Finding the Number of Coffee Orders

The owner of a small café found one day that the number of orders for tea was $\frac{1}{3}$ the number of orders for coffee. If the total number of orders for the two drinks was 76, how many orders were placed for coffee?

$$\text{Let} \quad x = \text{the number of orders for coffee}$$

$$\frac{1}{3}x = \text{the number of orders for tea.}$$

To set up an equation, use the fact that the total number of orders was 76.

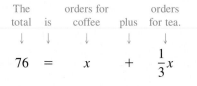

$$76 \quad = \quad x \quad + \quad \frac{1}{3}x$$

Now solve the equation.

$$76 = \frac{4}{3}x \qquad \text{Combine terms.}$$

$$\frac{3}{4}(76) = \frac{3}{4}\left(\frac{4}{3}x\right) \qquad \text{Multiply by } \frac{3}{4}.$$

$$57 = x \qquad \text{Combine terms.}$$

Solution set: {57}

There were 57 orders for coffee. The 57 coffee orders and the $(\frac{1}{3})(57) =$ 19 orders for tea give a total of $57 + 19 = 76$ orders, as required.

2. On one day of their vacation, Annie drove three times as far as Jim. Altogether they drove 84 miles that day. Find the number of miles driven by each.

◀◀ **WORK PROBLEM 2 AT THE SIDE.**

Sometimes it is necessary to find three unknown quantities in an applied problem. Frequently the three unknowns are compared in *pairs*. When this happens, it is usually easiest to let the variable represent the unknown found in both pairs. The next example illustrates this idea.

E X A M P L E 3   Dividing a Board into Pieces

Maria Gonzales is building a cabinet. The instructions require three pieces of shelving. The longest piece must be twice the length of the middle-sized piece, and the shortest piece must be 10 inches shorter than the middle-sized piece. She has a piece of shelving 70 inches long. How long will each of the three pieces be if she cuts them from this 70-inch piece?

ANSWERS

1. $9x + 5 = x - 19$; $\{-3\}$; -3

2. 21 miles for Jim; 63 miles for Annie

CONTINUED ON NEXT PAGE

Since the middle-sized piece appears in both pairs of comparisons, let x represent the length of the middle-sized piece. We have

$$x = \text{the length of the middle-sized piece}$$
$$2x = \text{the length of the longest piece}$$
$$x - 10 = \text{the length of the shortest piece.}$$

A sketch is helpful here.

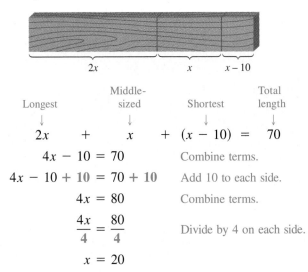

Longest	Middle-sized	Shortest	Total length
↓	↓	↓	↓

$$2x \quad + \quad x \quad + \quad (x - 10) \quad = \quad 70$$

$$4x - 10 = 70 \qquad \text{Combine terms.}$$
$$4x - 10 + 10 = 70 + 10 \qquad \text{Add 10 to each side.}$$
$$4x = 80 \qquad \text{Combine terms.}$$
$$\frac{4x}{4} = \frac{80}{4} \qquad \text{Divide by 4 on each side.}$$
$$x = 20$$

Solution set: {20}

The length of the middle-sized piece is 20 inches, the length of the longest piece is $2(20) = 40$ inches, and the length of the shortest piece is $20 - 10 = 10$ inches. Check to see that the sum of the three lengths is 70 inches.

WORK PROBLEM 3 AT THE SIDE. ▶▶

E X A M P L E 4   Analyzing a Gasoline/Oil Mixture

A lawn trimmer uses a mixture of gasoline and oil. For each ounce of oil, the mixture contains 16 ounces of gasoline. If the tank holds 68 ounces of the mixture, how many ounces of oil and how many ounces of gasoline does it require when it is full?

Let x = the number of ounces of oil required when full

$16x$ = the number of ounces of gasoline required when full.

Lawn trimmer   Gasoline   Oil

CONTINUED ON NEXT PAGE

3. A piece of pipe is 50 inches long. It is cut into three pieces. The longest piece is 10 inches more than the middle-sized piece, and the shortest piece measures 5 inches less than the middle-sized piece. Find the lengths of the three pieces.

4. At a meeting of the local stamp club, each member brought two non-members. If a total of 27 people attended, how many were members and how many were non-members?

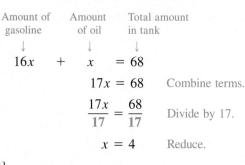

$$16x \quad + \quad x \quad = 68$$

$$17x = 68 \qquad \text{Combine terms.}$$

$$\frac{17x}{17} = \frac{68}{17} \qquad \text{Divide by 17.}$$

$$x = 4 \qquad \text{Reduce.}$$

Solution set: {4}

When the tank is full, it holds 4 ounces of oil and $16(4) = 64$ ounces of gasoline. This checks, since $4 + 64 = 68$.

◀◀ **WORK PROBLEM 4 AT THE SIDE.**

OBJECTIVE 2▶ The final examples of problem solving in this section deal with concepts from geometry. An angle can be measured by a unit called the **degree** (°). See Figure 1. Two angles whose sum is 90° are said to be **complementary** (kahm-pluh-MENT-uh-ree), or complements of each other. To remember this definition, think of the angles as *completing* a 90° angle, called a **right angle.**

Two angles whose sum is 180° are said to be **supplementary** (suh-pluh-MENT-uh-ree), or supplements of each other. One angle *supplements* the other to form a 180° angle, called a **straight angle.**

If x represents the degree measure of an angle, then

$90 - x$ represents the degree measure of its complement, and

$180 - x$ represents the degree measure of its supplement.

Angle of measure 1°

FIGURE 1

E X A M P L E   5   Finding the Measure of an Angle Given Information About Its Complement

Find the measure of an angle whose complement measures 27°.

Let the measure of the angle to be found be represented by x. Because an angle and its complement have measures that add to 90°,

$$x + 27 = 90$$
$$x + 27 - 27 = 90 - 27 \qquad \text{Subtract 27.}$$
$$x = 63.$$

Solution set: {63}

Since $63 + 27 = 90$ degrees, the answer is correct.

E X A M P L E   6   Finding the Measure of an Angle Given Information About Its Supplement

Find the measure of an angle whose measure is 25° less than the measure of its supplement.

Let x represent the measure of the unknown angle. Then $x + 25$ represents the measure of its supplement, since x must be 25 less than the

CONTINUED ON NEXT PAGE

measure of the supplement. The sum of the two angle measures must equal 180°.

$$x + x + 25 = 180$$
$$2x + 25 = 180 \qquad \text{Combine terms.}$$
$$2x + 25 - 25 = 180 - 25 \qquad \text{Subtract 25.}$$
$$2x = 155$$
$$x = 77.5$$

Solution set: {77.5}

The measure of the required angle is 77.5 degrees. The supplement measures $77.5 + 25 = 102.5$ degrees. Their sum is 180 degrees, so the answer checks.

WORK PROBLEM 5 AT THE SIDE. ▶▶

E X A M P L E   7   Finding the Measure of an Angle

Find the measure of an angle whose supplement is 10 degrees more than twice its complement.

Let $\qquad x =$ the degree measure of the angle;

$\qquad 90 - x =$ the degree measure of its complement;

$\qquad 180 - x =$ the degree measure of its supplement.

$$\begin{array}{ccccccc} \text{Supplement} & \text{is} & 10 & \text{more than} & \text{twice} & & \text{its complement.} \\ \downarrow & \downarrow & \downarrow & \downarrow & \downarrow & & \downarrow \\ 180 - x & = & 10 & + & 2 & \cdot & (90 - x) \end{array}$$

Solve the equation.

$$180 - x = 10 + 180 - 2x \qquad \text{Distributive property}$$
$$180 - x = 190 - 2x \qquad \text{Combine terms.}$$
$$180 - x + 2x = 190 - 2x + 2x \qquad \text{Add } 2x.$$
$$180 + x = 190 \qquad \text{Combine terms.}$$
$$180 + x - 180 = 190 - 180 \qquad \text{Subtract 180.}$$
$$x = 10$$

Solution set: {10}

The measure of the angle is 10 degrees. The complement of 10° is 80° and the supplement of 10° is 170°. 170° is equal to 10° more than twice 80° ($170 = 10 + 2(80)$ is true); therefore, the answer is correct.

WORK PROBLEM 6 AT THE SIDE. ▶▶

5. Find each of the following.

(a) The supplement of an angle that measures 92°

(b) The measure of an angle whose complement has twice its measure

6. Twice the complement of an angle is 30° less than its supplement. Find the measure of the angle.

ANSWERS

5. (a) 88° **(b)** 30°
6. 30°

Real World   *collaborative investigations*

Just How Is Windchill Factor Determined?

Meteorologists often refer to the windchill factor, which is a measure of the cooling effect that the wind has on a person's skin. It calculates the equivalent cooling temperature, if there were no wind. The table gives the windchill factor for various wind speeds and temperatures.

Wind/°F	40°	30°	20°	10°	0°	−10°	−20°	−30°	−40°	−50°
5 mph	37	27	16	6	−5	−15	−26	−36	−47	−57
10 mph	28	16	4	−9	−21	−33	−46	−58	−70	−83
15 mph	22	9	−5	−18	−36	−45	−58	−72	−85	−99
20 mph	18	4	−10	−25	−39	−53	−67	−82	−96	−110
25 mph	16	0	−15	−29	−44	−59	−74	−88	−104	−118
30 mph	13	−2	−18	−33	−48	−63	−79	−94	−109	−125
35 mph	11	−4	−20	−35	−49	−67	−82	−98	−113	−129
40 mph	10	−6	−21	−37	−53	−69	−85	−100	−116	−132

Source: Miller, A. and J. Thompson, *Elements of Meteorology*, Second Edition, Charles E. Merrill Publishing Co., 1975.

Suppose that we wish to determine the difference between two of these entries, and are interested only in the magnitude, or absolute value, of this difference. Then we subtract the two entries and find the absolute value. For example, the difference in windchill factors for wind at 20 miles per hour with a 20° F temperature and wind at 30 miles per hour with a 40° F temperature is $|-10° - 13°| = 23°$ F, or equivalently, $|13° - (-10°)| = 23°$ F.

FOR GROUP DISCUSSION

1. Suppose that it is 10° F outside and the wind picks up from 20 mph to 25 mph. By how much does the windchill factor decrease?

2. Suppose that a steady wind of 15 mph is blowing, and over a short period of time the windchill factor drops 18°. What were the starting and ending temperatures?

2.4 Exercises

1. List some words that will translate as "=" in an applied problem.

2. A problem requires finding the number of packages delivered in one day. Which one of the following would not be a reasonable answer?

(a) 7   (b) 0   (c) $6\frac{2}{3}$ (d) 15

3. Which one of the following would not be a reasonable answer to a problem that requires finding the number of pounds of coffee sold in a coffee shop?

(a) $14\frac{1}{2}$ (b) -9 (c) 8   (d) 10.25

4. In your own words, explain the general procedure for solving applied problems described in this section.

Solve the following problems. See Example 1.

5. If 1 is added to a number and this sum is doubled, the result is 5 more than the number. Find the number.

6. If 2 is subtracted from a number and this difference is tripled, the result is 4 more than the number. Find the number.

7. If 3 is added to twice a number and this sum is multiplied by 4, the result is the same as if the number is multiplied by 7 and 8 is added to the product. What is the number?

8. The sum of three times a number and 12 more than the number is the same as the difference between -6 and twice the number. What is the number?

Solve the following problems. See Examples 2 and 3.

9. The U.S. Senate has 100 members. After the 1990 election, there were 14 more Democrats than Republicans, with no other parties represented. How many members of each party were there in the Senate?

10. The total number of Democrats and Republicans in the U.S. House of Representatives in 1990 was 434. There were 100 fewer Republicans than Democrats. How many members of each party were there in the House of Representatives?

11. In his coaching career with the Boston Celtics, Red Auerbach had 558 more wins than losses. His total number of games coached was 1516. How many wins did Auerbach have?

12. In the first Super Bowl, played in 1966, Green Bay and Kansas City scored a total of 45 points. Green Bay won by 25 points. What was the score of the first Super Bowl?

13. Nagaraj Nanjappa has a strip of paper 39 inches long. He wants to cut it into two pieces so that one piece will be 9 inches shorter than the other. How long should each of the two pieces be?

14. The largest recorded dog was an English Mastiff named Zorba, who weighed 63 pounds more than an average lioness. The sum of the two animals' weights is 623 pounds. How much do the dog and the lioness each weigh? (*Source: The Guiness Book of Records,* 1996. Bantam Books, New York)

15. In one day, the newsroom received 13 tips of late-breaking news events. Murphy covered three times as many stories as Frank, while Frank covered two more than Corky. How many stories did each reporter cover?

16. In her job at the post office, Janie Quintana works a $6\frac{1}{2}$-hour day. She sorts mail, sells stamps, and does supervisory work. On one day, she sold stamps twice as long as she sorted mail, and sold stamps 1 hour longer than the time she spent doing supervisory work. How many hours did she spend at each task?

17. Venus is 31.2 million miles farther from the sun than Mercury, while Earth is 25.7 million miles farther from the sun than Venus. If the total of the distances for these three planets from the sun is 196.1 million miles, how far away from the sun is Mercury? (All distances given here are *mean* (*average*) distances.)

18. It is believed that Saturn has 5 more satellites (moons) than the known number of satellites for Jupiter, and 20 more satellites than the known number for Mars. If the total of these numbers is 41, how many satellites does Mars have?

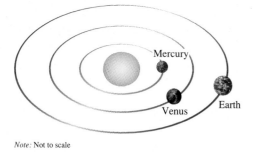

Note: Not to scale

19. During the 1996 baseball season, Atlanta Braves pitchers John Smoltz, Greg Maddux, and Mark Wohlers pitched a total of 576 innings. Maddux pitched $8\frac{2}{3}$ fewer innings than Smoltz, and $167\frac{2}{3}$ more innings than Wohlers. How many innings did each player pitch?

20. During their World Championship year of 1996, New York Yankees pitchers Dwight Gooden, David Cone, and Jeff Nelson pitched a total of 317 innings. Gooden pitched $96\frac{1}{3}$ innings more than Nelson, and Cone pitched $2\frac{1}{3}$ fewer innings than Nelson. How many innings did each player pitch?

21. The sum of the measures of the angles of any triangle is 180 degrees. In triangle *ABC*, angles *A* and *B* have the same measure, while the measure of angle *C* is 60 degrees larger than each of *A* and *B*. What are the measures of the three angles?

22. (See Exercise 21.) In triangle *ABC*, the measure of angle *A* is 141 degrees more than the measure of angle *B*. The measure of angle *B* is the same as the measure of angle *C*. Find the measure of each angle.

Solve the following problems. See Example 4.

23. The largest sheep ranch in the world is located in Australia. The number of sheep on the ranch is $\frac{8}{3}$ of the number of uninvited kangaroos grazing on the pastureland. Together, herds of these two animals number 88,000. How many sheep and how many kangaroos roam the ranch? (*Source: The Guiness Book of Records,* 1996. Bantam Books, New York)

24. The 1997 edition of *A Guide Book of United States Coins* lists the value of a Mint State-65 (uncirculated) 1950 Jefferson nickel minted at Denver as $\frac{4}{3}$ the value of a similar condition 1944 nickel minted at Philadelphia. Together the total value of the two coins is $14.00. What is the value of each coin?

25. A husky running the Iditarod (a thousand-mile race between Anchorage and Nome, Alaska) burns $5\frac{3}{8}$ calories in exertion for every 1 calorie burned in thermoregulation in extreme cold. According to one scientific study, a husky in top condition burns an amazing total of 11,200 calories per day. How many calories are burned for exertion and how many are burned for regulation of body temperature?

26. In 1988, a dairy in Alberta, Canada, created a sundae with approximately 1 pound of topping for every 83.2 pounds of ice cream. The total of the two ingredients weighed approximately 45,225 pounds. To the nearest tenth of a pound, how many pounds of ice cream and how many pounds of topping were there? (*Source: The Guiness Book of Records,* 1996. Bantam Books, New York)

Solve the following problems. See Examples 5 and 6.

27. Find the measure of an angle whose supplement is 3 times its measure.

28. Find the measure of an angle whose complement is 4 times its measure.

29. Find the measure of an angle whose supplement measures 38° less than three times its complement.

30. Find the measure of an angle whose supplement measures 39° more than twice its complement.

31. Find the measure of an angle such that the sum of the measures of its complement and its supplement is 160°.

32. Find the measure of an angle such that the difference between the measures of its supplement and three times its complement is 10°.

The following problems are a bit different from those described in the examples of this section. Solve these using the general method described. (In Exercises 33–36, consecutive integers and consecutive even integers are mentioned. Some examples of consecutive integers are 5, 6, 7, and 8; some examples of consecutive even integers are 4, 6, 8, and 10.)

33. If x represents an integer, then $x + 1$ represents the next larger consecutive integer. If the sum of two consecutive integers is 243, find the integers.

34. (See Exercise 33.) If the sum of two consecutive integers is -29, find the integers.

35. If x represents an even integer, then $x + 2$ represents the next larger even integer. Find two consecutive even integers such that the smaller added to three times the larger equals 86.

36. (See Exercise 35.) Find two consecutive even integers such that six times the smaller added to the larger gives a sum of 58.

Apply the ideas of this section to solve the problem based on the graph.

37. In 1991, the funding for Head Start programs increased by .50 billion dollars from the funding in 1990. In 1992, the increase was .25 billion dollars over the funding in 1991. For those three years the total funding was 5.6 billion dollars. How much was funded in each of these years? (*Source:* U.S. Department of Health and Human Services)

38. According to data provided by the National Safety Council, in 1992 the number of serious injuries per 100,000 participants in football, bicycling, and golf are illustrated in the graph. There were 800 more injuries in bicycling than in golf, and there were 1267 more in football than in bicycling. Altogether there were 3179 serious injuries per 100,000 participants. How many serious injuries were there in each sport?

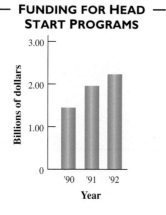

— FUNDING FOR HEAD —
START PROGRAMS

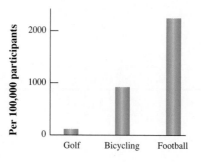

SERIOUS INJURIES IN 1992

2.5 *Formulas and Geometry Applications*

Many applied problems can be solved with a formula. For example, formulas exist for geometric figures such as squares and circles, for distance, for money earned on bank savings, and for converting English measurements to metric measurements. The formulas used in this book are given on the inside covers.

OBJECTIVE ▮▸ Given the values of all but one of the variables in a formula, we can find the value of the remaining variable by using the methods introduced in this chapter.

E X A M P L E I   **Using a Formula to Evaluate a Variable**

Find the value of the remaining variable in each of the following.

(a) $A = LW$; $A = 64$, $L = 10$

As shown in Figure 2, this formula gives the area of a rectangle with length L and width W. Substitute the given values into the formula and then solve for W.

$$A = LW$$
$$64 = 10W \qquad \text{Let } A = 64 \text{ and } L = 10.$$
$$6.4 = W \qquad \text{Divide by 10.}$$

Check that the width of the rectangle is 6.4.

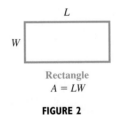

Rectangle
$A = LW$

FIGURE 2

(b) $A = \frac{1}{2}h(b + B)$; $A = 210$, $B = 27$, $h = 10$

This formula gives the area of a trapezoid with parallel sides of lengths b and B and distance h between the parallel sides. See Figure 3.

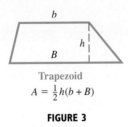

Trapezoid
$A = \frac{1}{2}h(b + B)$

FIGURE 3

Again, begin by substituting the given values into the formula.

$$A = \frac{1}{2}h(b + \boldsymbol{B})$$

$$210 = \frac{1}{2}(\boldsymbol{10})(b + \boldsymbol{27}) \qquad A = 210, B = 27, h = 10$$

CONTINUED ON NEXT PAGE

119

1. Find the value of the remaining variable in each of the following.

(a) $I = prt$; $I = \$246$, $r = .06$, $t = 2$

(b) $P = 2L + 2W$; $P = 126$, $W = 25$

2. A farmer has 800 meters of fencing material to enclose a rectangular field. The width of the field is 175 meters. Find the length of the field.

Now solve for b.

$$210 = 5(b + 27) \qquad \text{Multiply.}$$
$$210 = 5b + 135 \qquad \text{Distributive property}$$
$$210 - 135 = 5b + 135 - 135 \qquad \text{Subtract 135.}$$
$$75 = 5b \qquad \text{Combine terms.}$$
$$\frac{75}{5} = \frac{5b}{5} \qquad \text{Divide by 5.}$$
$$15 = b \qquad \text{Reduce.}$$

Solution set: $\{15\}$
Check that the length of the shorter parallel side, b, is 15.

◀◀ **WORK PROBLEM 1 AT THE SIDE.**

OBJECTIVE 2 As the next examples show, formulas are often used to solve applied problems. *It is a good idea to draw a sketch when a geometric figure is involved.* Example 2 uses the idea of *perimeter*. The **perimeter** (per-IM-it-er) of a figure is the sum of the lengths of its sides.

EXAMPLE 2   Finding the Width of a Rectangular Lot

A rectangular lot has perimeter 80 meters and length 25 meters. Find the width of the lot. See Figure 4.

We are told to find the width of the lot, so

let W = the width of the lot in meters.

The formula for the perimeter of a rectangle is

$$P = 2L + 2W.$$

Find the width by substituting 80 for P and 25 for L in the formula.

$$80 = 2(25) + 2W \qquad P = 80, L = 25$$

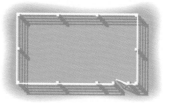

FIGURE 4

Solve the equation.

$$80 = 50 + 2W \qquad \text{Multiply.}$$
$$80 - 50 = 50 + 2W - 50 \qquad \text{Subtract 50.}$$
$$30 = 2W$$
$$15 = W \qquad \text{Divide by 2.}$$

Solution set: $\{15\}$
The width is 15 meters. Check this result. If the width is 15 meters and the length is 25 meters, the distance around the rectangular lot (perimeter) is $2(25) + 2(15) = 50 + 30 = 80$ meters, as required.

◀◀ **WORK PROBLEM 2 AT THE SIDE.**

The **area** of a geometric figure is a measure of the surface covered by the figure. Example 3 shows an application of area.

3. The area of a triangle is 120 square meters. The height is 24 meters. Find the length of the base of the triangle.

E X A M P L E 3   Finding the Height of a Triangular Sail

The area of a triangular sail of a sailboat is 126 square meters. The base of the sail is 21 meters. Find the height of the sail. See Figure 5.

Since we must find the height of the triangular sail,

let h = the height of the sail in meters.

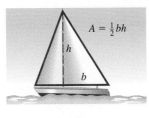

FIGURE 5

The formula for the area of a triangle is $A = \frac{1}{2}bh$, where A is the area, b is the base, and h is the height. Using the information given in the problem, substitute 126 for A and 21 for b in the formula.

$$A = \frac{1}{2}bh$$

$$126 = \frac{1}{2}(21)h \qquad A = 126, b = 21$$

Solve the equation.

$$126 = \frac{21}{2}h$$

$$\frac{2}{21}(126) = \frac{2}{21} \cdot \frac{21}{2}h \qquad \text{Multiply by } \frac{2}{21}.$$

$$12 = h \quad \text{or} \quad h = 12$$

Solution set: {12}
The height of the sail is 12 meters. Check to see that the values $A = 126$, $b = 21$, and $h = 12$ satisfy the formula for the area of a triangle.

WORK PROBLEM 3 AT THE SIDE. ▶▶

OBJECTIVE 3 Figure 6 shows two intersecting lines forming angles that are numbered ①, ②, ③, and ④. Angles ① and ③ lie "opposite" each other. They are called **vertical angles.** Another pair of vertical angles is ② and ④. In geometry, it is shown that the following property holds.

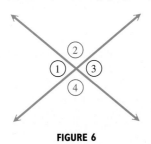

FIGURE 6

ANSWERS

3. 10 meters

4. Find the measures of each marked angle.

(a)

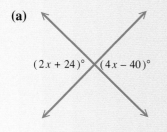

$(2x + 24)°$ $(4x - 40)°$

(b)

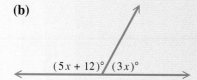

$(5x + 12)°$ $(3x)°$

Vertical Angles

Vertical angles have equal measures.

Now look at angles ① and ②. When their measures are added, we get the measure of a straight angle, which is 180°. There are three other such pairs of angles: ② and ③, ③ and ④, ① and ④.

The next example uses these ideas.

E X A M P L E   4   Finding Angle Measures

Refer to the appropriate figures in each part.

(a) Find the measure of each marked angle in Figure 7.

Since the marked angles are vertical angles, they have equal measures. Set $4x + 19$ equal to $6x - 5$ and solve.

$$4x + 19 = 6x - 5$$
$$-4x + 4x + 19 = -4x + 6x - 5 \qquad \text{Add } -4x.$$
$$19 = 2x - 5$$
$$19 + 5 = 2x - 5 + 5 \qquad \text{Add 5.}$$
$$24 = 2x$$
$$12 = x \qquad\qquad \text{Divide by 2.}$$

Solution set: {12}

Since $x = 12$, one angle has measure $4(12) + 19 = 67$ degrees. The other has the same measure, since $6(12) - 5 = 67$ as well. Each angle measures 67°.

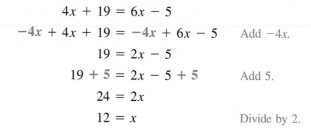

$(4x + 19)°$ $(6x - 5)°$

FIGURE 7

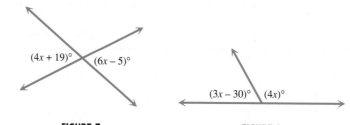

$(3x - 30)°$ $(4x)°$

FIGURE 8

(b) Find the measure of each marked angle in Figure 8.

The measures of the marked angles must add to 180° because together they form a straight angle. The equation to solve is

$$(3x - 30) + 4x = 180.$$
$$7x - 30 = 180 \qquad \text{Combine like terms.}$$
$$7x - 30 + 30 = 180 + 30 \qquad \text{Add 30.}$$
$$7x = 210$$
$$x = 30 \qquad\qquad \text{Divide by 7.}$$

Solution set: {30}

To find the measures of the angles, replace x with 30 in the two expressions.

$$3x - 30 = 3(30) - 30 = 90 - 30 = 60$$
$$4x = 4(30) = 120$$

The two angle measures are 60° and 120°.

◀◀ **WORK PROBLEM 4 AT THE SIDE.**

OBJECTIVE 4▶ Sometimes it is necessary to solve a large number of problems that use the same formula. For example, a surveying class might need to solve several problems that involve the formula for the area of a rectangle, $A = LW$. Suppose that in each problem the area (A) and the length (L) of a rectangle are given, and the width (W) must be found. Rather than solving for W each time the formula is used, it would be simpler to rewrite the *formula* so that it is solved for W. This process is called **solving for a specified variable** (SPESS-if-eyed VAIR-ee-uh-bul). As the following examples will show, solving a formula for a specified variable requires the same steps used earlier to solve equations with just one variable.

In solving a formula for a specified variable, we treat the specified variable as if it were the *only* variable in the equation, and treat the other variables as if they were numbers. We use the same steps to solve the equation for the specified variable that we have used to solve equations with just one variable.

5. (a) Solve $I = prt$ for t.

E X A M P L E 5   Solving for a Specified Variable

Solve $A = LW$ for W.

Think of undoing what has been done to W. Since W is multiplied by L, undo the multiplication by dividing both sides of $A = LW$ by L.

$$A = LW$$

$$\frac{A}{L} = \frac{LW}{L} \qquad \text{Divide by } L.$$

$$\frac{A}{L} = W \qquad \tfrac{L}{L} = 1;\ 1W = W$$

The formula is now solved for W.

(b) Solve $P = a + b + c$ for a.

WORK PROBLEM 5 AT THE SIDE. ▶▶

E X A M P L E 6   Solving for a Specified Variable

Solve $P = 2L + 2W$ for L.

We want to get L alone on one side of the equation. We begin by subtracting $2W$ on both sides.

$$P = 2L + 2W$$

$$P - 2W = 2L + 2W - 2W \qquad \text{Subtract } 2W.$$

$$P - 2W = 2L \qquad \text{Combine terms.}$$

$$\frac{P - 2W}{2} = \frac{2L}{2} \qquad \text{Divide by 2.}$$

$$\frac{P - 2W}{2} = L \qquad \tfrac{2}{2} = 1;\ 1L = L$$

The last step gives the formula solved for L, as required.

6. (a) Solve $A = p + prt$ for t.

E X A M P L E 7   Solving for a Specified Variable

Solve $F = \frac{9}{5}C + 32$ for C.

We need to isolate C on one side of the equation.

First undo the addition of 32 to $\frac{9}{5}C$ by subtracting 32 from both sides.

$$F = \frac{9}{5}C + 32$$

$$F - 32 = \frac{9}{5}C + 32 - 32 \qquad \text{Subtract 32.}$$

$$F - 32 = \frac{9}{5}C$$

Now multiply both sides by $\frac{5}{9}$. Use parentheses on the left.

$$\frac{5}{9}(F - 32) = \frac{5}{9} \cdot \frac{9}{5}C \qquad \text{Multiply by } \tfrac{5}{9}.$$

$$\frac{5}{9}(F - 32) = C$$

This last result is the formula for converting temperatures from Fahrenheit to Celsius.

◀◀ **WORK PROBLEM 6 AT THE SIDE.**

(b) Solve $Ax + By = C$ for y.

ANSWERS

6. (a) $t = \dfrac{A - p}{pr}$ **(b)** $y = \dfrac{C - Ax}{B}$

2.5 Exercises

1. In your own words, explain what is meant by the *perimeter* of a geometric figure.

2. If you don't remember or don't know the formula for the perimeter of a region with straight line sides, how could you find it?

3. Explain in your own words what is meant by the *area* of a geometric figure.

4. Look at the drawings of a rectangle and a trapezoid at the beginning of this section. How are the two figures alike? How do they differ?

Decide whether perimeter or area would be used to solve a problem concerning the measure of the quantity.

5. carpeting for a bedroom

6. sod for a lawn

7. fencing for a yard

8. baseboards for a living room

9. tile for a bathroom

10. fertilizer for a garden

11. determining the cost for replacing a linoleum floor with a wood floor

12. determining the cost for planting rye grass in a lawn for the winter

In the following exercises a formula is given, along with the values of all but one of the variables in the formula. Find the value of the variable that is not given. See Example 1.

13. $P = 2L + 2W$ (perimeter of a rectangle); $L = 6$, $W = 4$

14. $P = 2L + 2W$; $L = 8$, $W = 5$

15. $A = \frac{1}{2}bh$ (area of a triangle); $b = 10$, $h = 14$

16. $A = \frac{1}{2}bh$; $b = 8$, $h = 16$

17. $P = a + b + c$ (perimeter of a triangle); $P = 15$, $a = 3$, $b = 7$

18. $P = a + b + c$; $P = 12$, $a = 3$, $c = 5$

19. $d = rt$ (distance formula); $d = 100$, $t = 2.5$

20. $d = rt$; $d = 252$, $r = 45$

21. $I = prt$ (simple interest); $p = 5000$, $r = .025$, $t = 7$

22. $I = prt$; $p = 7500$, $r = .035$, $t = 6$

23. $C = 2\pi r$ (circumference of a circle); $C = 8.164$, $\pi = 3.14$*

24. $C = 2\pi r$; $C = 16.328$, $\pi = 3.14$

25. $A = \pi r^2$ (area of a circle); $r = 12$, $\pi = 3.14$

26. $A = \pi r^2$; $r = 4$, $\pi = 3.14$

---

*Actually, π is approximately equal to 3.14, not *exactly* equal to 3.14.

27. Explain why the formula for the area of a square is a special case of the formula for the area of a rectangle.

28. Complete the following: Triangles, rectangles, squares, parallelograms, and trapezoids are all examples of **polygons.** Perimeter is to a polygon as _____ is to a circle.

*The **volume** of a three-dimensional object is a measure of the space occupied by the object. For example, we would need to know the volume of a gasoline tank in order to know how many gallons of gasoline it would take to completely fill the tank. In each of the following exercises, a formula for the volume (V) of a three-dimensional object is given, along with values for the other variables. Evaluate V. See Example 1.*

29. $V = LWH$ (volume of a rectangular-sided box); $L = 12, W = 8, H = 4$

30. $V = LWH$; $L = 10, W = 5, H = 3$

31. $V = \frac{1}{3}Bh$ (volume of a pyramid); $B = 36, h = 4$

32. $V = \frac{1}{3}Bh$; $B = 12, h = 13$

33. $V = \frac{4}{3}\pi r^3$ (volume of a sphere); $r = 6, \pi = 3.14$

34. $V = \frac{4}{3}\pi r^3$; $r = 12, \pi = 3.14$

Use a formula to write an equation for each of the following applications, and use the problem-solving method of Section 2.4 to solve them. Formulas may be found on the inside covers of this book. See Examples 2 and 3.

35. The newspaper *The Constellation,* printed in 1859 in New York City as part of the Fourth of July celebration, had length 51 inches and width 35 inches. What was the perimeter? What was the area?

36. The *Daily Banner,* published in Roseberg, Oregon, in the nineteenth century, had page size 3 inches by 3.5 inches. What was the perimeter? What was the area?

37. The Skydome in Toronto, Canada, is the first stadium with a hard-shell, retractable roof. The steel dome is 630 feet in diameter. To the nearest foot, what is the circumference of this dome?

38. The largest drum ever constructed was played at the Royal Festival Hall in London in 1987. It had a diameter of 13 feet. What was the area of the circular face of the drum? (Use $\pi = 3.14$.)

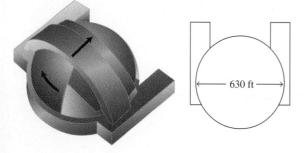

630 ft

39. The survey plat shown in the figure shows two lots that form a figure called a trapezoid. The measures of the parallel sides are 115.80 feet and 171.00 feet. The height of the trapezoid is 165.97 feet. Find the combined area of the two lots. Round your answer to the nearest hundredth of a square foot.

40. Lot *A* in the figure is in the shape of a trapezoid. The parallel sides measure 26.84 feet and 82.05 feet. The height of the trapezoid is 165.97 feet. Find the area of Lot *A*. Round your answer to the nearest hundredth of a square foot.

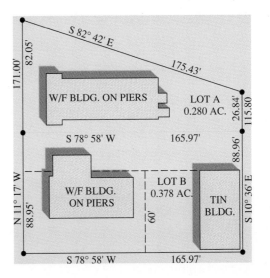

41. A color television set with a liquid crystal display was manufactured by Epson in 1985 and had dimensions 3 inches by $6\frac{3}{4}$ inches by $1\frac{1}{8}$ inches. What was the volume of this set?

42. The largest box of popcorn was filled by students in Jacksonville, Florida. The box was approximately 40 feet long, $20\frac{2}{3}$ feet wide, and 8 feet high. To the nearest cubic foot, what was the volume of the box?

Find the measure of each marked angle. See Example 4.

43. $(10x + 7)°$ $(7x + 3)°$

44. $(x + 1)°$ $(4x - 56)°$

45. $(3x + 45)°$ $(7x + 5)°$

46. $(5x - 129)°$ $(2x - 21)°$

47. $(11x - 37)°$ $(7x + 27)°$

48. $(10x + 15)°$ $(12x - 3)°$

Solve the formula for the specified variable. See Examples 5–7.

49. $d = rt$ for r

50. $d = rt$ for t

51. $A = LW$ for L

52. $V = LWH$ for H

53. $P = a + b + c$ for a

54. $P = a + b + c$ for b

55. $I = prt$ for p

56. $I = prt$ for r

57. $A = \dfrac{1}{2}bh$ for b

58. $A = \dfrac{1}{2}bh$ for h

59. $A = p + prt$ for r

60. $P = 2L + 2W$ for W

61. $V = \pi r^2 h$ for h

62. $V = \dfrac{1}{3}\pi r^2 h$ for h

63. $F = \dfrac{9}{5}C + 32$ for C

MATHEMATICAL CONNECTIONS (EXERCISES 64–66)

In many cases there are equally acceptable equivalent answers for a problem. Work Exercises 64–66 in order. They show how two seemingly different answers to a problem can both be correct.

64. Solve the formula $P = 2L + 2W$ for W using the following steps.
   (a) Subtract $2L$ from both sides.   **(b)** Divide both sides by 2.

65. Referring to the formula in Exercise 64, solve for W again using the following steps.
   (a) Divide each term on both sides by 2.   **(b)** Subtract L from both sides.

66. Compare the results in Exercises 64(b) and 65(b). Both are acceptable answers. To show that they are equivalent, give a justification for each step below.

   (a) $\dfrac{P}{2} - L = \dfrac{P}{2} - \dfrac{2}{2} \cdot L$

   (b) $\phantom{\dfrac{P}{2} - L} = \dfrac{P}{2} - \dfrac{2}{2} \cdot \dfrac{L}{1}$

   (c) $\phantom{\dfrac{P}{2} - L} = \dfrac{P}{2} - \dfrac{2L}{2}$

   (d) $\phantom{\dfrac{P}{2} - L} = \dfrac{P - 2L}{2}$

2.6 Ratio and Proportion; Percent

OBJECTIVE 1 Ratios provide a way of comparing two numbers or quantities using division. A **ratio** (RAY-shee-oh) is a quotient of two quantities with the same units.

The ratio of the number a to the number b is written

$$a \text{ to } b, \quad a:b, \quad \text{or} \quad \frac{a}{b}.$$

This last way of writing a ratio is most common in algebra.

Percents are ratios where the second number is always 100. For example 50% represents the ratio of 50 to 100, 27% represents the ratio of 27 to 100, and so on.

E X A M P L E 1   Writing a Word Phrase as a Ratio

Write a ratio for each word phrase.

(a) The ratio of 5 hours to 3 hours is

$$\frac{5 \text{ hours}}{3 \text{ hours}} = \frac{5}{3}.$$

(b) To find the ratio of 6 hours to 3 days, first convert 3 days to hours.

$$3 \text{ days} = 3 \cdot 24$$
$$= 72 \text{ hours}$$

The ratio of 6 hours to 3 days is thus

$$\frac{6 \text{ hours}}{3 \text{ days}} = \frac{6 \text{ hours}}{72 \text{ hours}} = \frac{6}{72} = \frac{1}{12}.$$

Note
Because the units of the two quantities in a ratio must be the same, ratios do not have units in their final forms.

WORK PROBLEM 1 AT THE SIDE. ▶▶

An example of the use of a ratio is in unit pricing, to see which size of an item offered in different sizes produces the best price per unit. To do this, set up the ratio of the price of the item to the number of units on the label. Then divide to obtain a decimal.

E X A M P L E 2   Finding the Price per Unit

The local supermarket charges the following prices for a popular brand of pancake syrup:

Size	Price
36-ounce	$3.89
24-ounce	$2.79
12-ounce	$1.89

Which size is the best buy? That is, which size has the lowest unit price?

— **CONTINUED ON NEXT PAGE**

OBJECTIVES

1 ▶ Write ratios.

2 ▶ Solve proportions.

3 ▶ Solve applied problems using proportions.

4 ▶ Find percentages and percents.

FOR EXTRA HELP

Tutorial   Tape 5   SSM, Sec. 2.6

1. Write each ratio.

(a) 9 women to 5 women

(b) 4 inches to 1 foot

ANSWERS

1. **(a)** $\frac{9}{5}$ **(b)** $\frac{4}{12} = \frac{1}{3}$

2. Paper towels sell for $.85 for 1 roll of 63 square feet and $2.89 for a package of 3 rolls of the same size. Find the best buy by finding the price per square foot in each case.

To find the best buy, write ratios comparing the price per size by the number of units (ounces here). The results in the following table were rounded to the nearest thousandth.

Size	Unit Cost (dollars per ounce)	
36-ounce	$\frac{\$3.89}{36} = \$.108$	← The best buy
24-ounce	$\frac{\$2.79}{24} = \$.116$	
12-ounce	$\frac{\$1.89}{12} = \$.158$	

Since the 36-ounce size produces the lowest price per unit, it would be the best buy. (Be careful: Sometimes the largest container is not the best buy.)

◀◀ **WORK PROBLEM 2 AT THE SIDE.**

OBJECTIVE 2   A ratio is used to compare two numbers or amounts. A **proportion** (proh-POR-shun) says that two ratios are equal. For example,

$$\frac{3}{4} = \frac{15}{20}$$

is a proportion that says that the ratios $\frac{3}{4}$ and $\frac{15}{20}$ are equal. In the proportion

$$\frac{a}{b} = \frac{c}{d}$$

$a, b, c,$ and d are the **terms** of the proportion. Beginning with the proportion

$$\frac{a}{b} = \frac{c}{d}$$

and multiplying both sides by the common denominator, bd, gives

$$bd \cdot \frac{a}{b} = bd \cdot \frac{c}{d}$$

$$\frac{b}{b}(d \cdot a) = \frac{d}{d}(b \cdot c) \qquad \text{Associative and commutative properties}$$

$$ad = bc. \qquad \text{Commutative and identity properties}$$

The products ad and bc are found by multiplying diagonally, as shown below.

For this reason, ad and bc are called **cross products.**

In our discussion below, we assume that no denominators are zero.

If $\frac{a}{b} = \frac{c}{d}$, then the cross products ad and bc are equal.
Also, if $ad = bc$, then $\frac{a}{b} = \frac{c}{d}$.

From the rule given above, if $\frac{a}{b} = \frac{c}{d}$ then $ad = bc$. However, if $\frac{a}{c} = \frac{b}{d}$, then $ad = cb$, or $ad = bc$. This means that the two proportions are equivalent, and

the proportion $\dfrac{a}{b} = \dfrac{c}{d}$ can always be written as $\dfrac{a}{c} = \dfrac{b}{d}$.

Sometimes one form is more convenient to work with than the other.

Four numbers are used in a proportion. If any three of these numbers are known, the fourth can be found.

E X A M P L E 3 Finding an Unknown in a Proportion

Find x in the proportion

$$\frac{5}{9} = \frac{x}{63}.$$

The cross products must be equal.

$5 \cdot 63 = 9 \cdot x$	Cross products
$315 = 9x$	Multiply.
$35 = x$	Divide by 9.

The solution set is $\{35\}$, and the value of x for which the proportion is true is 35.

WORK PROBLEM 3 AT THE SIDE. ▶▶

Caution
The cross product method cannot be used directly if there is more than one term on either side.

E X A M P L E 4 Solving an Equation Using Cross Products

Solve the equation

$$\frac{m - 2}{5} = \frac{m + 1}{3}.$$

Find the cross products.

$3(m - 2) = 5(m + 1)$	Be sure to use parentheses.
$3m - 6 = 5m + 5$	Distributive property
$3m = 5m + 11$	Add 6.
$-2m = 11$	Subtract $5m$.
$m = -\dfrac{11}{2}$	Divide by -2.

The solution set is $\left\{-\dfrac{11}{2}\right\}$.

WORK PROBLEM 4 AT THE SIDE. ▶▶

Note
Remember, when you set cross products equal to each other, you are really multiplying each ratio in the proportion by a common denominator.

OBJECTIVE 3 Proportions are useful in many practical applications.

E X A M P L E 5 Applying Proportions

A local drugstore is offering 3 packs of toothpicks for $.87. How much would be charged for 10 packs?

Let x represent the cost of 10 packs of toothpicks.

CONTINUED ON NEXT PAGE

---

3. Solve each equation.

(a) $\dfrac{y}{6} = \dfrac{35}{42}$

(b) $\dfrac{a}{24} = \dfrac{15}{16}$

4. Solve each equation.

(a) $\dfrac{z}{2} = \dfrac{z + 1}{3}$

(b) $\dfrac{p + 3}{3} = \dfrac{p - 5}{4}$

ANSWERS

3. (a) $\{5\}$ (b) $\left\{\dfrac{45}{2}\right\}$

4. (a) $\{2\}$ (b) $\{-27\}$

5. On a map, 12 inches represents 500 miles. How many miles would be represented by 30 inches?

Set up a proportion. One ratio in the proportion can involve the number of packs, and the other can involve the costs. Make sure that the corresponding numbers appear in the numerator and denominator.

$$\frac{\text{Cost of 3}}{\text{Cost of 10}} = \frac{3}{10}$$

$$\frac{.87}{x} = \frac{3}{10}$$

$$3x = .87(10) \qquad \text{Cross products}$$

$$3x = 8.7$$

$$x = 2.90 \qquad \text{Divide by 3.}$$

The 10 packs should cost $2.90. As we saw earlier, the proportion could also be written as

$$\frac{3}{.87} = \frac{10}{x},$$

which would give the same cross products.

◀◀ **WORK PROBLEM 5 AT THE SIDE.**

OBJECTIVE 4 We all see examples of the use of percent almost every day. We can use the techniques for solving proportions to solve percent problems. Recall, the decimal point is moved two places to the left to change a percent to a decimal number. Many calculators have a percent key that does this automatically.

We can solve percent problems by writing them as the proportion

$$\frac{\text{amount}}{\text{base}} = \frac{\text{percent}}{100} \quad \text{or} \quad \frac{a}{b} = \frac{p}{100}.$$

The amount, or **percentage** (per-CENT-ij), is compared to the base. Since *percent* means *per 100,* we compare the numerical value of the percent to 100. Thus, we write 50% as

$$\frac{p}{100} = \frac{50}{100}.$$

E X A M P L E   6   Finding Percentages

Solve each problem.

(a) Find 15% of 600.

Here, the base is 600, the percent is 15, and we must find the amount (or percentage).

$$\frac{a}{b} = \frac{p}{100}$$

$$\frac{a}{600} = \frac{15}{100}$$

$$100a = 600(15) \qquad \text{Cross multiply.}$$

$$a = \frac{600(15)}{100} \qquad \text{Divide by 100.}$$

$$a = 90$$

CONTINUED ON NEXT PAGE

(b) A video with a regular price of $18 is on sale this week at 22% off. Find the amount of the discount and the sale price of the video.

The discount is 22% of $18. We want to find a, given b is 18 and p is 22.

$$\frac{a}{b} = \frac{p}{100}$$

$$\frac{a}{18} = \frac{22}{100}$$

$$100a = 18(22) \qquad \text{Cross multiply.}$$

$$100a = 396$$

$$a = 3.96$$

Solution set: {3.96}

The amount of the discount on the video is $3.96, and the sale price is $18.00 − $3.96 = $14.04.

WORK PROBLEM 6 AT THE SIDE. ▶▶

To solve more involved percent problems, often we can translate the words to symbols and solve the equation just as we did other types of applied problems.

EXAMPLE 7 Solving an Applied Percent Problem

A newspaper ad offered a mountain bike at a sale price of $258. The regular price was $300. What percent of the regular price was the savings?

The savings amounted to $300 − $258 = $42. We can now restate the problem: What percent of 300 is 42? Substitute into the percent proportion. We have $a = 42$, $b = 300$, and p is to be found.

$$\frac{a}{b} = \frac{p}{100}$$

$$\frac{42}{300} = \frac{p}{100}$$

$$300p = 4200 \qquad \text{Cross multiply.}$$

$$\frac{300p}{300} = \frac{4200}{300} \qquad \text{Divide by 300.}$$

$$p = 14$$

Solution set: {14}

The sale price represented a 14% savings.

WORK PROBLEM 7 AT THE SIDE. ▶▶

6. Solve each problem.

(a) Find 20% of 70.

(b) Find the amount of discount on a television set with a regular price of $270 if the set is on sale at 25% off. Find the sale price of the set.

7. Answer each of the following.

(a) 90 is what percent of 270?

(b) The interest in one year on deposits of $11,000 was $682. What percent interest was paid?

Answers

6. (a) 14 **(b)** $67.50; $202.50

7. (a) $33\frac{1}{3}\%$ **(b)** 6.2%

The Changing Face of the United States

While the United States has long been considered a "melting pot" of cultures, it has never been so obvious as it is today. In particular, immigration is leading to a population boom in the West and South. According to a computer analysis of U.S. Census Bureau data by Paul Overberg, here are the projected population-percentage changes of racial groups and Hispanics in 1995–2000 in the most populous states.

State	White	Black	Asian	Native American	Hispanic
California	1.1%	.5%	18.3%	–2.3%	15.7%
Texas	6.5%	11.0%	25.5%	13.1%	13.6%
New York	–1.8%	3.4%	18.6%	5.8%	10.4%
Florida	6.5%	11.9%	22.5%	13.3%	22.3%
Pennsylvania	.3%	4.8%	22.5%	12.5%	19.7%
Illinois	1.0%	2.9%	18.2%	4.0%	16.2%
Ohio	.7%	5.6%	21.7%	0.0%	13.0%
Michigan	.5%	4.1%	23.5%	3.4%	12.0%
New Jersey	.6%	7.6%	27.3%	0.0%	16.5%
Georgia	7.5%	12.9%	26.8%	6.3%	26.0%

FOR GROUP DISCUSSION

1. It is expected that California, the nation's most populous state with 31.6 million people, will have 17.7 million more people by 2025, which represents a 56% growth. How was this percent obtained using these figures?

2. Based on the chart above, what is the fastest-growing racial group overall?

3. It is expected that people of Hispanic origin will account for 44% of the 72 million additional people in the United States in the year 2025. How many people is this?

4. As a class, discuss the problems that minorities face in mathematics courses.

2.6 Exercises

1. Which one of the following ratios is not the same as 3 to 4?
  (a) .75   **(c)** 4 to 3
  (b) 6 to 8   **(d)** 30 to 40

2. Give three ratios that are equivalent to 3 to 1.

Write a ratio for each word phrase. Write fractions in lowest terms. See Example 1.

3. 40 miles to 30 miles

4. 60 feet to 70 feet

5. 120 people to 90 people

6. 72 dollars to 220 dollars

7. 20 yards to 8 feet

8. 30 inches to 8 feet

9. 24 minutes to 2 hours

10. 16 minutes to 1 hour

11. 8 days to 40 hours

A supermarket was surveyed to find the prices charged for items in various sizes. Find the best buy (based on price per unit) for each of the following. See Example 2.

12. Trash bags
  20-count: $2.49
  30-count: $4.29

13. Black pepper
  4-ounce size: $1.57
  8-ounce size: $2.27

14. Spaghetti sauce
  $15\frac{1}{2}$-ounce size: $1.19
  32-ounce size: $1.69
  48-ounce size: $2.69

15. Breakfast cereal
  10-ounce size: $1.49
  13-ounce size: $1.85
  19-ounce size: $2.81

16. Tomato ketchup
  14-ounce size: $.93
  32-ounce size: $1.19
  44-ounce size: $2.19

17. Extra crunchy peanut butter
  12-ounce size: $1.49
  28-ounce size: $1.99
  40-ounce size: $3.99

18. Explain how percent and ratio are related.

19. Explain the distinction between *ratio* and *proportion*.

Solve each of the following. See Examples 3 and 4.

20. $\dfrac{k}{4} = \dfrac{175}{20}$

21. $\dfrac{x}{6} = \dfrac{18}{4}$

22. $\dfrac{49}{56} = \dfrac{z}{8}$

23. $\dfrac{z}{80} = \dfrac{20}{100}$

24. $\dfrac{3y - 2}{5} = \dfrac{6y - 5}{11}$

25. $\dfrac{2r + 8}{4} = \dfrac{3r - 9}{3}$

26. $\dfrac{5k + 1}{6} = \dfrac{3k - 2}{3}$ **27.** $\dfrac{2p + 7}{3} = \dfrac{p - 1}{4}$ **28.** $\dfrac{3m - 2}{5} = \dfrac{4 - m}{3}$

Solve each of the following problems involving proportion. See Example 5.

29. If 6 gallons of premium unleaded gasoline cost $9.72, how much would 9 gallons cost?

30. If sales tax on a $24 headset radio is $1.68, how much would the sales tax be on a pair of binoculars that costs $36?

31. Two slices of bacon contain 85 calories. How many calories are there in twelve slices of bacon?

32. Three ounces of liver contain 22 grams of protein. How many ounces of liver provide 121 grams of protein?

33. Biologists tagged 250 fish in Willow Lake on October 5. On a later date they found 7 tagged fish in a sample of 350. Estimate the total number of fish in Willow Lake to the nearest hundred.

34. On May 13, researchers at Argyle Lake tagged 420 fish. When they returned a few weeks later, their sample of 500 fish contained 9 that were tagged. Give an approximation of the fish population in Argyle Lake to the nearest hundred.

35. The distance between Singapore and Tokyo is 3300 miles. On a certain wall map, this distance is represented by 11 inches. The actual distance between Mexico City and Cairo is 7700 miles. How far apart are they on the same map?

36. The distance between Kansas City, Missouri, and Denver is 600 miles. On a certain wall map, this is represented by a length of 2.4 feet. On the map, how many feet would there be between Memphis and Philadelphia, two cities that are actually 1000 miles apart?

Answer each of the following. See Example 6.

37. What is 26% of 480?

38. What is 48.6% of 19?

39. What percent of 30 is 36?

40. What percent of 48 is 20?

41. 25% of what number is 150?

42. 12% of what number is 3600?

43. .392 is what percent of 28?

44. 78.84 is what percent of 292?

Work each of the following problems. Round all money amounts to the nearest dollar. See Examples 6(b) and 7.

45. According to a Knight-Ridder Newspapers report, several years ago the nation's "consumer-debt burden" was 16.4%. This means that the average American had consumer debts, such as credit card bills and auto loans, totaling 16.4% of his or her take-home pay. Suppose that Paul Eldersveld has a take-home pay of $3250 per month. What is 16.4% of his monthly take-home pay?

46. In a recent year General Motors announced that it would raise prices on its next year's vehicles by an average of 1.6%. If a certain vehicle had an original price of $10,526 and this price was raised 1.6%, what would the new price be?

47. At the 1996 Olympic Games in Atlanta, 20% of the medals won by Romania were gold medals. The team won a total of 20 medals. How many of the medals were *not* gold?

48. Forty percent of Florence Griffith Joyner's Olympic medals are not gold medals. She has 2 medals that are not gold. How many gold medals does she have?

49. An advertisement for a dot matrix printer gives a sale price of $150.00. The regular price is $180.00. What is the percent discount on this printer?

50. Jane Gunton bought a boat five years ago for $5000 and sold it this year for $2000. What percent of her original purchase price did she lose on the sale?

51. Quinhon Dac Ho earns $3200 per month. He wants to save 12% of this amount. How much will he save?

52. A family of four with a monthly income of $3800 plans to spend 8% of this amount on entertainment. How much will be spent on entertainment?

53. Here is a common business problem. If the sales tax rate is 6.5% and I have collected $3400 in sales tax, how much were my sales?

54. The 1916 dime minted in Denver is quite rare. The 1979 edition of *A Guide Book of United States Coins* listed its value in Extremely Fine condition as $625.00. The 1997 value had increased to $2400. What was the percent increase in the value of this coin?

At the start of play on September 16, 1996, the standings of the West division of the National League were as follows:

	Won	Lost
Los Angeles	84	65
San Diego	84	66
Colorado	79	71
San Francisco	60	89

"Winning percentage" is commonly expressed as a decimal rounded to the nearest thousandth. To find the winning percentage of a team, divide the number of wins by the total number of games played. Find the winning percentage of each of the following teams.

55. Los Angeles   **56.** San Diego   **57.** Colorado   **58.** San Francisco

59. Explain the difference between .5 and .5%.

60. Suppose that an item is discounted 10% and then marked up 10%. Is the final price the same as the starting price? If not, how do they compare?

MATHEMATICAL CONNECTIONS (EXERCISES 61–64)

In Section 2.3 we solved equations with fractions by first multiplying both sides by the common denominator. A proportion with a variable is this kind of equation. Work Exercises 61–64 in order. The steps justify the method of solving a proportion by cross products.

61. What is the LCD of the fractions in the equation $\frac{x}{6} = \frac{2}{5}$?

62. Solve the equation in Exercise 61 as follows.
(a) Multiply both sides by the LCD. What equation do you get?

(b) Solve the equation from part (a) by dividing both sides by the coefficient of x.

63. Solve the equation in Exercise 61 using cross products.

64. Compare your solution sets from Exercises 62 and 63. What do you notice?

KEY TERMS

2.1	**linear equation**	A linear equation is an equation that can be written in the form $Ax + B = C$, for real numbers A, B, and C, with $A \neq 0$.
	equivalent equations	Equations that have the same solution set are equivalent equations.
	identity	An identity is an equation that is true for all replacements of the variable.
2.4	**complementary angles**	Angles whose measures have a sum of $90°$ are complementary angles.
	right angle	A right angle measures $90°$.
	supplementary angles	Angles whose measures have a sum of $180°$ are supplementary angles.
	straight angle	A straight angle measures $180°$.
2.5	**vertical angles**	Vertical angles are angles formed by intersecting lines. They have the same measure.
2.6	**ratio**	A ratio is a quotient of two quantities with the same units.
	proportion	A proportion is a statement that two ratios are equal.
	cross products	The method of cross products provides a way of determining whether a proportion is true.
	percentage	A percentage is a part of the whole amount.

NEW SYMBOLS

{ }	set braces
a **to** b, $a : b$, **or** $\dfrac{a}{b}$	the ratio of a to b

---

QUICK REVIEW

Concepts	*Examples*
2.1 The Addition Property of Equality The same number may be added to (or subtracted from) each side of an equation without changing the solution.	$$x - 6 = 12$$ $$x - 6 + 6 = 12 + 6 \qquad \text{Add 6.}$$ $$x = 18 \qquad \text{Combine terms.}$$ The solution set is $\{18\}$.
2.2 The Multiplication Property of Equality Each side of an equation may be multiplied (or divided) by the same nonzero number without changing the solution.	$$\frac{3}{4}x = -9$$ $$\frac{4}{3} \cdot \frac{3}{4}x = \frac{4}{3}(-9) \qquad \text{Multiply by } \tfrac{4}{3}.$$ $$x = -12$$ The solution set is $\{-12\}$.

Concepts	Examples
2.3 More on Solving Linear Equations	Solve the equation $2x + 3(x + 1) = 38$.
1. Combine like terms to simplify each side.	**1.** $\quad 2x + 3x + 3 = 38 \qquad$ Distributive property $\quad\quad\quad 5x + 3 = 38 \qquad$ Combine like terms.
2. Get the variable term on one side, a number on the other.	**2.** $\quad 5x + 3 - 3 = 38 - 3 \quad$ Subtract 3. $\quad\quad\quad\quad 5x = 35 \qquad$ Combine terms.
3. Get the equation into the form $x = $ a number.	**3.** $\qquad \dfrac{5x}{5} = \dfrac{35}{5} \qquad$ Divide by 5. $\qquad\quad x = 7 \qquad$ Reduce.
4. Check by substituting the result into the original equation.	**4.** $\quad 2x + 3(x + 1) = 38 \qquad$ Check. $\quad 2(7) + 3(7 + 1) = 38 \quad$? Let $x = 7$. $\quad\quad\quad 14 + 24 = 38 \quad$? Add; then multiply. $\quad\quad\quad\quad\quad 38 = 38 \qquad$ True The solution set is $\{7\}$.
2.4 An Introduction to Applied Problems	One number is 5 more than another. Their sum is 21. What are the numbers?
1. Choose a variable to represent the unknown.	**1.** Let x be the smaller number.
2. Determine expressions for any other unknown quantities, using the variable. Draw figures or diagrams if they apply.	**2.** Let $x + 5$ be the larger number.
3. Translate the problem into an equation.	**3.** $x + (x + 5) = 21$
4. Solve the equation.	**4.** $\quad 2x + 5 = 21 \qquad$ Combine terms. $\quad 2x + 5 - 5 = 21 - 5 \quad$ Subtract 5. $\quad\quad\quad 2x = 16 \qquad$ Combine terms. $\quad\quad \dfrac{2x}{2} = \dfrac{16}{2} \qquad$ Divide by 2. $\quad\quad\quad x = 8 \qquad$ Reduce. Solution set: $\{8\}$
5. Answer the question asked in the problem.	**5.** The numbers are 8 and 13.
6. Check your solution by using the original words of the problem. Be sure that the answer is appropriate and makes sense.	**6.** 13 is 5 more than 8, and $8 + 13 = 21$. It checks.

Concepts	Examples
2.5 Formulas and Geometry Applications	
To find the values of one of the variables in a formula, given values for the others, substitute the known values into the formula.	Find L if $A = LW$, given that $A = 24$ and $W = 3$.

$$24 = L \cdot 3 \qquad\qquad A = 24,\ W = 3$$

$$\frac{24}{3} = \frac{L \cdot 3}{3} \qquad\qquad \text{Divide by 3.}$$

$$8 = L \qquad\qquad\qquad \text{Reduce.}$$

To solve a formula for one of the variables, isolate that variable by treating the other variables as numbers and using the steps for solving equations.	Solve $P = 2L + 2W$ for W.

$$P - 2L = 2L + 2W - 2L \qquad \text{Subtract } 2L.$$

$$P - 2L = 2W \qquad\qquad\qquad \text{Combine terms.}$$

$$\frac{P - 2L}{2} = \frac{2W}{2} \qquad\qquad \text{Divide by 2.}$$

$$\frac{P - 2L}{2} = W \quad \text{or} \quad W = \frac{P - 2L}{2}$$

2.6 Ratio and Proportion; Percent	
To write a ratio, express quantities in the same units.	4 feet to 8 inches

$$= 48 \text{ inches to } 8 \text{ inches} = \frac{48}{8} = \frac{6}{1}$$

To solve a proportion, use the method of cross products.	Solve $\dfrac{x}{12} = \dfrac{35}{60}$.

$$60x = 12 \cdot 35 \qquad \text{Cross products}$$

$$60x = 420 \qquad\qquad \text{Multiply.}$$

$$\frac{60x}{60} = \frac{420}{60} \qquad \text{Divide by 60.}$$

$$x = 7 \qquad\qquad\quad \text{Reduce.}$$

To solve a percent problem, use the proportion $$\frac{\text{amount}}{\text{base}} = \frac{\text{percent}}{100}.$$	Solution set: $\{7\}$

NUMBERS IN THE
Real World *collaborative investigations*

Calendar Mathematics: Number Patterns

One of the most common appearances of numbers in our everyday lives is in our calendar. There are many mathematical patterns in calendar months, and two samples are shown here.

NOVEMBER						
S	M	T	W	T	F	S
1	2	3	4	5	6	7
8	9	10	11	12	13	14
15	16	17	18	19	20	21
22	23	24	25	26	27	28
29	30					

DECEMBER						
S	M	T	W	T	F	S
		1	2	3	4	5
6	7	8	9	10	11	12
13	14	15	16	17	18	19
20	21	22	23	24	25	26
27	28	29	30	31		

FOR GROUP DISCUSSION

1. In the calendar for November, a three-by-three square of dates has been highlighted. Find the sum of these numbers, and divide by 9. Record your result. Then choose any other three-by-three square, find the sum of the numbers, divide by 9, and record your answer. What do you notice about the answers with respect to the original nine numbers added? Repeat to support your guess.

2. In the calendar for December, a rectangle of twelve numbers has been highlighted. Do each of the following:

   (a) Find the sum of the four numbers at the vertices. Then divide by 4 to find their average.

   (b) Find the sum of the three numbers on the interior of the rectangle. Then divide by 3 to find their average.

   (c) Find the sum of the twelve numbers that lie on the boundary of the rectangle. Then divide by 12 to find their average.

   (d) What do you notice about the results you found in parts (a)–(c)? Repeat for a similar set of numbers from the calendar. Does this pattern always hold true?

   Source: Adapted from *NCTM Student Math Notes*, January 1995, by Bonnie Litwiller and David Duncan.

[2.1–2.3] *Solve each equation. Check the solution.*

1. $x - 7 = 2$

2. $4r - 6 = 10$

3. $5x + 8 = 4x + 2$

4. $8t = 7t + \dfrac{3}{2}$

5. $(4r - 8) - (3r + 12) = 0$

6. $7(2x + 1) = 6(2x - 9)$

7. $-\dfrac{6}{5}y = -18$

8. $\dfrac{1}{2}r - \dfrac{1}{6}r + 3 = 2 + \dfrac{1}{6}r + 1$

9. $3x - (-2x + 6) = 4(x - 4) + x$

10. $.10(x + 80) + .20x = 14$

[2.4] *Solve each problem.*

11. If 7 is added to 5 times a number, the result is equal to 3 times the number. Find the number.

12. If 4 is subtracted from twice a number, the result is 36. Find the number.

13. The land area of Hawaii is 5213 square miles greater than that of Rhode Island. Together, the areas total 7637 square miles. What is the area of each state?

14. The height of Seven Falls in Colorado is $\frac{5}{2}$ the height (in feet) of Twin Falls in Idaho. The sum of the heights is 420 feet. Find the height of each.

15. The supplement of an angle measures 10 times the measure of its complement. What is the measure of the angle (in degrees)?

16. During a recent baseball season, Bret Saberhagen pitched 9 fewer innings than Jack Morris. Mark Langston pitched 6 more innings than Morris. How many innings did each pitch if their total number of innings pitched was 795?

[2.5] *A formula is given in each exercise, along with the values for all but one of the variables. Find the value of the variable that is not given.*

17. $A = \dfrac{1}{2}bh$; $A = 44$, $b = 8$

18. $A = \dfrac{1}{2}h(b + B)$; $b = 3$, $B = 4$, $h = 8$

19. $C = 2\pi r$; $C = 29.83$, $\pi = 3.14$

20. $V = \dfrac{4}{3}\pi r^3$; $r = 6$, $\pi = 3.14$

Solve the formula for the specified variable.

21. $A = LW$ for L

22. $A = \dfrac{1}{2}h(b + B)$ for h

Find the measure of each marked angle.

23.

24.

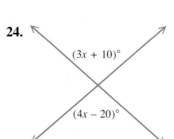

$(8x - 1)°$ $(3x - 6)°$

$(3x + 10)°$

$(4x - 20)°$

Solve each application of geometry.

25. A cinema screen in Indonesia has length 92.75 feet and width 70.5 feet. What is the perimeter? What is the area?

26. The Ziegfield Room in Reno, Nevada, has a circular turntable on which its show girls dance. The circumference of the turntable is 62.5 feet. What is the diameter of the turntable? What is the radius? What is its area? (Use $\pi = 3.14$.)

[2.6] *Write a ratio for each word phrase. Write fractions in lowest terms.*

27. 60 centimeters to 40 centimeters

28. 5 days to 2 weeks

29. 90 inches to 10 feet

30. 3 months to 3 years

Solve each proportion.

31. $\dfrac{p}{21} = \dfrac{5}{30}$

32. $\dfrac{5 + x}{3} = \dfrac{2 - x}{6}$

33. $\dfrac{y}{5} = \dfrac{6y - 5}{11}$

34. Explain how 40% can be expressed as a ratio of two whole numbers.

Solve each of the following problems involving proportion.

35. If 2 pounds of fertilizer will cover 150 square feet of lawn, how many pounds would be needed to cover 500 square feet?

36. If 8 ounces of medicine must be mixed with 20 ounces of water, how many ounces of medicine must be mixed with 90 ounces of water?

37. The tax on a $24.00 item is $2.04. How much tax would be paid on a $36.00 item?

38. The distance between two cities on a road map is 32 centimeters. The two cities are actually 150 kilometers apart. The distance on the map between two other cities is 80 centimeters. How far apart are these cities?

39. What is 23% of 76?

40. What percent of 12 is 21?

41. 6 is what percent of 18?

42. 36% of what number is 900?

43. Vinh must pay 6.5% sales tax on a new car. The cost of the car is $17,200. Find the amount of the tax.

44. An ad for steel-belted radials promises 15% better mileage when using them. Alexandra's import now gets 380 miles on a tank of gas. If she is to believe the ad, how many miles should she get per tank if she uses these tires?

MIXED REVIEW EXERCISES

Solve each of the following.

45. $\dfrac{y}{7} = \dfrac{y-5}{2}$

46. $I = prt$ for r

47. $.05x + .02x = 4.9$

48. $2 - 3(y - 5) = 4 + y$

49. $9x - (7x + 2) = 3x + (2 - x)$

50. $\dfrac{1}{3}s + \dfrac{1}{2}s + 7 = \dfrac{5}{6}s + 5 + 2$

Solve each problem.

51. Two-thirds of a number added to the number is 10. What is the number?

52. If three-fourths of a number is subtracted from twice the number, the result is 15. Find the number.

53. Buddy defeated Bob in an election. Buddy had twice as many votes as Bob, and together they had 1800 votes. How many votes did each of the candidates receive?

54. Norman and Janice Mello commute to work. Norman travels three times as far as Janice each day, and together they travel 112 miles. How far does each travel?

55. On a recent diet, Duc lost 18 pounds more than Hoa. Their total weight loss was 42 pounds. How much did Hoa lose?

56. Rick and Steve drove from different towns to a family reunion. Rick drove 43 miles farther than Steve. The two men drove a total of 293 miles. How far did Steve drive to the reunion?

57. A teacher noted that one week he had graded 32 more algebra tests than geometry tests. He had graded 102 tests altogether. How many geometry tests did he grade that week?

58. A parking lot attendant parked 63 cars one day. He parked 11 fewer large cars than small cars. How many small cars did he park?

59. The perimeter of a rectangle is 288 feet. The length is 4 feet longer than the width. Find the width.

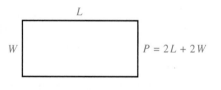

60. The perimeter of a triangle is 96 meters. One side is twice as long as another, and the third side is 30 meters long. What is the length of the longest side?

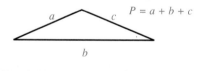

61. The area of a triangle is 182 square inches. The height is 14 inches. Find the length of the base.

62. The perimeter of a rectangle is 75 inches. The width is 17 inches. What is the length?

63. Find the measure of each marked angle.

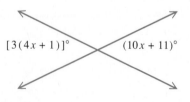

64. Nalima has grades of 82 and 96 on her first two English tests. What must she make on her third test so that her average will be at least 90?

Solve each equation and check the solution.

1. $3x - 7 = 11$

1. _____

2. $5x + 9 = 7x + 21$

2. _____

3. $2 - 3(y - 5) = 3 + (y + 1)$

3. _____

4. $2.3x + 13.7 = 1.3x + 2.9$

4. _____

5. $7 - (m - 4) = -3m + 2(m + 1)$

5. _____

6. $-\dfrac{4}{7}x = -12$

6. _____

7. $.06(x + 20) + .08(x - 10) = 4.6$

7. _____

8. $-8(2x + 4) = -4(4x + 8)$

8. _____

Solve each problem.

9. If 3 is subtracted from four times a number, the result is 10 less than five times the number. What is the number?

9. _____

10. The Golden Gate Bridge in San Francisco is 2605 feet longer than the Brooklyn Bridge. Together, their spans total 5795 feet. How long is each bridge?

10. _____

11. A piece of string is 40 centimeters long. It is cut into three pieces. The longest piece is 3 times as long as the middle-sized piece, and the shortest piece is 23 centimeters shorter than the longest piece. Find the lengths of the three pieces.

11. _____

12. _____

13. _____

14. _____

15. _____

16. _____

17. _____

18. _____

19. _____

20. _____

12. The formula for the perimeter of a rectangle is $P = 2L + 2W$. If $P = 116$ and $L = 40$, find W.

13. Solve the formula $P = 2L + 2W$ for W.

Find the measure of each marked angle.

14.

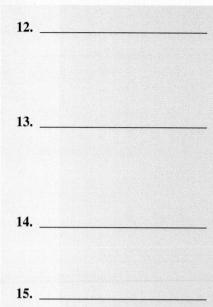

$(3x + 55)°$ $(7x - 25)°$

15.

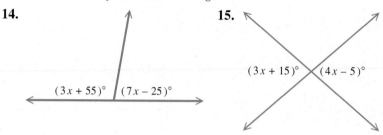

$(3x + 15)°$ $(4x - 5)°$

16. Find the measure of an angle if its supplement measures 10° more than three times its complement.

Solve each proportion.

17. $\dfrac{z}{8} = \dfrac{12}{16}$

18. $\dfrac{y + 5}{3} = \dfrac{y - 3}{4}$

19. Which is the better buy for processed cheese slices: 8 slices for $2.19 or 12 slices for $3.30?

20. The distance between Milwaukee and Boston is 1050 miles. On a certain map, this distance is represented by 42 inches. On the same map, Seattle and Cincinnati are 92 inches apart. What is the actual distance between Seattle and Cincinnati?

CUMULATIVE REVIEW EXERCISES CHAPTERS 1–2

Beginning with this chapter, each chapter in the text will conclude with a set of cumulative review exercises designed to cover the major topics from the beginning of the course, including review of basic arithmetic. This feature will allow the student to constantly review topics that have been introduced up to that point.

Write each fraction in lowest terms.

1. $\dfrac{15}{40}$

2. $\dfrac{108}{144}$

Work the following problems.

3. $\dfrac{5}{6} + \dfrac{1}{4} + \dfrac{7}{15}$

4. $16\dfrac{7}{8} - 3\dfrac{1}{10}$

5. $\dfrac{9}{8} \cdot \dfrac{16}{3}$

6. $\dfrac{3}{4} \div \dfrac{5}{8}$

7. $4.8 + 12.5 + 16.73$

8. $56.3 - 28.99$

9. $(67.8)(.45)$

10. $236.46 \div 4.2$

11. In making dresses, Earth Works uses $\frac{5}{8}$ yard of trim per dress. How many yards of trim would be used to make 56 dresses?

12. A cook wants to increase a recipe that serves 6 to make enough for 20 people. The recipe calls for $1\frac{1}{4}$ cups of cheese. How much cheese will be needed to serve 20?

13. One dog weighs $8\frac{1}{3}$ pounds, and another dog weighs $12\frac{5}{8}$ pounds. Find the total weight of both dogs.

14. A purchasing agent bought 3 desks at $211.40 each and 3 chairs for $195, $189.95, and $168.50. What was the final bill (without tax)?

Tell whether each of the following is true or false.

15. $\dfrac{8(7) - 5(6 + 2)}{3 \cdot 5 + 1} \geq 1$

16. $\dfrac{4(9 + 3) - 8(4)}{2 + 3 - 3} \geq 2$

Perform the indicated operations.

17. $-11 + 20 + (-2)$

18. $13 + (-19) + 7$

19. $9 - (-4)$

20. $-2(-5)(-4)$

21. $\dfrac{4 \cdot 9}{-3}$

22. $\dfrac{8}{7 - 7}$

23. $(-5 + 8) + (-2 - 7)$

24. $(-7 - 1)(-4) + (-4)$

25. $\dfrac{-3 - (-5)}{1 - (-1)}$

26. $\dfrac{6(-4) - (-2)(12)}{3^2 + 7^2}$

27. $\dfrac{(-3)^2 - (-4)(2^4)}{5 \cdot 2 - (-2)^3}$

28. $\dfrac{-2(5^3) - 6}{4^2 + 2(-5) + (-2)}$

Find the value of each expression when $x = -2$, $y = -4$, and $z = 3$.

29. $xz^3 - 5y^2$

30. $\dfrac{3x - y^3}{-4z}$

Name the property illustrated by each of the following examples.

31. $7(k + m) = 7k + 7m$

32. $3 + (5 + 2) = 3 + (2 + 5)$

33. $7 + (-7) = 0$

34. $3.5(1) = 3.5$

Simplify the following expressions by combining terms.

35. $4p - 6 + 3p - 8$

36. $-4(k + 2) + 3(2k - 1)$

Solve the following equations and check each solution.

37. $2r - 6 = 8$

38. $2(p - 1) = 3p + 2$

39. $4 - 5(a + 2) = 3(a + 1) - 1$

40. $2 - 6(z + 1) = 4(z - 2) + 10$

41. $-(m - 1) = 3 - 2m$

42. $\dfrac{y - 2}{3} = \dfrac{2y + 1}{5}$

43. $\dfrac{2x + 3}{5} = \dfrac{x - 4}{2}$

44. $\dfrac{2}{3}y + \dfrac{3}{4}y = -17$

Solve the formula for the indicated variable.

45. $P = a + b + c$ for c

46. $P = 4s$ for s

Solve the following problems.

47. The purchasing agent in Exercise 14 paid a sales tax of $6\frac{1}{4}\%$ on his purchase. What was the final bill, including tax?

48. A car has a price of $5000. For trading in her old car, Rosalie will get 25% off. Find the price of the car with the trade-in.

49. Margaret Westmoreland bought textbooks at the college bookstore for $52.47, including 6% sales tax. What did the books cost?

50. Louise Palla received a bill from her credit card company for $104.93. The bill included interest at $1\frac{1}{2}\%$ per month for one month and a $5.00 late charge. How much did her purchases amount to?

51. The perimeter of a rectangle is 98 centimeters. The width is 19 centimeters. Find the length.

52. The area of a triangle is 104 square inches. The base is 13 inches. Find the height.

Linear Inequalities and Absolute Value

3

3.1 Linear Inequalities in One Variable

Now we can extend our work with linear equations in Chapter 2 to *inequalities*. An **inequality** (in-ee-KWAHL-it-ee) says that two expressions are *not* equal. A **linear inequality in one variable** is an inequality such as

$$x + 5 < 2, \quad y - 3 \geq 5, \quad \text{or} \quad 2k + 5 > 10.$$

(Throughout this section the definitions and rules are given only for $<$, but they are also valid for $>$, $\leq$, and $\geq$.)

OBJECTIVE 1 We solve an inequality by finding all the numbers that make the inequality true. Usually, an inequality has an infinite number of solutions. These solutions are found, as are solutions of equations, by producing a series of simpler equivalent inequalities. **Equivalent inequalities** are inequalities with the same solution set. The inequalities in this chain of equivalent inequalities can be found with the **addition and multiplication properties of inequality.**

Addition Property of Inequality

For all real numbers a, b, and c, the inequalities

$$a < b \quad \text{and} \quad a + c < b + c$$

are equivalent. (The same number may be added to both sides of an inequality.)

When graphing the solution set on a number line, we use a parenthesis if the endpoint is not included and a square bracket if it is included.

EXAMPLE I   Solving a Linear Inequality

Solve each inequality, and graph the solution set.

(a) $x - 7 \leq -12$

$$x - 7 + 7 \leq -12 + 7 \qquad \text{Add 7.}$$
$$x \leq -5$$

The solution set is written in a shorthand form called *interval notation* as $(-\infty, -5]$, and is graphed in Figure 1.

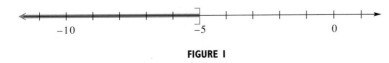

FIGURE I

CONTINUED ON NEXT PAGE

www.mathnotes.com

OBJECTIVES

1 ▶ Solve linear inequalities using the addition property.

2 ▶ Solve linear inequalities using the multiplication property.

3 ▶ Solve linear inequalities with three parts.

4 ▶ Solve applied problems with inequalities.

FOR EXTRA HELP

Tutorial   Tape 6   SSM, Sec. 3.1

1. Find the solution set of each inequality, and graph the solution set.

(a) $p + 6 \leq 8$

(b) $k - 5 \geq 1$

(c) $8y < 7y - 6$

(b) $3m > 2m + 14$

$$3m - 2m > 2m + 14 - 2m \qquad \text{Subtract } 2m.$$
$$m > 14$$

The solution set, $(14, \infty)$, is shown in Figure 2.

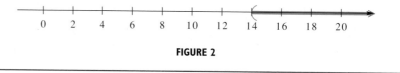

FIGURE 2

◀◀ **WORK PROBLEM I AT THE SIDE.**

OBJECTIVE 2 Solving an inequality such as $3x \leq 15$ requires dividing both sides by 3. This is done with the multiplication property of inequality, which is a little more involved than the corresponding property for equality. To see how this property works, start with the true statement

$$-2 < 5.$$

Multiply both sides by, say, 8.

$$-2(8) < 5(8) \qquad \text{Multiply by 8.}$$
$$-16 < 40 \qquad \text{True}$$

This gives a true statement. Start again with $-2 < 5$, and this time multiply both sides by -8.

$$-2(-8) < 5(-8) \qquad \text{Multiply by } -8.$$
$$16 < -40 \qquad \text{False}$$

The result, $16 < -40$, is false. To make it true, change the direction of the inequality symbol to get

$$16 > -40.$$

◀◀ **WORK PROBLEM 2 AT THE SIDE.**

2. Multiply both sides of each of the following by -5. Then insert the correct symbol, either $<$ or $>$, in the middle blank.

(a) $7 < 8$

$$-35\underline{\hspace{1cm}}-40$$

(b) $-1 > -4$

$$5\underline{\hspace{1cm}}\ \underline{\hspace{1cm}}$$

As these examples suggest, *multiplying* both sides of an inequality by a negative number requires that we reverse the direction of the inequality symbol. Because division is defined in terms of multiplication, we must also reverse the direction of the inequality symbol when *dividing* by a negative number.

Caution
Avoid the common error of forgetting to reverse the direction of the inequality sign when multiplying or dividing by a negative number!

The results of multiplying inequalities by positive and negative numbers are summarized here.

ANSWERS

1. (a) $(-\infty, 2]$

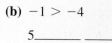

(b) $[6, \infty)$

(c) $(-\infty, -6)$

2. (a) $>$ **(b)** < 20

Multiplication Property of Inequality

For all real numbers a, b, and c, with $c \neq 0$,
the inequalities
$$a < b \quad \text{and} \quad ac < bc$$
are equivalent if $c > 0$;
and the inequalities
$$a < b \quad \text{and} \quad ac > bc$$
are equivalent if $c < 0$.

(Both sides of an inequality may be multiplied by a *positive* number without changing the direction of the inequality symbol. Multiplying [or dividing] by a *negative* number requires that we reverse the inequality symbol.)

3. Graph the solution set of each inequality.

(a) $2y < -10$

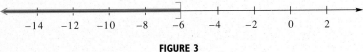

E X A M P L E 2   Solving a Linear Inequality

Solve each inequality, and graph the solution set.

(a) $5m \leq -30$

Use the multiplication property to divide both sides by 5. Since $5 > 0$, do *not* reverse the inequality symbol.

$$5m \leq -30$$

$$\frac{5m}{5} \leq \frac{-30}{5} \quad \text{Divide by 5.}$$

$$m \leq -6$$

The solution set, $(-\infty, -6]$, is graphed in Figure 3.

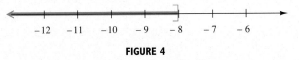

FIGURE 3

(b) $-7k \geq 8$

_____

(b) $-4k \geq 32$

Divide both sides by -4. Since $-4 < 0$, the inequality symbol must be reversed.

$$-4k \geq 32$$

$$\frac{-4k}{-4} \leq \frac{32}{-4} \quad \text{Change } \geq \text{ to } \leq.$$

$$k \leq -8$$

Figure 4 shows the graph of the solution set, $(-\infty, -8]$.

(c) $-9m < -81$

FIGURE 4

_____

WORK PROBLEM 3 AT THE SIDE. ▶▶

The steps required for solving a linear inequality are given below.

Solving a Linear Inequality

Step 1 Simplify each side of the inequality separately by combining like terms and clearing parentheses as needed.

Step 2 Use the addition property of inequality to write the inequality so that the variable is on one side.

Step 3 Use the multiplication property to write the inequality in the form $x < k$ or $x > k$.

Remember to change the direction of the inequality symbol **only** when multiplying or dividing both sides of an inequality by a **negative** number, and never otherwise.

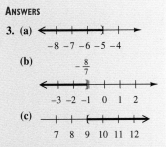

4. Graph the solution set of each inequality.

(a) $-y \leq 2$

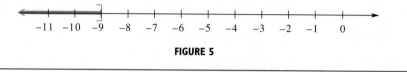

(b) $-z > -11$

(c) $4(x + 2) \geq 6x - 8$

(d) $5 - 3(m - 1)$
$\leq 2(m + 3) + 1$

E X A M P L E   3   Solving a Linear Inequality

Solve $-3(x + 4) + 2 \geq 8 - x$. Graph the solution set.

$-3x - 12 + 2 \geq 8 - x$	Clear parentheses.
$-3x - 10 \geq 8 - x$	Combine terms.
$-3x - 10 + x \geq 8 - x + x$	Add x.
$-2x - 10 \geq 8$	
$-2x - 10 + 10 \geq 8 + 10$	Add 10.
$-2x \geq 18$	
$\dfrac{-2x}{-2} \leq \dfrac{18}{-2}$	Divide by -2 and change the direction of the inequality symbol.
$x \leq -9$	

Notice that the inequality symbol was changed from $\geq$ to $\leq$ in the next to last step. Figure 5 shows the graph of the solution set, $(-\infty, -9]$.

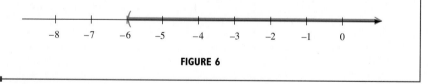

FIGURE 5

E X A M P L E   4   Solving a Linear Inequality

Solve $2 - 4(r - 3) < 3(5 - r) + 5$. Graph the solution set.

$2 - 4(r - 3) < 3(5 - r) + 5$	
$2 - 4r + 12 < 15 - 3r + 5$	Clear parentheses.
$14 - 4r < 20 - 3r$	Combine terms.
$14 - r < 20$	Add $3r$.
$-r < 6$	Subtract 14.

We want r, not $-r$. To get r, multiply both sides of the inequality by -1. Since -1 is negative, change the direction of the inequality symbol.

$-r < 6$	
$(-1)(-r) > (-1)(6)$	Multiply by -1 and change $<$ to $>$.
$r > -6$	

The solution set, $(-6, \infty)$, is graphed in Figure 6.

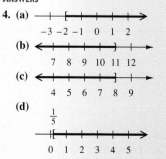

FIGURE 6

Note

In this text we will use the interval notation $(-\infty, \infty)$ if the solution set contains all real numbers.

◀◀ **WORK PROBLEM 4 AT THE SIDE.**

OBJECTIVE ▶ In further work in mathematics, it is sometimes necessary to work with an inequality such as $3 < x + 2 < 8$ where $x + 2$ is *between* 3 and 8. To solve a three-part inequality, we add or multiply all three parts, using the properties, as shown in the next example.

EXAMPLE 5 Solving a Three-Part Inequality

Solve the inequality $-2 \le 3k - 1 \le 5$, and graph the solution set.

Begin by adding 1 to each of the three parts to isolate the variable term in the middle.

$$-2 + 1 \le 3k - 1 + 1 \le 5 + 1 \qquad \text{Add 1.}$$

$$-1 \le 3k \le 6$$

$$\frac{-1}{3} \le \frac{3k}{3} \le \frac{6}{3} \qquad \text{Divide by 3.}$$

$$-\frac{1}{3} \le k \le 2$$

A graph of the solution, $[-\frac{1}{3}, 2]$, is shown in Figure 7.

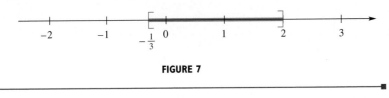

FIGURE 7

WORK PROBLEM 5 AT THE SIDE. ▶▶

The three-part inequality $-2 \le 3k - 1 \le 5$ can be written as the *compound* inequality

$$-2 \le 3k - 1 \quad \text{and} \quad 3k - 1 \le 5.$$

Compound inequalities are discussed further in the next section.

The types of solutions to be expected from solving linear equations or linear inequalities are shown below.

Solutions of Linear Equations and Inequalities

Equation or Inequality	Typical Solution Set	Graph of Solution Set
Linear equation $ax + b = c$	$\{p\}$	● p
Linear inequality $ax + b < c$	either $(-\infty, p)$ or (p, ∞)	) p
		(p
Three-part inequality $c < ax + b < d$	(p, q)	( ) p q

OBJECTIVE 4 There are several phrases that denote inequality. In addition to the familiar "is less than" and "is greater than" (which are examples of **strict** inequalities), the expressions "is no more than," "is at least," and others also denote inequalities. (These are called **nonstrict.**) Expressions like these sometimes appear in applied problems that we solve using inequalities. The following chart shows how these expressions are interpreted.

5. Graph the solution set of each inequality.

(a) $-3 \le x - 1 \le 7$

(b) $5 < 3x - 4 < 9$

ANSWERS

5. (a)

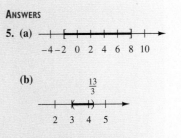

-4 -2 0 2 4 6 8 10

(b)

$\frac{13}{3}$

2 3 4 5

6. John can rent a car from Ames for $48 per day plus 10¢ per mile, or from Hughes at $40 per day plus 15¢ per mile. He plans to use the car for 3 days. What number of miles would make Hughes cost at most as much as Ames?

Word Expression	Interpretation	Word Expression	Interpretation
a **is at least** *b*	$a \geq b$	*a* **is at most** *b*	$a \leq b$
a **is no less than** *b*	$a \geq b$	*a* **is no more than** *b*	$a \leq b$

In Examples 6 and 7, we solve applied problems with inequalities.

E X A M P L E   6   Solving a Rental Problem

A rental company charges $15.00 to rent a chain saw, plus $2.00 per hour. Al Ghandi can spend no more than $35.00 to clear some logs from his yard. What is the *maximum* amount of time he can use the rented saw?

Let $h =$ the number of hours he can rent the saw. He must pay $15.00, plus $2.00h$, to rent the saw for h hours, and this amount must be *no more than* $35.00.

$$
\begin{array}{ccc}
\underbrace{\text{Cost of}}_{\text{renting}} & \underbrace{\text{is no}}_{\text{more than}} & \underbrace{\text{35 dollars}} \\
15 + 2h & \leq & 35 \\
15 + 2h - 15 & \leq & 35 - 15 \qquad \text{Subtract 15.} \\
2h & \leq & 20 \\
h & \leq & 10 \qquad \text{Divide by 2.}
\end{array}
$$

He can use the saw for a maximum of 10 hours. (Of course, he may use it for less time, as indicated by the inequality $h \leq 10$.)

◀◀ **WORK PROBLEM 6 AT THE SIDE.**

7. A student has grades of 92, 90, and 84 on his first three tests. What grade must the student make on his fourth test in order to keep an average of 90 or greater?

E X A M P L E   7   Solving a Grade-Averaging Problem

A student has grades of 88, 86, and 90 on her first three algebra tests. An average of 90 or above will earn an A in the class. What grade must the student make on her fourth test in order to have an A average?

Let x represent the score on the fourth test. Her average must be at least 90. To find the average of four numbers, add them and divide by 4.

$$\frac{88 + 86 + 90 + x}{4} \geq 90$$

$$\frac{264 + x}{4} \geq 90 \qquad \text{Add the scores.}$$

$$264 + x \geq 360 \qquad \text{Multiply by 4.}$$

$$x \geq 96 \qquad \text{Subtract 264.}$$

She must score *at least* 96 on her fourth test to keep an A average.

◀◀ **WORK PROBLEM 7 AT THE SIDE.**

ANSWERS

6. 480 miles or less

7. at least 94

3.1 Exercises

Match each inequality with the correct graph or interval notation.

_____ **1.** $x \le 3$

A.

_____ **2.** $x > 3$

B.

_____ **3.** $x < 3$

C. $(3, \infty)$

_____ **4.** $x \ge 3$

D. $(-\infty, 3]$

_____ **5.** $-3 \le x \le 3$

E. $(-3, 3)$

_____ **6.** $-3 < x < 3$

F. $[-3, 3]$

7. Explain how you will determine whether to use parentheses or brackets when graphing the solution set of an inequality.

8. When is it necessary to reverse the direction of the inequality sign when solving an inequality?

Solve each inequality, giving its solution set in both interval and graph forms. See Examples 1–4.

9. $5r \le -15$

10. $12m \le -36$

11. $4x + 1 \ge 21$

12. $5t + 2 \ge 52$

13. $\dfrac{3k - 1}{4} > 5$

14. $\dfrac{5z - 6}{8} < 8$

15. $-4x < 16$

16. $-2m > 10$

17. $-\dfrac{3}{4}r \ge 30$

18. $-\dfrac{2}{3}y \le 12$

19. $-\dfrac{3}{2}y \le -\dfrac{9}{2}$

20. $-\dfrac{2}{5}x \ge -\dfrac{4}{25}$

21. $-1.3m \geq -5.2$

22. $-2.5y \leq -1.25$

23. $\dfrac{2k - 5}{-4} > 5$

24. $\dfrac{3z - 2}{-5} < 6$

25. $y + 4(2y - 1) \geq y$

26. $m - 2(m - 4) \leq 3m$

27. $-(4 + r) + 2 - 3r < -14$

28. $-(9 + k) - 5 + 4k \geq 4$

29. $-3(z - 6) > 2z - 2$

30. $-2(y + 4) \leq 6y + 16$

MATHEMATICAL CONNECTIONS (EXERCISES 31–36)

Work Exercises 31–36 in order.

31. Solve the linear equation
$5(x + 3) - 2(x - 4) = 2(x + 7)$, and graph
the solution on a number line.

32. Solve the linear inequality
$5(x + 3) - 2(x - 4) > 2(x + 7)$, and graph
the solutions on a number line.

33. Solve the linear inequality
$5(x + 3) - 2(x - 4) < 2(x + 7)$, and graph
the solutions on a number line.

34. Graph all the solution sets of the equation and
inequalities in Exercises 31–33 on the same
number line. What set do you obtain?

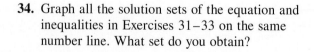

35. Based on the results of Exercises 31–33, complete the following using a conjecture (educated guess): The solution set of $-3(x + 2) = 3x + 12$ is $\{-3\}$, and the solution set of $-3(x + 2) < 3x + 12$ is $(-3, \infty)$. Therefore the solution set of $-3(x + 2) > 3x + 12$ is _____ .

36. Comment on the following statement: Equality is the boundary between less than and greater than.

Solve each inequality, giving its solution set in both interval and graph forms. See Example 5.

37. $-4 < x - 5 < 6$

38. $-1 < x + 1 < 8$

39. $-9 \leq k + 5 \leq 15$

40. $-4 \leq m + 3 \leq 10$

41. $-6 \leq 2z + 4 \leq 16$

42. $-15 < 3p + 6 < -12$

43. $-19 \leq 3x - 5 \leq 1$

44. $-16 < 3t + 2 < -10$

45. $-1 \leq \dfrac{2x - 5}{6} \leq 5$

46. $-3 \leq \dfrac{3m + 1}{4} \leq 3$

47. $4 \leq 5 - 9x < 8$

48. $4 \leq 3 - 2x < 8$

Everyday words and phrases like "exceed," "at least," and "fewer than" are closely related to mathematical inequality concepts. Based on the given graph, answer the questions in Exercises 49–52.

49. In which months did the percent of tornadoes exceed 7.7%?

50. In which months was the percent of tornadoes at least 12.9%?

——— **WHEN TORNADOES STRIKE** ———

December	2.5%
November	3.6%
October	3.0%
September	4.8%
August	7.7%
July	11.1%
June	20.7%
May	22.1%
April	12.9%
March	7.0%
February	2.8%
January	1.8%

Source: The USA Today Weather Book

51. The data used to determine the graph was based on the number of tornadoes sighted in the United States during the last twenty years. A total of 17,252 tornadoes were reported. In which months were fewer than 1500 reported?

52. How many more tornadoes occurred during March than October? (Use the total given in Exercise 51.)

Solve each problem. See Examples 6 and 7.

53. Margaret Westmoreland earned scores of 90 and 82 on her first two tests in English Literature. What score must she make on her third test to keep an average of 84 or greater?

54. Shannon d'Hemecourt scored 92 and 96 on her first two tests in Methods in Teaching Mathematics. What score must she make on her third test to keep an average of 90 or greater?

55. A couple wishes to rent a car for one day while on vacation. Ford Automobile Rental wants $15.00 per day and 14¢ per mile, while Chevrolet-For-A-Day wants $14.00 per day and 16¢ per mile. After how many miles would the price to rent the Chevrolet exceed the price to rent a Ford?

56. Jane and Terry Brandsma went to Mobile for a week. They needed to rent a car, so they checked out two rental firms. Avis wanted $28 per day, with no mileage fee. Downtown Toyota wanted $108 per week and 14¢ per mile. How many miles would they have to drive before the Avis price is less than the Toyota price?

A product will produce a profit only when the revenue (R) from selling the product exceeds the cost (C) of producing it. Find the smallest whole number of units x that must be sold for the business to show a profit for the item described.

57. Peripheral Visions, Inc. finds that the cost to produce x studio-quality videotapes is $C = 20x + 100$, while the revenue produced from them is $R = 24x$ (C and R in dollars).

58. Speedy Delivery finds that the cost to make x deliveries is $C = 3x + 2300$, while the revenue produced from them is $R = 5.50x$ (C and R in dollars).

3.2 Set Operations and Compound Inequalities

The words *and* and *or* are very important in interpreting certain kinds of equations and inequalities in algebra. They also occur in work with sets. In this section we study the use of these two words as they relate to sets and inequalities.

OBJECTIVE 1 We start by looking at the use of the word "and" with sets. The intersection of sets is defined below.

> For any two sets A and B, the **intersection** (IN-tur-sek-shun) of A and B, symbolized $A \cap B$, is defined as follows:
>
> $A \cap B = \{x \mid x$ is an element of A **and** x is an element of $B\}$.

EXAMPLE 1   Finding the Intersection of Two Sets

Let $A = \{1, 2, 3, 4\}$ and $B = \{2, 4, 6\}$. Find $A \cap B$.

The set $A \cap B$ contains those elements that belong to both A *and B:* the numbers 2 and 4. Therefore,

$$A \cap B = \{1, 2, 3, 4\} \cap \{2, 4, 6\}$$
$$= \{2, 4\}.$$

WORK PROBLEM 1 AT THE SIDE. ▶▶

OBJECTIVE 2 A **compound inequality** consists of two inequalities linked by a connective word such as *and* or *or*. Examples of compound inequalities are

$$x + 1 \leq 9 \quad \text{and} \quad x - 2 \geq 3$$

and

$$2x > 4 \quad \text{or} \quad 3x - 6 < 5.$$

To solve a compound inequality with the word *and,* we use the following steps.

Solving a Compound Inequality with *and*

Step 1 Solve each inequality in the compound inequality individually.

Step 2 Since the inequalities are joined with *and*, the solution will include all numbers that satisfy both solutions in Step 1 (the intersection of the solutions).

The next example shows how a compound inequality with *and* is solved.

EXAMPLE 2   Solving a Compound Inequality with *and*

Solve the compound inequality

$$x + 1 \leq 9 \quad \text{and} \quad x - 2 \geq 3.$$

Step 1 directs that we solve each inequality in the compound inequality individually.

$$x + 1 \leq 9 \qquad \text{and} \qquad x - 2 \geq 3$$
$$x + 1 - 1 \leq 9 - 1 \quad \text{and} \quad x - 2 + 2 \geq 3 + 2$$
$$x \leq 8 \qquad \text{and} \qquad x \geq 5$$

CONTINUED ON NEXT PAGE

1. Let $A = \{3, 4, 5, 6\}$ and $B = \{5, 6, 7\}$. Find $A \cap B$.

161

2. Graph the solution set of each compound inequality.

(a) $x < 10$ and $x > 2$

(b) $x + 3 < 1$ and
$x - 4 > -12$

3. Solve and graph.
$2x \geq x - 1$ and
$3x \geq 3 + 2x$

Now we apply Step 2. Because the inequalities are joined with the word *and*, the solution will include all numbers that satisfy both solutions in Step 1. Thus, the compound inequality is true whenever $x \leq 8$ and $x \geq 5$ are both true. The top graph in Figure 8 shows $x \leq 8$, and the middle graph shows $x \geq 5$. The bottom graph shows the numbers common to the first two graphs. As shown by this third graph, the solution consists of all numbers between 5 and 8, including both 5 and 8. This is the intersection of the two graphs, and we write it as $[5, 8]$.

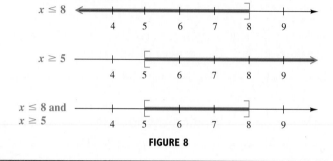

FIGURE 8

◀◀ WORK PROBLEM 2 AT THE SIDE.

E X A M P L E   3   Solving a Compound Inequality with *and*

Solve the compound inequality

$$-3x - 2 > 4 \quad \text{and} \quad 5x - 1 \leq -21.$$

Begin by solving each part separately.

$$-3x - 2 > 4 \quad \text{and} \quad 5x - 1 \leq -21$$
$$-3x > 6 \quad \text{and} \quad 5x \leq -20$$
$$x < -2 \quad \text{and} \quad x \leq -4$$

Now find all values of x that satisfy both conditions; that is, the real numbers that are less than -2 and also less than or equal to -4. As shown by the graphs in Figure 9, the solution set is $(-\infty, -4]$.

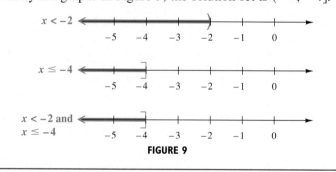

FIGURE 9

◀◀ WORK PROBLEM 3 AT THE SIDE.

E X A M P L E   4   Solving a Compound Inequality with *and*

Solve $x + 2 < 5$ and $x - 10 > 2$.

First solve each inequality separately.

$$x + 2 < 5 \quad \text{and} \quad x - 10 > 2$$
$$x < 3 \quad \text{and} \quad x > 12$$

ANSWERS

2. (a) ◀—+—[+++++]—+—▶
    0 2 4 6 8 10 12

(b) —+—[+++]—+—▶
    −10 −8 −6 −4 −2 0

3. —+++++[—+▶
    −1 0 1 2 3 4 5

CONTINUED ON NEXT PAGE ——

There is no number that is both less than 3 and greater than 12, so the given compound sentence has no solution (solution set is ∅). See Figure 10.

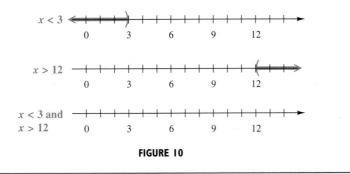

FIGURE 10

WORK PROBLEM 4 AT THE SIDE. ▶▶

4. Solve $x + 2 > 3$ and $2x + 1 < -3$.

OBJECTIVE ❸ We now discuss the union of two sets, which involves the use of the word "or."

For any two sets A and B, the **union** (YOON-yun) of A and B, symbolized $A \cup B$, is defined as follows:

$A \cup B = \{x \mid x \text{ is an element of } A \textbf{ or } x \text{ is an element of } B\}$.

E X A M P L E 5   Finding the Union of Two Sets

Find the union of the sets $A = \{1, 2, 3, 4\}$ and $B = \{2, 4, 6\}$.

Begin by listing all the elements of set A: 1, 2, 3, 4. Then list any additional elements from set B. In this case the elements 2 and 4 are already listed, so the only additional element is 6. Therefore,

$$A \cup B = \{1, 2, 3, 4\} \cup \{2, 4, 6\} = \{1, 2, 3, 4, 6\}.$$

The union consists of all elements in either A *or* B (or both).

5. Let $A = \{3, 4, 5, 6\}$ and $B = \{5, 6, 7\}$. Find $A \cup B$.

Notice in Example 5, that even though the elements 2 and 4 appeared in both sets A and B, they are only written once in $A \cup B$. It is not necessary to write them more than once in the union.

WORK PROBLEM 5 AT THE SIDE. ▶▶

OBJECTIVE ❹ To solve compound inequalities with the word *or*, we use the following steps.

Solving a Compound Inequality with *or*

Step 1 Solve each inequality in the compound inequality individually.

Step 2 Since the inequalities are joined with *or*, the solution will include all numbers that satisfy either one of the solutions in Step 1 (the union of the solutions).

The next examples show how to solve a compound inequality with *or*.

6. Graph each solution set.

(a) $x + 2 > 3$ or
$2x + 1 < -3$

(b) $y - 1 > 2$ or
$3y + 5 < 2y + 6$

EXAMPLE 6   Solving a Compound Inequality with *or*

Solve $6x - 4 < 2x$ or $-3x \leq -9$.

Solve each inequality separately:

$$6x - 4 < 2x \quad \text{or} \quad -3x \leq -9$$
$$4x < 4$$
$$x < 1 \quad \text{or} \quad x \geq 3.$$

The graphs of these results are shown in Figure 11. The third graph gives the combination of these two solutions. This final solution set is written

$$(-\infty, 1) \cup [3, \infty).$$

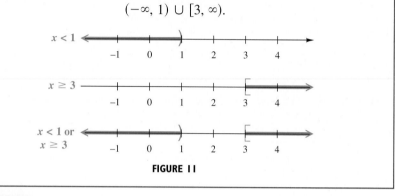

FIGURE 11

Caution
When interval notation is used to write the solution of Example 6, it *must* be written as

$$(-\infty, 1) \cup [3, \infty).$$

There is no short-cut way to write this solution.

◀◀ **WORK PROBLEM 6 AT THE SIDE.**

EXAMPLE 7   Solving a Compound Inequality with *or*

7. Solve
$3x - 2 \leq 13$ or
$x + 5 \geq 7$.

Solve $-4x + 1 \geq 9$ or $5x + 3 \geq -12$.

Solve each inequality separately.

$$-4x + 1 \geq 9 \quad \text{or} \quad 5x + 3 \geq -12$$
$$-4x \geq 8 \quad \text{or} \quad 5x \geq -15$$
$$x \leq -2 \quad \text{or} \quad x \geq -3$$

The graphs of these solutions are shown in Figure 12. As shown in the figure, every real number is a solution of the compound sentence because every real number satisfies at least one of the two inequalities.

The solution set is $(-\infty, \infty)$.

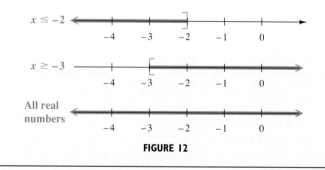

FIGURE 12

ANSWERS

6. (a)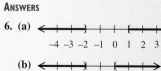

(b)

7. $(-\infty, \infty)$

◀◀ **WORK PROBLEM 7 AT THE SIDE.**

3.2 Exercises

Decide whether the statement is true or false. If it is false, explain why.

1. The union of the set of rational numbers and the set of irrational numbers is the set of real numbers.

2. The intersection of the set of rational numbers and the set of irrational numbers is the set $\{0\}$.

3. The union of the solution sets of $2x + 1 = 3$, $2x + 1 > 3$, and $2x + 1 < 3$ is the set of real numbers.

4. The intersection of the sets $\{x \mid x \geq 5\}$ and $\{x \mid x \leq 5\}$ is $\{5\}$.

5. The union of the sets $(-\infty, 6)$ and $(6, \infty)$ is $(-\infty, \infty)$.

6. The intersection of the sets $[6, \infty)$ and $(-\infty, 6]$ is $\emptyset$.

Let $A = \{1, 2, 3, 4, 5, 6\}$, $B = \{1, 3, 5\}$, $C = \{1, 6\}$, and $D = \{4\}$. Specify each of the following sets. See Examples 1 and 5.

7. $A \cap D$

8. $B \cap C$

9. $B \cap \emptyset$

10. $A \cap \emptyset$

11. $A \cup B$

12. $B \cup D$

13. $B \cup C$

14. $C \cup B$

15. $C \cup D$

16. $D \cup C$

17. Use the sets A, B, and C for Exercises 7–16 to show that $A \cap (B \cap C)$ is equal to $(A \cap B) \cap C$. This is true for any choices of sets. What property does this illustrate? (*Hint:* See Section 1.8.)

18. Repeat Exercise 17, showing that $A \cup (B \cup C)$ is equal to $(A \cup B) \cup C$. What property does this illustrate?

19. How can intersection be applied to a real-life situation?

20. A compound inequality uses one of the words *and* or *or*. Explain how you will determine whether to use *intersection* or *union* when graphing the solution set.

For each compound inequality, give the solution set in both interval and graph forms. See Examples 2, 3, and 4.

21. $x < 2$ and $x > -3$

22. $x < 5$ and $x > 0$

23. $x \le 2$ and $x \le 5$

24. $x \ge 3$ and $x \ge 6$

25. $x \le 3$ and $x \ge 6$

26. $x \le -1$ and $x \ge 3$

27. $x \ge -1$ and $x \ge 4$

28. $x \le -2$ and $x \le 2$

29. $x \ge -1$ and $x \le 3$

30. $x \ge 4$ and $x \le 5$

31. $x - 3 \le 6$ and $x + 2 \ge 7$

32. $x + 5 \le 11$ and $x - 3 \ge -1$

33. $3x - 4 \le 8$ and $4x - 1 \le 15$

—————————————▶

34. $7x + 6 \le 48$ and $-4x \ge -24$

—————————————▶

For each compound inequality, give the solution set in both interval and graph forms. See Examples 6 and 7.

35. $x \le 2$ or $x \ge 4$

—————————————▶

36. $x \le -5$ or $x \ge 6$

—————————————▶

37. $x \le 1$ or $x \le 8$

—————————————▶

38. $x \ge 1$ or $x \ge 8$

—————————————▶

39. $x \le 1$ or $x \ge 10$

—————————————▶

40. $x \le 2$ or $x \ge 9$

—————————————▶

41. $x \ge -2$ or $x \ge 5$

—————————————▶

42. $x \le -2$ or $x \le 6$

—————————————▶

43. $x \ge -2$ or $x \le 4$

—————————————▶

44. $x \ge 5$ or $x \le 7$

—————————————▶

45. $x + 2 > 7$ or $x - 1 < -6$

—————————————▶

46. $x + 1 > 3$ or $x + 4 < 2$

—————————————▶

47. $4x - 8 > 0$ or $4x - 1 < 7$

—————————————▶

48. $3x < x + 12$ or $3x - 8 > 10$

—————————————▶

For each compound inequality, decide whether intersection or union should be used. Then give the solution set in both interval and graph forms. See Examples 2, 3, 4, 6, and 7.

49. $x < -1$ and $x > -5$

50. $x > -1$ and $x < 7$

51. $x < 4$ or $x < -2$

52. $x < 5$ or $x < -3$

53. $x + 1 \geq 5$ and $x - 2 \leq 10$

54. $2x - 6 \leq -18$ and $2x \geq -18$

55. $-3x \leq -6$ or $-3x \geq 0$

56. $-8x \leq -24$ or $-5x \geq 15$

MATHEMATICAL CONNECTIONS (EXERCISES 57–60)

The figures represent the backyards of neighbors Luigi, Mario, Than, and Joe. Find the area and the perimeter of each yard. Suppose that each resident has 150 feet of fencing, and enough sod to cover 1400 square feet of lawn. Give the name or names of the residents whose yards satisfy the following descriptions.

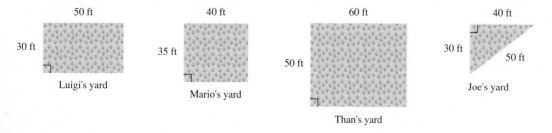

57. the yard can be fenced *and* the yard can be sodded

58. the yard can be fenced *and* the yard cannot be sodded

59. the yard cannot be fenced *and* the yard can be sodded

60. the yard cannot be fenced *and* the yard cannot be sodded

3.3 Absolute Value Equations and Inequalities

OBJECTIVES

1. Use the distance definition of absolute value.
2. Solve $|ax + b| = k, k > 0$.
3. Solve $|ax + b| < k$ and solve $|ax + b| > k, k > 0$.
4. Solve absolute value equations that involve rewriting.
5. Solve absolute value equations of the form
$$|ax + b| = |cx + d|.$$
6. Solve absolute value equations and inequalities that have only nonpositive constants on one side.

FOR EXTRA HELP

Tutorial   Tape 6   SSM, Sec. 3.3

OBJECTIVE 1 The absolute value of a number x, written $|x|$, represents the distance from x to 0 on the number line. For example, the solutions of $|x| = 4$ are 4 and -4, as shown in Figure 13.

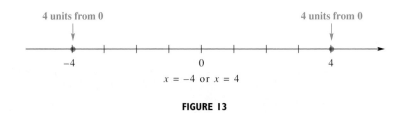

$$x = -4 \text{ or } x = 4$$

FIGURE 13

Because absolute value represents distance from 0, it is reasonable to interpret the solutions of $|x| > 4$ to be all numbers that are *more* than 4 units from 0. The set $(-\infty, -4) \cup (4, \infty)$ fits this description. Figure 14 shows the graph of the solution set of $|x| > 4$.

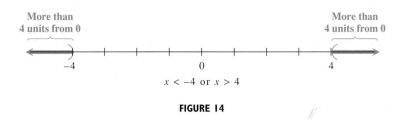

$$x < -4 \text{ or } x > 4$$

FIGURE 14

The solution set of $|x| < 4$ consists of all numbers that are *less* than 4 units from 0 on the number line. Another way of thinking of this is to think of all numbers *between* -4 and 4. This set of numbers is given by $(-4, 4)$, as shown in Figure 15.

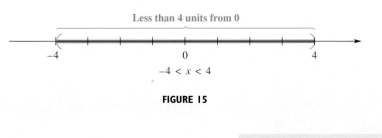

$$-4 < x < 4$$

FIGURE 15

WORK PROBLEM 1 AT THE SIDE. ▶▶

The equation and inequalities just described are examples of **absolute value equations and inequalities.** These are equations and inequalities that involve the absolute value of a variable expression. They generally take the form

$$|ax + b| = k, \qquad |ax + b| > k, \qquad \text{or} \qquad |ax + b| < k$$

where k is a positive number. We may solve them by rewriting them as compound equations or inequalities, as shown in the following summary.

1. Graph the solution set of each equation or inequality.

   (a) $|x| = 3$

   _____

   (b) $|x| > 3$

   _____

   (c) $|x| < 3$

   _____

ANSWERS

1. (a) $\begin{array}{c} \longleftrightarrow \\ -3\ -2\ -1\ \ 0\ \ 1\ \ 2\ \ 3 \end{array}$

   (b) $\begin{array}{c} \longleftrightarrow \\ -3\ -2\ -1\ \ 0\ \ 1\ \ 2\ \ 3 \end{array}$

   (c) $\begin{array}{c} \longleftrightarrow \\ -3\ -2\ -1\ \ 0\ \ 1\ \ 2\ \ 3 \end{array}$

2. Solve each equation, and graph the solution set.

(a) $|x + 2| = 3$

_____→

Solving Absolute Value Equations and Inequalities

Let k be a positive number, and p and q be two numbers.

1. To solve $|ax + b| = k$, solve the compound equation

$$ax + b = k \quad \text{or} \quad ax + b = -k.$$

The solution set is usually of the form $\{p, q\}$, with two numbers.

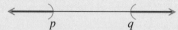

2. To solve $|ax + b| > k$, solve the compound inequality

$$ax + b > k \quad \text{or} \quad ax + b < -k.$$

The solution set is of the form $(-\infty, p) \cup (q, \infty)$, which consists of two separate intervals.

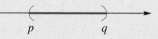

3. To solve $|ax + b| < k$, solve the compound inequality

$$-k < ax + b < k.$$

The solution set is of the form (p, q), a single interval.

(b) $|3x - 4| = 11$

_____→

Note

Some people prefer to write the compound statements in parts 1 and 2 of the summary as

$$ax + b = k \quad \text{or} \quad -(ax + b) = k$$

and

$$ax + b > k \quad \text{or} \quad -(ax + b) > k.$$

These forms are equivalent to those we give in the summary and produce the same results.

OBJECTIVE 2 The next example shows how we use a compound equation to solve a typical absolute value equation. Remember that because absolute value refers to distance from the origin, each absolute value equation will have two parts.

E X A M P L E 1   Solving an Absolute Value Equation

Solve $|2x + 1| = 7$.

For $|2x + 1|$ to equal 7, $2x + 1$ must be 7 units from 0 on the number line. This can happen only when $2x + 1 = 7$ or $2x + 1 = -7$. Solve this compound equation as follows.

$$2x + 1 = 7 \quad \text{or} \quad 2x + 1 = -7$$
$$2x = 6 \quad \text{or} \quad 2x = -8$$
$$x = 3 \quad \text{or} \quad x = -4$$

The solution set is $\{-4, 3\}$. Its graph is shown in Figure 16.

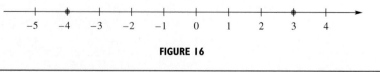

FIGURE 16

◀◀ WORK PROBLEM 2 AT THE SIDE.

OBJECTIVE 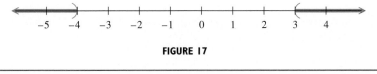 We now discuss how to solve absolute value inequalities.

┌ **E X A M P L E 2**   **Solving an Absolute Value Inequality with >**

Solve $|2x + 1| > 7$.

   As shown in the summary, this absolute value inequality should be rewritten as

$$2x + 1 > 7 \quad \text{or} \quad 2x + 1 < -7,$$

because $2x + 1$ must represent a number that is *more* than 7 units from 0 on the number line. Now solve the compound inequality.

$$2x + 1 > 7 \quad \text{or} \quad 2x + 1 < -7$$
$$2x > 6 \quad \text{or} \quad 2x < -8$$
$$x > 3 \quad \text{or} \quad x < -4$$

The solution set, $(-\infty, -4) \cup (3, \infty)$, is graphed in Figure 17. Notice that the graph consists of two intervals.

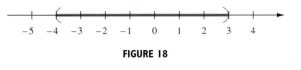

FIGURE 17

WORK PROBLEM 3 AT THE SIDE. ▶▶

┌ **E X A M P L E 3**   **Solving an Absolute Value Inequality with <**

Solve $|2x + 1| < 7$.

   The expression $2x + 1$ must represent a number that is less than 7 units from 0 on the number line. Another way of thinking of this is to realize that $2x + 1$ must be between -7 and 7. As the summary shows, this is written as the three-part inequality

$$-7 < 2x + 1 < 7.$$

We solved such inequalities in Section 3.1 by working with all three parts at the same time.

$$-7 < 2x + 1 < 7$$
$$-8 < 2x < 6 \qquad \text{Subtract 1 from each part.}$$
$$-4 < x < 3 \qquad \text{Divide each part by 2.}$$

The solution set is $(-4, 3)$, and the graph consists of a single interval as shown in Figure 18.

FIGURE 18

WORK PROBLEM 4 AT THE SIDE. ▶▶

3. Solve each inequality, and graph the solution set.

   (a) $|x + 2| > 3$

   _____

   (b) $|3x - 4| \geq 11$

   _____

4. Solve each inequality, and graph the solution set.

   (a) $|x + 2| < 3$

   _____

   (b) $|3x - 4| \leq 11$

   _____

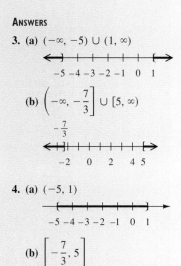

ANSWERS

3. (a) $(-\infty, -5) \cup (1, \infty)$

   (b) $\left(-\infty, -\dfrac{7}{3}\right] \cup [5, \infty)$

4. (a) $(-5, 1)$

   (b) $\left[-\dfrac{7}{3}, 5\right]$

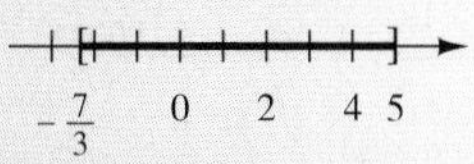

5. (a) Solve $|5a + 2| - 9 = -7$.

(b) Solve, and graph the solution set.

$$|m + 2| - 3 > 2$$

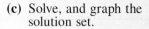

(c) Solve, and graph the solution set.

$$|3a + 2| + 4 \le 15$$

Look back at Figures 16, 17, and 18. These are the graphs of $|2x + 1| = 7$, $|2x + 1| > 7$, and $|2x + 1| < 7$. If we find the union of the three sets, we get the set of all real numbers. This is because for any value of x, $|2x + 1|$ will satisfy one and only one of the following: it is equal to 7, greater than 7, or less than 7.

Caution

When solving absolute value equations and inequalities of the types in Examples 1, 2, and 3, be sure to remember the following.

1. The methods described apply when the constant is alone on one side of the equation or inequality and is *positive*.
2. Absolute value equations and absolute value inequalities in the form $|ax + b| > k$ translate into "or" compound statements.
3. Absolute value inequalities in the form $|ax + b| < k$ translate into "and" compound statements, which may be written as three-part inequalities.
4. An "or" statement *cannot* be written in three parts. It would be incorrect to use

$$-7 > 2x + 1 > 7$$

in Example 2, because this would imply that $-7 > 7$, which is *false*.

OBJECTIVE Sometimes an absolute value equation or inequality is given in a form that requires some rewriting before it can be set up as a compound statement. The next example illustrates this process for an absolute value equation.

EXAMPLE 4   Solving an Absolute Value Equation Requiring Rewriting

Solve the equation $|x + 3| + 5 = 12$.

First get the absolute value alone on one side of the equals sign. Do this by subtracting 5 on each side.

$$|x + 3| + 5 - 5 = 12 - 5 \qquad \text{Subtract 5.}$$
$$|x + 3| = 7$$

Then use the method shown in Example 1.

$$x + 3 = 7 \quad \text{or} \quad x + 3 = -7$$
$$x = 4 \quad \text{or} \qquad x = -10$$

Check that the solution set is $\{4, -10\}$ by substituting in the original equation.

Solving an absolute value *inequality* requiring rewriting is done in a similar manner.

Caution

A common error in solving an absolute value equation such as the one in Example 4 is to forget about the absolute value symbols and solve $x + 3 + 5 = 12$. Do not make this error.

◀◀ **WORK PROBLEM 5 AT THE SIDE.**

OBJECTIVE The next example shows how to solve an equation with two absolute value expressions. For two expressions to have the same absolute value, they must either be equal or be negatives of each other.

ANSWERS

5. (a) $\left\{ -\dfrac{4}{5}, 0 \right\}$

(b) $(-\infty, -7) \cup (3, \infty)$

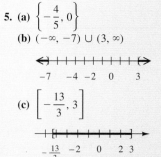

(c) $\left[-\dfrac{13}{3}, 3 \right]$

To solve an absolute value equation of the form

$$|ax + b| = |cx + d|,$$

solve the compound equation

$$ax + b = cx + d \quad \textbf{or} \quad ax + b = -(cx + d).$$

EXAMPLE 5   Solving an Equation with Two Absolute Values

Solve the equation $|z + 6| = |2z - 3|$.

This equation is satisfied either if $z + 6$ and $2z - 3$ are equal to each other, or if $z + 6$ and $2z - 3$ are negatives of each other. Thus, we have

$$z + 6 = 2z - 3 \quad \text{or} \quad z + 6 = -(2z - 3).$$

Solve each equation.

$$z + 6 = 2z - 3 \quad \text{or} \quad z + 6 = -2z + 3$$
$$9 = z \qquad\qquad 3z = -3$$
$$z = -1$$

The solution set is $\{9, -1\}$.

WORK PROBLEM 6 AT THE SIDE. ▶▶

OBJECTIVE ▶ When an absolute value equation or inequality involves a *negative* constant or *zero* alone on one side, simply use the properties of absolute value to solve. Keep in mind the following.

1. The absolute value of an expression can never be negative: $|a| \geq 0$ for all real numbers a.
2. The absolute value of an expression equals 0 only when the expression is equal to 0.

The next two examples illustrate these special cases.

EXAMPLE 6   Solving Special Cases of Absolute Value Equations

Solve each equation.

(a) $|5r - 3| = -4$

Since the absolute value of an expression can never be negative, there are no solutions for this equation. The solution set is $\emptyset$.

(b) $|7x - 3| = 0$

The expression $7x - 3$ will equal 0 *only* for the solution of the equation

$$7x - 3 = 0.$$

The solution of this equation is $\frac{3}{7}$. The solution set is $\left\{\frac{3}{7}\right\}$. It consists of only one element.

WORK PROBLEM 7 AT THE SIDE. ▶▶

---

6. Solve each equation.

(a) $|k - 1| = |5k + 7|$

(b) $|4r - 1| = |3r + 5|$

7. Solve.

(a) $|6x + 7| = -5$

(b) $\left|\dfrac{1}{4}x - 3\right| = 0$

ANSWERS

6. (a) $\{-1, -2\}$
   (b) $\left\{-\dfrac{4}{7}, 6\right\}$
7. (a) $\emptyset$ **(b)** $\{12\}$

8. Solve.

(a) $|x| > -1$

(b) $|y| < -5$

(c) $|k + 2| \leq 0$

E X A M P L E   7   **Solving Special Cases of Absolute Value Inequalities**

Solve each of the following inequalities.

(a) $|x| \geq -4$

The absolute value of a number is always nonnegative. For this reason, $|x| \geq -4$ is true for *all* real numbers. The solution set is $(-\infty, \infty)$.

(b) $|k + 6| < -2$

There is no number whose absolute value is less than -2, so this inequality has no solution. The solution set is $\emptyset$.

(c) $|m - 7| \leq 0$

The value of $|m - 7|$ will never be less than 0. However, $|m - 7|$ will equal 0 when $m = 7$. Therefore, the solution set is $\{7\}$.

◀◀ **WORK PROBLEM 8 AT THE SIDE.**

ANSWERS

8. (a) $(-\infty, \infty)$ **(b)** $\emptyset$ **(c)** $\{-2\}$

3.3 Exercises

Keeping in mind that the absolute value of a number can be interpreted as the distance between the graph of the number and 0 on the number line, match the absolute value equation or inequality with the graph of its solution set.

Choices   Choices

1. $|x| = 5$ _____   A.   **2.** $|x| = 9$ _____   A.

$|x| < 5$ _____   B. $|x| > 9$ _____   B.

$|x| > 5$ _____   C. $|x| \geq 9$ _____   C.

$|x| \leq 5$ _____   D. $|x| < 9$ _____   D.

$|x| \geq 5$ _____   E. $|x| \leq 9$ _____   E.

3. Explain when to use *and* and when to use *or* if you are solving an absolute value equation or inequality of the form $|ax + b| = k$, $|ax + b| < k$, or $|ax + b| > k$, where k is a positive number.

4. How many solutions will $|ax + b| = k$ have if **(a)** $k = 0$; **(b)** $k > 0$; **(c)** $k < 0$?

Solve each equation. See Example 1.

5. $|x| = 12$ **6.** $|k| = 14$ **7.** $|4x| = 20$ **8.** $|5x| = 30$

9. $|y - 3| = 9$ **10.** $|p - 5| = 13$ **11.** $|2x + 1| = 7$ **12.** $|2y + 3| = 19$

13. $|4r - 5| = 17$ **14.** $|5t - 1| = 21$ **15.** $|2y + 5| = 14$ **16.** $|2x - 9| = 18$

17. $\left|\dfrac{1}{2}x + 3\right| = 2$ **18.** $\left|\dfrac{2}{3}q - 1\right| = 5$ **19.** $\left|1 - \dfrac{3}{4}k\right| = 7$ **20.** $\left|2 - \dfrac{5}{2}m\right| = 14$

Solve each inequality, and graph the solution set. See Example 2.

21. $|x| > 3$

22. $|y| > 5$

23. $|k| \geq 4$

24. $|r| \geq 6$

25. $|t + 2| > 10$

26. $|r + 5| > 20$

27. $|3x - 1| \geq 8$

28. $|4x + 1| \geq 21$

29. $|3 - x| > 5$

30. $|5 - x| > 3$

31. The graph of the solution set of $|2x + 1| = 9$ is given below.

Without actually doing the algebraic work, graph the solution set of each inequality, referring to the graph above.

(a) $|2x + 1| < 9$ **(b)** $|2x + 1| > 9$

32. The graph of the solution set of $|3y - 4| < 5$ is given below.

Without actually doing the algebraic work, graph the solution set of each inequality, referring to the graph above.

(a) $|3y - 4| = 5$ **(b)** $|3y - 4| > 5$

Solve each inequality and graph its solution set. See Example 3. (Hint: Compare your answers to those in Exercises 21–30.)

33. $|x| \leq 3$

34. $|y| \leq 5$

35. $|k| < 4$

36. $|r| < 6$

37. $|t + 2| \leq 10$

38. $|r + 5| \leq 20$

39. $|3x - 1| < 8$

40. $|4x + 1| < 21$

41. $|3 - x| \leq 5$

42. $|5 - x| \leq 3$

Exercises 43–50 represent a sampling of the various types of absolute value equations and inequalities covered in Exercises 1–42. Decide which method of solution applies, find the solution set, and graph. See Examples 1, 2, and 3.

43. $|-4 + k| > 9$

44. $|-3 + t| > 8$

45. $|7 + 2z| = 5$

46. $|9 - 3p| = 3$

47. $|3r - 1| \leq 11$

48. $|2s - 6| \leq 6$

49. $|-6x - 6| \leq 1$

50. $|-2x - 6| \leq 5$

Solve each equation or inequality. Give the solution set in set notation for equations and in interval notation for inequalities. See Example 4.

51. $|x| - 1 = 4$

52. $|y| + 3 = 10$

53. $|x + 4| + 1 = 2$

54. $|y + 5| - 2 = 12$

55. $|2x + 1| + 3 > 8$

56. $|6x - 1| - 2 > 6$

57. $|x + 5| - 6 \le -1$

58. $|r - 2| - 3 \le 4$

Solve each equation. See Example 5.

59. $|3x + 1| = |2x + 4|$

60. $|7x + 12| = |x - 8|$

61. $\left| m - \dfrac{1}{2} \right| = \left| \dfrac{1}{2}m - 2 \right|$

62. $\left| \dfrac{2}{3}r - 2 \right| = \left| \dfrac{1}{3}r + 3 \right|$

63. $|6x| = |9x + 1|$

64. $|13y| = |2y + 1|$

65. $|2p - 6| = |2p + 11|$

66. $|3x - 1| = |3x + 9|$

Solve each equation or inequality. See Examples 6 and 7.

67. $|12t - 3| = -8$ **68.** $|13w + 1| = -3$ **69.** $|4x + 1| = 0$

70. $|6r - 2| = 0$ **71.** $|2q - 1| < -6$ **72.** $|8n + 4| < -4$

73. $|x + 5| > -9$ **74.** $|x + 9| > -3$ **75.** $|7x + 3| \leq 0$

76. $|4x - 1| \leq 0$ **77.** $|5x - 2| \geq 0$ **78.** $|4 + 7x| \geq 0$

79. $|10z + 7| > 0$ **80.** $|4x + 1| > 0$

MATHEMATICAL CONNECTIONS (EXERCISES 81–84)

The ten tallest buildings in Kansas City, Missouri, are listed along with their heights.

Building	Height (in feet)
One Kansas City Place	626
AT&T Town Pavilion	590
Hyatt Regency	504
Kansas City Power and Light	476
City Hall	443
Federal Office Building	413
Commerce Tower	402
City Center Square	402
Southwest Bell Telephone	394
Pershing Road Associates	352

Use this information to work through Exercises 81–84 in order.

81. To find the average of a group of numbers, we add the numbers and then divide by the number of items added. Use a calculator to find the average of the heights.

82. Let k represent the average height of these buildings. If a height x satisfies the inequality

$$|x - k| < t,$$

then the height is said to be within t feet of the average. Using your result from Exercise 81, list the buildings that are within 50 feet of the average.

83. Repeat Exercise 82, but find the buildings that are within 75 feet of the average.

84. (a) Write an absolute value inequality that describes the height of a building that is *not* within 75 feet of the average.
(b) Solve the inequality you wrote in part (a).
(c) Use the result of part (b) to find the buildings that are not within 75 feet of the average.
(d) Confirm that your answer to part (c) makes sense by comparing it with your answer to Exercise 83.

Students often have difficulty distinguishing between the various types of equations and inequalities introduced in this chapter and in Chapter 2. This section of miscellaneous equations and inequalities provides practice in solving all such types. Solve each equation or inequality.

1. $4z + 1 = 49$

2. $|m - 1| = 6$

3. $6q - 9 = 12 + 3q$

4. $3p + 7 = 9 + 8p$

5. $|a + 3| = -4$

6. $2m + 1 \leq m$

7. $8r + 2 \geq 5r$

8. $4(a - 11) + 3a = 20a - 31$

9. $2q - 1 = -7$

10. $|3q - 7| - 4 = 0$

11. $6z - 5 \leq 3z + 10$

12. $|5z - 8| + 9 \geq 7$

13. $9y - 3(y + 1) = 8y - 7$

14. $|y| \geq 8$

15. $9y - 5 \geq 9y + 3$

16. $13p - 5 > 13p - 8$

17. $|q| < 5.5$

18. $4z - 1 = 12 + z$

19. $\frac{2}{3}y + 8 = \frac{1}{4}y$

20. $-\frac{5}{8}y \geq -20$

21. $\frac{1}{4}p < -6$

22. $7z - 3 + 2z = 9z - 8z$

23. $\frac{3}{5}q - \frac{1}{10} = 2$

24. $|r - 1| < 7$

25. $r + 9 + 7r = 4(3 + 2r) - 3$ **26.** $6 - 3(2 - p) < 2(1 + p) + 3$ **27.** $|2p - 3| > 11$

28. $\dfrac{x}{4} - \dfrac{2x}{3} = -10$ **29.** $|5a + 1| \leq 0$ **30.** $5z - (3 + z) \geq 2(3z + 1)$

31. $-2 \leq 3x - 1 \leq 8$ **32.** $-1 \leq 6 - x \leq 5$ **33.** $|7z - 1| = |5z + 3|$

34. $|p + 2| = |p + 4|$ **35.** $|1 - 3x| \geq 4$ **36.** $\dfrac{1}{2} \leq \dfrac{2}{3} r \leq \dfrac{5}{4}$

37. $-(m + 4) + 2 = 3m + 8$ **38.** $\dfrac{p}{6} - \dfrac{3p}{5} = p - 86$ **39.** $-6 \leq \dfrac{3}{2} - x \leq 6$

40. $|5 - y| < 4$ **41.** $|y - 1| \geq -6$ **42.** $|2r - 5| = |r + 4|$

43. $8q - (1 - q) = 3(1 + 3q) - 4$ **44.** $8y - (y + 3) = -(2y + 1) - 12$

45. $|r - 5| = |r + 9|$ **46.** $|r + 2| < -3$

47. $2x + 1 > 5$ or $3x + 4 < 1$ **48.** $1 - 2x \geq 5$ and $7 + 3x \geq -2$

CHAPTER 3 SUMMARY

KEY TERMS

3.1 **linear inequality in one variable**
An inequality is linear in the variable x if it can be written in the form $ax + b < c$, $ax + b \leq c$, $ax + b > c$, or $ax + b \geq c$, where a, b, and c are real numbers, with $a \neq 0$.

equivalent inequalities
Equivalent inequalities are inequalities with the same solution set.

addition and multiplication properties of inequality
The same number may be added to (or subtracted from) both sides of an inequality to obtain an equivalent inequality. Both sides of an inequality may be multiplied or divided by the same positive number. If both sides are multiplied by or divided by a negative number, the inequality symbol must be reversed.

strict inequality
An inequality that involves $>$ or $<$ is called a strict inequality.

nonstrict inequality
An inequality that involves $\geq$ or $\leq$ is called a nonstrict inequality.

3.2 **intersection**
The intersection of two sets A and B is the set of elements that belong to both A and B.

union
The union of two sets A and B is the set of elements that belong to either A or B (or both).

compound inequality
A compound inequality is formed by joining two inequalities with a connective word, such as *and* or *or*.

3.3 **absolute value equation; absolute value inequality**
Absolute value equations and inequalities are equations and inequalities that involve the absolute value of a variable expression.

NEW SYMBOLS

$\cap$ set intersection
$\cup$ set union

$[3, \infty)$, $(-\infty, 2)$, $(-\infty, \infty)$ Examples of interval notation

QUICK REVIEW

Concepts	Examples
3.1 Linear Inequalities in One Variable	
Solving a Linear Inequality Simplify each side separately, combining like terms and removing parentheses. Use the addition property of inequality to get the variables on one side of the inequality sign and the numbers on the other. Combine like terms, and then use the multiplication property to change the inequality to the form $x < k$ or $x > k$.	Solve $3(x + 2) - 5x \leq 12$. $$3x + 6 - 5x \leq 12$$ $$-2x + 6 \leq 12$$ $$-2x \leq 6$$ $$\frac{-2x}{-2} \geq \frac{6}{-2}$$ $$x \geq -3$$
If an inequality is multiplied or divided by a *negative* number, the inequality symbol *must be reversed*.	The solution set is $[-3, \infty)$ and is graphed below.

Concepts	*Examples*
3.2 Set Operations and Compound Inequalities **Solving a Compound Inequality** Solve each inequality in the compound inequality individually. If the inequalities are joined with *and,* the solution is the intersection of the two individual solutions. If the inequalities are joined with *or,* the solution is the union of the two individual solutions.	Solve $x + 1 > 2$ and $2x < 6$. $\qquad x + 1 > 2 \quad$ and $\quad 2x < 6$ $\qquad\qquad x > 1 \quad$ and $\quad\; x < 3$ The solution set is $(1, 3)$. Solve $x \geq 4$ or $x \leq 0$. The solution set is $(-\infty, 0] \cup [4, \infty)$.
3.3 Absolute Value Equations and Inequalities Let k be a positive number. To solve $|ax + b| = k$, solve the compound equation $\qquad ax + b = k \quad$ or $\quad ax + b = -k$.	Solve $|x - 7| = 3$. $\qquad x - 7 = 3 \quad$ or $\quad x - 7 = -3$ $\qquad\quad x = 10 \qquad\qquad\quad x = 4$ The solution set is $\{4, 10\}$.
To solve $|ax + b| > k$, solve the compound inequality $\qquad ax + b > k \quad$ or $\quad ax + b < -k$.	Solve $|x - 7| > 3$. $\qquad x - 7 > 3 \quad$ or $\quad x - 7 < -3$ $\qquad\quad x > 10 \quad$ or $\qquad\quad x < 4$ The solution set is $(-\infty, 4) \cup (10, \infty)$.
To solve $|ax + b| < k$, solve the compound inequality $\qquad -k < ax + b < k$.	Solve $|x - 7| < 3$. $\qquad\qquad -3 < x - 7 < 3$ $\qquad\qquad\quad 4 < x < 10$ The solution set is $(4, 10)$.
To solve an absolute value equation of the form $\qquad |ax + b| = |cx + d|,$ solve the compound equation $\qquad ax + b = cx + d \quad$ or $\qquad ax + b = -(cx + d)$.	Solve $|x + 2| = |2x - 6|$. $\quad x + 2 = 2x - 6 \quad$ or $\quad x + 2 = -(2x - 6)$ $\qquad\quad x = 8 \qquad\qquad\qquad x + 2 = -2x + 6$ $\qquad\qquad\qquad\qquad\qquad\qquad 3x = 4$ $\qquad\qquad\qquad\qquad\qquad\qquad\; x = \dfrac{4}{3}$ The solution set is $\left\{\frac{4}{3}, 8\right\}$.

[3.1] *Solve each inequality. Give the solution set in both interval and graph forms.*

1. $-\dfrac{2}{3}k < 6$

2. $-5x - 4 \geq 11$

3. $\dfrac{6a + 3}{-4} < -3$

4. $\dfrac{9y + 5}{-3} > 3$

5. $5 - (6 - 4k) \geq 2k - 7$

6. $-6 \leq 2k \leq 24$

7. $8 \leq 3y - 1 < 14$

8. $-4 < 3 - 2k < 9$

9. To pass algebra, a student must have an average of at least 70% on five tests. On the first four tests, a student has grades of 75%, 79%, 64%, and 71%. What possible grades on the fifth test would guarantee a passing grade in the class?

10. While solving the inequality
$10x + 2(x - 4) < 12x - 13$, a student did all the work correctly and obtained the statement $-8 < -13$. The student did not know what to do at this point, because the variable "disappeared." How would you explain to the student the interpretation of this result?

[3.2] *In Exercises 11 and 12, let A* $= \{1, 3, 5, 7, 9\}$ *and let B* $= \{3, 6, 9, 12\}$.

11. Find $A \cap B$.

12. Find $A \cup B$.

Solve each compound inequality. Give the solution set in both interval and graph forms.

13. $x > 6$ and $x < 9$

14. $x + 4 > 12$ and $x - 2 < 12$

15. $x > 5$ or $x \leq -3$

16. $x \geq -2$ or $x < 2$

17. $x - 4 > 6$ and $x + 3 \leq 10$

18. $-5x + 1 \geq 11$ or $3x + 5 \geq 26$

Use the graphs to answer the questions in Exercises 19 and 20.

U.S. AIDS CASES REPORTED EACH YEAR

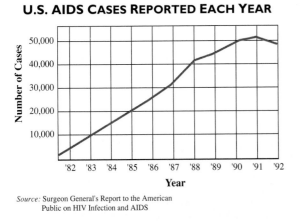

Source: Surgeon General's Report to the American
Public on HIV Infection and AIDS

**NEW AIDS CASES REPORTED AMONG
CHILDREN UNDER 13 YEARS OF AGE**

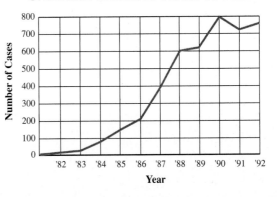

19. In which years did the number of U.S. AIDS cases exceed 30,000 *and* the new AIDS cases among children under thirteen years of age exceed 400?

20. In which years was the number of U.S. AIDS cases greater than 40,000 *or* the new AIDS cases among children under thirteen years of age less than 200?

[3.3] *Solve each absolute value equation.*

21. $|x| = 7$

22. $|y + 2| = 9$

23. $|3k - 7| = 8$

24. $|z - 4| = -12$

25. $|2k - 7| + 4 = 11$

26. $|4a + 2| - 7 = -3$

27. $|3p + 1| = |p + 2|$

28. $|2m - 1| = |2m + 3|$

Solve each absolute value inequality. Give the solution set in both interval and graph forms.

29. $|p| < 14$

30. $|-y + 6| \leq 7$

31. $|2p + 5| \leq 1$

    ⟶

32. $|x + 1| \geq -3$

    ⟶

33. $|5r - 1| > 9$

    ⟶

34. $|3k + 6| \geq 0$

    ⟶

MIXED REVIEW EXERCISES*

This set of exercises contains equations, inequalities, and applications from both Chapters 2 and 3.

Solve.

35. $(7 - 2k) + 3(5 - 3k) \geq k + 8$

36. $x < 5$ and $x \geq -4$

37. $-5(6p + 4) - 2p = -32p + 14$

38. The perimeter of a triangle is 34 inches. The middle side is twice as long as the shortest side. The longest side is 2 inches less than three times the shortest side. Find the lengths of the three sides.

39. $-5r \geq -10$ **40.** $|7x - 2| > 9$

41. $|2x - 10| = 20$ **42.** $|m + 3| \leq 13$

43. A square is such that if each side were increased by 4 inches, the perimeter would be 8 inches less than twice the perimeter of the original square. Find the length of a side of the original square.

44. In an election, one candidate received 151 more votes than the other. The total number of votes cast in the election was 1215. Find the number of votes received by each candidate.

---

* The order of exercises in this final group does not correspond to the order in which topics occur in the chapter. This random ordering should help you prepare for the chapter test in yet another way.

1. What is the special rule that must be remembered when multiplying or dividing both sides of an inequality by a negative number?

1. _____

Solve each inequality. Give the solution set in both interval and graph forms.

2. $4 - 6(x + 3) \leq -2 - 3(x + 6) + 3x$

2. _____→

3. $-\dfrac{4}{7}x > -16$

3. _____→

4. $-6 \leq \dfrac{4}{3}x - 2 \leq 2$

4. _____→

5. A student must have an average grade of 90 or greater on the three tests in Intermediate Algebra to earn a grade of A. A student had scores of 84 and 92 on her first two tests. What possible scores on her third test would assure her an A?

5. _____

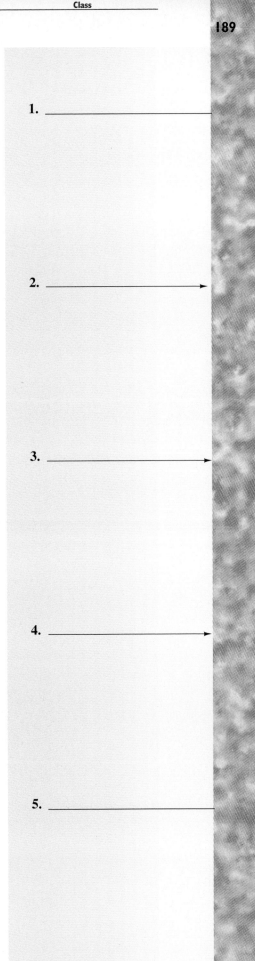

6. ──────────→

7. ──────────→

8. ──────────→

9. ──────────

10. ──────────

Solve each compound or absolute value inequality. Give the solution set in both interval and graph forms.

6. $3k \geq 6$ and $k - 4 < 5$

7. $|4x + 3| \leq 7$

8. $|5 - 6x| > 12$

Solve each absolute value equation.

9. $|3k - 2| + 1 = 8$

10. $|3 - 5x| = |2x + 8|$

Let $A = \{-8, -\frac{2}{3}, -\sqrt{6}, 0, \frac{4}{5}, 9, \sqrt{36}\}$. Simplify the elements of A as necessary, and then list the elements that belong to each set listed below.

1. Natural numbers

2. Whole numbers

3. Integers

4. Rational numbers

5. Irrational numbers

6. Real numbers

Add or subtract, as indicated.

7. $-\frac{4}{3} - \left(-\frac{2}{7}\right)$

8. $|-4| - |2| + |-6|$

9. $(-2)^4 + (-2)^3$

10. $|-4| - |4|$

Evaluate each of the following.

11. $(-3)^5$

12. $\left(\frac{6}{7}\right)^3$

13. $\frac{2}{3} + \frac{1}{2} - \frac{1}{6}$

14. -4^6

15. Which one of the following is not a real number: $-\sqrt{36}$ or $\sqrt{-36}$?

16. Which one of the following is undefined: $\dfrac{4 - 4}{4 + 4}$ or $\dfrac{4 + 4}{4 - 4}$?

Evaluate if $a = 2$, $b = -3$, and $c = 4$.

17. $-3a + 2b - c$

18. $-2b^2 - 4c$

19. $-8(a^2 + b^3)$

20. $\dfrac{3a^3 - b}{4 + 3c}$

Use the properties of real numbers to simplify each expression.

21. $-7r + 5 - 13r + 12$

22. $-(3k + 8) - 2(4k - 7) + 3(8k + 12)$

Identify the property of real numbers illustrated in each of the following.

23. $(a + b) + 4 = 4 + (a + b)$

24. $4x + 12x = (4 + 12)x$

25. $-9 + 9 = 0$

26. What is the reciprocal, or multiplicative inverse, of $-\frac{2}{3}$?

Solve each equation.

27. $-4x + 7(2x + 3) = 7x + 36$

28. $-\dfrac{3}{5}x + \dfrac{2}{3}x = 2$

29. $.06x + .03(100 + x) = 4.35$

30. $P = a + b + c$ for b

Solve each inequality. Give the solution set in both interval and graph forms.

31. $3 - 2(x + 7) \le -x + 3$

32. $-4 < 5 - 3x \le 0$

33. $2x + 1 > 5$ or $2 - x > 2$

34. $|-7k + 3| \ge 4$

According to figures provided by the Equal Employment Opportunity Commission, Bureau of Labor Statistics, the following are the median weekly earnings of full-time workers by occupation for men and women.

Occupation	Men	Women
Managerial and professional specialty	$753	$527
Mathematical and computer scientists	$923	$707
Waiters and waitresses	$281	$205
Bus drivers	$411	$321

Give the occupation that satisfies the description.

35. The median earnings for men are less than $900 *and* for women are greater than $500.

36. The median earnings for men are greater than $900 *or* for women are greater than $600.

Solve each problem.

37. How much pure alcohol should be added to 7 liters of 10% alcohol to increase the concentration to 30% alcohol?

38. A coin collection contains 29 coins. It consists of cents, nickels, and quarters. The number of quarters is 4 less than the number of nickels, and the face value of the collection is $2.69. How many of each denomination are there in the collection?

Linear Equations and Inequalities in Two Variables; Functions

4.1 The Rectangular Coordinate System

Graphs are widely used in the media. Newspapers and magazines, television, reports to stockholders, and newsletters all present information in the form of a graph. Figure 1(a) shows a **line graph** representing the federal debt (that is, the amount the government owes because it has borrowed to finance its purchasing) for selected years since 1980. The **bar graph** in Figure 1(b) gives the revenue of home-improvement retailers during the years 1990 through 1996. And the **circle graph,** or **pie chart,** in Figure 1(c) shows a breakdown of the sources of credit card fraud. Graphs are used so widely because they show a lot of information in a form that makes it easy to understand. As the saying goes, "A picture is worth a thousand words." In this section, we show how to graph equations of lines.

www.mathnotes.com

OBJECTIVES

1. Plot ordered pairs.
2. Find ordered pairs that satisfy a given equation.
3. Graph lines.
4. Find *x*- and *y*-intercepts.
5. Recognize equations of vertical or horizontal lines.

FOR EXTRA HELP

Tutorial   Tape 7   SSM, Sec. 4.1

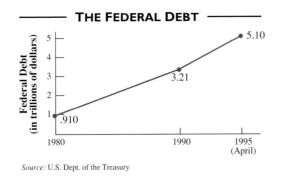

THE FEDERAL DEBT

Federal Debt (in trillions of dollars)

5.10
3.21
.910

1980   1990   1995 (April)

Source: U.S. Dept. of the Treasury

(a)

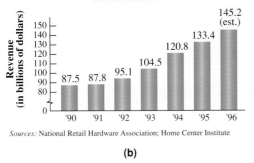

REVENUE OF HOME-IMPROVEMENT RETAILERS

Revenue (in billions of dollars)

145.2 (est.)
87.5 87.8 95.1 104.5 120.8 133.4

'90 '91 '92 '93 '94 '95 '96

Sources: National Retail Hardware Association; Home Center Institute

(b)

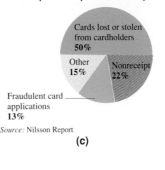

CREDIT CARD FRAUD

Lost or stolen cards account for half of the $15 million in fraud losses reported by banks each year.

Cards lost or stolen from cardholders **50%**

Other **15%**   Nonreceipt **22%**

Fraudulent card applications **13%**

Source: Nilsson Report

(c)

FIGURE 1

1. Plot the following points.

(a) $(-4, 2)$ **(b)** $(3, -2)$

(c) $(-5, -6)$ **(d)** $(4, 6)$

(e) $(-3, 0)$ **(f)** $(0, -5)$

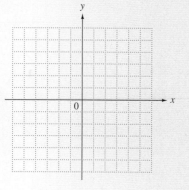

An **ordered pair** (OR-durd PAIR) is a pair of numbers written within parentheses in which the order of the numbers matters. By this definition, the ordered pairs $(2, 5)$ and $(5, 2)$ are different. The two numbers are called **components** of the ordered pair. It is customary for x to represent the first component and y the second component. We graph an ordered pair by using two perpendicular number lines that intersect at the zero points, as shown in Figure 2. The common zero point is called the **origin** (OR-ih-gin). The horizontal line, the **x-axis,** represents the first number in an ordered pair, and the vertical line, the **y-axis,** represents the second. The x-axis and the y-axis make up a **rectangular coordinate system.** It is also called the **Cartesian system,** named after the French mathematician René Descartes (1596–1650).

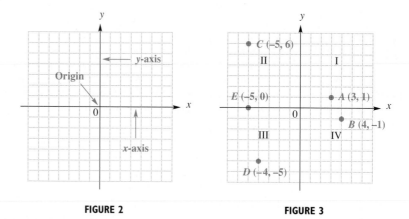

FIGURE 2   **FIGURE 3**

OBJECTIVE 1 To locate, or **plot,** the point on the graph that corresponds to the ordered pair $(3, 1)$, go three units from zero to the right along the x-axis, and then go one unit up parallel to the y-axis. The point corresponding to the ordered pair $(3, 1)$ is labeled A in Figure 3. The point $(4, -1)$ is labeled B, $(-5, 6)$ is labeled C, and $(-4, -5)$ is labeled D. Point E corresponds to $(-5, 0)$. The phrase "the point corresponding to the ordered pair $(2, 1)$" is often abbreviated as "the point $(2, 1)$." The numbers in the ordered pairs are called the **coordinates** (koh-OR-din-ets) of the corresponding point.

The four regions of the graph, shown in Figure 3, are called **quadrants** (KWAD-runts) **I, II, III,** and **IV,** reading counterclockwise from the upper right quadrant. The points on the x-axis and y-axis do not belong to any quadrant. For example, point E in Figure 3 belongs to no quadrant.

◀◀ **WORK PROBLEM I AT THE SIDE.**

OBJECTIVE 2 Each solution to an equation with two variables will include two numbers, one for each variable. To keep track of which number goes with which variable, we write the solutions as ordered pairs. For example, we can show that $(6, -2)$ is a solution of $2x + 3y = 6$ by substitution.

$$2x + 3y = 6$$
$$2(6) + 3(-2) = 6 \quad ?$$
$$12 - 6 = 6 \quad ?$$
$$6 = 6 \quad \text{True}$$

Because the pair of numbers $(6, -2)$ makes the equation true, it is a solution. On the other hand, because

$$2(5) + 3(1) = 10 + 3 = 13 \neq 6,$$

$(5, 1)$ is not a solution of the equation.

ANSWERS

1.

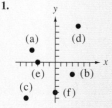

To find ordered pairs that satisfy an equation, we select any number for one of the variables, substitute it into the equation for that variable, and then solve for the other variable. For example, suppose we choose $x = 0$ in the equation $2x + 3y = 6$. Then, by substitution,

$$2x + 3y = 6$$

becomes

$$2(0) + 3y = 6 \qquad \text{Let } x = 0.$$
$$0 + 3y = 6$$
$$3y = 6$$
$$y = 2,$$

giving the ordered pair $(0, 2)$. Some other ordered pairs satisfying $2x + 3y = 6$ are $(6, -2)$, as shown above, and $(3, 0)$. Because every real number could be selected for one variable and would lead to a real number for the other variable, linear equations with two variables have an infinite number of solutions.

E X A M P L E 1   Completing Ordered Pairs

Complete the following ordered pairs for $2x + 3y = 6$.

(a) $(-3, \quad)$

We are given $x = -3$. Substitute into the equation.

$$2(-3) + 3y = 6 \qquad \text{Given } x = -3$$
$$-6 + 3y = 6$$
$$3y = 12$$
$$y = 4$$

The ordered pair is $(-3, 4)$.

(b) $(\quad , -4)$

Replace y with -4.

$$2x + 3y = 6$$
$$2x + 3(-4) = 6 \qquad \text{Given } y = -4$$
$$2x - 12 = 6$$
$$2x = 18$$
$$x = 9$$

The ordered pair is $(9, -4)$.

WORK PROBLEM 2 AT THE SIDE. ▶▶

OBJECTIVE 3▶ The **graph** of an equation is the set of points that correspond to all the ordered pairs that satisfy the equation. It gives a "picture" of the equation. Most equations with two variables have an infinite set of ordered pairs, so their graphs include an infinite number of points. To graph an equation, we plot a number of ordered pairs that satisfy the equation until we have enough points to suggest the shape of the graph. For example, to graph

2. (a) Complete the following ordered pairs for
$3x - 4y = 12$:
$(0, \quad), (\quad , 0), (\quad , -2),$
$(-4, \quad)$

(b) Find one other ordered pair that satisfies the equation.

ANSWERS

2. (a) $(0, -3)$, $(4, 0)$, $\left(\dfrac{4}{3}, -2\right)$,
$(-4, -6)$

(b) Many answers are possible; for
example, $\left(-6, -\dfrac{15}{2}\right)$.

3. Graph $3x - 4y = 12$. Use the points from Problem 2.

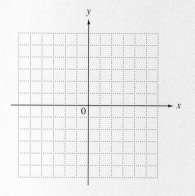

$2x + 3y = 6$, first graph all the ordered pairs found in Example 1 and Objective 2. These points are shown in Figure 4(a). The points appear to lie on a straight line. If all the ordered pairs that satisfy the equation $2x + 3y = 6$ were graphed, they would form a straight line. In fact, the graph of any first-degree equation in two variables is always a straight line. The graph of $2x + 3y = 6$ is the line shown in Figure 4(b).

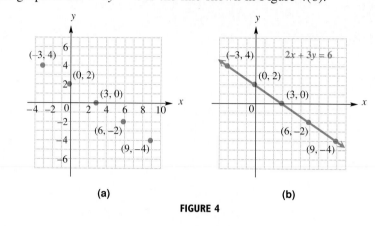

(a)   (b)

FIGURE 4

◀◀ **WORK PROBLEM 3 AT THE SIDE.**

OBJECTIVE 4 Because first-degree equations with two variables have straight-line graphs, they are called **linear equations in two variables.** (We discussed linear equations in one variable in Chapter 2.)

> **Standard Form of a Linear Equation**
>
> An equation that can be written in the **standard form**
> $$Ax + By = C$$
> is a linear equation.

A straight line is determined if any two different points on the line are known; finding two different points is enough to graph the line, but it is wise to find a third point as a check. Two points that are useful for graphing are the x- and y-intercepts. The x-**intercept** (IN-ter-sept) is the point (if any) where the line crosses the x-axis; likewise, the y-**intercept** is the point (if any) where the line crosses the y-axis. In Figure 4(b), the y-value of the point where the line crosses the x-axis is 0. Similarly, the x-value of the point where the line crosses the y-axis is 0. This suggests a method for finding the x- and y-intercepts.

> In the equation of a line, choose $y = 0$ and solve for x to find the x-intercept; choose $x = 0$ and solve for y to find the y-intercept.

E X A M P L E   2   Finding Intercepts

Find the x- and y-intercepts of $4x - y = -3$, and graph the equation.

To find the x-intercept, let $y = 0$.

$$4x - 0 = -3 \qquad \text{Let } y = 0.$$
$$4x = -3$$
$$x = -\frac{3}{4} \qquad x\text{-intercept is } \left(-\frac{3}{4}, 0\right)$$

CONTINUED ON NEXT PAGE

For the y-intercept, let $x = 0$.

$$4(0) - y = -3 \qquad \text{Let } x = 0.$$

$$-y = -3$$

$$y = 3 \qquad y\text{-intercept is } (0, 3)$$

We show the two intercepts in the form of a table, called a *table of values,* next to Figure 5. Plot the intercepts and draw a line through them to get the graph in Figure 5.

x	y
$-\frac{3}{4}$	0
0	3

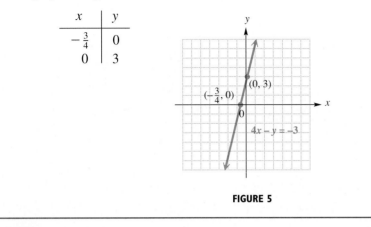

FIGURE 5

Note
While two points, such as the two intercepts in Figure 5, are sufficient to graph a straight line, it is a good idea to use a third point to guard against errors. Verify that $(-1, -1)$ also lies on the graph of $4x - y = -3$.

WORK PROBLEM 4 AT THE SIDE. ▶▶

OBJECTIVE 5▶ The next example shows that a graph may not have an x-intercept.

E X A M P L E   3   Graphing a Horizontal Line

Graph $y = 2$.

Writing $y = 2$ in standard form as $0x + 1y = 2$ shows that any value of x, including $x = 0$, gives $y = 2$, making the y-intercept $(0, 2)$. Every ordered pair that satisfies this equation has a y-coordinate of 2. Because y is always 2, there is no value of x corresponding to $y = 0$, and the graph has no x-intercept. The graph, shown in Figure 6, is a horizontal line.

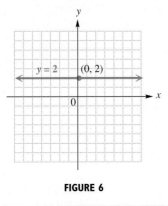

FIGURE 6

4. Find the intercepts, and graph $2x - y = 4$.

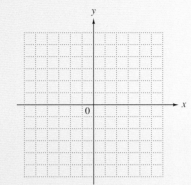

ANSWERS

4. x-intercept is $(2, 0)$; y-intercept is $(0, -4)$.

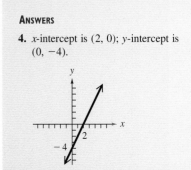

5. Find the intercepts and graph each line.

(a) $y + 4 = 0$

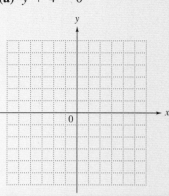

(b) $x = 2$

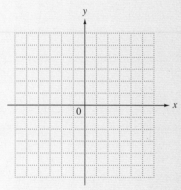

It is also possible for a graph to have no y-intercept, as in the next example.

E X A M P L E 4 Graphing a Vertical Line

Graph $x + 1 = 0$.

The standard form $1x + 0y = -1$ shows that *every* value of y leads to $x = -1$, so no value of y makes x equal to 0. The only way a straight line can have no y-intercept is to be vertical, as in Figure 7. Notice that every point on the line has $x = -1$, while all real numbers are used for the y-values.

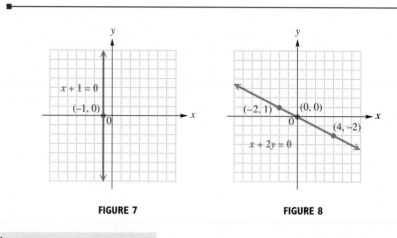

FIGURE 7   FIGURE 8

◀◀ WORK PROBLEM 5 AT THE SIDE.

The line graphed in the next example has both x-intercept and y-intercept at the origin.

E X A M P L E 5 Graphing a Line That Passes through the Origin

Graph $x + 2y = 0$.

Find the x-intercept by letting $y = 0$.

$$x + 2y = 0$$
$$x + 2(\mathbf{0}) = 0 \qquad \text{Let } y = 0.$$
$$x + 0 = 0$$
$$x = 0 \qquad x\text{-intercept is } (0, 0)$$

To find the y-intercept, let $x = 0$.

$$x + 2y = 0$$
$$\mathbf{0} + 2y = 0 \qquad \text{Let } x = 0.$$
$$y = 0 \qquad y\text{-intercept is } (0, 0)$$

Both intercepts are the same ordered pair, $(0, 0)$. Another point is needed to graph the line. Choose any number for x, say $x = 4$, and solve for y. (You could also choose any number for y to solve for x.)

$$x + 2y = 0$$
$$\mathbf{4} + 2y = 0 \qquad \text{Let } x = 4.$$
$$2y = -4$$
$$y = -2$$

This gives the ordered pair $(4, -2)$. These two points lead to the graph shown in Figure 8. As a check, verify that $(-2, 1)$ also lies on the line.

5. (a) no x-intercept; y-intercept is $(0, -4)$

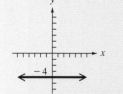

(b) no y-intercept; x-intercept is $(2, 0)$

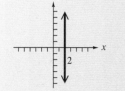

4.1 Exercises

Use the graph to answer the questions.

1. The graph indicates the percentage of all U.S. workers without any private or public health insurance. (*Source:* Census Bureau, Employee Benefit Research Institute)
   (a) Between which two years was the percentage approximately the same?
   (b) Between which two years was the increase the greatest?
   (c) In what year was the percent about 16.5%?

2. The graph indicates the production of handguns, rifles, and shotguns in the U.S. (*Source:* Bureau of Alcohol, Tobacco, and Firearms)
   (a) In what year between 1983 and 1992 was the production the greatest?
   (b) In what year was the production the least?
   (c) In what two successive years was the production less than the previous year's?

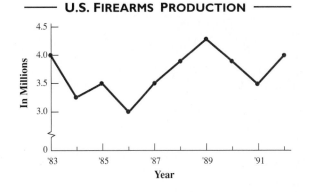

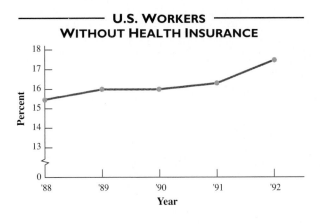

3. What is another name for the rectangular coordinate system? After whom is it named?

4. Observe the graphs in Exercises 1 and 2. If you were to use one of the other types of graphs mentioned in the opening paragraph of this section to depict the information given there, which one would you choose?

Fill in the blank with the correct response.

5. The point with coordinates (0, 0) is called the _____ of a rectangular coordinate system.

6. For any value of x, the point $(x, 0)$ lies on the _____-axis.

7. To find the x-intercept of a line, we let _____ equal 0 and solve for _____ .

8. The equation _____ = 4 has a horizontal
   $(x \text{ or } y)$
   line as its graph.

9. To graph a straight line, we must find a minimum of _____ points.

10. The point (_____, 4) is on the graph of $2x - 3y = 0$.

Name the quadrant, if any, in which each point is located.

11. _____ (a) (1, 6)
    _____ (b) (−4, −2)
    _____ (c) (−3, 6)
    _____ (d) (7, −5)
    _____ (e) (−3, 0)

12. _____ (a) (−2, −10)
    _____ (b) (4, 8)
    _____ (c) (−9, 12)
    _____ (d) (3, −9)
    _____ (e) (0, −8)

13. Use the given information to determine the possible quadrants in which the point (x, y) must lie.

(a) $xy > 0$ (b) $xy < 0$ (c) $\dfrac{x}{y} < 0$ (d) $\dfrac{x}{y} > 0$

14. What must be true about the coordinates of any point that lies on an axis?

Locate the following points on the rectangular coordinate system.

15. $(2, 3)$ **16.** $(-1, 2)$ **17.** $(-3, -2)$ **18.** $(1, -4)$

19. $(0, 5)$ **20.** $(-2, -4)$ **21.** $(-2, 4)$ **22.** $(3, 0)$

23. $(-2, 0)$ **24.** $(3, -3)$

In each exercise, complete the given ordered pairs for the equation, and then graph the equation. See Example 1.

25. $x - y = 3$
$(0, \_\_), (\_\_, 0)$
$(5, \_\_), (2, \_\_)$

26. $x - y = 5$
$(0, \_\_), (\_\_, 0)$
$(1, \_\_), (3, \_\_)$

27. $x + 2y = 5$
$(0, \_\_), (\_\_, 0)$
$(2, \_\_), (\_\_, 2)$

28. $x + 3y = -5$
$(0, \_\_), (\_\_, 0)$
$(1, \_\_), (\_\_, -1)$

29. $4x - 5y = 20$
$(0, \_\_), (\_\_, 0)$
$(2, \_\_), (\_\_, -3)$

30. $6x - 5y = 30$
$(0, \_\_), (\_\_, 0)$
$(3, \_\_), (\_\_, -2)$

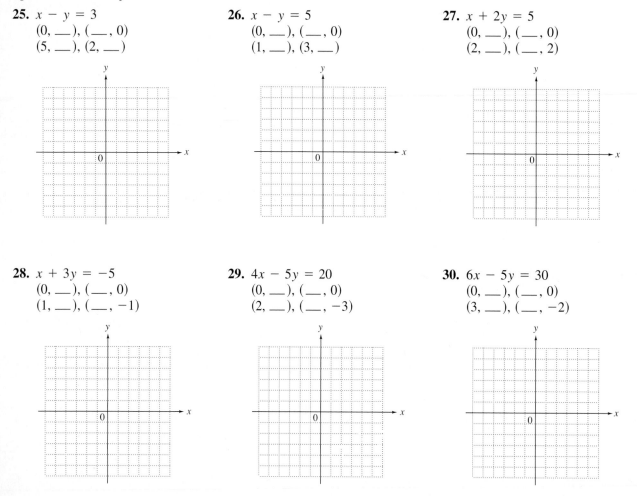

31. What is the equation of a line that coincides with the y-axis?

32. What is the equation of a line that coincides with the x-axis?

For each equation, find the x-intercept and the y-intercept, and graph. See Examples 2–5.

33. $2x + 3y = 12$ **34.** $5x + 2y = 10$ **35.** $x - 3y = 6$

36. $x - 2y = -4$ **37.** $3x - 7y = 9$ **38.** $5x + 6y = -10$

39. $y = 5$ **40.** $y = -3$ **41.** $x = 2$

42. $x = -3$ **43.** $x + 5y = 0$ **44.** $x - 3y = 0$

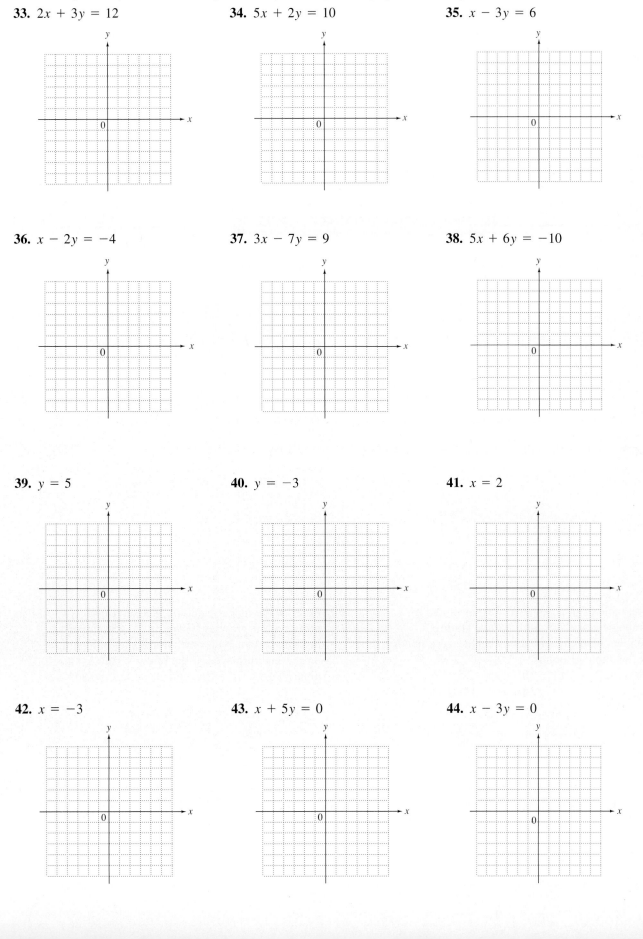

A linear equation can be used as a model to describe data in some cases. Exercises 45 and 46 are based on this idea.

45. Between the years 1980 and 1992, the linear model $y = 2.503x + 198.729$ approximated the winning speed for the Indianapolis 500 race. In the model, $x = 0$ corresponds to 1980, $x = 1$ to 1981, and so on, and y is in miles per hour. Use this model to approximate the speed of the 1988 winner, Rick Mears.

46. According to information provided by Families USA Foundation, the national average family health care cost in dollars between 1980 and 2000 (projected) can be approximated by the linear model $y = 382.75x + 1742$, where $x = 0$ corresponds to 1980 and $x = 20$ corresponds to 2000. Based on this model, what would be the expected national average health care cost in 1998?

INTERPRETING TECHNOLOGY (EXERCISES 47–50)

47. The screen shows the graph of one of the equations below, along with the coordinates of a point on the graph. Which one of the equations is it?
   (a) $x + 2y = 4$ (b) $-3x + 5y = 15$
   (c) $y = 4x - 2$ (d) $y = -2$

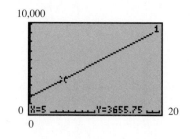

48. The screen shows the graph of one of the equations below. Two views of the graph are given, along with the intercepts. Which one of the equations is it?
   (a) $3x + 2y = 6$ (b) $-3x + 2y = 6$ (c) $-3x - 2y = 6$ (d) $3x - 2y = 6$

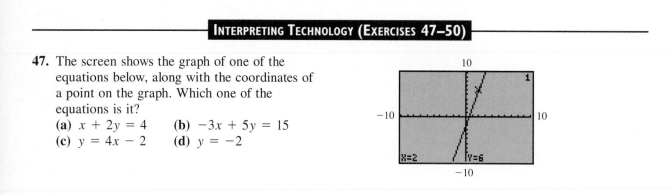

49. The table of points shown was generated by a graphing calculator with a *table* feature. Which one of the equations below corresponds to this table of points?
   (a) $y = 2x - 3$ (b) $y = -2x - 3$
   (c) $y = 2x + 3$ (d) $y = -2x + 3$

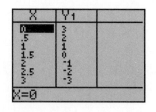

50. Refer to the model equation in Exercise 46. A portion of its graph is shown on the accompanying screen, along with the coordinates of a point on the line displayed at the bottom. How is this point interpreted in the context of the model?

4.2 *The Slope of a Line*

Slope is used in many ways in our everyday world. The slope of a hill (sometimes called the *grade*) is often given as a percent. For example, a 10% (or $\frac{10}{100} = \frac{1}{10}$) slope means the hill rises 1 unit for every 10 horizontal units. Stairs and roofs have slopes, too, as shown in Figure 9.

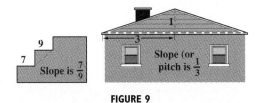

FIGURE 9

OBJECTIVES

1 ▶ Find the slope of a line, given two points on the line.

2 ▶ Find the slope of a line, given an equation of the line.

3 ▶ Graph a line, using its slope and a point on the line.

4 ▶ Use slopes to determine whether two lines are parallel, perpendicular, or neither.

5 ▶ Solve problems involving average rate of change.

FOR EXTRA HELP

Tutorial   Tape 7   SSM, Sec. 4.2

OBJECTIVE 1 ▶ To formulate a definition of the slope of a line, suppose (x_1, y_1) and (x_2, y_2) are two different points on a line as shown in Figure 10. (The notation x_1 (read "*x*-sub-one"), x_2, y_1, and y_2 represents specific *x*-values or *y*-values.) Then, as we move along the line from (x_1, y_1) to (x_2, y_2), the *y*-value changes from y_1 to y_2, an amount equal to $y_2 - y_1$. This vertical change is called the *rise*. As *y* changes from y_1 to y_2, the value of *x* changes from x_1 to x_2 by the amount $x_2 - x_1$. This horizontal change is called the *run*.

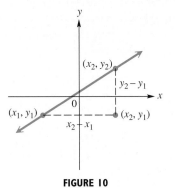

FIGURE 10

The ratio of the change in *y* to the change in *x* is called the **slope** of the line, with the letter *m* used for slope.

Slope

If $x_1 \neq x_2$, the slope of the line through the distinct points (x_1, y_1) and (x_2, y_2) is

$$m = \frac{\text{rise}}{\text{run}} = \frac{\text{change in } y}{\text{change in } x} = \frac{y_2 - y_1}{x_2 - x_1}.$$

The larger the absolute value of *m*, the steeper the corresponding line will be, because a larger value of *m* indicates a greater change in *y* as compared to the change in *x*.

EXAMPLE 1   Using the Definition of Slope

Find the slope of the line through the points $(2, -1)$ and $(-5, 3)$.
Let $(2, -1) = (x_1, y_1)$ and $(-5, 3) = (x_2, y_2)$; then

$$m = \frac{y_2 - y_1}{x_2 - x_1} = \frac{3 - (-1)}{-5 - 2} = \frac{4}{-7} = -\frac{4}{7}.$$

On the other hand, if the pairs are reversed, so that $(2, -1) = (x_2, y_2)$ and $(-5, 3) = (x_1, y_1)$, the slope is

$$m = \frac{-1 - 3}{2 - (-5)} = \frac{-4}{7} = -\frac{4}{7},$$

the same answer.

1. Find the slope of the line through each pair of points.

(a) $(-2, 7), (4, -3)$

Example 1 suggests that the slope is the same no matter which point we consider first. Also, using similar triangles from geometry, we can show that the slope is the same no matter which two different points on the line we choose.

> **Caution**
> In calculating the slope, be careful to subtract the y-values and the x-values in the *same* order.
>
> | Correct | Incorrect |
> | :-------------------------------: | :---------------------------------: |
> | $\dfrac{y_2 - y_1}{x_2 - x_1}$ or $\dfrac{y_1 - y_2}{x_1 - x_2}$ | $\dfrac{y_2 - y_1}{x_1 - x_2}$ or $\dfrac{y_1 - y_2}{x_2 - x_1}$ |

◀◀ **WORK PROBLEM 1 AT THE SIDE.**

OBJECTIVE 2 ▶ When an equation of a line is given, we can find the slope using the definition of slope by first finding two different points on the line.

E X A M P L E 2   Finding the Slope of a Line

Find the slope of the line $4x - y = 8$.

(b) $(1, 2), (8, 5)$

The intercepts can be used as the two different points needed to find the slope. Replace y with 0 to find that the x-intercept is $(2, 0)$; replace x with 0 to find that the y-intercept is $(0, -8)$. The slope is

$$m = \frac{-8 - 0}{0 - 2} = \frac{-8}{-2} = 4.$$

E X A M P L E 3   Finding the Slope of a Line

Find the slope of the following lines.

(a) $x = -3$

By inspection, $(-3, 5)$ and $(-3, -4)$ are two points that satisfy the equation $x = -3$. Use these two points to find the slope.

$$m = \frac{-4 - 5}{-3 - (-3)} = \frac{-9}{0}$$

(c) $(8, -2), (3, -2)$

Division by zero is undefined; therefore the slope is undefined. As shown in the previous section, the graph of an equation such as $x = -3$ is a vertical line (parallel to the y-axis). See Figure 11(a).

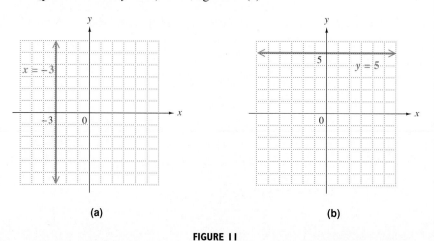

(a)   (b)

FIGURE 11

CONTINUED ON NEXT PAGE ─────

ANSWERS

1. (a) $-\dfrac{5}{3}$ **(b)** $\dfrac{3}{7}$ **(c)** 0

(b) $y = 5$

Find the slope by selecting two different points on the line, such as $(3, 5)$ and $(-1, 5)$, and using the definition of slope.

$$m = \frac{5 - 5}{3 - (-1)} = \frac{0}{4} = 0$$

The graph of $y = 5$ is the horizontal line in Figure 11(b).

Example 3 suggests the following generalization.

> The slope of a vertical line is undefined; the slope of a horizontal line is 0.

WORK PROBLEM 2 AT THE SIDE. ▶▶

OBJECTIVE 3 The following example shows how to graph a straight line by using the slope and one point on the line.

EXAMPLE 4 Using the Slope and a Point to Graph a Line

Graph the line that has slope $\frac{2}{3}$ and goes through the point $(-4, 2)$.

First locate the point $(-4, 2)$ on a graph (see Figure 12). Then, from the definition of slope,

$$m = \frac{\text{change in } y}{\text{change in } x} = \frac{2}{3},$$

move 2 units *up* in the y direction and then 3 units to the *right* in the x direction to locate another point on the graph, P. The line through $(-4, 2)$ and P is the required graph. An additional point Q, can be found by moving 2 units *down* and 3 units to the *left* because,

$$\frac{2}{3} = \frac{-2}{-3}.$$

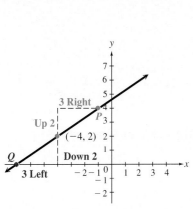

FIGURE 12

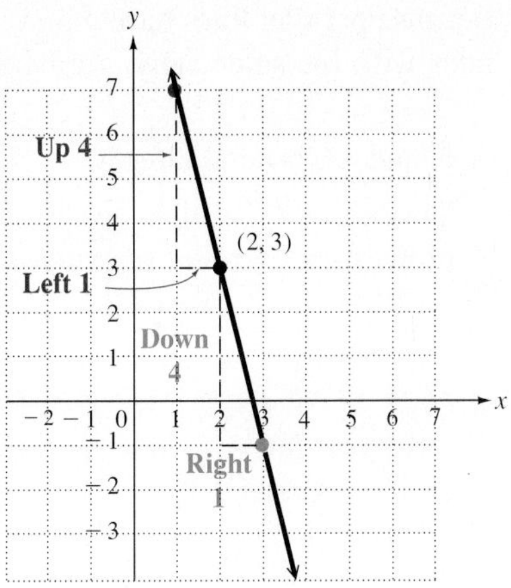

FIGURE 13

2. Find the slope of each line.

(a) $2x + y = 6$

(b) $3x - 4y = 12$

(c) $x = -6$

(d) $y + 5 = 0$

ANSWERS

2. (a) -2 **(b)** $\dfrac{3}{4}$ **(c)** undefined **(d)** 0

3. Graph the following lines.

(a) Through $(1, -3)$; $m = -\dfrac{3}{4}$

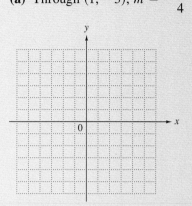

(b) Through $(-1, -4)$; $m = 2$

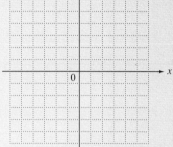

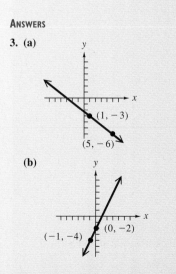
EXAMPLE 5   Using the Slope and a Point to Graph a Line

Graph the line through $(2, 3)$ with slope -4.

Start by locating the point $(2, 3)$ on the graph. Use the definition of slope to find a second point on the line, writing -4 as $\frac{-4}{1}$.

$$\text{Slope} = \frac{\text{change in } y}{\text{change in } x} = \frac{-4}{1}$$

Move *down* 4 units from $(2, 3)$ and then 1 unit to the *right*. Draw a line through this second point and $(2, 3)$, as in Figure 13. We could also write the slope as $\frac{4}{-1}$ and move 4 units *up* and 1 unit *left* to get another point.

◀◀ **WORK PROBLEM 3 AT THE SIDE.**

In Problem 3 at the side, the slope for part (b) is *positive*. As shown in the answer, the graph for this line goes up from left to right. The line in part (a) has a *negative* slope, and the graph goes down from left to right. This suggests the following conclusion.

> A positive slope indicates that the line goes up from left to right; a negative slope indicates that the line goes down from left to right.

Figure 14 shows lines of positive, zero, negative, and undefined slopes.

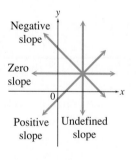

FIGURE 14

OBJECTIVE 4   The slopes of a pair of parallel or perpendicular lines are related in a special way. The slope of a line measures the steepness of a line. Because parallel lines have equal steepness, their slopes must be equal; also, lines with the same slope are parallel.

Slopes of Parallel Lines

Two nonvertical lines with the same slope are parallel; two nonvertical parallel lines have the same slope.

EXAMPLE 6   Determining Parallel Lines

Are the two lines L_1, through $(-2, 1)$ and $(4, 5)$, and L_2, through $(3, 0)$ and $(0, -2)$, parallel?

The slope of L_1 is

$$m_1 = \frac{5 - 1}{4 - (-2)} = \frac{4}{6} = \frac{2}{3}.$$

CONTINUED ON NEXT PAGE

The slope of L_2 is

$$m_2 = \frac{-2 - 0}{0 - 3} = \frac{-2}{-3} = \frac{2}{3}.$$

Because the slopes are equal, the two lines are parallel.

To see how the slopes of perpendicular lines are related, consider a nonvertical line with slope $\frac{a}{b}$. If this line is rotated 90°, the rise and run are exchanged, and the slope is $-\frac{b}{a}$ because the run is now negative. See Figure 15. Thus, the slopes of perpendicular lines have a product of -1 and are negative reciprocals of each other. For example, if the slopes of two lines are $\frac{3}{4}$ and $-\frac{4}{3}$, then the lines are perpendicular because $(\frac{3}{4})(-\frac{4}{3}) = -1$.

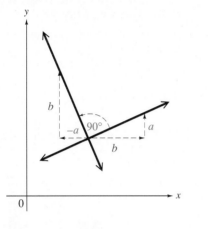

FIGURE 15

Slopes of Perpendicular Lines

If neither line is vertical, the slopes of perpendicular lines are negative reciprocals, so that their product is -1. Also, lines with slopes that are negative reciprocals are perpendicular.

EXAMPLE 7   Determining Perpendicular Lines

Are the lines with equations $2y = 3x - 6$ and $2x + 3y = -6$ perpendicular?

Find the slope of each line by first finding two points on the line. The points $(0, -3)$ and $(2, 0)$ are on the first line. The slope is

$$m_1 = \frac{0 - (-3)}{2 - 0} = \frac{3}{2}.$$

The second line goes through $(-3, 0)$ and $(0, -2)$ and has slope

$$m_2 = \frac{-2 - 0}{0 - (-3)} = -\frac{2}{3}.$$

Because these slopes are negative reciprocals, the product of the slopes is $(\frac{3}{2})(-\frac{2}{3}) = -1$, so the lines are perpendicular.

WORK PROBLEM 4 AT THE SIDE. ▶▶

---

4. Write *parallel, perpendicular,* or *neither* for each pair of two distinct lines.

a) The line through $(-1, 2)$ and $(3, 5)$ and the line through $(4, 7)$ and $(8, 10)$

(b) The line through $(5, -9)$ and $(3, 7)$ and the line through $(0, 2)$ and $(8, 3)$

(c) $2x - y = 4$ and $2x + y = 6$

(d) $3x + 5y = 6$ and $5x - 3y = 2$

Note

When deciding whether or not a pair of lines is parallel, in addition to checking for the same slope, be sure they are not the *same line*. The slopes of $4x + 2y = 6$ and $2x + y = 3$ are both -2; however, these equations describe the same line with y-intercept $(0, 3)$.

OBJECTIVE 5 We have seen how the slope of a line is the ratio of the change in y (vertical change) to the change in x (horizontal change). This idea can be extended to real-life situations as follows: the slope gives the average rate of change in y per unit of change in x, where the value of y depends on the value of x. The next example illustrates this idea of average rate of change. We assume a linear relationship between x and y.

E X A M P L E   8   Interpreting Slope as Average Rate of Change

The bar graph in Figure 16* shows the number of multimedia personal computers (PCs), in millions, in U.S. homes. Find the average rate of change in the number of multimedia PCs per year.

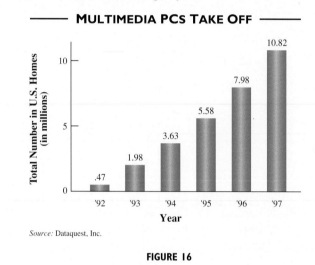

MULTIMEDIA PCS TAKE OFF

Source: Dataquest, Inc.

FIGURE 16

Since connecting the tops of the bars would closely approach a straight line, we can use the slope formula. We need two pairs of data. If we let 1992 represent 0, then 1993 represents 1, 1994 represents 2, and so on. Then the ordered pair for 1993 is (1, 1.98) and the pair for 1996 is (4, 7.98). The average rate of change in the number of multimedia PCs is found by using the slope formula.

$$\text{Average rate of change} = \frac{y_2 - y_1}{x_2 - x_1} = \frac{7.98 - 1.98}{4 - 1} = 2$$

The result, 2, indicates that the number of multimedia PCs increases by 2 million each year.

---

* Graph for Figure 16, "Multimedia PC's Take Off," from *The Wall Street Journal*, March 21, 1994. (The figures for 1995 to 1997 are estimates.) Reprinted by permission of The Wall Street Journal, Copyright © 1994 Dow Jones & Company, Inc. All Rights Reserved Worldwide.

4.2 Exercises

Use the given figure to determine the slope of the line segment described, by counting the number of units of "rise," the number of units of "run," and then finding the quotient.

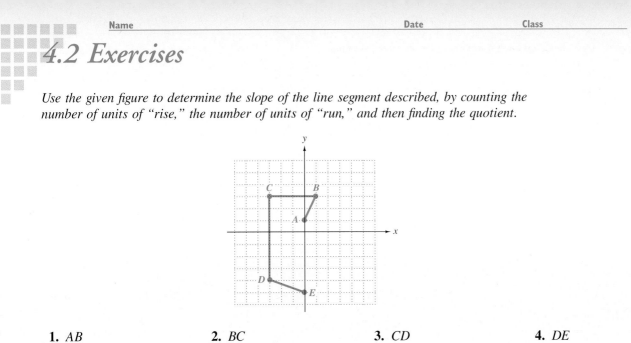

1. AB

2. BC

3. CD

4. DE

5. If a walkway rises 2 feet for every 10 feet of horizontal distance, which of the following expresses its slope (or grade)? (There may be more than one correct answer.)

    (a) .2   **(b)** $\dfrac{2}{10}$ **(c)** $\dfrac{1}{5}$ **(d)** 20%   **(e)** 5   **(f)** $\dfrac{20}{100}$ **(g)** 500%   **(h)** $\dfrac{10}{2}$

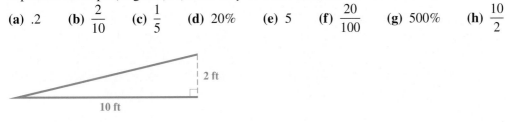

6. If the pitch of a roof is $\frac{1}{4}$, how many feet in the horizontal direction corresponds to a rise of 3 feet?

Find the slope of the line through each pair of points using the slope formula. See Example 1.

7. $(-2, -3)$ and $(-1, 5)$

8. $(-4, 3)$ and $(-3, -4)$

9. $(-4, 1)$ and $(2, 6)$

10. $(-3, -3)$ and $(5, 6)$

11. $(2, 4)$ and $(-4, 4)$

12. $(-6, 3)$ and $(2, 3)$

Graph the line with the given equation, and then find its slope based on the graph you have sketched. See Example 3.

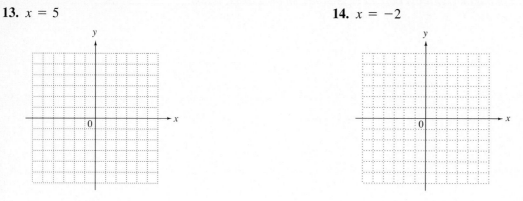

13. $x = 5$

14. $x = -2$

15. $y = 2$

16. $y = 6$

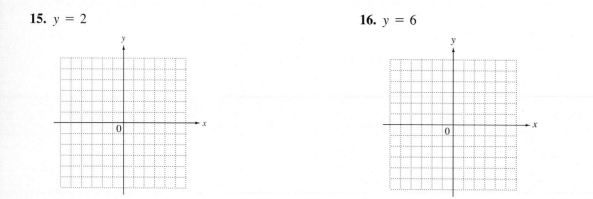

Based on the figure shown here, determine which line satisfies the given description.

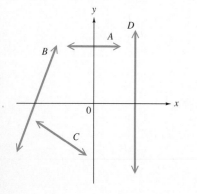

17. The line has positive slope.

18. The line has negative slope.

19. The line has slope 0.

20. The line has undefined slope.

Find the slope of each of the following lines, and sketch the graph. See Examples 1, 2, 4, and 5.

21. $x + 2y = 4$

22. $x + 3y = -6$

23. $-x + y = 4$

24. $-x + y = 6$

25. $6x + 5y = 30$

26. $3x + 4y = 12$

27. $5x - 2y = 10$

28. $4x - y = 4$

29. $y = 4x$

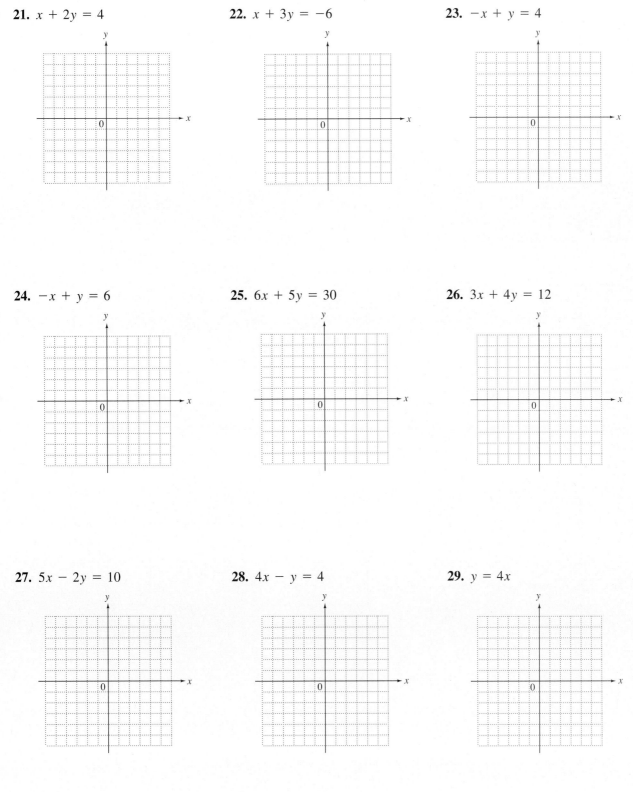

30. $y = -3x$

31. $y - 3 = 0$

32. $y + 5 = 0$

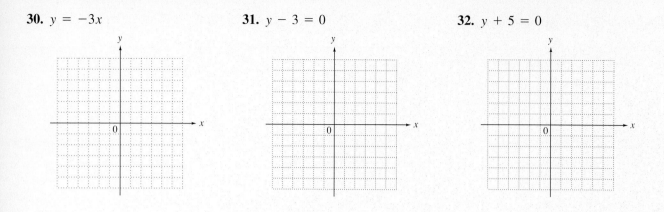

Use the methods shown in Examples 4 and 5 to graph each of the following lines.

33. Through $(-4, 2)$; $m = \dfrac{1}{2}$

34. Through $(-2, -3)$; $m = \dfrac{5}{4}$

35. Through $(0, -2)$; $m = -\dfrac{2}{3}$

36. Through $(0, -4)$; $m = -\dfrac{3}{2}$

37. Through $(-1, -2)$; $m = 3$

38. Through $(-2, -4)$; $m = 4$

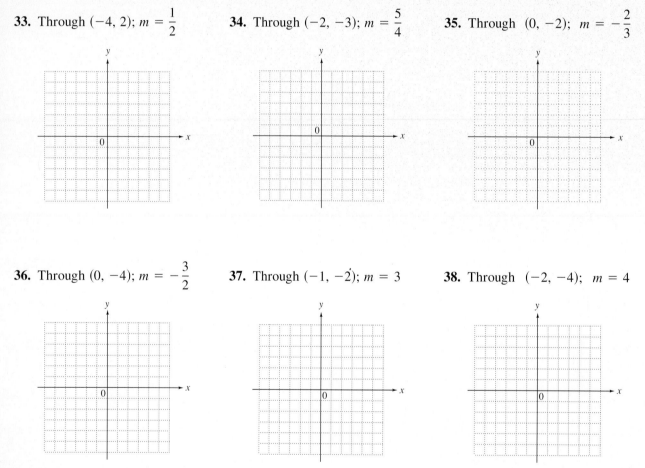

Decide whether the pair of lines is parallel, perpendicular, *or neither* parallel nor perpendicular. *See Examples 6 and 7.*

39. $2x + 5y = -7$ and $5x - 2y = 1$

40. $x + 4y = 7$ and $4x - y = 3$

41. The line through $(4, 6)$ and $(-8, 7)$ and the line through $(-5, 5)$ and $(7, 4)$

42. The line through $(15, 9)$ and $(12, -7)$ and the line through $(8, -4)$ and $(5, -20)$

43. $2x + y = 6$ and $x - y = 4$

44. $4x - 3y = 6$ and $3x - 4y = 2$

Use the concept of slope to solve the problem.

45. The upper deck at the new Comiskey Park in Chicago has produced, among other complaints, displeasure with its steepness. It's been compared to a ski jump. It is 160 feet from home plate to the front of the upper deck and 250 feet from home plate to the back. The top of the upper deck is 63 feet above the bottom. What is its slope?

46. When designing the new arena in Boston to replace the old Boston Garden, architects were careful to design the ramps leading up to the entrances so that circus elephants would be able to march up the ramps. The maximum grade (or slope) that an elephant will walk on is 13%. Suppose that such a ramp was constructed with a horizontal run of 150 feet. What would be the maximum vertical rise the architects could use?

Use the idea of average rate of change to solve the problem. See Example 8.

47. The graph shows how average monthly rates for cable television increased from 1980 to 1992. The graph can be approximated by a straight line.
   (a) Use the information provided for 1980 and 1992 to determine the average rate of change in price per year.
   (b) In your own words, explain how a *positive* rate of change affects the consumer in a situation such as the one illustrated by this graph.

48. Assuming a linear relationship, what is the average rate of change for cable industry revenues over the period from 1990 to 1992?

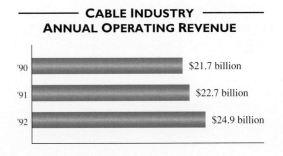

CABLE INDUSTRY
ANNUAL OPERATING REVENUE

'90 — $21.7 billion
'91 — $22.7 billion
'92 — $24.9 billion

Sources: Census Bureau; Paul Kagan Associates

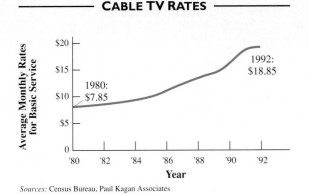

CABLE TV RATES

1980: $7.85
1992: $18.85

Sources: Census Bureau, Paul Kagan Associates

49. The 1993 Annual Report of AT&T cited the following figures concerning the global growth of international telephone calls: From 47.5 billion minutes in 1993, traffic was expected to rise to 60 billion minutes in 1995. Assuming a linear relationship, what was the average rate of change for this time period? (*Source:* TeleGeography, 1993, Washington, D.C.)

50. The market for international phone calls during the ten-year period from 1986 to 1995 is depicted in the accompanying bar graph. The tops of the bars approximate a straight line. Assuming that the traffic volume at the beginning of this period was 18 billion minutes and at the end was 60 billion minutes, what was the average rate of change for the ten-year period?

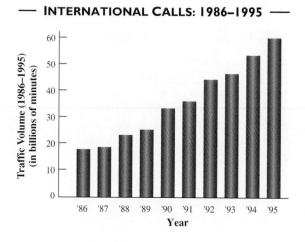

— INTERNATIONAL CALLS: 1986–1995 —

MATHEMATICAL CONNECTIONS (EXERCISES 51–56)

*In these exercises we examine a method of determining whether three points lie on the same straight line. (Such points are said to be **collinear**.) The points we consider are A(3, 1), B(6, 2), and C(9, 3). Work these exercises in order.*

51. Find the slope of segment AB.

52. Find the slope of segment BC.

53. Find the slope of segment AC.

54. If slope of AB = slope of BC = slope of AC, then A, B, and C are collinear. Use the results of Exercises 51–53 to show that this statement is satisfied.

55. Use the slope formula to determine whether the points $(1, -2)$, $(3, -1)$, and $(5, 0)$ are collinear.

56. Repeat Exercise 55 for the points $(0, 6)$, $(4, -5)$, and $(-2, 12)$.

4.3 Linear Equations in Two Variables

Many real-world situations can be described by straight-line graphs. In this section we see how to write a linear equation in such situations.

OBJECTIVE 1 We saw earlier that a straight line is a set of points in the plane such that the slope between any two points is the same. In Figure 17, point P is on the line through P_1 and P_2 if the slope of the line through points P_1 and P equals the slope of the line through points P and P_2. If these slopes are equal to m, then

$$\frac{y - y_1}{x - x_1} = \frac{y - y_2}{x - x_2} = m.$$

$$\frac{y - y_1}{x - x_1} = m$$

$$y - y_1 = m(x - x_1) \qquad \text{Multiply both sides by } x - x_1.$$

This last equation gives the *point-slope form* of the equation of the line, which shows the coordinates of a point (x_1, y_1) on the line and the slope of the line.

OBJECTIVES

1 ▸ Write the equation of a line, given its slope and a point on the line.

2 ▸ Write the equation of a line, given two points on the line.

3 ▸ Write the equation of a line, given its slope and *y*-intercept.

4 ▸ Find the slope and *y*-intercept of a line, given its equation.

5 ▸ Write the equation of a line parallel or perpendicular to a given line through a given point.

6 ▸ Apply concepts of linear equations to realistic examples.

FOR EXTRA HELP

Tutorial   Tape 7   SSM, Sec. 4.3

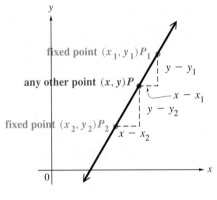

FIGURE 17

Point-slope Form

The **point-slope form** of the equation of a line is

$$\overset{\text{Slope}}{\underset{\text{Given point}}{y - y_1 = m(x - x_1).}}$$

Given a point on a line and the slope of the line, we can use the point-slope form to write the equation of the line.

E X A M P L E   1   Using the Point-slope Form to Find the Equation of a Line

Find an equation of the line with slope $\frac{1}{3}$ that goes through the point $(-2, 5)$.

Use the point-slope form of the equation of a line, with $(x_1, y_1) = (-2, 5)$ and $m = \frac{1}{3}$.

── CONTINUED ON NEXT PAGE

1. Write equations of the following lines in standard form.

   (a) Through $(-2, 7)$; $m = 3$

   $$y - y_1 = m(x - x_1) \qquad \text{Point-slope form}$$

   $$y - 5 = \frac{1}{3}[x - (-2)] \qquad y_1 = 5, \; m = \tfrac{1}{3}, \; x_1 = -2$$

   $$y - 5 = \frac{1}{3}(x + 2)$$

   $$3(y - 5) = 3 \cdot \frac{1}{3}(x + 2) \qquad \text{Multiply by 3.}$$

   $$3y - 15 = x + 2$$

   $$-x + 3y = 17 \qquad \text{Subtract } x; \text{ add 15.}$$

   $$x - 3y = -17 \qquad \text{Standard form}$$

   (b) Through $(1, 3)$; $m = -\dfrac{5}{4}$

◀◀ **WORK PROBLEM 1 AT THE SIDE.**

Notice that the point-slope form does not apply to a vertical line because the slope of a vertical line is undefined. A vertical line through the point (k, y), where k is a constant, has equation $x = k$.

A horizontal line has slope 0. From the point-slope form, the equation of a horizontal line through the point (x, k), where k is a constant, is

$$y - y_1 = m(x - x_1)$$

$$y - k = 0(x - x) \qquad \text{Let } m = 0, \; x_1 = x, \text{ and } y_1 = k.$$

$$y - k = 0$$

$$y = k.$$

2. Write equations of the following lines.

   (a) Through $(8, -2)$; $m = 0$

In summary, horizontal and vertical lines have special equations as follows.

> **Horizontal and Vertical Lines**
>
> If k is a constant, the vertical line through (k, y) has equation $x = k$, and the horizontal line through (x, k) has equation $y = k$.

◀◀ **WORK PROBLEM 2 AT THE SIDE.**

OBJECTIVE 2 ▶ When two points on a line are known, we can use the definition of slope to find the slope of the line. Then the point-slope form can be used to write an equation for the line.

   (b) The vertical line through $(3, 5)$

E X A M P L E   2   Finding the Equation of a Line Given Two Points

Find an equation of the line through the points $(-4, 3)$ and $(5, -7)$.

First find the slope by using the definition.

$$m = \frac{-7 - 3}{5 - (-4)} = -\frac{10}{9}$$

Either $(-4, 3)$ or $(5, -7)$ can be used as (x_1, y_1) in the point-slope form of the equation of a line. If we choose $(-4, 3)$, then $-4 = x_1$ and $3 = y_1$.

CONTINUED ON NEXT PAGE

ANSWERS

1. (a) $3x - y = -13$ (b) $5x + 4y = 17$
2. (a) $y = -2$ (b) $x = 3$

$$y - y_1 = m(x - x_1) \qquad \text{Point-slope form}$$

$$y - 3 = -\frac{10}{9}[x - (-4)] \qquad y_1 = 3,\ m = -\tfrac{10}{9},\ x_1 = -4$$

$$9(y - 3) = 9\left(-\frac{10}{9}\right)(x + 4) \qquad \text{Multiply by 9.}$$

$$9y - 27 = -10x - 40 \qquad \text{Distributive property}$$

$$10x + 9y = -13 \qquad \text{Standard form}$$

On the other hand, if $(5, -7)$ were used, the equation would be

$$y - (-7) = -\frac{10}{9}(x - 5) \qquad y_1 = -7,\ m = -\tfrac{10}{9},\ x_1 = 5$$

$$9y + 63 = -10x + 50 \qquad \text{Multiply by 9.}$$

$$10x + 9y = -13, \qquad \text{Standard form}$$

the same equation. Either way, the line through $(-4, 3)$ and $(5, -7)$ has equation $10x + 9y = -13$.

WORK PROBLEM 3 AT THE SIDE. ▶▶

OBJECTIVE 3 Suppose a line has slope m, and we know that the y-intercept is $(0, b)$. Using the slope-intercept form gives

$$y - y_1 = m(x - x_1)$$
$$y - b = m(x - 0)$$
$$y = mx + b.$$

When we solve the equation for y, the coefficient of x is the slope, m, and the constant is the y-value of the y-intercept, b. Because this form of the equation shows the slope and the y-intercept, it is called the *slope-intercept form*.

Slope-intercept Form

The equation of a line with slope m and y-intercept $(0, b)$ is written in **slope-intercept form** as

$$y = mx + b.$$

Slope y-intercept is $(0, b)$.

┌─ **E X A M P L E 3** **Using the Slope-intercept Form to Find the Equation of a Line**

Find an equation of the line with slope $-\frac{4}{5}$ and y-intercept $(0, -2)$.

Here $m = -\frac{4}{5}$ and $b = -2$. Substitute these values into the slope-intercept form.

$$y = mx + b \qquad \text{Slope-intercept form}$$

$$y = -\frac{4}{5}x - 2 \qquad m = -\tfrac{4}{5},\ b = -2$$

$$5y = -4x - 10 \qquad \text{Multiply by 5.}$$

$$4x + 5y = -10 \qquad \text{Standard form}$$

3. Write equations in standard form of the following lines.

(a) Through $(-1, 2)$ and $(5, 7)$

(b) Through $(-2, 6)$ and $(1, 4)$

ANSWERS

3. (a) $5x - 6y = -17$ **(b)** $2x + 3y = 14$

4. Write an equation in standard form for each line with the given slope and *y*-intercept.

(a) Slope 2; *y*-intercept $(0, -3)$

(b) Slope $-\frac{2}{3}$; *y*-intercept $(0, 0)$

(c) Slope 0; *y*-intercept $(0, 3)$

Note
The importance of the slope-intercept form of a linear equation cannot be overemphasized. First, every linear equation (of a nonvertical line) has a *unique* (one and only one) slope-intercept form. Second, at the end of this chapter, we will study linear *functions*. The slope-intercept form is necessary in specifying such functions.

◀◀ **WORK PROBLEM 4 AT THE SIDE.**

It is convenient to agree on the form for writing a linear equation. In Section 4.1, we defined *standard form* for a linear equation as

$$Ax + By = C.$$

In addition, from now on, let us agree that *A*, *B*, and *C* will be integers, with $A \geq 0$. For example, the final equation found in Example 3, $4x + 5y = -10$, is written in standard form.

Caution
The definition of "standard form" varies from one text to another. Any linear equation can be written in many different (all equally correct) forms. For example, the equation $2x + 3y = 8$ can be written as $2x = 8 - 3y$, $3y = 8 - 2x$, $x + \frac{3}{2}y = 4$, $4x + 6y = 16$, and so on. In addition to writing it in the form $Ax + By = C$ (with $A \geq 0$), let us agree that the form $2x + 3y = 8$ is preferred over any multiples of both sides, such as $4x + 6y = 16$.

OBJECTIVE 4 ▶ We can also use the slope-intercept form to determine the slope and *y*-intercept of a line from its equation by writing the equation in slope-intercept form.

5. Find the slope and the *y*-intercept of each line.

(a) $x + y = 2$

(b) $2x - 5y = 1$

E X A M P L E 4   Writing a Linear Equation in Slope-intercept Form

Write $3y + 2x = 9$ in slope-intercept form; then find the slope and *y*-intercept.

We solve for *y* to put the equation in slope-intercept form.

$$3y + 2x = 9$$
$$3y = -2x + 9 \quad \text{Subtract } 2x.$$
$$y = -\frac{2}{3}x + 3 \quad \text{Divide by 3.}$$

Slope ——↑   └── *y*-intercept is $(0, 3)$.

The slope-intercept form gives $-\frac{2}{3}$ for the slope, with *y*-intercept $(0, 3)$.

◀◀ **WORK PROBLEM 5 AT THE SIDE.**

OBJECTIVE 5 ▶ The previous section showed that parallel lines have the same slope, and perpendicular lines have slopes that are negative reciprocals. These results are used in the next two examples.

ANSWERS

4. (a) $2x - y = 3$ (b) $2x + 3y = 0$
   (c) $y = 3$
5. (a) $-1, (0, 2)$ (b) $\frac{2}{5}, \left(0, -\frac{1}{5}\right)$

┌─ **E X A M P L E 5**   **Finding the Equation of a Line Parallel to a Given Line**

Find an equation of the line going through the point $(-2, -3)$ and parallel to the line $2x + 3y = 6$.

The slope of a line parallel to the line $2x + 3y = 6$ can be found by solving for y.

$$2x + 3y = 6$$

$$3y = -2x + 6 \qquad \text{Subtract } 2x \text{ on both sides.}$$

$$y = -\frac{2}{3}x + 2 \qquad \text{Divide both sides by 3.}$$

The slope is given by the coefficient of x, so $m = -\frac{2}{3}$. This means that the line through $(-2, -3)$ and parallel to $2x + 3y = 6$ has slope $-\frac{2}{3}$. Use the point-slope form, with $(x_1, y_1) = (-2, -3)$ and $m = -\frac{2}{3}$ to write the equation.

$$y - y_1 = m(x - x_1) \qquad \text{Point-slope form}$$

$$y - (-3) = -\frac{2}{3}[x - (-2)] \qquad y_1 = -3, m = -\frac{2}{3}, x_1 = -2$$

$$y + 3 = -\frac{2}{3}(x + 2)$$

$$3(y + 3) = -2(x + 2) \qquad \text{Multiply by 3.}$$

$$3y + 9 = -2x - 4 \qquad \text{Distributive property}$$

$$2x + 3y = -13 \qquad \text{Standard form}$$

WORK PROBLEM 6 AT THE SIDE. ▶▶

┌─ **E X A M P L E 6**   **Finding the Equation of a Line Perpendicular to a Given Line**

Find an equation of the line perpendicular to $2x + 5y = 8$ and going through $(2, 3)$.

First, we find the slope of the line $2x + 5y = 8$ by solving for y.

$$5y = -2x + 8$$

$$y = -\frac{2}{5}x + \frac{8}{5}$$

The slope of this line is $-\frac{2}{5}$. Because the negative reciprocal of $-\frac{2}{5}$ is $\frac{5}{2}$, a line perpendicular to $2x + 5y = 8$ has slope $\frac{5}{2}$. We want an equation of a line with slope $\frac{5}{2}$ that goes through $(2, 3)$. Use the point-slope form.

$$y - 3 = \frac{5}{2}(x - 2) \qquad y_1 = 3, m = \frac{5}{2}, x_1 = 2$$

$$2(y - 3) = 5(x - 2) \qquad \text{Multiply by 2.}$$

$$2y - 6 = 5x - 10 \qquad \text{Distributive property}$$

$$-5x + 2y = -4 \qquad \text{Subtract } 5x; \text{ add 6.}$$

$$5x - 2y = 4 \qquad \text{Standard form}$$

WORK PROBLEM 7 AT THE SIDE. ▶▶

---

6. Write an equation in standard form for the line through $(5, 7)$ parallel to $2x - 5y = 15$.

7. Write equations in standard form of lines satisfying the given conditions.

(a) Through $(1, 6)$; perpendicular to $x + y = 9$

(b) Through $(-8, 3)$; perpendicular to $2x - 3y = 10$

Answers

6. $2x - 5y = -25$

7. (a) $x - y = -5$ **(b)** $3x + 2y = -18$

A summary of the forms of linear equations follows.

$Ax + By = C$	**Standard form**
(A, B, and C integers, neither A nor B equal to 0)	Slope is $-\dfrac{A}{B}$.
	x-intercept is $\left(\dfrac{C}{A}, 0\right)$.
	y-intercept is $\left(0, \dfrac{C}{B}\right)$.
$x = k$	**Vertical line** Slope is undefined. x-intercept is $(k, 0)$.
$y = k$	**Horizontal line** Slope is 0. y-intercept is $(0, k)$.
$y = mx + b$	**Slope-intercept form** Slope is m. y-intercept is $(0, b)$.
$y - y_1 = m(x - x_1)$	**Point-slope form** Slope is m. Line passes through (x_1, y_1).

OBJECTIVE 6 Suppose that it is time to fill up your car with gasoline. You drive into your local station and notice that the 89-octane gas that you will use is selling for \$1.20 per gallon. Experience has taught you that the final price you will pay can be determined by the number of gallons you buy multiplied by the price per gallon (in this case, \$1.20). As you pump the gas you observe two sets of numbers spinning by: one is the number of gallons you have pumped, and the other is the price you will pay for that number of gallons.

The table below shows how ordered pairs can be used to illustrate this situation.

Number of Gallons Pumped	*Price You Will Pay for this Number of Gallons*
0	\$0.00 = 0(\$0.00)
1	\$1.20 = 1(\$1.20)
2	\$2.40 = 2(\$1.20)
3	\$3.60 = 3(\$1.20)
4	\$4.80 = 4(\$1.20)

If we let x denote the number of gallons pumped, then the price y that we will pay can be found by the linear equation $y = 1.20x$, where y is in dollars. This is a simple realistic application of linear equations. Theoretically, there are infinitely many ordered pairs (x, y) that satisfy this equation, but in this application we are limited to nonnegative values for x, since we cannot have a negative number of gallons. The ordered pairs corresponding to the table above are (0, 0.00), (1, 1.20), (2, 2.40), (3, 3.60), and (4, 4.80). There is also a practical maximum value for x in this situation, that will vary from one car to another. What do you think determines this maximum value?

┌───

E X A M P L E   7   **Interpreting Ordered Pairs in an Application**

Name other ordered pairs that satisfy the equation $y = 1.20x$, choosing values for x that are not whole numbers. Interpret the ordered pair in the context of the gas-buying situation described earlier.

Let us arbitrarily choose x values of 1.5, 2.4, and 8.25. Substituting these values into $y = 1.20x$ gives the following:

When $x = 1.5$, $y = 1.20(1.5) = 1.80$.   Ordered pair: (1.5, 1.80)

When $x = 2.4$, $y = 1.20(2.4) = 2.88$.   Ordered pair: (2.4, 2.88)

When $x = 8.25$, $y = 1.20(8.25) = 9.90$.   Ordered pair: (8.25, 9.90)

These ordered pairs are interpreted as follows: When 1.5 gallons have been pumped, the meter shows that we owe $1.80. When 2.4 gallons have been pumped, we owe $2.88. And when 8.25 gallons have been pumped, the price is $9.90. ■

───┘

WORK PROBLEM 8 AT THE SIDE. ▶▶

Figure 18 shows the graphs of the eight ordered pairs found in Example 7 and the discussion preceding it. Notice that the points lie on a straight line. If we draw the line we obtain the graph of $y = 1.20x$. See Figure 19. Notice that we have used only nonnegative values for x in the graphs.

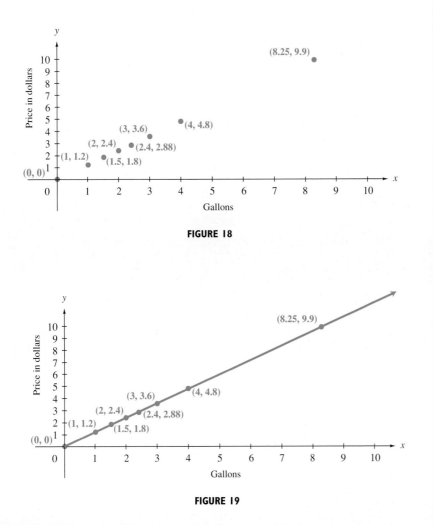

FIGURE 18

FIGURE 19

8. Repeat Example 7 for the x-value 5.5.

9. Suppose that a rental firm charges a flat rate of $15 plus $10 per day to rent a lawnmower. What equation defines the price y in dollars you will pay if you rent the lawnmower for x days?

In Example 7, the ordered pair $(0, 0)$ lies on the graph, meaning that the graph passes through the origin. If a realistic situation involves an initial charge plus a charge per unit of item purchased, the graph will not go through the origin. The next example illustrates this.

EXAMPLE 8 Determining an Equation and Graphing Ordered Pairs Satisfying It

Suppose that you can get a car wash at the gas station in Example 7 if you pay an additional $3.00.

(a) What is the equation that defines the price you will pay?

Since an additional $3.00 will be charged, you will pay $1.20x + 3.00$ dollars for x gallons of gasoline. Thus, if y represents the price, the equation is $y = 1.2x + 3$. (We deleted the unnecessary zeros here.)

(b) Graph the equation for x values between 0 and 10.

To graph this equation, we need only find two points, and possibly a third point as a check. Let's choose $x = 0$, $x = 2$, and $x = 10$:

When $x = 0$, $y = 1.2(0) + 3 = 3$.   Ordered pair: $(0, 3)$

When $x = 2$, $y = 1.2(2) + 3 = 5.4$.   Ordered pair: $(2, 5.4)$

When $x = 10$, $y = 1.2(10) + 3 = 15$.   Ordered pair: $(10, 15)$

The three ordered pairs are plotted and joined in Figure 20.

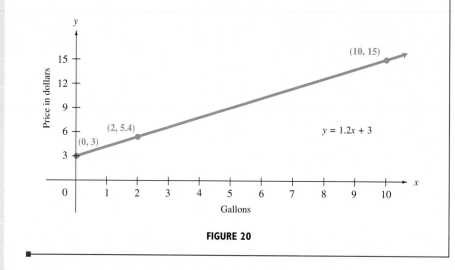

FIGURE 20

◀◀ WORK PROBLEM 9 AT THE SIDE.

4.3 Exercises

Match each equation with the graph that it most closely resembles. (Hint: Determining the signs of m and b will help you make your decision.)

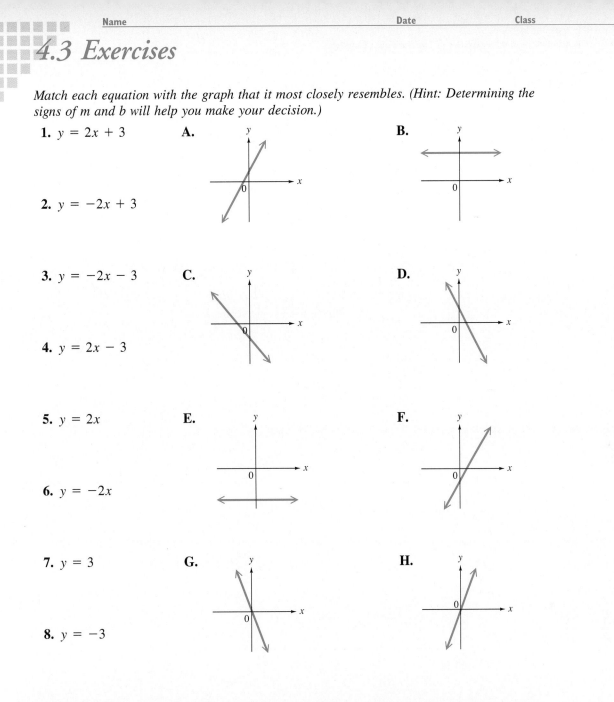

1. $y = 2x + 3$

A.

B.

2. $y = -2x + 3$

3. $y = -2x - 3$

C.

D.

4. $y = 2x - 3$

5. $y = 2x$

E.

F.

6. $y = -2x$

7. $y = 3$

G.

H.

8. $y = -3$

Write equations in standard form of lines satisfying the following conditions. See Example 1.

9. Through $(-2, 4)$; $m = -\dfrac{3}{4}$

10. Through $(-1, 6)$; $m = -\dfrac{5}{6}$

11. Through $(5, 8)$; $m = -2$

12. Through $(12, 10)$; $m = 1$

13. Through $(-5, 4)$; $m = \dfrac{1}{2}$

14. Through $(7, -2)$; $m = \dfrac{1}{4}$

15. Through $(-4, 12)$; horizontal

16. Through $(1, 5)$; horizontal

Write equations in standard form for the following lines. (Hint: What kind of line has undefined slope?)

17. Through $(9, 10)$; undefined slope

18. Through $(-2, 8)$; undefined slope

19. Through $(.5, .2)$; vertical

20. Through $\left(\frac{5}{8}, \frac{2}{9}\right)$; vertical

Write equations in standard form of the lines passing through the following pairs of points. See Example 2.

21. $(3, 4)$ and $(5, 8)$

22. $(5, -2)$ and $(-3, 14)$

23. $(6, 1)$ and $(-2, 5)$

24. $(-2, 5)$ and $(-8, 1)$

25. $\left(-\frac{2}{5}, \frac{2}{5}\right)$ and $\left(\frac{4}{3}, \frac{2}{3}\right)$

26. $\left(\frac{3}{4}, \frac{8}{3}\right)$ and $\left(\frac{2}{5}, \frac{2}{3}\right)$

27. $(2, 5)$ and $(1, 5)$

28. $(-2, 2)$ and $(4, 2)$

29. $(7, 6)$ and $(7, -8)$

30. $(13, 5)$ and $(13, -1)$

Write each equation in slope-intercept form; then give the slope of the line and the y-intercept. See Example 4.

	Equation	*Slope*	*y-intercept*
31. $5x + 2y = 20$	_____	_____	_____
32. $6x + 5y = 40$	_____	_____	_____

	Equation	*Slope*	*y-intercept*

33. $2x - 3y = 10$ _____   _____   _____

34. $4x - 3y = 7$ _____   _____   _____

Write equations in slope-intercept form of lines satisfying the following conditions. See Example 3.

35. $m = 5; b = 15$

36. $m = -2; b = 12$

37. $m = -\dfrac{2}{3}; b = \dfrac{4}{5}$

38. $m = -\dfrac{5}{8}; b = -\dfrac{1}{3}$

39. Slope $\dfrac{2}{5}$; y-intercept $(0, 5)$

40. Slope $-\dfrac{3}{4}$; y-intercept $(0, 7)$

Write equations in standard form of the lines satisfying the following conditions. See Examples 5 and 6.

41. Through $(7, 2)$; parallel to $3x - y = 8$

42. Through $(4, 1)$; parallel to $2x + 5y = 10$

43. Through $(-2, -2)$; parallel to $-x + 2y = 10$

44. Through $(-1, 3)$; parallel to $-x + 3y = 12$

45. Through $(8, 5)$; perpendicular to $2x - y = 7$

46. Through $(2, -7)$; perpendicular to $5x + 2y = 18$

47. Through $(-2, 7)$; perpendicular to $x = 9$

48. Through $(8, 4)$; perpendicular to $x = -3$

Write an equation in the form $y = ax$ for each of the following situations. Then give the three ordered pairs associated with the equation for x-values of 0, 5, and 10. For example, if a car travels 60 miles per hour, then after x hours it will have traveled $y = 60x$ miles. The three ordered pairs are (0, 0), (5, 300), and (10, 600). See Example 7.

49. x represents the number of hours traveling at 45 miles per hour, and y represents the distance traveled (in miles).

50. x represents the number of compact discs sold at $16 each, and y represents the total cost of the discs (in dollars).

51. x represents the number of gallons of gas sold at $1.30 per gallon, and y represents the total cost of the gasoline (in dollars).

52. x represents the number of days a videocassette is rented at $1.50 per day, and y represents the total charge for the rental (in dollars).

Write an equation in the form $y = ax + b$ for each of the following situations. Then give three ordered pairs associated with the equation for the x-values of 0, 5, and 10. For example, if it costs a flat rate of $.20 to make a certain long-distance call and the charge is $.10 per minute, then after x minutes the cost y in dollars is given by $y = .10x + .20$. The three ordered pairs are (0, .20), (5, .70), and (10, 1.20). See Example 8.

53. It costs a $15 flat fee to rent a chain saw, plus $3 per day starting with the first day. Let x represent the number of days rented, so that y represents the charge to the user (in dollars).

54. It costs a borrower $.05 per day for an overdue book, plus a flat $.50 charge for all books borrowed, to be contributed to a fund for building a new library. Let x represent the number of days the book is overdue, so that y represents the total fine to the tardy user.

55. A rental car costs $25.00 plus $.10 per mile. Let x represent the number of miles driven, so that y represents the total charge to the user.

56. It costs a flat fee of $450 to drill a well, plus $10 per foot. Let x represent the depth of the well, in feet, so that y represents the total cost of the well.

Write a linear equation and solve it in order to solve the problem.

57. Refer to Exercise 53. Suppose that the total charge is $69.00. For how many days was the saw rented?

58. Refer to Exercise 54. Suppose that the tardy user paid a $1.30 fine. How many days did this user keep the book past the due date?

59. Refer to Exercise 55. The renter paid $42.30. How many miles was the car driven?

60. Refer to Exercise 56. A certain well costs $6950. How deep is the well?

Assume that the situation described in the figure can be modeled by a straight-line graph, and use the information to find the y = mx + b form of the equation of the line.

61. The number of post offices in the United States has been declining in recent years. Use the information given on the bar graph for the years 1985 and 1993, letting $x = 0$ represent the year 1985 and letting y represent the number of post offices. (*Source:* U.S. Postal Service, Annual Report of the Postmaster General)

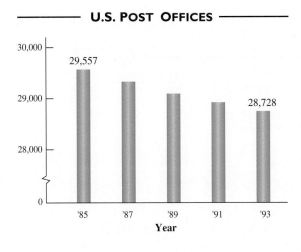

62. The number of motor vehicle deaths in the United States declined from 1990 to 1992 as seen in the accompanying bar graph. Use the information given, with $x = 0$ representing 1990 and y representing the number of deaths. (*Source:* National Safety Council)

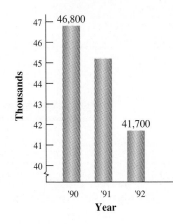

INTERPRETING TECHNOLOGY (EXERCISES 63–66)

63. The graphing calculator screen shows two lines. One is the graph of $y_1 = -2x + 3$ and the other is the graph of $y_2 = 3x - 4$. Which is which?

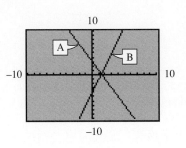

64. The graphing calculator screen shows two lines. One is the graph of $y_1 = 2x - 5$ and the other is the graph of $y_2 = 4x - 5$. Which is which?

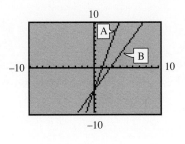

65. The graphing calculator screen shows a line with the coordinates of a point displayed at the bottom. The slope of the line is 3. What is the y-coordinate of the point on the line whose x-coordinate is 4?

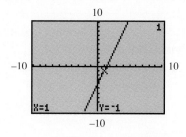

66. The table shown was generated by a graphing calculator. It gives several points that lie on the graph of a line. What is the slope of the line?

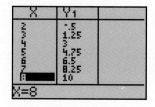

MATHEMATICAL CONNECTIONS (EXERCISES 67–72)

In Chapter 2 we learned how formulas can be applied to problem solving. In Exercises 67–72, we will see how the formula that relates the Celsius and Fahrenheit temperatures is derived. Work Exercises 67–72 in order.

67. There is a linear relationship between Celsius and Fahrenheit temperatures. When C = 0°,
F = _____°, and when C = 100°,
F = _____°.

68. Think of ordered pairs of temperatures (C, F), where C and F represent corresponding Celsius and Fahrenheit temperatures. The equation that relates the two scales has a straight-line graph that contains the two points determined in Exercise 67. What are these two points?

69. Find the slope of the line described in Exercise 68.

70. Now think of the point-slope form of the equation in terms of C and F, where C replaces x and F replaces y. Use the slope you found in Exercise 69 and one of the two points determined earlier, and find the equation that gives F in terms of C.

71. To obtain another form of the formula, use the equation you found in Exercise 70 and solve for C in terms of F.

72. The equation found in Exercise 70 is graphed on the graphing calculator screen shown here. Observe the display at the bottom, and interpret it in the context of this group of exercises.

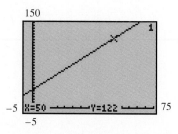

4.4 Linear Inequalities in Two Variables

In Chapter 3 we graphed linear inequalities with one variable on the number line. In this section we will graph linear inequalities in two variables on a rectangular coordinate system.

Linear Inequality

An inequality that can be written as

$$Ax + By < C \quad \text{or} \quad Ax + By > C,$$

where A, B, and C are real numbers and A and B are not both 0, is a **linear inequality in two variables.**

Also, $\leq$ and $\geq$ may replace $<$ and $>$ in the definition.

OBJECTIVE 1 A line divides the plane into three regions: the line itself and the two half-planes on either side of the line. Recall that the graphs of linear inequalities in one variable are intervals on the number line that sometimes include an endpoint. The graphs of linear inequalities in two variables are *regions* in the real number plane and may include a *boundary line*. The **boundary** (BOUN-dery) **line** for the inequality $Ax + By < C$ or $Ax + By > C$ is the graph of the *equation* $Ax + By = C$. To graph a linear inequality, we go through the following steps.

Graphing a Linear Inequality

Step 1 Draw the graph of the straight line that is the boundary. Make the line solid if the inequality involves $\leq$ or $\geq$; make the line dashed if the inequality involves $<$ or $>$.

Step 2 Choose any point not on the line as a test point.

Step 3 Shade the region that includes the test point if the test point satisfies the original inequality; otherwise, shade the region on the other side of the boundary line.

E X A M P L E 1   Graphing a Linear Inequality

Graph the solutions of $3x + 2y \geq 6$.

Graph the linear inequality $3x + 2y \geq 6$ by first graphing the straight line $3x + 2y = 6$. The graph of this line, the boundary of the graph of the inequality, is shown in Figure 21.

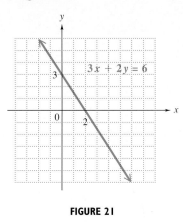

FIGURE 21

— **CONTINUED ON NEXT PAGE**

OBJECTIVES

1 Graph linear inequalities.

2 Graph the intersection of two linear inequalities.

3 Graph the union of two linear inequalities.

4 Graph first-degree absolute value inequalities.

FOR EXTRA HELP

Tutorial   Tape 8   SSM, Sec. 4.4

1. Graph the solutions of each inequality.

(a) $x + y \leq 4$

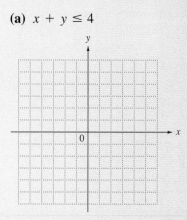

(b) $3x + y \geq 6$

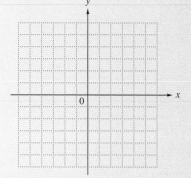

The graph of $3x + 2y \geq 6$ includes the points of the line $3x + 2y = 6$, and either the points *above* the line $3x + 2y = 6$ or the points *below* that line. To decide which side belongs to the graph, first select any point not on the line $3x + 2y = 6$. The origin, $(0, 0)$, is often a good choice. Substitute the values from the test point $(0, 0)$ for x and y in the inequality $3x + 2y > 6$:

$$3(0) + 2(0) > 6$$
$$0 > 6. \qquad \text{False}$$

Because the result is false, $(0, 0)$ does not satisfy the inequality, and the solutions include all points on the other side of the line. This region is shaded in Figure 22.

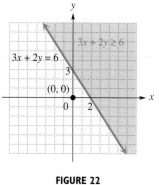

FIGURE 22

◀◀ **WORK PROBLEM 1 AT THE SIDE.**

If the inequality is written in the form $y > mx + b$ or $y < mx + b$, the inequality symbol indicates which half-plane to shade.

If $y > mx + b$, shade **above** the boundary line;

if $y < mx + b$, shade **below** the boundary line.

E X A M P L E 2   Graphing an Inequality

Graph the solutions of $x - 3y < 4$.

First graph the boundary line, shown in Figure 23. The points of the boundary line do not belong to the inequality $x - 3y < 4$ (because the inequality symbol is $<$, not $\leq$). For this reason, the line is dashed. Now solve the inequality for y.

$$x - 3y < 4$$
$$-3y < -x + 4$$
$$y > \frac{x}{3} - \frac{4}{3} \qquad \begin{array}{l}\text{Multiply by } -\frac{1}{3}; \\ \text{change the inequality.}\end{array}$$

Because of the *is greater than* symbol, we should shade *above* the line. As a check, we can choose any point not on the line, say $(1, 2)$, and substitute for x and y in the original inequality.

$$1 - 3(2) < 4$$
$$-5 < 4 \qquad \text{True}$$

This result agrees with our decision to shade above the line. The solutions, graphed in Figure 23, include only those points in the shaded half-plane (not those on the line).

ANSWERS

1. (a)

(b)

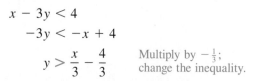

CONTINUED ON NEXT PAGE

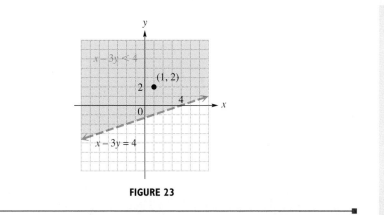

FIGURE 23

WORK PROBLEM 2 AT THE SIDE.

OBJECTIVE 2 In Section 3.2 we discussed how the words "and" and "or" are used with compound inequalities. In that section, the inequalities were in a single variable. Those ideas can be extended to include inequalities in two variables. If a pair of inequalities is joined with the word "and," it is interpreted as the intersection of the solutions of the inequalities. The graph of the intersection of two or more inequalities is the region of the plane where all points satisfy all of the inequalities at the same time.

EXAMPLE 3 Graphing the Intersection of Two Inequalities

Graph the intersection of the solutions of $2x + 4y \geq 5$ and $x \geq 1$.

To begin, graph each of the two inequalities $2x + 4y \geq 5$ and $x \geq 1$ separately. The graph of $2x + 4y \geq 5$ is shown in Figure 24(a), and the graph of $x \geq 1$ is shown in Figure 24(b). The graph of the intersection is the region common to both graphs, as shown in Figure 24(c).

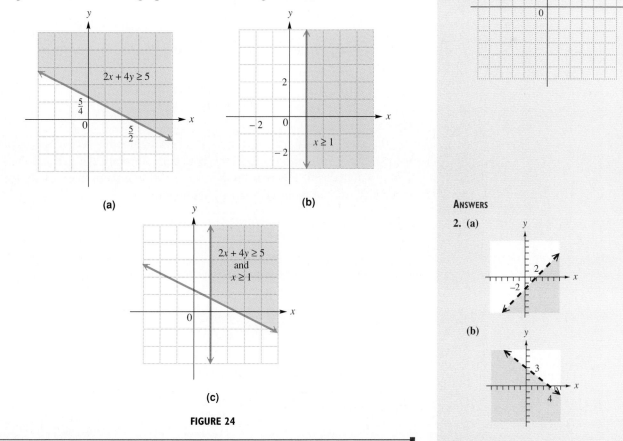

FIGURE 24

2. Graph the solutions of each inequality.

(a) $x - y > 2$

(b) $3x + 4y < 12$

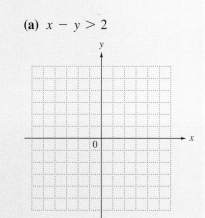

ANSWERS

2. (a)

(b)

3. Graph the intersection of the solutions of the inequalities:
$x - y \le 4$ and $x \ge -2$.

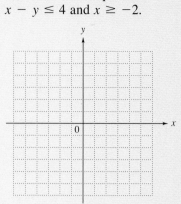

4. Graph the union of the solutions of the inequalities:
$7x - 3y < 21$ or $x > 2$.

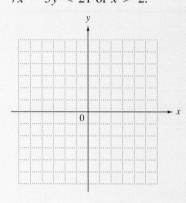

◀◀ **WORK PROBLEM 3 AT THE SIDE.**

OBJECTIVE ▶ When two inequalities are joined by the word "or," we must find the union of the graphs of the inequalities. The graph of the union of two inequalities includes all of the points that satisfy either inequality.

E X A M P L E   4   Graphing the Union of Two Inequalities

Graph the union of the solutions of $2x + 4y \ge 5$ or $x \ge 1$.

The graphs of the two inequalities are shown in Figures 24(a) and 24(b). The graph of the union is shown in Figure 25.

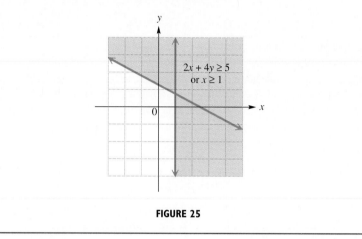

FIGURE 25

◀◀ **WORK PROBLEM 4 AT THE SIDE.**

OBJECTIVE ▶ An absolute value inequality should first be written without absolute value bars as shown in Chapter 3. Then the methods given in the previous examples can be used to graph the inequality.

E X A M P L E   5   Graphing an Absolute Value Inequality

Graph the solutions of $|x| \le 4$.

Rewrite $|x| \le 4$ as $-4 \le x \le 4$. The topic of this section is linear inequalities in *two* variables, so the boundary lines are the vertical lines $x = -4$ and $x = 4$. Points from the region between these lines satisfy the inequality, so that region is shaded. See Figure 26.

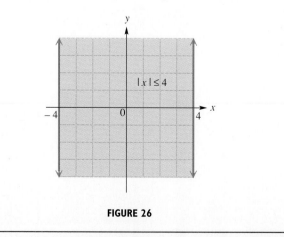

FIGURE 26

ANSWERS

3.

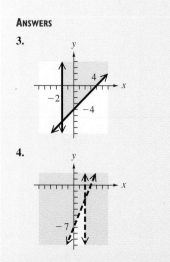

4.

E X A M P L E 6 Graphing an Absolute Value Inequality

Graph the solutions of $|y + 2| > 3$.

As shown in Section 3.3, the equation of the boundary, $|y + 2| = 3$, can be rewritten as

$$y + 2 = 3 \quad \text{or} \quad y + 2 = -3$$
$$y = 1 \quad \text{or} \quad y = -5.$$

This shows that $y = 1$ and $y = -5$ are boundary lines. Checking points from each of the three regions determined by the horizontal boundary lines gives the graph shown in Figure 27.

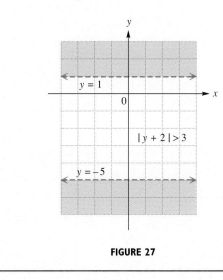

FIGURE 27

WORK PROBLEM 5 AT THE SIDE. ▶▶

5. Graph the solutions of each absolute value inequality.

(a) $|x| \geq 4$

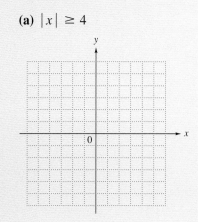

(b) $|x - 2| < 1$

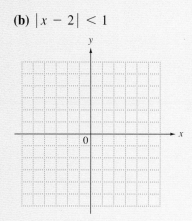

ANSWERS

5. (a)

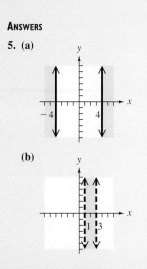

(b)

NUMBERS IN THE
Real World

a graphing calculator minicourse

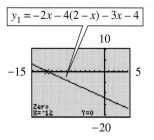

$y_1 = -4x + 7$

$y_1 = -2x - 4(2 - x) - 3x - 4$

Lesson 1: Solution of Linear Equations

A graphing calculator can be used to solve a linear equation of the form $y = 0$. First, we observe that the graph of $y = mx + b$ is a straight line, and if $m \neq 0$, the x-intercept of the graph is the solution of the equation $mx + b = 0$. For a simple example, suppose we want to solve $-4x + 7 = 0$ graphically. We begin by graphing $y_1 = -4x + 7$ in a viewing window that shows the x-intercept. Then we use the capability of the calculator to find the x-intercept. As seen in the first figure to the left, the x-intercept of the graph, which is also the solution or root of the equation, is 1.75. (For convenience, we will refer to the number as the x-intercept rather than the point.) Therefore, the solution set of $-4x + 7 = 0$ is $\{1.75\}$. This can easily be verified using strictly algebraic methods.

If we want to solve a more complicated linear equation, such as

$$-2x - 4(2 - x) = 3x + 4,$$

we begin by writing it as an equivalent equation with 0 on one side. If we subtract $3x$ and subtract 4 from both sides, we get

$$-2x - 4(2 - x) - 3x - 4 = 0.$$

Then we graph $y_1 = -2x - 4(2 - x) - 3x - 4$, and find the x-intercept. Notice that the viewing window must be altered from the one shown in the first figure, because the x-intercept does not lie in the interval $[-10, 10]$. As seen in the second figure, the x-intercept of the line, and thus the solution or root of the equation, is -12. The solution set is $\{-12\}$.

GRAPHING CALCULATOR EXPLORATIONS

1. Solve the equation $-2x - 4(2 - x) = 3x + 4$ using traditional algebraic methods, and show that the solution set is $\{-12\}$, supporting the graphical approach used in the second figure.

2. Use traditional algebraic methods to solve $2(x - 5) + 3x = 0$. Then use a graphing calculator to support your result by showing that the x-intercept of $y_1 = 2(x - 5) + 3x$ corresponds to your solution.

3. Use traditional algebraic methods to solve $6x - 4(3 - 2x) = 5(x - 4) - 10$. Then use a graphing calculator to support your result by showing that the x-intercept of $y_1 = 6x - 4(3 - 2x) - 5(x - 4) + 10$ corresponds to your solution.

4. Using traditional algebraic methods, we can show that the solution of

$$\frac{x + 7}{6} + \frac{2x - 8}{2} = -4 \text{ is } -1.$$

Graph $y_1 = \dfrac{x + 7}{6} + \dfrac{2x - 8}{2} + 4$, being careful to use parentheses around the numerators as you make the keystrokes. Show that the x-intercept of the graph is -1.

4.4 Exercises

In each statement, fill in the first blank with either solid *or* dashed. *Fill in the second blank with* above *or* below.

1. The boundary of the graph of $y \leq -x + 2$ will be a _____ line, and the shading will be _____ the line.

2. The boundary of the graph of $y < -x + 2$ will be a _____ line, and the shading will be _____ the line.

3. The boundary of the graph of $y > -x + 2$ will be a _____ line, and the shading will be _____ the line.

4. The boundary of the graph of $y \geq -x + 2$ will be a _____ line, and the shading will be _____ the line.

Graph each linear inequality in two variables. See Examples 1 and 2.

5. $x + y \leq 2$

6. $x + y \leq -3$

7. $4x - y < 4$

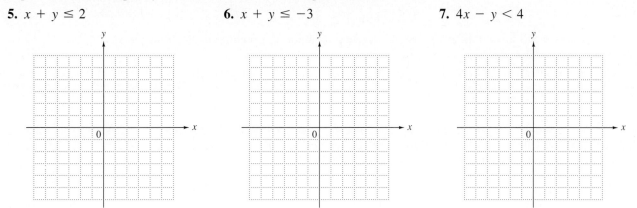

8. $3x - y < 3$

9. $x + 3y \geq -2$

10. $x + 4y \geq -3$

11. $x + y > 0$

12. $x + 2y > 0$

13. $x - 3y \leq 0$

14. $x - 5y \leq 0$

15. $y < x$

16. $y \leq 4x$

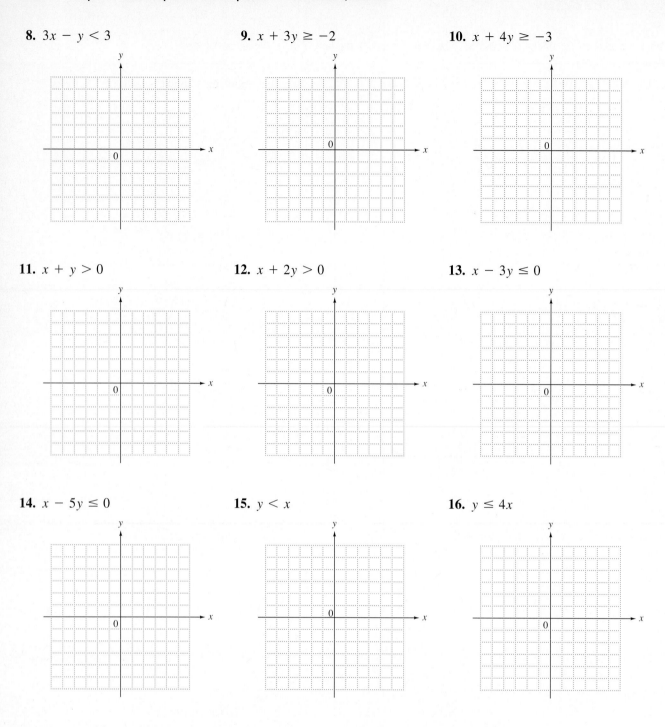

17. Explain how you will know whether to use intersection or union in graphing inequalities like those found in Examples 3 and 4 in this section.

18. Explain how you will know whether to use "and" or "or" in graphing absolute value inequalities like those found in Examples 5 and 6 in this section.

Graph the intersection or union, as appropriate, of the solutions of the following pairs of linear inequalities. See Examples 3 and 4.

19. $x + y \leq 1$ and $x \geq 1$ **20.** $x - y \geq 2$ and $x \geq 3$ **21.** $2x - y \geq 2$ and $y < 4$

22. $3x - y \geq 3$ and $y < 3$ **23.** $x + y > -5$ and $y < -2$ **24.** $6x - 4y < 10$ and $y > 2$

25. $x - y \geq 1$ or $y \geq 2$ **26.** $x + y \leq 2$ or $y \geq 3$ **27.** $x - 2 > y$ or $x < 1$

28. $x + 3 < y$ or $x > 3$ **29.** $3x + 2y < 6$ or $x - 2y > 2$ **30.** $x - y \geq 1$ or $x + y \leq 4$

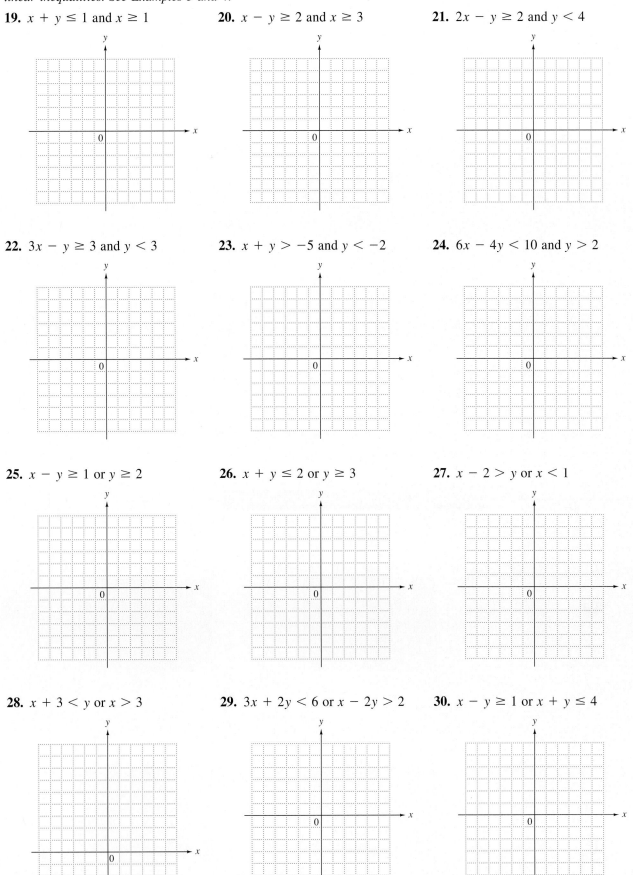

Graph each of the following inequalities involving absolute value. See Examples 5 and 6.

31. $|x| \geq 3$

32. $|y| < 5$

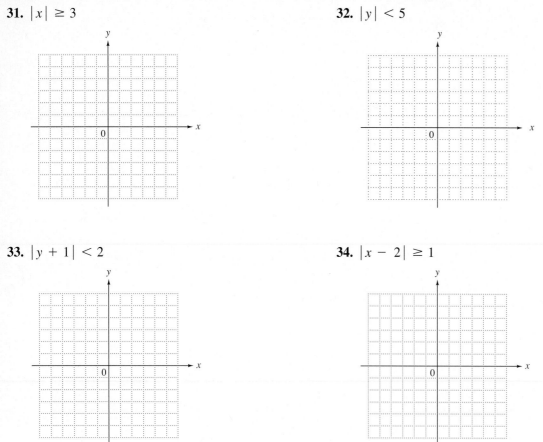

33. $|y + 1| < 2$

34. $|x - 2| \geq 1$

INTERPRETING TECHNOLOGY (EXERCISES 35–38)

Match each inequality with its calculator-generated graph. (Hint: Use the slope, y-intercept, and inequality symbol in making your choice.)

_____ **35.** $y \leq 3x - 6$ **A.**

_____ **36.** $y \geq 3x - 6$ **B.**

_____ **37.** $y \leq -3x - 6$ **C.**

_____ **38.** $y \geq -3x - 6$ **D.**

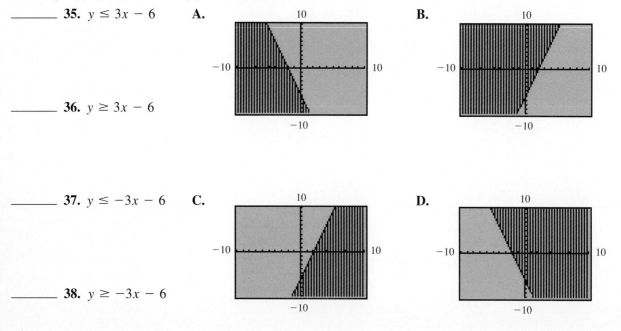

4.5 An Introduction to Functions

One of the most important concepts in algebra is that of a *function* (FUNK-shun). In this section we will examine functions, and see how they can be expressed in several forms. The table below indicates the number of people on Earth for various years from 1800 through 2050. These figures come from The Population Reference Bureau, Inc., and figures after 1987 are projections.

Year	Population in Billions
1800	1
1927	2
1960	3
1975	4
1987	5
1998	6
2008	7
2019	8
2032	9
2050	10

The information in the table can be expressed as a set of ordered pairs, where the first entry in the ordered pair is the year, and the second entry is the population in billions. For example, three such ordered pairs are (1800, 1), (1960, 3), and (1998, 6).

It has been shown that under certain conditions, the Fahrenheit temperature determines the number of times a cricket will chirp per minute. If x represents the temperature and y represents the number of chirps per minute, the equation $y = .25x + 40$ gives the relationship between them. For example, when $x = 40$, $y = 50$, meaning that at 40°F the cricket chirps 50 times per minute. This can be represented by the ordered pair (40, 50). Some other such ordered pairs are (20, 45) and (60, 55).

A third way of representing data that leads to ordered pairs is a graph. The graph in Figure 28 shows the public debt of the United States for the fiscal years 1990 through 1994.

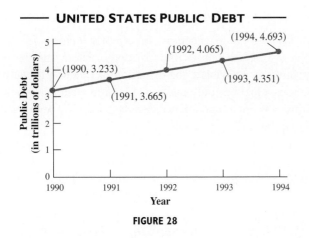

UNITED STATES PUBLIC DEBT

(1994, 4.693)
(1992, 4.065)
(1990, 3.233)
(1993, 4.351)
(1991, 3.665)

Public Debt (in trillions of dollars)

Year

FIGURE 28

If x represents the year and y the public debt in trillions of dollars, one ordered pair describing their relationship is (1990, 3.233).

OBJECTIVES

1 ▸ Identify a function.
2 ▸ Find the domain and range of a function.
3 ▸ Use the vertical line test.
4 ▸ Use $f(x)$ notation.
5 ▸ Write a linear equation as a linear function.

FOR EXTRA HELP

Tutorial   Tape 8   SSM, Sec. 4.5

1. (a) Name two other ordered pairs in the world population example given in the table.

In each example we have given, the second entry in the ordered pair is dependent on the first. That is, the population is dependent on the year, the number of chirps is dependent on the temperature, and the public debt is dependent on the year. In general, when the value of y depends on the value of x, y is called the **dependent variable** (DEE-pen-dent VAIR-ee-uh-bul) and x is called the **independent variable** (IN-dih-pen-dent VAIR-ee-uh-bul).

◀◀ **WORK PROBLEM 1 AT THE SIDE.**

Objective 1▶ We are now ready to introduce the function concept. First, we will examine a more general concept, the relation. Since related quantities can be represented by ordered pairs (as we have just seen), we define a *relation* (ree-LAY-shun) as follows.

Relation

A **relation** is a set of ordered pairs of real numbers.

(b) Name two other ordered pairs in the cricket example given by the equation.

In each of the examples of relations just discussed, for every value of x, there was one and only one value of y. A relation of this type has an important role in algebra, and it is called a *function*.

Function

A **function** is a relation in which, for each value of the first component of the ordered pairs, there is exactly one value of the second component.

(c) Name two other ordered pairs in the public debt example given in Figure 28.

E X A M P L E 1   Identifying a Function Expressed as a Set of Ordered Pairs

Determine whether each of the following relations is a function.

$$F = \{(1, 2), (-2, 5), (3, -1)\}$$
$$G = \{(-2, 1), (-1, 0), (0, 1), (1, 2), (2, 2)\}$$
$$H = \{(-4, 1), (-2, 1), (-2, 0)\}$$

Relations F and G are functions, because for each x-value, there is only one y-value. Notice that in G, the last two ordered pairs have the same y-value. This does not violate the definition of function because each first component (x-value) has only one second component (y-value).

Relation H is not a function because the last two ordered pairs have the same x-value, but different y-values.

In addition to sets of ordered pairs, tables, equations, and graphs, a function can be represented by a mapping, as described in Example 2.

E X A M P L E 2   Recognizing a Function Expressed as a Mapping

A function can also be expressed as a correspondence or *mapping* from one set to another. The mapping in Figure 29 is a function that assigns to a state its population (in millions) expected by the year 2000.

Answers

(Answers will vary.)
1. (a) (1927, 2), (1975, 4)
   (b) (44, 51), (48, 52)
   (c) (1991, 3.665), (1992, 4.065)

CONTINUED ON NEXT PAGE ───

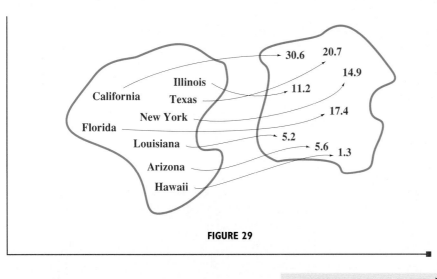

FIGURE 29

WORK PROBLEM 2 AT THE SIDE. ▶▶

Objective 2▶ The set of all first components (*x*-values) of the ordered pairs of a relation is called the **domain** (doh-MAIN) of the relation, and the set of all second components (*y*-values) is called the **range.** Another way of interpreting this is that the domain is the set of inputs, and the range is the set of outputs. The domain of function F in Example 1 is $\{1, -2, 3\}$; the range is $\{2, 5, -1\}$. Also, the domain of function G is $\{-2, -1, 0, 1, 2\}$, and the range is $\{0, 1, 2\}$. Domains and ranges can also be defined in terms of independent and dependent variables.

Domain and Range

In a relation, the set of all values of the independent variable (*x*) is the **domain;** the set of all values of the dependent variable (*y*) is the **range.**

E X A M P L E 3   Finding Domains and Ranges

Give the domain and range of each function.

(a) $\{(3, -1), (4, 2), (0, 5)\}$

The domain, the set of *x*-values, is $\{3, 4, 0\}$; the range is the set of *y*-values, $\{-1, 2, 5\}$.

(b) The function in Figure 29

The domain is {Illinois, Texas, California, New York, Florida, Louisiana, Arizona, Hawaii} and the range is $\{1.3, 5.2, 5.6, 11.2, 14.9, 17.4, 20.7, 30.6\}$.

WORK PROBLEM 3 AT THE SIDE. ▶▶

2. Which are functions?

(a) $\{(1, 2), (2, 4), (3, 3), (4, 2)\}$

(b) $\{(0, 3), (-1, 2), (-1, 3)\}$

(c)
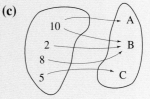

(d)
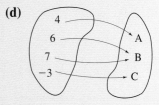

3. Give the domain and range of each function.

(a) $\{(1, 2), (2, 4), (3, 3), (4, 2)\}$

(b)
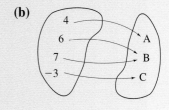

4. Give the domain and range of each relation from its graph.

(a)

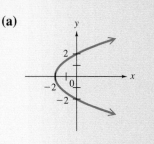

EXAMPLE 4 **Finding Domains and Ranges from Graphs**

Three relations are graphed in Figure 30. Give the domain and range of each.

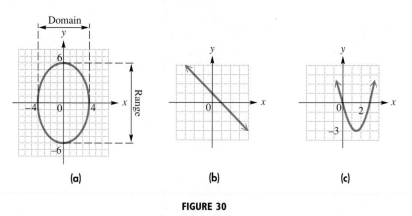

FIGURE 30

(a) In Figure 30(a), the *x*-values of the points on the graph include all numbers between -4 and 4, inclusive. The *y*-values include all numbers between -6 and 6, inclusive. Using interval notation,

the domain is $[-4, 4]$; the range is $[-6, 6]$.

(b) In Figure 30(b), the arrowheads indicate that the line extends indefinitely left and right, as well as up and down. Therefore, both the domain and the range are the set of all real numbers, written $(-\infty, \infty)$.

(c) In Figure 30(c), the arrowheads indicate that the graph extends indefinitely left and right, as well as upward. The domain is $(-\infty, \infty)$. Because there is a least *y*-value, -3, the range includes all numbers greater than or equal to -3, written $[-3, \infty)$.

◄◄ **WORK PROBLEM 4 AT THE SIDE.**

Relations are often defined by equations such as $y = 2x + 3$ and $y^2 = x$. Sometimes we need to determine the domain of a relation from its equation. The domain of a relation is assumed to be all real numbers that produce real numbers for the dependent variable when substituted for the independent variable. For example, because any real number can be used as a replacement for *x* in $y = 2x + 3$, the domain of this function is the set of real numbers. As another example, the function defined by $y = \frac{1}{x}$ has all real numbers except 0 as a domain, because *y* is undefined if $x = 0$. In general, the domain of a function defined by an algebraic expression is all real numbers except those numbers that lead to division by zero or an even root of a negative number.

(b)

EXAMPLE 5 **Identifying a Function from an Equation or Inequality**

For each of the following decide whether it defines a function, and give the domain.

(a) $y = \sqrt{2x - 1}$

For any choice of *x* in the domain, there is exactly one corresponding value for *y*, so this equation defines a function. We saw earlier that the square root of a negative number is not a real number, so we must have

CONTINUED ON NEXT PAGE

$$2x - 1 \geq 0$$
$$2x \geq 1$$
$$x \geq \frac{1}{2}.$$

The domain is $\left[\frac{1}{2}, \infty\right)$.

(b) $y^2 = x$

The ordered pairs $(16, 4)$ and $(16, -4)$ both satisfy this equation. Because one value of x, 16, corresponds to two values of y, 4 and -4, this equation does not define a function. If $y^2 = x$, then $y = \sqrt{x}$ or $y = -\sqrt{x}$, which shows that two values of y correspond to each positive value of x. Because x is equal to the square of y, the values of x must always be nonnegative. The domain is $[0, \infty)$.

(c) $y \leq x - 1$

By definition, y is a function of x if a value of x leads to exactly one value of y. In this example, a particular value of x, say 1, corresponds to many values of y. The ordered pairs $(1, 0)$, $(1, -1)$, $(1, -2)$, $(1, -3)$, and so on, all satisfy the inequality. For this reason, this inequality does not define a function. Any number can be used for x, so the domain is the set of real numbers, $(-\infty, \infty)$.

(d) $y = \dfrac{5}{x - 1}$

Given any value of x, we find y by subtracting 1, then dividing the result into 5. This process produces exactly one value of y for each x-value, so this equation defines a function. The domain includes all real numbers except those that make the denominator zero. We find these numbers by setting the denominator equal to zero and solving for x.

$$x - 1 = 0$$
$$x = 1$$

Thus, the domain includes all real numbers except 1. In interval notation this is written as

$$(-\infty, 1) \cup (1, \infty).$$

Caution
The parentheses used to represent an ordered pair are also used to represent an open interval. In general, there is no confusion between these symbols because the context of the discussion tells us whether we are discussing ordered pairs or open intervals.

WORK PROBLEM 5 AT THE SIDE. ▶▶

OBJECTIVE ③ In a function each value of x leads to only one value of y, so any vertical line drawn through the graph of a function would intersect the graph in at most one point. This is the **vertical line test for a function.**

Vertical Line Test

If a vertical line intersects the graph of a relation in more than one point, then the relation is not a function.

5. Decide whether or not each equation or inequality defines a function, and give the domain.

(a) $y = 6x + 12$

(b) $y \leq 4x$

(c) $y = -\sqrt{3x - 2}$

(d) $y^2 = 25x$

ANSWERS
5. (a) yes; $(-\infty, \infty)$ **(b)** no; $(-\infty, \infty)$
  (c) yes; $\left[\frac{2}{3}, \infty\right)$
  (d) no; $[0, \infty)$

6. Which of the following graphs represent functions?

(a)

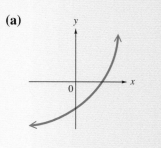

(b)

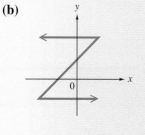

(c)

For example, the graph shown in Figure 31(a) is not the graph of a function, because a vertical line intersects the graph in more than one point. The graph of Figure 31(b) does represent a function. Any vertical line will cross the graph at most once.

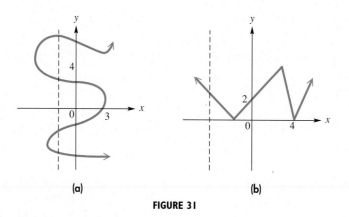

(a)   (b)

FIGURE 31

◀◀ **WORK PROBLEM 6 AT THE SIDE.**

OBJECTIVE 4 To say that y is a function of x means that for each value of x from the domain of the function, there is exactly one value of y. To emphasize that y *is a function of x,* or that y depends on x, it is common to write

$$y = f(x),$$

with $f(x)$ read "f of x." (In this special notation, the parentheses do not indicate multiplication.) The letter f stands for *function.* (Other letters, such as g and h are often used as well.) For example, if $y = 9x - 5$, we emphasize that y is a function of x by writing $y = 9x - 5$ as

$$f(x) = 9x - 5.$$

We can use this **function notation** to simplify certain statements. For example, if $y = 9x - 5$, then replacing x with 2 gives

$$y = 9 \cdot 2 - 5$$
$$= 18 - 5$$
$$= 13.$$

The statement "if $x = 2$, then $y = 13$" is abbreviated with function notation as

$$f(2) = 13.$$

We read this as "f of 2 equals 13." Also, $f(0) = 9 \cdot 0 - 5 = -5$, and $f(-3) = -32$.

These ideas and the symbols used to represent them can be explained as follows.

Name of the function   Defining expression

$$y = f(x) = 9x - 5$$

Value of the function   Name of the independent variable

Caution

The symbol $f(x)$ *does not* indicate "f times x," but represents the y-value for the indicated x-value. As shown above, $f(2)$ is the y-value that corresponds to the x-value 2.

ANSWERS

6. (a) and (c) are the graphs of functions.

E X A M P L E 6   Using Function Notation

Let $f(x) = -x^2 + 5x - 3$. Find the following.

(a) $f(2)$

Replace x with 2.

$$f(2) = -2^2 + 5 \cdot 2 - 3 = -4 + 10 - 3 = 3$$

(b) $f(-1)$

$$f(-1) = -(-1)^2 + 5(-1) - 3 = -1 - 5 - 3 = -9$$

(c) $f(q)$

Replace x with q.

$$f(q) = -q^2 + 5q - 3$$

The replacement of one variable with another is important in later courses.

WORK PROBLEM 7 AT THE SIDE. ▶▶

OBJECTIVE 5 By the vertical line test, linear equations (except for the type where $x = k$) define functions because their graphs are non-vertical straight lines.

Linear Function

A function that can be written in the form

$$f(x) = mx + b$$

is a **linear function.**

As mentioned earlier, this form of the equation is the slope-intercept form, where m is the slope, $(0, b)$ is the y-intercept, and $f(x)$ is just another name for y.

E X A M P L E 7   Graphing a Linear Function

Graph $f(x) = -2x + 4$. Give the domain and the range.

To graph a linear function $f(x) = mx + b$, we simply replace $f(x)$ with y and then graph the linear equation $y = mx + b$. The graph of $f(x) = -2x + 4$ is a line with slope -2 and y-intercept $(0, 4)$, as shown in Figure 32. The domain and the range are both $(-\infty, \infty)$.

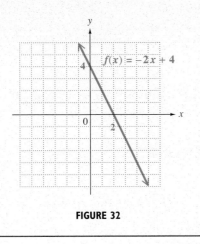

FIGURE 32

WORK PROBLEM 8 AT THE SIDE. ▶▶

---

7. Find $f(-3)$ and $f(p)$.

(a) $f(x) = 6x - 2$

(b) $f(x) = \dfrac{-3x + 5}{2}$

(c) $f(x) = \dfrac{1}{6}x^2 - 1$

8. Graph the linear function $f(x) = 3x + 1$. Give the domain and the range.

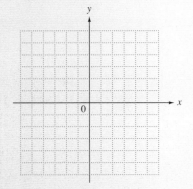

ANSWERS

7. (a) -20; $6p - 2$

  (b) 7; $\dfrac{-3p + 5}{2}$

  (c) $\dfrac{1}{2}$; $\dfrac{1}{6}p^2 - 1$

8.

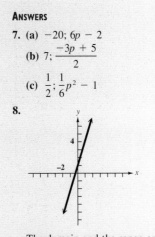

The domain and the range are both $(-\infty, \infty)$.

NUMBERS IN THE *Real World*

a graphing calculator minicourse

Lesson 2: Solution of Linear Inequalities

In Lesson 1 we observed that the *x*-intercept of the graph of the line $y = mx + b$ is the solution of the equation $mx + b = 0$. We can extend our observations to consider the solution of the associated inequalities $mx + b > 0$ and $mx + b < 0$. The solution set of $mx + b > 0$ is the set of all *x*-values for which the graph of $y = mx + b$ is *above* the *x*-axis. (We consider points above because the symbol is >.) On the other hand, the solution set of $mx + b < 0$ is the set of all *x*-values for which the graph of $y = mx + b$ is *below* the *x*-axis. (We consider points below because the symbol is <.) Therefore, once we know the solution set of the equation and have the graph of the line, we can determine the solution sets of the corresponding inequalities.

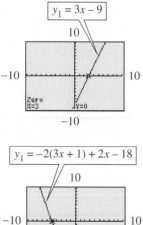

In the first figure to the left, we see that the *x*-intercept of $y_1 = 3x - 9$ is 3. Therefore, by the concepts of Lesson 1, the solution set of $3x - 9 = 0$ is {3}. Because the graph of y_1 lies above the *x*-axis for *x*-values greater than 3, the solution set of $3x - 9 > 0$ is $(3, \infty)$. Because the graph lies below the *x*-axis for *x*-values less than 3, the solution set of $3x - 9 < 0$ is $(-\infty, 3)$.

Suppose that we wish to solve the equation $-2(3x + 1) = -2x + 18$, and the associated inequalities $-2(3x + 1) > -2x + 18$ and $-2(3x + 1) < -2x + 18$. We begin by considering the equation rewritten so that the right side is equal to 0: $-2(3x + 1) + 2x - 18 = 0$. Graphing $y_1 = -2(3x + 1) + 2x - 18$ yields the *x*-intercept -5, as shown in the second figure. The first inequality listed is equivalent to $y_1 > 0$. Because the line lies *above* the *x*-axis for *x*-values less than -5, the solution set of $-2(3x + 1) > -2x + 18$ is $(-\infty, -5)$. Because the line lies *below* the *x*-axis for *x*-values greater than -5, the solution set of $-2(3x + 1) < -2x + 18$ is $(-5, \infty)$.

GRAPHING CALCULATOR EXPLORATIONS

1. Refer to the first figure in Lesson 1 on page 234. Use the graph to solve each inequality:
   (a) $-4x + 7 > 0$ (b) $-4x + 7 < 0$.

2. The graph below is that of $y_1 = 3(5 - 2x)$. Use the graph and the display to give the solution set of each of the following:
   (a) $3(5 - 2x) = 0$ (b) $3(5 - 2x) > 0$ (c) $3(5 - 2x) \leq 0$.

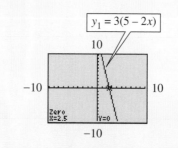

246

4.5 Exercises

For the various methods of representing a function shown in Exercises 1–3, give the three ordered pairs with the first entries 1990, 1992, and 1993.

1. Commissioned Officers in the U.S. Army

Year	Number
1990	746,220
1992	661,391
1993	590,324
1994	553,627
1995	521,036

Sources: Department of the Army, U.S. Dept. of Defense

2. $y = 18x - 35{,}753$ gives an approximation for Medicare costs, in billions of dollars, where x is the year and y is the cost.

3. PERSONAL CONSUMPTION EXPENDITURES IN THE UNITED STATES

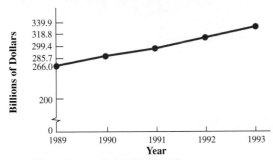

Sources: Bureau of Economic Analysis; U.S. Dept. of Commerce

4. Explain what is meant by each of the following terms.
   (a) relation   **(c)** range of a relation
   (b) domain of a relation   **(d)** function

5. Describe the use of the vertical line test.

6. Give an example of a relation that is not a function, having domain $\{-3, 2, 6\}$ and range $\{4, 6\}$. (There are many possible correct answers.)

Decide whether the relation is a function, and give the domain and the range of the relation.
Use the vertical line test in Exercises 13–16. See Examples 1–4.

7. {(5, 1), (3, 2), (4, 9), (7, 3)}

8. {(8, 0), (5, 4), (9, 3), (3, 9)}

9. {(2, 4), (0, 2), (2, 6)}

10. {(9, −2), (−3, 5), (9, 1)}

11.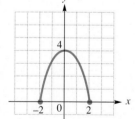

12.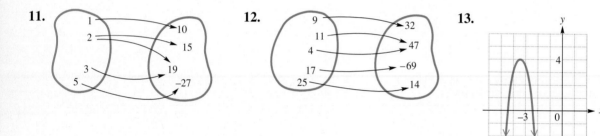

13.

14.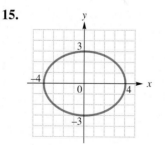

15.

16.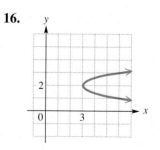

Decide whether the given equation defines y as a function of x. Give the domain. Identify any linear functions. See Example 5.

17. $y = x^2$ **18.** $y = x^3$ **19.** $x = y^6$ **20.** $x = y^4$

21. $x + y < 4$ **22.** $x - y < 3$ **23.** $y = \sqrt{x}$ **24.** $y = -\sqrt{x}$

25. $xy = 1$ **26.** $xy = -3$ **27.** $y = 2x - 6$ **28.** $y = -6x + 8$

29. $y = \sqrt{4x + 2}$ **30.** $y = \sqrt{9 - 2x}$ **31.** $y = \dfrac{2}{x - 9}$ **32.** $y = \dfrac{-7}{x - 16}$

Let f(x) = -3x + 4 and g(x) = -x² + 4x + 1. Find each of the following. See Example 6.

33. $f(0)$ **34.** $f(-3)$ **35.** $g(-2)$ **36.** $g(10)$

37. $f(p)$ **38.** $g(k)$ **39.** $f(-x)$ **40.** $g(-x)$

41. $f(x + 2)$ **42.** $g(2p)$ **43.** $f(g(1))$ **44.** $g(f(1))$

45. Compare the answers to Exercises 43 and 44. Do you think that $f(g(x))$ is, in general, equal to $g(f(x))$?

46. Make up two linear functions f and g such that $f(g(2)) = 4$. (There are many ways to do this.)

47. Fill in the blanks with the correct responses. The equation $2x + y = 4$ has a straight _____ as its graph. One point that lies on the graph is (3, _____). If we solve the equation for y and use function notation, we have a _____ function $f(x) = $ _____. For this function, $f(3) = $ _____, meaning that the point (_____, _____) lies on the graph of the function.

48. Which one of the following defines a linear function?

(a) $y = \dfrac{x - 5}{4}$ (b) $y = \sqrt[3]{x}$

(c) $y = x^2$ (d) $y = \sqrt{x}$

Sketch the graph of the linear function. Give the domain and range. See Example 7.

49. $f(x) = -2x + 5$ **50.** $g(x) = 4x - 1$ **51.** $h(x) = \dfrac{1}{2}x + 2$

52. $F(x) = -\dfrac{1}{4}x + 1$

53. $G(x) = 2$

54. $H(x) = -3x$

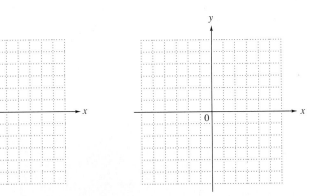

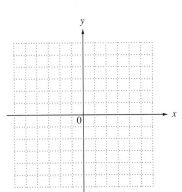

55. The linear function $f(x) = -123x + 29{,}685$ provides a model for the number of post offices in the United States from 1984 to 1990, where $x = 0$ corresponds to 1984, $x = 1$ corresponds to 1985, and so on. Use this model to give the approximate number of post offices during the given year. (*Source:* U.S. Postal Service, Annual Report of the Postmaster General)
  (a) 1985 **(b)** 1987 **(c)** 1990

56. The linear function $f(x) = 1650x + 3817$ provides a model for the United States defense budget for the decade of the 1980s, where $x = 0$ corresponds to 1980, $x = 1$ corresponds to 1981, and so on, with $f(x)$ representing the budget in millions of dollars. Use this model to approximate the defense budget during the given year. (*Source:* U.S. Office of Management and Budget)
  (a) 1983 **(b)** 1985 **(c)** 1988

INTERPRETING TECHNOLOGY (EXERCISES 57–62)

---

57. Refer to the linear function in Exercise 55. The graphing calculator screen shows a portion of the graph of $y = f(x)$ with the coordinates of a point on the graph displayed at the bottom of the screen. Interpret the meaning of the display in the context of Exercise 55.

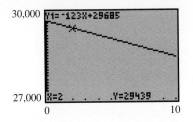

58. Refer to the linear function in Exercise 56. The graphing calculator screen shows a portion of the graph of $y = f(x)$ with the coordinates of a point on the graph displayed at the bottom of the screen. Interpret the meaning of the display in the context of Exercise 56.

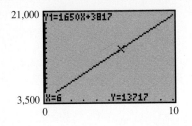

59. The graphing calculator screen shows the graph of a linear function $y = f(x)$, along with the display of coordinates of a point on the graph. Use function notation to write what the display indicates.

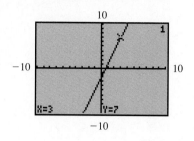

60. The table was generated by a graphing calculator for a linear function $y = f(x)$. Use the table to answer the following questions.
- **(a)** What is $f(2)$?
- **(b)** If $f(x) = -3.7$, what is the value of x?
- **(c)** What is the slope of the line?
- **(d)** What is the y-intercept of the line?
- **(e)** Find the expression for $f(x)$.

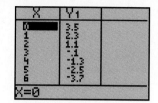

61. The two screens show the graph of the same linear function $y = f(x)$. Find the expression for $f(x)$.

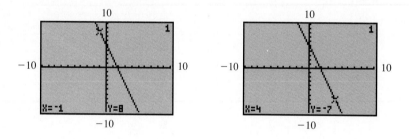

62. The formula for converting Celsius to Fahrenheit is $F = 1.8C + 32$. If we graph $y = f(x) = 1.8x + 32$ on a graphing calculator screen, we obtain the accompanying picture. (We used the interval $[-50, 50]$ on the x-axis and $[-75, 50]$ on the y-axis.) The point $(-40, -40)$ lies on the graph, as indicated by the display. Interpret the meaning of this in the context of this exercise.

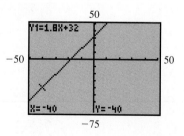

4.6 An Application of Functions: Variation

Certain types of functions are very common, especially in the physical sciences. These are functions where y depends on a multiple of x, or y depends on a number divided by x. In such situations, y is said to *vary directly* as x (in the first case) or *vary inversely* as x (in the second case). For example, the period of a pendulum varies directly as the square root of the length of the pendulum and inversely as the square root of the acceleration due to gravity. In this section we discuss several types of variation.

OBJECTIVE 1 The circumference of a circle is a function of the radius, defined by the formula $C = 2\pi r$, where r is the radius of the circle. As the formula shows, the circumference is always a constant multiple of the radius (C is always found by multiplying r by the constant 2π). Because of this, the circumference is said to *vary directly* as the radius.

Direct Variation

y varies directly as x if there exists some constant k such that

$$y = kx.$$

Also, y is said to be **proportional to** x. The number k is called the **constant of variation.** In direct variation, for $k > 0$, as the value of x increases, the value of y also increases. Similarly, as x decreases, y decreases.

The direct variation equation defines a linear function. In applications, functions are often defined by variation equations. For example, if Tom earns $8 per hour, his wages vary directly as, or are proportional to, the number of hours he works. If y represents his total wages and x the number of hours he has worked, then

$$y = 8x.$$

Here k, the constant of variation, is 8. We can see that the constant of variation is the slope of the line $y = 8x$. The graph of a function that is an example of a direct variation will always be a line through the origin.

OBJECTIVE 2 The following examples show how to find the value of the constant k.

EXAMPLE 1 Finding the Constant of Variation

Steven is paid an hourly wage. One week he worked 43 hours and was paid $795.50. Find the constant of variation and the direct variation equation.

Let h represent the number of hours he works and T represent his corresponding pay. Then, T varies directly as h, and

$$T = kh.$$

Here k represents Steven's hourly wage. Since $T = 795.50$ when $h = 43$,

$$795.50 = 43k$$

$$k = 18.50. \quad \text{Use a calculator.}$$

His hourly wage is $18.50. Thus, T and h are related by

$$T = 18.50h.$$

OBJECTIVES

1 ▶ Write an equation expressing direct variation.

2 ▶ Find the constant of variation, and solve direct variation problems.

3 ▶ Solve inverse variation problems.

4 ▶ Solve joint variation problems.

5 ▶ Solve combined variation problems.

FOR EXTRA HELP

Tutorial   Tape 8   SSM, Sec. 4.6

1. Vicki is paid a daily wage. One month she worked 17 days and earned $1334.50. Find the constant of variation and write a direct variation equation.

EXAMPLE 2 Finding the Constant of Variation

Suppose y varies directly as z, and $y = 50$ when $z = 100$. Find k and the equation relating y and z.

Because y varies directly as z,

$$y = kz,$$

for some constant k. We know that $y = 50$ when $z = 100$. Substituting these values into the equation $y = kz$ gives

$$y = kz$$

$$50 = k \cdot \mathbf{100}. \qquad \text{Let } y = 50 \text{ and } z = 100.$$

Now solve for k.

$$k = \frac{50}{100} = \frac{1}{2}$$

The variables y and z are related by the equation

$$y = \frac{1}{2}z.$$

2. In Example 2, find y for each of the following values of z.

   (a) $z = 80$

   (b) $z = 6$

◀◀ **WORK PROBLEMS I AND 2 AT THE SIDE.**

EXAMPLE 3 Solving a Direct Variation Problem

Power consumption is measured in kilowatt-hours (kwh). The charge to customers varies directly as the number of hours of consumption. If it costs $76.50 to use 850 kwh, how much will 1000 kwh cost?

If c represents the cost and h is the number of kilowatt-hours, then $c = kh$ for some constant k. Because 850 kwh cost $76.50, let $c = 76.50$ and $h = 850$ in the equation $c = kh$ to find k.

$$c = k\mathbf{h}$$

$$\mathbf{76.50} = k(\mathbf{850})$$

$$k = \frac{76.50}{850}$$

$$k = .09$$

3. It costs $52 to use 800 kwh of electricity. How much would the following kilowatt-hours cost?

   (a) 1000

For 1000 kwh,

$$c = (.09)(1000)$$

$$c = 90.$$

It will cost $90 to use 1000 kilowatts of power.

   (b) 650

◀◀ **WORK PROBLEM 3 AT THE SIDE.**

The direct variation equation $y = kx$ defines a linear function. However, other kinds of variation are defined by nonlinear functions. Often, one variable is directly proportional to a *power* of another variable.

Direct Variation as a Power

y varies directly as the nth power of x if there exists a real number k such that

$$y = kx^n.$$

ANSWERS

1. $k = 78.50$; Let E represent her earnings for d days. Then $E = 78.50d$.
2. (a) 40
   (b) 3
3. (a) $65 (b) $42.25

An example of direct variation as a power is the formula for the area of a circle, $A = \pi r^2$. Here, π is the constant of variation, and the area varies directly as the square of the radius.

E X A M P L E 4   Solving a Direct Variation Problem

The distance a body falls from rest varies directly as the square of the time it falls (disregarding air resistance). If an object falls 64 feet in 2 seconds, how far will it fall in 8 seconds?

If d represents the distance the object falls, and t the time it takes to fall, then d is a function of t, and

$$d = kt^2$$

for some constant k. To find the value of k, use the fact that the object falls 64 feet in 2 seconds.

$$d = kt^2 \qquad \text{Formula}$$
$$64 = k(2)^2 \qquad \text{Let } d = 64 \text{ and } t = 2.$$
$$k = 16 \qquad \text{Find } k.$$

Using 16 for k, the variation equation becomes

$$d = 16t^2.$$

Now let $t = 8$ to find the number of feet the object will fall in 8 seconds.

$$d = 16(8)^2 \qquad \text{Let } t = 8.$$
$$= 1024$$

The object will fall 1024 feet in 8 seconds.

WORK PROBLEM 4 AT THE SIDE. ▶▶

OBJECTIVE ▶ 3 Another type of variation is *inverse variation*. If $k > 0$, with inverse variation, as one variable increases, the other decreases.

Inverse Variation

y varies inversely as x if there exists a real number k such that

$$y = \frac{k}{x}.$$

Also, **y varies inversely as the nth power of x** if there exists a real number k such that

$$y = \frac{k}{x^n}.$$

Notice that the inverse variation equations also define functions. Because x is in the denominator, these functions are called *rational functions*.

---

4. The area of a circle varies directly as the square of its radius. A circle with a radius of 3 inches has an area of 28.278 square inches.

  (a) Find k and write a variation equation.

  (b) What is the area of a circle with a radius of 4.1 inches?

5. If the temperature is constant, the volume of a gas varies inversely as the pressure. For a certain gas, the volume is 10 cubic centimeters when the pressure is 6 kilograms per square centimeter.

(a) Find the variation equation.

(b) Find the volume when the pressure is 12 kilograms per square centimeter.

E X A M P L E 5 Solving an Inverse Variation Problem

The weight of an object above the earth varies inversely as the square of its distance from the center of the earth. A space vehicle in an elliptical orbit has a maximum distance from the center of the earth (apogee) of 6700 miles. Its minimum distance from the center of the earth (perigee) is 4090 miles. See Figure 33. If an astronaut in the vehicle weighs 57 pounds at its apogee, what does the astronaut weigh at its perigee?

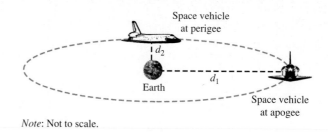

Space vehicle at perigee

d_2

Earth

d_1

Space vehicle at apogee

Note: Not to scale.

FIGURE 33

If w is the weight and d is the distance from the center of the earth, then

$$w = \frac{k}{d^2}$$

for some constant k. At the apogee the astronaut weighs 57 pounds, and the distance from the center of the earth is 6700 miles. Use these values to find k.

$$57 = \frac{k}{(6700)^2} \quad \text{Let } w = 57 \text{ and } d = 6700.$$

$$k = 57(6700)^2$$

Then the weight at the perigee with $d = 4090$ miles is

$$w = \frac{57(6700)^2}{(4090)^2} \approx 153 \text{ pounds.}$$

Note
The approximate answer in Example 5, 153 pounds, was obtained by using a calculator. A calculator will often be helpful in performing operations required in variation problems.

◀◀ WORK PROBLEM 5 AT THE SIDE.

OBJECTIVE 4▶ It is common for the value of one variable to depend on the values of several others. For example, if the value of one variable varies directly as the product of the values of several other variables (perhaps raised to powers), the first variable is said to **vary jointly** as the others. The next example illustrates joint variation.

E X A M P L E 6 Solving a Joint Variation Problem

The strength of a rectangular beam varies jointly as its width and the square of its depth. If the strength of a beam 2 inches wide and 10 inches deep is 1000 pounds per square inch, what is the strength of a beam 4 inches wide and 8 inches deep?

ANSWERS

5. **(a)** $V = \dfrac{60}{P}$ **(b)** 5 cubic centimeters

CONTINUED ON NEXT PAGE

If S represents the strength, w the width, and d the depth, then,

$$S = kwd^2$$

for some constant k. Because $S = 1000$ if $w = 2$ and $d = 10$,

$$1000 = k(2)(10)^2. \qquad \text{$S = 1000$, $w = 2$, and $d = 10$.}$$

Solving this equation for k gives

$$1000 = k \cdot 2 \cdot 100$$
$$1000 = 200k,$$

or $\qquad\qquad\qquad\qquad k = 5,$

so that $\qquad\qquad\qquad S = 5wd^2.$

Now find S when $w = 4$ and $d = 8$.

$$S = 5(4)(8)^2 \qquad \text{$w = 4$ and $d = 8$.}$$
$$= 1280$$

The strength of the beam is 1280 pounds per square inch.

OBJECTIVE 5 There are many combinations of direct and inverse variation. The final example shows a typical **combined variation** problem.

E X A M P L E   7   Solving a Combined Variation Problem

The body-mass index, or BMI, is used by physicians to assess a person's level of fatness.* The BMI varies directly as an individual's weight in pounds and inversely as the square of the individual's height in inches. A person who weighs 118 pounds and is 64 inches tall has a BMI of 20. (The BMI is rounded to the nearest whole number.) Find the BMI of a person who weighs 165 pounds with a height of 70 inches.

Let B represent the BMI, w the weight, and h the height. Then

$$B = \frac{kw}{h^2}. \quad \begin{array}{l} \leftarrow \text{ BMI varies directly as the weight.} \\ \leftarrow \text{ BMI varies inversely as the square of the height.} \end{array}$$

To find k, let $B = 20$, $w = 118$, and $h = 64$.

$$20 = \frac{k(118)}{64^2}$$

$$k = \frac{20(64^2)}{118} \quad \text{Multiply by 64^2; divide by 118.}$$

$$= 694 \qquad \text{Use a calculator.}$$

Now find B when $k = 694$, $w = 165$, and $h = 70$.

$$B = \frac{694(165)}{70^2} = 23 \quad \text{(rounded)}$$

The required BMI is 23. A BMI from 20 through 26 is considered desirable.

WORK PROBLEM 6 AT THE SIDE. ▶▶

---

* From *Reader's Digest,* October 1993.

---

6. The maximum load that a cylindrical column with a circular cross section can hold varies directly as the fourth power of the diameter of the cross section and inversely as the square of the height. A 9-meter column 1 meter in diameter will support 8 metric tons. How many metric tons can be supported by a column 12 meters high and $\frac{2}{3}$ meter in diameter?

ANSWERS

6. $\dfrac{8}{9}$ metric ton

Numbers in the *Real World*

a graphing calculator minicourse

Lesson 3: Tables

In addition to the obvious capability of graphing functions, graphing calculators are also capable of generating tables of values. For example, suppose that we are interested in finding a table of x- and y-values for the linear equation $2x + y = 6$. We begin by solving the equation for y to get $y = -2x + 6$. This expression for y is then used to define y_1. We can set up the table by directing the calculator to start (TblStart) at 0 and have an increment (ΔTbl) of 1. (See the figure at the left below.) Then the table can be viewed, as seen in the middle figure. We can scroll up or down to find other values automatically. Another option is to have the calculator ask for specific x-values; it will then return the corresponding y-value. (See the figure at the right.) The x-values -100, -45.7, π, 3.33, $\sqrt{23}$, 100, and 10^7 have been entered, and the y-values returned.

 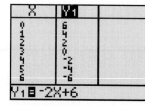 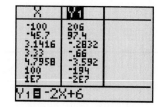

Some graphing calculator models allow the user to see the graph of a function and a table of values on the same screen. Two such screens are seen below for the function $y_1 = -2x + 6$.

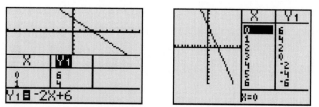

GRAPHING CALCULATOR EXPLORATIONS

1. Use a graphing calculator to generate a table of values for $y = -3x + 7$, starting at $x = 0$ and having increment 1.

2. Use a graphing calculator to generate a table of values for $4x - y = 6$, starting at $x = -5$ and having increment 2. (*Hint:* You will have to solve for y first.)

3. Use the table function of a graphing calculator to find the value of y when $x = 2$ for the equation $3x - 6y = 12$.

4.6 Exercises

Solve each of the following. See Examples 2–6.

1. If x varies directly as y, and $x = 9$ when $y = 3$, find x when $y = 12$.

2. If x varies directly as y, and $x = 10$ when $y = 7$, find y when $x = 50$.

3. If z varies inversely as w, and $z = 10$ when $w = .5$, find z when $w = 8$.

4. If t varies inversely as s, and $t = 3$ when $s = 5$, find s when $t = 5$.

5. Assume p varies jointly as q and r^2, and $p = 200$ when $q = 2$ and $r = 3$. Find p when $q = 5$ and $r = 2$.

6. Assume f varies jointly as g^2 and h, and $f = 50$ when $g = 4$ and $h = 2$. Find f when $g = 3$ and $h = 6$.

7. For $k > 0$, if y varies directly as x, when x increases, y _____ , and when x decreases, y _____ .

8. For $k > 0$, if y varies inversely as x, when x increases, y _____ , and when x decreases, y _____ .

Solve each problem involving variation. See Examples 1–7.

9. Todd bought 8 gallons of gasoline and paid $8.79. To the nearest tenth of a cent, what is the price of gasoline per gallon?

10. Melissa gives horseback rides at Shadow Mountain Ranch. A 2.5-hour ride costs $50.00. What is the price per hour?

11. The weight of an object on Earth is directly proportional to the weight of that same object on the moon. A 200-pound astronaut would weigh 32 pounds on the moon. How much would a 50-pound dog weigh on the moon?

12. In the study of electricity, the resistance of a conductor of uniform cross-sectional area is directly proportional to its length. Suppose that the resistance of a certain type of copper wire is .640 ohm per 1000 feet. What is the resistance of 2500 feet of the wire?

13. The frequency of a vibrating string varies inversely as its length. That is, a longer string vibrates fewer times in a second than a shorter string. Suppose a piano string 2 feet long vibrates 250 cycles per second. What frequency would a string 5 feet long have?

14. The current in a simple electrical circuit is inversely proportional to the resistance. If the current is 20 amperes (an *ampere* is a unit for measuring current) when the resistance is 5 ohms, find the current when the resistance is 7.5 ohms.

15. The illumination produced by a light source varies inversely as the square of the distance from the source. If the illumination produced 4 meters from a light source is 48 footcandles, find the illumination produced 16 meters from the same source.

16. The force with which the earth attracts an object above the earth's surface varies inversely with the square of the object's distance from the center of the earth. If an object 4000 miles from the center of the earth is attracted with a force of 160 pounds, find the force of attraction on an object 6000 miles from the center of the earth.

17. For a given interest rate, simple interest varies jointly as principal and time. If $2000 left in an account for 4 years earned interest of $280, how much interest would be earned in 6 years?

18. The collision impact of an automobile varies jointly as its mass and the square of its speed. Suppose a 2000-pound car traveling at 55 miles per hour has a collision impact of 6.1. What is the collision impact of the same car at 65 miles per hour?

19. The amount of water emptied by a pipe varies directly as the square of the diameter of the pipe. For a certain constant water flow, a pipe emptying into a canal will allow 200 gallons of water to escape in an hour. The diameter of the pipe is 6 inches. How much water would a 12-inch pipe empty into the canal in an hour, assuming the same water flow?

20. The number of long distance phone calls between two cities in a certain time period varies jointly as the populations of the cities, p_1 and p_2, and inversely as the distance between them. If 80,000 calls are made between two cities 400 miles apart, with populations of 70,000 and 100,000, how many calls are made between cities with populations of 50,000 and 75,000 that are 250 miles apart?

21. Ken Griffey, Jr. weighs 205 pounds and is 6 feet, 3 inches tall. Use the information given in Example 7 to find his body-mass index.

22. A body-mass index from 27 through 29 carries a slight risk of weight-related health problems, while one of 30 or more indicates a great increase in risk. Use your own height and weight and the information in Example 7 to determine whether you are at risk.

INTERPRETING TECHNOLOGY (EXERCISES 23–24)

23. The graphing calculator screen shows a portion of the graph of a function $y = f(x)$ that satisfies the conditions for direct variation. What is $f(36)$?

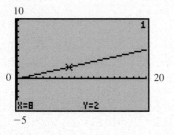

24. The accompanying table of points was generated by a graphing calculator. The points lie on the graph of a function $y_1 = f(x)$ that satisfies the conditions for direct variation. What is $f(36)$?

X	Y₁
0	0
1	1.5
2	3
3	4.5
4	6
5	7.5
6	9

X=0

MATHEMATICAL CONNECTIONS (EXERCISES 25–30)

A routine activity such as pumping gasoline can be related to many of the concepts studied in this chapter. Suppose that premium unleaded costs $1.25 per gallon. Work Exercises 25–30 in order.

25. 0 gallons of gasoline cost $0.00, while 1 gallon costs $1.25. Represent these two pieces of information as ordered pairs of the form (gallons, price).

26. Use the information from Exercise 25 to find the slope of the line on which the two points lie.

27. Write the slope-intercept form of the equation of the line on which the two points lie.

28. Using function notation, if $f(x) = ax + b$ represents the line from Exercise 27, what are the values of a and b?

29. How does the value of a from Exercise 28 relate to gasoline in this situation? With relationship to the line, what do we call this number?

30. Why does the equation from Exercise 28 satisfy the conditions for direct variation? In the context of variation, what do we call the value of a?

4.1	**line graph, bar graph, circle graph, pie chart**	These are various methods of depicting data using pictorial representations.
	ordered pair	An ordered pair is a pair of numbers written in parentheses in which the order of the numbers matters.
	origin	When two number lines intersect at a right angle, the origin is the common zero point.
	rectangular (Cartesian) coordinate system	Two number lines that intersect at a right angle at their zero points form a rectangular coordinate system, also called the Cartesian system.
	x-axis	The horizontal number line in a rectangular coordinate system is called the x-axis.
	y-axis	The vertical number line in a rectangular coordinate system is called the y-axis.
	plot	To plot an ordered pair is to locate it on a rectangular coordinate system.
	coordinates	The numbers in an ordered pair are called the coordinates of the corresponding point.
	quadrant	A quadrant is one of the four regions in the plane determined by a rectangular coordinate system.
	linear equation in two variables	A first-degree equation with two variables is a linear equation in two variables.
	standard form	A linear equation is in standard form when written as $Ax + By = C$, with $A \geq 0$, and A, B, and C integers.
	x-intercept	The point where a line crosses the x-axis is the x-intercept.
	y-intercept	The point where a line crosses the y-axis is the y-intercept.
4.2	**slope**	The ratio of the change in y compared to the change in x along a line is the slope of the line.
4.4	**linear inequality in two variables**	A first-degree inequality with two variables is a linear inequality in two variables.
	boundary line	In the graph of a linear inequality, the boundary line separates the region that satisfies the inequality from the region that does not satisfy the inequality.
4.5	**dependent variable**	If the quantity y depends on x, then y is called the dependent variable in an equation relating x and y.
	independent variable	If y depends on x, then x is the independent variable in an equation relating x and y.
	relation	A relation is a set of ordered pairs of real numbers.
	function	A function is a set of ordered pairs in which each value of the first component, x, corresponds to exactly one value of the second component, y.
	domain	The domain of a relation is the set of first components (x-values) of the ordered pairs of the relation.
	range	The range of a relation is the set of second components (y-values) of the ordered pairs of the relation.

graph of a relation	The graph of a relation is the graph of the ordered pairs of the relation.
vertical line test	The vertical line test says that if a vertical line cuts the graph of a relation in more than one point, then the relation is not a function.
linear function	A function that can be written in the form $f(x) = mx + b$ is a linear function.

New Symbols

(a, b)	ordered pair
x_1	a specific value of the variable x (read "x sub one")
m	slope
$f(x)$	function of x (read "f of x")

Quick Review

Concepts	*Examples*
4.1 The Rectangular Coordinate System	
Finding Intercepts To find the x-intercept, let $y = 0$. To find the y-intercept, let $x = 0$.	The graph of $2x + 3y = 12$ has x-intercept $(6, 0)$ and y-intercept $(0, 4)$.
4.2 The Slope of a Line	
Slope $$m = \frac{\text{rise}}{\text{run}} = \frac{\text{change in } y}{\text{change in } x} = \frac{y_2 - y_1}{x_2 - x_1}$$	For $2x + 3y = 12$, $$m = \frac{4 - 0}{0 - 6} = -\frac{2}{3}.$$
A vertical line has undefined slope.	$x = 3$ has undefined slope.
A horizontal line has 0 slope.	$y = -5$ has $m = 0$.
Distinct parallel lines have equal slopes.	$\begin{array}{ll} y = 2x + 5 & 4x - 2y = 6 \\ m = 2 & m = 2 \end{array}$ Lines are **parallel**.
The slopes of perpendicular lines are negative reciprocals with a product of -1.	$\begin{array}{ll} y = 3x - 1 & x + 3y = 4 \\ m = 3 & m = -\dfrac{1}{3} \end{array}$ Lines are **perpendicular**.

Concepts	Examples
4.3 Linear Equations in Two Variables	
Standard form $Ax + By = C$	$2x - 5y = 8$
Vertical line $x = k$	$x = -1$
Horizontal line $y = k$	$y = 4$
Slope-intercept form $$y = mx + b$$	$y = 2x + 3$ $m = 2$, y-intercept is $(0, 3)$.
Point-slope form $$y - y_1 = m(x - x_1)$$	$y - 3 = 4(x - 5)$ $(5, 3)$ is on the line, $m = 4$.

| **4.4 Linear Inequalities in Two Variables** | |
| **Graphing a Linear Inequality** | |

Step 1   Draw the graph of the line that is the boundary. Make the line solid if the inequality involves ≤ or ≥; make the line dashed if the inequality involves < or >.

Step 2   Choose any point not on the line as a test point.

Step 3   Shade the region that includes the test point if the test point satisfies the original inequality; otherwise, shade the region on the other side of the boundary line.

Graph $2x - 3y \le 6$.
Draw the graph of $2x - 3y = 6$. Use a solid line because the symbol ≤ is used.

Choose $(1, 2)$.
$$2(1) - 3(2) = 2 - 6 \le 6 \quad \text{True}$$
Shade the side of the line that includes $(1, 2)$.

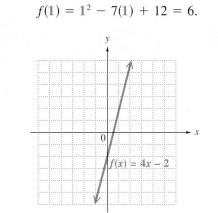

4.5 An Introduction to Functions

To evaluate a function using function notation (that is, $f(x)$ notation) for a given value of x, substitute the value wherever x appears.

If $f(x) = x^2 - 7x + 12$, then
$$f(1) = 1^2 - 7(1) + 12 = 6.$$

To graph the linear function $f(x) = mx + b$, replace $f(x)$ with y and graph $y = mx + b$, as in Section 4.2.

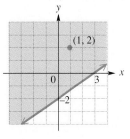

Concepts	Examples
4.6 An Application of Functions: Variation If there is a constant of variation k such that: $y = kx^n$, then y varies directly as, or is proportional to, x^n; $y = \dfrac{k}{x^n}$, then y varies inversely as x^n.	Area of a circle **varies directly as** the square of the radius. $$A = kr^2$$ Pressure **varies inversely as** volume. $$P = \dfrac{k}{V}$$

CHAPTER 4 REVIEW EXERCISES

[4.1] *Complete the given ordered pairs for each equation, and then graph the equation.*

1. $3x + 2y = 6$
$(0, \underline{}), (\underline{}, 0), (2, \underline{}), (\underline{}, -2)$

2. $x - y = 6$
$(2, \underline{}), (\underline{}, -3), (1, \underline{}), (\underline{}, -2)$

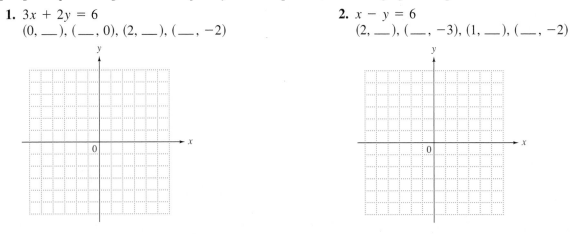

Find the x- and y-intercepts, and graph each of the following equations.

3. $4x + 3y = 12$

4. $5x + 7y = 15$

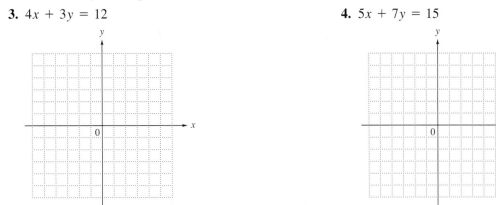

[4.2] *Find the slope for each line in Exercises 5–10.*

5. Through $(-1, 2)$ and $(4, -6)$

6. $y = 2x + 3$

7. $-3x + 4y = 5$

8. $y = 4$

9. A line parallel to $3y = -2x + 5$

10. A line perpendicular to $3x - y = 6$

Tell whether the line has positive, negative, zero, or undefined slope.

11.   **12.**   **13.**   **14.**

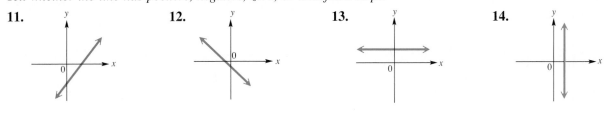

[4.3] *Write an equation for each line. In Exercises 21–24, write the equations in standard form.*

15. Slope $\dfrac{3}{5}$; y-intercept $(0, -8)$ **16.** Slope $-\dfrac{1}{3}$; y-intercept $(0, 5)$ **17.** Slope 0; y-intercept $(0, 12)$

18. Undefined slope; through $(2, 7)$ **19.** Horizontal; through $(-1, 4)$ **20.** Vertical; through $(.3, .6)$

21. Through $(2, -5)$ and $(1, 4)$ **22.** Through $(-3, -1)$ and $(2, 6)$

23. Parallel to $4x - y = 3$ and through $(6, -2)$ **24.** Perpendicular to $2x - 5y = 7$ and through $(0, 1)$

Solve each problem.

25. The national average for family health care cost in dollars between 1980 and 2000 (projected) can be approximated by the linear equation

$$y = 382.75x + 1742$$

where $x = 0$ corresponds to 1980 and $x = 20$ corresponds to the year 2000.
(a) What would be the national average for family health care cost in 1999 according to this model?
(b) In what year was the cost $3273?
(c) Graph this linear equation model.

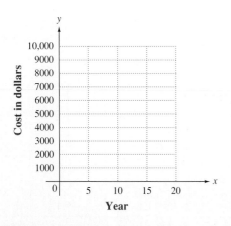

26. For the years 1980 through 1992, the average winning speed in miles per hour in the Indianapolis 500 can be approximated by the linear equation

$$y = 2.503x + 198.729,$$

where $x = 0$ corresponds to 1980, $x = 12$ corresponds to 1992, and y is in miles per hour.
(a) Based on this model, what would have been Al Unser's winning speed in 1987?
(b) Unser's actual winning speed in 1987 was 215.390 miles per hour. By how much does your answer in part (a) differ from this? Why do you think there is a discrepancy?

[4.4] *Graph the solution of each inequality.*

27. $3x - 2y \leq 12$ **28.** $5x - y > 6$ **29.** $x \geq 2$

30. $2x + y \leq 1$ and $x \geq 2y$ **31.** $x - 2y < 4$ or $x + y < 3$ **32.** $|x - 1| < 4$

[4.5] *Give the domain and range of each relation. Identify any functions.*

33. $\{(-4, 2), (-4, -2),$
    $(1, 5), (1, -5)\}$

34.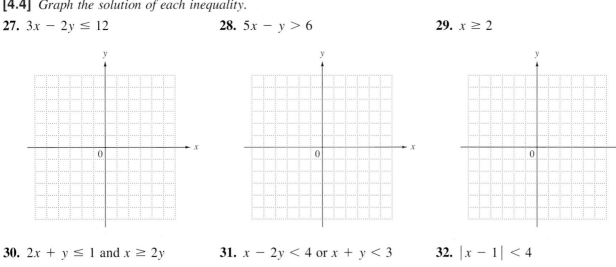

35.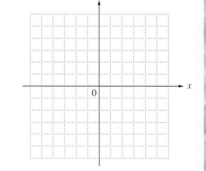

Given $f(x) = -2x^2 + 3x - 6,$ *find each of the following.*

36. $f(0)$ **37.** $f(3)$ **38.** $f[f(0)]$ **39.** $f(2p)$

Determine whether the equation defines y as a function of x. Identify any linear functions. Give the domain in each case.

40. $y = 3x - 3$ **41.** $y < x + 2$ **42.** $y = |x - 4|$

43. $y = \sqrt{4x + 7}$ **44.** $x = y^2$ **45.** $y = \dfrac{7}{x - 36}$

46. Graph the linear function $f(x) = -\dfrac{3}{2}x + \dfrac{7}{2}$.

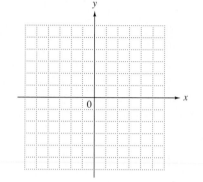

[4.6] *Solve each problem involving variation.*

47. m varies directly as p^2 and inversely as q, and $m = 32$ when $p = 8$ and $q = 10$. Find q when $p = 12$ and $m = 48$.

48. x varies jointly as y and z and inversely as $\sqrt{w}$. If $x = 12$ when $y = 3$, $z = 8$, and $w = 36$, find y when $x = 12$, $z = 4$, and $w = 25$.

49. The resistance in ohms of a platinum wire temperature sensor varies directly as the temperature in *degrees Kelvin* (°K). If the resistance is .646 ohm at a temperature of 190°K, find the resistance at a temperature of 250°K.

50. For the subject in a photograph to appear in the same perspective in the photograph as in real life, the viewing distance must be properly related to the amount of enlargement. For a particular camera, the viewing distance varies directly as the amount of enlargement. A picture taken with this camera that is enlarged 5 times should be viewed from a distance of 250 millimeters. Suppose a print 8.6 times the size of the negative is made. From what distance should it be viewed?

51. A meteorite approaching the earth has velocity inversely proportional to the square root of its distance from the center of the earth. If the velocity is 5 kilometers per second when the distance is 8100 kilometers from the center of the earth, find the velocity at a distance of 6400 kilometers.

52. The period of a pendulum varies directly as the square root of the length of the pendulum and inversely as the square root of the acceleration due to gravity. Find the period when the length is 4 feet and the acceleration due to gravity is 32 feet per second per second, if the period is 1.06π seconds when the length is 9 feet and the acceleration due to gravity is 32 feet per second per second.

CHAPTER 4 TEST

1. Find the slope of the line through $(6, 4)$ and $(-4, -1)$.

1. _____

For each line, find the slope and the x- and y-intercepts.

2. $3x - 2y = 13$

2. _____

3. $y = 5$

3. _____

4. Describe the graph of a line with undefined slope in a rectangular coordinate system.

4. _____

Write the equation of each line in standard form.

5. Through $(-3, 14)$; horizontal

5. _____

6. Through $(4, -1)$; $m = -5$

6. _____

7. Through $(-7, 2)$;
   (a) parallel to $3x + 5y = 6$;
   (b) perpendicular to $y = 2x$

7. **(a)** _____

   (b) _____

Graph each of the following.

8. $4x - 3y = -12$

8.

9. $y - 2 = 0$

9.

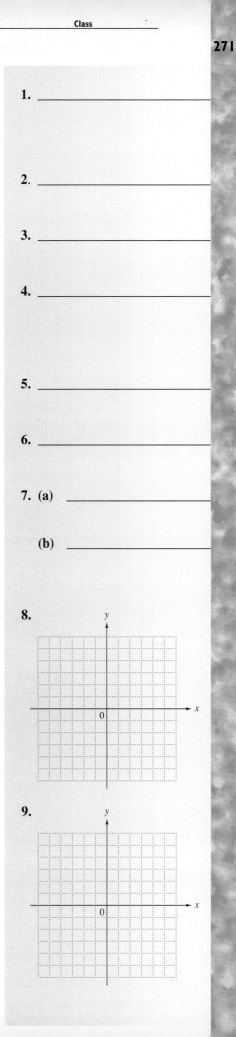

10.

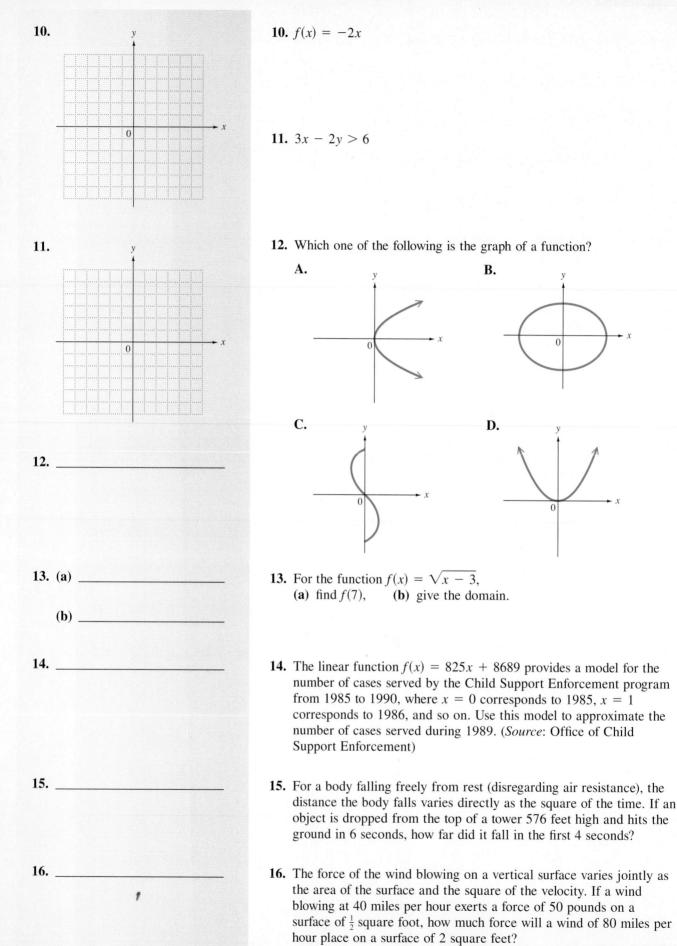

11.

12. _____

13. (a) _____

(b) _____

14. _____

15. _____

16. _____

10. $f(x) = -2x$

11. $3x - 2y > 6$

12. Which one of the following is the graph of a function?

A.

B.

C.

D.

13. For the function $f(x) = \sqrt{x - 3}$,
(a) find $f(7)$,   **(b)** give the domain.

14. The linear function $f(x) = 825x + 8689$ provides a model for the number of cases served by the Child Support Enforcement program from 1985 to 1990, where $x = 0$ corresponds to 1985, $x = 1$ corresponds to 1986, and so on. Use this model to approximate the number of cases served during 1989. (*Source*: Office of Child Support Enforcement)

15. For a body falling freely from rest (disregarding air resistance), the distance the body falls varies directly as the square of the time. If an object is dropped from the top of a tower 576 feet high and hits the ground in 6 seconds, how far did it fall in the first 4 seconds?

16. The force of the wind blowing on a vertical surface varies jointly as the area of the surface and the square of the velocity. If a wind blowing at 40 miles per hour exerts a force of 50 pounds on a surface of $\frac{1}{2}$ square foot, how much force will a wind of 80 miles per hour place on a surface of 2 square feet?

CUMULATIVE REVIEW EXERCISES CHAPTERS 1–4

Decide which of the following are true.

1. $5 \cdot 6 \geq |32 - 20|$ **2.** $5 - |-4| \leq 9$ **3.** $-4(4 - 8) \geq |-20|$

Perform each operation.

4. $-|-2| - 4 + |-3| + 7$ **5.** $(-.8)^2$ **6.** $(-3) - (-1)^2$ **7.** $-\dfrac{2}{3}\left(-\dfrac{12}{5}\right)$

Use the properties of real numbers to simplify.

8. $-2(m - 3)$ **9.** $-(-4m + 3)$ **10.** $3x^2 - 4x + 4 + 9x - x^2$

Write in interval notation.

11. $\{x \mid x > 2\}$ **12.** $\{x \mid x \leq 1\}$ **13.** $\{x \mid -3 < x \leq 5\}$

14. Is $\sqrt{\dfrac{-2 + 4}{-5}}$ a real number?

Evaluate if $p = -4$, $q = -2$, and $r = 5$.

15. $-3(2q - 3p)$ **16.** $8r^2 + q^2$ **17.** $|p|^3 - |q^3|$

18. $\dfrac{\sqrt{r}}{-p + 2q}$ **19.** $\dfrac{5p + 6r^2}{p^2 + q - 1}$

Solve.

20. $2z - 5 + 3z = 4 - (z + 2)$ **21.** $\dfrac{3a - 1}{5} + \dfrac{a + 2}{2} = -\dfrac{3}{10}$ **22.** $-\dfrac{4}{3}d \geq -5$

23. $3 - 2(m + 3) < 4m$ **24.** $2k + 4 < 10$ and $3k - 1 > 5$ **25.** $2k + 4 > 10$ or $3k - 1 < 5$

26. $|5x + 3| = 13$ **27.** $|x + 2| < 9$ **28.** $|2y - 5| \geq 9$

29. Twelve more than twice a number is equal to four times the number. Find the number.

30. Ms. Bell must take at least 30 units of a certain medication each day. She can get the medication from white pills or yellow pills, each of which contains 3 units of the drug. To provide other benefits, she needs to take twice as many of the yellow pills as white pills. Find the smallest number of white pills that will satisfy these requirements.

31. Complete the ordered pairs (0,), (, 0), and (2,) for the equation $3x - 4y = 12$.

32. Graph $-4x + 2y = 8$ and give the intercepts.

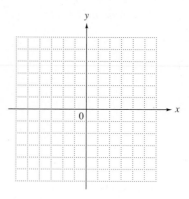

Find the slope of each line described.

33. Through $(-5, 8)$ and $(-1, 2)$

34. Perpendicular to $4x - 3y = 12$

In Exercises 35–37, write an equation in standard form for each line.

35. Slope $-\dfrac{3}{4}$; y-intercept $(0, -1)$

36. Horizontal; through $(2, -2)$

37. Through $(4, -3)$ and $(1, 1)$

38. For the function $f(x) = -4x + 10$,
(a) what is the domain?
(b) what is $f(-3)$?

Use the graph to answer the questions in Exercises 39 and 40.

39. What is the slope of the line segment joining the points for 1988 and 1991?

40. Which one of the two line segments shown has a greater slope?

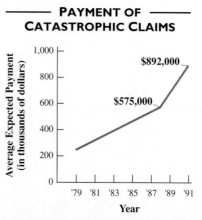

Systems of Linear Equations

5.1 Solving Systems of Linear Equations by Graphing

When a number of equations in several variables are considered simultaneously, we have what is known as a system of equations. A **system** (SISS-tem) **of linear equations** consists of two or more linear equations with the same variables. Examples of systems of two linear equations with two variables are shown below.

$$2x + 3y = 4 \qquad x + 3y = 1 \qquad x - y = 1$$
$$3x - y = -5 \qquad -y = 4 - 2x \qquad y = 3$$

In the system on the right, think of $y = 3$ as an equation in two variables by writing it as $0x + y = 3$.

OBJECTIVE ▶ Applications often require solving a system of equations. The **solution of a system** of linear equations includes all the ordered pairs that make both equations true at the same time.

www.mathnotes.com

OBJECTIVES

▶ Decide whether a given ordered pair is a solution of a system.

▶ Solve linear systems by graphing.

▶ Identify systems with no solutions or with an infinite number of solutions.

FOR EXTRA HELP

Tutorial   Tape 9   SSM, Sec. 5.1

─ **E X A M P L E I**   **Determining Whether an Ordered Pair Is a Solution**

Is $(4, -3)$ a solution of the following systems?

(a) $x + 4y = -8$
$3x + 2y = 6$

Decide whether or not $(4, -3)$ is a solution of the system by substituting 4 for x and -3 for y in each equation.

$x + 4y = -8$	$3x + 2y = 6$
$4 + 4(-3) = -8$?	$3(4) + 2(-3) = 6$?
$4 + (-12) = -8$? Multiply.	$12 + (-6) = 6$? Multiply.
$-8 = -8$ True	$6 = 6$ True

Because $(4, -3)$ satisfies both equations, it is a solution of the system.

(b) $2x + 5y = -7$
$3x + 4y = 2$

Again, substitute 4 for x and -3 for y in both equations.

$2x + 5y = -7$	$3x + 4y = 2$
$2(4) + 5(-3) = -7$?	$3(4) + 4(-3) = 2$?
$8 + (-15) = -7$? Multiply.	$12 + (-12) = 2$? Multiply.
$-7 = -7$ True	$0 = 2$ False

The ordered pair $(4, -3)$ is not a solution because it does not satisfy the second equation.

1. Decide whether the given ordered pair is a solution of the system.

(a) $(2, 5)$
$$3x - 2y = -4$$
$$5x + y = 15$$

◀◀ **WORK PROBLEM I AT THE SIDE.**

OBJECTIVE ▶ Several methods of solving a system of two linear equations in two variables are discussed in this chapter. The **solution set of a system** is the set of all ordered pairs that satisfy the system. One way to find the solution set of a system of two linear equations is to graph both equations on the same axes. The graph of each line shows points whose coordinates satisfy the equation of that line. The coordinates of any point where the lines intersect give a solution of the system. Because two different straight lines can intersect at no more than one point, there can never be more than one solution for such a system.

E X A M P L E 2 Solving Systems by Graphing

Solve each system of equations by graphing both equations on the same axes.

(a) $2x + 3y = 4$
$$3x - y = -5$$

As shown in Chapter 4, we can graph these two equations by plotting points for each line.

$2x + 3y = 4$			$3x - y = -5$	
x	y		x	y
0	$\frac{4}{3}$		0	5
2	0		$-\frac{5}{3}$	0
-2	$\frac{8}{3}$		-2	-1

The lines in Figure 1 suggest that the graphs intersect at the point $(-1, 2)$. Check this by substituting -1 for x and 2 for y in both equations. Because $(-1, 2)$ satisfies both equations, the solution set of this system is $\{(-1, 2)\}$.

(b) $2x + y = 0$
$$4x - 3y = 10$$

Find the solution set of the system by graphing the two lines on the same axes. As suggested by Figure 2, the solution set is $\{(1, -2)\}$, the point at which the graphs of the two lines intersect. Check by substituting 1 for x and -2 for y in both equations of the system.

(b) $(1, -2)$
$$x - 3y = 7$$
$$4x + y = 5$$

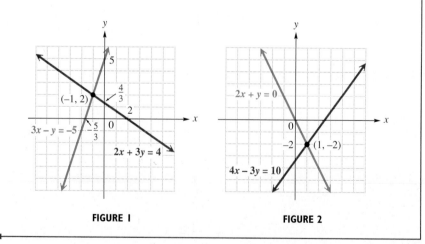

FIGURE I

FIGURE 2

Caution
A difficulty with the graphing method of solution is that it may not be possible to determine from the graph the exact coordinates of the point that represents the solution. For this reason, algebraic methods of solution are explained later in this chapter. The graphing method does, however, show geometrically how solutions are found.

WORK PROBLEM 2 AT THE SIDE. ▶▶

OBJECTIVE ▶ Sometimes the graphs of the two equations in a system either do not intersect at all or are the same line, as in the systems in Example 3.

E X A M P L E 3 **Solving Special Systems**

Solve each system by graphing.

(a) $2x + y = 2$
   $2x + y = 8$

The graphs of these lines are shown in Figure 3. The two lines are parallel with equal slopes and have no points in common. For a system whose equations lead to graphs with no points in common, we will write $\emptyset$, to indicate that the solution set has no elements.

(b) $2x + 5y = 1$
   $6x + 15y = 3$

The graphs of these two equations are the same line. See Figure 4. Here the second equation can be obtained by multiplying each side of the first equation by 3. In this case, every point on the line is a solution of the system, and the solution is an infinite number of ordered pairs. We will write "infinite number of solutions" to indicate this situation.

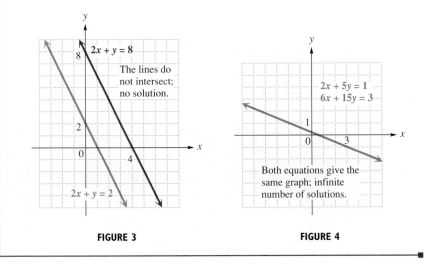

FIGURE 3   **FIGURE 4**

Each system in Example 2 has a solution. A system with a solution is called a **consistent** (kuhn-SISS-tent) **system.** A system of equations with no solutions, such as the one in Example 3(a), is called an **inconsistent** (IN-kuhn-siss-tent) **system.** The equations in Example 2 are independent equations. **Independent** (IN-di-pen-dent) **equations** have different graphs. The equations of the system in Example 3(b) have the same graph. Because they are different forms of the same equation, these equations are called **dependent** (DEE-pen-dent or di-PEN-dent) **equations.**

---

2. Solve each system of equations by graphing both equations on the same axes. Check your answers.

(a) $5x - 3y = 9$
   $x + 2y = 7$
(One of the lines is already graphed.)

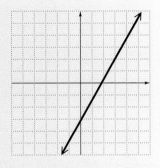

(b) $x + y = 4$
   $2x - y = -1$

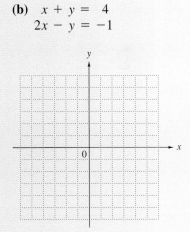

3. Solve each system of equations by graphing both equations on the same axes.

(a) $3x - y = 4$
$6x - 2y = 12$
(One of the lines is already graphed.)

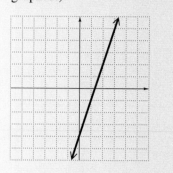

(b) $-x + 3y = 2$
$2x - 6y = -4$

◀◀ **WORK PROBLEM 3 AT THE SIDE.**

Examples 2 and 3 show the three cases that may occur in a system of two equations with two unknowns.

1. The graphs intersect at exactly one point, which gives the (single) solution of the system. The system is consistent, and the equations are independent.

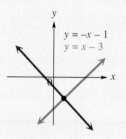

2. The graphs are parallel lines, so there is no solution and the solution set is ∅. The system is inconsistent.

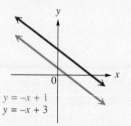

3. The graphs are the same line. The solution set is an infinite set of ordered pairs. The equations are dependent.

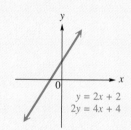

ANSWERS

3. (a) ∅
  (b) infinite number of solutions

5.1 *Exercises*

Fill in the blanks with the correct responses.

1. When two or more linear equations with the same variables are considered together, we have a(n) _____ of linear equations.

2. Any ordered pair that makes both equations true in a system is called a(n) _____ of the system.

3. A system of equations that has a solution is called a(n) _____ system.

4. A system of equations with no solutions is called a(n) _____ system.

Decide whether the given ordered pair is a solution of the given system. See Example 1.

5. $(2, -3)$
$$x + y = -1$$
$$2x + 5y = 19$$

6. $(4, 3)$
$$x + 2y = 10$$
$$3x + 5y = 3$$

7. $(-1, -3)$
$$3x + 5y = -18$$
$$4x + 2y = -10$$

8. $(-9, -2)$
$$2x - 5y = -8$$
$$3x + 6y = -39$$

9. $(7, -2)$
$$4x = 26 - y$$
$$3x = 29 + 4y$$

10. $(9, 1)$
$$2x = 23 - 5y$$
$$3x = 24 + 3y$$

11. $(6, -8)$
$$-2y = x + 10$$
$$3y = 2x + 30$$

12. $(-5, 2)$
$$5y = 3x + 20$$
$$3y = -2x - 4$$

13. Which one of the ordered pairs below could possibly be a solution of the system graphed? Why is it the only valid choice?
(a) $(2, 2)$
(b) $(-2, 2)$
(c) $(-2, -2)$
(d) $(2, -2)$

14. Which one of the ordered pairs below could possibly be a solution of the system graphed? Why is it the only valid choice?
(a) $(2, 0)$
(b) $(0, 2)$
(c) $(-2, 0)$
(d) $(0, -2)$

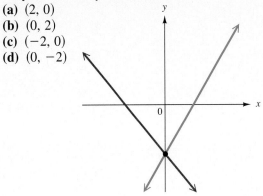

Solve each system of equations by graphing both equations on the same axes. See Example 2.

15. $x - y = 2$
$x + y = 6$

16. $x - y = 3$
$x + y = -1$

17. $x + y = 4$
$y - x = 4$

18. $x + y = -5$
$x - y = 5$

19. $x - 2y = 6$
$x + 2y = 2$

20. $2x - y = 4$
$4x + y = 2$

21. $3x - 2y = -3$
$-3x - y = -6$

22. $2x - y = 4$
$2x + 3y = 12$

23. $2x - 3y = -6$
$y = -3x + 2$

24. $-3x + y = -3$
$y = x - 3$

25. $3x - 4y = 24$
$y = -\dfrac{3}{2}x + 3$

26. $3x - 2y = 12$
$y = -4x + 5$

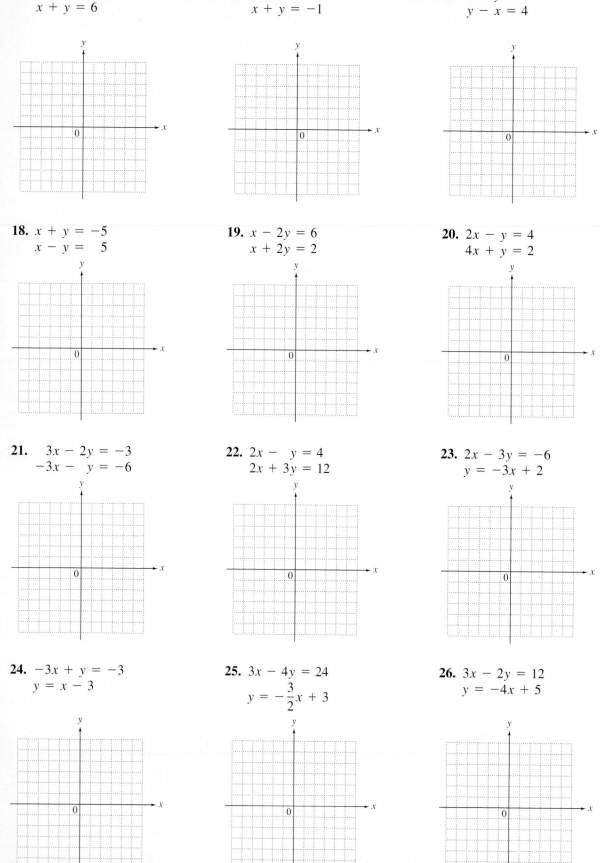

27. Explain why a system of two linear equations cannot have exactly two solutions.

28. Write each equation in the following systems in slope-intercept form. Use what you learned in Chapter 4 about slope and the y-intercept to describe the graphs of these equations.

(a) $3x + 2y = 6$ 　　(b) $2x - y = 4$ 　(c) $x - 3y = 5$
　　$-2y = 3x - 5$ 　　　$x = .5y + 2$ 　　　$2x + y = 8$

Explain how you can use this information to determine the number of solutions of a system of equations.

Solve each system by graphing. If the two equations produce parallel lines, write ∅. If the two equations produce the same line, write infinite number of solutions. *See Example 3.*

29. $x + 2y = 6$
　　$2x + 4y = 8$

30. $2x - y = 6$
　　$6x - 3y = 12$

31. $-2x + y = -4$
　　$4x = 2y + 8$

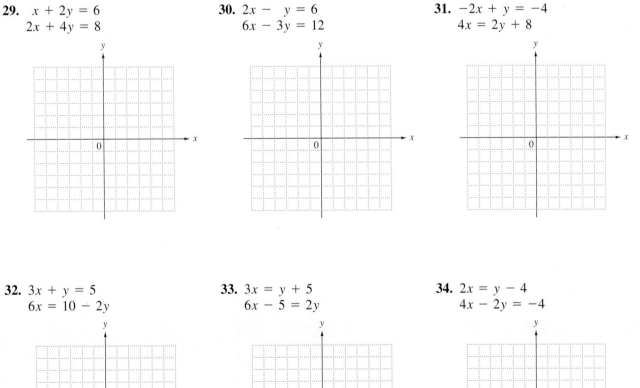

32. $3x + y = 5$
　　$6x = 10 - 2y$

33. $3x = y + 5$
　　$6x - 5 = 2y$

34. $2x = y - 4$
　　$4x - 2y = -4$

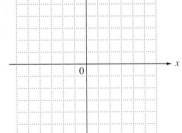

INTERPRETING TECHNOLOGY (EXERCISES 35–37)

Graphing calculators can be used to find the point of intersection of two lines. Use the display at the bottom of the screen to find the values of x and y that form the ordered pair that is the solution of the given system.

35. $x + 4y = -8$
$3x + 2y = 6$

36. $3x - y = 5$
$2x + y = 10$

37. $y = -.5x + 6$
$y = \quad x + 12$

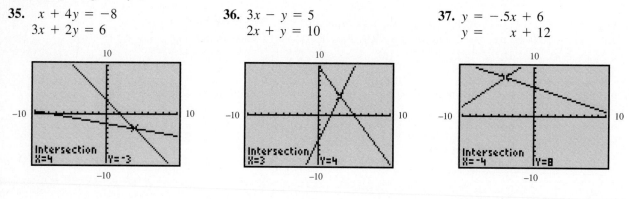

38. Explain one of the drawbacks of solving a system of equations graphically.

39. If the two lines that are the graphs of the equations in a system are parallel, how many solutions does the system have? If the two lines coincide, how many solutions does the system have?

40. Find a system of equations with the solution set $\{(-2, 3)\}$, and show the graph.

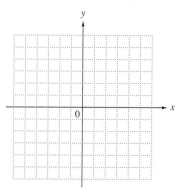

41. Graph the system

$$2x + 3y = 6$$
$$x - 3y = 5.$$

Can you check your answer? What is the problem?

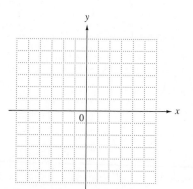

5.2 Solving Systems of Linear Equations by Addition

Graphing to solve a system of equations has a serious drawback: It is difficult to accurately find a solution such as $(\frac{1}{3}, -\frac{5}{6})$ from a graph.

OBJECTIVE ▶ An algebraic method that depends on the addition property of equality can be used to solve systems. As mentioned earlier, adding the same quantity to each side of an equation results in equal sums.

$$\text{If } A = B, \quad \text{then} \quad A + C = B + C.$$

This addition can be taken a step further. Adding *equal* quantities, rather than the *same* quantity, to both sides of an equation also results in equal sums.

$$\text{If } A = B \quad \text{and} \quad C = D, \quad \text{then} \quad A + C = B + D.$$

The use of the addition property to solve systems is called the **addition method** for solving systems of equations. For most systems, this method is more efficient than graphing.

> **Note**
>
> When using the addition method, the idea is to eliminate one of the variables. To do this, one of the variables must have coefficients that are opposites. Keep this in mind throughout the examples in this section.

EXAMPLE 1 Using the Addition Method

Use the addition method to solve the system

$$x + y = 5$$
$$x - y = 3.$$

Each equation in this system is a statement of equality, so, as discussed above, the sum of the right-hand sides equals the sum of the left-hand sides. Adding in this way gives

$$(x + y) + (x - y) = 5 + 3.$$

Combine terms to get

$$2x = 8 \quad \text{Combine terms.}$$
$$x = 4. \quad \text{Divide by 2.}$$

The result, $x = 4$, gives the x-value of the solution of the given system. Find the y-value of the solution by substituting 4 for x in either of the two equations in the system.

WORK PROBLEM 1 AT THE SIDE. ▶▶

The solution found at the side, $(4, 1)$, can be checked by substituting 4 for x and 1 for y into both equations in the given system.

Check: $\quad x + y = 5 \qquad\qquad x - y = 3$
$\qquad\qquad 4 + 1 = 5 \quad ? \qquad\quad 4 - 1 = 3 \quad ?$
$\qquad\qquad\quad 5 = 5 \quad \text{True} \qquad\quad 3 = 3 \quad \text{True}$

Because both results are true, the solution set of the given system is $\{(4, 1)\}$. The two equations are graphed in Figure 5 on the next page. Notice that the point of intersection is $(4, 1)$, as indicated by the solution of the system using the addition method.

CONTINUED ON NEXT PAGE

OBJECTIVES

▶ Solve linear systems by addition.

▶ Multiply one or both equations of a system so the addition method can be used.

▶ Write equations in the proper form to use the addition method.

▶ Solve linear systems having parallel lines as their graphs.

▶ Solve linear systems having the same line as their graphs.

FOR EXTRA HELP

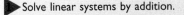

Tutorial   Tape 9   SSM, Sec. 5.2

1. **(a)** Substitute $x = 4$ in the equation $x + y = 5$ to find the value of y.

   (b) Give the solution set of the system.

ANSWERS

1. **(a)** $y = 1$ **(b)** $\{(4, 1)\}$

283

2. Solve each system by the addition method. Check each solution.

(a) $x + y = 8$
$x - y = 2$

(b) $3x - y = 7$
$2x + y = 3$

3. Solve each system by the addition method. Check each solution.

(a) $2x - y = 2$
$4x + y = 10$

(b) $8x - 5y = 32$
$4x + 5y = 4$

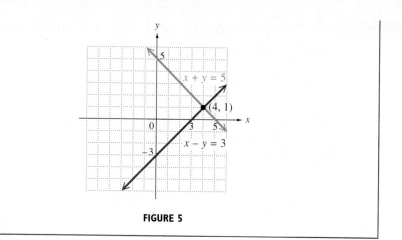

FIGURE 5

Caution
A system is not completely solved until you find values for *both* x and y. Do not make the mistake of finding the value of only one variable.

◀◀ **WORK PROBLEM 2 AT THE SIDE.**

EXAMPLE 2 **Using the Addition Method**

Solve the system

$$-2x + y = -11$$
$$5x - y = 26.$$

As above, add left-hand sides and add right-hand sides. This may be done most easily by drawing a line under the second equation and adding vertically. (Like terms must be placed in columns.)

$$\begin{aligned} -2x + y &= -11 \\ \underline{5x - y} &= \underline{26} \\ 3x &= 15 \qquad \text{Add in columns.} \\ x &= 5 \qquad \text{Divide by 3.} \end{aligned}$$

Substitute 5 for x in either of the original equations. We choose the first.

$$\begin{aligned} -2x + y &= -11 \\ -2(5) + y &= -11 \qquad \text{Let } x = 5. \\ -10 + y &= -11 \qquad \text{Multiply.} \\ y &= -1 \qquad \text{Add 10.} \end{aligned}$$

The solution set is $\{(5, -1)\}$. Check this solution by substituting 5 for x and -1 for y in both of the original equations.

◀◀ **WORK PROBLEM 3 AT THE SIDE.**

OBJECTIVE ▶ In both earlier examples, the addition step eliminated a variable. Sometimes one or both equations in a system must be multiplied by some number before the addition step will eliminate a variable.

EXAMPLE 3 **Multiplying Before Using the Addition Method**

Solve the system

$$x + 3y = 7 \qquad\qquad \textbf{(1)}$$
$$2x + 5y = 12. \qquad\qquad \textbf{(2)}$$

CONTINUED ON NEXT PAGE

Adding the two equations gives $3x + 8y = 19$, which does not help to solve the system. However, if each side of equation (1) is first multiplied by -2 using the multiplication property of equality, the terms with the variable x will drop out after adding.

$$-2(x + 3y) = -2(7)$$
$$-2x - 6y = -14 \qquad (3)$$

Now add equations (3) and (2).

$$
\begin{array}{ll}
-2x - 6y = -14 & (3) \\
\underline{2x + 5y = 12} & (2) \\
-y = -2 & \text{Add.} \\
y = 2 & \text{Multiply by } -1.
\end{array}
$$

Substituting into equation (1) gives

$$
\begin{array}{ll}
x + 3y = 7 & \\
x + 3(2) = 7 & \text{Let } y = 2. \\
x + 6 = 7 & \text{Multiply.} \\
x = 1. & \text{Subtract 6.}
\end{array}
$$

The solution set of this system is $\{(1, 2)\}$. Check that this ordered pair satisfies both of the original equations.

> **WORK PROBLEM 4 AT THE SIDE.** ▶▶

E X A M P L E 4 Multiplying Twice Before Using the Addition Method

Solve the system

$$
\begin{array}{ll}
2x + 3y = -15 & (1) \\
5x + 2y = 1. & (2)
\end{array}
$$

Here we must use the multiplication property of equality with both equations instead of just one. Multiply by numbers that will cause the coefficients of x (or of y) in the two equations to be additive inverses of each other. For example, multiply each side of equation (1) by 5, and each side of equation (2) by -2.

$$
\begin{array}{ll}
10x + 15y = -75 & \text{Multiply (1) by 5.} \\
\underline{-10x - 4y = -2} & \text{Multiply (2) by } -2. \\
11y = -77 & \text{Add.} \\
y = -7 & \text{Divide by 11.}
\end{array}
$$

Substituting -7 for y in either equation (1) or (2) gives $x = 3$. The solution set of the system is $\{(3, -7)\}$. Check this solution.

The same result would have been obtained by multiplying each side of equation (1) by 2 and each side of equation (2) by -3. This process would eliminate the y terms so that the value of x would have been found first.

> **WORK PROBLEM 5 AT THE SIDE.** ▶▶

OBJECTIVE ▶ Before a system can be solved by the addition method, the two equations of the system must have like terms in the same positions. When this is not the case, the terms should first be rearranged, as the next example shows. This example also shows an alternative way to get the second number when finding the solution of a system.

---

4. Solve each system by the addition method. Check each solution.

    (a) $x - 3y = -7$
             $3x + 2y = 23$

    (b) $8x + 2y = 2$
            $3x - y = 6$

5. Solve each system of equations. Check each solution.

    (a) $4x - 5y = -18$
          $3x + 2y = -2$

    (b) $6x + 7y = 4$
          $5x + 8y = -1$

ANSWERS

4. (a) $\{(5, 4)\}$ **(b)** $\{(1, -3)\}$

5. (a) $\{(-2, 2)\}$ **(b)** $\{(3, -2)\}$

6. Solve each system of equations.

(a) $5x = 7 + 2y$
$5y = 5 - 3x$

(b) $3y = 8 + 4x$
$6x = 9 - 2y$

EXAMPLE 5 Rearranging Terms Before Using the Addition Method

Solve the system

$$4x = 9 - 3y \qquad (1)$$
$$5x - 2y = 8. \qquad (2)$$

Rearrange the terms in equation (1) so that the like terms can be aligned in columns. Add $3y$ to each side to get the following system.

$$4x + 3y = 9 \qquad (3)$$
$$5x - 2y = 8 \qquad (2)$$

Let us eliminate y by multiplying each side of equation (3) by 2 and each side of equation (2) by 3, and then adding.

$$
\begin{array}{ll}
8x + 6y = 18 & \text{Multiply by 2.} \\
\underline{15x - 6y = 24} & \text{Multiply by 3.} \\
23x \quad\quad = 42 & \text{Add.}
\end{array}
$$

$$x = \frac{42}{23} \qquad \text{Divide by 23.}$$

Substituting $\frac{42}{23}$ for x in one of the given equations would give y, but the arithmetic involved would be messy. Instead, solve for y by starting again with the original equations and eliminating x. Do this by multiplying each side of equation (3) by 5 and each side of equation (2) by -4, and then adding.

$$
\begin{array}{ll}
20x + 15y = 45 & \text{Multiply by 5.} \\
\underline{-20x + 8y = -32} & \text{Multiply by }-4. \\
23y = 13 & \text{Add.}
\end{array}
$$

$$y = \frac{13}{23} \qquad \text{Divide by 23.}$$

The solution set is $\{(\frac{42}{23}, \frac{13}{23})\}$.

When the value of the first variable is a fraction, the method used in Example 5 prevents errors that often occur when working with fractions. (Of course, this method could be used in solving any system of equations.)

◀◀ **WORK PROBLEM 6 AT THE SIDE.**

The solution set of a linear system of equations having exactly one solution can be found by the addition method. A summary of the steps is given below.

Solving a Linear System

Step 1 Write both equations of the system in the form $Ax + By = C.$

Step 2 If necessary, multiply one or both equations by appropriate numbers so that the coefficients of x (or y) are negatives of each other.

Step 3 Add the two equations to get an equation with only one variable.

Step 4 Solve the equation from Step 3.

Step 5 Substitute the solution from Step 4 into either of the original equations.

Step 6 Solve the resulting equation from Step 5 for the remaining variable.

Step 7 Check the answer.

ANSWERS

6. (a) $\left\{ \left(\frac{45}{31}, \frac{4}{31} \right) \right\}$ **(b)** $\left\{ \left(\frac{11}{26}, \frac{42}{13} \right) \right\}$

OBJECTIVE ▶ In Section 5.1 some of the systems had equations with graphs that were parallel lines (from an inconsistent system), while the equations of other systems had graphs that were the same line (dependent equations). These systems can also be solved with the addition method.

E X A M P L E   6   Using Addition to Solve an Inconsistent System

Solve by the addition method.

$$2x + 4y = 5$$
$$4x + 8y = -9$$

Multiply each side of $2x + 4y = 5$ by -2 and then add.

$$\begin{array}{r} -4x - 8y = -10 \\ 4x + 8y = -9 \\ \hline 0 = -19 \quad \text{False} \end{array}$$

The false statement $0 = -19$ shows that the given system is self-contradictory. *It has no solution and the solution set is ∅.* This means that the graphs of the equations of this system are parallel lines, as shown in Figure 6. Since this system has no solution, it is inconsistent.

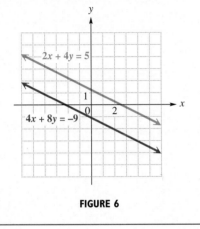

FIGURE 6

WORK PROBLEM 7 AT THE SIDE. ▶▶

OBJECTIVE ▶ The next example shows the result of using the addition method when the equations of the system are dependent, with the graphs of the equations in the system the same line.

E X A M P L E   7   Using Addition to Solve a System of Dependent Equations

Solve by the addition method.

$$3x - y = 4$$
$$-9x + 3y = -12$$

Multiply each side of the first equation by 3 and then add the two equations to get

$$\begin{array}{r} 9x - 3y = 12 \\ -9x + 3y = -12 \\ \hline 0 = 0. \quad \text{True} \end{array}$$

This result means that every solution of one equation is also a solution of the other, so the system has an infinite number of solutions: all the ordered

— **CONTINUED ON NEXT PAGE**

7. Solve each system by the addition method.

(a) $4x + 3y = 10$
$2x + \dfrac{3}{2}y = 12$

(b) $-2x - 4y = -1$
$5x + 10y = 15$

ANSWERS

7. (a) ∅ **(b)** ∅

8. Solve each system by the addition method.

(a) $6x + 3y = 9$
$\quad\ -8x - 4y = -12$

pairs corresponding to points that lie on the common graph. As mentioned in Section 5.1, the equations in this system are dependent. In the answers at the back of this book, a solution of such a system of dependent equations is indicated by *infinite number of solutions*. A graph of the equations of this system is shown in Figure 7.

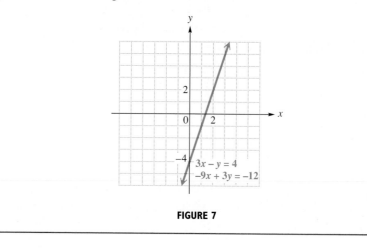

$3x - y = 4$
$-9x + 3y = -12$

FIGURE 7

◀◀ **WORK PROBLEM 8 AT THE SIDE.**

(b) $4x - 6y = 10$
$\quad -10x + 15y = -25$

One of three situations may occur when the addition method is used to solve a linear system of equations.

1. The result of the addition step is a statement such as $x = 2$ or $y = -3$. The solution set will contain only one ordered pair. The graphs of the equations of the system will intersect at exactly one point.

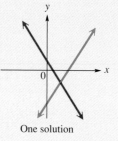

One solution

2. The result of the addition step is a false statement, such as $0 = 4$. In this case, the graphs are parallel lines, and the solution set is $\emptyset$.

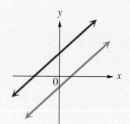

No solution

3. The result of the addition step is a true statement, such as $0 = 0$. The graphs of the equations of the system are the same line, and an infinite number of ordered pairs are solutions. These ordered pairs must satisfy the equation of the line.

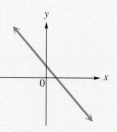

Infinite number of solutions

ANSWERS
8. (a) infinite number of solutions
(b) infinite number of solutions

5.2 Exercises

Answer true or false for each of the following statements. If false, tell why.

1. The ordered pair (0, 0) *must* be a solution of a system of the form

$$Ax + By = 0$$
$$Cx + Dy = 0.$$

2. To eliminate the *y* terms in the system

$$2x + 12y = 7$$
$$3x + 4y = 1,$$

we should multiply the bottom equation by 3 and then add.

3. The system

$$x + y = 1$$
$$x + y = 2$$

has no solutions.

4. The ordered pair (4, −5) cannot be a solution of a system that contains the equation $5x − 4y = 0$.

Solve each system by the addition method. Check your answers. See Examples 1 and 2.

5. $x + y = 2$
   $2x − y = −5$

6. $3x − y = −12$
   $x + y = 4$

7. $2x + y = −5$
   $x − y = 2$

8. $2x + y = −15$
   $−x − y = 10$

9. $3x + 2y = 0$
   $−3x − y = 3$

10. $5x − y = 5$
   $−5x + 2y = 0$

11. $6x − y = −1$
   $−6x + 5y = 17$

12. $6x + y = 9$
   $−6x + 3y = 15$

Solve each system by the addition method. Check your answers. See Example 3.

13. $2x - y = 12$
$3x + 2y = -3$

14. $x + y = 3$
$-3x + 2y = -19$

15. $x + 3y = 19$
$2x - y = 10$

16. $4x - 3y = -19$
$2x + y = 13$

17. $x + 4y = 16$
$3x + 5y = 20$

18. $2x + y = 8$
$5x - 2y = -16$

19. $5x - 3y = -20$
$-3x + 6y = 12$

20. $4x + 3y = -28$
$5x - 6y = -35$

21. $2x - 8y = 0$
$4x + 5y = 0$

22. $3x - 15y = 0$
$6x + 10y = 0$

Solve each system by the addition method. Check your answers. See Examples 4 and 5.

23. $3x + 5y = 7$
$5x + 4y = -10$

24. $2x + 3y = 13$
$5x + 2y = -6$

25. $2x + 3y = 0$
$7y - 29 = 5x$

26. $2x + 9y = 44$
$6y - 61 = 5x$

27. $24x + 12y = -7$
$16x - 17 = 18y$

28. $9x + 4y = -3$
$6x + 7 = -6y$

29. $3x = 3 + 2y$
$-\dfrac{4}{3}x + y = \dfrac{1}{3}$

30. $3x = 27 + 2y$
$x - \dfrac{7}{2}y = -25$

Use the addition method to solve each system. See Examples 6 and 7.

31. $x + y = 7$
$x + y = -3$

32. $x - y = 4$
$x - y = -3$

33. $-x + 3y = 4$
$-2x + 6y = 8$

34. $6x - 2y = 24$
$-3x + y = -12$

35. $5x - 2y = 3$
$10x - 4y = 5$

36. $3x - 5y = 1$
$6x - 10y = 4$

37. $6x + 3y = 0$
$-18x - 9y = 0$

38. $3x - 5y = 0$
$9x - 15y = 0$

MATHEMATICAL CONNECTIONS (EXERCISES 39–44)

In this group of exercises we show the connections between solving linear equations in one variable and solving a linear system of two equations in two variables. Work through Exercises 39–44 in order.

39. Solve the linear equation $\frac{1}{2}x + 4 = 3x - 1$ using the methods described in Chapter 2.

40. Check your solution for the equation in Exercise 39 by substituting back into the original equation. What is the value that you get for both the left and right sides after you make this substitution?

41. Graph the linear system

$$y = \frac{1}{2}x + 4$$
$$y = 3x - 1$$

and find the solution set of the system.

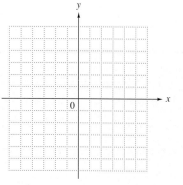

42. How does the *x*-coordinate of the solution of the system in Exercise 41 compare to the solution of the linear equation in Exercise 39?

43. How does the *y*-coordinate of the solution of the system in Exercise 41 compare to the value you obtained in the check in Exercise 40?

44. Based on your observations in Exercises 39–43, fill in the blanks with the correct responses.

The solution of the linear equation $\frac{2}{3}x + 3 = -x + 8$ is 3. When we substitute 3 back into the equation, we get the value _____ on both sides, verifying that 3 is indeed the solution. Now if we graph the system

$$y = \frac{2}{3}x + 3$$
$$y = -x + 8,$$

the solution set of the system is $\{($_____ , _____$)\}$.

5.3 Solving Systems of Linear Equations by Substitution

OBJECTIVE ▶ The graphical method and the addition method for solving systems of linear equations were discussed in the preceding sections. A third method, the **substitution method,** is particularly useful for solving systems where one equation is solved, or can be solved quickly, for one of the variables.

OBJECTIVES

▶ Solve linear systems by substitution.

▶ Solve linear systems with fractions as coefficients.

FOR EXTRA HELP

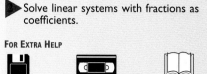

Tutorial   Tape 9   SSM, Sec. 5.3

E X A M P L E 1   Using the Substitution Method

Solve the system

$$3x + 5y = 26$$
$$y = 2x.$$

The second of these two equations is already solved for y. This equation says that $y = 2x$. Substituting $2x$ for y in the first equation gives

$$3x + 5y = 26$$
$$3x + 5(2x) = 26 \qquad \text{Let } y = 2x.$$
$$3x + 10x = 26 \qquad \text{Multiply.}$$
$$13x = 26 \qquad \text{Combine terms.}$$
$$x = 2. \qquad \text{Divide by 13.}$$

Because $x = 2$, we find y from the equation $y = 2x$ by substituting 2 for x.

$$y = 2(2) = 4 \qquad \text{Let } x = 2.$$

Check that the solution set of the given system is $\{(2, 4)\}$.

WORK PROBLEM 1 AT THE SIDE. ▶▶

E X A M P L E 2   Using the Substitution Method

Use substitution to solve the system

$$2x + 5y = 7$$
$$x = -1 - y.$$

The second equation gives x in terms of y. Substitute $-1 - y$ for x in the first equation.

$$2x + 5y = 7$$
$$2(-1 - y) + 5y = 7 \qquad \text{Let } x = -1 - y.$$
$$-2 - 2y + 5y = 7 \qquad \text{Distributive property}$$
$$-2 + 3y = 7 \qquad \text{Combine terms.}$$
$$3y = 9 \qquad \text{Add 2.}$$
$$y = 3 \qquad \text{Divide by 3.}$$

To find x, substitute $y = 3$ in the equation $x = -1 - y$ to get $x = -1 - 3 = -4$. Check that the solution set of the given system is $\{(-4, 3)\}$.

1. Solve by the substitution method.

(a) $3x + 5y = 69$
$\qquad\; y = 4x$

(b) $-x + 4y = 26$
$\qquad\qquad y = -3x$

ANSWERS

1. **(a)** $\{(3, 12)\}$ **(b)** $\{(-2, 6)\}$

2. Solve each system by substitution. Check each solution.

(a) $3x - 4y = -11$
$x = y - 2$

EXAMPLE 3 Using the Substitution Method

Use substitution to solve the system

$$x = 5 - 2y$$
$$2x + 4y = 6.$$

Substitute $5 - 2y$ for x in the second equation.

$$2x + 4y = 6$$
$$2(5 - 2y) + 4y = 6 \qquad \text{Let } x = 5 - 2y.$$
$$10 - 4y + 4y = 6 \qquad \text{Distributive property}$$
$$10 = 6 \qquad \text{False}$$

As shown in the last section, this false result means that the equations in the system have graphs that are parallel lines. The system is inconsistent and the solution set is $\emptyset$.

◀◀ **WORK PROBLEM 2 AT THE SIDE.**

EXAMPLE 4 Using the Substitution Method

Use substitution to solve the system

$$2x + 3y = 8$$
$$-4x - 2y = 0.$$

To use the substitution method, one of the equations must be solved for one of the variables. Let us choose the first equation of the system, $2x + 3y = 8$, and solve for x.

To get x alone on one side, subtract $3y$ from each side.

$$2x + 3y = 8$$
$$2x = 8 - 3y \qquad \text{Subtract } 3y.$$

(b) $8x - y = 4$
$y = 8x + 4$

Now divide each side by 2.

$$x = \frac{8 - 3y}{2} \qquad \text{Divide by 2.}$$

Substitute this value for x in the second equation of the system.

$$-4x - 2y = 0$$
$$-4\left(\frac{8 - 3y}{2}\right) - 2y = 0 \qquad \text{Let } x = \frac{8 - 3y}{2}.$$
$$-2(8 - 3y) - 2y = 0 \qquad \text{Divide } -4 \text{ by 2.}$$
$$-16 + 6y - 2y = 0 \qquad \text{Distributive property}$$
$$-16 + 4y = 0 \qquad \text{Combine terms.}$$
$$4y = 16 \qquad \text{Add 16.}$$
$$y = 4 \qquad \text{Divide by 4.}$$

Find x by letting $y = 4$ in $x = \dfrac{8 - 3y}{2}$.

$$x = \frac{8 - 3(4)}{2} \qquad \text{Let } y = 4.$$
$$x = \frac{8 - 12}{2} \qquad \text{Multiply.}$$
$$x = -2 \qquad \text{Subtract and divide.}$$

CONTINUED ON NEXT PAGE

Check:

$$2x + 3y = 8 \qquad\qquad -4x - 2y = 0$$

$$2(-2) + 3(4) = 8 \quad ? \qquad\qquad -4(-2) - 2(4) = 0 \quad ?$$

$$-4 + 12 = 8 \quad ? \qquad\qquad 8 - 8 = 0 \quad ?$$

$$8 = 8 \qquad \text{True} \qquad\qquad 0 = 0 \qquad \text{True}$$

The solution set of the given system is $\{(-2, 4)\}$.

WORK PROBLEM 3 AT THE SIDE. ▶▶

E X A M P L E 5 Using the Substitution Method

Use substitution to solve the system

$$2x = 4 - y \qquad\qquad (1)$$

$$6 + 3y + 4x = 16 - x. \qquad\qquad (2)$$

Start by simplifying the second equation by adding x and subtracting 6 on each side. This gives the simplified system

$$2x = 4 - y \qquad\qquad (1)$$

$$5x + 3y = 10. \qquad\qquad (3)$$

For the substitution method, one of the equations must be solved for either x or y. Because the coefficient of y in equation (1) is -1, we avoid fractions by solving this equation for y.

$$2x = 4 - y \qquad\qquad (1)$$

$$2x - 4 = -y \qquad \text{Subtract 4.}$$

$$-2x + 4 = y \qquad \text{Multiply by } -1.$$

Now substitute $-2x + 4$ for y in equation (3).

$$5x + 3y = 10$$

$$5x + 3(-2x + 4) = 10 \qquad \text{Let } y = -2x + 4.$$

$$5x - 6x + 12 = 10 \qquad \text{Distributive property}$$

$$-x + 12 = 10 \qquad \text{Combine terms.}$$

$$-x = -2 \qquad \text{Subtract 12.}$$

$$x = 2 \qquad \text{Multiply by } -1.$$

Since $y = -2x + 4$ and $x = 2$,

$$y = -2(2) + 4 = 0,$$

and the solution set is $\{(2, 0)\}$.

Check:

$$2x = 4 - y \qquad\qquad 6 + 3y + 4x = 16 - x$$

$$2(2) = 4 - 0 \quad ? \qquad\qquad 6 + 3(0) + 4(2) = 16 - 2 \quad ?$$

$$4 = 4 \qquad \text{True} \qquad\qquad 6 + 0 + 8 = 14 \qquad \text{True}$$

WORK PROBLEM 4 AT THE SIDE. ▶▶

OBJECTIVE ▶ When a system includes equations with fractions as coefficients, we eliminate the fractions by multiplying each side by a common denominator. Then we solve the resulting system.

---

3. Solve each system by substitution. Check each solution.

(a) $x + 4y = -1$
$2x - 5y = 11$

(b) $2x + 5y = 4$
$x + \ y = -1$

4. Solve each system by substitution. First simplify where necessary.

(a) $\qquad x = 5 - 3y$
$2x + 3 = 5x - 4y + 14$

(b) $5x - y = -14 + 2x + y$
$7x + 9y + 4 = 3x + 8y$

5. Verify that the same solution is found if the addition method is used to solve the system of equations (3) and (4) in Example 6.

6. Solve the following system by any method. First clear all fractions.

$$\frac{2}{3}x + \frac{1}{2}y = 6$$

$$\frac{1}{2}x - \frac{3}{4}y = 0$$

E X A M P L E   6   Using the Substitution Method with Fractions as Coefficients

Solve the following system by the substitution method.

$$3x + \frac{1}{4}y = 2 \qquad\qquad (1)$$

$$\frac{1}{2}x + \frac{3}{4}y = -\frac{5}{2} \qquad\qquad (2)$$

Clear equation (1) of fractions by multiplying each side by 4.

$$4\left(3x + \frac{1}{4}y\right) = 4(2) \qquad \text{Multiply by 4.}$$

$$4(3x) + 4\left(\frac{1}{4}y\right) = 4(2) \qquad \text{Distributive property}$$

$$12x + y = 8 \qquad\qquad (3)$$

Now clear equation (2) of fractions by multiplying each side by the common denominator 4.

$$4\left(\frac{1}{2}x + \frac{3}{4}y\right) = 4\left(-\frac{5}{2}\right) \qquad \text{Multiply by 4.}$$

$$4\left(\frac{1}{2}x\right) + 4\left(\frac{3}{4}y\right) = 4\left(-\frac{5}{2}\right) \qquad \text{Distributive property}$$

$$2x + 3y = -10 \qquad\qquad (4)$$

The given system of equations has been simplified as follows.

$$12x + y = 8 \qquad\qquad (3)$$
$$2x + 3y = -10 \qquad\qquad (4)$$

Let us solve this system by the substitution method. Equation (3) can be solved for y by subtracting $12x$ from each side.

$$12x + y = 8$$
$$y = -12x + 8 \qquad \text{Subtract } 12x.$$

Now substitute the result for y in equation (4).

$$2x + 3(-12x + 8) = -10 \qquad \text{Let } y = -12x + 8.$$
$$2x - 36x + 24 = -10 \qquad \text{Distributive property}$$
$$-34x = -34 \qquad \text{Combine terms; subtract 24.}$$
$$x = 1 \qquad \text{Divide by } -34.$$

Using $x = 1$ in $y = -12x + 8$ gives $y = -12(1) + 8 = -4$. The solution set is $\{(1, -4)\}$. Check by substituting 1 for x and -4 for y in both of the original equations.

◀◀ **WORK PROBLEMS 5 AND 6 AT THE SIDE.**

ANSWERS

5. The solution is the same.

6. $\{(6, 4)\}$

1. A student solves the system

$$5x - y = 15$$
$$7x + y = 21$$

and finds that $x = 3$, which is the correct value for x. The student gives the solution as "$x = 3$." Is this correct? Explain.

2. A student solves the system

$$x + y = 4$$
$$2x + 2y = 8$$

and obtains the equation $0 = 0$. The student gives the solution set as $\{(0, 0)\}$. Is this correct? Explain.

3. Professor Brandsma gave the following item on a test in algebra:
Solve the system

$$3x - y = 13$$
$$2x + 5y = 20$$

by the substitution method.
One student worked the problem by solving first for y in the first equation. Another student worked it by solving first for x in the second equation. Both students got the correct solution, $(5, 2)$. Which student, do you think, had less work to do? Explain.

4. Which one of the following systems would be easier to solve using the substitution method? Why?

$$5x - 3y = 7 \qquad\qquad 7x + 2y = 4$$
$$2x + 8y = 3 \qquad\qquad\quad y = -3x$$

Solve each system by the substitution method. Check each solution. See Examples 1–4.

5. $3x + 2y = 27$
   $x = y + 4$

6. $4x + 3y = -5$
   $x = y - 3$

7. $3x + 5y = 14$
   $x - 2y = -10$

8. $5x + 2y = -1$
   $2x - y = -13$

9. $3x + 4 = -y$
   $2x + y = 0$

10. $2x - 5 = -y$
   $x + 3y = 0$

11. $7x + 4y = 13$
     $x + y = 1$

12. $3x - 2y = 19$
     $x + y = 8$

13. $3x - y = 5$
     $y = 3x - 5$

14. $4x - y = -3$
     $y = 4x + 3$

15. $6x - 8y = 6$
     $-3x + 2y = -2$

16. $3x + 2y = 6$
     $-6x + 4y = -8$

17. $2x + 8y = 3$
     $x = 8 - 4y$

18. $2x + 10y = 3$
     $x = 1 - 5y$

19. $12x - 16y = 8$
     $3x = 4y + 2$

20. $6x + 9y = 6$
     $2x = 2 - 3y$

In Exercises 21 and 22, (a) solve the system by the addition method, (b) then solve the system by the substitution method, (c) and finally tell which method you prefer for that particular system, and why.

21. $4x - 3y = -8$
     $x + 3y = 13$

22. $2x + 5y = 0$
     $x = -3y + 1$

Solve each system either by the addition method or the substitution method. First simplify equations where necessary. Check each solution. See Example 5.

23. $4 + 4x - 3y = 34 + x$
     $4x = -y - 2 + 3x$

24. $5x - 4y = 42 - 8y - 2$
     $2x + y = x + 1$

25. $2x - 8y + 3y + 2 = 5y + 16$
     $8x - 2y = 4x + 28$

26. $7x - 9 + 2y - 8 = -3y + 4x + 13$
     $4y - 8x = -8 + 9x + 32$

27. $\begin{aligned} -2x + 3y &= 12 + 2y \\ 2x - 5y + 4 &= -8 - 4y \end{aligned}$

28. $\begin{aligned} 2x + 5y &= 7 + 4y - x \\ 5x + 3y + 8 &= 22 - x + y \end{aligned}$

29. $\begin{aligned} 5x + y &= 12 - x - 7y \\ 3x + 2y &= 10 - 6x - 10y \end{aligned}$

30. $\begin{aligned} -2x + 3y &= 7 - 5x - y \\ -4x + 2y &= 1 - 10x - 6y \end{aligned}$

Exercises 31 and 32 refer to the system

$$\frac{1}{3}x - \frac{1}{2}y = 7$$

$$\frac{1}{6}x + \frac{1}{3}y = 0.$$

31. One student solved the system by multiplying both equations by 6 to clear fractions, and another student multiplied by 12. Assuming they do all other work correctly, should they both get the same answer?

32. One student solved the system and wrote as his answer "$x = 12$," while another solved it and wrote as her answer "$y = -6$." Who, if either, was correct? Why?

Solve each system either by the addition method or the substitution method. First clear all fractions. Check each solution. See Example 6.

33. $\begin{aligned} x + \frac{1}{3}y &= y - 2 \\ \frac{1}{4}x + y &= x + y \end{aligned}$

34. $\begin{aligned} \frac{5}{3}x + 2y &= \frac{1}{3} + y \\ 3x - 3 + \frac{y}{3} &= -2 + 2x \end{aligned}$

35. $\begin{aligned} \frac{x}{6} + \frac{y}{6} &= 2 \\ -\frac{1}{2}x - \frac{1}{3}y &= -8 \end{aligned}$

36. $\dfrac{x}{2} - \dfrac{y}{3} = 9$

$\dfrac{x}{5} - \dfrac{y}{4} = 5$

37. $\dfrac{x}{3} - \dfrac{3y}{4} = -\dfrac{1}{2}$

$\dfrac{x}{6} + \dfrac{y}{8} = \dfrac{3}{4}$

38. $\dfrac{x}{5} + 2y = \dfrac{16}{5}$

$\dfrac{3x}{5} + \dfrac{y}{2} = -\dfrac{7}{5}$

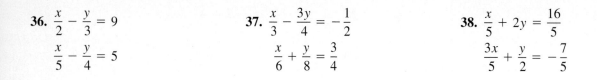

MATHEMATICAL CONNECTIONS (EXERCISES 39–42)

A system of linear equations can be used to model the cost and the revenue of a business. Work Exercises 39–42 in order.

39. Suppose that you start a business manufacturing and selling bicycles, and it costs you $5000 to get started. You determine that each bicycle will cost $400 to manufacture. Explain why the linear equation $y_1 = 400x + 5000$ gives your *total* cost to manufacture x bicycles (y_1 in dollars).

40. You decide to sell each bike for $600. What expression in x represents the revenue you will take in if you sell x bikes? Write an equation using y_2 to express your revenue when you sell x bikes (y_2 in dollars).

41. Form a system from the two equations in Exercises 39 and 40 and then solve the system.

42. The value of x from Exercise 41 is the number of bikes it takes to *break even*. Fill in the blanks: When _____ bikes are sold, the break-even point is reached. At that point, you have spent _____ dollars and taken in _____ dollars.

5.4 Linear Systems of Equations in Three Variables

OBJECTIVES

1. ▶ Solve linear systems with three equations and three unknowns by the elimination method.

2. ▶ Solve linear systems with three equations and three unknowns where some of the equations have missing terms.

3. ▶ Solve linear systems with three equations and three unknowns that are inconsistent or that include dependent equations.

FOR EXTRA HELP

Tutorial   Tape 10   SSM, Sec. 5.4

A solution of an equation in three variables, such as $2x + 3y - z = 4$, is called an **ordered triple** and is written (x, y, z). For example, the ordered triple $(1, 1, 1)$ is a solution of the equation, because

$$2(1) + 3(1) - (1) = 2 + 3 - 1 = 4.$$

Verify that another solution of this equation is $(10, -3, 7)$.

In the rest of this chapter, the term *linear equation* is extended to equations of the form $Ax + By + Cz + \cdots + Dw = K$, where not all the coefficients $A, B, C, \ldots, D$ equal zero. For example, $2x + 3y - 5z = 7$ and $x - 2y - z + 3u - 2w = 8$ are linear equations, the first with three variables and the second with five variables.

In this section we discuss the solution of a system of linear equations in three variables such as

$$4x + 8y + z = 2$$
$$x + 7y - 3z = -14$$
$$2x - 3y + 2z = 3.$$

Theoretically, a system of this type can be solved by graphing. However, the graph of a linear equation with three variables is a *plane* and not a line. Since the graph of each equation of the system is a plane, which requires three-dimensional graphing, this method is not practical. However, it does illustrate the number of solutions possible for such systems, as Figure 8 shows.

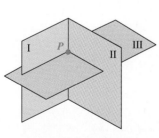

(a) A single solution

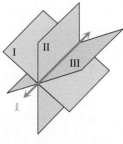

(b) Points of a line in common

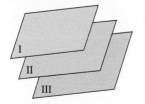

(c) No points in common

(d) All points in common

FIGURE 8

301

Figure 8 illustrates the following cases.

Graphs of Linear Systems in Three Variables

1. The three planes may meet at a single, common point that is the solution of the system. See Figure 8(a).
2. The three planes may have the points of a line in common so that the infinite set of points that satisfy the equation of the line is the solution of the system. See Figure 8(b).
3. The planes may have no points common to all three, in which case there is no solution for the system. See Figure 8(c).
4. The three planes may coincide, in which case the solution of the system is the set of all points on a plane. See Figure 8(d).

OBJECTIVE ▶ Because a graphic solution of a system of three equations in three variables is impractical, these systems can be solved with an extension of the elimination method, summarized as follows.

Solving Linear Systems in Three Variables

Step 1 Use the elimination method to eliminate any variable from any two of the given equations. The result is an equation in two variables.

Step 2 Eliminate the *same* variable from any *other* two equations. The result is an equation in the same two variables as in Step 1.

Step 3 Use the elimination or the substitution method to eliminate a variable from the two equations in two variables that result from Steps 1 and 2. The result is an equation in one variable that gives the value of that variable.

Step 4 Substitute the value of the variable found in Step 3 into either of the equations in two variables to find the value of the second variable.

Step 5 Use the values of the two variables from Steps 3 and 4 to find the value of the third variable by substituting into any of the original equations.

EXAMPLE 1   Solving a System in Three Variables

Solve the system

$$4x + 8y + z = 2 \tag{1}$$
$$x + 7y - 3z = -14 \tag{2}$$
$$2x - 3y + 2z = 3. \tag{3}$$

As before, the elimination method involves eliminating a variable from the sum of two equations. To begin, choose equations (1) and (2) and eliminate z by multiplying both sides of equation (1) by 3 and then adding the result to equation (2).

$$
\begin{array}{rl}
12x + 24y + 3z = 6 & \text{3 times both sides of equation (1)} \\
\underline{x + 7y - 3z = -14} & \tag{2} \\
13x + 31y \phantom{{}+ 3z} = -8 & \tag{4}
\end{array}
$$

CONTINUED ON NEXT PAGE

Equation (4) has only two variables, x and y. To get another equation without z, multiply both sides of equation (1) by -2 and add the result to equation (3). It is important at this point to eliminate the *same variable, z.*

$$
\begin{array}{ll}
-8x - 16y - 2z = -4 & \text{-2 times both sides of equation (1)} \\
\underline{2x - 3y + 2z = 3} & \text{(3)} \\
-6x - 19y = -1 & \text{(5)}
\end{array}
$$

Now solve the system of equations (4) and (5) for x and y.

WORK PROBLEM 1 AT THE SIDE. ▶▶

As shown in Problem 1 at the side, the solution of the system of equations (4) and (5) is $x = -3$ and $y = 1$. To find z, substitute -3 for x and 1 for y in equation (1). (Any of the three given equations could be used.)

$$
\begin{array}{ll}
4x + 8y + z = 2 & \text{(1)} \\
4(-3) + 8(1) + z = 2 & \\
z = 6 &
\end{array}
$$

The ordered triple $(-3, 1, 6)$ is the only solution of the system. Check that the solution satisfies all three equations of the system so the solution set is $\{(-3, 1, 6)\}$.

WORK PROBLEM 2 AT THE SIDE. ▶▶

OBJECTIVE ▶ When one or more of the equations of a system has a missing term, one elimination step can be omitted.

┌ **E X A M P L E 2   Solving a System of Equations with Missing Terms**

Solve the system

$$
\begin{array}{ll}
6x - 12y = -5 & \text{(6)} \\
8y + z = 0 & \text{(7)} \\
9x - z = 12. & \text{(8)}
\end{array}
$$

Since equation (8) is missing the variable y, one way to begin the solution is to eliminate y again with equations (6) and (7). Multiply both sides of equation (6) by 2 and both sides of equation (7) by 3, and then add.

$$
\begin{array}{ll}
12x - 24y = -10 & \text{2 times both sides of equation (6)} \\
\underline{24y + 3z = 0} & \text{3 times both sides of equation (7)} \\
12x + 3z = -10 & \text{(9)}
\end{array}
$$

Use this result, together with equation (8), to eliminate z. Multiply both sides of equation (8) by 3. This gives

$$
\begin{array}{ll}
27x - 3z = 36 & \text{3 times both sides of equation (8)} \\
\underline{12x + 3z = -10} & \text{(9)} \\
39x = 26 &
\end{array}
$$

$$
x = \frac{26}{39} = \frac{2}{3}.
$$

CONTINUED ON NEXT PAGE

---

1. Solve the system of equations for x and y.

$$
\begin{array}{l}
13x + 31y = -8 \\
-6x - 19y = -1
\end{array}
$$

2. Solve the system.

$$
\begin{array}{l}
x + y + z = 2 \\
x - y + 2z = 2 \\
-x + 2y - z = 1
\end{array}
$$

3. Solve the system.

$$x - y = 6$$
$$2y + 5z = 1$$
$$3x - 4z = 8$$

4. Solve each system.

(a) $3x - 5y + 2z = 1$
$5x + 8y - z = 4$
$-6x + 10y - 4z = 5$

(b) $7x - 9y + 2z = 0$
$y + z = 0$
$8x - z = 0$

Substitution into equation (8) gives

$$9x - z = 12 \qquad \text{(8)}$$
$$9\left(\frac{2}{3}\right) - z = 12 \qquad x = \tfrac{2}{3}$$
$$6 - z = 12$$
$$z = -6.$$

Substitution of -6 for z in equation (7) gives

$$8y + z = 0 \qquad \text{(7)}$$
$$8y - 6 = 0 \qquad z = -6$$
$$8y = 6$$
$$y = \frac{3}{4}.$$

Check in each of the original equations of the system to verify that the solution set of the system is $\{(\tfrac{2}{3}, \tfrac{3}{4}, -6)\}$.

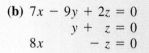 **WORK PROBLEM 3 AT THE SIDE.**

OBJECTIVE ▶ Linear systems with three variables may be inconsistent or may include dependent equations. The next two examples illustrate these cases.

E X A M P L E 3 Solving an Inconsistent System with Three Variables

Solve the following system.

$$2x - 4y + 6z = 5 \qquad \text{(10)}$$
$$-x + 3y - 2z = -1 \qquad \text{(11)}$$
$$x - 2y + 3z = 1 \qquad \text{(12)}$$

Eliminate x by adding equations (11) and (12) to get the equation

$$y + z = 0.$$

Now to eliminate x again, multiply both sides of equation (12) by -2 and add the result to equation (10).

$$\begin{array}{r} -2x + 4y - 6z = -2 \\ \underline{2x - 4y + 6z = 5} \\ 0 = 3 \quad \text{False} \end{array}$$

The resulting false statement indicates that equations (10) and (12) have no common solution; the system is inconsistent and the solution set is ∅. The graph of the equations of the system would show at least two of the planes parallel to one another.

> **Note**
> If you get a false statement from the addition step, as in Example 3, you do not need to go any further with the solution. Since two of the three planes are parallel, it is not possible for the three planes to have any common points.

◀◀ **WORK PROBLEM 4 AT THE SIDE.**

─ E X A M P L E   4   **Solving a System of Dependent Equations with Three Variables**

Solve the system.

$$2x - 3y + 4z = 8 \tag{13}$$

$$-x + \frac{3}{2}y - 2z = -4 \tag{14}$$

$$6x - 9y + 12z = 24 \tag{15}$$

Multiplying both sides of equation (13) by 3 gives equation (15). Multiplying both sides of equation (14) by -6 also results in equation (15). Because of this, the three equations are dependent. All three equations have the same graph. We will continue to indicate this as "infinite number of solutions," as we did for systems in two variables.

WORK PROBLEM 5 AT THE SIDE. ▶▶

We can extend the method discussed in this section to solve larger systems. For example, to solve a system of four equations in four variables, eliminate the same variable from three pairs of equations to get a system of three equations in three unknowns. Then proceed as shown above.

5. Solve the system.

$$x - y + z = 4$$
$$-3x + 3y - 3z = -12$$
$$2x - 2y + 2z = 8$$

NUMBERS IN THE
Real World *a graphing calculator minicourse*

Lesson 4: Solving a Linear System of Equations

We have learned algebraic methods of solving linear systems. Consider the system

$$5x - 2y = 4$$
$$2x + 3y = 13$$

can be solved by elimination. As shown there, the solution set of the system is $\{(2, 3)\}$, meaning that the graphs of the two lines intersect at the point $(2, 3)$.

This can be supported with a graphing calculator by graphing the two lines and using the capabilities of the calculator to find the point of intersection. However, we must first solve each equation for y so that the equations can be entered, because this is a requirement for most graphing calculators. If we solve $5x - 2y = 4$ for y, we get $y = \frac{5}{2}x - 2$, and solving $2x + 3y = 13$ for y yields $y = -\frac{2}{3}x + \frac{13}{3}$. If we enter these equations as y_1 and y_2, respectively, we get the graphics shown in the figure at the left. The display at the bottom of the figure indicates that the point of intersection is $(2, 3)$, as determined algebraically.

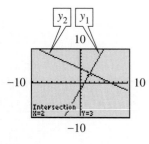

GRAPHING CALCULATOR EXPLORATIONS

1. Consider the system $4x - 3y = 7$, $3x - 2y = 6$.

  (a) Solve it by elimination.

  (b) Solve $4x - 3y = 7$ for y, and graph it as y_1.

  (c) Solve $3x - 2y = 6$ for y, and graph it as y_2.

  (d) Use your calculator to find the point of intersection of the two lines. It should be the same as the one you found algebraically in part (a).

2. Consider the system $2x - y = 3$, $6x - 3y = 9$.

  (a) Solve $2x - y = 3$ for y, and graph it as y_1.

  (b) Solve $6x - 3y = 9$ for y, and graph it as y_2.

  (c) Discuss how the display supports the algebraic conclusion that the system has infinitely many solutions.

3. Consider the system $x + 3y = 4$, $-2x - 6y = 3$.

  (a) Solve $x + 3y = 4$ for y, and graph it as y_1.

  (b) Solve $-2x - 6y = 3$ for y, and graph it as y_2.

  (c) Discuss how the display supports the algebraic conclusion that the system has no solutions.

5.4 *Exercises*

1. Explain what the following statement means: The solution set of the system

$$2x + y + z = 3$$
$$3x - y + z = -2 \quad \text{is } \{(-1, 2, 3)\}.$$
$$4x - y + 2z = 0$$

2. Write a system of three linear equations in three variables that has the solution set {(3, 1, 2)}. Then solve the system. (*Hint:* Start with the solution and make up three equations that are satisfied by the solution. There are many ways to do this.)

Solve each of the following systems of equations. See Example 1.

3. $3x + 2y + z = 8$
$2x - 3y + 2z = -16$
$x + 4y - z = 20$

4. $-3x + y - z = -10$
$-4x + 2y + 3z = -1$
$2x + 3y - 2z = -5$

5. $2x + 5y + 2z = 0$
$4x - 7y - 3z = 1$
$3x - 8y - 2z = -6$

6. $5x - 2y + 3z = -9$
$4x + 3y + 5z = 4$
$2x + 4y - 2z = 14$

7. $x + y - z = -2$
$2x - y + z = -5$
$-x + 2y - 3z = -4$

8. $x + 2y + 3z = 1$
$-x - y + 3z = 2$
$-6x + y + z = -2$

Solve each system of equations. See Example 2.

9. $2x - 3y + 2z = -1$
$\quad x + 2y + \ z = 17$
$\quad\quad\ 2y - \ z = 7$

10. $2x - \ y + 3z = 6$
$\quad\ x + 2y - \ z = 8$
$\quad\quad\ 2y + \ z = 1$

11. $4x + 2y - 3z = 6$
$\quad\ x - 4y + \ z = -4$
$\quad -x \quad\quad + 2z = 2$

12. $2x + 3y - 4z = 4$
$\quad\ x - 6y + \ z = -16$
$\quad -x \quad\quad + 3z = 8$

13. $2x + \ y \quad\quad = 6$
$\quad\quad\ 3y - 2z = -4$
$\quad 3x \quad\quad - 5z = -7$

14. $\quad 4x - 8y \quad\quad = -7$
$\quad\quad\quad 4y + z = 7$
$\quad -8x \quad\quad + z = -4$

15. Using your immediate surroundings, give an example of three planes that
   (a) intersect in a single point;
   (b) do not intersect;
   (c) intersect in infinitely many points.

16. Suppose that a system has infinitely many ordered pair solutions of the form

$$(x, y, z) \text{ such that } x + y + 2z = 1.$$

Give three specific ordered pairs that are solutions of the system.

Solve each of the following systems of equations. See Examples 1, 3, and 4.

17. $\quad 2x + 2y - 6z = 5$
$\quad -3x + \ y - \ z = -2$
$\quad -x - \ y + 3z = 4$

18. $-2x + \ 5y + \ z = -3$
$\quad\ 5x + 14y - \ z = -11$
$\quad\ 7x + \ 9y - 2z = -5$

19. $-5x + 5y - 20z = -40$
$x - y + 4z = 8$
$3x - 3y + 12z = 24$

20. $x + 4y - z = 3$
$-2x - 8y + 2z = -6$
$3x + 12y - 3z = 9$

21. $2x + y - z = 6$
$4x + 2y - 2z = 12$
$-x - \frac{1}{2}y + \frac{1}{2}z = -3$

22. $2x - 8y + 2z = -10$
$-x + 4y - z = 5$
$\frac{1}{8}x - \frac{1}{2}y + \frac{1}{8}z = -\frac{5}{8}$

23. $x + y - 2z = 0$
$3x - y + z = 0$
$4x + 2y - z = 0$

24. $2x + 3y - z = 0$
$x - 4y + 2z = 0$
$3x - 5y - z = 0$

MATHEMATICAL CONNECTIONS (EXERCISES 25–30)

Suppose that on a distant planet a function of the form

$$f(x) = ax^2 + bx + c \quad (a \neq 0)$$

describes the height in feet of a projectile x seconds after it has been projected upward. Work through Exercises 25–30 in order to see how this can be related to a system of three equations in three variables a, b, and c.

25. After 1 second, the height of a certain projectile is 128 feet. Thus, $f(1) = 128$. Use this information to find one equation in the variables a, b, and c. (*Hint:* Substitute 1 for x and 128 for $f(x)$.)

26. After 1.5 seconds, the height is 140 feet. Find a second equation in a, b, and c.

27. After 3 seconds, the height is 80 feet. Find a third equation in a, b, and c.

28. Write a system of three equations in a, b, and c, based on your answers in Exercises 25–27. Solve the system.

29. What is the function f for this particular projectile?

30. What was the initial height of the projectile? (*Hint:* Find $f(0)$.)

Systems involving more than three equations can be solved by extending the methods shown in this section. Solve each of the following systems.

31.
$$\begin{aligned}
x + y + z - w &= 5 \\
2x + y - z + w &= 3 \\
x - 2y + 3z + w &= 18 \\
x + y - z - 2w &= -8
\end{aligned}$$

32.
$$\begin{aligned}
3x + y - z + 2w &= 9 \\
x + y + 2z - w &= 10 \\
x - y - z + 3w &= -2 \\
x - y + z - w &= 6
\end{aligned}$$

5.5 Applications of Linear Systems

Many applied problems involve more than one unknown quantity. Although some problems with two unknowns can be solved using just one variable, many times it is easier to use two variables. To solve a problem with two unknowns, we write two equations that relate the unknown quantities. We can then solve the system formed by the pair of equations using the methods given at the beginning of this chapter.

OBJECTIVES

1. Solve problems requiring values of unknown quantities using two variables.
2. Solve money problems using two variables.
3. Solve mixture problems using two variables.
4. Solve problems about distance, rate, and time using two variables.
5. Solve problems with three unknowns using a system of three equations.

FOR EXTRA HELP

Tutorial   Tape 10   SSM, Sec. 5.5

Solving Problems with More than One Variable

Step 1 **Determine what you are to find.** Assign a variable for each unknown and *write down* what it represents.

Step 2 **Write down other information.** If appropriate, draw a figure or a diagram and label it using the variables from Step 1. Use a chart to summarize the information.

Step 3 **Write a system of equations.** Write as many equations as there are unknowns.

Step 4 **Solve the system.**

Step 5 **Answer the question(s).** Be sure you have answered all questions posed.

Step 6 **Check.** Check your solution(s) in the original problem. Be sure your answer makes sense.

OBJECTIVE 1 The next example shows how to write a system of equations to solve a problem that requires finding two unknown values.

EXAMPLE 1 Solving a Problem to Find Unknown Values

The length of the foundation of a rectangular house is to be 6 meters more than its width. Find the length and width of the house if the perimeter must be 48 meters.

Begin by sketching a rectangle to represent the foundation of the house. Let

$$x = \text{the length}$$

and

$$y = \text{the width}.$$

See Figure 9.

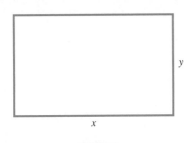

FIGURE 9

The length, x, is 6 meters more than the width, y. Therefore,

$$x = 6 + y.$$

CONTINUED ON NEXT PAGE

311

1. Solve the system shown in Example 1 to find the width and the length.

2. Write a system of equations to solve the following problem. Do not solve it.

   The perimeter of a rectangle is 76 inches. If the width were doubled, it would be 13 inches more than the length. Find the width and length.

3. (a) Jamilla bought 4 pounds of peaches and 2 pounds of apricots, paying $5. Later, she bought 7 pounds of peaches and 3 pounds of apricots for $8.25. Find the cost per pound for each fruit.

   (b) A cashier has $1260 in tens and twenties, with a total of 98 bills. How many of each type are there?

The formula for the perimeter of a rectangle is $P = 2L + 2W$. Here $P = 48$, $L = x$, and $W = y$, so

$$48 = 2x + 2y.$$

The length and width can now be found by solving the system

$$x = 6 + y$$
$$48 = 2x + 2y.$$

◄◄ **WORK PROBLEMS 1 AND 2 AT THE SIDE.**

OBJECTIVE 2 ▶ Another type of problem that often leads to a system of equations is one about different amounts of money.

E X A M P L E 2   Solving a Problem about Money

For an art project Kay bought 8 pieces of poster board and 3 marker pens for $6.50. She later needed 2 pieces of poster board and 2 pens. These items cost $3.00. Find the cost of 1 marker pen and 1 sheet of poster board.

Let x represent the cost of a piece of poster board and y represent the cost of a pen. For the first purchase, $8x$ represents the cost of the pieces of poster board and $3y$ the cost of the pens. The total cost was $6.50, so

$$8x + 3y = 6.50.$$

For the second purchase,

$$2x + 2y = 3.00.$$

To solve the system, multiply both sides of the second equation by -4 and add the result to the first equation.

$$
\begin{array}{rcr}
8x + 3y = & 6.50 \\
-8x - 8y = & -12.00 \\
\hline
-5y = & -5.50 \\
y = & 1.10
\end{array}
$$

By substituting 1.10 for y in either of the equations, verify that $x = .40$. Kay paid $.40 for a piece of poster board and $1.10 for a pen.

Note
In Example 2, x and y represented costs in *dollars* because the right side of each equation was in dollars, so the left side had to agree. Therefore, $x = .40$ represents $.40, not .40¢.

◄◄ **WORK PROBLEM 3 AT THE SIDE.**

OBJECTIVE 3 ▶ For many mixture problems it is natural to use more than one variable and a system of equations.

E X A M P L E 3   Solving a Mixture Problem

How many ounces each of 5% hydrochloric acid and 20% hydrochloric acid must be combined to get 10 ounces of solution that is 12.5% hydrochloric acid?

CONTINUED ON NEXT PAGE ─┘

ANSWERS

1. width = 9; length = 15 (Both answers are in meters.)
2.   Let x = width
          y = length.
     $2x + 2y = 76$
        $2x = y + 13$
3. (a) $.75 for peaches, $1 for apricots
   (b) 70 tens, 28 twenties

Let x represent the number of ounces of 5% solution and y represent the number of ounces of 20% solution. A table summarizes the information from the problem.

Kind of Solution	Ounces of Solution	Ounces of Acid
5%	x	$.05x$
20%	y	$.20y$
12.5%	**10**	**$(.125)10$**

When the x ounces of 5% solution and the y ounces of 20% solution are combined, the total number of ounces is 10, so that

$$x + y = 10. \qquad \textbf{(1)}$$

The ounces of acid in the 5% solution $(.05x)$ plus the ounces of acid in the 20% solution $(.20y)$ should equal the total ounces of acid in the mixture, which is $(.125)10$. That is,

$$.05x + .20y = (.125)10. \qquad \textbf{(2)}$$

Eliminate x by first multiplying both sides of equation (2) by 100 to clear it of decimals and then multiplying both sides of equation (1) by -5.

$$
\begin{array}{rl}
5x + 20y = 125 & \text{100 times both sides of equation (2)} \\
\underline{-5x - 5y = -50} & \text{-5 times both sides of equation (1)} \\
15y = 75 & \\
y = 5 &
\end{array}
$$

Because $y = 5$ and $x + y = 10$, x is also 5, so the desired mixture will require 5 ounces of the 5% solution and 5 ounces of the 20% solution.

WORK PROBLEM 4 AT THE SIDE. ▶▶

OBJECTIVE ▶ Constant rate applications require the distance formula, $d = rt$, where d is distance, r is rate (or speed), and t is time. These applications often lead naturally to a system of equations, as in the next example.

E X A M P L E   4   Solving a Motion Problem

A car travels 250 kilometers in the same time that a truck travels 225 kilometers. If the speed of the car is 8 kilometers per hour faster than the speed of the truck, find both speeds.

A table is useful to organize the information in problems about distance, rate, and time. Fill in the given information for each vehicle (in this case, distance) and use variables for the unknown speeds (rates) as follows.

	d	r	t
Car	250	x	
Truck	225	y	

The table shows nothing about time. Get an expression for time by solving the distance formula, $d = rt$, for t.

$$\frac{d}{r} = t$$

The two times can be written as $\frac{250}{x}$ and $\frac{225}{y}$.

— **CONTINUED ON NEXT PAGE**

4. (a) A grocer has some $4 per pound coffee and some $8 per pound coffee, which he will mix to make 50 pounds of $5.60 per pound coffee. How many pounds of each should be used?

(b) Some 40% ethyl alcohol solution is to be mixed with some 80% solution to get 200 liters of a 50% mixture. How many liters of each should be used?

5. A train travels 600 miles in the same time that a truck travels 520 miles. Find the speed of each vehicle if the train's average speed is 8 miles per hour faster than the truck's.

The problem states that the car travels 8 kilometers per hour faster than the truck. Since the two speeds are x and y,

$$x = y + 8.$$

Both vehicles travel for the same time, so

$$\frac{250}{x} = \frac{225}{y}.$$

This is not a linear equation. However, multiplying both sides by xy gives

$$250y = 225x,$$

which is linear. Now solve the system.

$$x = y + 8 \tag{3}$$
$$250y = 225x \tag{4}$$

The substitution method can be used. Replace x with $y + 8$ in equation (4).

$$250y = 225(y + 8) \qquad \text{Let } x = y + 8.$$
$$250y = 225y + 1800 \qquad \text{Distributive property}$$
$$25y = 1800 \qquad \text{Subtract } 225y.$$
$$y = 72 \qquad \text{Divide by 25.}$$

Because $x = y + 8$, the value of x is $72 + 8 = 80$. It is important to check the solution in the original problem since one of the equations had variable denominators. Checking verifies that the speeds are 80 kilometers per hour for the car and 72 kilometers per hour for the truck.

◀◀ **WORK PROBLEM 5 AT THE SIDE.**

OBJECTIVE ▶ Some applications involve three unknowns. When three variables are used, three equations are necessary to solve the problem. We can then use the methods of the previous section to solve the system. The next two examples illustrate this.

E X A M P L E 5   Solving a Problem about Food Prices

Joe Schwartz bought apples, hamburger, and milk at the grocery store. Apples cost $.70 a pound, hamburger was $1.50 a pound, and milk was $.80 a quart. He bought twice as many pounds of apples as hamburger. The number of quarts of milk was one more than the number of pounds of hamburger. If his total bill was $8.20, how much of each item did he buy?

First choose variables to represent the three unknowns.

Let x = the number of pounds of apples;
      y = the number of pounds of hamburger;
      z = the number of quarts of milk.

Next, use the information in the problem to write three equations. Since Joe bought twice as many pounds of apples as hamburger,

$$x = 2y$$

or

$$x - 2y = 0.$$

CONTINUED ON NEXT PAGE

ANSWERS

5. The train travels at 60 miles per hour and the truck at 52 miles per hour.

The number of quarts of milk amounted to one more than the number of pounds of hamburger, so

$$z = 1 + y$$

or

$$-y + z = 1.$$

Multiplying the cost of each item by the amount of that item and adding gives the total bill.

$$.70x + 1.50y + .80z = 8.20$$

Multiply both sides of this equation by 10 to clear it of decimals.

$$7x + 15y + 8z = 82$$

Use the method shown in the previous section to solve the system

$$x - 2y = 0$$
$$-y + z = 1$$
$$7x + 15y + 8z = 82.$$

Verify that the solution is (4, 2, 3). Now go back to the statements defining the variables to decide what the numbers of the solution represent. Doing this shows that Joe bought 4 pounds of apples, 2 pounds of hamburger, and 3 quarts of milk.

WORK PROBLEM 6 AT THE SIDE. ▶▶

Business problems involving production sometimes require the solution of a system of equations. The final example shows how to set up such a system.

E X A M P L E 6   Solving a Business Production Problem

A company produces three color television sets, models X, Y, and Z. Each model X set requires 2 hours of electronics work, 2 hours of assembly time, and 1 hour of finishing time. Each model Y requires 1, 3, and 1 hours of electronics, assembly, and finishing time, respectively. Each model Z requires 3, 2, and 2 hours of the same work, respectively. There are 100 hours available for electronics, 100 hours available for assembly, and 65 hours available for finishing per week. How many of each model should be produced each week if all available time must be used?

Let x = the number of model X produced per week;
y = the number of model Y produced per week;
z = the number of model Z produced per week.

A chart is useful for organizing the information in a problem of this type.

	Each Model X	*Each Model Y*	*Each Model Z*	*Totals*
Hours of electronics work	2	1	3	100
Hours of assembly time	2	3	2	100
Hours of finishing time	1	1	2	65

— **CONTINUED ON NEXT PAGE**

6. A department store has three kinds of perfume: cheap, better, and best. It has 10 more bottles of cheap than better, and 3 fewer bottles of the best than better. Each bottle of cheap costs $8, better costs $15, and best costs $32. The total value of all the perfume is $589. How many bottles of each are there?

7. A paper mill makes newsprint, bond, and copy machine paper. Each ton of newsprint requires 3 tons of recycled paper and 1 ton of wood pulp. Each ton of bond requires 2 tons of recycled paper, 4 tons of wood pulp, and 3 tons of rags. A ton of copy machine paper requires 2 tons of recycled paper, 3 tons of wood pulp, and 2 tons of rags. The mill has 4200 tons of recycled paper, 5800 tons of wood pulp, and 3900 tons of rags. How much of each kind of paper can be made from these supplies?

The x model X sets require $2x$ hours of electronics, the y model Y sets require $1y$ (or y) hours of electronics, and the z model Z sets require $3z$ hours of electronics. Since 100 hours are available for electronics,

$$2x + y + 3z = 100.$$

Similarly, from the fact that 100 hours are available for assembly,

$$2x + 3y + 2z = 100,$$

and the fact that 65 hours are available for finishing leads to the equation

$$x + y + 2z = 65.$$

The system

$$2x + y + 3z = 100$$
$$2x + 3y + 2z = 100$$
$$x + y + 2z = 65$$

may be solved to find $x = 15$, $y = 10$, and $z = 20$. The company should produce 15 model X, 10 model Y, and 20 model Z sets per week.

Notice the advantage of setting up the chart in Example 6. By reading across, we can easily determine the coefficients and the constants in the system.

◀◀ **WORK PROBLEM 7 AT THE SIDE.**

5.5 Exercises

For each application in this exercise set, select variables to represent the unknown quantities, write equations using the variables, and solve the resulting systems. The applications in Exercises 1–34 require solving systems with two variables. Exercises 35–42 require solving systems with three variables.

Solve each problem. See Example 1.

1. In a recent year, the number of daily newspapers in Texas was 52 more than the number of daily newspapers in Florida. Together the two states had a total of 134 dailies. How many did each state have? (*Source:* Editor & Publisher International Yearbook)

2. In the United States, the number of nuclear power plants in the South exceeds that of the Midwest by 11. Together these two areas of the country have a total of 73 nuclear power plants. How many plants does each region have? (*Source:* U.S. Energy Information Administration)

3. Find the measures of the angles marked *x* and *y*. (*Hint: x + y = 90 and y + 3x + 10 = 180.*)

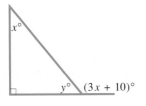

4. Find the measures of the angles marked *x* and *y*. (*Hint: x + y = 180 and x − 20 = y.*)

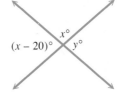

5. Pete and Monica measured the perimeter of a tennis court and found that it was 42 feet longer than it was wide and had a perimeter of 228 feet. What were the length and the width of the tennis court?

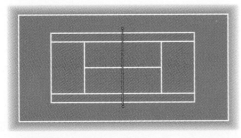

6. Michael and Penny found that the width of their basketball court was 44 feet less than the length. If the perimeter was 288 feet, what were the length and the width of their court?

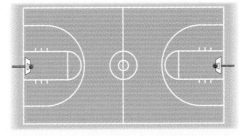

Solve each problem. See Examples 2 and 6.

7. Valencia Community College has decided to supply its mathematics labs with color monitors. A trip to the local electronics outlet provides the following information: 4 CGA monitors and 6 VGA monitors can be purchased for $4600, while 6 CGA monitors and 4 VGA monitors will cost $4400. What are the prices of a single CGA monitor and a single VGA monitor?

8. The Texas Rangers Gift Shop will sell 5 New Era baseball caps and 3 Diamond Collection jerseys for $430, or 3 New Era caps and 4 Diamond Collection jerseys for $500. What are the prices of a single cap and a single jersey?

9. A factory makes use of two basic machines, *A* and *B*, which turn out two different products, yarn and thread. Each unit of yarn requires 1 hour on machine *A* and 2 hours on machine *B*, while each unit of thread requires 1 hour on *A* and 1 hour on *B*. Machine *A* runs 8 hours per day, while machine *B* runs 14 hours per day. How many units each of yarn and thread should the factory make to keep its machines running at capacity?

10. The Doors Company makes personal computers. It has found that each standard model requires 4 hours to manufacture electronics and 2 hours for the case. The top-of-the-line model requires 5 hours for the electronics and 1.5 hours for the case. On a particular production run, the company has available 200 hours in the electronics department and 76 hours in the cabinet department. How many of each model can be made?

11. Theodis bought 2 kilograms of dark clay and 3 kilograms of light clay, paying $22 for the clay. He later needed 1 kilogram of dark clay and 2 kilograms of light clay, costing $13 altogether. How much did he pay for each type of clay?

12. Laronda wants to grow two types of algae, green and brown. She has 15 kilograms of nutrient X and 26 kilograms of nutrient Y. A vat of green algae needs 2 kilograms of nutrient X and 3 kilograms of nutrient Y, while a vat of brown algae needs 1 kilogram of nutrient X and 2 kilograms of nutrient Y. How many vats of each type of algae should she grow in order to use all the nutrients?

The formulas p = br (percentage = base × rate) and I = prt (simple interest = principal × rate × time) are used in the applications found in Exercises 17–28. To prepare for the use of these formulas, answer the questions in Exercises 13 and 14.

13. If a container of liquid contains 32 ounces of solution, what is the number of ounces of pure acid if the given solution contains the following acid concentrations?
   (a) 10%   **(b)** 25%   **(c)** 40%   **(d)** 50%

14. If $4000 is invested in an account paying simple annual interest, how much interest will be earned during the first year at the following rates?
   (a) 2%   **(b)** 3%   **(c)** 4%   **(d)** 3.5%

15. If a pound of oranges costs $.89, how much will *x* pounds cost?

16. If a ticket to a movie costs $5.50, and *y* tickets are sold, how much money is collected from the sale?

Solve each problem. See Examples 2 and 3.

17. How many gallons each of 25% alcohol and 35% alcohol should be mixed to get 20 gallons of 32% alcohol?

Kind	Amount	Pure Alcohol
25%	x	
35%	y	
32%	20	

18. How many liters each of 15% acid and 33% acid should be mixed to get 120 liters of 21% acid?

Kind	Amount	Pure Acid
15%	x	
33%	y	
21%	120	

19. Pure acid is to be added to a 10% acid solution to obtain 54 liters of a 20% acid solution. What amounts of each should be used?

20. A truck radiator holds 36 liters of fluid. How much pure antifreeze must be added to a mixture that is 4% antifreeze in order to fill the radiator with a mixture that is 20% antifreeze?

21. A party mix is made by adding nuts that sell for $2.50 a kilogram to a cereal mixture that sells for $1 a kilogram. How much of each should be added to get 30 kilograms of a mix that will sell for $1.70 a kilogram?

Ingredient	Amount	Value of the Ingredients
Nuts	x	$2.50x$
Cereal	y	$1.00y$
Mixed	30	$1.70(30)$

22. A popular fruit drink is made by mixing fruit juices. Such a mixture with 50% juice is to be mixed with another mixture that is 30% juice to get 200 liters of a mixture that is 45% juice. How much of each should be used?

Ingredient	Amount	Amount of Pure Juice
50% juice	x	$.50x$
30% juice	y	$.30y$
Mixed	200	$.45(200)$

23. Tickets to a production of *King Lear* at Delgado Community College cost $5.00 for general admission or $4.00 with a student ID. If 184 people paid to see a performance and $812 was collected, how many of each type of admission were sold?

24. Carol Britz plans to mix pecan clusters that sell for $3.60 per pound with chocolate truffles that sell for $7.20 per pound to get a mixture that she can sell in Valentine boxes for $4.95 per pound. How much of the $3.60 clusters and the $7.20 truffles should she use to create 80 pounds of the mix?

25. Cliff Morris has been saving dimes and quarters. He has 94 coins in all. If the total value is $19.30, how many dimes and how many quarters does he have?

26. A teller at the Bank of New Roads received a checking account deposit in twenty-dollar bills and fifty-dollar bills. She received a total of 70 bills, and the amount of the deposit was $3200. How many of each denomination were deposited?

27. A total of $3000 is invested, part at 2% simple interest and part at 4%. If the total annual return from the two investments is $100, how much is invested at each rate?

Rate	Amount	Interest
2%	x	
4%	y	
Total	$3000	$100

28. An investor must invest a total of $15,000 in two accounts, one paying 4% annual simple interest, and the other 3%. If he wants to earn $550 annual interest, how much should he invest at each rate?

Rate	Amount	Interest
4%	x	
3%	y	
Total	$15,000	$550

The formula d = rt (distance = rate × time) is used in the applications found in Exercises 31–34. To prepare for the use of this formula, answer the questions in Exercises 29 and 30.

29. If the speed of a killer whale is 25 miles per hour, and the whale swims for y hours, how many miles does the whale travel?

30. If the speed of a boat in still water is 20 miles per hour, and the speed of the current of a river is x miles per hour, what is the speed of the boat
  (a) going upstream (against the current) and
  (b) going downstream (with the current)?

Solve each problem. See Example 4.

31. A freight train and an express train leave towns 390 kilometers apart, traveling toward one another. The freight train travels 30 kilometers per hour slower than the express train. They pass one another 3 hours later. What are their speeds?

32. A train travels 150 kilometers in the same time that a plane covers 400 kilometers. If the speed of the plane is 20 kilometers per hour less than 3 times the speed of the train, find both speeds.

33. Braving blizzard conditions on the planet
Hoth, Luke Skywalker sets out at top speed in
his snow speeder for a rebel base 3600 miles
away. He travels into a steady headwind, and
makes the trip in 2 hours. Returning, he finds
that the trip back, still at top speed but now
with a tailwind, takes only 1.5 hours. Find the
top speed of Luke's snow speeder and the
speed of the wind.

34. Traveling for three hours into a steady headwind, a
plane flies 1650 miles. The pilot determines that
flying *with* the same wind for two hours, he could
make a trip of 1300 miles. What is the speed of
the plane and the speed of the wind?

*Solve each problem involving three unknowns. See Examples 5 and 6. (In Exercises 35–38,
remember that the sum of the measures of the angles of a triangle is 180°.)*

35. In the figure shown, $z = x + 10$ and
$x + y = 100$. Determine a third equation
involving x, y, and z, and then find the measures
of the three angles.

36. In the figure shown, x is 10 less than y and 20
less than z. Write a system of equations, and
find the measures of the three angles.

37. In a certain triangle, the measure of the
second angle is 10° more than three times
the first. The third angle measure is equal to
the sum of the measures of the other two. Find
the measures of the three angles.

38. The measure of the largest angle of a triangle is
12° less than the sum of the measures of the other
two. The smallest angle measures 58° less than the
largest. Find the measures of the angles.

39. The perimeter of a triangle is 70 centimeters.
The longest side is 4 centimeters less than the
sum of the other two sides. Twice the shortest
side is 9 centimeters less than the longest side.
Find the length of each side of the triangle.

40. The perimeter of a triangle is 56 inches. The
longest side measures 4 inches less than the sum of
the other two sides. Three times the shortest side is
4 inches more than the longest side. Find the
lengths of the three sides.

41. A Mardi Gras trinket manufacturer supplies three wholesalers, A, B, and C. The output from a day's production is 320 cases of trinkets. She must send wholesaler A three times as many cases as she sends B, and she must send wholesaler C 160 cases less than she provides A and B together. How many cases should she send to each wholesaler to distribute the entire day's production to them?

42. A motorcycle manufacturer produces three different models: the Avalon, the Durango, and the Roadripper. Production restrictions require them to make, on a monthly basis, 10 more Roadrippers than the total of the other models, and twice as many Durangos as Avalons. The shop must produce a total of 490 cycles per month. How many cycles of each type should be made per month?

MATHEMATICAL CONNECTIONS (EXERCISES 43–47)

Thus far in this text we have studied only linear *equations. In later chapters we will study the graphs of other kinds of equations. One such graph is a* circle. *We will see that an equation of the form*

$$x^2 + y^2 + ax + by + c = 0$$

may have a circle as its graph. It is a fact from geometry that given three noncollinear points (that is, points that do not all lie on the same straight line), there will be a circle that contains them. For example, the points $(4, 2)$, $(-5, -2)$, *and* $(0, 3)$ *lie on the circle whose equation is*

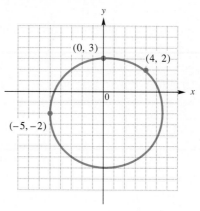

$$x^2 + y^2 - \frac{7}{5}x + \frac{27}{5}y - \frac{126}{5} = 0.$$

The circle is shown in the figure.

Work Exercises 43–47 in order, so that the equation of the circle passing through the points $(2, 1)$, $(-1, 0)$, *and* $(3, 3)$ *can be found.*

43. Let $x = 2$ and $y = 1$ in the equation $x^2 + y^2 + ax + by + c = 0$ to find an equation in a, b, and c.

44. Let $x = -1$ and $y = 0$ to find a second equation in a, b, and c.

45. Let $x = 3$ and $y = 3$ to find a third equation in a, b, and c.

46. Solve the system of equations formed by your answers in Exercises 43–45 to find the values of a, b, and c. What is the equation of the circle?

47. Explain why the graph of a circle is not that of a function.

5.6 Determinants

The topics of *matrices* and *determinants* are studied in many branches of mathematics. While matrices are important mathematical entities in their own right, we will not study their applications in this text. However, we will examine the uses of determinants.

An ordered array of numbers within square brackets, such as

$$\text{Rows} \begin{array}{c} \rightarrow \\ \rightarrow \end{array} \overset{\text{Columns}}{\underset{\downarrow \quad \downarrow \quad \downarrow}{\begin{bmatrix} 2 & 3 & 5 \\ 7 & 1 & 2 \end{bmatrix}}},$$

is called a **matrix** (MAY-triks). Matrices (the plural of matrix) are named according to the number of horizontal rows and vertical columns they contain. The number of rows is given first and the number of columns second, so that the matrix above is "two by three," written 2×3. A matrix with the same number of rows as columns, such as

$$\begin{bmatrix} -1 & 0 \\ 1 & -2 \end{bmatrix} \quad \text{and} \quad \begin{bmatrix} 8 & -1 & -3 \\ 2 & 1 & 6 \\ 0 & 5 & -3 \end{bmatrix},$$

is called a **square matrix.**

Every square matrix is associated with a real number called its **determinant** (dee-TERM-in-ent). A determinant is symbolized by the entries of the matrix placed between two vertical lines, such as

$$\begin{vmatrix} 2 & 3 \\ 7 & 1 \end{vmatrix} \quad \text{or} \quad \begin{vmatrix} 7 & 4 & 3 \\ 0 & 1 & 5 \\ -6 & 0 & -1 \end{vmatrix}.$$

Determinants are also named according to the number of rows and columns they contain. For example, the first determinant above is a 2×2 (read "two by two") determinant and the second is a 3×3 determinant.

OBJECTIVE 1 The value of the 2×2 determinant

$$\begin{vmatrix} a & b \\ c & d \end{vmatrix}$$

is defined as follows.

Value of a 2 × 2 Determinant

$$\begin{vmatrix} a & b \\ c & d \end{vmatrix} = ad - bc.$$

EXAMPLE 1   Evaluating a 2 × 2 Determinant

Evaluate the determinant.

$$\begin{vmatrix} -1 & -3 \\ 4 & -2 \end{vmatrix}$$

Here $a = -1$, $b = -3$, $c = 4$, and $d = -2$. Using these values, the determinant equals

$$\begin{vmatrix} -1 & -3 \\ 4 & -2 \end{vmatrix} = (-1)(-2) - (-3)(4)$$

$$= 2 + 12 = 14.$$

OBJECTIVES

1. Evaluate 2×2 determinants.

2. Use expansion by minors about the first column to evaluate 3×3 determinants.

3. Use expansion by minors about any row or column to evaluate determinants.

4. Evaluate larger determinants.

FOR EXTRA HELP

Tutorial   Tape 10   SSM, Sec. 5.6

1. Evaluate each determinant.

(a) $\begin{vmatrix} -4 & 6 \\ 2 & 3 \end{vmatrix}$

◀◀ **WORK PROBLEM 1 AT THE SIDE.**

A 3×3 determinant can be evaluated in a similar way.

Value of a 3 × 3 Determinant

$$\begin{vmatrix} a_1 & b_1 & c_1 \\ a_2 & b_2 & c_2 \\ a_3 & b_3 & c_3 \end{vmatrix} = \begin{aligned} &(a_1 b_2 c_3 + b_1 c_2 a_3 + c_1 a_2 b_3) \\ &- (a_3 b_2 c_1 + b_3 c_2 a_1 + c_3 a_2 b_1) \end{aligned}$$

This rule for evaluating a 3×3 determinant is difficult to remember. A method for calculating a 3×3 determinant that is easier to use is based on the rule above. Rearranging terms and using the distributive property gives

$$\begin{vmatrix} a_1 & b_1 & c_1 \\ a_2 & b_2 & c_2 \\ a_3 & b_3 & c_3 \end{vmatrix} = \begin{aligned} &a_1(b_2 c_3 - b_3 c_2) - a_2(b_1 c_3 - b_3 c_1) \\ &+ a_3(b_1 c_2 - b_2 c_1). \end{aligned} \qquad (1)$$

(b) $\begin{vmatrix} 3 & -1 \\ 0 & 2 \end{vmatrix}$

Each of the quantities in parentheses represents a 2×2 determinant that is the part of the 3×3 determinant remaining when the row and column of the multiplier are eliminated, as shown below.

$$a_1(b_2 c_3 - b_3 c_2) \quad \begin{vmatrix} a_1 & b_1 & c_1 \\ a_2 & b_2 & c_2 \\ a_3 & b_3 & c_3 \end{vmatrix}$$

$$a_2(b_1 c_3 - b_3 c_1) \quad \begin{vmatrix} a_1 & b_1 & c_1 \\ a_2 & b_2 & c_2 \\ a_3 & b_3 & c_3 \end{vmatrix}$$

$$a_3(b_1 c_2 - b_2 c_1) \quad \begin{vmatrix} a_1 & b_1 & c_1 \\ a_2 & b_2 & c_2 \\ a_3 & b_3 & c_3 \end{vmatrix}$$

These 2×2 determinants are called **minors** of the elements in the 3×3 determinant. In the determinant above, the minors of a_1, a_2, and a_3 are, respectively,

$$\begin{vmatrix} b_2 & c_2 \\ b_3 & c_3 \end{vmatrix}, \quad \begin{vmatrix} b_1 & c_1 \\ b_3 & c_3 \end{vmatrix}, \quad \begin{vmatrix} b_1 & c_1 \\ b_2 & c_2 \end{vmatrix}.$$

(c) $\begin{vmatrix} -2 & 5 \\ 1 & 5 \end{vmatrix}$

OBJECTIVE A 3×3 determinant can be evaluated by multiplying each element in the first column by its minor and combining the products as indicated in equation (1). This is called the **expansion of the determinant by minors** (ex-PAN-shun of the dee-TERM-in-ent by MY-ners) about the first column.

EXAMPLE 2 Evaluating a 3 × 3 Determinant

Evaluate the determinant using expansion by minors about the first column.

$$\begin{vmatrix} 1 & 3 & -2 \\ -1 & -2 & -3 \\ 1 & 1 & 2 \end{vmatrix}$$

In this determinant, $a_1 = 1$, $a_2 = -1$, and $a_3 = 1$. Multiply each of these numbers by its minor, and combine the three terms using the definition. Notice that the second term in the definition is *subtracted*.

$$\begin{vmatrix} 1 & 3 & -2 \\ -1 & -2 & -3 \\ 1 & 1 & 2 \end{vmatrix} = 1\begin{vmatrix} -2 & -3 \\ 1 & 2 \end{vmatrix} - (-1)\begin{vmatrix} 3 & -2 \\ 1 & 2 \end{vmatrix} + 1\begin{vmatrix} 3 & -2 \\ -2 & -3 \end{vmatrix}$$

$$= 1[(-2)(2) - (-3)(1)] + 1[(3)(2) - (-2)(1)]$$
$$+ 1[(3)(-3) - (-2)(-2)]$$
$$= 1(-1) + 1(8) + 1(-13)$$
$$= -1 + 8 - 13$$
$$= -6$$

WORK PROBLEM 2 AT THE SIDE. ▶▶

OBJECTIVE 3 To get equation (1) we could have rearranged terms in the definition of the determinant and used the distributive property to factor out the three elements of the second or third columns or of any of the three rows. Therefore, expanding by minors about any row or any column results in the same value for a 3×3 determinant. To determine the correct signs for the terms of other expansions, the following **array of signs** is helpful.

Array of Signs for a 3 × 3 Determinant

$$\begin{array}{ccc} + & - & + \\ - & + & - \\ + & - & + \end{array}$$

The signs alternate for each row and column beginning with a $+$ in the first row, first column position. For example, if the expansion is to be about the second column, the first term would have a minus sign associated with it, the second term a plus sign, and the third term a minus sign.

┌─ **E X A M P L E   3   Evaluating a 3 × 3 Determinant**

Evaluate the determinant of Example 2 using expansion by minors about the second column.

$$\begin{vmatrix} 1 & 3 & -2 \\ -1 & -2 & -3 \\ 1 & 1 & 2 \end{vmatrix}$$

$$= -3\begin{vmatrix} -1 & -3 \\ 1 & 2 \end{vmatrix} + (-2)\begin{vmatrix} 1 & -2 \\ 1 & 2 \end{vmatrix} - 1\begin{vmatrix} 1 & -2 \\ -1 & -3 \end{vmatrix}$$

$$= -3(1) - 2(4) - 1(-5)$$
$$= -3 - 8 + 5$$
$$= -6$$

As expected, the result is the same as in Example 2.

WORK PROBLEM 3 AT THE SIDE. ▶▶

2. Expand by minors about the first column.

(a) $\begin{vmatrix} 0 & -1 & 0 \\ 2 & 4 & 2 \\ 3 & 1 & 5 \end{vmatrix}$

(b) $\begin{vmatrix} 2 & 1 & 4 \\ -3 & 0 & 2 \\ -2 & 1 & 5 \end{vmatrix}$

3. Evaluate each determinant using expansion by minors about the second column.

(a) $\begin{vmatrix} 2 & 1 & 3 \\ -1 & 0 & 4 \\ 2 & 4 & 3 \end{vmatrix}$

(b) $\begin{vmatrix} 5 & -1 & 2 \\ 0 & 4 & 3 \\ -1 & 2 & 0 \end{vmatrix}$

ANSWERS

2. (a) 4 **(b)** −5

3. (a) −33 **(b)** −19

4. Evaluate.

$$\begin{vmatrix} 1 & 0 & 2 & 0 \\ 3 & 0 & 0 & 4 \\ 0 & -1 & 1 & 0 \\ 2 & 0 & -1 & 0 \end{vmatrix}$$

OBJECTIVE ▶ The method of expansion by minors can be extended to evaluate larger determinants, such as 4×4 or 5×5. For a larger determinant, the sign array also is extended. For example, the signs for a 4×4 determinant are arranged as follows.

Array of Signs for a 4 × 4 Determinant

$$\begin{array}{cccc} + & - & + & - \\ - & + & - & + \\ + & - & + & - \\ - & + & - & + \end{array}$$

EXAMPLE 4   Evaluating a 4 × 4 Determinant

Evaluate the determinant below.

$$\begin{vmatrix} -1 & -2 & 3 & 2 \\ 0 & 1 & 4 & -2 \\ 3 & -1 & 4 & 0 \\ 2 & 1 & 0 & 3 \end{vmatrix}$$

The work can be reduced by choosing a row or column with zeros, say the fourth row. Expand by minors about the fourth row using the elements of the fourth row and the signs from the fourth row of the sign array, as shown below. The minors are 3×3 determinants.

$$\begin{vmatrix} -1 & -2 & 3 & 2 \\ 0 & 1 & 4 & -2 \\ 3 & -1 & 4 & 0 \\ 2 & 1 & 0 & 3 \end{vmatrix} = -2 \begin{vmatrix} -2 & 3 & 2 \\ 1 & 4 & -2 \\ -1 & 4 & 0 \end{vmatrix} + 1 \begin{vmatrix} -1 & 3 & 2 \\ 0 & 4 & -2 \\ 3 & 4 & 0 \end{vmatrix}$$

$$- 0 \begin{vmatrix} -1 & -2 & 2 \\ 0 & 1 & -2 \\ 3 & -1 & 0 \end{vmatrix} + 3 \begin{vmatrix} -1 & -2 & 3 \\ 0 & 1 & 4 \\ 3 & -1 & 4 \end{vmatrix}$$

Now evaluate each 3×3 determinant.

$$= -2(6) + 1(-50) - 0 + 3(-41)$$
$$= -185$$

◀◀ **WORK PROBLEM 4 AT THE SIDE.**

Each of the four 3×3 determinants in Example 4 is evaluated by expansion of three 2×2 minors. Thus, a great deal of work is needed to evaluate a 4×4 or larger determinant. However, such large determinants can be evaluated quickly with the aid of a computer. Many graphing calculators also have this capability.

ANSWERS

4. 20

5.6 Exercises

Decide whether the statement is true or false.

1. A matrix is an array of numbers, while a determinant is just a number.

2. A square matrix has the same number of rows as columns.

3. The determinant $\begin{vmatrix} a & b \\ c & d \end{vmatrix}$ is equal to $ad - bc$.

4. The value of $\begin{vmatrix} 0 & 0 \\ x & y \end{vmatrix}$ is zero for any replacements for x and y.

5. The value of $\begin{vmatrix} a & a \\ a & a \end{vmatrix}$ is a for any replacement for a.

6. $\begin{vmatrix} a & b \\ c & d \end{vmatrix} = \begin{vmatrix} b & a \\ c & d \end{vmatrix}$ for any replacements for the variables.

Evaluate the following determinants. See Example 1.

7. $\begin{vmatrix} -2 & 5 \\ -1 & 4 \end{vmatrix}$

8. $\begin{vmatrix} 3 & -6 \\ 2 & -2 \end{vmatrix}$

9. $\begin{vmatrix} 1 & -2 \\ 7 & 0 \end{vmatrix}$

10. $\begin{vmatrix} -5 & -1 \\ 1 & 0 \end{vmatrix}$

11. $\begin{vmatrix} 0 & 4 \\ 0 & 4 \end{vmatrix}$

12. $\begin{vmatrix} 8 & -3 \\ 0 & 0 \end{vmatrix}$

Evaluate the following determinants using expansion by minors about the first column. See Example 2.

13. $\begin{vmatrix} -1 & 2 & 4 \\ -3 & -2 & -3 \\ 2 & -1 & 5 \end{vmatrix}$

14. $\begin{vmatrix} 2 & -3 & -5 \\ 1 & 2 & 2 \\ 5 & 3 & -1 \end{vmatrix}$

15. $\begin{vmatrix} 1 & 0 & -2 \\ 0 & 2 & 3 \\ 1 & 0 & 5 \end{vmatrix}$

16. $\begin{vmatrix} 2 & -1 & 0 \\ 0 & -1 & 1 \\ 1 & 2 & 0 \end{vmatrix}$
  17. $\begin{vmatrix} 1 & 0 & 0 \\ 0 & 1 & 0 \\ 0 & 0 & 1 \end{vmatrix}$
  18. $\begin{vmatrix} 0 & 0 & 1 \\ 0 & 1 & 0 \\ 1 & 0 & 0 \end{vmatrix}$

Evaluate the following determinants by expansion about any row or column. (Hint: If possible, choose a row or column with zeros.) See Example 3.

19. $\begin{vmatrix} 4 & 4 & 2 \\ 1 & -1 & -2 \\ 1 & 0 & 2 \end{vmatrix}$
  20. $\begin{vmatrix} 3 & -1 & 2 \\ 1 & 5 & -2 \\ 0 & 2 & 0 \end{vmatrix}$
  21. $\begin{vmatrix} 2 & 0 & 1 \\ -1 & 0 & 2 \\ 5 & 0 & 4 \end{vmatrix}$

22. $\begin{vmatrix} 2 & -4 & 0 \\ 3 & -5 & 0 \\ 6 & -7 & 0 \end{vmatrix}$
  23. $\begin{vmatrix} -6 & 3 & 5 \\ -3 & 2 & 2 \\ 0 & 0 & 0 \end{vmatrix}$
  24. $\begin{vmatrix} 0 & 0 & 0 \\ 4 & 0 & -2 \\ 2 & -1 & 3 \end{vmatrix}$

25. $\begin{vmatrix} 3 & 5 & -2 \\ 1 & -4 & 1 \\ 3 & 1 & -2 \end{vmatrix}$
  26. $\begin{vmatrix} 1 & 3 & 2 \\ 3 & -1 & -2 \\ 1 & 10 & 20 \end{vmatrix}$
  27. $\begin{vmatrix} 1 & 3 & -2 \\ -1 & 4 & 5 \\ 2 & 6 & -4 \end{vmatrix}$

Evaluate the following. Expand by minors about the second row. See Example 4.

28. $\begin{vmatrix} 1 & 4 & 2 & 0 \\ 0 & 2 & 0 & -1 \\ 3 & -1 & 2 & 0 \\ 1 & 4 & -1 & 3 \end{vmatrix}$
  29. $\begin{vmatrix} 4 & 1 & 0 & 2 \\ 1 & 0 & 0 & -2 \\ 3 & 4 & 1 & -3 \\ -2 & 1 & 1 & -1 \end{vmatrix}$

30. $\begin{vmatrix} 3 & -5 & 1 & 9 \\ 0 & 5 & 2 & 0 \\ 2 & -1 & -1 & -1 \\ -4 & 2 & 2 & 2 \end{vmatrix}$ 　　　　　**31.** $\begin{vmatrix} 4 & 2 & 2 & 2 \\ 1 & -1 & 0 & 0 \\ 2 & 1 & 1 & 1 \\ 0 & 0 & -3 & -2 \end{vmatrix}$

32. Consider the following statement: For every square matrix, there is one and only one determinant associated with the matrix.

　　Explain how this statement illustrates the concept of function.

MATHEMATICAL CONNECTIONS (EXERCISES 33–36)

Recall the formula for slope and the point-slope form of the equation of a line, as found in Chapter 4. Use these formulas in working Exercises 33–36 in order, so that you can see how a determinant can be used in writing the equation of a line.

33. Write the expression for the slope of a line passing through the points (x_1, y_1) and (x_2, y_2).

34. Using the expression from Exercise 33 as m, and the point (x_1, y_1), write the point-slope form of the equation of the line.

35. Using the equation obtained in Exercise 34, multiply both sides by $x_2 - x_1$, and write the equation so that 0 is on the right side.

36. Consider the *determinant equation*

$$\begin{vmatrix} x & y & 1 \\ x_1 & y_1 & 1 \\ x_2 & y_2 & 1 \end{vmatrix} = 0.$$

Expand by minors on the left and show that this determinant equation yields the same result that you obtained in Exercise 35.

INTERPRETING TECHNOLOGY (EXERCISES 37–40)

In Example 2 we showed how to evaluate the determinant

$$\begin{vmatrix} 1 & 3 & -2 \\ -1 & -2 & -3 \\ 1 & 1 & 2 \end{vmatrix}$$

by expanding by minors. Modern graphing calculators have the capability of finding determinants at the stroke of a key. The display on the left is a graphing calculator-generated depiction of the matrix with the same entries as shown in Example 2, and the display on the right shows that its determinant is indeed −6.

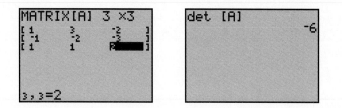

Predict the display that the calculator would give if the determinant of the matrix shown were evaluated.

37.

38.

39.

40.

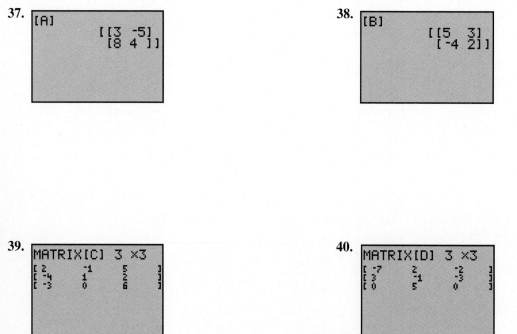

5.7 Solution of Linear Systems of Equations by Determinants—Cramer's Rule

OBJECTIVES

1 Understand the derivation of Cramer's rule.

2 Apply Cramer's rule to a linear system with two equations and two unknowns.

3 Apply Cramer's rule to a linear system with three equations and three unknowns.

4 Determine when Cramer's rule does not apply.

FOR EXTRA HELP

Tutorial   Tape 11   SSM, Sec. 5.7

In the previous section we discussed how to evaluate determinants. In this section we will show how determinants can be used to solve systems of equations. They will be important in the development of Cramer's rule, stated later in this section.

OBJECTIVE 1 We begin by using the elimination method to solve the general system of two equations in two variables,

$$a_1x + b_1y = c_1 \qquad (1)$$
$$a_2x + b_2y = c_2. \qquad (2)$$

The result will be a formula that can be used directly for any system of two equations with two unknowns. To get this general solution, eliminate y and solve for x by first multiplying both sides of equation (1) by b_2 and both sides of equation (2) by $-b_1$. Add these results and solve for x.

$$
\begin{array}{ll}
a_1b_2x + b_1b_2y = c_1b_2 & b_2 \text{ times both sides of equation (1)} \\
-a_2b_1x - b_1b_2y = -c_2b_1 & -b_1 \text{ times both sides of equation (2)} \\
\hline
(a_1b_2 - a_2b_1)x = c_1b_2 - c_2b_1 &
\end{array}
$$

$$x = \frac{c_1b_2 - c_2b_1}{a_1b_2 - a_2b_1}, \quad a_1b_2 - a_2b_1 \neq 0$$

(What happens if $a_1b_2 - a_2b_1 = 0$? We will have more to say about this later in this section.)

To solve for y, multiply both sides of equation (1) by $-a_2$ and both sides of equation (2) by a_1 and add.

$$
\begin{array}{ll}
-a_1a_2x - a_2b_1y = -a_2c_1 & -a_2 \text{ times both sides of equation (1)} \\
a_1a_2x + a_1b_2y = a_1c_2 & a_1 \text{ times both sides of equation (2)} \\
\hline
(a_1b_2 - a_2b_1)y = a_1c_2 - a_2c_1 &
\end{array}
$$

$$y = \frac{a_1c_2 - a_2c_1}{a_1b_2 - a_2b_1}, \quad a_1b_2 - a_2b_1 \neq 0$$

Both numerators and the common denominator of these values for x and y can be written as determinants because

$$a_1c_2 - a_2c_1 = \begin{vmatrix} a_1 & c_1 \\ a_2 & c_2 \end{vmatrix},$$

$$c_1b_2 - c_2b_1 = \begin{vmatrix} c_1 & b_1 \\ c_2 & b_2 \end{vmatrix},$$

and

$$a_1b_2 - a_2b_1 = \begin{vmatrix} a_1 & b_1 \\ a_2 & b_2 \end{vmatrix}.$$

Using these results, the solutions for x and y become

$$x = \frac{\begin{vmatrix} c_1 & b_1 \\ c_2 & b_2 \end{vmatrix}}{\begin{vmatrix} a_1 & b_1 \\ a_2 & b_2 \end{vmatrix}} \quad \text{and} \quad y = \frac{\begin{vmatrix} a_1 & c_1 \\ a_2 & c_2 \end{vmatrix}}{\begin{vmatrix} a_1 & b_1 \\ a_2 & b_2 \end{vmatrix}}, \quad \begin{vmatrix} a_1 & b_1 \\ a_2 & b_2 \end{vmatrix} \neq 0.$$

For convenience, denote the three determinants in the solution as

$$\begin{vmatrix} a_1 & b_1 \\ a_2 & b_2 \end{vmatrix} = D, \quad \begin{vmatrix} c_1 & b_1 \\ c_2 & b_2 \end{vmatrix} = D_x, \quad \begin{vmatrix} a_1 & c_1 \\ a_2 & c_2 \end{vmatrix} = D_y.$$

Notice that the elements of D are the four coefficients of the variables in the given system; the elements of D_x are obtained by replacing the

1. Solve by Cramer's rule.

(a) $x + y = 5$
     $x - y = 1$

(b) $2x - 3y = -26$
     $3x + 4y = 12$

(c) $4x - 5y = -8$
     $3x + 7y = -6$

coefficients of x by the respective constants; the elements of D_y are obtained by replacing the coefficients of y by the respective constants.

     These results are summarized as **Cramer's rule.**

Cramer's Rule for 2 × 2 Systems

Given the system

$$a_1x + b_1y = c_1$$
$$a_2x + b_2y = c_2 \quad \text{with} \quad a_1b_2 - a_2b_1 = D \neq 0,$$

then

$$x = \frac{\begin{vmatrix} c_1 & b_1 \\ c_2 & b_2 \end{vmatrix}}{\begin{vmatrix} a_1 & b_1 \\ a_2 & b_2 \end{vmatrix}} = \frac{D_x}{D} \quad \text{and} \quad y = \frac{\begin{vmatrix} a_1 & c_1 \\ a_2 & c_2 \end{vmatrix}}{\begin{vmatrix} a_1 & b_1 \\ a_2 & b_2 \end{vmatrix}} = \frac{D_y}{D}.$$

OBJECTIVE 2▶ To use Cramer's rule to solve a system of equations, find the three determinants, D, D_x, and D_y and then write the necessary quotients for x and y.

Caution
As indicated above, Cramer's rule does not apply if $D = a_1b_2 - a_2b_1$ is 0. When $D = 0$, the system is inconsistent or has dependent equations. For this reason, it is a good idea to evaluate D first.

E X A M P L E   I   **Using Cramer's Rule for a 2 × 2 System**

Use Cramer's rule to solve the system

$$5x + 7y = -1$$
$$6x + 8y = 1.$$

     By Cramer's rule, $x = \dfrac{D_x}{D}$ and $y = \dfrac{D_y}{D}$. As mentioned above, it is a good idea to find D first, since if $D = 0$, Cramer's rule does not apply. If $D \neq 0$, then find D_x and D_y.

$$D = \begin{vmatrix} 5 & 7 \\ 6 & 8 \end{vmatrix} = 5(8) - 6(7) = -2;$$

$$D_x = \begin{vmatrix} -1 & 7 \\ 1 & 8 \end{vmatrix} = (-1)8 - 7(1) = -15;$$

$$D_y = \begin{vmatrix} 5 & -1 \\ 6 & 1 \end{vmatrix} = 5(1) - (-1)6 = 11.$$

From Cramer's rule,

$$x = \frac{D_x}{D} = \frac{-15}{-2} = \frac{15}{2},$$

and

$$y = \frac{D_y}{D} = \frac{11}{-2} = -\frac{11}{2}.$$

The solution set is $\{(\frac{15}{2}, -\frac{11}{2})\}$, as can be verified by checking in the given system.

OBJECTIVE ▶3▶ In a similar manner, Cramer's rule can be applied to systems of three equations with three variables.

2. Find D_y and D_z.

Cramer's Rule for 3 × 3 Systems

Given the system

$$a_1x + b_1y + c_1z = d_1$$
$$a_2x + b_2y + c_2z = d_2$$
$$a_3x + b_3y + c_3z = d_3$$

with

$$D_x = \begin{vmatrix} d_1 & b_1 & c_1 \\ d_2 & b_2 & c_2 \\ d_3 & b_3 & c_3 \end{vmatrix}, \quad D_y = \begin{vmatrix} a_1 & d_1 & c_1 \\ a_2 & d_2 & c_2 \\ a_3 & d_3 & c_3 \end{vmatrix},$$

$$D_z = \begin{vmatrix} a_1 & b_1 & d_1 \\ a_2 & b_2 & d_2 \\ a_3 & b_3 & d_3 \end{vmatrix}, \quad D = \begin{vmatrix} a_1 & b_1 & c_1 \\ a_2 & b_2 & c_2 \\ a_3 & b_3 & c_3 \end{vmatrix} \neq 0,$$

then

$$x = \frac{D_x}{D}, \quad y = \frac{D_y}{D}, \quad z = \frac{D_z}{D}.$$

E X A M P L E   2 **Using Cramer's Rule for a 3 × 3 System**

Use Cramer's rule to solve the system

$$x + y - z + 2 = 0$$
$$2x - y + z + 5 = 0$$
$$x - 2y + 3z - 4 = 0.$$

To use Cramer's rule, we must first rewrite the system in the form

$$x + y - z = -2$$
$$2x - y + z = -5$$
$$x - 2y + 3z = 4.$$

Expand by minors about row 1 to find D.

$$D = \begin{vmatrix} 1 & 1 & -1 \\ 2 & -1 & 1 \\ 1 & -2 & 3 \end{vmatrix} = 1\begin{vmatrix} -1 & 1 \\ -2 & 3 \end{vmatrix} - 1\begin{vmatrix} 2 & 1 \\ 1 & 3 \end{vmatrix} + (-1)\begin{vmatrix} 2 & -1 \\ 1 & -2 \end{vmatrix}$$

$$= 1(-1) - 1(5) - 1(-3) = -3$$

Expanding D_x by minors about row 1 gives

$$D_x = \begin{vmatrix} -2 & 1 & -1 \\ -5 & -1 & 1 \\ 4 & -2 & 3 \end{vmatrix}$$

$$= -2\begin{vmatrix} -1 & 1 \\ -2 & 3 \end{vmatrix} - 1\begin{vmatrix} -5 & 1 \\ 4 & 3 \end{vmatrix} + (-1)\begin{vmatrix} -5 & -1 \\ 4 & -2 \end{vmatrix}$$

$$= -2(-1) - 1(-19) - 1(14) = 7.$$

WORK PROBLEM 2 AT THE SIDE. ▶▶

CONTINUED ON NEXT PAGE

3. Solve by Cramer's rule.

(a) $x + y + z = 2$
$ 2x - z = -3$
$ y + 2z = 4$

Using the results for D and D_x and the results from Problem 2 at the side, apply Cramer's rule to get

$$x = \frac{D_x}{D} = \frac{7}{-3} = -\frac{7}{3}, \quad y = \frac{D_y}{D} = \frac{-22}{-3} = \frac{22}{3},$$

$$z = \frac{D_z}{D} = \frac{-21}{-3} = 7.$$

The solution set is $\{(-\frac{7}{3}, \frac{22}{3}, 7)\}$.

◀◀ WORK PROBLEM 3 AT THE SIDE.

OBJECTIVE ▶ As mentioned earlier, Cramer's rule does not apply when $D = 0$. The next example illustrates this case.

EXAMPLE 3 Determining When Cramer's Rule Does Not Apply

If possible, use Cramer's rule to solve the following system.

$$2x - 3y + 4z = 8$$
$$6x - 9y + 12z = 24$$
$$x + 2y - 3z = 5$$

(b) $3x - 2y + 4z = 5$
$ 4x + y + z = 14$
$ x - y - z = 1$

Find D, D_x, D_y, and D_z. Here

$$D = \begin{vmatrix} 2 & -3 & 4 \\ 6 & -9 & 12 \\ 1 & 2 & -3 \end{vmatrix} = 2 \begin{vmatrix} -9 & 12 \\ 2 & -3 \end{vmatrix} - 6 \begin{vmatrix} -3 & 4 \\ 2 & -3 \end{vmatrix} + 1 \begin{vmatrix} -3 & 4 \\ -9 & 12 \end{vmatrix}$$

$$= 2(3) - 6(1) + 1(0)$$

$$= 0.$$

Since $D = 0$ here, Cramer's rule does not apply and we must use another method to solve the system. Multiplying the first equation on both sides by 3 shows that the first two equations have the same solutions, so this system has dependent equations and has infinitely many solutions.

4. Solve by Cramer's rule if applicable.

$$x - y + z = 6$$
$$3x + 2y + z = 4$$
$$2x - 2y + 2z = 14$$

◀◀ WORK PROBLEM 4 AT THE SIDE.

Cramer's rule can be extended to 4×4 or larger systems. See a standard college algebra text for details.

5.7 Exercises

1. For the system

$$8x - 4y = 8$$
$$x + 3y = 22,$$

$D_x = 112$, $D_y = 168$, and $D = 28$. What is the solution set of the system?

2. For the system

$$x + 3y - 6z = 7$$
$$2x - y + z = 1$$
$$x + 2y + 2z = -1,$$

the solution set is $\{(1, 0, -1)\}$ and $D = -43$. Find the values of D_x, D_y, and D_z.

Use Cramer's rule to solve each of the following linear systems in two variables. See Example 1.

3. $3x + 5y = -5$
    $-2x + 3y = 16$

4. $5x + 2y = -3$
    $4x - 3y = -30$

5. $3x + 2y = 3$
    $2x - 4y = 2$

6. $7x - 2y = 6$
    $4x - 5y = 15$

7. $8x + 3y = 1$
    $6x - 5y = 2$

8. $3x - y = 9$
    $2x + 5y = 8$

Use Cramer's rule where applicable to solve the following linear systems. If Cramer's rule does not apply, say so. See Examples 2 and 3.

9. $2x + 3y + 2z = 15$
    $x - y + 2z = 5$
    $x + 2y - 6z = -26$

10. $x - y + 6z = 19$
     $3x + 3y - z = 1$
     $x + 9y + 2z = -19$

11. $2x + 2y + z = 10$
     $4x - y + z = 20$
     $-x + y - 2z = -5$

12. $x + 3y - 4z = -12$
     $3x + y - z = -5$
     $5x - y + z = -3$

13. $2x - 3y + 4z = 8$
     $6x - 9y + 12z = 24$
     $-4x + 6y - 8z = -16$

14. $7x + y - z = 4$
     $2x - 3y + z = 2$
     $-6x + 9y - 3z = -6$

15. $3x + 5z = 0$
     $2x + 3y = 1$
     $-y + 2z = -11$

16. $-x + 2y = 4$
     $3x + y = -5$
     $2x + z = -1$

17. $x - 3y = 13$
     $2y + z = 5$
     $-x + z = -7$

18.
$$-5x - y \qquad = -10$$
$$3x + 2y + z = -3$$
$$-y - 2z = -13$$

19.
$$3x + 2y \qquad - w = 0$$
$$2x \qquad + z + 2w = 5$$
$$x + 2y - z \qquad = -2$$
$$2x - y + z + w = 2$$

20.
$$x + 2y - z + w = 8$$
$$2x - y \qquad - w = 12$$
$$y + 3z \qquad = 11$$
$$x \qquad - z - w = 4$$

21. Make up a system of two equations in two variables having consecutive integers as coefficients and constants. For example, one such system would be

$$2x + 3y = 4$$
$$5x + 6y = 7.$$

Now solve the system using Cramer's rule. Repeat, using six different consecutive integers. Compare the two solution sets. What do you notice?

MATHEMATICAL CONNECTIONS (EXERCISES 22–24)

In this section we have seen how determinants can be used to solve systems of equations. In the Mathematical Connections in the exercises for Section 5.6, we saw how a determinant can be used to write the equation of a line given two points on a line. Here, we show how a determinant can be used to find the area of a triangle if we know the coordinates of its vertices.

Suppose that $A(x_1, y_1)$, $B(x_2, y_2)$, and $C(x_3, y_3)$ are the coordinates of the vertices of triangle ABC in the coordinate plane. Then it can be shown that the area of the triangle is given by the absolute value of

$$\frac{1}{2} \begin{vmatrix} x_1 & y_1 & 1 \\ x_2 & y_2 & 1 \\ x_3 & y_3 & 1 \end{vmatrix}.$$

Work Exercises 22–24 in order.

22. Sketch triangle ABC in the coordinate plane, given that the coordinates of A are $(0, 0)$, of B are $(-3, -4)$, and of C are $(2, -2)$.

23. Write the determinant expression as described above that gives the area of triangle ABC described in Exercise 22.

24. Evaluate the absolute value of the determinant expression in Exercise 23 to find the area.

5.1	**system of linear equations**	A system of linear equations consists of two or more linear equations with the same variables.
	solution of a system	The solution of a system of linear equations includes all the ordered pairs that make all the equations of the system true at the same time.
	solution set of a system	The set of all solutions of a system forms its solution set.
	consistent system	A system of equations with a solution is a consistent system.
	inconsistent system	An inconsistent system of equations is a system with no solutions.
	independent equations	Equations of a system that have different graphs are called independent equations.
	dependent equations	Equations of a system that have the same graph (because they are different forms of the same equation) are called dependent equations.
5.6	**matrix**	A matrix is a rectangular array of numbers, consisting of horizontal rows and vertical columns.
	square matrix	A square matrix is a matrix that has the same number of rows as columns.
	determinant	Associated with every square matrix is a real number called its determinant, symbolized by the entries of the matrix between two vertical lines.
	expansion by minors	A method of evaluating a 3 × 3 or larger determinant is called expansion by minors.

NEW SYMBOLS

(x, y, z) ordered triple

$\begin{bmatrix} a & b \\ c & d \end{bmatrix}$ 2 × 2 matrix

$\begin{vmatrix} a & b & c \\ d & e & f \\ g & h & i \end{vmatrix}$ 3 × 3 determinant

$\begin{vmatrix} a & b \\ c & d \end{vmatrix}$ 2 × 2 determinant

QUICK REVIEW

Concepts	Examples
5.1 Solving Systems of Linear Equations by Graphing An ordered pair is a solution of a system if it makes all equations of the system true at the same time. If the graphs of the equations of a system are both sketched on the same axes, the points of intersection, if any, are solutions of the system. The set of all solutions is the solution set.	Is $(4, -1)$ a solution of the following system? $$x + y = 3$$ $$2x - y = 9$$ Yes, because $4 + (-1) = 3$, and $2(4) - (-1) = 9$ are both true. $\{(3, 2)\}$ is the solution set of the system $$x + y = 5$$ $$2x - y = 4.$$

Concepts	Examples
5.2 Solving Systems of Linear Equations by Addition	Solve by addition.
Step 1 Write both equations in the form $Ax + By = C$.	$$x + 3y = 7$$ $$3x - y = 1$$
Step 2 If necessary, multiply one or both equations by appropriate numbers so that the coefficients of x (or y) are negatives of each other.	Multiply the top equation by -3 to eliminate the x terms.
Step 3 Add the equations to get an equation with only one variable.	$\begin{array}{r} -3x - 9y = -21 \\ \underline{3x - y = 1} \\ -10y = -20 \quad \text{Add.} \end{array}$
Step 4 Solve the equation from Step 3.	$y = 2 \qquad \text{Divide by } -10.$
Step 5 Substitute the solution from Step 4 into either of the original equations.	Substitute to get the value of x.
Step 6 Solve the resulting equation from Step 5 for the remaining variable.	$$x + 3y = 7$$ $$x + 3(2) = 7 \qquad \text{Let } y = 2.$$ $$x + 6 = 7 \qquad \text{Multiply.}$$ $$x = 1 \qquad \text{Subtract 6.}$$
Step 7 Check the answer.	The solution set is $\{(1, 2)\}$.
If the result of the addition step is a false statement, such as $0 = 4$, the graphs are parallel lines, and there is *no solution* for the system. The solution set is $\emptyset$.	$\begin{array}{r} x - 2y = 6 \\ \underline{-x + 2y = -2} \\ \mathbf{0 = 4} \qquad \text{No solution; solution set is } \emptyset. \end{array}$
If the result is a true statement, such as $0 = 0$, the graphs are the *same line,* and an infinite number of ordered pairs are solutions.	$\begin{array}{r} x - 2y = 6 \\ \underline{-x + 2y = -6} \\ \mathbf{0 = 0} \qquad \text{Infinite number of solutions} \end{array}$
5.3 Solving Systems of Linear Equations by Substitution	Solve by substitution.
Solve one equation for one variable, and substitute the expression into the other equation to get an equation in one variable. Solve the equation, and then substitute the solution into either of the original equations to obtain the other variable. Check the answer.	$$x + 2y = -5 \qquad \textbf{(1)}$$ $$y = -2x - 1 \qquad \textbf{(2)}$$ Substitute $-2x - 1$ for y in equation (1). $$x + 2(-2x - 1) = -5$$ Solve to get $x = 1$. To find y, let $x = 1$ in equation (2): $y = -2(1) - 1 = -3$. The solution set is $\{(1, -3)\}$.

Concepts	Examples
5.4 Linear Systems of Equations in Three Variables	
Solving Linear Systems in Three Variables	Solve the system
Step 1 Use the elimination method to eliminate any variable from any two of the given equations. The result is an equation in two variables.	$$x + 2y - z = 6$$ $$x + y + z = 6$$ $$2x + y - z = 7.$$ Add the first and second equations; z is eliminated and the result is $2x + 3y = 12$.
Step 2 Eliminate the *same* variable from any *other* two equations. The result is an equation in the same two variables as in Step 1.	Eliminate z again by adding the second and third equations to get $3x + 2y = 13$. Now solve the system $$2x + 3y = 12 \qquad (*)$$ $$3x + 2y = 13.$$
Step 3 Use the elimination method to eliminate a second variable from the two equations in two variables that result from Steps 1 and 2. The result is an equation in one variable that gives the value of that variable.	To eliminate x, multiply the top equation by -3 and the bottom equation by 2. $$-6x - 9y = -36$$ $$\underline{6x + 4y = 26}$$ $$-5y = -10$$ $$y = 2$$
Step 4 Substitute the value of the variable found in Step 3 into either of the equations in two variables to find the value of the second variable.	Let $y = 2$ in equation $(*)$. $$2x + 3(2) = 12$$ $$2x + 6 = 12$$ $$2x = 6$$ $$x = 3$$
Step 5 Use the values of the two variables from Steps 3 and 4 to find the value of the third variable by substituting into any of the original equations.	Let $y = 2$ and $x = 3$ in any of the original equations to find $z = 1$. The solution set is $\{(3, 2, 1)\}$.
5.5 Applications of Linear Systems	
To solve an applied problem with two (three) unknowns, write two (three) equations that relate the unknowns. Then solve the system.	The perimeter of a rectangle is 18 feet. The length is 3 feet more than twice the width. Find the dimensions of the rectangle.
	Let x represent the length and y represent the width. From the perimeter formula, one equation is $2x + 2y = 18$. From the problem, another equation is $x = 3 + 2y$. Now solve the system $$2x + 2y = 18$$ $$x = 3 + 2y.$$
	The solution of the system is $(7, 2)$. Therefore, the length is 7 feet and the width is 2 feet.

Concepts	*Examples*

5.6 Determinants

Value of a 2 × 2 Determinant

$$\begin{vmatrix} a & b \\ c & d \end{vmatrix} = ad - bc.$$

Determinants larger than 2 × 2 are evaluated by expansion by minors about a column or row.

Array of Signs for a 3 × 3 Determinant

$$\begin{array}{ccc} + & - & + \\ - & + & - \\ + & - & + \end{array}$$

$$\begin{vmatrix} 3 & 4 \\ -2 & 6 \end{vmatrix} = (3)(6) - (4)(-2) = 26$$

Evaluate the following determinant by expanding about the second column.

$$\begin{vmatrix} 2 & -3 & -2 \\ -1 & -4 & -3 \\ -1 & 0 & 2 \end{vmatrix} = 3(-5) + (-4)(2) - (0)(-8)$$

$$= -15 - 8 + 0$$

$$= -23$$

5.7 Solution of Linear Systems of Equations by Determinants—Cramer's Rule

Cramer's Rule for 2 × 2 Systems

Given the system

$$a_1 x + b_1 y = c_1$$
$$a_2 x + b_2 y = c_2$$

with $a_1 b_2 - a_2 b_1 = D \neq 0$, then

$$x = \frac{\begin{vmatrix} c_1 & b_1 \\ c_2 & b_2 \end{vmatrix}}{\begin{vmatrix} a_1 & b_1 \\ a_2 & b_2 \end{vmatrix}} = \frac{D_x}{D}$$

and

$$y = \frac{\begin{vmatrix} a_1 & c_1 \\ a_2 & c_2 \end{vmatrix}}{\begin{vmatrix} a_1 & b_1 \\ a_2 & b_2 \end{vmatrix}} = \frac{D_y}{D}.$$

For a 3 × 3 system, D is the determinant of the matrix of coefficients; D_x is found by replacing the coefficients of x in D with the constants; D_y is found by replacing the coefficients of y in D with the constants; D_z is found by replacing the coefficients of z in D with the constants.

Solve using Cramer's rule.

$$x - 2y = -1$$
$$2x + 5y = 16$$

$$x = \frac{\begin{vmatrix} -1 & -2 \\ 16 & 5 \end{vmatrix}}{\begin{vmatrix} 1 & -2 \\ 2 & 5 \end{vmatrix}} = \frac{-5 + 32}{5 + 4} = \frac{27}{9} = 3$$

$$y = \frac{\begin{vmatrix} 1 & -1 \\ 2 & 16 \end{vmatrix}}{\begin{vmatrix} 1 & -2 \\ 2 & 5 \end{vmatrix}} = \frac{16 + 2}{5 + 4} = \frac{18}{9} = 2$$

The solution set is $\{(3, 2)\}$.

Solve using Cramer's rule.

$$3x + 2y + z = -5$$
$$x - y + 3z = -5$$
$$2x + 3y + z = 0$$

Using expansion by minors, it can be shown that $D_x = 45$, $D_y = -30$, $D_z = 0$, and $D = -15$. Therefore,

$$x = \frac{D_x}{D} = \frac{45}{-15} = -3,$$

$$y = \frac{D_y}{D} = \frac{-30}{-15} = 2,$$

$$z = \frac{D_z}{D} = \frac{0}{-15} = 0.$$

The solution set is $\{(-3, 2, 0)\}$.

[5.1] *Decide whether the given ordered pair is a solution of the given system.*

1. (3, 4)
$$4x - 2y = 4$$
$$5x + y = 19$$

2. (1, −3)
$$5x + 3y = -4$$
$$2x - 3y = 11$$

3. (−5, 2)
$$x - 4y = -13$$
$$2x + 3y = 4$$

4. (0, 1)
$$3x + 8y = 8$$
$$4x - 3y = 3$$

Solve each system by graphing.

5. $x + y = 4$
$2x - y = 5$

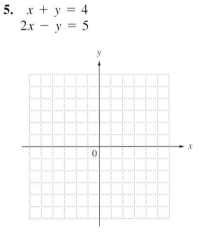

6. $x - 2y = 4$
$2x + y = -2$

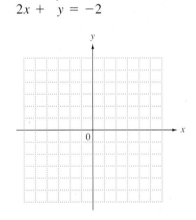

7. $x - 2 = 2y$
$2x - 4y = 4$

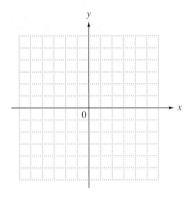

8. $2x + 4 = 2y$
$y - x = -3$

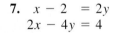

9. Why would a system of linear equations having the solution set $\{(-\frac{1}{2}, \frac{2}{3})\}$ be difficult to solve using the graphing method?

10. When a student was asked to determine whether the ordered pair $(1, -2)$ is a solution of the system

$$x + y = -1$$
$$2x + y = 4,$$

he answered "yes." His reasoning was that the ordered pair satisfies the equation $x + y = -1$; that is, $1 + (-2) = -1$ is true. Why is the student's answer wrong?

[5.2] *Solve each system by the addition method.*

11. $2x - y = 13$
$\quad\ x + y = \ 8$

12. $3x - \ y = -13$
$\quad\ x - 2y = -1$

13. $5x + 4y = -7$
$\quad\ 3x - 4y = -17$

14. $-4x + 3y = \ \ 25$
$\quad\ \ \ 6x - 5y = -39$

15. $3x - 4y = 9$
$\quad\ 6x - 8y = 18$

16. $\quad\ 2x + \ y = 3$
$\quad -4x - 2y = 6$

17. Only one of the following systems does not require that we multiply one or both equations by a constant in order to solve the system by the addition method. Which one is it?

(a) $-4x + 3y = 7$
$\quad\ \ \ 3x - 4y = 4$

(b) $\quad\ 5x + 8y \ \ = 13$
$\quad\ 12x + 24y = 36$

(c) $2x + 3y = 5$
$\quad\ x - 3y = 12$

(d) $\quad x + 2y = 9$
$\quad 3x - \ \ y = 6$

18. For the system

$$2x + 12y = 7$$
$$3x + \ \ 4y = 1,$$

if we were to multiply the first (top) equation by -3, by what number would we have to multiply the second (bottom) equation in order to
(a) eliminate the x terms when solving by the addition method?
(b) eliminate the y terms when solving by the addition method?

[5.3] *Solve each system by the substitution method.*

19. $3x + y = 7$
    $x = 2y$

20. $2x - 5y = -19$
    $y = x + 2$

21. $4x + 5y = 44$
    $x + 2 = 2y$

22. $5x + 15y = 3$
    $x + 3y = 2$

Solve each system by any method. First simplify equations, and clear them of fractions where necessary.

23. $2x + 3y = -5$
    $3x + 4y = -8$

24. $6x - 9y = 0$
    $2x - 3y = 0$

25. $2x + y - x = 3y + 5$
    $y + 2 = x - 5$

26. $5x - 3 + y = 4y + 8$
    $2y + 1 = x - 3$

27. $\dfrac{x}{2} + \dfrac{y}{3} = 7$

$\dfrac{x}{4} + \dfrac{2y}{3} = 8$

28. $\dfrac{3x}{4} - \dfrac{y}{3} = \dfrac{7}{6}$

$\dfrac{x}{2} + \dfrac{2y}{3} = \dfrac{5}{3}$

[5.4] *Solve the following systems of equations using the addition method.*

29. $2x + 3y - z = -16$
  $x + 2y + 2z = -3$
  $-3x + y + z = -5$

30. $3x - y - z = -8$
  $4x + 2y + 3z = 15$
  $-6x + 2y + 2z = 10$

31. $4x - y \quad\;\; = 2$
  $3y + z = 9$
  $x \quad\;\; + 2z = 7$

32. $x + 5y - 3z = 0$
  $2x + 6y + z = 0$
  $3x - y + 4z = 0$

[5.5] *Solve the following problems by writing a system of equations.*

33. A rectangular table top is 2 feet longer than it is wide, and its perimeter is 20 feet. Find the length and the width of the table top.

34. On an 8-day business trip, Clay rented a car for $32 per day at weekday rates and $19 per day at weekend rates. If his total rental bill was $217, how many days did he rent at each rate?

35. A plane flies 560 miles in 1.75 hours traveling with the wind. The return trip later against the same wind takes the plane 2 hours. Find the speed of the plane and the speed of the wind.

36. Sweet's Candy Store is offering a special mix for Valentine's Day. Latoya Sweet will mix some $2-a-pound candy with some $1-a-pound candy to get 100 pounds of mix that she will sell at $1.30 a pound. How many pounds of each should she use?

37. The sum of the measures of the angles of a triangle is 180°. One angle measures 10° less
than the sum of the other two. The measure of
the middle-sized angle is the average of the
other two. Find the measures of the three
angles.

38. David Zerangue sells real estate. On three recent sales, he made 10% commission,
6% commission, and 5% commission. His total commissions on these sales were $17,000,
and he sold property worth $280,000. If the
5% sale amounted to the sum of the other two,
what were the three sales' prices?

39. The manager of a candy store wants to feature
a special Easter candy mixture of jelly beans,
small chocolate eggs, and marshmallow
chicks. She plans to make 15 pounds of mix to
sell at $1 a pound. Jelly beans sell for $.80 a
pound, chocolate eggs for $2 a pound, and
marshmallow chicks for $1 a pound. She will
use twice as many pounds of jelly beans as
eggs and chicks combined, and fives times as
many pounds of jelly beans as chocolate eggs.
How many pounds of each candy should she
use?

40. Gil Troutman has a collection of tropical fish.
For each fish, he paid either $20, $40, or $65.
The number of $40 fish is one less than twice
the number of $20 fish. If there are 29 fish in
all worth $1150, how many of each kind of
fish are in the collection?

[5.6] *Evaluate each of the following determinants.*

41. $\begin{vmatrix} 2 & -9 \\ 8 & 4 \end{vmatrix}$

42. $\begin{vmatrix} 7 & 0 \\ 5 & -3 \end{vmatrix}$

43. $\begin{vmatrix} 2 & 10 & 4 \\ 0 & 1 & 3 \\ 0 & 6 & -1 \end{vmatrix}$

44. $\begin{vmatrix} 0 & 0 & 0 \\ 0 & 2 & 5 \\ -1 & 3 & 6 \end{vmatrix}$

45. $\begin{vmatrix} 0 & 0 & 2 \\ 2 & 1 & 0 \\ -1 & 0 & 0 \end{vmatrix}$

46. $\begin{vmatrix} 1 & 3 & -2 \\ 2 & 6 & -4 \\ 5 & 0 & 1 \end{vmatrix}$

[5.7]

47. Under what conditions can a system *not* be solved using Cramer's rule?

Use Cramer's rule to solve the following systems of equations.

48. $3x - 4y = 5$
$2x + y = 8$

49. $-4x + 3y = -12$
$2x + 6y = 15$

50. $4x + y + z = 11$
$x - y - z = 4$
$y + 2z = 0$

51. $-x + 3y - 4z = 2$
$2x + 4y + z = 3$
$3x - z = 9$

MIXED REVIEW EXERCISES

Solve by any method.

52. $\dfrac{2}{3}x + \dfrac{1}{6}y = \dfrac{19}{2}$

$\dfrac{1}{3}x - \dfrac{2}{9}y = 2$

53. $2x + 5y - z = 12$
$-x + y - 4z = -10$
$-8x - 20y + 4z = 31$

54. $x = 7y + 10$
$2x + 3y = 3$

55. $x + 4y = 17$
$-3x + 2y = -9$

56. $-7x + 3y = 12$
$5x + 2y = 8$

57. $2x - 5y = 8$
$3x + 4y = 10$

58. To make a 10% acid solution for chemistry class, Xavier wants to mix some 5% solution with 10 liters of 20% solution. How many liters of 5% solution should he use?

59. The sum of the three angles of a triangle is 180°. The largest angle is twice the measure of the smallest, and the third angle measures 10° less than the largest. Find the measures of the three angles.

CHAPTER 5 TEST

1. Use a graph to solve the system

$x + y = 7$
$x - y = 5$.

1. _____

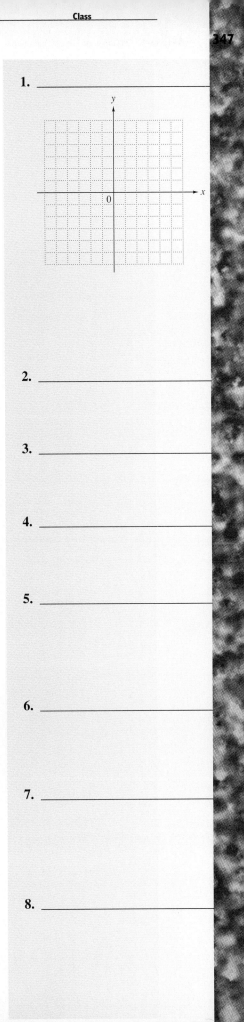

Solve each system using elimination.

2. $3x + y = 12$
$2x - y = 3$

2. _____

3. $-5x + 2y = -4$
$6x + 3y = -6$

3. _____

4. $3x + 4y = 8$
$8y = 7 - 6x$

4. _____

5. $3x + 5y + 3z = 2$
$6x + 5y + z = 0$
$3x + 10y - 2z = 6$

5. _____

Solve each system using substitution.

6. $2x - 3y = 24$
$y = -\dfrac{2}{3}x$

6. _____

7. $12x - 5y = 8$
$3x = \dfrac{5}{4}y + 2$

7. _____

Solve each problem by writing a system of equations.

8. In an election, one candidate received 45 more votes than the other. The total number of votes cast in the election was 405. Find the number of votes received by each candidate.

8. _____

9. _____

9. A chemist needs 12 liters of a 40% alcohol solution. She must mix a 20% solution and a 50% solution. How many liters of each will be required to obtain what she needs?

10. _____

10. A local electronics store will sell 7 AC adaptors and 2 rechargeable flashlights for $86, or 3 AC adaptors and 4 rechargeable flashlights for $84. What is the price of a single AC adaptor and a single rechargeable flashlight?

11. _____

11. The owner of a tea shop wants to mix three kinds of tea to make 100 ounces of a mixture that will sell for $.83 an ounce. He uses Orange Pekoe, which sells for $.80 an ounce, Irish Breakfast, for $.85 an ounce, and Earl Grey, for $.95 an ounce. If he wants to use twice as much Orange Pekoe as Irish Breakfast, how much of each kind of tea should he use?

Evaluate each of the following determinants.

12. _____

12. $\begin{vmatrix} 6 & -3 \\ 5 & -2 \end{vmatrix}$

13. _____

13. $\begin{vmatrix} 4 & 1 & 0 \\ -2 & 7 & 3 \\ 0 & 5 & 2 \end{vmatrix}$

Solve each of the following systems by Cramer's rule.

14. _____

14. $3x - y = -8$
$2x + 6y = 3$

15. _____

15. $x + y + z = 4$
$-2x + z = 5$
$3y + z = 9$

Evaluate.

1. $(-3)^4$ **2.** -3^4 **3.** $-(-3)^4$ **4.** $\sqrt{.49}$

5. $-\sqrt{.49}$ **6.** $\sqrt{-.49}$ **7.** $\sqrt{16}$ **8.** $-\sqrt{16}$

Evaluate if $x = -4$, $y = 3$, and $z = 6$.

9. $|2x| + 3y - z^3$ **10.** $-5(x^3 - y^3)$ **11.** $\dfrac{2x^2 - x + z}{3y - z}$

Solve each equation.

12. $7(2x + 3) - 4(2x + 1) = 2(x + 1)$ **13.** $|6x - 8| = 4$

14. $ax + by = cx + d$ for x **15.** $.04x + .06(x - 1) = 1.04$

Solve each inequality.

16. $\dfrac{2}{3}y + \dfrac{5}{12}y \leq 20$ **17.** $|3x + 2| \leq 4$ **18.** $|12t + 7| \geq 0$

Solve each problem.

19. Two cars start from points 420 miles apart and travel toward each other. They meet after 3.5 hours. Find the average speed of each car if one travels 30 miles per hour slower than the other.

20. A triangle has an area of 42 square meters. The base is 14 meters long. Find the height of the triangle.

21. A jar contains only pennies, nickels, and dimes. The number of dimes is 1 more than the number of nickels, and the number of pennies is 6 more than the number of nickels. How many of each denomination can be found in the jar, if the total value is $4.80?

22. Two angles of a triangle have the same measure. The measure of the third angle is 4° less than twice the measure of each of the equal angles. Find the measures of the three angles.

In Exercises 23–28, point A has coordinates $(-2, 6)$ and point B has coordinates $(4, -2)$.

23. What is the equation of the horizontal line through *A*?

24. What is the equation of the vertical line through *B*?

25. What is the slope of AB?

26. What is the slope of a line perpendicular to line AB?

27. What is the standard form of the equation of line AB?

28. Write the equation of the line in the form of a linear function.

29. Find the standard form of the equation of the line with x-intercept $(-3, 0)$ and y-intercept $(0, 5)$.

30. Graph the linear function whose graph has slope $\frac{2}{3}$ and passes through the point $(-1, -3)$.

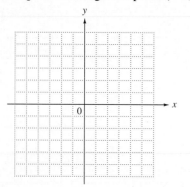

31. Graph the inequality $-3x - 2y \le 6$.

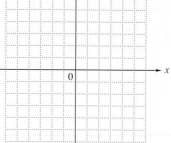

Solve by any method.

32. $-2x + 3y = -15$
$4x - y = 15$

33. $x + y + z = 10$
$x - y - z = 0$
$-x + y - z = -4$

34. Evaluate.

$$\begin{vmatrix} 1 & 2 & 3 \\ 0 & 5 & 1 \\ -1 & 0 & 4 \end{vmatrix}$$

In Exercises 35–36, solve the problem using a system of equations. Use the method of your choice: elimination, substitution, or Cramer's rule.

35. Mabel Johnston bought apples and oranges at DeVille's Grocery. She bought 6 pounds of fruit. Oranges cost $.90 per pound, while apples cost $.70 per pound. If she spent a total of $5.20, how many pounds of each kind of fruit did she buy?

36. Alexis has inherited $80,000 from her aunt. She invests part of the money in a video rental firm that produces a return of 7% per year, and divides the rest equally between a tax-free bond at 6% a year and a money market fund at 4% a year. Her annual return on these investments is $5200. How much is invested in each?

The graph shows a company's costs to produce computer parts and the revenue from the sale of computer parts.

37. At what production level does the cost equal the revenue? What is the revenue at that point?

38. Profit is revenue less cost. Estimate the profit on the sale of 1100 parts.

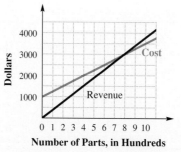

Exponents and Polynomials

6

6.1 The Product Rule and Power Rules for Exponents

OBJECTIVE 1 In Chapter 1 we used exponents to write repeated products. Recall that in the expression 5^2, the number 5 is called the *base* and 2 is called the *exponent* or *power*. The expression 5^2 is called an *exponential expression*. Usually we do not write a quantity with an exponent of 1; however, sometimes it is convenient to do so. In general, for any quantity a, $a = a^1$.

EXAMPLE 1 Using Exponents

Write $3 \cdot 3 \cdot 3 \cdot 3 \cdot 3$ in exponential form, and evaluate the exponential expression.

Since 3 occurs as a factor five times, the base is 3 and the exponent is 5. The exponential expression is 3^5, read "3 to the fifth power" or simply "3 to the fifth." The value is

$$3^5 = 3 \cdot 3 \cdot 3 \cdot 3 \cdot 3 = 243.$$

WORK PROBLEM 1 AT THE SIDE. ▶▶

EXAMPLE 2 Evaluating an Exponential Expression

Evaluate each exponential expression. Name the base and the exponent.

	Base	Exponent
(a) $5^4 = 5 \cdot 5 \cdot 5 \cdot 5 = 625$	5	4
(b) $-5^4 = -1 \cdot 5^4 = -1 \cdot (5 \cdot 5 \cdot 5 \cdot 5) = -625$	5	4
(c) $(-5)^4 = (-5)(-5)(-5)(-5) = 625$	-5	4

Caution

It is important to understand the difference between parts (b) and (c) of Example 2. In -5^4 the lack of parentheses shows that the exponent 4 refers only to the base 5, and not -5; in $(-5)^4$ the parentheses show that the exponent 4 refers to the base -5. In summary, $-a^n$ and $(-a)^n$ are not always the same.

Expression	Base	Exponent	Example
$-a^n$	a	n	$-3^2 = -(3 \cdot 3) = -9$
$(-a)^n$	$-a$	n	$(-3)^2 = (-3)(-3) = 9$

www.mathnotes.com

OBJECTIVES

1 ▶ Use exponents.

2 ▶ Use the product rule for exponents.

3 ▶ Use the rule $(a^m)^n = a^{mn}$.

4 ▶ Use the rule $(ab)^m = a^m b^m$.

5 ▶ Use the rule $\left(\dfrac{a}{b}\right)^m = \dfrac{a^m}{b^m}$.

6 ▶ Use combinations of the rules for exponents.

FOR EXTRA HELP

Tutorial   Tape 11   SSM, Sec. 6.1

1. Write $2 \cdot 2 \cdot 2 \cdot 2$ in exponential form, and evaluate.

351

2. Evaluate each exponential expression. Name the base and the exponent.

(a) $(-2)^5$ (b) -2^5

(c) -4^2 (d) $(-4)^2$

3. Find each product by the product rule, if possible.

(a) $8^2 \cdot 8^5$

(b) $(-7)^5 \cdot (-7)^3$

(c) $y^3 \cdot y$

(d) $4^2 \cdot 3^5$

(e) $6^4 + 6^2$

◀◀ WORK PROBLEM 2 AT THE SIDE.

OBJECTIVE 2 By the definition of exponential expressions,

$$2^4 \cdot 2^3 = \overbrace{(2 \cdot 2 \cdot 2 \cdot 2)}^{4\ \text{factors}}\overbrace{(2 \cdot 2 \cdot 2)}^{3\ \text{factors}}$$
$$= \underbrace{2 \cdot 2 \cdot 2 \cdot 2 \cdot 2 \cdot 2 \cdot 2}_{4 + 3 = 7\ \text{factors}}$$
$$= 2^7.$$

Also,

$$6^2 \cdot 6^3 = (6 \cdot 6)(6 \cdot 6 \cdot 6)$$
$$= 6 \cdot 6 \cdot 6 \cdot 6 \cdot 6$$
$$= 6^5.$$

Generalizing from these examples, $2^4 \cdot 2^3 = 2^{4+3} = 2^7$ and $6^2 \cdot 6^3 = 6^5$. In each case, adding the exponents gives the exponent of the product, suggesting the **product rule for exponents.**

Product Rule for Exponents

For any positive integers m and n, $\quad a^m \cdot a^n = a^{m+n}$. (Keep the same base and add the exponents.)
Example: $6^2 \cdot 6^5 = 6^{2+5} = 6^7$

EXAMPLE 3 Using the Product Rule

Use the product rule for exponents to find each product, if possible.

(a) $6^3 \cdot 6^5 = 6^{3+5} = 6^8$ by the product rule.

(b) $(-4)^7(-4)^2 = (-4)^{7+2} = (-4)^9$ by the product rule.

(c) $x^2 \cdot x = x^2 \cdot x^1 = x^{2+1} = x^3$

(d) $m^4 \cdot m^3 = m^{4+3} = m^7$

(e) The product rule does not apply to the product $2^3 \cdot 3^2$ because the bases are different.

$$2^3 \cdot 3^2 = 8 \cdot 9 = 72.$$

(f) The product rule does not apply to $2^3 + 2^4$ because it is a *sum*, not a *product*.

$$2^3 + 2^4 = 8 + 16 = 24$$

Caution
The bases must be the same before we can apply the product rule for exponents.

◀◀ WORK PROBLEM 3 AT THE SIDE.

ANSWERS
2. (a) -32; -2; 5 (b) -32; 2; 5
(c) -16; 4; 2 (d) 16; -4; 2
3. (a) 8^7 (b) $(-7)^8$ (c) y^4
(d) cannot use the product rule
(e) cannot use the product rule

E X A M P L E 4   Using the Product Rule

Multiply $2x^3$ and $3x^7$.

Since $2x^3$ means $2 \cdot x^3$ and $3x^7$ means $3 \cdot x^7$, we use the associative and commutative properties and the product rule to get

$$2x^3 \cdot 3x^7 = 2 \cdot 3 \cdot x^3 \cdot x^7 = 6x^{10}.$$

Caution
Be sure you understand the difference between *adding* and *multiplying* exponential expressions. For example,

$$8x^3 + 5x^3 = 13x^3$$

but $\qquad (8x^3)(5x^3) = 8 \cdot 5 \cdot x^{3+3} = 40x^6.$

WORK PROBLEM 4 AT THE SIDE. ▶▶

OBJECTIVE 3 We can simplify an expression such as $(8^3)^2$ with the product rule for exponents, as follows.

$$(8^3)^2 = (8^3)(8^3) = 8^{3+3} = 8^6$$

The product of the exponents, 3 and 2, in $(8^3)^2$ gives the exponent in 8^6. As another example,

$$(5^2)^3 = 5^2 \cdot 5^2 \cdot 5^2$$
$$= 5^{2+2+2}$$
$$= 5^6,$$

and $2 \cdot 3 = 6$. These examples suggest power rule (a) for exponents.

Power Rule (a) for Exponents

For any positive integers m and n, $\qquad (a^m)^n = a^{mn}$.
(Raise a power to a power by multiplying exponents.)
Example: $(3^2)^4 = 3^{2 \cdot 4} = 3^8$

E X A M P L E 5   Using Power Rule (a)

Use power rule (a) for exponents to simplify each expression.

(a) $(2^5)^3 = 2^{5 \cdot 3} = 2^{15}$

(b) $(5^7)^2 = 5^{7 \cdot 2} = 5^{14}$

(c) $(x^2)^5 = x^{2 \cdot 5} = x^{10}$

(d) $(n^3)^2 = n^{3 \cdot 2} = n^6$

WORK PROBLEM 5 AT THE SIDE. ▶▶

OBJECTIVE 4 The properties studied in Chapter 1 can be used to develop two more rules for exponents. Using the definition of an exponential expression and the commutative and associative properties, we can evaluate the expression $(4x)^3$ as shown below.

$$(4x)^3 = (4x)(4x)(4x) \qquad \text{Definition of exponent}$$
$$= 4 \cdot 4 \cdot 4 \cdot x \cdot x \cdot x \qquad \text{Commutative and associative properties}$$
$$= 4^3x^3 \qquad \text{Definition of exponent}$$

This example suggests power rule (b) for exponents.

---

4. Multiply.

(a) $5m^2 \cdot 2m^6$

(b) $3p^5 \cdot 9p^4$

(c) $-7p^5 \cdot (3p^8)$

5. Simplify each expression.

(a) $(5^3)^4$

(b) $(6^2)^5$

(c) $(3^2)^4$

(d) $(a^6)^5$

6. Simplify.

(a) $5(mn)^3$

(b) $(3a^2b^4)^5$

(c) $(5m^2)^3$

Power Rule (b) for Exponents

For any positive integer m, $(ab)^m = a^m b^m$.
(Raise a product to a power by raising each factor to the power.)
Example: $(2p)^5 = 2^5 p^5$

E X A M P L E   6   Using Power Rule (b)

Use power rule (b) to simplify each expression.

(a) $(3xy)^2 = 3^2 x^2 y^2$

$\qquad\qquad = 9x^2 y^2$

(b) $9(pq)^2 = 9(p^2 q^2)$ Power rule (b)

$\qquad\qquad = 9p^2 q^2$ Multiply.

(c) $5(2m^2 p^3)^4 = 5[2^4 (m^2)^4 (p^3)^4]$ Power rule (b)

$\qquad\qquad\quad = 5(2^4 m^8 p^{12})$ Power rule (a)

$\qquad\qquad\quad = 5 \cdot 2^4 m^8 p^{12}$

$\qquad\qquad\quad = 80 m^8 p^{12}$ $5 \cdot 2^4 = 5 \cdot 16 = 80$

◀◀ **WORK PROBLEM 6 AT THE SIDE.**

7. Simplify. Assume all variables represent nonzero real numbers.

(a) $\left(\dfrac{5}{2}\right)^4$

(b) $\left(\dfrac{p}{q}\right)^2$

(c) $\left(\dfrac{r}{t}\right)^3$

OBJECTIVE 5 Since the quotient $\frac{a}{b}$ can be written as $a \cdot \frac{1}{b}$, we can use power rule (b), together with some of the properties of real numbers, to get power rule (c) for exponents.

Power Rule (c) for Exponents

For any positive integer m, $\left(\dfrac{a}{b}\right)^m = \dfrac{a^m}{b^m}$ $(b \neq 0)$.

(Raise a quotient to a power by raising both the numerator and the denominator to the power.)
Example: $\left(\dfrac{5}{3}\right)^2 = \dfrac{5^2}{3^2}$

E X A M P L E   7   Using Power Rule (c)

Simplify each expression.

(a) $\left(\dfrac{2}{3}\right)^5 = \dfrac{2^5}{3^5}$

(b) $\left(\dfrac{m}{n}\right)^4 = \dfrac{m^4}{n^4}, \; n \neq 0$

◀◀ **WORK PROBLEM 7 AT THE SIDE.**

The rules for exponents discussed in this section are now summarized. These rules are basic to the study of algebra and should be *memorized*.

ANSWERS

6. (a) $5m^3 n^3$ (b) $3^5 a^{10} b^{20}$ (c) $5^3 m^6$

7. (a) $\dfrac{5^4}{2^4}$ (b) $\dfrac{p^2}{q^2}$ (c) $\dfrac{r^3}{t^3}$

Rules for Exponents

For positive integers m and n:		*Examples*
Product rule	$a^m \cdot a^n = a^{m+n}$	$6^2 \cdot 6^5 = 6^{2+5} = 6^7$
Power rules (a)	$(a^m)^n = a^{mn}$	$(3^2)^4 = 3^{2 \cdot 4} = 3^8$
(b)	$(ab)^m = a^m b^m$	$(2p)^5 = 2^5 p^5$
(c)	$\left(\dfrac{a}{b}\right)^m = \dfrac{a^m}{b^m} \quad (b \neq 0)$	$\left(\dfrac{5}{3}\right)^2 = \dfrac{5^2}{3^2}$

OBJECTIVE 6 ▶ As shown in the next example, more than one rule may be needed to simplify an expression.

─**E X A M P L E 8**   **Using Combinations of Rules**

Simplify each expression.

(a) $\left(\dfrac{2}{3}\right)^2 \cdot 2^3$

$$\left(\dfrac{2}{3}\right)^2 \cdot 2^3 = \dfrac{2^2}{3^2} \cdot \dfrac{2^3}{1} \qquad \text{Power rule (c)}$$

$$= \dfrac{2^2 \cdot 2^3}{3^2 \cdot 1} \qquad \text{Multiply fractions.}$$

$$= \dfrac{2^5}{3^2} \qquad \text{Product rule}$$

(b) $(5x)^3(5x)^4$

$$(5x)^3(5x)^4 = (5x)^7 \qquad \text{Product rule}$$

$$= 5^7 x^7 \qquad \text{Power rule (b)}$$

(c) $(2x^2 y^3)^4 (3xy^2)^3$

$$(2x^2 y^3)^4(3xy^2)^3 = 2^4(x^2)^4(y^3)^4 \cdot 3^3 x^3 (y^2)^3 \qquad \text{Power rule (b)}$$

$$= 2^4 \cdot 3^3 x^8 y^{12} x^3 y^6 \qquad \text{Power rule (a)}$$

$$= 16 \cdot 27 x^{11} y^{18} \qquad \text{Product rule}$$

$$= 432 x^{11} y^{18}$$

■

Caution

Refer to Example 8(c). Notice that

$$(2x^2 y^3)^4 \neq (2 \cdot 4) x^{2 \cdot 4} y^{3 \cdot 4}.$$

Do not multiply the coefficient 2 and the exponent 4.

WORK PROBLEM 8 AT THE SIDE. ▶▶

8. Simplify.

(a) $(2m)^3(2m)^4$

(b) $\left(\dfrac{5k^3}{3}\right)^2$

(c) $\left(\dfrac{1}{5}\right)^4 (2x)^2$

(d) $(3xy^2)^3(x^2 y)^4$

ANSWERS

8. (a) $2^7 m^7$ **(b)** $\dfrac{5^2 k^6}{3^2}$

(c) $\dfrac{2^2 x^2}{5^4}$ **(d)** $3^3 x^{11} y^{10}$

Real World
collaborative investigations

Do You Have the Knack for Percent?

Do you have the "knack?" We are all born with different abilities. That's what makes us unique individuals. Some people have musical talents, while others can't carry a tune. Some people can repair automobiles, sew, or build things while others can't do any of these. And some people have a knack for percents, while others are completely befuddled by them. Do you have the knack?

For example, suppose that you need to compute 20% of 50. You have the knack if you use one of these methods:

1. You think "Well, 20% means $\frac{1}{5}$, and to find $\frac{1}{5}$ of something I divide by 5, so 50 divided by 5 is 10. The answer is 10."

2. You think "20% is twice 10%, and to find 10% of something I move the decimal point one place to the left. So, 10% of 50 is 5, and 20% is twice 5, or 10. The answer is 10."

If you don't have the knack, you probably search for a calculator whenever you need to compute a percent, and hope like crazy that you'll work it correctly. Keep in mind one thing, however: There's nothing to be ashamed of if you don't have the knack. The methods explained in this section allow you to learn how to compute percents using tried-and-true mathematical methods. And just because you don't have the knack, it doesn't mean that you can't succeed in mathematics or can't learn other concepts. One ability does not necessarily assure success in another seemingly similar area. Case in point: The nineteenth-century German Zacharias Dase was a lightning-swift calculator, and could do things like multiply 79,532,853 by 93,758,479 in his head in less than one minute, but had no concept of theoretical mathematics!

FOR GROUP DISCUSSION

1. Suppose you are in a restaurant and want to leave a 15% tip for the server. Discuss the various ways of determining 15% of the bill.

2. A television reporter once asked a professional wrist-wrestler what percent of his sport was physical and what percent was mental. The athlete responded "I would say it's 50% physical and 90% mental." Comment on this response.

3. We often hear claims such as "She gave 110% effort." Comment on this claim. Do you think that this is actually possible?

6.1 Exercises

Write each expression using exponents. See Example 1.

1. $3 \cdot 3 \cdot 3 \cdot 3 \cdot 3 \cdot 3 \cdot 3$

2. $(-6)(-6)(-6)(-6)$

3. $(-2)(-2)(-2)(-2)(-2)$

4. $t \cdot t \cdot t \cdot t \cdot t \cdot t$

5. $w \cdot w \cdot w \cdot w \cdot w \cdot w$

6. $\left(\frac{1}{2}\right)\left(\frac{1}{2}\right)\left(\frac{1}{2}\right)\left(\frac{1}{2}\right)\left(\frac{1}{2}\right)\left(\frac{1}{2}\right)$

7. $\left(-\frac{1}{4}\right)\left(-\frac{1}{4}\right)\left(-\frac{1}{4}\right)\left(-\frac{1}{4}\right)\left(-\frac{1}{4}\right)$

8. $(-8p)(-8p)$

9. $(-7x)(-7x)(-7x)(-7x)$

10. Explain how the expressions $(-3)^4$ and -3^4 are different.

11. Explain how the expressions $(5x)^3$ and $5x^3$ are different.

Identify the base and the exponent for each exponential expression. In Exercises 12–15, also evaluate the expression. See Example 2.

12. 3^5

13. 2^7

14. $(-3)^5$

15. $(-2)^7$

16. $(-6x)^4$

17. $(-8x)^4$

18. $-6x^4$

19. $-8x^4$

20. Explain why the product rule does not apply to the expression $5^2 + 5^3$. Then evaluate the expression by finding the individual powers and then adding the results.

INTERPRETING TECHNOLOGY (EXERCISES 21–22)

---

21. Two graphing calculator screens are shown here. What do you think the calculator notation $a \wedge b$ means?

```
(-6)(-6)(-6)(-6)
           1296
(-6)^4
           1296
-6^4
          -1296
```

```
(-7)(-7)(-7)(-7)
           2401
7^4
           2401
-7^4
          -2401
```

22. Use your answer to Exercise 21 to evaluate $4 \wedge 3$.

23. What exponent is understood on the base x in the expression xy^2?

24. How are the expressions 3^2, 5^3, and 7^4 read?

Use the product rule to simplify each expression. Write each answer in exponential form. See Examples 3 and 4.

25. $5^2 \cdot 5^6$

26. $3^6 \cdot 3^7$

27. $4^2 \cdot 4^7 \cdot 4^3$

28. $5^3 \cdot 5^8 \cdot 5^2$

29. $(-7)^3(-7)^6$

30. $(-9)^8(-9)^5$

31. $t^3 \cdot t^8 \cdot t^{13}$

32. $n^5 \cdot n^6 \cdot n^9$

33. $(-8r^4)(7r^3)$

34. $(10a^7)(-4a^3)$

35. $(-6p^5)(-7p^5)$

36. $(-5w^8)(-9w^8)$

37. Explain why the product rule does not apply to the expression $3^2 \cdot 4^3$. Then evaluate the expression by finding the individual powers and multiplying the results.

38. Repeat Exercise 37 for the expression $(-3)^3 \cdot (-2)^5$.

In each of the following exercises, first add the given terms. Then start over and multiply them.

39. $5x^4, 9x^4$

40. $8t^5, 3t^5$

41. $-7a^2, 2a^2, 10a^2$

42. $6x^3, 9x^3, -2x^3$

Use the power rules for exponents to simplify each expression. Write each answer in exponential form. See Examples 5–7.

43. $(4^3)^2$

44. $(8^3)^6$

45. $(t^4)^5$

46. $(y^6)^5$

47. $(7r)^3$

48. $(11x)^4$

49. $(5xy)^5$

50. $(9pq)^6$

51. $(-5^2)^6$

52. $(-9^4)^8$

53. $(-8^3)^5$

54. $(-7^5)^7$

55. $8(qr)^3$

56. $4(vw)^5$

57. $\left(\dfrac{1}{2}\right)^3$

58. $\left(\dfrac{1}{3}\right)^5$

59. $\left(\dfrac{a}{b}\right)^3 (b \neq 0)$

60. $\left(\dfrac{r}{t}\right)^4 (t \neq 0)$

61. $\left(\dfrac{9}{5}\right)^8$

62. $\left(\dfrac{12}{7}\right)^3$

Use a combination of the rules of exponents introduced in this section to simplify each expression. See Example 8.

63. $\left(\dfrac{5}{2}\right)^3 \cdot \left(\dfrac{5}{2}\right)^2$

64. $\left(\dfrac{3}{4}\right)^5 \cdot \left(\dfrac{3}{4}\right)^6$

65. $\left(\dfrac{9}{8}\right)^3 \cdot 9^2$

66. $\left(\dfrac{8}{5}\right)^4 \cdot 8^3$

67. $(2x)^9(2x)^3$

68. $(6y)^5(6y)^8$

69. $(-6p)^4(-6p)$

70. $(-13q)^3(-13q)$

71. $(6x^2y^3)^5$

72. $(5r^5t^6)^7$

73. $(x^2)^3(x^3)^5$

74. $(y^4)^5(y^3)^5$

75. $(2w^2x^3y)^2(x^4y)^5$

76. $(3x^4y^2z)^3(yz^4)^5$

77. $(-r^4s)^2(-r^2s^3)^5$

78. $(-ts^6)^4(-t^3s^5)^3$

79. $\left(\dfrac{5a^2b^5}{c^6}\right)^3 (c \neq 0)$

80. $\left(\dfrac{6x^3y^9}{z^5}\right)^4 (z \neq 0)$

81. A student tried to simplify $(10^2)^3$ as 1000^6. Is this correct? If not, how is it simplified using the product rule for exponents?

82. Explain why $(3x^2y^3)^4$ is *not* equivalent to $(3 \cdot 4)x^8y^{12}$.

6.2 Integer Exponents and the Quotient Rule

OBJECTIVES

1. Use zero as an exponent.
2. Use negative numbers as exponents.
3. Use the quotient rule for exponents.
4. Use combinations of rules.

FOR EXTRA HELP

Tutorial   Tape 11   SSM, Sec. 6.2

OBJECTIVE 1 In the previous section we studied the product rule for exponents. In all of our work, exponents were positive integers. To develop meanings for exponents other than positive integers (such as 0 and negative integers), we want to define them in such a way that rules for exponents are the same, regardless of the kind of number used for the exponents.

Suppose we want to find a meaning for an expression such as

$$6^0,$$

where 0 is used as an exponent. If we were to multiply this by 6^2, for example, we would want the product rule to still be valid. Therefore, we would have

$$6^0 \cdot 6^2 = 6^{0+2} = 6^2.$$

So multiplying 6^2 by 6^0 should give 6^2. Because 6^0 is acting as if it were 1 here, we should define 6^0 to equal 1. This is the definition for 0 used as an exponent with any nonzero base.

Definition of Zero Exponent

For any nonzero real number a, $a^0 = 1$.
Example: $17^0 = 1$

E X A M P L E 1   Using Zero Exponents
Evaluate each exponential expression.

(a) $60^0 = 1$

(b) $(-60)^0 = 1$

(c) $-60^0 = -(1) = -1$

(d) $y^0 = 1$, if $y \neq 0$

(e) $-r^0 = -1$, if $r \neq 0$

Caution

Notice the difference between parts (b) and (c) of Example 1. In Example 1(b) the base is -60 and the exponent is 0. Any nonzero base raised to a zero exponent is 1. But in Example 1(c), the base is 60. Then $60^0 = 1$, and $-60^0 = -1$.

WORK PROBLEM 1 AT THE SIDE. ▶▶

OBJECTIVE 2 Now let us consider how we can define negative integers as exponents. Suppose that we want to give a meaning to

$$6^{-2}$$

so that the product rule is still valid. If we multiply 6^{-2} by 6^2, we get

$$6^{-2} \cdot 6^2 = 6^{-2+2} = 6^0 = 1.$$

The expression 6^{-2} is acting as if it were the reciprocal of 6^2, because their product is 1. The reciprocal of 6^2 may be written $\frac{1}{6^2}$, leading us to define 6^{-2} as $\frac{1}{6^2}$. This is a particular case of the definition of negative exponents.

1. Evaluate.

(a) 28^0

(b) $(-16)^0$

(c) -7^0

(d) $m^0, m \neq 0$

(e) $-p^0, p \neq 0$

Definition of Negative Exponents

For any nonzero real number a and any integer n, $\quad a^{-n} = \dfrac{1}{a^n}$.

Example: $\quad 3^{-2} = \dfrac{1}{3^2}$

By definition, a^{-n} and a^n are reciprocals, since

$$a^n \cdot a^{-n} = a^n \cdot \frac{1}{a^n} = 1.$$

Since $1^n = 1$, the definition of a^{-n} also can be written

$$a^{-n} = \frac{1}{a^n} = \frac{1^n}{a^n} = \left(\frac{1}{a}\right)^n.$$

For example,

$$6^{-3} = \left(\frac{1}{6}\right)^3 \quad \text{and} \quad \left(\frac{1}{3}\right)^{-2} = 3^2.$$

E X A M P L E 2   Using Negative Exponents

Simplify by writing with positive exponents.

(a) $3^{-2} = \dfrac{1}{3^2} = \dfrac{1}{9}$

(b) $5^{-3} = \dfrac{1}{5^3} = \dfrac{1}{125}$

(c) $\left(\dfrac{1}{2}\right)^{-3} = 2^3 = 8$ $\frac{1}{2}$ and 2 are reciprocals.

As shown above, we can change the base to its reciprocal if we also change the sign of the exponent.

(d) $\left(\dfrac{2}{5}\right)^{-4} = \left(\dfrac{5}{2}\right)^4$ $\frac{2}{5}$ and $\frac{5}{2}$ are reciprocals.

(e) $4^{-1} - 2^{-1} = \dfrac{1}{4} - \dfrac{1}{2} = \dfrac{1}{4} - \dfrac{2}{4} = -\dfrac{1}{4}$

Apply the exponents first, then subtract.

(f) $p^{-2} = \dfrac{1}{p^2}, p \neq 0$

(g) $\dfrac{1}{x^{-4}}, x \neq 0$

$$\frac{1}{x^{-4}} = \frac{1^{-4}}{x^{-4}} \qquad 1^{-4} = 1$$

$$= \left(\frac{1}{x}\right)^{-4} \qquad \text{Power rule (c)}$$

$$= x^4 \qquad \tfrac{1}{x} \text{ and } x \text{ are reciprocals.}$$

Caution

A negative exponent does not indicate a negative number; negative exponents lead to reciprocals.

Expression	**Example**	
a^{-n}	$3^{-2} = \dfrac{1}{3^2} = \dfrac{1}{9}$	Not negative
$-a^{-n}$	$-3^{-2} = -\dfrac{1}{3^2} = -\dfrac{1}{9}$	Negative

WORK PROBLEM 2 AT THE SIDE. ▶▶

The definition of negative exponents allows us to move factors in a fraction if we also change the signs of the exponents. For example,

$$\frac{2^{-3}}{3^{-4}} = \frac{\dfrac{1}{2^3}}{\dfrac{1}{3^4}} = \frac{1}{2^3} \cdot \frac{3^4}{1} = \frac{3^4}{2^3}$$

so that

$$\frac{2^{-3}}{3^{-4}} = \frac{3^4}{2^3}.$$

Changing from Negative to Positive Exponents

For any nonzero numbers a and b, and any integers m and n,

$$\frac{a^{-m}}{b^{-n}} = \frac{b^n}{a^m}.$$

Example: $\dfrac{3^{-5}}{2^{-4}} = \dfrac{2^4}{3^5}$

E X A M P L E 3 **Changing from Negative to Positive Exponents**

Write with only positive exponents. Assume all variables represent nonzero real numbers.

(a) $\dfrac{4^{-2}}{5^{-3}} = \dfrac{5^3}{4^2}$

(b) $\dfrac{m^{-5}}{p^{-1}} = \dfrac{p^1}{m^5} = \dfrac{p}{m^5}$

(c) $\dfrac{a^{-2}b}{3d^{-3}} = \dfrac{bd^3}{3a^2}$

(d) $x^3 y^{-4} = \dfrac{x^3 y^{-4}}{1} = \dfrac{x^3}{y^4}$

WORK PROBLEM 3 AT THE SIDE. ▶▶

2. Write with positive exponents.

(a) 4^{-3}

(b) 6^{-2}

(c) $\left(\dfrac{2}{3}\right)^{-2}$

(d) $2^{-1} + 5^{-1}$

(e) $m^{-5}, m \neq 0$

(f) $\dfrac{1}{z^{-4}}, z \neq 0$

3. Write with only positive exponents. Assume all variables represent nonzero real numbers.

(a) $\dfrac{7^{-1}}{5^{-4}}$

(b) $\dfrac{x^{-3}}{y^{-2}}$

(c) $\dfrac{4h^{-5}}{m^{-2}k}$

(d) $p^2 q^{-5}$

ANSWERS

2. (a) $\dfrac{1}{4^3}$ (b) $\dfrac{1}{6^2}$ (c) $\left(\dfrac{3}{2}\right)^2$

(d) $\dfrac{1}{2} + \dfrac{1}{5} = \dfrac{7}{10}$ (e) $\dfrac{1}{m^5}$ (f) z^4

3. (a) $\dfrac{5^4}{7}$ (b) $\dfrac{y^2}{x^3}$ (c) $\dfrac{4m^2}{h^5 k}$ (d) $\dfrac{p^2}{q^5}$

4. Simplify. Write answers with positive exponents.

(a) $\dfrac{5^{11}}{5^8}$

(b) $\dfrac{4^7}{4^{10}}$

(c) $\dfrac{6^{-5}}{6^{-2}}$

(d) $\dfrac{8^4 \cdot m^9}{8^5 \cdot m^{10}}, m \neq 0$

OBJECTIVE ▶3 What about the quotient of two exponential expressions with the same base? We know that

$$\frac{6^5}{6^3} = \frac{6 \cdot 6 \cdot 6 \cdot 6 \cdot 6}{6 \cdot 6 \cdot 6} = 6^2.$$

Notice that $5 - 3 = 2$. Also,

$$\frac{6^2}{6^4} = \frac{6 \cdot 6}{6 \cdot 6 \cdot 6 \cdot 6} = \frac{1}{6^2} = 6^{-2}.$$

Here, $2 - 4 = -2$. These examples suggest the quotient rule for exponents.

Quotient Rule for Exponents

For any nonzero real number a and any integers m and n,

$$\frac{a^m}{a^n} = a^{m-n}.$$

(Keep the base and subtract the exponents.)

Example: $\dfrac{5^8}{5^4} = 5^{8-4} = 5^4$

E X A M P L E 4 Using the Quotient Rule for Exponents

Simplify, using the quotient rule for exponents. Write answers with positive exponents.

(a) $\dfrac{5^8}{5^6} = 5^{8-6} = 5^2$

(b) $\dfrac{4^2}{4^9} = 4^{2-9} = 4^{-7} = \dfrac{1}{4^7}$

(c) $\dfrac{5^{-3}}{5^{-7}} = 5^{-3-(-7)} = 5^4$

(d) $\dfrac{q^5}{q^{-3}} = q^{5-(-3)} = q^8, q \neq 0$

(e) $\dfrac{3^2 x^5}{3^4 x^3} = \dfrac{3^2}{3^4} \cdot \dfrac{x^5}{x^3} = 3^{2-4} \cdot x^{5-3}$

$$= 3^{-2} x^2 = \dfrac{x^2}{3^2}, x \neq 0$$

Sometimes numerical expressions with small exponents, such as 3^2, are evaluated. Doing that would give the result as $\frac{x^2}{9}$.

◀◀ **WORK PROBLEM 4 AT THE SIDE.**

Since exponential expressions with negative exponents can be written with positive exponents, the rules for exponents also are true for negative exponents.

The definitions and rules for exponents given in this section and the previous one are summarized below.

Definitions and Rules for Exponents

For any integers m and n: *Examples*

Product rule	$a^m \cdot a^n = a^{m+n}$	$7^4 \cdot 7^5 = 7^9$
Zero exponent	$a^0 = 1 \quad (a \neq 0)$	$(-3)^0 = 1$
Negative exponent	$a^{-n} = \dfrac{1}{a^n} \quad (a \neq 0)$	$5^{-3} = \dfrac{1}{5^3}$
Quotient rule	$\dfrac{a^m}{a^n} = a^{m-n} \ (a \neq 0)$	$\dfrac{2^2}{2^5} = 2^{2-5} = 2^{-3} = \dfrac{1}{2^3}$
Power rules (a)	$(a^m)^n = a^{mn}$	$(4^2)^3 = 4^6$
(b)	$(ab)^m = a^m b^m$	$(3k)^4 = 3^4 k^4$
(c)	$\left(\dfrac{a}{b}\right)^m = \dfrac{a^m}{b^m} \ (b \neq 0)$	$\left(\dfrac{2}{3}\right)^{-2} = \dfrac{2^{-2}}{3^{-2}} = \dfrac{3^2}{2^2}$
Negative to positive exponent	$\dfrac{a^{-m}}{b^{-n}} = \dfrac{b^n}{a^m}$	$\dfrac{2^{-4}}{5^{-3}} = \dfrac{5^3}{2^4}$

OBJECTIVE 4 As shown in the next example, we may sometimes need to use more than one rule to simplify an expression.

EXAMPLE 5   Using a Combination of Rules

Use a combination of the rules for exponents to simplify each expression. Assume all variables represent nonzero real numbers.

(a) $\dfrac{(4^2)^3}{4^5}$

$$\dfrac{(4^2)^3}{4^5} = \dfrac{4^6}{4^5} \qquad \text{Power rule (a)}$$
$$= 4^{6-5} \qquad \text{Quotient rule}$$
$$= 4^1 = 4$$

(b) $(2x)^3(2x)^2$

$$(2x)^3(2x)^2 = (2x)^5 \qquad \text{Product rule}$$
$$= 2^5 x^5 \text{ or } 32x^5 \qquad \text{Power rule (b)}$$

(c) $\left(\dfrac{2x^3}{5}\right)^{-4}$

By the definition of a negative exponent and the power rules,

$$\left(\dfrac{2x^3}{5}\right)^{-4} = \left(\dfrac{5}{2x^3}\right)^4 \qquad \text{Change the base to its reciprocal, and change the sign of the exponent.}$$
$$= \dfrac{5^4}{2^4 x^{12}}.\qquad \text{Power rules}$$

(d) $\left(\dfrac{3x^{-2}}{4^{-1}y^3}\right)^{-3} = \dfrac{3^{-3}x^6}{4^3 y^{-9}} \qquad \text{Power rules}$

$$= \dfrac{x^6 y^9}{4^3 \cdot 3^3} \qquad \text{Negative to positive exponent rule}$$

CONTINUED ON NEXT PAGE

5. Simplify. Assume all variables represent nonzero real numbers.

(a) $12^5 \cdot 12^{-7} \cdot 12^6$

(e) $\dfrac{(4m)^{-3}}{(3m)^{-4}} = \dfrac{4^{-3}m^{-3}}{3^{-4}m^{-4}}$ Power rule (b)

$= \dfrac{3^4 m^4}{4^3 m^3}$ Negative to positive exponent rule

$= \dfrac{3^4 m^{4-3}}{4^3}$ Quotient rule

$= \dfrac{3^4 m}{4^3}$

Note
Since the steps can be done in several different orders, there are many equally good ways to simplify problems like Examples 5(d) and 5(e).

(b) $y^{-2} \cdot y^5 \cdot y^{-8}$

◀◀ **WORK PROBLEM 5 AT THE SIDE.**

(c) $\dfrac{(6x)^{-1}}{(3x^2)^{-2}}$

(d) $\dfrac{3^9 \cdot (x^2 y)^{-2}}{3^3 \cdot x^{-4} y}$

ANSWERS

5. (a) 12^4 **(b)** $\dfrac{1}{y^5}$ **(c)** $\dfrac{3x^3}{2}$ **(d)** $\dfrac{3^6}{y^3}$

6.2 Exercises

Each of the following expressions is equal to either 0, 1, or −1. Decide which is correct.
See Example 1.

1. $(-4)^0$ **2.** $(-10)^0$ **3.** -9^0 **4.** -5^0

5. $(-2)^0 - 2^0$ **6.** $(-8)^0 - 8^0$ **7.** $(5x)^0 \quad (x \neq 0)$ **8.** $\left(\dfrac{3}{y}\right)^0 \quad (y \neq 0)$

Evaluate each expression. See Examples 1 and 2.

9. $7^0 + 9^0$ **10.** $8^0 + 6^0$ **11.** 4^{-3} **12.** 5^{-4}

13. $\left(\dfrac{1}{2}\right)^{-4}$ **14.** $\left(\dfrac{1}{3}\right)^{-3}$ **15.** $\left(\dfrac{6}{7}\right)^{-2}$ **16.** $\left(\dfrac{2}{3}\right)^{-3}$

17. $(-3)^{-4}$ **18.** $(-4)^{-3}$ **19.** $5^{-1} + 3^{-1}$ **20.** $6^{-1} + 2^{-1}$

Decide whether each of the following is positive, negative, or zero.

21. $(-2)^{-3}$ **22.** $(-3)^{-2}$ **23.** -2^4 **24.** -3^6

25. $\left(\dfrac{1}{4}\right)^{-2}$ **26.** $\left(\dfrac{1}{5}\right)^{-2}$ **27.** $1 - 5^0$ **28.** $1 - 7^0$

MATHEMATICAL CONNECTIONS (EXERCISES 29–32)

---

*In Objective 1, we used the product rule to motivate the definition of a zero exponent. We can
also use the quotient rule. To see this, work Exercises 29–32 in order.*

29. Consider the expression $\frac{25}{25}$. What is its simplest form?

30. Write the quotient in Exercise 29 using the fact that $25 = 5^2$.

31. Apply the quotient rule for exponents to your answer for Exercise 30. Give the answer as
a power of 5.

32. Because your answers for Exercises 29 and 31 both represent $\frac{25}{25}$, they must be equal.
Write this equality. What definition does it support?

Use the quotient rule to simplify each expression. Write each expression with positive exponents. Assume that all variables represent nonzero real numbers. See Examples 2, 3, and 4.

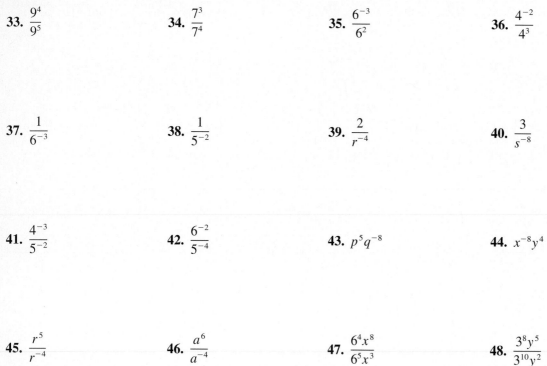

33. $\dfrac{9^4}{9^5}$

34. $\dfrac{7^3}{7^4}$

35. $\dfrac{6^{-3}}{6^2}$

36. $\dfrac{4^{-2}}{4^3}$

37. $\dfrac{1}{6^{-3}}$

38. $\dfrac{1}{5^{-2}}$

39. $\dfrac{2}{r^{-4}}$

40. $\dfrac{3}{s^{-8}}$

41. $\dfrac{4^{-3}}{5^{-2}}$

42. $\dfrac{6^{-2}}{5^{-4}}$

43. $p^5 q^{-8}$

44. $x^{-8} y^4$

45. $\dfrac{r^5}{r^{-4}}$

46. $\dfrac{a^6}{a^{-4}}$

47. $\dfrac{6^4 x^8}{6^5 x^3}$

48. $\dfrac{3^8 y^5}{3^{10} y^2}$

Use a combination of the rules for exponents to simplify each expression. Write answers with only positive exponents. Assume that all variables represent nonzero real numbers. See Example 5.

49. $\dfrac{(7^4)^3}{7^9}$

50. $\dfrac{(5^3)^2}{5^2}$

51. $x^{-3} \cdot x^5 \cdot x^{-4}$

52. $y^{-8} \cdot y^5 \cdot y^{-2}$

53. $\dfrac{(3x)^{-2}}{(4x)^{-3}}$

54. $\dfrac{(2y)^{-3}}{(5y)^{-4}}$

55. $\left(\dfrac{x^{-1}y}{z^2}\right)^{-2}$

56. $\left(\dfrac{p^{-4}q}{r^{-3}}\right)^{-3}$

57. $(6x)^4(6x)^{-3}$

58. $(10y)^9(10y)^{-8}$

59. $\dfrac{(m^7 n)^{-2}}{m^{-4} n^3}$

60. $\dfrac{(m^8 n^{-4})^2}{m^{-2} n^5}$

61. $\dfrac{5x^{-3}}{(4x)^2}$

62. $\dfrac{-3k^5}{(2k)^2}$

63. $\left(\dfrac{2p^{-1}q}{3^{-1}m^2}\right)^2$

64. $\left(\dfrac{4xy^2}{x^{-1}y}\right)^{-2}$

6.3 An Application of Exponents: Scientific Notation

OBJECTIVES

1. Express numbers in scientific notation.

2. Convert numbers in scientific notation to numbers without exponents.

3. Use scientific notation in calculations.

FOR EXTRA HELP

Tutorial   Tape 12   SSM, Sec. 6.3

OBJECTIVE 1 One example of the use of exponents comes from science. The numbers occurring in science are often extremely large (such as the distance from the earth to the sun, 93,000,000 miles) or extremely small (the wavelength of yellow-green light is approximately .0000006 meter). Because of the difficulty of working with many zeros, scientists often express such numbers with exponents. Each number is written as $a \times 10^n$, where $1 \leq |a| < 10$ and n is an integer. This form is called **scientific notation.** There is always one nonzero digit before the decimal point. For example, 35 is written 3.5×10^1, or 3.5×10; 56,200 is written 5.62×10^4, since

$$56{,}200 = 5.62 \times 10{,}000 = 5.62 \times 10^4,$$

and .09 is written as 9×10^{-2}.

The steps involved in writing a number in scientific notation are given below. For negative numbers, follow these steps using the absolute value of the number; then make the result negative.

Writing a Number in Scientific Notation

Step 1 Move the decimal point to the right of the first nonzero digit.

Step 2 Count the number of places you moved the decimal point.

Step 3 The number of places in Step 2 is the absolute value of the exponent on 10.

Step 4 The exponent on 10 is positive if you made the number smaller in Step 1; the exponent is negative if you made the number larger in Step 1.

EXAMPLE 1 Using Scientific Notation

Write each number in scientific notation.

(a) 93,000,000

The number will be written in scientific notation as 9.3×10^n. To find the value of n, first compare 9.3 with 93,000,000. Since 9.3 is *smaller* than 93,000,000, we must multiply by a *positive* power of 10 so the product 9.3×10^n will equal the larger number.

Move the decimal point to follow the first nonzero digit. Count the number of places the decimal point was moved.

$$9.3\,000\,000 \qquad \text{7 places}$$

Since the decimal point was moved 7 places, and since n is positive, $93{,}000{,}000 = 9.3 \times 10^7$.

(b) $463{,}000{,}000{,}000{,}000 = 4.63\,000\,000\,000\,000 \qquad \text{14 places}$

$$= 4.63 \times 10^{14}$$

(c) $-302{,}100 = -3.021 \times 10^5$

(d) .00462

Move the decimal point to the right of the first nonzero digit and count the number of places the decimal point was moved.

$$004.62 \qquad \text{3 places}$$

CONTINUED ON NEXT PAGE

1. Write each number in scientific notation.

(a) 63,000

(b) 5,870,000

(c) .0571

(d) .000062

2. Write without exponents.

(a) 4.2×10^3

(b) 8.7×10^5

(c) 6.42×10^{-3}

3. Simplify, and write without exponents.

(a) $(2.6 \times 10^4)(2 \times 10^{-6})$

(b) $\dfrac{4.8 \times 10^2}{2.4 \times 10^{-3}}$

Because 4.62 is *larger* than .00462, the exponent must be *negative*.

$$.00462 = 4.62 \times 10^{-3}$$

(e) $.0000762 = 7.62 \times 10^{-5}$

◀◀ WORK PROBLEM I AT THE SIDE.

OBJECTIVE 2 To convert a number written in scientific notation to a number without exponents, work in reverse. Multiplying a number by a positive power of 10 will make the number larger; multiplying by a negative power of 10 will make the number smaller.

E X A M P L E   2   Writing Numbers without Exponents

Write each number without exponents.

(a) 6.2×10^3

Since the exponent is positive, make 6.2 larger by moving the decimal point 3 places to the right.

$$6.2 \times 10^3 = 6.200 = 6200$$

(b) $4.283 \times 10^5 = 4.28300 = 428,300$ Move 5 places to the right.

(c) $-9.73 \times 10^{-2} = -09.73 = -.0973$ Move 2 places to the left.

As these examples show, the exponent tells the number of places and the direction that the decimal point is moved.

◀◀ WORK PROBLEM 2 AT THE SIDE.

OBJECTIVE 3 The next example shows how scientific notation can be used with products and quotients.

E X A M P L E   3   Multiplying and Dividing with Scientific Notation

Write each product or quotient without exponents.

(a) $(6 \times 10^3)(5 \times 10^{-4})$

$(6 \times 10^3)(5 \times 10^{-4})$

$= (6 \times 5)(10^3 \times 10^{-4})$ Commutative and associative properties

$= 30 \times 10^{-1}$ Product rule for exponents

$= 30. = 3$ Write without exponents.

(b) $\dfrac{6 \times 10^{-5}}{2 \times 10^3} = \dfrac{6}{2} \times \dfrac{10^{-5}}{10^3} = 3 \times 10^{-8} = .00000003$

◀◀ WORK PROBLEM 3 AT THE SIDE.

Calculators usually have a key labeled EE or EXP for scientific notation. See An Introduction to Scientific Calculators at the front of this book for more information.

6.3 Exercises

Write the numbers (other than dates) mentioned in the following statements in scientific notation.

1. How is it that the average CEO in Japan receives an income of $300,000, while the average CEO in the United States earns $2.8 million? (Andrew Zimbalist, *Baseball and Billions*. New York: Basic Books, 1992, p. 78.)

2. In 1995, women made up about 22 percent or 5.4 million, of all golfers. (From "Driving Force," an article in the *Sacramento Bee,* October 6, 1996, Sec. D.)

3. In 1995 sales of imported stemmed roses increased to 752 million, while sales of U.S.-produced roses decreased to 394 million. (Knight-Ridder Tribune graphic in the *Sacramento Bee,* Sept. 22, 1996, Sec. E, p. 3.)

4. The number of engineers employed in the U.S. during 1993 was 1.67 million. (Engineering Workforce Commission)

Determine whether or not the given number is written in scientific notation as defined in Objective 1. If it is not, write it as such.

5. 4.56×10^3 6. 7.34×10^5 7. 5,600,000   8. 34,000

9. $.8 \times 10^2$ 10. $.9 \times 10^3$ 11. .004   12. .0007

13. Explain in your own words what it means for a number to be written in scientific notation.

14. Explain how to multiply a number by a positive power of ten. Then explain how to multiply a number by a negative power of ten.

Write each number in scientific notation. See Example 1.

15. 5,876,000,000   16. 9,994,000,000   17. 82,350   18. 78,330

19. .000007   20. .0000004   21. $-.00203$ 22. $-.0000578$

Write each number without exponents. See Example 2.

23. 7.5×10^5 24. 8.8×10^6 25. 5.677×10^{12} 26. 8.766×10^9

27. -6.21×10^0 28. -8.56×10^0 29. 7.8×10^{-4}

30. 8.9×10^{-5} 31. 5.134×10^{-9} 32. 7.123×10^{-10}

Perform the indicated operations, and write the answers in scientific notation and then without exponents. See Example 3.

33. $(2 \times 10^8) \times (3 \times 10^3)$

34. $(4 \times 10^7) \times (3 \times 10^3)$

35. $(5 \times 10^4) \times (3 \times 10^2)$

36. $(8 \times 10^5) \times (2 \times 10^3)$

37. $(3.15 \times 10^{-4}) \times (2.04 \times 10^8)$

38. $(4.92 \times 10^{-3}) \times (2.25 \times 10^7)$

Perform the indicated operations, and write the answers in scientific notation. See Example 3.

39. $\dfrac{9 \times 10^{-5}}{3 \times 10^{-1}}$

40. $\dfrac{12 \times 10^{-4}}{4 \times 10^{-3}}$

41. $\dfrac{8 \times 10^3}{2 \times 10^2}$

42. $\dfrac{5 \times 10^4}{1 \times 10^3}$

43. $\dfrac{2.6 \times 10^{-3} \times 7.0 \times 10^{-1}}{2 \times 10^2 \times 3.5 \times 10^{-3}}$

44. $\dfrac{9.5 \times 10^{-1} \times 2.4 \times 10^4}{5 \times 10^3 \times 1.2 \times 10^{-2}}$

If the number in the statement is written in scientific notation, write it without exponents. If it is written without exponents, write it in scientific notation. See Examples 1 and 2.

45. The number of possible hands in contract bridge is about 6.35×10^{11}.

46. If there are forty numbers to choose from in a lottery, and a player must choose six different ones, the player has about 3.84×10^6 ways to make a choice.

47. Quentin Tarantino's black comedy, *Pulp Fiction*, boosted John Travolta's movie career by grossing a little over $45,500,000.

48. Whitney Houston's starring debut film, *The Bodyguard*, grossed $415,000,000 worldwide.

49. The body of a 150-pound person contains about 2.3×10^{-4} pounds of copper and about 6×10^{-3} pounds of iron.

50. The mean distance from Venus to the sun is about 6.7×10^6 miles.

6.4 Addition and Subtraction of Polynomials

OBJECTIVES

1. Identify terms and coefficients.
2. Add like terms.
3. Know the vocabulary for polynomials.
4. Add polynomials.
5. Subtract polynomials.
6. Apply the rules and definitions for polynomials to multivariable polynomials.

FOR EXTRA HELP

Tutorial   Tape 12   SSM, Sec. 6.4

OBJECTIVE 1 Recall that in an expression such as

$$4x^3 + 6x^2 + 5x + 8,$$

the quantities that are added, $4x^3$, $6x^2$, $5x$, and 8 are called *terms*. In the term $4x^3$, the number 4 is called the *numerical coefficient*, or simply the *coefficient* (koh-uh-FISH-ent), of x^3. In the same way, 6 is the coefficient of x^2 in the term $6x^2$, 5 is the coefficient of x in the term $5x$, and 8 is the coefficient in the term 8. A constant term, like 8 in the polynomial above, can be thought of as $8x^0$, where x^0 is defined to equal 1, as shown in Section 6.2.

EXAMPLE 1 Identifying Coefficients

Name the coefficient of each term in these expressions.

(a) $4x^3$

The coefficient is 4.

(b) $x - 6x^4$

The coefficient of x is 1 because $x = 1 \cdot x$. The coefficient of x^4 is -6, since $x - 6x^4$ can be written as the sum $x + (-6x^4)$.

(c) $5 - v^3$

The coefficient of the term 5 is 5 since $5 = 5v^0$. By writing $5 - v^3$ as a sum, $5 + (-v^3)$, or $5 + (-1v^3)$, the coefficient of v^3 can be identified as -1.

> **WORK PROBLEM 1 AT THE SIDE.** ▶▶

1. Name the coefficient of each term in these expressions.

(a) $3m^2$

OBJECTIVE 2 Recall from Section 1.9 that *like terms* have exactly the same combination of variables, with the same exponents on the variables. Only the coefficients may be different. Examples of like terms are

$$19m^5 \quad \text{and} \quad 14m^5;$$
$$6y^9, \quad -37y^9, \quad \text{and} \quad y^9;$$
$$3pq^2, \quad -2pq^2, \quad \text{and} \quad 4pq^2.$$

By the distributive property, we add like terms by adding their coefficients.

(b) $2x^2 - x$

EXAMPLE 2 Adding Like Terms

Simplify each expression by adding like terms.

(a) $-4x^3 + 6x^3 = (-4 + 6)x^3$ Distributive property

$$= 2x^3$$

(b) $9x^6 - 14x^6 + x^6 = (9 - 14 + 1)x^6 = -4x^6$

(c) $12m^2 + 5m + 4m^2 = (12 + 4)m^2 + 5m$

$$= 16m^2 + 5m$$

(d) $3x^2y + 4x^2y - x^2y = (3 + 4 - 1)x^2y = 6x^2y$

(c) $x + 8$

ANSWERS

1. (a) 3 **(b)** 2; -1 **(c)** 1; 8

371

2. Add like terms.

  (a) $5x^4 + 7x^4$

  (b) $9pq + 3pq - 2pq$

  (c) $r^2 + 3r + 5r^2$

  (d) $8t + 6w$

3. Choose all descriptions that apply for each of the expressions in parts (a)–(d).

  (1) Polynomial

  (2) Polynomial written in descending order

  (3) Not a polynomial

  (a) $3m^5 + 5m^2 - 2m + 1$

  (b) $2p^4 + p^6$

  (c) $\dfrac{1}{x} + 2x^2 + 3$

  (d) $x - 3$

In Example 2(c), it is not possible to add $16m^2$ and $5m$. These two terms are unlike because the exponents on the variables are different. *Unlike terms* have different variables or different exponents on the same variables.

◄◄ **WORK PROBLEM 2 AT THE SIDE.**

OBJECTIVE **3** Polynomials are basic to algebra. A **polynomial** (pah-luh-NOH-mee-ul) **in** x is a term or the sum of a finite number of terms of the form ax^n, for any real number a and any whole number n. For example,

$$16x^8 - 7x^6 + 5x^4 - 3x^2 + 4$$

is a polynomial in x (here 4 can be written as $4x^0$). This polynomial is written in **descending** (dee-SEND-ing) **powers** of the variable, because the exponents on x decrease from left to right. On the other hand,

$$2x^3 - x^2 + \frac{4}{x}, \quad 5\sqrt{x} - 2x, \quad \text{and} \quad x^{-2} + 4 - x^2$$

are not polynomials in x. The first expression is not a polynomial because $\frac{4}{x} = 4x^{-1}$ is not a *product, ax^n,* for a whole number n. The second expression is not a polynomial because of the term $5\sqrt{x}$, and the third expression is not a polynomial because of the negative exponent on x. (Of course, a polynomial could be defined using any variable, or variables, and not just x.)

◄◄ **WORK PROBLEM 3 AT THE SIDE.**

The **degree of a term with one variable** is the exponent on the variable. For example, $3x^4$ has degree 4, $6x^{17}$ has degree 17, $5x$ has degree 1, and -7 has degree 0 (since -7 can be written as $-7x^0$). The **degree of a polynomial** is the highest degree in any nonzero term of the polynomial. For example, $3x^4 - 5x^2 + 6$ is degree 4, the polynomial $5x + 7$ is degree 1, and 3 (or $3x^0$) is degree 0.

Three types of polynomials are very common and are given special names. A polynomial with exactly three terms is called a **trinomial** (TRY-noh-mee-ul). (*Tri-* means "three," as in *tri*angle.) Examples are

$$9m^3 - 4m^2 + 6, \quad \frac{19}{3}y^2 + \frac{8}{3}y + 5, \quad \text{and} \quad -3m^5 - 9m^2 + 2.$$

A polynomial with exactly two terms is called a **binomial** (BY-noh-mee-ul). (*Bi-* means "two," as in *bi*cycle.) Examples are

$$-9x^4 + 9x^3, \quad 8m^2 + 6m, \quad \text{and} \quad 3m^5 - 9m^2.$$

A polynomial with only one term is called a **monomial** (MAH-noh-mee-ul). (*Mon(o)-* means "one," as in *mono*rail.) Examples are

$$9m, \quad -6y^5, \quad a^2, \quad \text{and} \quad 6.$$

EXAMPLE 3   Classifying Polynomials

For each polynomial, first simplify if possible by combining like terms. Then give the degree and tell whether the polynomial is a monomial, a binomial, a trinomial, or none of these.

(a) $2x^3 + 5$

The polynomial cannot be simplified. The degree is 3. The polynomial is a binomial.

CONTINUED ON NEXT PAGE

(b) $4x - 5x + 2x$

Add like terms to simplify: $4x - 5x + 2x = x$. The degree is 1. The simplified polynomial is a monomial.

WORK PROBLEM 4 AT THE SIDE. ▶▶

OBJECTIVE ▶4▶ Polynomials may be added, subtracted, multiplied, and divided. Polynomial addition and subtraction are explained in the rest of this section.

Adding Polynomials

To add two polynomials, add like terms.

E X A M P L E 4   **Adding Polynomials Horizontally**

(a) Add $6x^3 - 4x^2 + 3$ and $-2x^3 + 7x^2 - 5$.

Write the sum.

$$(6x^3 - 4x^2 + 3) + (-2x^3 + 7x^2 - 5)$$

Rewrite this sum with the parentheses removed.

$$6x^3 + (-4x^2) + 3 + (-2x^3) + 7x^2 + (-5)$$

Place like terms together.

$$6x^3 + (-2x^3) + (-4x^2) + 7x^2 + \mathbf{3} + \mathbf{(-5)}$$

Combine like terms to get

$$4x^3 + 3x^2 + (-2) \quad \text{or} \quad 4x^3 + 3x^2 - 2.$$

(b) Add $(2x^2 - 4x + 3)$ and $(x^3 + 5x)$.

$$(2x^2 - 4x + 3) + (x^3 + 5x) = 2x^2 - 4x + 3 + x^3 + 5x$$
$$= x^3 + 2x^2 + x + 3 \quad \text{Combine like terms.}$$

WORK PROBLEM 5 AT THE SIDE. ▶▶

The polynomials in Example 4 also could be added vertically, as shown in the next example. This form of addition is used in multiplication of polynomials.

E X A M P L E 5   **Adding Polynomials Vertically**

Add $6x^3 - 4x^2 + 3$ and $-2x^3 + 7x^2 - 5$.

Write like terms in columns.

$$6x^3 - 4x^2 + 3$$
$$\underline{-2x^3 + 7x^2 - 5}$$

Now add, column by column.

$$\begin{array}{ccc} 6x^3 & -4x^2 & 3 \\ \underline{-2x^3} & \underline{7x^2} & \underline{-5} \\ 4x^3 & 3x^2 & -2 \end{array}$$

Add the three sums together.

$$4x^3 + 3x^2 + (-2) = 4x^3 + 3x^2 - 2$$

This is the same answer found in Example 4(a).

4. For each polynomial, first simplify if possible. Then give the degree and tell whether the polynomial is a monomial, binomial, trinomial, or none of these.

(a) $3x^2 + 2x - 4$

(b) $x^3 + 4x^3$

(c) $x^8 - x^7 + 2x^8$

5. Find each sum.

(a) $(2x^4 - 6x^2 + 7)$
$\quad + (-3x^4 + 5x^2 + 2)$

(b) $(3x^2 + 4x + 2)$
$\quad + (6x^3 - 5x - 7)$

ANSWERS

4. (a) degree 2; trinomial **(b)** degree 3; monomial (simplify to $5x^3$) **(c)** degree 8; binomial (simplify to $3x^8 - x^7$)

5. (a) $-x^4 - x^2 + 9$
(b) $6x^3 + 3x^2 - x - 5$

6. Add each pair of polynomials.

(a) $4x^3 - 3x^2 + 2x$
$\underline{6x^3 + 2x^2 - 3x}$

(b) $x^2 - 2x + 5$
$\underline{4x^2 + 3x - 2}$

7. Subtract, and check your answers by addition.

(a) $(14y^3 - 6y^2 + 2y - 5)$
$- (2y^3 - 7y^2 - 4y + 6)$

(b) $\left(\dfrac{7}{2}y^2 - \dfrac{11}{3}y + 8\right)$
$- \left(-\dfrac{3}{2}y^2 + \dfrac{4}{3}y + 6\right)$

◀◀ **WORK PROBLEM 6 AT THE SIDE.**

OBJECTIVE 5 Earlier, the difference $x - y$ was defined as $x + (-y)$. (We find the difference $x - y$ by adding x and the opposite of y.) For example,

$$7 - 2 = 7 + (-2) = 5$$

and

$$-8 - (-2) = -8 + 2 = -6.$$

A similar method is used to subtract polynomials.

Subtracting Polynomials

To subtract polynomials, change all the signs of the second polynomial and add the result to the first polynomial.

EXAMPLE 6 Subtracting Polynomials

Subtract $(5x - 2) - (3x - 8)$.

By the definition of subtraction, we should change the signs of the second polynomial and then add the two polynomials.

$$(5x - 2) - (3x - 8) = (5x - 2) + (-3x + 8)$$
$$= 5x - 2 - 3x + 8$$
$$= 2x + 6$$

EXAMPLE 7 Subtracting Polynomials

Subtract $6x^3 - 4x^2 + 2$ from $11x^3 + 2x^2 - 8$.

Start with

$$(11x^3 + 2x^2 - 8) - (6x^3 - 4x^2 + 2).$$

Change all the signs on the second polynomial and add like terms.

$$(11x^3 + 2x^2 - 8) + (-6x^3 + 4x^2 - 2)$$
$$= 11x^3 + 2x^2 - 8 - 6x^3 + 4x^2 - 2$$
$$= 5x^3 + 6x^2 - 10$$

We can check a subtraction problem by using the fact that if $a - b = c$, then $a = b + c$. For example, $6 - 2 = 4$. Check by writing $6 = 2 + 4$, which is correct. Check the polynomial subtraction above by adding $6x^3 - 4x^2 + 2$ and $5x^3 + 6x^2 - 10$. Since the sum is $11x^3 + 2x^2 - 8$, the subtraction was performed correctly.

◀◀ **WORK PROBLEM 7 AT THE SIDE.**

Subtraction also can be done in columns. We will use vertical subtraction in polynomial division in Section 6.8.

ANSWERS
6. (a) $10x^3 - x^2 - x$ **(b)** $5x^2 + x + 3$
7. (a) $12y^3 + y^2 + 6y - 11$
   (b) $5y^2 - 5y + 2$

EXAMPLE 8 Subtracting Polynomials Vertically

Use the method of subtracting by columns to find

$$(14y^3 - 6y^2 + 2y - 5) - (2y^3 - 7y^2 - 4y + 6).$$

Step 1 Arrange like terms in columns.

$$14y^3 - 6y^2 + 2y - 5$$
$$2y^3 - 7y^2 - 4y + 6$$

Step 2 Change all signs in the second row, and then add (*Step 3*).

$$14y^3 - 6y^2 + 2y - 5$$
$$\underline{-2y^3 + 7y^2 + 4y - 6} \quad \text{All signs changed}$$
$$12y^3 + y^2 + 6y - 11 \quad \text{Add.}$$

Either the horizontal or the vertical method may be used for adding and subtracting polynomials.

WORK PROBLEM 8 AT THE SIDE. ▶▶

EXAMPLE 9 Adding and Subtracting More than Two Polynomials

Perform the indicated operations to simplify the expression

$$(4 - x + 3x^2) - (2 - 3x + 5x^2) + (8 + 2x - 4x^2).$$

Since additions and subtractions are performed from left to right, we subtract first to get

$$(4 - x + 3x^2) - (2 - 3x + 5x^2) + (8 + 2x - 4x^2)$$
$$= (4 - x + 3x^2) + (-2 + 3x - 5x^2) + (8 + 2x - 4x^2).$$

Now add the first two polynomials, then add the sum and the third polynomial.

$$= (2 + 2x - 2x^2) + (8 + 2x - 4x^2)$$
$$= 10 + 4x - 6x^2$$

WORK PROBLEM 9 AT THE SIDE. ▶▶

OBJECTIVE 6 A **multivariable** (muhl-tih-VAIR-ee-uh-bul) **polynomial** has more than one variable. Some examples are $a + ab + b$ and $6x^2y + xy - 5x$. Most of the definitions and rules in this section also apply to multivariable polynomials. The **degree of a term with more than one variable** is the sum of the exponents on the variables. Thus, the term $2x^2y$ has degree $2 + 1 = 3$, and the polynomial $6x^2y + xy - 5$ has degree 3.

EXAMPLE 10 Determining the Degree of a Multivariable Polynomial

Give the degree of each polynomial.

(a) $7m^3n + 5m^2n - 8mn$

The degree of each term is $3 + 1 = 4$, $2 + 1 = 3$, and $1 + 1 = 2$, respectively. Thus, the degree of the polynomial is 4, the largest of these three numbers.

— **CONTINUED ON NEXT PAGE**

8. Subtract, using the method of subtracting by columns.

$$(14y^3 - 6y^2 + 2y)$$
$$- (2y^3 - 7y^2 + 6)$$

9. Perform the indicated operations.

$$(6p^4 - 8p^3 + 2p - 1)$$
$$- (-7p^4 + 6p^2 - 12)$$
$$+ (p^4 - 3p + 8)$$

ANSWERS
8. $12y^3 + y^2 + 2y - 6$
9. $14p^4 - 8p^3 - 6p^2 - p + 19$

10. Add or subtract. Give the degree of the answer.

(a) $(3mn + 2m - 4n)$
$+ (-mn + 4m + n)$

(b) $-x^2y + 5xy^3 - 4y + x^3y^2$

The four terms have the following degrees, respectively: $2 + 1 = 3$, $1 + 3 = 4$, 1, and $3 + 2 = 5$. The polynomial has degree 5.

Multivariable polynomials are added and subtracted by combining like terms, just as we did with single variable polynomials.

EXAMPLE 11   Adding and Subtracting Multivariable Polynomials

Add or subtract as indicated, and give the degree of the answer.

(a) $(4a + 2ab - b) + (3a - ab + b)$

$$(4a + 2ab - b) + (3a - ab + b)$$
$$= 4a + 2ab - b + 3a - ab + b$$
$$= 7a + ab$$

The polynomial $7a + ab$ has degree 2.

(b) $(2x^2y + 3xy + y^2) - (3x^2y - xy - 2y^2)$

$$(2x^2y + 3xy + y^2) - (3x^2y - xy - 2y^2)$$
$$= 2x^2y + 3xy + y^2 - 3x^2y + xy + 2y^2$$
$$= -x^2y + 4xy + 3y^2$$

The polynomial $-x^2y + 4xy + 3y^2$ has degree 3.

◄◄ WORK PROBLEM 10 AT THE SIDE.

(b) $(5p^2q^2 - 4p^2 + 2q)$
$- (2p^2q^2 - p^2 - 3q)$

6.4 Exercises

Tell whether each statement is true always, sometimes, *or* never.

1. A polynomial is a binomial.

2. A polynomial is a trinomial.

3. A trinomial is a polynomial.

4. A binomial is a polynomial.

5. A trinomial is a binomial.

6. A binomial is a trinomial.

7. A monomial is a polynomial.

8. A polynomial is a monomial.

For each of the following, determine the number of terms in the polynomial, and name the coefficients of the terms. See Example 1.

9. $6x^4$

10. $-9y^5$

11. t^4

12. s^7

13. $-19r^2 - r$

14. $2y^3 - y$

15. $x + 8x^2$

16. $v - 2v^3$

In each polynomial, add like terms whenever possible. Write the result in descending powers of the variable. See Example 2.

17. $-3m^5 + 5m^5$

18. $-4y^3 + 3y^3$

19. $2r^5 + (-3r^5)$

20. $-19y^2 + 9y^2$

21. $.2m^5 - .5m^2$

22. $-.9y + .9y^2$

23. $-3x^5 + 2x^5 - 4x^5$

24. $6x^3 - 8x^3 + 9x^3$

25. $-4p^7 + 8p^7 + 5p^9$

26. $-3a^8 + 4a^8 - 3a^2$

27. $-4y^2 + 3y^2 - 2y^2 + y^2$

28. $3r^5 - 8r^5 + r^5 + 2r^5$

For each polynomial, first simplify, if possible, and write it in descending powers of the variable. Then give the degree of the resulting polynomial, and tell whether it is a monomial, a binomial, a trinomial, or none of these. See Example 3.

29. $6x^4 - 9x$

30. $7t^3 - 3t$

31. $5m^4 - 3m^2 + 6m^5 - 7m^3$

32. $6p^5 + 4p^3 - 8p^4 + 10p^2$

33. $\frac{5}{3}x^4 - \frac{2}{3}x^4 + \frac{1}{3}x^2 - 4$

34. $\frac{4}{5}r^6 + \frac{1}{5}r^6 - r^4 + \frac{2}{5}r$

35. $.8x^4 - .3x^4 - .5x^4 + 7$

36. $1.2t^3 - .9t^3 - .3t^3 + 9$

If an object is thrown upward under certain conditions, its height in feet is given by the trinomial $-16x^2 + 60x + 80$, *where x is in seconds. If we evaluate the polynomial for a specific value of x, we will get one and only one value as a result. This idea is basic to the study of* functions, *one of the most important concepts in mathematics. Work Exercises 37–40 in order.*

37. If gasoline costs \$1.25 per gallon, then the monomial $1.25x$ gives the cost, in dollars, of x gallons. Evaluate this monomial for $x = 4$, then use the result to fill in the blanks. If _____ gallons are purchased, the cost is _____ .

38. If it costs \$15 to rent a chain saw plus \$2 per day, the binomial $2x + 15$ gives the cost, in dollars, to rent the chain saw for x days. Evaluate this polynomial for $x = 6$ and use the result to fill in the blanks. If the saw is rented for _____ days, the cost is _____ .

39. Evaluate the polynomial $-16x^2 + 60x + 80$, given in the directions above, for $x = 2.5$ and use the result to fill in the blanks. If _____ seconds have elapsed, the height of the object is _____ feet.

40. A certain viral infection causes a fever that typically lasts 6 days. A patient's temperature in °F on day x is given by the trinomial

$$-\frac{2}{3}x^2 + \frac{14}{3}x + 96.$$

Evaluate the trinomial for $x = 3$ and fill in the blanks. On day _____ , the patient has a temperature of _____ °F.

Add or subtract as indicated. See Examples 5 and 8.

41. Add.

$3m^2 + 5m$
$2m^2 - 2m$

42. Add.

$4a^3 - 4a^2$
$6a^3 + 5a^2$

43. Subtract.

$12x^4 - x^2$
$8x^4 + 3x^2$

44. Subtract.

$13y^5 - y^3$
$7y^5 + 5y^3$

45. Add.

$\dfrac{2}{3}x^2 + \dfrac{1}{5}x + \dfrac{1}{6}$
$\dfrac{1}{2}x^2 - \dfrac{1}{3}x + \dfrac{2}{3}$

46. Add.

$\dfrac{4}{7}y^2 - \dfrac{1}{5}y + \dfrac{7}{9}$
$\dfrac{1}{3}y^2 - \dfrac{1}{3}y + \dfrac{2}{5}$

47. Subtract.

$12m^3 - 8m^2 + 6m + 7$
$-3m^3 + 5m^2 - 2m - 4$

48. Subtract.

$5a^4 - 3a^3 + 2a^2 - a + 6$
$-6a^4 + a^3 - a^2 + a - 1$

Perform the indicated operations. See Examples 4, 6, 7, and 9.

49. $(2r^2 + 3r - 12) + (6r^2 + 2r)$

50. $(3r^2 + 5r - 6) + (2r - 5r^2)$

51. $(8m^2 - 7m) - (3m^2 + 7m - 6)$

52. $(x^2 + x) - (3x^2 + 2x - 1)$

53. $(16x^3 - x^2 + 3x) + (-12x^3 + 3x^2 + 2x)$

54. $(-2b^6 + 3b^4 - b^2) + (b^6 + 2b^4 + 2b^2)$

55. $(7y^4 + 3y^2 + 2y) - (18y^5 - 5y^3 + y)$

56. $(8t^5 + 3t^3 + 5t) - (19t^4 - 6t^2 + t)$

57. $[(8m^2 + 4m - 7) - (2m^3 - 5m + 2)] - (m^2 + m + 1)$

58. $[(9b^3 - 4b^2 + 3b + 2) - (-2b^3 - 3b^2 + b)] - (8b^3 + 6b + 4)$

Find the perimeter of each of the following geometric figures.

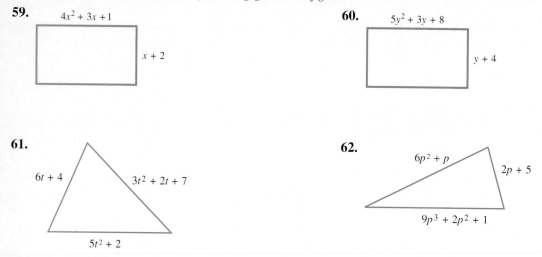

59. $4x^2 + 3x + 1$

  $x + 2$

60. $5y^2 + 3y + 8$

  $y + 4$

61.

$6t + 4$ $3t^2 + 2t + 7$

  $5t^2 + 2$

62. $6p^2 + p$ $2p + 5$

  $9p^3 + 2p^2 + 1$

63. Subtract $9x^2 - 3x + 7$ from $-2x^2 - 6x + 4$.

64. Subtract $-5w^3 + 5w^2 - 7$ from $6w^3 + 8w + 5$.

65. Explain why the degree of the term 3^4 is not 4. What is its degree?

66. Can the sum of two polynomials in x, both of degree 3, be of degree 2? If so, give an example.

Add or subtract as indicated. Give the degree of the answer. See Examples 10 and 11.

67. $(9a^2b - 3a^2 + 2b) + (4a^2b - 4a^2 - 3b)$

68. $(4xy^3 - 3x + y) + (5xy^3 + 13x - 4y)$

69. $(2c^4d + 3c^2d^2 - 4d^2) - (c^4d + 8c^2d^2 - 5d^2)$

70. $(3k^2h^3 + 5kh + 6k^3h^2) - (2k^2h^3 - 9kh + k^3h^2)$

71. Subtract. $9m^3n - 5m^2n^2 + 4mn^2$
$\phantom{\text{Subtract.}}$ $\underline{-3m^3n + 6m^2n^2 + 8mn^2}$

72. Subtract. $12r^5t + 11r^4t^2 - 7r^3t^3$
$\phantom{\text{Subtract.}}$ $\underline{-8r^5t + 10r^4t^2 + 3r^3t^3}$

6.5 Multiplication of Polynomials

OBJECTIVE 1 As shown earlier, we find the product of two monomials by using the rules for exponents and the commutative and associative properties. For example,

$$(6x^3)(4x^4) = 6 \cdot 4 \cdot x^3 \cdot x^4 = 24x^7.$$

Also,

$$(-8m^6)(-9n^6) = (-8)(-9)(m^6)(n^6) = 72m^6n^6.$$

To find the product of a monomial and a polynomial with more than one term, we use the distributive property and then the method shown above.

┌─ **E X A M P L E 1** **Multiplying a Monomial and a Polynomial**

Use the distributive property to find each product.

(a) $4x^2(3x + 5)$

$$4x^2(3x + 5) = (4x^2)(3x) + (4x^2)(5) \quad \text{Distributive property}$$
$$= 12x^3 + 20x^2 \quad \text{Multiply monomials.}$$

(b) $-8m^3(4m^3 + 3m^2 + 2m - 1)$
$$= (-8m^3)(4m^3) + (-8m^3)(3m^2)$$
$$+ (-8m^3)(2m) + (-8m^3)(-1) \quad \text{Distributive property}$$
$$= -32m^6 - 24m^5 - 16m^4 + 8m^3 \quad \text{Multiply monomials.}$$

WORK PROBLEM 1 AT THE SIDE. ▶▶

OBJECTIVE 2 The distributive property is used also to find the product of any two polynomials. For example, to find the product of the polynomials $x + 1$ and $x - 4$, think of $x + 1$ as a single quantity and use the distributive property as follows.

$$(x + 1)(x - 4) = (x + 1)\,x + (x + 1)(-4)$$

Now use the distributive property to find $(x + 1)x$ and $(x + 1)(-4)$.

$$(x + 1)x + (x + 1)(-4) = x(x) + 1(x) + x(-4) + 1(-4)$$
$$= x^2 + x + (-4x) + (-4)$$
$$= x^2 - 3x - 4$$

┌─ **E X A M P L E 2** **Multiplying Two Binomials**
Find the product $(2x + 1)(3x + 5)$.

$$(2x + 1)(3x + 5) = (2x + 1)(3x) + (2x + 1)(5)$$
$$= (2x)(3x) + (1)(3x) + (2x)(5) + (1)(5)$$
$$= 6x^2 + 3x + 10x + 5$$
$$= 6x^2 + 13x + 5$$

WORK PROBLEM 2 AT THE SIDE. ▶▶

1. Find each product.

(a) $5m^3(2m + 7)$

(b) $2x^4(3x^2 + 2x - 5)$

(c) $-4y^2(3y^3 + 2y^2 - 4y + 8)$

2. Find each product.

(a) $(4x + 3)(2x + 1)$

(b) $(3k - 2)(2k + 1)$

(c) $(m + 5)(3m - 4)$

ANSWERS

1. (a) $10m^4 + 35m^3$
   (b) $6x^6 + 4x^5 - 10x^4$
   (c) $-12y^5 - 8y^4 + 16y^3 - 32y^2$
2. (a) $8x^2 + 10x + 3$
   (b) $6k^2 - k - 2$
   (c) $3m^2 + 11m - 20$

381

3. Multiply.

(a) $(m^3 - 2m + 1)$
      $\cdot (2m^2 + 4m + 3)$

A rule for multiplying any two polynomials is given below.

> **Multiplying Polynomials**
>
> Multiply two polynomials by multiplying each term of the second polynomial by each term of the first polynomial and adding the products.

E X A M P L E   3   Multiplying Any Two Polynomials

Find the product $(3y^2 + 2y - 1)(y^3 + y^2 - 5)$.

Multiply each term in the first polynomial by each term in the second polynomial, then add any like terms.

$$(3y^2 + 2y - 1)(y^3 + y^2 - 5)$$
$$= 3y^2(y^3) + 3y^2(y^2) + 3y^2(-5) + 2y(y^3) + 2y(y^2)$$
$$\quad + 2y(-5) - 1(y^3) - 1(y^2) - 1(-5)$$
$$= 3y^5 + 3y^4 - 15y^2 + 2y^4 + 2y^3 - 10y$$
$$\quad - y^3 - y^2 + 5$$
$$= 3y^5 + 5y^4 + y^3 - 16y^2 - 10y + 5$$

◀◀ **WORK PROBLEM 3 AT THE SIDE.**

OBJECTIVE 3▶ We can also multiply two polynomials by writing one polynomial above the other.

(b) $(6p^2 + 2p - 4)(3p^2 - 5)$

E X A M P L E   4   Multiplying Polynomials Vertically

Multiply $2x^2 + 4x + 1$ by $3x + 5$.

Start with

$$2x^2 + 4x + 1$$
$$\underline{\qquad 3x + 5.}$$

It is not necessary to line up like terms in columns, because any terms may be multiplied (not just like terms). Begin by multiplying each of the terms in the top row by 5.

Step 1

$$2x^2 + \ 4x + 1$$
$$\underline{\qquad\quad 3x + 5}$$
$$10x^2 + 20x + 5 \qquad \leftarrow \text{Product of 5 and } 2x^2 + 4x + 1$$

Notice how this process is similar to multiplication of whole numbers. Now multiply each term in the top row by $3x$. Be careful to place the like terms in columns, since the final step will involve addition (as in multiplying two whole numbers).

Step 2

$$2x^2 + \ 4x + 1$$
$$\underline{\qquad\quad 3x + 5}$$
$$10x^2 + 20x + 5$$
$$6x^3 + 12x^2 + \ 3x \qquad\qquad \leftarrow \text{Product of } 3x \text{ and } 2x^2 + 4x + 1$$

ANSWERS

3. (a) $2m^5 + 4m^4 - m^3 - 6m^2 - 2m + 3$
  (b) $18p^4 + 6p^3 - 42p^2 - 10p + 20$

CONTINUED ON NEXT PAGE

Step 3 Add like terms.

$$2x^2 + 4x + 1$$
$$\underline{3x + 5}$$
$$10x^2 + 20x + 5$$
$$\underline{6x^3 + 12x^2 + 3x}$$
$$6x^3 + 22x^2 + 23x + 5$$

The product is $6x^3 + 22x^2 + 23x + 5$.

WORK PROBLEM 4 AT THE SIDE. ▶▶

E X A M P L E 5   Multiplying Polynomials Vertically

Find the product of $3p - 5q$ and $2p + 7q$.

$$3p - 5q$$
$$\underline{2p + 7q}$$
$$21pq - 35q^2 \quad \leftarrow 7q(3p - 5q)$$
$$\underline{6p^2 - 10pq} \quad \leftarrow 2p(3p - 5q)$$
$$6p^2 + 11pq - 35q^2$$

E X A M P L E 6   Multiplying Polynomials Vertically

Find the product of $4m^3 - 2m^2 + 4m$ and $\frac{1}{2}m^2 + \frac{5}{2}$.

$$4m^3 - 2m^2 + 4m$$
$$\underline{\frac{1}{2}m^2 + \frac{5}{2}}$$
$$10m^3 - 5m^2 + 10m \quad \text{Terms of top row multiplied by } \tfrac{5}{2}$$
$$\underline{2m^5 - m^4 + 2m^3} \quad \text{Terms of top row multiplied by } \tfrac{1}{2}m^2$$
$$2m^5 - m^4 + 12m^3 - 5m^2 + 10m$$

WORK PROBLEM 5 AT THE SIDE. ▶▶

OBJECTIVE 4 To find higher powers of binomials, such as $(x + 5)^3$, we use the definition of exponent and perform repeated multiplication.

E X A M P L E 7   Squaring a Polynomial

Find $(3n^2 + 5n - 1)^2$.

By definition, $(3n^2 + 5n - 1)^2 = (3n^2 + 5n - 1)(3n^2 + 5n - 1)$. Use the vertical method of multiplication.

$$3n^2 + 5n - 1$$
$$\underline{3n^2 + 5n - 1}$$
$$-3n^2 - 5n + 1$$
$$15n^3 + 25n^2 - 5n$$
$$\underline{9n^4 + 15n^3 - 3n^2}$$
$$9n^4 + 30n^3 + 19n^2 - 10n + 1$$

4. Find each product.

(a) $4k - 6$
$\underline{2k + 5}$

(b) $3x^2 + 4x - 5$
$\underline{x + 4}$

5. Find each product.

(a) $2m + 3p$
$\underline{5m - 4p}$

(b) $k^3 - k^2 + k + 1$
$\underline{\frac{2}{3}k - \frac{1}{3}}$

(c) $a^3 + 3a - 4$
$\underline{2a^2 + 6a + 5}$

ANSWERS

4. (a) $8k^2 + 8k - 30$
(b) $3x^3 + 16x^2 + 11x - 20$
5. (a) $10m^2 + 7mp - 12p^2$
(b) $\frac{2}{3}k^4 - k^3 + k^2 + \frac{1}{3}k - \frac{1}{3}$
(c) $2a^5 + 6a^4 + 11a^3 + 10a^2 - 9a - 20$

6. Find each product.

(a) $(2x^2 - 3x + 4)^2$

E X A M P L E 8 **Cubing a Binomial**

Find $(x + 5)^3$.

Since $(x + 5)^3 = (x + 5)(x + 5)(x + 5)$, the first step is to find the product $(x + 5)(x + 5)$.

$$(x + 5)(x + 5) = x^2 + 5x + 5x + 25$$
$$= x^2 + 10x + 25$$

Now multiply this result by $x + 5$.

$$(x + 5)(x^2 + 10x + 25) = x^3 + 10x^2 + 25x + 5x^2 + 50x + 125$$
$$= x^3 + 15x^2 + 75x + 125$$

(b) $(m + 1)^3$

E X A M P L E 9 **Finding the Fourth Power of a Binomial**

Find $(2y - 3)^4$.

One way to proceed is to note that $(2y - 3)^4 = (2y - 3)^2(2y - 3)^2$.

$$(2y - 3)^4 = (2y - 3)^2(2y - 3)^2$$
$$= (4y^2 - 6y - 6y + 9)(4y^2 - 6y - 6y + 9)$$
$$= (4y^2 - 12y + 9)(4y^2 - 12y + 9)$$

Multiplying vertically will help to keep like terms together so they can be combined.

$$
\begin{array}{r}
4y^2 - 12y + 9 \\
4y^2 - 12y + 9 \\
\hline
36y^2 - 108y + 81 \\
-48y^3 + 144y^2 - 108y \\
16y^4 - 48y^3 + 36y^2 \\
\hline
16y^4 - 96y^3 + 216y^2 - 216y + 81
\end{array}
$$

◄◄ **WORK PROBLEM 6 AT THE SIDE.**

(c) $(3k - 2)^4$

1. In multiplying a monomial by a polynomial, such as in $4x(3x^2 + 7x^3) = 4x(3x^2) + 4x(7x^3)$, the first property that is used is the _____ property.

2. In multiplying two monomials, we use (but do not always show) the commutative and associative properties. In each item below, name the property that has been used.

$$(5x^3)(6x^5) = (5x^3 \cdot 6)x^5 \qquad \underline{\hspace{3cm}}$$
$$= (5 \cdot 6x^3)x^5 \qquad \underline{\hspace{3cm}}$$
$$= (5 \cdot 6)(x^3 \cdot x^5) \qquad \underline{\hspace{3cm}}$$
$$= 30x^8$$

Find each product. See Example 1.

3. $-2m(3m + 2)$

4. $-5p(6 + 3p)$

5. $\dfrac{3}{4}p(8 - 6p + 12p^3)$

6. $\dfrac{4}{3}x(3 + 2x + 5x^3)$

7. $2y^5(3 + 2y + 5y^4)$

8. $2m^4(3m^2 + 5m + 6)$

Find each binomial product. See Examples 2 and 5.

9. $(n - 2)(n + 3)$

10. $(r - 6)(r + 8)$

11. $(4r + 1)(2r - 3)$

12. $(5x + 2)(2x - 7)$

13. $(3x + 2)(3x - 2)$

14. $(7x + 3)(7x - 3)$

15. $(3q + 1)(3q + 1)$

16. $(4w + 7)(4w + 7)$

17. $(3t + 4s)(2t + 5s)$

18. $(8v + 5w)(2v + 3w)$

19. $(-.3t + .4)(t + .6)$

20. $(-.5x + .9)(x - .2)$

21. Perform the following multiplications: $(x + 4)(x - 4)$; $(y + 2)(y - 2)$; $(r + 7)(r - 7)$. Observe your answers, and explain the pattern that can be found in the answers.

22. Repeat Exercise 21 for the following: $(x + 4)(x + 4)$; $(y - 2)(y - 2)$; $(r + 7)(r + 7)$.

Find each product. See Examples 3, 4, and 6.

23. $(6x + 1)(2x^2 + 4x + 1)$

24. $(9y - 2)(8y^2 - 6y + 1)$

25. $(4m + 3)(5m^3 - 4m^2 + m - 5)$

26. $(y + 4)(3y^3 - 2y^2 + y + 3)$

27. $(5x^2 + 2x + 1)(x^2 - 3x + 5)$

28. $(2m^2 + m - 3)(m^2 - 4m + 5)$

Find each product. See Examples 7, 8, and 9.

29. $(x + 12)^2$

30. $(y + 11)^2$

31. $(3t + 1)^2$

32. $(4y + 3)^2$

33. $(5x - 2y)^2$

34. $(6p - 5s)^2$

35. $(h - 5)^3$

36. $(n - 3)^3$

37. $(r + 1)^4$

38. $(k - 1)^4$

39. $(3x^2 + x - 4)^2$

40. $(4y^2 - y + 3)^2$

MATHEMATICAL CONNECTIONS (EXERCISES 41–46)

Work Exercises 41–46 in order. Refer to the figure as necessary.

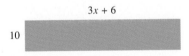

3x + 6

10

41. Find a polynomial that represents the area of the rectangle.

42. Suppose you know that the area of the rectangle is 600 square yards. Use this information and the polynomial from Exercise 41 to write an equation and solve it for x.

43. What are the dimensions of the rectangle (assume units are all in yards)?

44. Suppose the rectangle represents a lawn and its costs $3.50 per square yard to lay sod on the lawn. How much will it cost to sod the entire lawn?

45. Use the result of Exercise 43 to find the perimeter of the lawn.

46. Again, suppose the rectangle represents a lawn and it costs $9.00 per yard to fence the lawn. How much will it cost to fence the lawn?

6.6 Products of Binomials

OBJECTIVES

1. Multiply binomials by the FOIL method.
2. Square binomials.
3. Find the product of the sum and difference of two terms.

FOR EXTRA HELP

Tutorial   Tape 13   SSM, Sec. 6.6

OBJECTIVE 1 We can use the methods introduced in the last section to find the product of any two polynomials. They are the only practical methods for multiplying polynomials with three or more terms. However, many of the polynomials to be multiplied are binomials, with only two terms, so in this section we discuss a shortcut that eliminates the need to write all the steps. To develop this shortcut, let us first multiply $x + 3$ and $x + 5$ using the distributive property.

$$(x + 3)(x + 5) = (x + 3)x + (x + 3)5$$
$$= (x)(x) + (3)(x) + (x)(5) + (3)(5)$$
$$= x^2 + 3x + 5x + 15$$
$$= x^2 + 8x + 15$$

The first term in the second line, $(x)(x)$, is the product of the first terms of the two binomials.

$$(x + 3)(x + 5) \qquad \text{Multiply the \textbf{first} terms: } (x)(x).$$

The term $(x)(5)$ is the product of the first term of the first binomial and the last term of the second binomial. This is the **outer product.**

$$(x + 3)(x + 5) \qquad \text{Multiply the \textbf{outer} terms: } (x)(5).$$

The term $(3)(x)$ is the product of the last term of the first binomial and the first term of the second binomial. The product of these middle terms is called the **inner product.**

$$(x + 3)(x + 5) \qquad \text{Multiply the \textbf{inner} terms: } (3)(x).$$

Finally, $(3)(5)$ is the product of the last terms of the two binomials.

$$(x + 3)(x + 5) \qquad \text{Multiply the \textbf{last} terms: } (3)(5).$$

In the third step of the multiplication above, the inner product and the outer product are added. This step should be performed mentally, so that the three terms of the answer can be written without extra steps as

$$(x + 3)(x + 5) = x^2 + 8x + 15.$$

> **WORK PROBLEM I AT THE SIDE.** ▶▶

A summary of these steps is given below. This procedure is sometimes called the **FOIL method,** which comes from the initial letters of the words *First, Outer, Inner,* and *Last.*

Multiplying Binomials by the FOIL Method

Step 1 Multiply the two **F**irst terms of the binomials to get the first term of the answer.

Step 2 Find the **O**uter product and the **I**nner product and add them (mentally, if possible) to get the middle term of the answer.

Step 3 Multiply the two **L**ast terms of the binomials to get the last term of the answer.

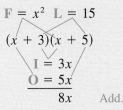

$$F = x^2 \quad L = 15$$
$$(x + 3)(x + 5)$$
$$I = 3x$$
$$\underline{O = 5x}$$
$$8x \qquad \text{Add.}$$

---

1. For the product $(2p - 5)(3p + 7)$, find the following.

(a) Product of first terms

(b) Outer product

(c) Inner product

(d) Product of last terms

(e) Complete product in simplified form

2. Use the FOIL method to find each product.

(a) $(m + 4)(m - 3)$

E X A M P L E 1   Using the FOIL Method

Use the FOIL method to find the product $(x + 8)(x - 6)$.

Step 1 **F** Multiply the **first** terms.

$$x(x) = x^2$$

Step 2 **O** Find the product of the **outer** terms.

$$x(-6) = -6x$$

I Find the product of the **inner** terms.

$$8(x) = 8x$$

Add the outer and inner products mentally.

$$-6x + 8x = 2x$$

Step 3 **L** Multiply the **last** terms.

$$8(-6) = -48$$

(b) $(y + 7)(y + 2)$

The product of $x + 8$ and $x - 6$ is found by adding the terms found in the three steps above, so

$$(x + 8)(x - 6) = x^2 + 2x - 48.$$

As a shortcut, this product can be found in the following manner.

  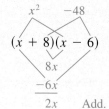

◀◀ **WORK PROBLEM 2 AT THE SIDE.**

(c) $(r - 8)(r - 5)$

E X A M P L E 2   Using the FOIL Method

Multiply $9x - 2$ and $3x + 1$.

First	$(9x - 2)(3x + 1)$	$27x^2$
Outer	$(9x - 2)(3x + 1)$	$9x$
Inner	$(9x - 2)(3x + 1)$	$-6x$
Last	$(9x - 2)(3x + 1)$	-2

$$\hspace{3cm} \text{F} \quad \text{O} \quad \text{I} \quad \text{L}$$
$$(9x - 2)(3x + 1) = 27x^2 + 9x - 6x - 2$$
$$= 27x^2 + 3x - 2$$

E X A M P L E 3 Using the FOIL Method

Find the following products.

$$\begin{array}{cccc} \text{F} & \text{O} & \text{I} & \text{L} \end{array}$$

(a) $(2k + 5y)(k + 3y) = (2k)(k) + (2k)(3y) + (5y)(k) + (5y)(3y)$

$$= 2k^2 + 6ky + 5ky + 15y^2$$

$$= 2k^2 + 11ky + 15y^2$$

(b) $(7p + 2q)(3p - q) = 21p^2 - pq - 2q^2$

WORK PROBLEM 3 AT THE SIDE. ▶▶

Objective ▶ Certain special types of binomial products occur so often that the form of the answers should be memorized. For example, to find the square of a binomial quickly, use the method shown in Example 4.

E X A M P L E 4 Squaring a Binomial

Find $(2m + 3)^2$.

Squaring $2m + 3$ by the FOIL method gives

$$(2m + 3)(2m + 3) = 4m^2 + 6m + 6m + 9 = 4m^2 + 12m + 9.$$

The result has the square of both the first and the last terms of the binomial:

$$4m^2 = (2m)^2 \quad \text{and} \quad 9 = 3^2.$$

The middle term is twice the product of the two terms of the binomial, that is,

$$12m = 2(6m) = 2(2m)(3).$$

This example suggests the following rule.

Square of a Binomial

The square of a binomial is a trinomial made up of the square of the first term, plus twice the product of the two terms, plus the square of the last term of the binomial. For a and b,

$$(a + b)^2 = a^2 + 2ab + b^2.$$

Also, $$(a - b)^2 = a^2 - 2ab + b^2.$$

E X A M P L E 5 Squaring Binomials

Use the formula to square each binomial.

$$(a - b)^2 = a^2 - 2 \cdot a \cdot b + b^2$$

(a) $(5z - 1)^2 = (5z)^2 - 2(5z)(1) + (1)^2$

$$= 25z^2 - 10z + 1$$

Recall that $(5z)^2 = 5^2z^2 = 25z^2$.

(b) $(3b + 5r)^2 = (3b)^2 + 2(3b)(5r) + (5r)^2$

$$= 9b^2 + 30br + 25r^2$$

CONTINUED ON NEXT PAGE

---

3. Find each product.

(a) $(4k - 1)(2k + 3)$

(b) $(6m + 5)(m - 4)$

(c) $(8y + 3)(2y + 1)$

(d) $(3r + 2t)(3r + 4t)$

Answers

3. (a) $8k^2 + 10k - 3$
(b) $6m^2 - 19m - 20$
(c) $16y^2 + 14y + 3$
(d) $9r^2 + 18rt + 8t^2$

4. Find each square by using the pattern for the square of a binomial.

(a) $(t + u)^2$

(b) $(2m - p)^2$

(c) $(4p + 3q)^2$

(d) $(5r - 6s)^2$

(e) $\left(3k - \dfrac{1}{2}\right)^2$

5. Find each product by using the pattern for the sum and difference of two terms.

(a) $(6a + 3)(6a - 3)$

(b) $(10m + 7)(10m - 7)$

(c) $(7p + 2q)(7p - 2q)$

(d) $\left(3r - \dfrac{1}{2}\right)\left(3r + \dfrac{1}{2}\right)$

(c) $(2a - 9x)^2 = 4a^2 - 36ax + 81x^2$

(d) $\left(4m + \dfrac{1}{2}\right)^2 = (4m)^2 + 2(4m)\left(\dfrac{1}{2}\right) + \left(\dfrac{1}{2}\right)^2$

$$= 16m^2 + 4m + \dfrac{1}{4}$$

Caution
A common error in squaring a binomial is forgetting the middle term of the product. In general,
$$(a + b)^2 \neq a^2 + b^2.$$

◄◄ **WORK PROBLEM 4 AT THE SIDE.**

OBJECTIVE 3 Binomial products of the form $(a + b)(a - b)$ also occur frequently. In these products, one binomial is the sum of two terms, and the other is the difference of the same two terms. As an example, the product of $x + 2$ and $x - 2$ is
$$(x + 2)(x - 2) = x^2 - 2x + 2x - 4$$
$$= x^2 - 4.$$

As this example suggests, the product of $a + b$ and $a - b$ is the difference between two squares.

Product of the Sum and Difference of Two Terms
$$(a + b)(a - b) = a^2 - b^2$$

E X A M P L E 6   **Finding the Product of the Sum and Difference of Two Terms**

Find each product.

$(a \quad + \quad b)(a \quad - \quad b)$
$\downarrow \qquad \downarrow \quad \downarrow \qquad \downarrow$

(a) $(5m + 3)(5m - 3)$

Use the pattern for the sum and difference of two terms.
$$(5m + 3)(5m - 3) = (5m)^2 - 3^2$$
$$= 25m^2 - 9$$

(b) $(4x + y)(4x - y) = (4x)^2 - y^2$
$$= 16x^2 - y^2$$

(c) $\left(z - \dfrac{1}{4}\right)\left(z + \dfrac{1}{4}\right) = z^2 - \dfrac{1}{16}$

◄◄ **WORK PROBLEM 5 AT THE SIDE.**

The product formulas of this section will be very useful in later work, particularly in Chapter 7 on factoring. Therefore, it is important to memorize these formulas and practice using them.

ANSWERS

4. (a) $t^2 + 2tu + u^2$
(b) $4m^2 - 4mp + p^2$
(c) $16p^2 + 24pq + 9q^2$
(d) $25r^2 - 60rs + 36s^2$
(e) $9k^2 - 3k + \dfrac{1}{4}$

5. (a) $36a^2 - 9$ **(b)** $100m^2 - 49$
(c) $49p^2 - 4q^2$ **(d)** $9r^2 - \dfrac{1}{4}$

6.6 Exercises

1. Consider the product $(2x + 3)(x - 5)$.
   (a) What is the product of the first terms, $2x(x)$? _____
   (b) What is the product of the outer terms, $2x(-5)$? _____
   (c) What is the product of the inner terms, $3(x)$? _____
   (d) What is the product of the last terms, $3(-5)$? _____
   (e) What is the sum of the outer and inner products found in parts (b) and (c)? _____
   (f) Write the complete product, which is a trinomial, using your results in
   parts (a), (e), and (d). _____

2. Repeat Exercise 1 for the product $(3y - 8)(2y + 5)$.

Find each product using the FOIL method. See Examples 1–3.

3. $(r + 1)(r + 3)$ **4.** $(p + 2)(p + 3)$ **5.** $(w - 4)(w - 6)$

6. $(y - 3)(y - 7)$ **7.** $(x - .3)(x + .8)$ **8.** $(y + .7)(y + .4)$

9. $\left(x - \dfrac{2}{3}\right)\left(x + \dfrac{1}{4}\right)$ **10.** $\left(z - \dfrac{5}{6}\right)\left(z + \dfrac{3}{4}\right)$ **11.** $(9x + 2)(3x + 7)$

12. $(6x + 7)(2x + 3)$ **13.** $(3m + 7)(3m + 5)$ **14.** $(7x - 1)(7x - 8)$

15. $(3 - 2x)(5 - 3x)$ **16.** $(4 - 3r)(5 - 2r)$ **17.** $(-5 + 6z)(3 - z)$

18. $(-4 + 5w)(5 - w)$ **19.** $\left(-\dfrac{8}{3} + 3k\right)\left(-\dfrac{2}{3} - k\right)$ **20.** $\left(-\dfrac{5}{4} + 2r\right)\left(-\dfrac{3}{4} - r\right)$

21. $(3w + 2z)(9w - z)$ **22.** $(2x - y)(5x + 3y)$ **23.** $(5y - 3z)(6y + 5z)$

24. $(4x + 5y)(7x - y)$ **25.** $(-.8p + .3s)(.2p + s)$ **26.** $(.7h + k)(-.4h + .3k)$

27. Consider the square $(2x + 3)^2$.
   (a) What is the square of the first term, $(2x)^2$? _____
   (b) What is twice the product of the two terms, $2(2x)(3)$? _____
   (c) What is the square of the last term, 3^2? _____
   (d) Write the final product, which is a trinomial, using your results in parts (a)–(c). _____

28. Repeat Exercise 27 for the square $(3x - 2)^2$.

Find each square. See Examples 4 and 5.

29. $(a - c)^2$ **30.** $(p - y)^2$ **31.** $(p + 2)^2$ **32.** $(r + 5)^2$

33. $(4x - 3)^2$ **34.** $(5y + 2)^2$ **35.** $(.8t + .7s)^2$ **36.** $(.7z - .3w)^2$

37. $\left(5x + \dfrac{2}{5}y\right)^2$ **38.** $\left(6m - \dfrac{4}{5}n\right)^2$

39. Consider the product $(7x + 3y)(7x - 3y)$.
   (a) What is the product of the first terms, $(7x)(7x)$? _____
   (b) Multiply the outer terms, $(7x)(-3y)$. Then multiply the inner terms, $(3y)(7x)$. Add the results. What is this sum? _____
   (c) What is the product of the last terms, $(3y)(-3y)$? _____
   (d) Write the complete product using your answers in parts (a) and (c). _____
   Why is the sum found in part (b) omitted here?

40. Repeat Exercise 39 for the product $(5x + 7y)(5x - 7y)$.

Find the following products. See Example 6.

41. $(q + 2)(q - 2)$ **42.** $(x + 8)(x - 8)$ **43.** $(2w + 5)(2w - 5)$ **44.** $(3z + 8)(3z - 8)$

45. $(10x + 3y)(10x - 3y)$ **46.** $(13r + 2z)(13r - 2z)$ **47.** $(2x^2 - 5)(2x^2 + 5)$ **48.** $(9y^2 - 2)(9y^2 + 2)$

49. $\left(7x + \dfrac{3}{7}\right)\left(7x - \dfrac{3}{7}\right)$ **50.** $\left(9y + \dfrac{2}{3}\right)\left(9y - \dfrac{2}{3}\right)$

MATHEMATICAL CONNECTIONS (EXERCISES 51–54)

To understand how the special product $(a + b)^2 = a^2 + 2ab + b^2$ can be applied to a purely numerical problem, work Exercises 51–54 in order.

51. Evaluate 35^2 using either traditional paper-and-pencil methods or a calculator.

52. The number 35 can be written as $30 + 5$. Therefore, $35^2 = (30 + 5)^2$. Use the special product for squaring a binomial with $a = 30$ and $b = 5$ to write an expression for $(30 + 5)^2$. Do not simplify at this time.

53. Use the rule for order of operations to simplify the expression you found in Exercise 52.

54. How do the answers in Exercises 51 and 53 compare?

6.7 Division of a Polynomial by a Monomial

OBJECTIVE 1 We add two fractions with a common denominator as follows.

$$\frac{a}{c} + \frac{b}{c} = \frac{a+b}{c}.$$

Looking at this statement in reverse gives us a rule for dividing a polynomial by a monomial.

Dividing a Polynomial by a Monomial

To divide a polynomial by a monomial, divide each term of the polynomial by the monomial:

$$\frac{a+b}{c} = \frac{a}{c} + \frac{b}{c} \quad (c \neq 0).$$

The quotient rule for exponents is used to reduce each fraction as shown in the following examples.

EXAMPLE 1   Dividing a Polynomial by a Monomial

Divide $5m^5 - 10m^3$ by $5m^2$.

Use the rule above, with $+$ replaced by $-$.

$$\frac{5m^5 - 10m^3}{5m^2} = \frac{5m^5}{5m^2} - \frac{10m^3}{5m^2} = m^3 - 2m$$

Check by multiplication.

$$5m^2(m^3 - 2m) = 5m^5 - 10m^3$$

Because division by 0 is undefined, the quotient

$$\frac{5m^5 - 10m^3}{5m^2}$$

is undefined if $m = 0$. In the rest of the chapter, we assume that no denominators are 0.

WORK PROBLEM 1 AT THE SIDE. ▶▶

EXAMPLE 2   Dividing a Polynomial by a Monomial

Divide: $\dfrac{16a^5 - 12a^4 + 8a^2}{4a^3}$.

Divide each term of $16a^5 - 12a^4 + 8a^2$ by $4a^3$.

$$\frac{16a^5 - 12a^4 + 8a^2}{4a^3} = \frac{16a^5}{4a^3} - \frac{12a^4}{4a^3} + \frac{8a^2}{4a^3}$$

$$= 4a^2 - 3a + \frac{2}{a}$$

CONTINUED ON NEXT PAGE

OBJECTIVE

1 ▶ Divide a polynomial by a monomial.

FOR EXTRA HELP

Tutorial   Tape 13   SSM, Sec. 6.7

1. Divide.

(a) $\dfrac{6p^4 + 18p^7}{3p^2}$

(b) $\dfrac{12m^6 + 18m^5 + 30m^4}{6m^2}$

(c) $(18r^7 - 9r^2) \div (3r)$

ANSWERS

1. (a) $2p^2 + 6p^5$
   (b) $2m^4 + 3m^3 + 5m^2$
   (c) $6r^6 - 3r$

2. Divide.

(a) $\dfrac{20x^4 - 25x^3 + 5x}{5x^2}$

(b) $\dfrac{50m^4 - 30m^3 + 20m}{10m^3}$

3. Divide.

(a) $\dfrac{8y^7 - 9y^6 - 11y - 4}{y^2}$

(b) $\dfrac{12p^5 + 8p^4 + 3p^3 - 5p^2}{3p^3}$

(c) $\dfrac{45x^4 + 30x^3 - 60x^2}{-15x^2}$

The quotient is not a polynomial because of the expression $\frac{2}{a}$, which has a variable in the denominator. While the sum, difference, and product of two polynomials are always polynomials, the quotient of two polynomials may not be.

Again, check by multiplying.

$$4a^3\left(4a^2 - 3a + \frac{2}{a}\right) = 4a^3(4a^2) - 4a^3(3a) + 4a^3\left(\frac{2}{a}\right)$$
$$= 16a^5 - 12a^4 + 8a^2$$

◀◀ WORK PROBLEM 2 AT THE SIDE.

E X A M P L E 3   Dividing a Polynomial by a Monomial
Divide.

$$\frac{12x^4 - 7x^3 + 4x}{4x} = \frac{12x^4}{4x} - \frac{7x^3}{4x} + \frac{4x}{4x}$$
$$= 3x^3 - \frac{7x^2}{4} + 1$$

Check by multiplication.

Caution
In Example 3, notice that the quotient $\frac{4x}{4x} = 1$. It is a common error to leave the 1 out of the answer. Checking by multiplication will show that the answer $3x^3 - \frac{7}{4}x^2$ is not correct.

E X A M P L E 4   Dividing a Polynomial by a Monomial
Divide the polynomial

$$180y^{10} - 150y^8 + 120y^6 - 90y^4 + 100y$$

by the monomial $-30y^2$.
Using the methods of this section,

$$\frac{180y^{10} - 150y^8 + 120y^6 - 90y^4 + 100y}{-30y^2}$$
$$= \frac{180y^{10}}{-30y^2} - \frac{150y^8}{-30y^2} + \frac{120y^6}{-30y^2} - \frac{90y^4}{-30y^2} + \frac{100y}{-30y^2}$$
$$= -6y^8 + 5y^6 - 4y^4 + 3y^2 - \frac{10}{3y}$$

To check, multiply this answer and $-30y^2$.

◀◀ WORK PROBLEM 3 AT THE SIDE.

ANSWERS

2. (a) $4x^2 - 5x + \dfrac{1}{x}$ (b) $5m - 3 + \dfrac{2}{m^2}$

3. (a) $8y^5 - 9y^4 - \dfrac{11}{y} - \dfrac{4}{y^2}$

(b) $4p^2 + \dfrac{8p}{3} + 1 - \dfrac{5}{3p}$

(c) $-3x^2 - 2x + 4$

6.7 Exercises

1. Explain why the division problem $\dfrac{16m^3 - 12m^2}{4m}$ can be performed using the methods of this section, while the division problem $\dfrac{4m}{16m^3 - 12m^2}$ cannot.

2. If the area of a rectangle is $24x^6$ and the width is $4x^2$, what is the length?

3. Evaluate $\dfrac{5y + 6}{2}$ when $y = 2$. Evaluate $5y + 3$ when $y = 2$. Does $\dfrac{5y + 6}{2}$ equal $5y + 3$?

4. Evaluate $\dfrac{10r + 7}{5}$ when $r = 1$. Evaluate $2r + 7$ when $r = 1$. Does $\dfrac{10r + 7}{5}$ equal $2r + 7$?

Divide each polynomial by 2m. See Examples 1–4.

5. $10m^5 - 16m^4 + 8m^3$ 6. $6m^5 - 4m^3 + 2m^2$ 7. $8m^5 - 4m^3 + 4m^2$

8. $8m^4 - 4m^3 + 6m^2$ 9. $m^5 - 4m^2 + 8$ 10. $m^3 + m^2 + 6$

Divide each polynomial by $3x^2$. See Examples 1–4.

11. $12x^5 - 9x^4 + 6x^3$ 12. $24x^6 - 12x^5 + 30x^4$ 13. $3x^2 + 15x^3 - 27x^4$ 14. $3x^2 - 18x^4 + 30x^5$

15. $36x + 24x^2 + 6x^3$ 16. $9x - 12x^2 + 9x^3$ 17. $4x^4 + 3x^3 + 2x$ 18. $5x^4 - 6x^3 + 8x$

Perform each division. See Examples 1–4.

19. $\dfrac{27r^4 - 36r^3 - 6r^2 + 3r - 2}{3r}$ 20. $\dfrac{8k^4 - 12k^3 - 2k^2 - 2k - 3}{2k}$

21. $\dfrac{2m^5 - 6m^4 + 8m^2}{-2m^3}$ 22. $\dfrac{6r^5 - 8r^4 + 10r^2}{-2r^4}$

23. $(20a^4 - 15a^5 + 25a^3) \div (5a^4)$

24. $(16y^5 - 8y^2 + 12y) \div (4y^2)$

25. $(120x^{11} - 60x^{10} + 140x^9 - 100x^8) \div (10x^{12})$

26. $(120x^{12} - 84x^9 + 60x^8 - 36x^7) \div (12x^9)$

27. The quotient in Exercise 17 is $\dfrac{4x^2}{3} + x + \dfrac{2}{3x}$. Notice how the third term is written with x in the denominator. Would $\frac{2}{3}x$ be an acceptable form for this term? Explain why or why not. Is $\frac{4}{3}x^2$ an acceptable form for the first term? Why or why not?

28. If the area of a rectangle is represented by $12x^2 - 4x + 2$ and the width is $2x$, what expression represents the length?

29. What polynomial, when divided by $5x^3$, yields $3x^2 - 7x + 7$ as a quotient?

30. The quotient of a certain polynomial and $-12y^3$ is $6y^3 - 5y^2 + 2y - 3 + \frac{7}{y}$. Find the polynomial.

MATHEMATICAL CONNECTIONS (EXERCISES 31–34)

Our system of numeration is called a decimal system. It is based on powers of ten. In a whole number such as 2846, each digit is understood to represent the number of powers of ten for its place value. The 2 represents two thousands (2×10^3), the 8 represents eight hundreds (8×10^2), the 4 represents four tens (4×10^1), and the 6 represents six ones (or units) (6×10^0). In expanded form we write

$$2846 = (2 \times 10^3) + (8 \times 10^2) + (4 \times 10^1) + (6 \times 10^0).$$

Keeping this information in mind, work Exercises 31–34 in order.

31. Divide 2846 by 2, using paper-and-pencil methods: $2\overline{)2846}$.

32. Write your answer in Exercise 31 in expanded form.

33. Use the methods of this section to divide the polynomial $2x^3 + 8x^2 + 4x + 6$ by 2.

34. Compare your answers in Exercises 32 and 33. How are they similar? How are they different? For what value of x does the answer in Exercise 33 equal the answer in Exercise 32?

6.8 The Quotient of Two Polynomials

OBJECTIVE 1 A method of "long division" is used to divide a polynomial by a polynomial (other than a monomial). This method is similar to the method of long division used for two whole numbers. For comparison, the division of whole numbers is shown alongside the division of polynomials. The polynomial must be in descending order.

OBJECTIVE

1 Divide a polynomial by a polynomial.

FOR EXTRA HELP

Tutorial   Tape 13   SSM, Sec. 6.8

Division of Whole Numbers	Division of Polynomials

Step 1

Divide 27 into 6696.

$$27 \overline{)6696}$$

Divide $2x + 3$ into $8x^3 - 4x^2 - 14x + 15$.

$$2x + 3 \overline{)8x^3 - 4x^2 - 14x + 15}$$

Step 2

27 divides into 66 **2** times; $2 \cdot 27 = \mathbf{54}$.

$$\begin{array}{r} 2 \\ 27 \overline{)6696} \\ 54 \end{array}$$

$2x$ divides into $8x^3$ **$4x^2$** times; $4x^2(2x + 3) = \mathbf{8x^3 + 12x^2}$.

$$\begin{array}{r} 4x^2 \\ 2x + 3 \overline{)8x^3 - 4x^2 - 14x + 15} \\ 8x^3 + 12x^2 \end{array}$$

Step 3

Subtract; then bring down the next digit.

$$\begin{array}{r} 2 \\ 27 \overline{)6696} \\ 54 \downarrow \\ \hline 129 \end{array}$$

Subtract; then bring down the next term.

$$\begin{array}{r} 4x^2 \\ 2x + 3 \overline{)8x^3 - 4x^2 - 14x + 15} \\ 8x^3 + 12x^2 \quad \downarrow \\ \hline -16x^2 - 14x \end{array}$$

(To subtract two polynomials, change the sign of the second and then add.)

Step 4

27 divides into 129 **4** times; $4 \cdot 27 = \mathbf{108}$.

$$\begin{array}{r} 24 \\ 27 \overline{)6696} \\ 54 \\ \hline 129 \\ 108 \end{array}$$

$2x$ divides into $-16x^2$ **$-8x$** times; $-8x(2x + 3) = \mathbf{-16x^2 - 24x}$.

$$\begin{array}{r} 4x^2 - 8x \\ 2x + 3 \overline{)8x^3 - 4x^2 - 14x + 15} \\ 8x^3 + 12x^2 \\ \hline -16x^2 - 14x \\ -16x^2 - 24x \end{array}$$

Step 5

Subtract; then bring down the next digit.

$$\begin{array}{r} 24 \\ 27 \overline{)6696} \\ 54 \\ \hline 129 \\ 108 \\ \hline 216 \end{array}$$

Subtract; then bring down the next term.

$$\begin{array}{r} 4x^2 - 8x \\ 2x + 3 \overline{)8x^3 - 4x^2 - 14x + 15} \\ 8x^3 + 12x^2 \\ \hline -16x^2 - 14x \\ -16x^2 - 24x \\ \hline 10x + 15 \end{array}$$

Step 6

27 divides into 216 **8** times;
8 · 27 = **216**.

$$
\begin{array}{r}
24\underline{8} \\
27\overline{)6696} \\
54 \\
\overline{129} \\
108 \\
\overline{216} \\
\underline{216}
\end{array}
$$

6696 divided by 27 is 248.
There is no remainder.

2x divides into 10x **5** times;
5(2x + 3) = **10x + 15.**

$$
\begin{array}{r}
4x^2 - 8x + 5 \\
2x + 3\overline{)8x^3 - 4x^2 - 14x + 15} \\
\underline{8x^3 + 12x^2} \\
-16x^2 - 14x \\
\underline{-16x^2 - 24x} \\
10x + 15 \\
\underline{10x + 15}
\end{array}
$$

$8x^3 - 4x^2 - 14x + 15$ divided by
$2x + 3$ is $4x^2 - 8x + 5$. There is
no remainder.

Step 7

Check by multiplication.

$$27 \cdot 248 = 6696$$

Check by multiplication.

$$(2x + 3)(4x^2 - 8x + 5)$$
$$= 8x^3 - 4x^2 - 14x + 15$$

E X A M P L E I Dividing a Polynomial by a Polynomial

Divide $4x^3 - 4x^2 + 5x - 8$ by $2x - 1$.

$$
\begin{array}{r}
2x^2 - x + 2 \\
2x - 1\overline{)4x^3 - 4x^2 + 5x - 8} \\
\underline{4x^3 - 2x^2} \\
-2x^2 + 5x \\
\underline{-2x^2 + x} \\
4x - 8 \\
\underline{4x - 2} \\
-6 \quad \leftarrow \text{Remainder}
\end{array}
$$

Step 1 2x divides into $4x^3$ **$2x^2$** times; $2x^2(2x - 1) = 4x^3 - 2x^2$.

Step 2 Subtract; bring down the next term.

Step 3 2x divides into $-2x^2$ **$-x$** times; $-x(2x - 1) = -2x^2 + x$.

Step 4 Subtract; bring down the next term.

Step 5 2x divides into 4x **2** times; 2(2x − 1) = 4x − 2.

Step 6 Subtract. The remainder is −6. Thus 2x − 1 divides into
$4x^3 - 4x^2 + 5x - 8$ with a quotient of $2x^2 - x + 2$ and a remainder of −6. Write the remainder as the numerator of a fraction that has 2x − 1 as its denominator. The answer is not a
polynomial because of the remainder.

$$\frac{4x^3 - 4x^2 + 5x - 8}{2x - 1} = 2x^2 - x + 2 + \frac{-6}{2x - 1}$$

CONTINUED ON NEXT PAGE

Step 7   Check by multiplication.

$$(2x - 1)\left(2x^2 - x + 2 + \frac{-6}{2x - 1}\right)$$

$$= (2x - 1)(2x^2) + (2x - 1)(-x) + (2x - 1)(2)$$

$$+ (2x - 1)\left(\frac{-6}{2x - 1}\right)$$

$$= 4x^3 - 2x^2 - 2x^2 + x + 4x - 2 - 6$$

$$= 4x^3 - 4x^2 + 5x - 8$$

WORK PROBLEM 1 AT THE SIDE. ▶▶

E X A M P L E   2   **Dividing into a Polynomial with Missing Terms**

Divide $x^3 - 1$ by $x - 1$.

Here the polynomial $x^3 - 1$ is missing the x^2 term and the x term. When terms are missing, use 0 as the coefficient for the missing terms. (Zero acts as a placeholder here, just as it does in our number system.)

$$x^3 - 1 = x^3 + 0x^2 + 0x - 1$$

Now divide.

$$
\begin{array}{r}
x^2 + x + 1 \\
x - 1 \overline{) x^3 + 0x^2 + 0x - 1} \\
\underline{x^3 - x^2} \\
x^2 + 0x \\
\underline{x^2 - x} \\
x - 1 \\
\underline{x - 1} \\
0
\end{array}
$$

The remainder is 0. The quotient is $x^2 + x + 1$. Check by multiplication.

$$(x^2 + x + 1)(x - 1) = x^3 - 1$$

WORK PROBLEM 2 AT THE SIDE. ▶▶

E X A M P L E   3   **Dividing by a Polynomial with Missing Terms**

Divide $x^4 + 2x^3 + 2x^2 - x - 1$ by $x^2 + 1$.

Since $x^2 + 1$ has a missing x term, write it as $x^2 + 0x + 1$. Then go through the division process as follows.

$$
\begin{array}{r}
x^2 + 2x + 1 \\
x^2 + 0x + 1 \overline{) x^4 + 2x^3 + 2x^2 - x - 1} \\
\underline{x^4 + 0x^3 + x^2} \\
2x^3 + x^2 - x \\
\underline{2x^3 + 0x^2 + 2x} \\
x^2 - 3x - 1 \\
\underline{x^2 + 0x + 1} \\
-3x - 2
\end{array}
$$

CONTINUED ON NEXT PAGE

---

1. Divide.

(a) $(x^3 + x^2 + 4x - 6) \div (x - 1)$

(b) $\dfrac{p^3 - 2p^2 - 5p + 9}{p + 2}$

2. Divide.

(a) $\dfrac{r^2 - 5}{r + 4}$

(b) $(x^3 - 8) \div (x - 2)$

ANSWERS

1. (a) $x^2 + 2x + 6$

   (b) $p^2 - 4p + 3 + \dfrac{3}{p + 2}$

2. (a) $r - 4 + \dfrac{11}{r + 4}$

   (b) $x^2 + 2x + 4$

3. Divide.

(a)
$$(2x^4 + 3x^3 - x^2 + 6x + 5)$$
$$\div (x^2 - 1)$$

When the result of subtracting $(-3x - 2$, in this case) is a polynomial of smaller degree than the divisor $(x^2 + 0x + 1)$, that polynomial is the remainder. Write the answer as

$$x^2 + 2x + 1 + \frac{-3x - 2}{x^2 + 1}.$$

Verify by multiplication that this is the correct quotient.

◀◀ **WORK PROBLEM 3 AT THE SIDE.**

(b)
$$\frac{2m^5 + m^4 + 6m^3 - 3m^2 - 18}{m^2 + 3}$$

ANSWERS

3. (a) $2x^2 + 3x + 1 + \dfrac{9x + 6}{x^2 - 1}$

  (b) $2m^3 + m^2 - 6$

6.8 Exercises

1. In the division problem $(4x^4 + 2x^3 - 14x^2 + 19x + 10) \div (2x + 5) = 2x^3 - 4x^2 + 3x + 2$, which polynomial is the divisor? Which is the quotient?

2. When dividing one polynomial by another, how do you know when to stop dividing?

3. In dividing $12m^2 - 20m + 3$ by $2m - 3$, what is the first step?

4. In the division in Exercise 3, what is the second step?

Perform each division. See Example 1.

5. $\dfrac{x^2 - x - 6}{x - 3}$

6. $\dfrac{m^2 - 2m - 24}{m - 6}$

7. $\dfrac{2y^2 + 9y - 35}{y + 7}$

8. $\dfrac{2y^2 + 9y + 7}{y + 1}$

9. $\dfrac{p^2 + 2p + 20}{p + 6}$

10. $\dfrac{x^2 + 11x + 16}{x + 8}$

11. $(r^2 - 8r + 15) \div (r - 3)$

12. $(t^2 + 2t - 35) \div (t - 5)$

13. $\dfrac{4a^2 - 22a + 32}{2a + 3}$

14. $\dfrac{9w^2 + 6w + 10}{3w - 2}$

15. $\dfrac{8x^3 - 10x^2 - x + 3}{2x + 1}$

16. $\dfrac{12t^3 - 11t^2 + 9t + 18}{4t + 3}$

MATHEMATICAL CONNECTIONS (EXERCISES 17–20)

We can find the value of a polynomial in x for a given value of x by substituting that number for x. Surprisingly, we can accomplish the same thing by division. For example, to find the value of $2x^2 - 4x + 3$ for $x = -3$, we would divide $x - (-3)$ into $2x^2 - 4x + 3$. The remainder will give the value of the polynomial for $x = -3$. Work Exercises 17–20 in order.

17. Find the value of $2x^2 - 4x + 3$ for $x = -3$ by substitution.

18. Divide $2x^2 - 4x + 3$ by $x + 3$. Give the remainder.

19. Compare your answers to Exercises 17 and 18. What do you notice?

20. Choose another polynomial and evaluate it both ways at some value of the variable. Do the answers agree?

Perform each division. See Examples 2 and 3.

21. $\dfrac{3y^3 + y^2 + 2}{y + 1}$

22. $\dfrac{2r^3 - 6r - 36}{r - 3}$

23. $\dfrac{3k^3 - 4k^2 - 6k + 10}{k^2 - 2}$

24. $\dfrac{5z^3 - z^2 + 10z + 2}{z^2 + 2}$

25. $(x^4 - x^2 - 2) \div (x^2 - 2)$

26. $(r^4 + 2r^2 - 3) \div (r^2 - 1)$

27. $\dfrac{6p^4 - 15p^3 + 14p^2 - 5p + 10}{3p^2 + 1}$

28. $\dfrac{6r^4 - 10r^3 - r^2 + 15r - 8}{2r^2 - 3}$

29. $\dfrac{2x^5 + 9x^4 + 8x^3 + 10x^2 + 14x + 5}{2x^2 + 3x + 1}$

30. $\dfrac{4t^5 - 11t^4 - 6t^3 + 5t^2 - t + 3}{4t^2 + t - 3}$

31. $\dfrac{x^4 - 1}{x^2 - 1}$

32. $\dfrac{y^3 + 1}{y + 1}$

KEY TERMS

6.3	scientific notation	A number written as $a \times 10^n$, where $1 \leq |a| < 10$ and n is an integer, is in scientific notation.
6.4	polynomial	A polynomial is a term or the sum of a finite number of terms.
	descending powers	A polynomial is written in descending powers if the degrees of its terms are in decreasing order.
	degree of a term	The degree of a term with one variable is the exponent on the variable.
	degree of a polynomial	The degree of a polynomial in one variable is the highest exponent found in any term of the polynomial.
	trinomial	A trinomial is a polynomial with three terms.
	binomial	A binomial is a polynomial with two terms.
	monomial	A monomial is a polynomial with one term.

NEW SYMBOLS

x^{-n} x to the negative n

QUICK REVIEW

Concepts	Examples
6.1 The Product Rule and Power Rules for Exponents	
For any integers m and n:	
Product rule $a^m \cdot a^n = a^{m+n}$	$2^4 \cdot 2^5 = 2^9$
Power rules (a) $(a^m)^n = a^{mn}$	$(3^4)^2 = 3^8$
(b) $(ab)^m = a^m b^m$	$(6a)^5 = 6^5 a^5$
(c) $\left(\dfrac{a}{b}\right)^m = \dfrac{a^m}{b^m}$ $(b \neq 0)$	$\left(\dfrac{2}{3}\right)^4 = \dfrac{2^4}{3^4}$
6.2 Integer Exponents and the Quotient Rule	
If $a \neq 0$, for integers m and n:	
Zero exponent $a^0 = 1$	$15^0 = 1$
Negative exponent $a^{-n} = \dfrac{1}{a^n}$	$5^{-2} = \dfrac{1}{5^2} = \dfrac{1}{25}$
Quotient rule $\dfrac{a^m}{a^n} = a^{m-n}$	$\dfrac{4^8}{4^3} = 4^5$
$\dfrac{a^{-m}}{b^{-n}} = \dfrac{b^n}{a^m}$ $\qquad \left(\dfrac{a}{b}\right)^{-m} = \left(\dfrac{b}{a}\right)^m$	$\dfrac{6^{-2}}{7^{-3}} = \dfrac{7^3}{6^2}$ $\qquad \left(\dfrac{5}{3}\right)^{-4} = \left(\dfrac{3}{5}\right)^4$
6.3 An Application of Exponents: Scientific Notation	
To write a number in scientific notation (as $a \times 10^n$), move the decimal point to the right of the first nonzero digit. If the decimal point is moved n places, and this makes the number smaller, n is positive; otherwise, n is negative. If the decimal point is not moved, n is 0.	$247 = 2.47 \times 10^2$ $.0051 = 5.1 \times 10^{-3}$ $4.8 = 4.8 \times 10^0$

Concepts	Examples
6.4 Addition and Subtraction of Polynomials **Addition:** Add like terms. **Subtraction:** Change the signs of the terms in the second polynomial and add to the first polynomial.	Add: $\begin{array}{r} 2x^2 + 5x - 3 \\ 5x^2 - 2x + 7 \\ \hline 7x^2 + 3x + 4 \end{array}$ $(2x^2 + 5x - 3) - (5x^2 - 2x + 7)$ $= (2x^2 + 5x - 3) + (-5x^2 + 2x - 7)$ $= -3x^2 + 7x - 10$
6.5 Multiplication of Polynomials Multiply each term of the first polynomial by each term of the second polynomial. Then add like terms.	Multiply: $\begin{array}{r} 3x^3 - 4x^2 + 2x - 7 \\ 4x + 3 \\ \hline 9x^3 - 12x^2 + 6x - 21 \\ 12x^4 - 16x^3 + 8x^2 - 28x \\ \hline 12x^4 - 7x^3 - 4x^2 - 22x - 21 \end{array}$
6.6 Products of Binomials **FOIL Method** *Step 1* Multiply the two first terms to get the first term of the answer. *Step 2* Find the outer product and the inner product and mentally add them, when possible, to get the middle term of the answer. *Step 3* Multiply the two last terms to get the last term of the answer. **Square of a Binomial** $(a + b)^2 = a^2 + 2ab + b^2$ $(a - b)^2 = a^2 - 2ab + b^2$ **Product of the Sum and Difference of Two Terms** $(a + b)(a - b) = a^2 - b^2$	Find $(2x + 3)(5x - 4)$. $$2x(5x) = 10x^2$$ $$2x(-4) + 3(5x) = 7x$$ $$3(-4) = -12$$ $$(2x + 3)(5x - 4) = 10x^2 + 7x - 12$$ $$(3x + 1)^2 = 9x^2 + 6x + 1$$ $$(2m - 5n)^2 = 4m^2 - 20mn + 25n^2$$ $$(4a + 3)(4a - 3) = 16a^2 - 9$$
6.7 Division of a Polynomial by a Monomial Divide each term of the polynomial by the monomial: $$\frac{a + b}{c} = \frac{a}{c} + \frac{b}{c}$$	Divide: $\dfrac{4x^3 - 2x^2 + 6x - 8}{2x}$ $$= 2x^2 - x + 3 - \frac{4}{x}$$
6.8 The Quotient of Two Polynomials Use "long division."	Divide: $3x + 4\,\overline{\smash{)}\,6x^2 - 7x - 21}$ quotient $2x - 5 + \dfrac{-1}{3x + 4}$ $\begin{array}{r} 6x^2 + 8x \\ \hline -15x - 21 \\ -15x - 20 \\ \hline -1 \end{array}$

CHAPTER 6 REVIEW EXERCISES

[6.1] *Use the product rule to simplify each expression. Write each answer in exponential form.*

1. $4^3 \cdot 4^8$ **2.** $(-5)^6(-5)^5$ **3.** $(-8x^4)(9x^3)$ **4.** $(2x^2)(5x^3)(x^9)$

Use the power rules to simplify each expression. Write each answer in exponential form.

5. $(19x)^5$ **6.** $(-4y)^7$ **7.** $5(pt)^4$ **8.** $\left(\dfrac{7}{5}\right)^6$

Use a combination of rules to simplify each expression. Leave answers in exponential form.

9. $(3x^2y^3)^3$ **10.** $(t^4)^8(t^2)^5$ **11.** $(6x^2z^4)^2(x^3yz^2)^4$

12. Explain why the product rule for exponents does not apply to the expression $7^2 + 7^4$.

[6.2] *Evaluate each expression.*

13. $5^0 + 8^0$ **14.** 2^{-5} **15.** $\left(\dfrac{6}{5}\right)^{-2}$ **16.** $4^{-2} - 4^{-1}$

Simplify. Write each answer in exponential form, using only positive exponents. Assume all variables are nonzero.

17. $\dfrac{6^{-3}}{6^{-5}}$ **18.** $\dfrac{x^{-7}}{x^{-9}}$ **19.** $\dfrac{p^{-8}}{p^4}$ **20.** $\dfrac{r^{-2}}{r^{-6}}$

21. $(2^4)^2$ **22.** $(9^3)^{-2}$ **23.** $(5^{-2})^{-4}$

24. $(8^{-3})^4$ **25.** $\dfrac{(m^2)^3}{(m^4)^2}$ **26.** $\dfrac{y^4 \cdot y^{-2}}{y^{-5}}$

27. $\dfrac{r^9 \cdot r^{-5}}{r^{-2} \cdot r^{-7}}$ **28.** $(-5m^3)^2$ **29.** $(2y^{-4})^{-3}$

30. $\dfrac{ab^{-3}}{a^4b^2}$ **31.** $\dfrac{(6r^{-1})^2 \cdot (2r^{-4})}{r^{-5}(r^2)^{-3}}$ **32.** $\dfrac{(2m^{-5}n^2)^3(3m^2)^{-1}}{m^{-2}n^{-4}(m^{-1})^2}$

[6.3] *Write each number in scientific notation.*

33. 48,000,000   **34.** 28,988,000,000   **35.** .000065   **36.** .0000000824

Write each number without exponents.

37. 2.4×10^4 **38.** 7.83×10^7 **39.** 8.97×10^{-7} **40.** 9.95×10^{-12}

Perform the indicated operation and write the answer without exponents.

41. $(2 \times 10^{-3}) \times (4 \times 10^5)$ **42.** $\dfrac{8 \times 10^4}{2 \times 10^{-2}}$ **43.** $\dfrac{12 \times 10^{-5} \times 5 \times 10^4}{4 \times 10^3 \times 6 \times 10^{-2}}$ **44.** $\dfrac{2.5 \times 10^5 \times 4.8 \times 10^{-4}}{7.5 \times 10^8 \times 1.6 \times 10^{-5}}$

[6.4] *For each polynomial, first simplify, if possible, and write it in descending powers of the variable. Then give the degree of the resulting polynomial, and tell whether it is a monomial, a binomial, a trinomial, or none of these.*

45. $9m^2 + 11m^2 + 2m^2$

46. $-4p + p^3 - p^2 + 8p + 2$

47. $2r^4 - r^3 + 8r^4 + r^3 - 6r^4 + 8r^5$

48. $12a^5 + 19a^4 + 8a^3 + 2a^3 - 9a^4 + 3a^5$

Add or subtract as indicated.

49. Add.
$$\begin{aligned} -2a^3 + 5a^2 \\ \underline{-3a^3 - a^2} \end{aligned}$$

50. Add.
$$\begin{aligned} 4r^3 - 8r^2 + 6r \\ \underline{-2r^3 + 5r^2 + 3r} \end{aligned}$$

51. Subtract.
$$\begin{aligned} 6y^2 - 8y + 2 \\ \underline{-5y^2 + 2y - 7} \end{aligned}$$

52. Subtract.
$$\begin{aligned} -12k^4 - 8k^2 + 7k - 5 \\ \underline{k^4 + 7k^2 + 11k + 1} \end{aligned}$$

53. $(2m^3 - 8m^2 + 4) + (8m^3 + 2m^2 - 7)$

54. $(-5y^2 + 3y + 11) + (4y^2 - 7y + 15)$

55. $(6p^2 - p - 8) - (-4p^2 + 2p + 3)$

56. $(12r^4 - 7r^3 + 2r^2) - (5r^4 - 3r^3 + 2r^2 + 1)$

[6.5] *Find each product.*

57. $5x(2x + 14)$ **58.** $-3p^3(2p^2 - 5p)$ **59.** $(m - 9)(m + 2)$ **60.** $(3k - 6)(2k + 1)$

61. $(3r - 2)(2r^2 + 4r - 3)$ **62.** $(2y + 3)(4y^2 - 6y + 9)$ **63.** $(r + 2)^3$

64. Explain why $(a + b)^2$ is not equal to $a^2 + b^2$.

[6.6] *Find each product.*

65. $(3k + 1)(2k + 3)$ **66.** $(a + 3b)(2a - b)$ **67.** $(6k - 3q)(2k - 7q)$

68. $(a + 4)^2$ **69.** $(3p - 2)^2$ **70.** $(2r + 5s)^2$

71. $(6m - 5)(6m + 5)$ **72.** $(2z + 7)(2z - 7)$ **73.** $(5a + 6b)(5a - 6b)$ **74.** $(2x^2 + 5)(2x^2 - 5)$

[6.7] *Perform each division.*

75. $\dfrac{-15y^4}{-9y^2}$ **76.** $\dfrac{-12x^3y^2}{6xy}$

77. $\dfrac{6y^4 - 12y^2 + 18y}{-6y}$ **78.** $\dfrac{2p^3 - 6p^2 + 5p}{2p^2}$

79. $(5x^{13} - 10x^{12} + 20x^7 - 35x^5) \div (-5x^4)$ **80.** $(-10m^4n^2 + 5m^3n^3 + 6m^2n^4) \div (5m^2n)$

[6.8] *Perform each division.*

81. $(2r^2 + 3r - 14) \div (r - 2)$ **82.** $\dfrac{12m^2 - 11m - 10}{3m - 5}$

83. $\dfrac{10a^3 + 5a^2 - 14a + 9}{5a^2 - 3}$

84. $\dfrac{2k^4 + 4k^3 + 9k^2 - 8}{2k^2 + 1}$

MIXED REVIEW EXERCISES

Simplify. Perform the indicated operations. Write with positive exponents. Assume that no denominators are equal to zero.

85. $19^0 - 3^0$

86. $(3p)^4(3p^{-7})$

87. 7^{-2}

88. $(-7 + 2k)^2$

89. $\dfrac{2y^3 + 17y^2 + 37y + 7}{2y + 7}$

90. $\left(\dfrac{6r^2s}{5}\right)^4$

91. $-m^5(8m^2 + 10m + 6)$

92. $\left(\dfrac{1}{2}\right)^{-5}$

93. $(25x^2y^3 - 8xy^2 + 15x^3y) \div (5x)$

94. $(6r^{-2})^{-1}$

95. $(2x + y)^3$

96. $2^{-1} + 4^{-1}$

97. $(a + 2)(a^2 - 4a + 1)$

98. $(5y^3 - 8y^2 + 7) - (-3y^3 + y^2 + 2)$

99. $(2r + 5)(5r - 2)$

100. $(12a + 1)(12a - 1)$

Evaluate each expression.

1. 5^{-4}

2. $(-3)^0 + 4^0$

3. $4^{-1} + 3^{-1}$

Simplify, and write each answer using only positive exponents. Assume that variables represent nonzero numbers.

4. $\dfrac{8^{-1} \cdot 8^4}{8^{-2}}$

5. $\dfrac{(x^{-3})^{-2}(x^{-1}y)^2}{(xy^{-2})^2}$

Write each number in scientific notation.

6. (a) 344,000,000,000
   (b) .00000557

Write each number without exponents.

7. (a) 2.96×10^7
   (b) 6.07×10^{-8}

For each polynomial, combine like terms when possible, and write the polynomial in descending powers of the variable. Give the degree of the simplified polynomial. Decide whether the simplified polynomial is a monomial, binomial, trinomial, or none of these.

8. $5x^2 + 8x - 12x^2$

9. $13n^3 - n^2 + n^4 + 3n^4 - 9n^2$

Perform the indicated operations.

10. $(5t^4 - 3t^2 + 7t + 3) - (t^4 - t^3 + 3t^2 + 8t + 3)$

11. $(2y^2 - 8y + 8) + (-3y^2 + 2y + 3) - (y^2 + 3y - 6)$

12. Subtract.
$$9t^3 - 4t^2 + 2t + 2$$
$$9t^3 + 8t^2 - 3t - 6$$

1. _____

2. _____

3. _____

4. _____

5. _____

6. (a) _____

  (b) _____

7. (a) _____

  (b) _____

8. _____

9. _____

10. _____

11. _____

12. _____

13. _____

13. $3x^2(-9x^3 + 6x^2 - 2x + 1)$

14. _____

14. $(t - 8)(t + 3)$

15. _____

15. $(4x + 3y)(2x - y)$

16. _____

16. $(5x - 2y)^2$

17. _____

17. $(10v + 3w)(10v - 3w)$

18. _____

18. $(2r - 3)(r^2 + 2r - 5)$

19. _____

19. $(x + 1)^3$

20. _____

20. $\dfrac{8y^3 - 6y^2 + 4y + 10}{2y}$

21. _____

21. $(-9x^2y^3 + 6x^4y^3 + 12xy^3) \div (3xy)$

22. _____

22. $\dfrac{2x^2 + x - 36}{x - 4}$

23. _____

23. $(3x^3 - x + 4) \div (x - 2)$

24. _____

24. What polynomial expression represents the area of this square?

$3x + 9$

25. _____

25. Give an example of this situation: the sum of two fourth degree polynomials in x is a third degree polynomial in x.

1. True or false: The number $-.75$ is both an integer and a rational number.

2. Simplify $-3(4^2 - 7) + 2$.

3. Evaluate $xy^2 - 3x$ if $x = 2$ and $y = -3$.

4. Add: $-113\dfrac{2}{3} + 56\dfrac{1}{4}$.

5. Subtract: $-18 - (-34)$.

6. Multiply: $(-4)(-3)(25)(-1)$.

7. Divide: $\dfrac{12 - 14}{-6}$.

8. What property justifies the statement $3 + (9 + 4) = (9 + 4) + 3$?

9. Simplify: $3(2t - 4) - 5(t^2 - 3t + 8)$.

10. Solve: $5 - (2x + 4) = x + 1$.

11. Express in words, using x as the variable: 3 more than twice the product of a number and 5.

12. Solve: $-\dfrac{3}{4}r = 36$.

13. A farmer has 34 more pigs than goats, with 98 animals in all. Find the number of pigs and the number of goats on this farm.

14. Solve: $A = LW$ for W.

15. In a mixture of concrete, there are 3 pounds of cement mix for every 1 pound of gravel. If the mixture contains a total of 248 pounds of these two ingredients, how many pounds of gravel are there?

16. What percent of 12 is 36?

17. Solve: $3x + 3 \geq -2x - 1$.

18. If $A = \{3, 6, 9\}$ and $B = \{6, 9, 12\}$, find $A \cap B$.

19. Solve: $2x > 4$ and $3x \leq 30$.

20. Solve: $|4x - 7| = 3$.

21. Solve: $|2x + 5| > 3$.

22. If a point has coordinates $(-3, -5)$, in which quadrant does it lie?

23. What is the slope of the line joining the points $(-7, 3)$ and $(4, 6)$?

24. What is the slope of the line with equation $3x - 5y = 3$?

25. Graph $x - 3y = 6$.

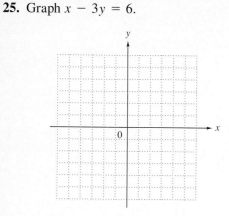

26. Graph $x - 3y \leq 6$.

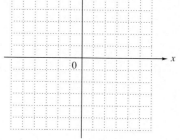

27. If $f(x) = -x^2 + 3x + 2$, find $f(9)$.

28. If p varies directly as q, and $p = 9$ when $q = 6$, find p when $q = 15$.

29. Solve the system by addition.

$$3x + 8y = 14$$
$$x - y = 1$$

30. Solve the system by substitution.

$$9x - 3y = -3$$
$$y = x + 3$$

31. Solve by any method.

$$x + 2y + z = 8$$
$$2x - y + 3z = 15$$
$$-x + 3y - 3z = -11$$

32. At the Mandeville Nut Shop, 6 pounds of peanuts and 12 pounds of cashews cost $60, while 3 pounds of peanuts and 4 pounds of cashews cost $22. Find the cost of each kind of nut.

33. Evaluate the determinant: $\begin{vmatrix} 4 & -3 \\ 2 & -6 \end{vmatrix}$.

34. Evaluate the determinant: $\begin{vmatrix} 5 & 4 & 0 \\ -1 & 2 & 3 \\ 1 & 0 & 5 \end{vmatrix}$.

35. Evaluate $8^{-1} + 3^{-2}$.

36. Write in scientific notation: $-.0000000588$.

37. Simplify: $(-3x^3 - 4x^2 + 5x - 6) + 2(4x^3 - 9x^2 + 8x + 5)$.

38. Multiply: $(5r - 6)(3r^2 + 2r + 7)$.

39. Divide: $\dfrac{25w^2 + 10w - 15}{5w}$.

40. Divide $4x^3 - 2x^2 + 6x - 8$ by $x - 1$.

Factoring

7.1 Factors; The Greatest Common Factor

Factoring is the opposite of multiplication; to **factor** a number means to write it as the product of two or more numbers. The product is called the **factored form** of the number. Now we extend the idea of factoring to any polynomial.

OBJECTIVE 1 An integer that is a factor of two or more integers is a **common factor** of those integers. For example, 6 is a common factor of 18 and 24 because 6 is a factor of both 18 and 24. Other common factors of 18 and 24 are 1, 2, and 3. The **greatest common factor** of a list of integers is the largest common factor of those integers. This means 6 is the greatest common factor of 18 and 24, since it is the largest of their common factors.

EXAMPLE 1 Finding the Greatest Common Factor for Numbers

Find the greatest common factor for each group of numbers.

(a) 30, 45

First write each number in prime factored form.

$$30 = 2 \cdot 3 \cdot 5 \qquad 45 = 3^2 \cdot 5$$

Use each prime the *least* number of times it appears in all the factored forms. There is no 2 in the prime factored form of 45, so there will be no 2 in the greatest common factor. The least number of times 3 appears in all the factored forms is 1; the least number of times 5 appears is also 1. From this, the greatest common factor is

$$3^1 \cdot 5^1 = 3 \cdot 5 = 15.$$

(b) 72, 120, 432

Find the prime factored form of each number.

$$72 = 2^3 \cdot 3^2 \quad 120 = 2^3 \cdot 3 \cdot 5 \quad 432 = 2^4 \cdot 3^3$$

The least number of times 2 appears in all the factored forms is 3, and the least number of times 3 appears is 1. There is no 5 in the prime factored form of either 72 or 432, so the greatest common factor is

$$2^3 \cdot 3 = 24.$$

CONTINUED ON NEXT PAGE

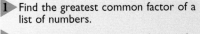
www.mathnotes.com

OBJECTIVES

1 Find the greatest common factor of a list of numbers.

2 Find the greatest common factor of a list of variable terms.

3 Factor out the greatest common factor.

4 Factor by grouping.

FOR EXTRA HELP

Tutorial   Tape 14   SSM, Sec. 7.1

1. Find the greatest common factor for each group of numbers.

(a) 30, 20, 15

(b) 42, 28, 35

(c) 12, 18, 26, 32

(d) 10, 15, 21

(c) 10, 11, 14

Write the prime factored form of each number.

$$10 = 2 \cdot 5 \quad 11 = 11 \quad 14 = 2 \cdot 7$$

There are no primes common to all three numbers, so the greatest common factor is 1.

◀◀ **WORK PROBLEM 1 AT THE SIDE.**

OBJECTIVE 2 The greatest common factor can also be found for a list of variable terms. For example, the terms $x^4, x^5, x^6,$ and x^7 have x^4 as the greatest common factor because each of these terms can be written with x^4 as a factor.

$$x^4 = 1 \cdot x^4, \quad x^5 = x \cdot x^4, \quad x^6 = x^2 \cdot x^4, \quad x^7 = x^3 \cdot x^4$$

Note
The exponent on a variable in the greatest common factor is the *smallest* exponent that appears on that variable in all the terms.

E X A M P L E   2   Finding Greatest Common Factors for Variable Terms

Find the greatest common factor for each list of terms.

(a) $21m^7, -18m^6, 45m^8, -24m^5$

$$21m^7 = 3 \cdot 7 \cdot m^7$$
$$-18m^6 = -1 \cdot 2 \cdot 3^2 \cdot m^6$$
$$45m^8 = 3^2 \cdot 5 \cdot m^8$$
$$-24m^5 = -1 \cdot 2^3 \cdot 3 \cdot m^5$$

First, 3 is the greatest common factor of the coefficients 21, -18, 45, and -24. The smallest exponent on m is 5, so the greatest common factor of the terms is $3m^5$.

(b) $x^4y^2, x^7y^5, x^3y^7, y^{15}$

$$x^4y^2 = x^4 \cdot y^2$$
$$x^7y^5 = x^7 \cdot y^5$$
$$x^3y^7 = x^3 \cdot y^7$$
$$y^{15} = y^{15}$$

There is no x in the last term, y^{15}, so x will not appear in the greatest common factor. There is a y in each term, however, and 2 is the smallest exponent on y. The greatest common factor is y^2.

(c) $-a^2b, -ab^2$

Write $-a^2b$ as $-1a^2b$ and $-ab^2$ as $-1ab^2$. The factors of -1 are -1 and 1. Since $1 > -1$, the greatest common factor is $1ab$ or ab.

Note
In a group of terms with negatives, many times a negative common factor is preferable (even though it is not the greatest common factor). In Example 2(c), for instance, we might prefer $-ab$ as the common factor. In factoring exercises, either answer will be acceptable.

ANSWERS

1. (a) 5 **(b)** 7 **(c)** 2 **(d)** 1

In summary, we find the greatest common factor of a list of terms as follows.

> **Finding The Greatest Common Factor**
>
> *Step 1* Write each number in prime factored form.
>
> *Step 2* List each prime number that is a factor of every number in the list.
>
> *Step 3* Use as exponents on the prime factors the *smallest* exponent from the prime factored forms. (If a prime does not appear in one of the prime factored forms, it cannot appear in the greatest common factor.)
>
> *Step 4* Multiply together the primes from Step 3. If there are no primes left after Step 3, the greatest common factor is 1.

WORK PROBLEM 2 AT THE SIDE. ▶▶

OBJECTIVE 3 The idea of a greatest common factor can be used to write a polynomial in factored form. For example, the polynomial

$$3m + 12$$

consists of the two terms $3m$ and 12. The greatest common factor for these two terms is 3. Write $3m + 12$ so that each term is a product with 3 as one factor.

$$3m + 12 = 3 \cdot m + 3 \cdot 4$$

Now use the distributive property.

$$3m + 12 = 3 \cdot m + 3 \cdot 4 = 3(m + 4)$$

The factored form of $3m + 12$ is $3(m + 4)$. This process is called **factoring out the greatest common factor.**

> **Caution**
> Notice that the polynomial $3m + 12$ is *not* in factored form when written as
>
> $$3 \cdot m + 3 \cdot 4.$$
>
> The *terms* are factored, but the polynomial is not. The factored form of $3m + 12$ is the *product*
>
> $$3(m + 4).$$

E X A M P L E   3   Factoring Out the Greatest Common Factor

Factor out the greatest common factor.

(a) $20m^5 + 10m^4 + 15m^3$

   The greatest common factor for the terms of this polynomial is $5m^3$.

$$20m^5 + 10m^4 + 15m^3$$
$$= (5m^3)(4m^2) + (5m^3)(2m) + (5m^3)3 \qquad \text{Factor each term.}$$
$$= 5m^3(4m^2 + 2m + 3) \qquad \text{Factor out } 5m^3.$$

Check this work by multiplying $5m^3$ and $4m^2 + 2m + 3$. You should get the original polynomial as your answer.

CONTINUED ON NEXT PAGE

2. Find the greatest common factor for each list of terms.

(a) $6m^4, 9m^2, 12m^5$

(b) $-12p^5, -18q^4$

(c) y^4z^2, y^6z^8, z^9

(d) $12p^{11}, 17q^5$

ANSWERS

2. (a) $3m^2$ **(b)** 6 **(c)** z^2 **(d)** 1

3. Factor out the greatest common factor.

(a) $10y^5 - 8y^4 + 6y^2$

(b) $m^7 + m^9$

(c) $8p^5q^2 + 16p^6q^3 - 12p^4q^7$

(d) $13x^2 - 27$

(e) $r(t - 4) + 5(t - 4)$

4. Factor by grouping.

(a) $pq + 5q + 2p + 10$

(b) $x^2 + 7x + 2x + 14$

(b) $x^5 + x^3 = (x^3)x^2 + (x^3)1 = x^3(x^2 + 1)$

(c) $20m^7p^2 - 36m^3p^4 = 4m^3p^2(5m^4 - 9p^2)$

(d) $a(a + 3) + 4(a + 3)$

The binomial $a + 3$ is the greatest common factor here.

$$a(a + 3) + 4(a + 3) = (a + 3)(a + 4)$$

> **Caution**
> Be careful to avoid the common error of leaving out the 1 in a problem like Example 3(b). Always be sure that the factored form can be multiplied out to give the original polynomial.

◀◀ **WORK PROBLEM 3 AT THE SIDE.**

OBJECTIVE 4 Common factors are used in **factoring by grouping,** explained in the next example.

EXAMPLE 4 Factoring by Grouping

Factor by grouping.

(a) $2x + 6 + ax + 3a$

The first two terms have a common factor of 2, and the last two terms have a common factor of a.

$$2x + 6 + ax + 3a = 2(x + 3) + a(x + 3)$$

The expression is still not in factored form because it is the *sum* of two terms. Now, however, $x + 3$ is a common factor and can be factored out.

$$2x + 6 + ax + 3a = 2(x + 3) + a(x + 3) = (x + 3)(2 + a)$$

The final result is in factored form because it is a *product*. Note that the goal in factoring by grouping is to get a common factor, $x + 3$ here, so that the last step is possible.

(b) $m^2 + 6m + 2m + 12 = m(m + 6) + 2(m + 6)$ Factor each group.
$$= (m + 6)(m + 2)$$ Factor out the common binomial factor.

◀◀ **WORK PROBLEM 4 AT THE SIDE.**

Use these steps when factoring by grouping.

> **Factoring by Grouping**
>
> *Step 1* Group the terms so that each group has a common factor.
>
> *Step 2* Use the distributive property to factor each group of terms.
>
> *Step 3* If possible, factor a common binomial factor from the results of Step 2.
>
> *Step 4* If Step 2 does not result in a common binomial factor, try grouping the terms of the original polynomial in a different way.

ANSWERS

3. **(a)** $2y^2(5y^3 - 4y^2 + 3)$
   (b) $m^7(1 + m^2)$
   (c) $4p^4q^2(2p + 4p^2q - 3q^5)$
   (d) no common factor (except 1)
   (e) $(t - 4)(r + 5)$
4. **(a)** $(p + 5)(q + 2)$
   (b) $(x + 7)(x + 2)$

E X A M P L E 5   **Rearranging Terms Before Factoring by Grouping**

Factor by grouping.

(a) $10x^2 - 12y^2 + 15xy - 8xy$

Factoring out the common factor of 2 from the first two terms and the common factor of xy terms from the last two terms gives

$$10x^2 - 12y^2 + 15xy - 8xy = 2(5x^2 - 6y^2) + xy(15 - 8).$$

This did not lead to a common factor, so we try rearranging the terms. There is usually more than one way to do this. Let's try

$$10x^2 - 8xy - 12y^2 + 15xy,$$

grouping the first two terms and the last two terms as follows.

$$
\begin{aligned}
10x^2 - 8xy - 12y^2 + 15xy &= 2x(5x - 4y) + 3y(-4y + 5x) \\
&= 2x(5x - 4y) + 3y(5x - 4y) \\
&= (5x - 4y)(2x + 3y)
\end{aligned}
$$

(b) $2xy + 12 - 3y - 8x$

We need to rearrange these terms to get two groups that each have a common factor. Trial and error suggests the following grouping.

$$
\begin{aligned}
2xy + 12 - 3y - 8x &= (2xy - 3y) + (-8x + 12) \\
&= y(\mathbf{2x - 3}) - 4(\mathbf{2x - 3}) \qquad \text{Factor each group.} \\
&= (2x - 3)(y - 4) \qquad \text{Factor out the common binomial factor.}
\end{aligned}
$$

Since the quantities in parentheses in the second step must be the same, we factored out -4 rather than 4.

■

WORK PROBLEM 5 AT THE SIDE. ▶▶

Caution
Note the careful use of signs in Example 5(b). Sign errors often occur when grouping with negative signs.

---

5. Factor by grouping.

(a)
$$6y^2 - 20w^2 + 15yw - 8yw$$

(b) $9mn - 4 + 12m - 3n$

Number Patterns in Our World

In the 1997 film *Contact*, Jodie Foster plays a scientist who makes contact with extraterrestrial life. The initial signal received is a pattern of prime numbers: 2, 3, 5, 7, 11, ... Greek mathematicians of 2500 years ago studied geometric patterns in numbers. The beauty of number patterns will no doubt be studied by humans as long as they inhabit the earth, and it is a good bet that if intelligent life exists elsewhere in the universe, it too knows of the universality of mathematical number patterns.

Consider the following geometric arrangements.

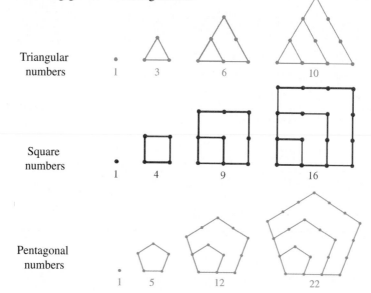

Triangular numbers   1   3   6   10

Square numbers   1   4   9   16

Pentagonal numbers   1   5   12   22

Collectively, these numbers are called **figurate numbers.** The figurate numbers possess numerous interesting patterns. Every square number greater than 1 is the sum of two consecutive triangular numbers. (For example, 9 = 3 + 6 and 25 = 10 + 15.) Every pentagonal number can be represented as the sum of a square number and a triangular number. (For example, 5 = 4 + 1 and 12 = 9 + 3.)

FOR GROUP DISCUSSION

Figurate number	1st	2nd	3rd	4th	5th	6th	7th	8th
Triangular	1	3	6	10	15	21		
Square	1	4	9	16	25			
Pentagonal	1	5	12	22				
Hexagonal	1	6	15					
Heptagonal	1	7						
Octagonal	1							

By increasing the number of sides of the arrangements depicting the figurate numbers, we obtain the *hexagonal, heptagonal, octagonal* numbers, and so on. With the accompanying table reproduced on the chalkboard, discuss the many patterns that can be found, and complete the table through the eighth column.

418

7.1 Exercises

1. Is 3 the greatest common factor of 18, 24, and 42? If not, what is?

2. Is pq the greatest common factor of pq^2, p^2q, and p^2q^2? If not, what is?

3. Give an example of three numbers whose greatest common factor is 7.

4. How would you respond to the statement "25 and 36 have no common factor"?

Find the greatest common factor for each group of numbers. See Example 1.

5. 12, 16

6. 18, 24

7. 40, 20, 4

8. 50, 30, 5

9. 18, 24, 36, 48

10. 15, 30, 45, 75

11. 4, 9, 12

12. 9, 16, 24

Find the greatest common factor for each list of terms. See Example 2.

13. $16y$, 24

14. $18w$, 27

15. $30x^3$, $40x^6$, $50x^7$

16. $60z^4$, $70z^8$, $90z^9$

17. $-x^4y^3$, $-xy^2$

18. $-a^4b^5$, $-a^3b$

19. $42ab^3$, $-36a$, $90b$, $-48ab$

20. $45c^3d$, $75c$, $90d$, $-105cd$

21. Is $-xy$ a common factor of $-x^4y^3$ and $-xy^2$? If so, what is the other factor that when multiplied by $-xy$ gives $-x^4y^3$?

22. Is $-a^5b^2$ a common factor of $-a^4b^5$ and $-a^3b$? Explain why or why not.

Complete the factoring.

23. $9m^4 = 3m^2(\quad)$

24. $12p^5 = 6p^3(\quad)$

25. $-8z^9 = -4z^5(\quad)$

26. $-15k^{11} = -5k^8(\quad)$

27. $6m^4n^5 = 3m^3n(\quad)$

28. $27a^3b^2 = 9a^2b(\quad)$

29. $-14x^4y^3 = 2xy(\quad)$

30. $-16m^3n^3 = 4mn^2(\quad)$

Factor out the greatest common factor for each expression. See Example 3.

31. $65y^{10} + 35y^6$

32. $100a^5 + 16a^3$

33. $11w^3 - 100$

34. $13z^5 - 80$

35. $8m^2n^3 + 24m^2n^2$

36. $19p^2y - 38p^2y^3$

37. $13y^8 + 26y^4 - 39y^2$

38. $5x^5 + 25x^4 - 20x^3$

39. $45q^4p^5 + 36qp^6 + 81q^2p^3$

40. $125a^3z^5 + 60a^4z^4 - 85a^5z^2$

41. $a^2(2a + b) + b^2(2a + b)$

42. $3x(x^2 + 5) - y(x^2 + 5)$

Factor by grouping. See Example 4.

43. $p^2 + 4p + 3p + 12$

44. $m^2 + 2m + 5m + 10$

45. $a^2 - 2a + 5a - 10$

46. $y^2 - 6y + 4y - 24$

47. $7z^2 + 14z - az - 2a$

48. $5m^2 + 15mp - 2mp - 6p^2$

49. $18r^2 + 12ry - 3xr - 2xy$

50. $8s^2 - 4st + 6sy - 3yt$

51. $3a^3 + 3ab^2 + 2a^2b + 2b^3$

52. $4x^3 + 3x^2y + 4xy^2 + 3y^3$

53. $1 - a + ab - b$

54. $6 - 3x - 2y + xy$

55. $16m^3 - 4m^2p^2 - 4mp + p^3$

56. $10t^3 - 2t^2s^2 - 5ts + s^3$

MATHEMATICAL CONNECTIONS (EXERCISES 57–60)

In many cases, the choice of which pairs of terms to group when factoring by grouping can be done in different ways to get the correct answer. To see this for Example 5(b), work Exercises 57–60 in order.

57. Factor the polynomial from Example 5(b), $2xy + 12 - 3y - 8x$, by rearranging the terms as follows: $2xy - 8x - 3y + 12$. What properties from Section 1.8 allow this?

58. To prepare to factor the rearranged polynomial by grouping, group the first two terms and the last two terms. Then factor each group.

59. Is your result from Exercise 58 in factored form? If not, why?

60. If your answer to Exercise 59 is *no*, factor the polynomial. Is the result the same as the one shown for Example 5(b)?

---

61. Refer to Exercise 53. The answer given in the back of the book is $(1 - a)(1 - b)$. A student factored this same polynomial and got the result $(a - 1)(b - 1)$.
   (a) Is this student's answer correct?
   (b) If your answer to part (a) is *yes*, explain why these two seemingly different answers are both acceptable.

62. Explain how you, while tutoring someone in introductory algebra, could make up a polynomial like those in Exercises 43–56 that could be factored by grouping.

7.2 Factoring Trinomials

Using FOIL, we find the product of the binomials $k - 3$ and $k + 1$ is

$$(k - 3)(k + 1) = k^2 - 2k - 3.$$

In this section, we show how to *factor* the polynomial $k^2 - 2k - 3$, that is, write it as the product $(k - 3)(k + 1)$. This product is called the *factored form* of $k^2 - 2k - 3$. The discussion of factoring in this section is limited to trinomials where the coefficient of the squared term is 1.

OBJECTIVE 1 ▶ When factoring polynomials with only integer coefficients, we use only integers for the numerical coefficients. For example, we can factor $x^2 + 5x + 6$ by finding integers m and n such that

$$x^2 + 5x + 6 = (x + m)(x + n).$$

To find these integers m and n, we first use FOIL to multiply the two binomials:

$$(x + m)(x + n) = x^2 + nx + mx + mn.$$

By the distributive property,

$$x^2 + nx + mx + mn = x^2 + (n + m)x + mn.$$

Comparing this result with $x^2 + 5x + 6$ shows that we must find integers m and n having a sum of 5 and a product of 6.

Product of m and n is 6.

$$x^2 + 5x + 6 = x^2 + (n + m)x + mn$$

Sum of m and n is 5.

Because many pairs of integers have a sum of 5, it is best to begin by listing those pairs of integers whose product is 6. Both 5 and 6 are positive, so we need to consider only pairs in which both integers are positive.

WORK PROBLEM 1 AT THE SIDE. ▶▶

From Problem 1 at the side, we see that the numbers 1 and 6 and the numbers 2 and 3 both have a product of 6, but only the pair 2 and 3 has a sum of 5. So 2 and 3 are the required integers, and

$$x^2 + 5x + 6 = (x + 2)(x + 3).$$

We can check by multiplying the binomials, using FOIL. Make sure that the sum of the outer and inner products produces the correct middle term.

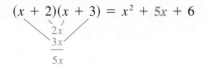

$$(x + 2)(x + 3) = x^2 + 5x + 6$$

This method of factoring can be used only for trinomials having the coefficient of the squared term equal to 1. Methods for factoring other trinomials will be given in the next section.

EXAMPLE 1 Factoring a Trinomial with Only Positive Terms

Factor $m^2 + 9m + 14$.

Look for two integers whose product is 14 and whose sum is 9. List the pairs of integers whose products are 14. Then examine the sums.

CONTINUED ON NEXT PAGE

OBJECTIVES

1 ▶ Factor trinomials with a coefficient of 1 for the squared term.

2 ▶ Factor such polynomials after factoring out the greatest common factor.

FOR EXTRA HELP

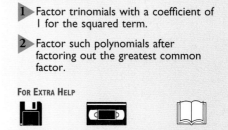

Tutorial   Tape 14   SSM, Sec. 7.2

1. (a) List all pairs of positive integers whose product is 6.

(b) Find the pair from part (a) whose sum is 5.

2. Complete the given lists of numbers; then factor
$y^2 + 12y + 20$.

Factors of 20	Sum of Factors
20, 1	20 + 1 = 21
10, __	10 + __ = __
5, __	5 + __ = __

3. Factor each trinomial.

(a) $a^2 - 9a - 22$

(b) $r^2 - 6r - 16$

ANSWERS
2. 2; 2; 12; 4; 4; 9; $(y + 10)(y + 2)$
3. (a) $(a - 11)(a + 2)$
   (b) $(r - 8)(r + 2)$

Again, only positive integers are needed because all signs in $m^2 + 9m + 14$ are positive.

Factors of 14	Sum of Factors
14, 1	14 + 1 = 15
7, 2	7 + 2 = **9** Sum is 9.

From the list, 7 and 2 are the required integers, since $7 \cdot 2 = 14$ and $7 + 2 = 9$. Thus

$$m^2 + 9m + 14 = (m + 2)(m + 7).$$

Note
In Example 1, the answer also could have been written $(m + 7)(m + 2)$. Because of the commutative property of multiplication, the order of the factors does not matter.

◀◀ **WORK PROBLEM 2 AT THE SIDE.**

E X A M P L E   2   Factoring a Trinomial with Two Negative Terms

Factor $p^2 - 2p - 15$.

Find two integers whose product is -15 and whose sum is -2. If these numbers do not come to mind right away, find them (if they exist) by listing all the pairs of integers whose product is -15. Because the last term, -15, is negative, we need pairs of integers with different signs.

Factors of -15	Sum of Factors
15, -1	15 + (-1) = 14
5, -3	5 + (-3) = 2
-15, 1	-15 + 1 = -14
-5, 3	-5 + 3 = -2 Sum is -2.

The required integers are -5 and 3, and

$$p^2 - 2p - 15 = (p - 5)(p + 3).$$

◀◀ **WORK PROBLEM 3 AT THE SIDE.**

As shown in the next example, some trinomials cannot be factored using only integer coefficients. We call such trinomials **prime polynomials.**

E X A M P L E   3   Factoring a Trinomial with One Negative Term

Factor each trinomial.

(a) $x^2 - 5x + 12$

Since the middle term is negative, the sum must be negative. The last term is positive, so both factors must be negative to give a positive product and a negative sum. First, list all pairs of negative integers whose product is 12. Then examine the sums.

Factors of 12	Sum of Factors
$-12, -1$	$-12 + (-1) = -13$
$-6, -2$	$-6 + (-2) = -8$
$-3, -4$	$-3 + (-4) = -7$

CONTINUED ON NEXT PAGE

None of the pairs of integers has a sum of -5. Because of this, the trinomial $x^2 - 5x + 12$ *cannot be factored using only integer coefficients,* showing that it is a *prime polynomial.*

(b) $k^2 - 8k + 11$

There is no pair of integers whose product is 11 and whose sum is -8, so $k^2 - 8k + 11$ is a prime polynomial.

> WORK PROBLEM 4 AT THE SIDE. ▶▶

We can now summarize the procedure for factoring a trinomial of the form $x^2 + bx + c$.

Factoring $x^2 + bx + c$

Find two integers whose product is c and whose sum is b.

1. Both integers must be positive if b and c are positive.
2. Both integers must be negative if c is positive and b is negative.
3. One integer must be positive and one must be negative if c is negative.

EXAMPLE 4 Factoring a Trinomial with Two Variables

Factor $z^2 - 2bz - 3b^2$.

Look for two expressions whose product is $-3b^2$ and whose sum is $-2b$. The expressions are $-3b$ and b, so

$$z^2 - 2bz - 3b^2 = (z - 3b)(z + b).$$

> WORK PROBLEM 5 AT THE SIDE. ▶▶

OBJECTIVE 2 The trinomial in the next example does not have a coefficient of 1 for the squared term. (In fact, there is no squared term.) A preliminary step must be taken before using the steps discussed above.

EXAMPLE 5 Factoring a Trinomial with a Common Factor

Factor $4x^5 - 28x^4 + 40x^3$.

First, factor out the greatest common factor, $4x^3$.

$$4x^5 - 28x^4 + 40x^3 = 4x^3(x^2 - 7x + 10)$$

Now factor $x^2 - 7x + 10$. The integers -5 and -2 have a product of 10 and a sum of -7. The complete factored form is

$$4x^5 - 28x^4 + 40x^3 = 4x^3(x - 5)(x - 2).$$

Caution
When factoring, always remember to look for a common factor first. Do not forget to include the common factor as part of the answer. Multiplying out the factored form should always give the original polynomial. (This is a good way to check your answer.)

> WORK PROBLEM 6 AT THE SIDE. ▶▶

---

4. Factor each trinomial, where possible.

(a) $r^2 - 3r - 4$

(b) $m^2 - 2m + 5$

5. Factor each trinomial.

(a) $b^2 - 3ab - 4a^2$

(b) $r^2 - 6rs + 8s^2$

6. Factor each trinomial as completely as possible.

(a) $2p^3 + 6p^2 - 8p$

(b) $3x^4 - 15x^3 + 18x^2$

ANSWERS

4. (a) $(r - 4)(r + 1)$
    (b) prime
5. (a) $(b - 4a)(b + a)$
    (b) $(r - 4s)(r - 2s)$
6. (a) $2p(p + 4)(p - 1)$
    (b) $3x^2(x - 3)(x - 2)$

NUMBERS IN THE
Real World *collaborative investigations*

Using Number Properties in Identification

In order to have a consistent numbering system for published books, the International Standard Book Number (ISBN) was established. A book can be identified completely by its ISBN. It is usually composed of 10 digits with hyphens appearing at various positions. For example, the ISBN of *Mathematical Ideas*, 8th edition, by Heeren, Hornsby, and Miller, is 0-673-99893-2. The 0 is for the United States and a few other English-speaking countries. The 673 is associated with the publisher, Addison-Wesley, while the next 5 digits are the book number assigned to the text. The final digit is a check digit between 0 and 10 with X representing 10. The reason for a check is to avoid errors such as transposing digits when copying an ISBN number. The check digit helps to determine if a number is a valid ISBN. To determine if it is valid one can complete the table.

0		
6	6	6
7	13	19
3	16	35
9	25	60
9	34	94
8	42	136
9	51	187
3	54	241
2	56	297

Bar-code scanners not only translate the bar code to a sequence of numbers, they are also programmed to use properties of our numeration system to check that the code is valid.

In the first column write down the 10-digit ISBN. In the second column calculate the running total of the numbers in the first column. Similarly, in the third column calculate the total of the numbers in the second column. If the final sum in the lower right corner of the table is evenly divisible by 11 (297 ÷ 11 = 27), it is considered a valid ISBN. Try changing one number in 0-673-99893-2 or transposing two adjacent digits and then test the validity of the ISBN.

FOR GROUP DISCUSSION

1. Apply the process described above to the ISBN number of this book.

2. Determine whether the following ISBN numbers are valid.
   (a) *The Beauty of Fractals*, by H. O. Peitgen and P. H. Richter, 3-540-15851-0
   (b) *Women in Science*, by Vivian Gornick, 0-671-41738-3
   (c) *Beyond Numeracy*, by John Allen Paulos, 0-394-58640-9
   (d) *Iron John*, by Robert Bly, 0-201-51720-X

Source: Hamming, R., *Coding and Information Theory*, Prentice-Hall, Englewood Cliffs, New Jersey, 1986.

7.2 Exercises

1. In factoring a trinomial in x as $(x + a)(x + b)$, what must be true of a and b, if the coefficient of the last term of the trinomial is negative?

2. In Exercise 1, what must be true of a and b if the coefficient of the last term is positive?

3. What is meant by a *prime polynomial?*

4. How can you check your work when factoring a trinomial? Does the check ensure that the trinomial is completely factored?

In Exercises 5–10, list all pairs of integers with the given product. Then find the pair whose sum is given. See the charts in Examples 1–3.

5. Product: 12   Sum: 7   **6.** Product: 18   Sum: 9

7. Product: -24 Sum: -5 **8.** Product: -36 Sum: -16

9. Product: 27   Sum: 28   **10.** Product: 32   Sum: 33

11. Which one of the following is the correct factored form of $x^2 - 12x + 32$?
     (a) $(x - 8)(x + 4)$ **(b)** $(x + 8)(x - 4)$ **(c)** $(x - 8)(x - 4)$ **(d)** $(x + 8)(x + 4)$

12. What would be the first step in factoring $2x^3 + 8x^2 - 10x$?

Complete the factoring.

13. $x^2 + 15x + 44 = (x + 4)(\quad)$ **14.** $r^2 + 15r + 56 = (r + 7)(\quad)$

15. $x^2 - 9x + 8 = (x - 1)(\quad)$ **16.** $t^2 - 14t + 24 = (t - 2)(\quad)$

17. $y^2 - 2y - 15 = (y + 3)(\quad)$ **18.** $t^2 - t - 42 = (t + 6)(\quad)$

19. $x^2 + 9x - 22 = (x - 2)(\quad)$ **20.** $x^2 + 6x - 27 = (x - 3)(\quad)$

21. $y^2 - 7y - 18 = (y + 2)(\quad)$ **22.** $y^2 - 2y - 24 = (y + 4)(\quad)$

Factor completely. If a polynomial cannot be factored, write prime. *See Examples 1–3.*

23. $y^2 + 9y + 8$ **24.** $a^2 + 9a + 20$ **25.** $b^2 + 8b + 15$

26. $x^2 + 6x + 8$ **27.** $m^2 + m - 20$ **28.** $p^2 + 4p - 5$

29. $y^2 - 8y + 15$ **30.** $y^2 - 6y + 8$ **31.** $r^2 - r - 30$

32. $q^2 - q - 42$ **33.** $t^2 - 8t + 16$ **34.** $s^2 - 10s + 25$

Factor completely. See Examples 4 and 5.

35. $r^2 + 3ra + 2a^2$ **36.** $x^2 + 5xa + 4a^2$ **37.** $t^2 - tz - 6z^2$

38. $a^2 - ab - 12b^2$ **39.** $x^2 + 4xy + 3y^2$ **40.** $p^2 + 9pq + 8q^2$

41. $v^2 - 11vw + 30w^2$ **42.** $v^2 - 11vx + 24x^2$ **43.** $4x^2 + 12x - 40$

44. $5y^2 - 5y - 30$ **45.** $2t^3 + 8t^2 + 6t$ **46.** $3t^3 + 27t^2 + 24t$

47. $2x^6 + 8x^5 - 42x^4$ **48.** $4y^5 + 12y^4 - 40y^3$ **49.** $m^3n - 10m^2n^2 + 24mn^3$

50. $y^3z + 3y^2z^2 - 54yz^3$

51. Use the FOIL method from Section 6.6 to show that $(2x + 4)(x - 3) = 2x^2 - 2x - 12$. Why, then, is it incorrect to completely factor $2x^2 - 2x - 12$ as $(2x + 4)(x - 3)$?

52. Why is it incorrect to completely factor $3x^2 + 9x - 12$ as the product $(x - 1)(3x + 12)$?

53. What polynomial can be factored to give $(a + 9)(a + 4)$?

54. What polynomial can be factored to give $(y - 7)(y + 3)$?

7.3 *More on Factoring Trinomials*

OBJECTIVES

1. ► Factor trinomials by grouping when the coefficient of the squared term is not 1.

2. ► Factor trinomials using FOIL.

FOR EXTRA HELP

Tutorial   Tape 14   SSM, Sec. 7.3

Trinomials such as $2x^2 + 7x + 6$, in which the coefficient of the squared term is *not* 1, are factored with extensions of methods we used in the previous sections.

OBJECTIVE 1 ► Recall that a trinomial such as $m^2 + 3m + 2$ is factored by finding two numbers whose product is 2 and whose sum is 3. To factor $2x^2 + 7x + 6$, we look for two integers whose product is $2 \cdot 6 = 12$ and whose sum is 7.

$$
\begin{array}{c}
\text{Sum is 7.} \\
2x^2 + 7x + 6 \\
\text{Product } 2 \cdot 6 = 12.
\end{array}
$$

By considering the pairs of positive integers whose product is 12, the necessary integers are found to be 3 and 4. We use these integers to write the middle term, $7x$, as $7x = 3x + 4x$. With this, the trinomial $2x^2 + 7x + 6$ becomes

$$2x^2 + 7x + 6 = 2x^2 + \underbrace{3x + 4x}_{7x = 3x + 4x} + 6.$$

$$= x(2x + 3) + 2(2x + 3) \qquad \text{Factor each group.}$$

Must be same

$$2x^2 + 7x + 6 = (2x + 3)(x + 2) \qquad \text{Factor out } 2x + 3.$$

In the example above, we could have written $7x$ as $4x + 3x$. Check that the resulting factoring by grouping would give the same answer. Check the factorization by finding the product of $2x + 3$ and $x + 2$.

E X A M P L E 1   Factoring Trinomials by Grouping

Factor the trinomial.

(a) $6r^2 + r - 1$

We must find two integers with a product of $6(-1) = -6$ and a sum of 1.

$$
\begin{array}{c}
\text{Sum is 1.} \\
6r^2 + r - 1 = 6r^2 + 1r - 1 \\
\text{Product is } 6(-1) = -6.
\end{array}
$$

The integers are -2 and 3. We write the middle term, $+r$, as $-2r + 3r$, so that

$$6r^2 + r - 1 = 6r^2 - 2r + 3r - 1.$$

CONTINUED ON NEXT PAGE

1. Factor each trinomial by grouping.

(a) $2m^2 + 7m + 3$

(b) $5p^2 - 2p - 3$

(c) $15k^2 - k - 2$

Factor by grouping on the right-hand side.

$6r^2 + r - 1 = 6r^2 - 2r + 3r - 1$
$\qquad\qquad = 2r(3r - 1) + 1(3r - 1)$ The binomials must be the same.
$\qquad\qquad = (3r - 1)(2r + 1)$

(b) $12z^2 - 5z - 2$

Look for two integers whose product is $12(-2) = -24$ and whose sum is -5. The required integers are 3 and -8, and

$12z^2 - 5z - 2 = 12z^2 + 3z - 8z - 2$ $\qquad -5z = 3z - 8z$
$\qquad\qquad = 3z(4z + 1) - 2(4z + 1)$ Group terms and factor each group.
$\qquad\qquad = (4z + 1)(3z - 2).$ Factor out $4z + 1$.

(c) $10m^2 + mn - 3n^2$

Two integers whose product is $10(-3) = -30$ and whose sum is 1 are -5 and 6. Rewrite the trinomial with four terms.

$10m^2 + mn - 3n^2 = 10m^2 - 5mn + 6mn - 3n^2$ $\quad mn = -5mn + 6mn$
$\qquad\qquad = 5m(2m - n) + 3n(2m - n)$ Group terms and factor each group.
$\qquad\qquad = (2m - n)(5m + 3n)$ Factor out the common factor.

◀◀ **WORK PROBLEM 1 AT THE SIDE.**

2. Factor each polynomial as completely as possible.

(a) $4x^2 - 2x - 30$

(b) $18p^4 + 63p^3 + 27p^2$

(c) $6a^2 + 3ab - 18b^2$

E X A M P L E   2   Factoring a Trinomial with a Common Factor

Factor $28x^5 - 58x^4 - 30x^3$.

First factor out the greatest common factor, $2x^3$.

$$28x^5 - 58x^4 - 30x^3 = 2x^3(14x^2 - 29x - 15)$$

Now, to factor the trinomial $14x^2 - 29x - 15$, we must find two integers whose product is $(14)(-15) = -210$ and whose sum is -29. Factoring 210 into its prime factors gives

$$210 = 2 \cdot 3 \cdot 5 \cdot 7.$$

By combining these prime factors in pairs in different ways and using one positive and one negative (to get -210) we find the factors 6 and -35 have the correct sum. Now we can rewrite the given trinomial and factor it.

$$28x^5 - 58x^4 - 30x^3 = 2x^3(14x^2 + 6x - 35x - 15)$$
$$= 2x^3[(14x^2 + 6x) - (35x + 15)]$$
$$= 2x^3[2x(7x + 3) - 5(7x + 3)]$$
$$= 2x^3[(7x + 3)(2x - 5)]$$
$$= 2x^3(7x + 3)(2x - 5)$$

Caution
Do not forget to include the common factor in the final result.

◀◀ **WORK PROBLEM 2 AT THE SIDE.**

ANSWERS

1. (a) $(2m + 1)(m + 3)$
   (b) $(5p + 3)(p - 1)$
   (c) $(5k - 2)(3k + 1)$
2. (a) $2(2x + 5)(x - 3)$
   (b) $9p^2(2p + 1)(p + 3)$
   (c) $3(2a - 3b)(a + 2b)$

OBJECTIVE **2** ▶ The rest of this section shows an alternative method of factoring trinomials in which the coefficient of the squared term is not 1. This method uses trial and error. In the next example, the alternative method is used to factor $2x^2 + 7x + 6$, the same trinomial factored at the beginning of this section.

To factor $2x^2 + 7x + 6$ by trial and error, we must use FOIL backwards. We want to write $2x^2 + 7x + 6$ as the product of two binomials.

$$2x^2 + 7x + 6 = (\quad)(\quad)$$

The product of the two first terms of the binomials is $2x^2$. The possible factors of $2x^2$ are $2x$ and x or $-2x$ and $-x$. Since all terms of the trinomial are positive, only positive factors should be considered. Thus, we have

$$2x^2 + 7x + 6 = (2x\quad)(x\quad).$$

The product of the two last terms, 6, can be factored as $6 \cdot 1$, $1 \cdot 6$, $2 \cdot 3$, or $3 \cdot 2$. Try each pair to find the pair that gives the correct middle term.

WORK PROBLEM 3 AT THE SIDE. ▶▶

In part (b) at the side, since $2x + 6 = 2(x + 3)$, the binomial $2x + 6$ has a common factor of 2, while $2x^2 + 7x + 6$ has no common factor other than 1. The product $(2x + 6)(x + 1)$ cannot be correct. (Part (c) also has one factor with a common factor.)

> **Note**
> If the original polynomial has no common factor, then none of its binomial factors will either.

Now try the numbers 2 and 3 as factors of 6. Because of the common factor of 2 in $2x + 2$, $(2x + 2)(x + 3)$ will not work. Try $(2x + 3)(x + 2)$.

$$\underset{\underset{\underset{7x}{4x}}{3x}}{(2x + 3)(x + 2)} = 2x^2 + 7x + 6 \qquad \text{Correct}$$

Add.

Finally, we see that $2x^2 + 7x + 6$ factors as

$$2x^2 + 7x + 6 = (2x + 3)(x + 2).$$

Check by multiplying $2x + 3$ and $x + 2$.

┌─ **E X A M P L E   3   Factoring a Trinomial with All Terms Positive Using FOIL**

Factor $8p^2 + 14p + 5$.

The number 8 has several possible pairs of factors, but 5 has only 1 and 5 or -1 and -5. For this reason, it is easier to begin by considering the factors of 5. Ignore the negative factors since all coefficients in the trinomial are positive. If $8p^2 + 14p + 5$ can be factored, the factors will have the form

$$(\quad + 5)(\quad + 1).$$

└─ **CONTINUED ON NEXT PAGE**

---

3. Decide whether each factored form is correct or incorrect for $2x^2 + 7x + 6$.

(a) $(2x + 1)(x + 6)$

(b) $(2x + 6)(x + 1)$

(c) $(2x + 2)(x + 3)$

4. Factor each trinomial.

(a) $2p^2 + 9p + 9$

(b) $6p^2 + 19p + 10$

(c) $8x^2 + 14x + 3$

The possible pairs of factors of $8p^2$ are $8p$ and p, or $4p$ and $2p$. Try various combinations, checking the middle term in each case.

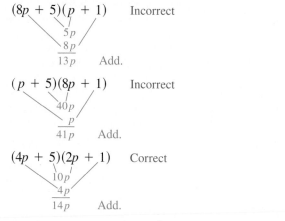

$(8p + 5)(p + 1)$ Incorrect
$5p$
$8p$
$13p$ Add.

$(p + 5)(8p + 1)$ Incorrect
$40p$
p
$41p$ Add.

$(4p + 5)(2p + 1)$ Correct
$10p$
$4p$
$14p$ Add.

Since $14p$ is the correct middle term, $8p^2 + 14p + 5$ factors as $(4p + 5)(2p + 1)$.

◀◀ **WORK PROBLEM 4 AT THE SIDE.**

E X A M P L E   4   Factoring a Trinomial with a Negative Middle Term Using FOIL

Factor $6x^2 - 11x + 3$.

Since 3 has only 1 and 3 or -1 and -3 as factors, it is better here to begin by factoring 3. The last term of the trinomial $6x^2 - 11x + 3$ is positive and the middle term has a negative coefficient, so only negative factors should be considered. Try -3 and -1 as factors of 3:

$$(\quad - 3)(\quad - 1).$$

The factors of $6x^2$ may be either $6x$ and x, or $2x$ and $3x$. Try $2x$ and $3x$.

$(2x - 3)(3x - 1)$ Correct
$-9x$
$-2x$
$-11x$ Add.

These factors give the correct middle term, so

$$6x^2 - 11x + 3 = (2x - 3)(3x - 1).$$

E X A M P L E   5   Factoring a Trinomial with a Negative Last Term Using FOIL

Factor $8x^2 + 6x - 9$.

The integer 8 has several possible pairs of factors, as does -9. Since the last term is negative, one positive factor and one negative factor of -9 are needed. Since the coefficient of the middle term is small, it is wise to avoid large factors such as 8 or 9. Let us try 4 and 2 as factors of 8, and 3 and -3 as factors of -9, and check the middle term.

$(4x + 3)(2x - 3)$ Incorrect
$6x$
$-12x$
$-6x$ Add.

CONTINUED ON NEXT PAGE

Let us try exchanging 3 and -3.

$$(4x - 3)(2x + 3) \qquad \text{Correct}$$

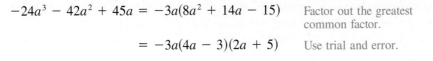

$$\begin{array}{r} -6x \\ 12x \\ \hline 6x \quad \text{Add.} \end{array}$$

This time we got the correct middle term, so

$$8x^2 + 6x - 9 = (4x - 3)(2x + 3).$$

WORK PROBLEM 5 AT THE SIDE. ▶▶

EXAMPLE 6 Factoring a Trinomial with Two Variables

Factor $12a^2 - ab - 20b^2$.

There are several pairs of factors of $12a^2$, including $12a$ and a, $6a$ and $2a$, and $3a$ and $4a$, just as there are many possible pairs of factors of $-20b^2$, including $-20b$ and b, $10b$ and $-2b$, $-10b$ and $2b$, $4b$ and $-5b$, and $-4b$ and $5b$. Once again, since the desired middle term is small, we should avoid the larger factors. Let us try as factors $6a$ and $2a$ and $4b$ and $-5b$.

$$(6a + 4b)(2a - 5b)$$

This cannot be correct, as mentioned before, since $6a + 4b$ has a common factor while the given trinomial has none. Let us try $3a$ and $4a$ with $4b$ and $-5b$.

$$(3a + 4b)(4a - 5b) = 12a^2 + ab - 20b^2 \qquad \text{Incorrect}$$

Here the middle term has the wrong sign, so we change the signs in the factors.

$$(3a - 4b)(4a + 5b) = 12a^2 - ab - 20b^2 \qquad \text{Correct}$$

EXAMPLE 7 Factoring a Trinomial with a Negative Common Factor

Factor $-24a^3 - 42a^2 + 45a$.

The common factor could be $3a$ or $-3a$. If we factor out $-3a$, the first term of the trinomial factor will be positive, which makes it easier to factor.

$$-24a^3 - 42a^2 + 45a = -3a(8a^2 + 14a - 15) \qquad \text{Factor out the greatest common factor.}$$

$$= -3a(4a - 3)(2a + 5) \qquad \text{Use trial and error.}$$

WORK PROBLEM 6 AT THE SIDE. ▶▶

5. Factor each trinomial.

(a) $6x^2 + 5x - 4$

(b) $6m^2 - 11m - 10$

(c) $4x^2 - 3x - 7$

6. Factor each trinomial.

(a) $2x^2 - 5xy - 3y^2$

(b) $8a^2 + 2ab - 3b^2$

(c) $-12x^3 + 16x^2y + 3xy^2$

ANSWERS

5. (a) $(3x + 4)(2x - 1)$
(b) $(2m - 5)(3m + 2)$
(c) $(4x - 7)(x + 1)$
6. (a) $(2x + y)(x - 3y)$
(b) $(4a + 3b)(2a - b)$
(c) $-x(6x + y)(2x - 3y)$

NUMBERS IN THE

Real World collaborative investigations

The Influence of Spanish Coinage on Stock Prices

Did you ever wonder why fractions used in stock-market reports have denominators that are powers of 2, such as $\frac{1}{2}$, $\frac{3}{4}$ and $\frac{5}{8}$? During the early years of the United States, prior to the minting of its own coinage, the Spanish eight-reales coin, also known as the Spanish milled dollar, circulated freely in the states. Its fractional parts, the four reales, two reales, and one real, were known as *pieces of eight*, and described as such in pirate and treasure lore. When the New York Stock Exchange was founded in 1792, it chose to use the Spanish milled dollar as its price basis, rather than the decimal base as proposed by Thomas Jefferson that same year.

In the September 1997 issue of *COINage*, Tom Delorey's article "The End of 'Pieces of Eight'," gives the following account:

> As the Spanish dollar and its fractions continued to be legal tender in America alongside the decimal coins until 1857, there was no urgency to change the system—and by the time the Spanish-American money was withdrawn in 1857, pricing stocks in eighths of a dollar—and no less—was a tradition carved in stone. Being somewhat a conservative organization, the NYSE saw no need to fix what was not broken.

At this writing, there is movement to begin reporting stock prices in decimal values. (Delorey gives an excellent account of how this will affect buying and selling.) While old habits die hard, it seems that even this one is finally coming to an end.

FOR GROUP DISCUSSION

1. Have you ever heard this old cheer? "Two bits, four bits, six bits, a dollar. All for the (home team), stand up and holler." The term **two bits** refers to 25 cents. Discuss how this cheer is based on the Spanish eight-reales coin.

2. Have you ever heard "If it ain't broke, don't fix it?" Discuss how this might apply to a real-life situation. How do you think the public will react to the decimalization of stock prices?

7.3 Exercises

In Exercises 1–4, give the first steps in factoring the trinomial $2x^2 + x - 21$ by grouping. Fill in the blanks.

1. Find the product of _____ and _____ .

2. Find factors of _____ that have a sum of _____ .

3. Write x as _____ + _____ .

4. Factor the polynomial _____ by grouping.

Decide which is the correct factored form of the given polynomial.

5. $2x^2 - x - 1$
   (a) $(2x - 1)(x + 1)$ (b) $(2x + 1)(x - 1)$

6. $3a^2 - 5a - 2$
   (a) $(3a + 1)(a - 2)$ (b) $(3a - 1)(a + 2)$

7. $4y^2 + 17y - 15$
   (a) $(y + 5)(4y - 3)$ (b) $(2y - 5)(2y + 3)$

8. $12c^2 - 7c - 12$
   (a) $(6c - 2)(2c + 6)$ (b) $(4c + 3)(3c - 4)$

9. $4k^2 + 13mk + 3m^2$
   (a) $(4k + m)(k + 3m)$ (b) $(4k + 3m)(k + m)$

10. $2x^2 + 11x + 12$
    (a) $(2x + 3)(x + 4)$ (b) $(2x + 4)(x + 3)$

Complete the factoring.

11. $6a^2 + 7ab - 20b^2 = (3a - 4b)(\qquad)$

12. $9m^2 - 3mn - 2n^2 = (3m + n)(\qquad)$

13. $2x^2 + 6x - 8 = 2(\qquad)$
    $= 2(\qquad)(\qquad)$

14. $3x^2 - 9x - 30 = 3(\qquad)$
    $= 3(\qquad)(\qquad)$

15. $4z^3 - 10z^2 - 6z = 2z(\qquad)$
    $= 2z(\qquad)(\qquad)$

16. $15r^3 - 39r^2 - 18r = 3r(\qquad)$
    $= 3r(\qquad)(\qquad)$

17. For the polynomial $12x^2 + 7x - 12$, 2 is not a common factor. Explain why the binomial $2x - 6$, then, cannot be a factor of the polynomial.

18. Explain how the signs of the last terms of the two binomial factors of a trinomial are determined.

Factor completely. Use either method described in this section. See Examples 1–7.

19. $2x^2 + 7x + 3$

20. $3y^2 + 13y + 4$

21. $3a^2 + 10a + 7$

22. $7r^2 + 8r + 1$

23. $4r^2 + r - 3$

24. $4r^2 + 3r - 10$

25. $15m^2 + m - 2$

26. $6x^2 + x - 1$

27. $8m^2 - 10m - 3$

28. $12s^2 + 11s - 5$

29. $20x^2 + 11x - 3$

30. $20x^2 - 28x - 3$

31. $21m^2 + 13m + 2$

32. $38x^2 + 23x + 2$

33. $20y^2 + 39y - 11$

34. $10x^2 + 11x - 6$

35. $6b^2 + 7b + 2$ **36.** $6w^2 + 19w + 10$ **37.** $24x^2 - 42x + 9$ **38.** $48b^2 - 74b - 10$

39. $40m^2q + mq - 6q$ **40.** $15a^2b + 22ab + 8b$ **41.** $2m^3 + 2m^2 - 40m$ **42.** $3x^3 + 12x^2 - 36x$

43. $15n^4 - 39n^3 + 18n^2$ **44.** $24a^4 + 10a^3 - 4a^2$ **45.** $18x^5 + 15x^4 - 75x^3$

46. $32z^5 - 20z^4 - 12z^3$ **47.** $15x^2y^2 - 7xy^2 - 4y^2$ **48.** $14a^2b^3 + 15ab^3 - 9b^3$

49. $12p^2 + 7pq - 12q^2$ **50.** $6m^2 - 5mn - 6n^2$ **51.** $25a^2 + 25ab + 6b^2$

52. $6x^2 - 5xy - y^2$ **53.** $6a^2 - 7ab - 5b^2$ **54.** $25g^2 - 5gh - 2h^2$

55. $6m^6n + 7m^5n^2 + 2m^4n^3$ **56.** $12k^3q^4 - 4k^2q^5 - kq^6$ **57.** $5 - 6x + x^2$

58. $7 + 8x + x^2$ **59.** $16 + 16x + 3x^2$ **60.** $18 + 65x + 7x^2$

If a trinomial has a negative coefficient for the squared term, such as $-2x^2 + 11x - 12$, *it may be easier to factor by first factoring out the common factor* -1:

$$-2x^2 + 11x - 12 = -1(2x^2 - 11x + 12)$$
$$= -1(2x - 3)(x - 4).$$

Use this method to factor the trinomials in Exercises 61–66.

61. $-x^2 - 4x + 21$ **62.** $-x^2 + x + 72$ **63.** $-3x^2 - x + 4$

64. $-5x^2 + 2x + 16$ **65.** $-2a^2 - 5ab - 2b^2$ **66.** $-3p^2 + 13pq - 4q^2$

MATHEMATICAL CONNECTIONS (EXERCISES 67–72)

One of the most common problems that beginning algebra students face is this: if an answer obtained doesn't look exactly like the one given in the back of the book, is it necessarily incorrect? Very often there are several different equivalent forms of an answer that are all correct. Work Exercises 67–72 in order, so that you can see how and why this is possible for factoring problems.

67. Factor the integer 35 as the product of two prime numbers.

68. Factor the integer 35 as the product of the negatives of two prime numbers.

69. Verify the following factorization: $6x^2 - 11x + 4 = (3x - 4)(2x - 1)$.

70. Verify the following factorization: $6x^2 - 11x + 4 = (4 - 3x)(1 - 2x)$.

71. Compare the two valid factorizations in Exercises 69 and 70. How do the factors in each case compare?

72. Suppose you know that the correct factorization of a particular trinomial is $(7t - 3)(2t - 5)$. Based on your observations in Exercises 69–71, what is another valid factorization?

7.4 *Special Factorizations*

By reversing the rules for multiplication of binomials that we learned in the last chapter, we get three rules for factoring polynomials in certain forms.

OBJECTIVE The formula for the product of the sum and difference of the same two terms is

$$(a + b)(a - b) = a^2 - b^2.$$

Reversing this rule leads to the following factorization.

Factoring a Difference of Two Squares

$$a^2 - b^2 = (a + b)(a - b)$$

For example,

$$m^2 - 16 = m^2 - 4^2 = (m + 4)(m - 4).$$

E X A M P L E 1   Factoring a Difference of Squares

Factor each binomial that is the difference of two squares.

$$\underset{\downarrow \quad\quad \downarrow \quad\quad\quad \downarrow \quad \downarrow \quad\quad \downarrow \quad \downarrow}{a^2 \;-\; b^2 \;=\; (a \;+\; b)(\,a \;-\; b)}$$

(a) $x^2 - 49 = x^2 - 7^2 = (x + 7)(x - 7)$

(b) $z^2 - \dfrac{9}{16} = z^2 - \left(\dfrac{3}{4}\right)^2 = \left(z + \dfrac{3}{4}\right)\left(z - \dfrac{3}{4}\right)$

(c) $y^2 - m^2 = (y + m)(y - m)$

(d) $p^2 + 16$

    Since $p^2 + 16$ is the *sum* of two squares, it is not equal to $(p + 4)(p - 4)$. Also, using FOIL,

$$(p - 4)(p - 4) = p^2 - 8p + 16 \neq p^2 + 16$$

and

$$(p + 4)(p + 4) = p^2 + 8p + 16 \neq p^2 + 16$$

so $p^2 + 16$ is a prime polynomial.

Caution
As Example 1(d) suggests, after any common factor is removed, the sum of two squares cannot be factored.

E X A M P L E 2   Factoring a Difference of Squares

Factor each binomial that is the difference of two squares.

$$\underset{\downarrow \quad\quad\quad \downarrow \quad\quad\quad \downarrow \quad \downarrow \quad\quad \downarrow \quad \downarrow}{a^2 \quad\; -\; b^2 \;=\; (a \;+\; b)(\,a \quad -\; b)}$$

(a) $25m^2 - 16 = (5m)^2 - 4^2 = (5m + 4)(5m - 4)$

(b) $49z^2 - 64 = (7z)^2 - (8)^2 = (7z + 8)(7z - 8)$

WORK PROBLEM 1 AT THE SIDE. ▶▶

OBJECTIVES

1 Factor the difference of two squares.

2 Factor a perfect square trinomial.

FOR EXTRA HELP

Tutorial   Tape 15   SSM, Sec. 7.4

1. Factor if possible.

(a) $p^2 - 100$

(b) $9m^2 - 49$

(c) $64a^2 - 25$

(d) $x^2 + y^2$

ANSWERS

1. (a) $(p + 10)(p - 10)$
   (b) $(3m + 7)(3m - 7)$
   (c) $(8a + 5)(8a - 5)$
   (d) prime

2. Factor.

(a) $50r^2 - 32$

(b) $27y^2 - 75$

(c) $k^4 - 49$

(d) $81r^4 - 16$

E X A M P L E   3   Factoring More Involved Differences of Squares

Factor completely.

$$a^2 \;-\; b^2 \;=\; (a \;+\; b)(\,a \;-\; b)$$

(a) $9x^2 - 4z^2 = (3x)^2 - (2z)^2 = (3x + 2z)(3x - 2z)$

(b) $81y^2 - 36$

$$81y^2 - 36 = 9(9y^2 - 4) \qquad \text{Factor out 9.}$$
$$= 9(3y + 2)(3y - 2) \qquad \text{Difference of squares}$$

(c) $p^4 - 36 = (p^2)^2 - 6^2 = (p^2 + 6)(p^2 - 6)$

Neither $p^2 + 6$ nor $p^2 - 6$ can be factored further.

(d) $m^4 - 16 = (m^2)^2 - 4^2$

$$= (m^2 + 4)(m^2 - 4) \qquad \text{Difference of squares}$$
$$= (m^2 + 4)(m + 2)(m - 2) \qquad \text{Difference of squares}$$

Caution
A common error is to forget to factor the difference of two squares a second time when several steps are required, as in Example 3(d).

◀◀ **WORK PROBLEM 2 AT THE SIDE.**

OBJECTIVE 2 The expressions 144, $4x^2$, and $81m^6$ are called perfect squares because

$$144 = 12^2, \quad 4x^2 = (2x)^2, \quad \text{and} \quad 81m^6 = (9m^3)^2.$$

A **perfect square trinomial** is a trinomial that is the square of a binomial. As an example, $x^2 + 8x + 16$ is a perfect square trinomial because it is the square of the binomial $x + 4$:

$$x^2 + 8x + 16 = (x + 4)^2.$$

For a trinomial to be a perfect square, two of its terms must be perfect squares. For this reason, $16x^2 + 4x + 15$ is not a perfect square trinomial because only the term $16x^2$ is a perfect square.

On the other hand, even if two of the terms are perfect squares, the trinomial may not be a perfect square trinomial. For example, $x^2 + 6x + 36$ has two perfect square terms, but it is not a perfect square trinomial. (Try to find a binomial that can be squared to give $x^2 + 6x + 36$.)

We can multiply to see that the square of a binomial gives the following perfect square trinomials.

Factoring Perfect Square Trinomials

$$a^2 + 2ab + b^2 = (a + b)^2$$
$$a^2 - 2ab + b^2 = (a - b)^2$$

The middle term of a perfect square trinomial is always twice the product of the two terms in the squared binomial. (This was shown in Section 6.6.) Use this to check any attempt to factor a trinomial that appears to be a perfect square.

ANSWERS
2. (a) $2(5r + 4)(5r - 4)$
(b) $3(3y + 5)(3y - 5)$
(c) $(k^2 + 7)(k^2 - 7)$
(d) $(9r^2 + 4)(3r + 2)(3r - 2)$

E X A M P L E   4   Factoring a Perfect Square Trinomial

Factor $x^2 + 10x + 25$.

The term x^2 is a perfect square, and so is 25. We can try to factor the trinomial as

$$x^2 + 10x + 25 = (x + 5)^2.$$

To check this we take twice the product of the two terms in the squared binomial.

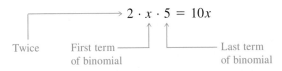

$$2 \cdot x \cdot 5 = 10x$$

Twice   First term —⌐   ⌐— Last term
of binomial   of binomial

Since $10x$ is the middle term of the trinomial, the trinomial is a perfect square and can be factored as $(x + 5)^2$.

WORK PROBLEM 3 AT THE SIDE. ▶▶

E X A M P L E   5   Factoring Perfect Square Trinomials

Factor each perfect square trinomial.

(a) $x^2 - 22x + 121$

The first and last terms are perfect squares ($121 = 11^2$). Check to see whether the middle term of $x^2 - 22x + 121$ is twice the product of the first and last terms of the binomial $(x - 11)$.

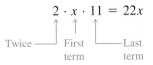

$$2 \cdot x \cdot 11 = 22x$$

Twice —⌐ First └─ Last
         term   term

Since twice the product of the first and last terms of the binomial is the middle term, $x^2 - 22x + 121$ is a perfect square trinomial and

$$x^2 - 22x + 121 = (x - 11)^2.$$

(b) $9m^2 - 24m + 16 = (3m)^2 - 2(3m)(4) + 4^2 = (3m - 4)^2$

Twice —⌐ First └─ Last
         term   term

(c) $25y^2 + 20y + 16$

The first and last terms are perfect squares.

$$25y^2 = (5y)^2 \quad \text{and} \quad 16 = 4^2$$

Twice the product of the first and last terms of the binomial $5y + 4$ is

$$2 \cdot 5y \cdot 4 = 40y$$

which is not the middle term of $25y^2 + 20y + 16$. This polynomial is not a perfect square. In fact, the polynomial cannot be factored even with the methods of Section 7.3; it is a prime polynomial.

3. Factor each trinomial that is a perfect square.

(a) $p^2 + 14p + 49$

(b) $m^2 + 8m + 16$

(c) $x^2 + 2x + 1$

4. Factor each trinomial that is a perfect square.

(a) $p^2 - 18p + 81$

(b) $16a^2 + 56a + 49$

(c) $121p^2 + 110p + 100$

(d) $64x^2 - 48x + 9$

Note
The sign of the second term in the squared binomial is always the same as the sign of the middle term in the trinomial. Also, the first and last terms of a perfect square trinomial must be *positive*, because they are squares. For example, the polynomial $x^2 - 2x - 1$ cannot be a perfect square because the last term is negative.

◀◀ **WORK PROBLEM 4 AT THE SIDE.**

The methods of factoring discussed in this section are summarized here. These rules should be memorized.

Special Factorizations

Difference of two squares	$a^2 - b^2 = (a + b)(a - b)$
Perfect square trinomials	$a^2 + 2ab + b^2 = (a + b)^2$
	$a^2 - 2ab + b^2 = (a - b)^2$

ANSWERS

4. (a) $(p - 9)^2$ **(b)** $(4a + 7)^2$
   (c) not a perfect square trinomial
   (d) $(8x - 3)^2$

7.4 Exercises

1. In order to develop the factoring techniques described in this section, complete the following list of squares.

   $1^2 =$ _____ $2^2 =$ _____ $3^2 =$ _____ $4^2 =$ _____ $5^2 =$ _____

   $6^2 =$ _____ $7^2 =$ _____ $8^2 =$ _____ $9^2 =$ _____ $10^2 =$ _____

   $11^2 =$ _____ $12^2 =$ _____ $13^2 =$ _____ $14^2 =$ _____ $15^2 =$ _____

   $16^2 =$ _____ $17^2 =$ _____ $18^2 =$ _____ $19^2 =$ _____ $20^2 =$ _____

2. In order to use the factoring techniques described in this section, you will sometimes need to recognize fourth powers of integers. Complete the following list of fourth powers.

   $1^4 =$ _____ $2^4 =$ _____ $3^4 =$ _____ $4^4 =$ _____ $5^4 =$ _____

3. The following powers of x are all perfect squares: x^2, x^4, x^6, x^8, x^{10}. Based on this observation, we may make a conjecture (an educated guess) that if the power of a variable is divisible by _____ (with 0 remainder) then we have a perfect square.

4. Is the conjecture of Exercise 3 true in all cases? Why or why not?

Factor each binomial completely. Use your answers in Exercises 1 and 2 as necessary. See Examples 1–3.

5. $y^2 - 25$ 6. $t^2 - 16$ 7. $9r^2 - 4$ 8. $4x^2 - 9$

9. $36m^2 - \dfrac{16}{25}$ 10. $100b^2 - \dfrac{4}{49}$ 11. $36x^2 - 16$ 12. $32a^2 - 8$

13. $196p^2 - 225$ 14. $361q^2 - 400$ 15. $16r^2 - 25a^2$ 16. $49m^2 - 100p^2$

17. $100x^2 + 49$ 18. $81w^2 + 16$ 19. $p^4 - 49$ 20. $r^4 - 25$

21. $x^4 - 1$ 22. $y^4 - 16$ 23. $p^4 - 256$ 24. $16k^4 - 1$

25. When a student was directed to factor $x^4 - 81$ completely, his teacher did not give him full credit when he answered $(x^2 + 9)(x^2 - 9)$. The student argued that because his answer does indeed give $x^4 - 81$ when multiplied out, he should be given full credit. Was the teacher justified in her grading of this item? Why or why not?

26. The binomial $4x^2 + 16$ is a sum of two squares that *can* be factored. How is this binomial factored? When can the sum of two squares be factored?

Factor each trinomial completely. It may be necessary to factor out the greatest common factor first. See Examples 4 and 5.

27. $w^2 + 2w + 1$

28. $p^2 + 4p + 4$

29. $x^2 - 8x + 16$

30. $x^2 - 10x + 25$

31. $t^2 + t + \dfrac{1}{4}$

32. $m^2 + \dfrac{2}{3}m + \dfrac{1}{9}$

33. $x^2 - 1.0x + .25$

34. $y^2 - 1.4y + .49$

35. $2x^2 + 24x + 72$

36. $3y^2 - 48y + 192$

37. $16x^2 - 40x + 25$

38. $36y^2 - 60y + 25$

39. $49x^2 - 28xy + 4y^2$

40. $4z^2 - 12zw + 9w^2$

41. $64x^2 + 48xy + 9y^2$

42. $9t^2 + 24tr + 16r^2$

43. $-50h^2 + 40hy - 8y^2$

44. $-18x^2 - 48xy - 32y^2$

45. In the polynomial $9y^2 + 14y + 25$, the first and last terms are perfect squares. Can the polynomial be factored? If it can, factor it. If it cannot, explain why it is not a perfect square trinomial.

46. Find the value of b so that $100a^2 + ba + 9 = (10a + 3)^2$ is true.

MATHEMATICAL CONNECTIONS (EXERCISES 47–50)

We have seen that multiplication and factoring are reverse processes. We know that multiplication and division are also related: to check a division problem, we multiply the quotient by the divisor to get the dividend. To see how we can use this idea, work Exercises 47–50 in order.

47. Factor $10x^2 + 11x - 6$.

48. Use long division to divide $10x^2 + 11x - 6$ by $2x + 3$.

49. Could we have predicted the result in Exercise 48 from the result in Exercise 47? Explain your answer.

50. Divide $x^3 - 1$ by $x - 1$. Use your answer to factor $x^3 - 1$.

SUMMARY EXERCISES ON FACTORING*

As you factor a polynomial, these questions will help you decide on a suitable factoring technique.

> **Factoring a Polynomial**
> 1. Is there a common factor? If so, factor it out.
> 2. How many terms are in the polynomial?
>   *Two terms:* Check to see whether it is the difference of two squares.
>   *Three terms:* Is it a perfect square trinomial? If the trinomial is not a perfect square, check to see whether the coefficient of the squared term is 1. If so, use the method of Section 7.2. If the coefficient of the squared term of the trinomial is not 1, use the general factoring methods of Section 7.3.
>   *Four terms:* Try to factor the polynomial by grouping.
> 3. Can any factors be factored further? If so, factor them.

Factor as completely as possible.

1. $32m^9 + 16m^5 + 24m^3$

2. $2m^2 - 10m - 48$

3. $14k^3 + 7k^2 - 70k$

4. $9z^2 + 64$

5. $6z^2 + 31z + 5$

6. $m^2 - 3mn - 4n^2$

7. $49z^2 - 16y^2$

8. $100n^2r^2 + 30nr^3 - 50n^2r$

9. $16x + 20$

10. $2m^2 + 5m - 3$

11. $10y^2 - 7yz - 6z^2$

12. $y^4 - 16$

13. $m^2 + 2m - 15$

14. $6y^2 - 5y - 4$

15. $32z^3 + 56z^2 - 16z$

16. $15y + 5$

17. $z^2 - 12z + 36$

18. $9m^2 - 64$

19. $y^2 - 4yk - 12k^2$

20. $16z^2 - 8z + 1$

21. $6y^2 - 6y - 12$

22. $72y^3z^2 + 12y^2 - 24y^4z^2$

23. $p^2 - 17p + 66$

24. $a^2 + 17a + 72$

---

*This exercise set includes all kinds of factoring methods. The exercises are randomly mixed to give you practice at deciding which method should be used.

25. $k^2 + 9$

26. $108m^2 - 36m + 3$

27. $z^2 - 3za - 10a^2$

28. $45a^3b^5 - 60a^4b^2 + 75a^6b^4$

29. $4k^2 - 12k + 9$

30. $a^2 - 3ab - 28b^2$

31. $16r^2 + 24rm + 9m^2$

32. $3k^2 + 4k - 4$

33. $3k^3 - 12k^2 - 15k$

34. $a^4 - 625$

35. $16k^2 - 48k + 36$

36. $8k^2 - 10k - 3$

37. $36y^6 - 42y^5 - 120y^4$

38. $8p^2 + 23p - 3$

39. $5z^3 - 45z^2 + 70z$

40. $8k^2 - 2kh - 3h^2$

41. $54m^2 - 24z^2$

42. $4k^2 - 20kz + 25z^2$

43. $6a^2 + 10a - 4$

44. $15h^2 + 11hg - 14g^2$

45. $m^2 - 81$

46. $10z^2 - 7z - 6$

47. $125m^4 - 400m^3n + 195m^2n^2$

48. $9y^2 + 12y - 5$

49. $m^2 - 4m + 4$

50. $27p^{10} - 45p^9 - 252p^8$

51. $24k^4p + 60k^3p^2 + 150k^2p^3$

52. $10m^2 + 25m - 60$

53. $12p^2 + pq - 6q^2$

54. $k^2 - 64$

55. $64p^2 - 100m^2$

56. $2m^2 + 7mn - 15n^2$

57. $100a^2 - 81y^2$

58. $8a^2 + 23ab - 3b^2$

59. $a^2 + 8a + 16$

60. $4y^2 - 25$

7.5 Solving Quadratic Equations by Factoring

In this section we introduce **quadratic** (kwah-DRAD-ik) **equations,** which are equations that contain a squared term and no terms of higher degree.

Quadratic Equations

Quadratic equations can be written in the form

$$ax^2 + bx + c = 0$$

where a, b, and c are real numbers, with $a \neq 0$.

The form $ax^2 + bx + c = 0$ is the **standard form** of a quadratic equation. For example,

$$x^2 + 5x + 6 = 0, \quad 2a^2 - 5a = 3, \quad \text{and} \quad y^2 = 4$$

are all quadratic equations but only $x^2 + 5x + 6 = 0$ is in standard form.

> **WORK PROBLEM I AT THE SIDE.** ▶▶

OBJECTIVE ▶ Some quadratic equations can be solved by factoring. A more general method for solving those equations that cannot be solved by factoring is given in Chapter 10. We use the **zero-factor property** to solve a quadratic equation by factoring.

Zero-Factor Property

If a and b are real numbers and if $ab = 0$, then $a = 0$ or $b = 0$.

In other words, if the product of two numbers is zero, then at least one of the numbers must be zero. One number *must* be 0, but both *may* be 0.

EXAMPLE I Using the Zero-Factor Property

Solve the equation $(x + 3)(2x - 1) = 0$.

The product $(x + 3)(2x - 1)$ is equal to zero. By the zero-factor property, the only way that the product of these two factors can be zero is if at least one of the factors is zero. Therefore, either $x + 3 = 0$ or $2x - 1 = 0$. Solve each of these two linear equations as in Chapter 2.

$$
\begin{array}{lll}
x + 3 = 0 & 2x - 1 = 0 \\
x = -3 & 2x = 1 & \text{Add 1 to both sides.} \\
& x = \dfrac{1}{2} & \text{Divide by 2.}
\end{array}
$$

The given equation $(x + 3)(2x - 1) = 0$ has two solutions, $x = -3$ and $x = \frac{1}{2}$. Check these answers by substituting -3 for x in the original equation, $(x + 3)(2x - 1) = 0$. Then start over and substitute $\frac{1}{2}$ for x.

If $x = -3$, then

$$(-3 + 3)[2(-3) - 1] = 0 \quad ?$$
$$0(-7) = 0. \quad \text{True}$$

If $x = \frac{1}{2}$, then

$$\left(\frac{1}{2} + 3\right)\left(2 \cdot \frac{1}{2} - 1\right) = 0 \quad ?$$
$$\frac{7}{2}(1 - 1) = 0 \quad ?$$
$$\frac{7}{2} \cdot 0 = 0. \quad \text{True}$$

Both -3 and $\frac{1}{2}$ result in true equations, so the solution set is $\{-3, \frac{1}{2}\}$.

1. Write each quadratic equation in standard form.

(a) $x^2 - 3x = 4$

(b) $y^2 = 9y - 8$

2. Solve each equation. Check your answers.

(a) $(x - 5)(x + 2) = 0$

(b) $(3x - 2)(x + 6) = 0$

(c) $x(5x + 3) = 0$

3. Solve each equation.

(a) $m^2 - 3m - 10 = 0$

(b) $x^2 - 7x = 0$

(c) $r^2 + 2r = 8$

◀◀ **WORK PROBLEM 2 AT THE SIDE.**

In Example 1 the equation to be solved was presented with the polynomial in factored form. If the polynomial in an equation is not already factored, first make sure that the equation is in standard form. Then factor.

E X A M P L E 2   Solving a Quadratic Equation

Solve each equation.

(a) $x^2 - 5x = -6$

First, rewrite the equation in standard form by adding 6 to both sides.

$$x^2 - 5x + 6 = -6 + 6 \qquad \text{Add 6.}$$
$$x^2 - 5x + 6 = 0$$

Now factor $x^2 - 5x + 6$. Find two numbers whose product is 6 and whose sum is -5. These two numbers are -2 and -3, so the equation becomes

$$(x - 2)(x - 3) = 0. \qquad \text{Factor.}$$
$$x - 2 = 0 \quad \text{or} \quad x - 3 = 0 \qquad \text{Zero-factor property}$$
$$x = 2 \quad \text{or} \qquad x = 3 \qquad \text{Solve each equation.}$$

Check the solution set $\{2, 3\}$ by substituting first 2 and then 3 for x in the original equation.

(b) $y^2 = y + 20$

$$y^2 = y + 20$$
$$y^2 - y - 20 = 0 \qquad \text{Subtract } y \text{ and 20.}$$
$$(y - 5)(y + 4) = 0 \qquad \text{Factor.}$$
$$y - 5 = 0 \quad \text{or} \quad y + 4 = 0 \qquad \text{Zero-factor property}$$
$$y = 5 \quad \text{or} \qquad y = -4 \qquad \text{Solve each equation.}$$

Check the solution set $\{5, -4\}$ by substituting in the original equation.

Note
The word "or" as used in Example 2 means "one or the other or both."

◀◀ **WORK PROBLEM 3 AT THE SIDE.**

In summary, we go through the following steps to solve quadratic equations by factoring.

Solving a Quadratic Equation by Factoring

Step 1 Write the equation in standard form: all terms on one side of the equals sign, with 0 on the other side.

Step 2 Factor completely.

Step 3 Set each factor with a variable equal to 0, and solve the resulting equations.

Step 4 Check each solution in the original equation.

ANSWERS

2. (a) $\{5, -2\}$ (b) $\left\{\dfrac{2}{3}, -6\right\}$ (c) $\left\{0, -\dfrac{3}{5}\right\}$

3. (a) $\{-2, 5\}$ (b) $\{0, 7\}$ (c) $\{-4, 2\}$

┌─ **E X A M P L E 3**   **Solving a Quadratic Equation with a Common Factor**

Solve the equation $4p^2 + 40 = 26p$.

Subtract $26p$ from each side and write in descending powers to get

$$4p^2 - 26p + 40 = 0.$$

$$2(2p^2 - 13p + 20) = 0 \qquad \text{Factor out 2.}$$

$$2(2p - 5)(p - 4) = 0 \qquad \text{Factor the trinomial.}$$

$$2p - 5 = 0 \quad \text{or} \quad p - 4 = 0 \qquad \text{Zero-factor property}$$

$$2p = 5 \quad \text{or} \qquad p = 4$$

$$p = \frac{5}{2}$$

The solution set is $\{\frac{5}{2}, 4\}$. Check by substituting in the original equation. ∎

Caution

A common error is to include the common factor 2 as a solution in Example 3.

WORK PROBLEM 4 AT THE SIDE. ▶▶

┌─ **E X A M P L E 4**   **Solving Quadratic Equations**

Solve each equation.

(a) $16m^2 - 25 = 0$

$$16m^2 - 25 = 0$$

$$(4m + 5)(4m - 5) = 0 \qquad \text{Factor.}$$

$$4m + 5 = 0 \quad \text{or} \quad 4m - 5 = 0 \qquad \text{Zero-factor property}$$

$$4m = -5 \quad \text{or} \qquad 4m = 5$$

$$m = -\frac{5}{4} \quad \text{or} \qquad m = \frac{5}{4}$$

You should check the two solutions, $-\frac{5}{4}$ and $\frac{5}{4}$, using the original equation.

(b) $k(2k + 5) = 3$

$$k(2k + 5) = 3$$

$$2k^2 + 5k = 3 \qquad \text{Multiply.}$$

$$2k^2 + 5k - 3 = 0 \qquad \text{Subtract 3.}$$

$$(2k - 1)(k + 3) = 0 \qquad \text{Factor.}$$

$$2k - 1 = 0 \quad \text{or} \quad k + 3 = 0 \qquad \text{Zero-factor property}$$

$$2k = 1$$

$$k = \frac{1}{2} \quad \text{or} \qquad k = -3$$

Check that the solution set is $\{\frac{1}{2}, -3\}$.

└─ **CONTINUED ON NEXT PAGE**

4. Solve each equation.

(a) $10a^2 - 5a - 15 = 0$

(b) $4x^2 - 2x = 42$

ANSWERS

4. (a) $\left\{-1, \frac{3}{2}\right\}$ **(b)** $\left\{-3, \frac{7}{2}\right\}$

5. Solve each equation.

(a) $49m^2 - 9 = 0$

(b) $p(4p + 7) = 2$

(c) $m^2 = 3m$

6. Solve each equation.

(a) $r^3 - 16r = 0$

(b) $x^3 - 3x^2 - 18x = 0$

(c) $y^2 = 2y$

First write the equation in standard form.

$$y^2 - 2y = 0 \qquad \text{Standard form}$$
$$y(y - 2) = 0 \qquad \text{Factor.}$$
$$y = 0 \quad \text{or} \quad y - 2 = 0 \qquad \text{Zero-factor property}$$
$$y = 2$$

The solution set is $\{0, 2\}$.

> **Caution**
>
> In Example 4(b) the zero-factor property could not be used to solve the equation as given because of the 3 on the right. Remember that the zero-factor property applies only to a product that equals 0.
>
> In Example 4(c) it is tempting to begin by dividing both sides of the equation by y to get $y = 2$. Note that we do not get the other solution, 0, by this method.
>
> We *may* divide both sides of an equation by a *nonzero* real number, however. For instance, in Example 3 we could have divided both sides by 2 to begin. This would not have affected the solutions.

◀◀ **WORK PROBLEM 5 AT THE SIDE.**

OBJECTIVE 2 The zero-factor property can also be used to solve equations that result in more than two factors, as shown in Example 5. (These equations are not quadratic equations. Why not?)

E X A M P L E   5   Solving Equations with More Than Two Factors

Solve the equation $6z^3 - 6z = 0$.

$$6z^3 - 6z = 0$$
$$6z(z^2 - 1) = 0 \qquad \text{Factor out } 6z.$$
$$6z(z + 1)(z - 1) = 0 \qquad \text{Factor } z^2 - 1.$$

By an extension of the zero-factor property, this product can equal 0 only if at least one of the factors is 0. Write and solve three equations, one for each factor with a variable.

$$6z = 0 \quad \text{or} \quad z + 1 = 0 \quad \text{or} \quad z - 1 = 0$$
$$z = 0 \quad \text{or} \qquad z = -1 \quad \text{or} \qquad z = 1$$

Check the solution set $\{0, -1, 1\}$ by substituting in the original equation.

◀◀ **WORK PROBLEM 6 AT THE SIDE.**

ANSWERS

5. (a) $\left\{-\dfrac{3}{7}, \dfrac{3}{7}\right\}$ **(b)** $\left\{-2, \dfrac{1}{4}\right\}$ **(c)** $\{0, 3\}$

6. (a) $\{-4, 0, 4\}$ **(b)** $\{-3, 0, 6\}$

┌─
E X A M P L E 6 **Solving Equations with More Than Two Factors**

Solve the equation $(2x - 1)(x^2 - 9x + 20) = 0$.

$$(2x - 1)(x^2 - 9x + 20) = 0$$

$$(2x - 1)(x - 5)(x - 4) = 0 \qquad \text{Factor } x^2 - 9x + 20.$$

$$2x - 1 = 0 \quad \text{or} \quad x - 5 = 0 \quad \text{or} \quad x - 4 = 0 \qquad \text{Zero-factor property}$$

$$x = \frac{1}{2} \quad \text{or} \qquad x = 5 \quad \text{or} \qquad x = 4$$

The solution set is $\{\frac{1}{2}, 5, 4\}$. Check each solution.
─┘

WORK PROBLEM 7 AT THE SIDE. ▶▶

Caution
In Example 6, it would be unproductive to begin by multiplying the two factors together. Keep in mind the zero-factor property requires the product of two or more factors equal to zero. Always consider first whether an equation is given in the appropriate form for the zero-factor property.

7. Solve each equation.

(a) $(m + 3)(m^2 - 11m + 10) = 0$

(b) $(2x + 5)(4x^2 - 9) = 0$

NUMBERS IN THE
Real World *collaborative investigations*

Using Computers to Break Codes

Finding prime factors of extremely large numbers has been considered a mere computer exercise—interesting for the improved methods of working with computers, but of no value in its own right. This has changed in recent years with new methods of computer coding in which very large numbers are used in an attempt to provide unbreakable codes for computer data. Just as fast as these numbers are used, other people try to find prime factors of them so that the code can be broken.

An article in *Science News*, May 7, 1994 ("Team Sieving Cracks a Huge Number," p. 292) reported that a method known as RSA was used to factor a 129-digit number. It was accomplished by more than 600 computers working together. The number and its two factors appear at left.

114,381,625,757,888,867,669,235,
779,976,146,612,010,218,296,721,
242,362,562,561,842,935,706,935,
245,733,897,830,597,123,563,958,
705,058,989,075,147,599,290,026,
879,543,541
= 3,490,529,510,847,650,949,147,
849,619,903,898,133,417,764,638,
493,387,843,990,820,577
× 32,769,132,993,266,709,549,961,
988,190,834,461,413,177,642,967,
992,942,539,798,288,533.

FOR GROUP DISCUSSION

1. A number of the form $2^n - 1$, where n is prime, may be either prime or composite. At one time it was believed that all such numbers were prime, until it was shown that for $n = 11$, the result is a composite number. What is the value of $2^{11} - 1$? Factor it into its prime factors.

2. A **Smith number** is a composite number that satisfies this property: When the digits of its prime factors are added, the result is the same as the sum of the digits of the number itself. Show that 4,937,775 is a Smith number. (*Hint:* 65,837 is one of its prime factors.)

3. Comment on the following story:

   In 1903, long before computers, the mathematician F. N. Cole presented before a meeting of the American Mathematical Society his discovery of a factorization of the number $2^{67} - 1$. He walked up to the chalkboard, raised 2 to the 67th power, and then subtracted 1. Then he moved over to another part of the board and multiplied out 193,707,721 × 761,838,257,287. The two calculations agreed, and Cole received a standing ovation for a presentation that did not include a single word.

7.5 Exercises

Fill in the blanks.

1. A quadratic equation is an equation that can be put in the form _____ = 0.

2. If a quadratic equation is in standard form, to solve the equation we should begin by _____ the polynomial.

Solve each equation and check your answer. See Example 1.

3. $(x + 5)(x - 2) = 0$

4. $(x - 1)(x + 8) = 0$

5. $(2m - 7)(m - 3) = 0$

6. $(6k + 5)(k + 4) = 0$

7. $2x(3x - 4) = 0$

8. $6y(4y + 9) = 0$

9. Students often become confused as to how to handle a constant, such as 2 in the equation $2x(3x - 4) = 0$ of Exercise 7. How would you explain to someone how to solve this equation, and how to handle the constant 2?

10. The zero-factor property can be extended to more than two factors. For example, to solve $(x - 4)(x + 3)(2x - 7) = 0$, we would set each factor equal to zero and solve three equations. Find the solutions to this equation.

11. Why do you think that 9 is called a *double solution* of the equation $(x - 9)^2 = 0$?

12. Write an equation with the two solutions 5 and $-\frac{4}{3}$.

Solve each equation, and check your answer. See Examples 2–4.

13. $y^2 + 3y + 2 = 0$

14. $p^2 + 8p + 7 = 0$

15. $y^2 - 3y + 2 = 0$

16. $r^2 - 4r + 3 = 0$

17. $x^2 = 24 - 5x$

18. $t^2 = 2t + 15$

19. $x^2 = 3 + 2x$

20. $m^2 = 4 + 3m$

21. $z^2 = -2 - 3z$

22. $p^2 = 2p + 3$

23. $m^2 + 8m + 16 = 0$

24. $b^2 - 6b + 9 = 0$

25. $3x^2 + 5x - 2 = 0$

26. $6r^2 - r - 2 = 0$

27. $6p^2 = 4 - 5p$

28. $6x^2 = 4 + 5x$

29. $9s^2 + 12s = -4$

30. $36x^2 + 60x = -25$

31. $y^2 - 9 = 0$

32. $m^2 - 100 = 0$

33. $16k^2 - 49 = 0$

34. $4w^2 - 9 = 0$

35. $n^2 = 121$

36. $x^2 = 400$

37. What is wrong with this reasoning in solving the equation in Exercise 35? "To solve $n^2 = 121$, I must find a number that, when multiplied by itself, gives 121. Because $11^2 = 121$, the solution of the equation is 11."

38. What is wrong with this reasoning in solving $x^2 = 7x$? "To solve $x^2 = 7x$, first divide both sides by x to get $x = 7$. Therefore, the solution is 7."

Solve each equation, and check your answer. See Examples 4, 5, and 6.

39. $x^2 = 7x$

40. $t^2 = 9t$

41. $6r^2 = 3r$

42. $10y^2 = -5y$

43. $g(g - 7) = -10$

44. $r(r - 5) = -6$

45. $z(2z + 7) = 4$

46. $b(2b + 3) = 9$

47. $2(y^2 - 66) = -13y$

48. $3(t^2 + 4) = 20t$

49. $3x(x + 1) = (2x + 3)(x + 1)$

50. $2k(k + 3) = (3k + 1)(k + 3)$

51. $(2r + 5)(3r^2 - 16r + 5) = 0$

52. $(3m + 4)(6m^2 + m - 2) = 0$

53. $(2x + 7)(x^2 + 2x - 3) = 0$

54. $(x + 1)(6x^2 + x - 12) = 0$

55. $9y^3 - 49y = 0$

56. $16r^3 - 9r = 0$

7.6 Applications of Quadratic Equations

We are now ready to use factoring to solve quadratic equations that arise from applied problems. Many problems in this section will require a formula from the inside covers. The general approach is the same as in Chapter 2. We still follow the six steps listed in Section 2.4.

OBJECTIVE 1 We begin with a geometry problem.

OBJECTIVES

1 Solve applied problems about geometric figures.

2 Solve applied problems about moving objects.

3 Solve applied problems about consecutive integers.

4 Solve applied problems using the Pythagorean formula.

FOR EXTRA HELP

Tutorial   Tape 15   SSM, Sec. 7.6

EXAMPLE 1   Solving an Area Problem

The Goldsteins are planning to plant a rectangular garden in their yard. The width of the garden will be 4 feet less than its length, and they want it to have an area of 96 square feet. Find the length and width of the garden.

Let x = the length of the garden,

$x - 4$ = the width (the width is 4 less than the length).

See Figure 1. The area of a rectangle is given by the formula

$$\text{Area} = LW = \text{Length} \times \text{Width}.$$

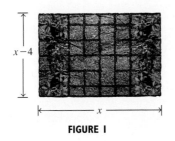

FIGURE 1

Substitute 96 for the area, x for the length, and $x - 4$ for the width in the formula.

$A = LW$

$96 = x(x - 4)$ Let $A = 96$, $L = x$, $W = x - 4$.

$96 = x^2 - 4x$ Distributive property

$0 = x^2 - 4x - 96$ Subtract 96 from both sides.

$0 = (x - 12)(x + 8)$ Factor.

$x - 12 = 0$ or $x + 8 = 0$ Zero-factor property

$x = 12$ or $x = -8$

The solutions are $x = 12$ and $x = -8$. Because a rectangle cannot have a side of negative length, discard the solution -8. Then the length of the garden will be 12 feet, and the width will be $12 - 4 = 8$ feet. As a check, the width is 4 feet less than the length and the area is $8 \cdot 12 = 96$ square feet, as required.

Caution
In an applied problem, always be careful to check solutions against physical facts.

1. (a) The length of a room is 2 meters more than the width. The area of the floor is 48 square meters. Find the length and width of the room.

(b) The length of each side of a square is increased by 4 inches. The sum of the areas of the original square and the larger square is 106 square inches. What is the length of a side of the original square?

WORK PROBLEM 1 AT THE SIDE. ▶▶

ANSWERS

1. (a) length: 8 meters; width: 6 meters
   (b) 5 inches

2. The number of impulses fired after a nerve has been stimulated is given by the function $I(x) = -x^2 + 2x + 60$, where x is in milliseconds after the stimulation. When will 45 impulses occur? Do you get two solutions? Why is only one answer given?

OBJECTIVE ▶2▶ The next problem involves the relationship between the height of a ball and the time since it was hit.

EXAMPLE 2   Finding the Height of a Ball

A tennis player has a 180-feet-per-second (125 miles per hour) serve. If he serves straight up in the air, the height of the ball in feet at time t in seconds is given by the function

$$h(x) = -16t^2 + 180t + 6.$$

How long will it take for the ball to reach a height of 206 feet?

A height of 206 feet gives $h(x) = 206$, so the equation becomes

$$206 = -16t^2 + 180t + 6.$$

To solve the equation, we should begin by subtracting 206 from both sides to get 0 on one side. Then we factor the trinomial. For convenience we reverse the sides of the equation.

$$-16t^2 + 180t + 6 = 206$$

$$-16t^2 + 180t - 200 = 0 \qquad \text{Subtract 206.}$$

$$4t^2 - 45t + 50 = 0 \qquad \text{Divide by } -4.$$

$$(4t - 5)(t - 10) = 0 \qquad \text{Factor.}$$

$$4t - 5 = 0 \quad \text{or} \quad t - 10 = 0 \qquad \text{Zero-factor property}$$

$$t = \frac{5}{4} \quad \text{or} \qquad t = 10$$

Solution set: $\{\frac{5}{4}, 10\}$

Since we found two answers, the ball will be 206 feet above the ground twice (once on its way up and once on its way down) at $\frac{5}{4}$ seconds and at 10 seconds.

◀◀ **WORK PROBLEM 2 AT THE SIDE.**

OBJECTIVE ▶3▶ **Consecutive** (kuhn-SEK-yoo-tiv) **integers** are integers that are next to each other on a number line, such as 5 and 6, or -11 and -10. **Consecutive odd integers** are odd integers that are next to each other, such as 21 and 23, or -17 and -15. **Consecutive even integers** are defined similarly; for example, 4 and 6 are consecutive even integers, as are -10 and -8.

The following list may be helpful in working with consecutive integers. Here x represents the first of the integers.

Consecutive Integers	
Two consecutive integers	$x, x + 1$
Three consecutive integers	$x, x + 1, x + 2$
Two consecutive odd integers	$x, x + 2$
Three consecutive even integers	$x, x + 2, x + 4$

In the next example we show how quadratic equations can occur in work with consecutive integers.

EXAMPLE 3   Solving a Consecutive Integer Problem

The product of two consecutive odd integers is 1 less than five times their sum. Find the integers.

$$\text{Let} \qquad s = \text{the small integer}$$
$$s + 2 = \text{the next larger odd integer.}$$

Because the problem mentions consecutive *odd* integers, use $s + 2$ for the larger of the two integers. According to the problem, the product is 1 less than five times the sum.

$$
\underset{\begin{array}{c}\text{The}\\\text{product}\\\downarrow\end{array}}{s(s + 2)}
\;\underset{\begin{array}{c}\text{is}\\\downarrow\end{array}}{=}\;
\underset{\begin{array}{c}\text{five times}\\\text{the sum}\\\downarrow\end{array}}{5(s + s + 2)}
\;\underset{\begin{array}{c}\text{less 1.}\\\downarrow\end{array}}{-\;1}
$$

Simplify this equation and solve it.

$s^2 + 2s = 5s + 5s + 10 - 1$	Distributive property
$s^2 + 2s = 10s + 9$	Combine terms.
$s^2 - 8s - 9 = 0$	Subtract $10s$ and 9.
$(s - 9)(s + 1) = 0$	Factor.
$s - 9 = 0 \qquad \text{or} \qquad s + 1 = 0$	Zero-factor property
$s = 9 \qquad \text{or} \qquad s = -1$	

The solution set is $\{9, -1\}$. We need to find two consecutive odd integers.

If $s = 9$ is the first, then $s + 2 = 11$ is the second.

If $s = -1$ is the first, then $s + 2 = 1$ is the second.

Check that two pairs of integers satisfy the problem: 9 and 11 or -1 and 1.

WORK PROBLEM 3 AT THE SIDE. ▶▶

OBJECTIVE 4 ▶ The next example requires the **Pythagorean** (puh-THAG-uh-REE-un) **formula** from geometry.

Pythagorean Formula

If a right triangle (a triangle with a 90° angle) has longest side of length c and two other sides of lengths a and b, then

$$a^2 + b^2 = c^2.$$

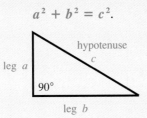

The longest side is called the **hypotenuse,** and the two shorter sides are the **legs** of the triangle.

---

3. (a) The product of two consecutive even integers is 4 more than two times their sum. Find the integers.

(b) Find three consecutive odd integers such that the product of the smallest and largest is 16 more than the middle integer.

4. The hypotenuse of a right triangle is 3 inches longer than the longer leg. The shorter leg is 3 inches shorter than the longer leg. Find the lengths of the sides of the triangle.

EXAMPLE 4   Using the Pythagorean Formula

Ed and Mark leave their office, with Ed traveling north and Mark traveling east. When Mark is 1 mile farther than Ed from the office, the distance between them is 2 miles more than Ed's distance from the office. Find their distances from the office and the distance between them.

Let x represent Ed's distance from the office, $x + 1$ represent Mark's distance from the office, and $x + 2$ represent the distance between them. Place these on a right triangle, as in Figure 2.

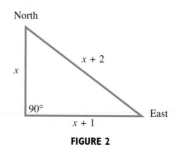

FIGURE 2

Substitute into the Pythagorean formula.

$$a^2 + b^2 = c^2$$
$$x^2 + (x + 1)^2 = (x + 2)^2$$

Because $(x + 1)^2 = x^2 + 2x + 1$, and $(x + 2)^2 = x^2 + 4x + 4$, the equation becomes

$$x^2 + x^2 + 2x + 1 = x^2 + 4x + 4.$$

$$x^2 - 2x - 3 = 0 \qquad \text{Standard form}$$
$$(x - 3)(x + 1) = 0 \qquad \text{Factor.}$$
$$x - 3 = 0 \quad \text{or} \quad x + 1 = 0 \qquad \text{Zero-factor property}$$
$$x = 3 \quad \text{or} \qquad x = -1$$

Solution set: $\{3, -1\}$

Since -1 cannot represent a distance, 3 is the only possible answer. Ed's distance is 3 miles, Mark's distance is 4 miles, and the distance between them is 5 miles. Check that $3^2 + 4^2 = 5^2$.

Caution

When solving a problem involving the Pythagorean formula, be sure that the expressions for the sides are properly placed.

$$\text{leg}^2 + \text{leg}^2 = \text{hypotenuse}^2$$

◀◀ **WORK PROBLEM 4 AT THE SIDE.**

7.6 Exercises

Complete these statements which review the key steps for solving applied problems first introduced in Chapter 2.

1. Read the problem carefully, choose _____ to represent _____ and write it down.

2. Write down a mathematical _____ for any other unknown quantities.

3. Translate the problem into _____ .

4. Solve the _____ and answer the _____ in the problem.

In Exercises 5 and 6 a figure and a corresponding geometric formula are given. (These and other geometric formulas may be found on the inside covers of this book.)
(a) Write an equation using the formula and the given information.
(b) Solve the equation, giving only solutions that make sense in the problem.
(c) Use the solution to find the indicated dimensions of the figure.

5.

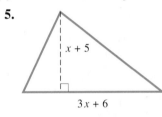

$x + 5$

$3x + 6$

Area of a triangle: $A = \frac{1}{2}bh$

The area of this triangle is 60 square units. Find its base and its height.

6.

4

x

$x + 2$

Volume of a rectangular Chinese box: $V = LWH$

The volume of this box is 192 cubic units. Find its length and its width.

Solve each problem. Check your answer to be sure that it is reasonable. Refer to the formulas on the inside covers. See Examples 1 and 2.

7. The length of a VHS videocassette shell is 3 inches more than its width. The area of the rectangular top side of the shell is 28 square inches. Find the length and the width of the videocassette shell.

8. A plastic box that holds a standard audiocassette has a length 4 centimeters longer than its width. The area of the rectangular top of the box is 77 square centimeters. Find the length and the width of the box.

9. The dimensions of a certain IBM computer monitor screen are such that its length is 3 inches more than its width. If the length were increased by 1 inch while the width remained the same, the area would increase by 8 square inches. What are the dimensions of the screen?

10. The keyboard of the computer mentioned in Exercise 9 is 11 inches longer than it is wide. If both its length and width were increased by 2 inches, the area of the top of the keyboard would increase by 58 square inches. What are the length and the width of the keyboard?

11. A square mirror has sides measuring 2 feet less than the sides of a square painting. If the difference between their areas is 32 square feet, find the lengths of the sides of the mirror and the painting.

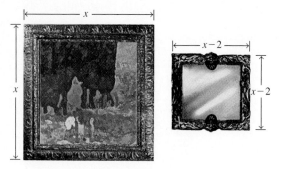

12. The sides of one square have a length 3 meters more than the sides of a second square. If the area of the larger square is subtracted from 4 times the area of the smaller square, the result is 36 square meters. What are the lengths of the sides of each square?

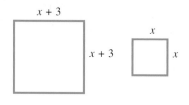

13. If an object is propelled upward from a height of s feet at an initial velocity of v feet per second, then its height after t seconds is given by the function

$$h(t) = -16t^2 + vt + s,$$

where h is in feet. For example, if the object is propelled from a height of 48 feet with an initial velocity of 32 feet per second, its height is given by

$$h(t) = -16t^2 + 32t + 48.$$

(a) After how many seconds is the height 64 feet? (*Hint:* Let $h(t) = 64$ and solve.)

(b) After how many seconds is the height 60 feet?

(c) After how many seconds does the object hit the ground? (*Hint:* When the object hits the ground, $h(t) = 0$.)

(d) The quadratic equation from part (c) has two solutions, yet only one of them is appropriate for answering the question. Why is this so?

14. If an object is propelled upward from ground level with an initial velocity of 64 feet per second, its height in feet t seconds later is given by the function $h(t) = -16t^2 + 64t$. Use this information in parts (a)–(d).

(a) After how many seconds is the height 48 feet?

(b) It can be shown using concepts developed in other courses that the object reaches its maximum height 2 seconds after it is propelled. What is this maximum height?

(c) After how many seconds does the object hit the ground?

(d) The quadratic equation from part (c) has two solutions, yet only one of them is appropriate for answering the question. Why is this so?

15. If an object is dropped, the distance d it falls in t seconds (disregarding air resistance) is given by

$$d(t) = \frac{1}{2}gt^2,$$

where g is approximately 32 feet per second per second. Find the distance an object would fall in the following times.
(a) 4 seconds
(b) 8 seconds

16. Refer to the formula in Exercise 15. How long would it take an object to fall from the top of the building described below? Use a calculator as necessary, and round your answer to the nearest tenth of a second.
(a) Navarre Building, New York City
512 feet
(b) One Canada Square, London   800 feet

Solve each problem. See Example 3.

17. The product of two consecutive integers is 11 more than their sum. Find the integers.

18. The product of two consecutive integers is 4 less than 4 times their sum. Find the integers.

19. Find three consecutive odd integers such that 3 times the sum of all three is 18 more than the product of the smaller two.

20. Find three consecutive odd integers such that the sum of all three is 42 less than the product of the larger two.

21. Find three consecutive even integers such that the sum of the squares of the smaller two is equal to the square of the largest.

22. Find three consecutive even integers such that the square of the sum of the smaller two is equal to twice the largest.

Use the Pythagorean formula to solve the following problems. See Example 4.

23. Wei-Jen works due north of home. Her husband Alan works due east. They leave for work at the same time. By the time Wei-Jen is 5 miles from home, the distance between them is one mile more than Alan's distance from home. How far from home is Alan?

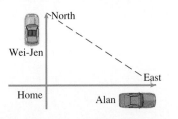

24. Two cars left an intersection at the same time. One traveled north. The other traveled 14 miles farther, but to the east. How far apart were they then, if the distance between them was 4 miles more than the distance traveled east?

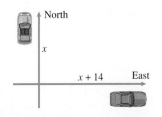

25. A ladder is leaning against a building. The distance from the bottom of the ladder to the building is 4 feet less than the length of the ladder. How high up the side of the building is the top of the ladder if that distance is 2 feet less than the length of the ladder?

26. A lot has the shape of a right triangle with one leg 2 meters longer than the other. The hypotenuse is two meters less than twice the length of the shorter leg. Find the length of the shorter leg.

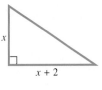

Solve Exercises 27 and 28 using the formula for the volume of a pyramid,
$V = \frac{1}{3}Bh$, *where B is the area of the base.*

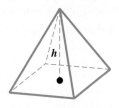

27. The volume of a pyramid is 32 cubic meters. Suppose the numerical value of the height is 10 meters less than the numerical value of the area of the base. Find the height and the area of the base.

28. Suppose a pyramid has a rectangular base whose width is 3 centimeters less than the length. If the height is 8 centimeters and the volume is 144 cubic centimeters, find the dimensions of the base.

Find a polynomial representing the area of each shaded region.

29.

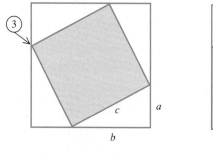

30.

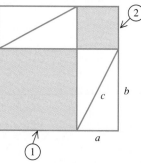

MATHEMATICAL CONNECTIONS (EXERCISES 31–34)

One of the many known proofs of the Pythagorean formula is based on the figures shown. Refer to the appropriate figure and answer Exercises 31–34 in order.

FIGURE A   **FIGURE B**

31. What is an expression for the area of the dark square labeled ③ in Figure A?

32. The five regions in Figure A are equal in area to the six regions in Figure B. What is an expression for the area of the square labeled ① in Figure B?

33. What is an expression for the area of the square labeled ② in Figure B?

34. Represent this statement using algebraic expressions: The sum of the areas of the dark regions in Figure B is equal to the area of the dark region in Figure A. What does this equation represent?

7.1	**factor**	An expression A is a factor of an expression B if B can be divided by A with zero remainder.
	factored form	An expression is in factored form when it is written as a product.
	greatest common factor	The greatest common factor is the largest quantity that is a factor of each of a group of quantities.
7.2	**prime polynomial**	A prime polynomial is a polynomial that cannot be factored.
7.4	**perfect square trinomial**	A perfect square trinomial is a trinomial that can be factored as the square of a binomial.
7.5	**quadratic equation**	A quadratic equation is an equation that can be written in the form $ax^2 + bx + c = 0$, with $a \neq 0$.
	standard form	The form $ax^2 + bx + c = 0$ is the standard form of a quadratic equation.

QUICK REVIEW

Concepts	*Examples*
7.1 Factors; The Greatest Common Factor	
Finding the Greatest Common Factor	Find the greatest common factor of $4x^2y$, $-6x^2y^3$, $2xy^2$.
1. Include the largest numerical factor of every term.	$$4x^2y = 2^2 \cdot x^2 \cdot y$$ $$-6x^2y^3 = -1 \cdot 2 \cdot 3 \cdot x^2 \cdot y^3$$ $$2xy^2 = 2 \cdot x \cdot y^2$$
2. Include each variable that is a factor of every term raised to the smallest exponent that appears in a term.	The greatest common factor is $2xy$.
Factoring by Grouping	Factor $2a^2 + 2ab + a + b$.
1. Group the terms so that each group has a common factor.	$= (2a^2 + 2ab) + (a + b)$
2. Factor out the greatest common factor in each group.	$= 2a(a + b) + 1(a + b)$
3. Factor a common binomial factor from the result of Step 2.	$= (a + b)(2a + 1)$
4. If Step 3 cannot be performed, try a different grouping.	
7.2 Factoring Trinomials	
To factor $x^2 + bx + c$, find m and n such that $mn = c$ and $m + n = b$.	Factor $x^2 + 6x + 8$.
$$\begin{array}{c} mn = c \\ \downarrow \\ x^2 + bx + c \\ \nwarrow \\ m + n = b \end{array}$$	$$\begin{array}{c} mn = 8 \\ \downarrow \\ x^2 + 6x + 8 \\ \nwarrow \\ m + n = 6 \end{array}$$ $$m = 2 \quad \text{and} \quad n = 4$$
Then $x^2 + bx + c = (x + m)(x + n)$.	$x^2 + 6x + 8 = (x + 2)(x + 4)$.

Concepts	Examples
7.3 More on Factoring Trinomials	Factor $3x^2 + 14x - 5$.
To factor $ax^2 + bx + c$:	
By Grouping	$$mn = 3 \cdot -5 = -15$$
Find m and n:	$$m + n = 14$$
	$$m = -1,\ n = 15$$
$$\overbrace{ax^2 + bx + c}^{\substack{mn = ac \\ m + n = b}}$$	
Then factor $ax^2 + mx + nx + b$ by grouping.	$3x^2 - x + 15x - 5 = x(3x - 1) + 5(3x - 1)$ $\qquad = (3x - 1)(x + 5)$
By Trial and Error	By trial and error,
Use FOIL backwards.	$$3x^2 + 14x - 5 = (3x - 1)(x + 5).$$
7.4 Special Factorizations	
$$a^2 - b^2 = (a + b)(a - b)$$	$$4x^2 - 9 = (2x + 3)(2x - 3)$$
$$a^2 + 2ab + b^2 = (a + b)^2$$	$$9x^2 + 6x + 1 = (3x + 1)^2$$
$$a^2 - 2ab + b^2 = (a - b)^2$$	$$4x^2 - 20x + 25 = (2x - 5)^2$$
7.5 Solving Quadratic Equations by Factoring	
Zero-Factor Property	
If a and b are real numbers and if $ab = 0$, then $a = 0$ or $b = 0$.	If $(x - 2)(x + 3) = 0$, then $x - 2 = 0$ or $x + 3 = 0$.
Solving a Quadratic Equation by Factoring	Solve $2x^2 = 7x + 15$.
1. Write in standard form.	**1.** $2x^2 - 7x - 15 = 0$
2. Factor.	**2.** $(2x + 3)(x - 5) = 0$
3. Use the zero-factor property.	**3.** $2x + 3 = 0 \quad$ or $\quad x - 5 = 0$ $\qquad 2x = -3 \qquad\qquad x = 5$ $\qquad x = -\dfrac{3}{2}$
	Solution set: $\{-\frac{3}{2}, 5\}$
4. Check.	**4.** Both solutions satisfy the original equation.
7.6 Applications of Quadratic Equations	
Pythagorean Formula	In a right triangle with legs measuring 8 meters and 15 meters, the square of the hypotenuse is equal to
In a right triangle, the square of the hypotenuse equals the sum of the squares of the legs.	$$8^2 + 15^2 = 289$$
$$a^2 + b^2 = c^2$$	Therefore, the hypotenuse measures 17 meters, since $289 = 17^2$.

In a right triangle, hypotenuse c, leg a, leg b, with $90°$ angle.

[7.1] _Factor out the greatest common factor or factor by grouping._

1. $7t + 14$

2. $60z^3 + 30z$

3. $35x^3 + 70x^2$

4. $2xy - 8y + 3x - 12$

5. $100m^2n^3 - 50m^3n^4 + 150m^2n^2$

6. $6y^2 + 9y + 4y + 6$

[7.2] _Factor completely._

7. $x^2 + 5x + 6$

8. $y^2 - 13y + 40$

9. $q^2 + 6q - 27$

10. $r^2 - r - 56$

11. $r^2 - 4rs - 96s^2$

12. $p^2 + 2pq - 120q^2$

13. $8p^3 - 24p^2 - 80p$

14. $3x^4 + 30x^3 + 48x^2$

15. $m^2 - 3mn - 18n^2$

16. $y^2 - 8yz + 15z^2$

17. $p^7 - p^6q - 2p^5q^2$

18. $3r^5 - 6r^4s - 45r^3s^2$

19. $x^2 + x + 1$

20. $3x^2 + 6x + 6$

[7.3]

21. In order to begin factoring $6r^2 - 5r - 6$, what are the possible first terms of the two binomial factors, if we consider only positive integer coefficients?

22. What is the first step you would use to factor $2z^3 + 9z^2 - 5z$?

Factor completely.

23. $2k^2 - 5k + 2$

24. $3r^2 + 11r - 4$

25. $6r^2 - 5r - 6$

26. $10z^2 - 3z - 1$

27. $8v^2 + 17v - 21$

28. $24x^5 - 20x^4 + 4x^3$

29. $-6x^2 + 3x + 30$

30. $10r^3s + 17r^2s^2 + 6rs^3$

[7.4]

31. Which one of the following is the difference of two squares?
   (a) $32x^2 - 1$ **(b)** $4x^2y^2 - 25z^2$ **(c)** $x^2 + 36$ **(d)** $25y^3 - 1$

32. Which one of the following is a perfect square trinomial?
   (a) $x^2 + x + 1$ **(b)** $y^2 - 4y + 9$ **(c)** $4x^2 + 10x + 25$ **(d)** $x^2 - 20x + 100$

Factor completely.

33. $n^2 - 49$

34. $25b^2 - 121$

35. $49y^2 - 25w^2$

36. $144p^2 - 36q^2$

37. $x^2 + 100$

38. $z^2 + 10z + 25$

39. $r^2 - 12r + 36$

40. $9t^2 - 42t + 49$

41. $16m^2 + 40mn + 25n^2$

42. $54x^3 - 72x^2 + 24x$

[7.5] *Solve each equation and check the solutions.*

43. $(4t + 3)(t - 1) = 0$

44. $(x + 7)(x - 4)(x + 3) = 0$

45. $x(2x - 5) = 0$

46. $z^2 + 4z + 3 = 0$

47. $m^2 - 5m + 4 = 0$

48. $x^2 = -15 + 8x$

49. $3z^2 - 11z - 20 = 0$

50. $81t^2 - 64 = 0$

51. $y^2 = 8y$

52. $n(n - 5) = 6$

53. $t^2 - 14t + 49 = 0$

54. $t^2 = 12(t - 3)$

55. $(5z + 2)(z^2 + 3z + 2) = 0$

56. $x^2 = 9$

[7.6] *Solve each problem.*

57. The length of a rug is 6 feet more than the width. The area is 40 square feet. Find the length and width of the rug.

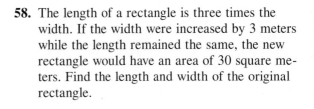

58. The length of a rectangle is three times the width. If the width were increased by 3 meters while the length remained the same, the new rectangle would have an area of 30 square meters. Find the length and width of the original rectangle.

59. The length of a rectangle is 2 centimeters more than the width. The area is numerically 44 more than the perimeter. Find the length and width of the rectangle.

60. The volume of a box is to be 120 cubic meters. The width of the box is to be 4 meters, and the height 1 meter less than the length. Find the length and height of the box.

61. Two cars left an intersection at the same time. One traveled west, and the other traveled 14 miles less, but to the south. How far apart were they then, if the distance between them was 16 miles more than the distance traveled south?

62. The surface area S of a box is given by

$$S = 2WH + 2WL + 2LH.$$

A treasure chest from a sunken galleon has dimensions as shown in the figure. Its surface area is 650 square feet. Find its width.

63. The product of two consecutive integers is 29 more than their sum. What are the integers?

64. The sides of a right triangle have lengths (in inches) that are consecutive even integers. What are the lengths of the sides?

If an object is thrown straight up with an initial velocity of 128 feet per second, its height after t seconds is

$$h(t) = 128t - 16t^2.$$

Find the height of the object after the following periods of time.

65. 1 second

66. 2 seconds

67. 4 seconds

68. For the object described above, when does it return to the ground?

69. A 9-inch by 12-inch picture is to be placed on a cardboard mat so that there is an equal border around the picture. The area of the finished mat and picture is to be 208 square inches. How wide will the border be?

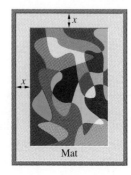

Mat

70. A box is made from a 12-centimeter by 10-centimeter piece of cardboard by cutting a square from each corner and folding up the sides. The area of the bottom of the box is to be 48 square centimeters. Find the length of a side of the cutout squares.

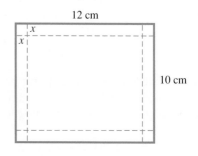

MIXED REVIEW EXERCISES

Factor completely.

71. $z^2 - 11zx + 10x^2$

72. $3k^2 + 11k + 10$

73. $15m^2 + 20mp - 12mp - 16p^2$

74. $y^4 - 625$

75. $6m^3 - 21m^2 - 45m$

76. $24ab^3c^2 - 56a^2bc^3 + 72a^2b^2c$

77. $25a^2 + 15ab + 9b^2$

78. $12x^2yz^3 + 12xy^2z - 30x^3y^2z^4$

79. $2a^5 - 8a^4 - 24a^3$

80. $12r^2 + 18rq - 10rq - 15q^2$

81. $100a^2 - 9$

82. $49t^2 + 56t + 16$

Solve.

83. $t(t - 7) = 0$

84. $x(x + 3) = 10$

85. $25x^2 + 20x + 4 = 0$

86. A lot is shaped like a right triangle. The hypotenuse is 3 meters longer than the longer leg. The longer leg is 6 meters longer than twice the length of the shorter leg. Find the lengths of the sides of the lot.

87. A pyramid has a rectangular base with a length that is 2 meters more than the width. The height of the pyramid is 6 meters, and its volume is 48 cubic meters. Find the length and width of the base.

88. The product of the smaller two of three consecutive integers is equal to 23 plus the largest. Find the integers.

89. The sum of two consecutive even integers is 34 less than their product. Find the integers.

90. The floor plan for a house is a rectangle with length 7 meters more than its width. The area is 170 square meters. Find the width and length of the house.

91. The triangular sail of a schooner has an area of 30 square meters. The height of the sail is 4 meters more than the base. Find the base of the sail.

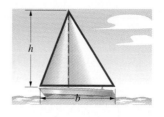

CHAPTER 7 TEST

1. Which one of the following is the correct completely factored form of $2x^2 - 2x - 24$?
   (a) $(2x + 6)(x - 4)$ **(b)** $(x + 3)(2x - 8)$
   (c) $2(x + 4)(x - 3)$ **(d)** $2(x + 3)(x - 4)$

1. _____

Factor each polynomial as completely as possible.

2. $12x^2 - 30x$

2. _____

3. $2m^3n^2 + 3m^3n - 5m^2n^2$

3. _____

4. $x^2 - 5x - 24$

4. _____

5. $x^2 - 9x + 14$

5. _____

6. $2x^2 + x - 3$

6. _____

7. $6x^2 - 19x - 7$

7. _____

8. $3x^2 - 12x - 15$

8. _____

9. $10z^2 - 17z + 3$

9. _____

10. $t^2 + 2t + 3$

10. _____

11. $x^2 + 36$

11. _____

12. $y^2 - 49$

12. _____

13. $9y^2 - 64$

13. _____

14. $x^2 + 16x + 64$

14. _____

15. $4x^2 - 28xy + 49y^2$

15. _____

16. $-2x^2 - 4x - 2$

16. _____

17. $6t^4 + 3t^3 - 108t^2$

17. _____

18. $4r^2 + 10rt + 25t^2$

18. _____

19. _____

19. $4t^3 + 32t^2 + 64t$

20. _____

20. $x^4 - 81$

21. _____

21. Why is $(p + 3)(p + 3)$ *not* the correct factorization of $p^2 + 9$?

Solve each equation.

22. _____

22. $(x + 3)(x - 9) = 0$

23. _____

23. $2r^2 - 13r + 6 = 0$

24. _____

24. $25x^2 - 4 = 0$

25. _____

25. $x(x - 20) = -100$

26. _____

26. $t^2 = 3t$

27. _____

27. Why isn't "$x = \frac{2}{3}$" the correct response to "Solve the equation $x^2 = \frac{4}{9}$"?

Solve each problem.

28. _____

28. The length of a rectangular flower bed is 3 feet less than twice its width. The area of the bed is 54 square feet. Find the dimensions of the flower bed.

29. _____

29. A carpenter needs to cut a brace to support a wall stud, as shown in the figure. The brace should be 7 feet less than three times the length of the stud. If the brace will be anchored on the floor 15 feet away from the stud, how long should the brace be?

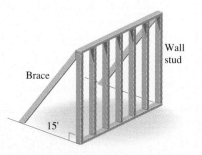

30. _____

30. Find two consecutive integers such that the square of the sum of the two integers is 11 more than the smaller integer.

1. Simplify: $\dfrac{2(3) - 1}{4(5 + 3) + 4}$.

2. Find the value of the expression $2x + 3y$ when $x = 5$ and $y = 4$.

3. True or false: Some whole numbers are not rational numbers.

4. Simplify: $\dfrac{3(-6) - 4(2)}{5^2 + 1}$.

5. Which property of real numbers justifies the statement $5(t + r) = 5t + 5r$?

6. Simplify: $-4(5r + 3) - 6[2 + 3(r - 9)]$.

7. Solve: $\dfrac{1}{3}(p + 18) + 2p + 3 = 3p + 9$.

8. The largest of three consecutive integers is added to twice the middle integer to obtain a sum that is 18 more than the smallest. What are the three consecutive integers?

9. The formula for the volume of a pyramid is $V = (1/3)Bh$, where B is the area of the base and h is the height. Find the volume of a pyramid with base area 100 square feet and height 150 feet.

10. A recipe for green salad for 35 people calls for 9 heads of lettuce. How many heads of lettuce would be needed for 105 people?

11. Solve: $x + 4(2x - 1) \geq x$.

12. Solve the following compound inequality.

$$6x - 4 < 2x \quad \text{or} \quad -4x \leq -12$$

13. Solve: $|3 - 2x| \leq 5$.

14. Solve: $|5 - 6x| = |x - 1|$.

15. The graph of $x = 4$ in the rectangular coordinate plane is a _____ line.
    (horizontal/vertical)

16. A line that slants from upper left to lower right has a slope that is a _____ number.
    (positive/negative)

17. Sketch the graph of $y = -2x + 5$.

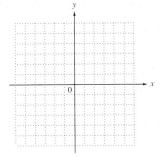

18. Sketch the graph of $y \leq -2x + 5$.

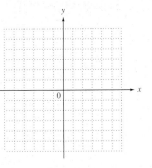

19. If $f(x) = 3x^2 - 2x + 6$ and $g(x) = 4x - 12$, find $f(3) - g(2)$.

20. Suppose that f varies jointly as g^2 and h, and $f = 50$ when $g = 4$ and $h = 2$. Find f when $g = 3$ and $h = 6$.

21. Solve the system.

$$4x + 8y + \ z = 2$$
$$3x + 4y - \ z = -11$$
$$2x - 3y + 2z = 3$$

22. Margaret Maggio has a collection of pennies, dimes, and quarters. The number of dimes is one less than twice the number of pennies. If there are 27 coins in all worth a total of $4.20, how many of each denomination does she have?

23. Find the value of the determinant.

$$\begin{vmatrix} 2 & -5 & 6 \\ 0 & 2 & 1 \\ -4 & 0 & 3 \end{vmatrix}$$

24. Solve by Cramer's rule.

$$4x - 3y = -19$$
$$2x + \ y = \ 13$$

25. Write with only positive exponents.

$$(4x^{-2}y^3z^{-1})(x^{-3}y^2z^{-4})^{-2}$$

26. Write in standard notation.

$$4.79 \times 10^{-4}$$

27. Subtract.

$$\begin{array}{r} -6x^3 - 4x^2 + 3x - 9 \\ \underline{12x^3 + 7x^2 - 2x + 3} \end{array}$$

28. Expand the binomial.

$$(4x^2 - 3y)^3$$

29. Divide $27x^3 + 64$ by $3x + 4$.

30. Factor: $24x^6 - 12x^5 - 30x^3$.

31. Factor: $20m^2 + 17m + 3$.

32. Factor: $81x^4 - 16y^4$.

33. Factor: $4x^2 - 20xy + 25y^2$.

34. Solve: $(3x + 2)(x - 4)(2x + 9) = 0$.

35. Solve: $3(x^2 + 4) - 20x = 0$.

36. If a right triangle has one leg measuring 15 inches and hypotenuse measuring 25 inches, what is the length of the other leg?

37. Solve for t: $at^2 = 36$, $a \neq 0$.

38. The product of two consecutive integers is 55 more than their sum. Find the integers.

39. The distance traveled by an object is given by the function f, where $f(t) = 16t^2$. Here t is in seconds and $f(t)$ is in feet. After how many seconds has the object traveled 144 feet?

40. (Refer to the formula in Exercise 9.) Suppose a pyramid has a rectangular base whose width is 4 inches less than its length. If the height is 48 inches and the volume is 720 cubic inches, find the length of the base.

Rational Expressions

8.1 The Fundamental Property of Rational Expressions

The quotient of two integers (with denominator not zero) is called a rational number. In the same way, the quotient of two polynomials with denominator not equal to zero is called a **rational expression.** Our work with rational expressions will require much of what we learned in Chapters 6 and 7 on polynomials and factoring, as well as the rules for fractions from basic arithmetic.

> **A rational expression** is an expression of the form
>
> $$\frac{P}{Q},$$
>
> where P and Q are polynomials, with $Q \neq 0$.

Examples of rational expressions include

$$\frac{-6x}{x^3 + 8}, \quad \frac{9x}{y + 3}, \quad \text{and} \quad \frac{2m^3}{8}.$$

OBJECTIVE 1 ▶ A quotient with a zero denominator is *not* a rational expression because division by zero is undefined. For that reason, be careful when substituting a number in the denominator of a rational expression. For example, in

$$\frac{8x^2}{x - 3}$$

the variable x can take on any value except 3. When $x = 3$ the denominator becomes $3 - 3 = 0$, making the quotient undefined.

EXAMPLE 1 Finding Values That Make a Rational Expression Undefined

Find any values of the variable for which the following are undefined.

(a) $\dfrac{p + 5}{3p + 2}$

CONTINUED ON NEXT PAGE

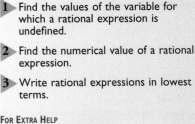

www.mathnotes.com

OBJECTIVES

1 ▶ Find the values of the variable for which a rational expression is undefined.

2 ▶ Find the numerical value of a rational expression.

3 ▶ Write rational expressions in lowest terms.

FOR EXTRA HELP

Tutorial   Tape 16   SSM, Sec. 8.1

1. Find all values for which the following rational expressions are undefined.

(a) $\dfrac{x + 2}{x - 5}$

(b) $\dfrac{3r}{r^2 + 6r + 8}$

(c) $\dfrac{-5m}{m^2 + 4}$

This rational expression is undefined for any value of p that makes the denominator equal to zero. We find these values by solving the equation $3p + 2 = 0$ to get

$$3p = -2 \quad \text{or} \quad p = -\dfrac{2}{3}.$$

Since $p = -\frac{2}{3}$ will make the denominator zero, the given expression is undefined for $-\frac{2}{3}$.

(b) $\dfrac{9m^2}{m^2 - 5m + 6}$

Solve the equation $m^2 - 5m + 6 = 0$.

$(m - 2)(m - 3) = 0$ Factor.

$m - 2 = 0 \quad \text{or} \quad m - 3 = 0$ Set each factor equal to 0.

$m = 2 \quad \text{or} \quad m = 3$

The original fraction is undefined for $m = 2$ and for $m = 3$.

(c) $\dfrac{2r}{r^2 + 1}$

This denominator cannot equal zero for any value of r because r^2 is always greater than or equal to zero, and adding 1 makes the sum greater than zero. Thus, there are no values for which this rational expression is undefined.

◀◀ **WORK PROBLEM 1 AT THE SIDE.**

OBJECTIVE ▶2 The next example shows how to find the numerical value of a rational expression for a given value of the variable.

E X A M P L E   2   **Evaluating a Rational Expression**

Find the numerical value of $\dfrac{3x + 6}{2x - 4}$ for each of the following values of x.

(a) $x = 1$

$$\dfrac{3x + 6}{2x - 4} = \dfrac{3(1) + 6}{2(1) - 4} \quad \text{Let } x = 1.$$

$$= \dfrac{9}{-2} = -\dfrac{9}{2}$$

(b) $x = 2$

$$\dfrac{3x + 6}{2x - 4} = \dfrac{3(2) + 6}{2(2) - 4} = \dfrac{6 + 6}{4 - 4} = \dfrac{12}{0} \quad \text{Let } x = 2.$$

Substituting 2 for x makes the denominator zero, so the expression is undefined when $x = 2$.

2. Find the value of each rational expression when $x = 3$.

(a) $\dfrac{x}{2x + 1}$

(b) $\dfrac{2x + 6}{x - 3}$

◀◀ **WORK PROBLEM 2 AT THE SIDE.**

ANSWERS

1. (a) 5 (b) $-4, -2$ (c) never undefined

2. (a) $\dfrac{3}{7}$ (b) undefined

OBJECTIVE 3 A rational expression represents a number for each value of the variable that does not make the denominator zero. For this reason, the properties of rational numbers from basic arithmetic also apply to rational expressions. For example, the **fundamental property of rational expressions** permits rational expressions to be written in lowest terms. A rational expression is in **lowest terms** when there are no common factors in the numerator and the denominator (except 1).

Fundamental Property of Rational Expressions

If $\frac{P}{Q}$ is a rational expression and if K represents any factor where $K \neq 0$, then

$$\frac{PK}{QK} = \frac{P}{Q}.$$

This property is based on the identity property of multiplication:

$$\frac{PK}{QK} = \frac{P}{Q} \cdot \frac{K}{K} = \frac{P}{Q} \cdot 1 = \frac{P}{Q}.$$

The next example shows how to write both a rational number and a rational expression in lowest terms. Notice the similarity in the procedures.

EXAMPLE 3 Writing in Lowest Terms

Write in lowest terms.

(a) $\frac{30}{72}$

Begin by factoring.

$$\frac{30}{72} = \frac{2 \cdot 3 \cdot 5}{2 \cdot 2 \cdot 2 \cdot 3 \cdot 3}$$

Group any factors common to the numerator and denominator.

$$\frac{30}{72} = \frac{5 \cdot (2 \cdot 3)}{2 \cdot 2 \cdot 3 \cdot (2 \cdot 3)}$$

Use the fundamental property.

$$\frac{30}{72} = \frac{5}{2 \cdot 2 \cdot 3} = \frac{5}{12}$$

(b) $\frac{14k^2}{2k^3}$

Write k^2 as $k \cdot k$.

$$\frac{14k^2}{2k^3} = \frac{2 \cdot 7 \cdot k \cdot k}{2 \cdot k \cdot k \cdot k}$$

$$\frac{14k^2}{2k^3} = \frac{7(2 \cdot k \cdot k)}{k(2 \cdot k \cdot k)}$$

$$\frac{14k^2}{2k^3} = \frac{7}{k}$$

WORK PROBLEM 3 AT THE SIDE. ▶▶

EXAMPLE 4 Writing in Lowest Terms

Write $\frac{3x - 12}{5x - 20}$ in lowest terms.

Begin by factoring both numerator and denominator. Then use the fundamental property of rational expressions.

$$\frac{3x - 12}{5x - 20} = \frac{3(x - 4)}{5(x - 4)} = \frac{3}{5}$$

3. Use the fundamental property of rational expressions to write the following rational expressions in lowest terms.

(a) $\frac{5x^4}{15x^2}$

(b) $\frac{6p^3}{2p^2}$

ANSWERS

3. (a) $\frac{x^2}{3}$ (b) $3p$

4. Write each rational expression in lowest terms.

(a) $\dfrac{4y + 2}{6y + 3}$

| **Caution** |

Although x appears in both the numerator and denominator in Example 4, and 12 and 20 have a common factor of 4, the fundamental property cannot be used before factoring because $3x$, $5x$, 12, and 20 are *terms*, not *factors*. Terms are *added* or *subtracted;* factors are *multiplied* or *divided.* For example,

$$\frac{6 + 2}{3 + 2} = \frac{8}{5}, \quad \text{not} \quad \frac{6}{3} + \frac{2}{2} = 2 + 1 = 3.$$

Also, $\quad \dfrac{2x + 3}{4x + 6} = \dfrac{2x + 3}{2(2x + 3)} = \dfrac{1}{2}, \quad$ but $\quad \dfrac{x^2 + 6}{x + 3} \neq x + 2.$

◄◄ **WORK PROBLEM 4 AT THE SIDE.**

(b) $\dfrac{8p + 8q}{5p + 5q}$

E X A M P L E 5 Writing in Lowest Terms

Write $\dfrac{m^2 + 2m - 8}{2m^2 - m - 6}$ in lowest terms.

Always begin by factoring both numerator and denominator, if possible. Then use the fundamental property to remove common factors.

$$\frac{m^2 + 2m - 8}{2m^2 - m - 6} = \frac{(m + 4)(m - 2)}{(2m + 3)(m - 2)} = \frac{m + 4}{2m + 3}$$

◄◄ **WORK PROBLEM 5 AT THE SIDE.**

5. Write each rational expression in lowest terms.

(a) $\dfrac{x^2 + 4x + 4}{4x + 8}$

In Example 4, the rational expression $\dfrac{3x - 12}{5x - 20}$ is restricted to values of x not equal to 4. For this reason

$$\frac{3x - 12}{5x - 20} = \frac{3}{5} \quad \text{for} \quad x \neq 4.$$

Similarly, in Example 5,

$$\frac{m^2 + 2m - 8}{2m^2 - m - 6} = \frac{m + 4}{2m + 3} \quad \text{for} \quad m \neq -\frac{3}{2} \text{ or } 2.$$

From now on we will assume that such restrictions are understood without actually writing them down.

(b) $\dfrac{a^2 - b^2}{a^2 + 2ab + b^2}$

E X A M P L E 6 Writing in Lowest Terms

Write $\dfrac{x - y}{y - x}$ in lowest terms.

At first glance, there does not seem to be any way in which $x - y$ and $y - x$ can be factored to get a common factor. However, $y - x$ can be factored as

$$y - x = -1(-y + x) = -1(x - y).$$

With these factors, use the fundamental property to simplify the rational expression.

$$\frac{x - y}{y - x} = \frac{1(x - y)}{-1(x - y)} = \frac{1}{-1} = -1$$

ANSWERS

4. (a) $\dfrac{2}{3}$ (b) $\dfrac{8}{5}$

5. (a) $\dfrac{x + 2}{4}$ (b) $\dfrac{a - b}{a + b}$

In Example 6, notice that $y - x$ is the negative or opposite of $x - y$. A general rule for this situation follows.

A fraction with numerator and denominator that are negatives equals -1.

Caution

Although x and y appear in both the numerator and denominator in Example 6, it is not possible to use the fundamental property right away because they are *terms*, not *factors*. Terms are *added*, while factors are *multiplied*.

E X A M P L E   7   Writing in Lowest Terms

Write each rational expression in lowest terms.

(a) $\dfrac{2 - m}{m - 2}$

Since $2 - m$ is the negative of $m - 2$ (or $-2 + m$),

$$\frac{2 - m}{m - 2} = -1.$$

(b) $\dfrac{3 + r}{3 - r}$

The quantity $3 - r$ *is not* the negative of $3 + r$. This rational expression cannot be written in simpler form.

WORK PROBLEM 6 AT THE SIDE. ▶▶

6. Write each rational expression in lowest terms.

(a) $\dfrac{5 - y}{y - 5}$

(b) $\dfrac{m - n}{n - m}$

(c) $\dfrac{9 - k}{9 + k}$

ANSWERS

6. (a) -1 **(b)** -1
   (c) cannot be written in simpler form

Numbers in the Real World

Real World *collaborative investigations*

The Dangers of Ultraviolet Rays

Sunbathing is a popular pastime, but excessive exposure to ultraviolet light can be responsible for both skin and eye damage. Ultraviolet light from the sun is responsible for both the tanning and burning of exposed skin. Only about 6 percent of solar radiation reaching the earth is in the form of ultraviolet light. The intensity of ultraviolet light is affected by seasonal changes in the ozone layer, cloud cover, and time of day. The table shows maximum ultraviolet intensity, measured in milliwatts per square meter for various latitudes and dates.

Latitude	Mar. 12	June 21	Sept. 21	Dec. 21
0°	325	254	325	272
10°	311	275	280	220
20°	249	292	256	143
30°	179	248	182	80
40°	99	199	127	34
50°	57	143	75	13

47.5° Seattle

20° Hawaii

If a student from Seattle, located at a latitude of 47.5°, spends spring break in Hawaii with a latitude of 20°, the sun's ultraviolet rays in Hawaii will be approximately 249/57 ≈ 4.37 times more intense than in Seattle. Hawaii's spring sun is approximately 249/143 ≈ 1.74 times stronger than Seattle's most intense sun in the summer. Because ultraviolet light is scattered through reflection, clouds and shade do not stop sunburn completely. A person sitting in the shade can receive 40 percent of the ultraviolet light occurring in sunlit areas. Thin clouds transmit 80 percent of ultraviolet light while a person swimming $1\frac{1}{2}$ feet below the surface of the water also receives 80 percent of the sun's ultraviolet radiation. So, even sitting in the shade our Seattle student will still feel a more intense sun in Hawaii.

FOR GROUP DISCUSSION

1. How much more intense is the sun's ultraviolet rays in Hawaii than Seattle on June 21?

2. Repeat Exercise 1 for September 21.

3. Repeat Exercise 1 for December 21.

8.2 Multiplication and Division of Rational Expressions

OBJECTIVES

1. Multiply rational expressions.
2. Find reciprocals.
3. Divide rational expressions.

FOR EXTRA HELP

Tutorial   Tape 16   SSM, Sec. 8.2

OBJECTIVE 1 The product of two fractions is found by multiplying the numerators and multiplying the denominators. Rational expressions are multiplied in the same way.

Multiplying Rational Expressions

The product of the rational expressions P/Q and R/S is

$$\frac{P}{Q} \cdot \frac{R}{S} = \frac{PR}{QS}.$$

In words: Multiply the numerators and multiply the denominators.

The next example shows the multiplication of both two rational numbers and two rational expressions. This parallel discussion lets you compare the steps.

EXAMPLE 1 Multiplying Rational Expressions

Multiply. Write answers in lowest terms.

(a) $\dfrac{3}{10} \cdot \dfrac{5}{9}$ | (b) $\dfrac{6}{x} \cdot \dfrac{x^2}{12}$

Find the product of the numerators and the denominators.

$$\frac{3}{10} \cdot \frac{5}{9} = \frac{3 \cdot 5}{10 \cdot 9} \qquad\qquad \frac{6}{x} \cdot \frac{x^2}{12} = \frac{6 \cdot x^2}{x \cdot 12}$$

Factor the numerator and denominator to identify any common factors. Then use the fundamental property to write each product in lowest terms.

$$\frac{3}{10} \cdot \frac{5}{9} = \frac{3 \cdot 5}{2 \cdot 5 \cdot 3 \cdot 3} = \frac{1}{6} \qquad\quad \frac{6}{x} \cdot \frac{x^2}{12} = \frac{6 \cdot x \cdot x}{2 \cdot 6 \cdot x} = \frac{x}{2}$$

Notice in the second step above that the products were left in factored form because common factors are needed to write the product in lowest terms.

> **WORK PROBLEM 1 AT THE SIDE.** ▶▶

EXAMPLE 2 Multiplying Rational Expressions

Find the product of $\dfrac{x + y}{2x}$ and $\dfrac{x^2}{(x + y)^2}$.

Use the definition of multiplication.

$$\frac{x + y}{2x} \cdot \frac{x^2}{(x + y)^2} = \frac{(x + y)x^2}{2x(x + y)^2} \qquad \text{Multiply numerators.}$$
$$\text{Multiply denominators.}$$

$$= \frac{(x + y)x \cdot x}{2x(x + y)(x + y)} \qquad \text{Factor; identify common factors.}$$

$$= \frac{x}{2(x + y)} \qquad\qquad \text{Lowest terms}$$

> **WORK PROBLEM 2 AT THE SIDE.** ▶▶

1. Multiply.

(a) $\dfrac{3m^2}{2} \cdot \dfrac{10}{m}$

(b) $\dfrac{8p^2 q}{3} \cdot \dfrac{9}{q^2 p}$

2. Multiply.

(a) $\dfrac{a + b}{5} \cdot \dfrac{30}{2(a + b)}$

(b) $\dfrac{3(p - q)}{p} \cdot \dfrac{q}{2(p - q)}$

ANSWERS

1. (a) $15m$ (b) $\dfrac{24p}{q}$

2. (a) 3 (b) $\dfrac{3q}{2p}$

3. Multiply.

(a)

$$\frac{x^2 + 7x + 10}{3x + 6} \cdot \frac{6x - 6}{x^2 + 2x - 15}$$

(b)

$$\frac{m^2 + 4m - 5}{m + 5} \cdot \frac{m^2 + 8m + 15}{m - 1}$$

4. Find the reciprocal.

(a) $\dfrac{6b^5}{3r^2b}$

(b) $\dfrac{t^2 - 4t}{t^2 + 2t - 3}$

E X A M P L E   3   Multiplying Rational Expressions

Find the product of $\dfrac{x^2 + 3x}{x^2 - 3x - 4}$ and $\dfrac{x^2 - 5x + 4}{x^2 + 2x - 3}$.

Use the definition of multiplication. Factor the numerators and denominators wherever possible. Use the fundamental property to write the product in lowest terms.

$$\frac{x^2 + 3x}{x^2 - 3x - 4} \cdot \frac{x^2 - 5x + 4}{x^2 + 2x - 3} = \frac{(x^2 + 3x)(x^2 - 5x + 4)}{(x^2 - 3x - 4)(x^2 + 2x - 3)}$$

$$= \frac{x(x + 3)(x - 4)(x - 1)}{(x - 4)(x + 1)(x + 3)(x - 1)}$$

$$= \frac{x}{x + 1}$$

■

◀◀ **WORK PROBLEM 3 AT THE SIDE.**

The *reciprocal* of the fraction c/d is d/c. In arithmetic, we learn that the product of two reciprocals is 1. To divide by the nonzero fraction c/d, we multiply by the reciprocal of c/d. Division of rational expressions is defined in the same way.

> **Dividing Rational Expressions**
>
> If P/Q and R/S are any two rational expressions, with $R/S \neq 0$, then
>
> $$\frac{P}{Q} \div \frac{R}{S} = \frac{P}{Q} \cdot \frac{S}{R} = \frac{PS}{QR}.$$
>
> In words: Multiply the first rational expression by the reciprocal of the second rational expression.

OBJECTIVE 2▶ The reciprocal of a rational expression is found by inverting the fraction. For example, the reciprocal of $\dfrac{2x - 1}{x - 5}$ is $\dfrac{x - 5}{2x - 1}$.

E X A M P L E   4   Finding the Reciprocal of Rational Expressions

Find the reciprocal of each rational expression.

(a) $\dfrac{4p^3}{9q}$

Invert the rational expression. The reciprocal is $\dfrac{9q}{4p^3}$.

(b) $\dfrac{k^2 - 9}{k^2 - k - 20}$

The reciprocal is $\dfrac{k^2 - k - 20}{k^2 - 9}$.

■

◀◀ **WORK PROBLEM 4 AT THE SIDE.**

ANSWERS

3. (a) $\dfrac{2(x - 1)}{x - 3}$ **(b)** $(m + 5)(m + 3)$

4. (a) $\dfrac{3r^2b}{6b^5}$ **(b)** $\dfrac{t^2 + 2t - 3}{t^2 - 4t}$

OBJECTIVE ▶3▶ The next example shows the division of two rational numbers and the division of two rational expressions.

E X A M P L E 5 **Dividing Rational Expressions**

Divide. Write answers in lowest terms.

(a) $\dfrac{5}{8} \div \dfrac{7}{16}$

(b) $\dfrac{y}{y+3} \div \dfrac{4y}{y+5}$

Multiply the first expression and the reciprocal of the second.

$$\dfrac{5}{8} \div \dfrac{7}{16} = \dfrac{5}{8} \cdot \dfrac{\mathbf{16}}{\mathbf{7}} \qquad \begin{array}{l}\text{Reciprocal}\\\text{of } \frac{7}{16}\end{array}$$

$$= \dfrac{5 \cdot 16}{8 \cdot 7}$$

$$= \dfrac{5 \cdot 8 \cdot 2}{8 \cdot 7}$$

$$= \dfrac{10}{7}$$

$$\dfrac{y}{y+3} \div \dfrac{4y}{y+5}$$

$$= \dfrac{y}{y+3} \cdot \dfrac{\mathbf{y+5}}{\mathbf{4y}} \qquad \begin{array}{l}\text{Reciprocal}\\\text{of } \frac{4y}{y+5}\end{array}$$

$$= \dfrac{y(y+5)}{(y+3)(4y)}$$

$$= \dfrac{y+5}{4(y+3)}$$

> **WORK PROBLEM 5 AT THE SIDE.** ▶▶

E X A M P L E 6 **Dividing Rational Expressions**

Divide: $\dfrac{(3m)^2}{(2p)^3} \div \dfrac{6m^3}{16p^2}$.

$$\dfrac{(3m)^2}{(2p)^3} \div \dfrac{6m^3}{16p^2} = \dfrac{(3m)^2}{(2p)^3} \cdot \dfrac{16p^2}{6m^3} \qquad \text{Multiply by the reciprocal.}$$

$$= \dfrac{(3m)(3m)}{(2p)(2p)(2p)} \cdot \dfrac{16p^2}{6m^3} \qquad \text{Factor.}$$

$$= \dfrac{9 \cdot 16m^2 p^2}{8 \cdot 6 p^3 m^3} \qquad \begin{array}{l}\text{Multiply numerators.}\\\text{Multiply denominators.}\end{array}$$

$$= \dfrac{3}{mp} \qquad \text{Lowest terms}$$

> **WORK PROBLEM 6 AT THE SIDE.** ▶▶

E X A M P L E 7 **Dividing Rational Expressions**

Divide: $\dfrac{x^2 - 4}{(x+3)(x-2)} \div \dfrac{(x+2)(x+3)}{-2x}$.

First, use the definition of division.

$$\dfrac{x^2-4}{(x+3)(x-2)} \div \dfrac{(x+2)(x+3)}{-2x}$$

$$= \dfrac{x^2-4}{(x+3)(x-2)} \cdot \dfrac{-2x}{(x+2)(x+3)} \qquad \begin{array}{l}\text{Multiply by the reciprocal of}\\\text{second expression.}\end{array}$$

CONTINUED ON NEXT PAGE

---

5. Divide.

(a) $\dfrac{r}{r-1} \div \dfrac{3r}{r+4}$

(b) $\dfrac{6x-4}{3} \div \dfrac{15x-10}{9}$

6. Divide.

(a) $\dfrac{5a^2 b}{2} \div \dfrac{10ab^2}{8}$

(b) $\dfrac{(3t)^2}{w} \div \dfrac{3t^2}{5w^4}$

ANSWERS

5. (a) $\dfrac{r+4}{3(r-1)}$ **(b)** $\dfrac{6}{5}$

6. (a) $\dfrac{2a}{b}$ **(b)** $15w^3$

7. Divide.

(a)

$$\frac{y^2 + 4y + 3}{y + 3} \div \frac{y^2 - 4y - 5}{y - 3}$$

(b) $\dfrac{4x(x + 3)}{2x + 1} \div \dfrac{-x^2(x + 3)}{4x^2 - 1}$

8. Divide.

(a) $\dfrac{ab - a^2}{a^2 - 1} \div \dfrac{b - a}{a^2 + 2a + 1}$

(b) $\dfrac{x^2 - y^2}{x^2 - 1} \div \dfrac{x^2 + 2xy + y^2}{x^2 + x}$

Next, be sure all numerators and all denominators are factored.

$$= \frac{(x + 2)(x - 2)}{(x + 3)(x - 2)} \cdot \frac{-2x}{(x + 2)(x + 3)}$$

Now multiply numerators and denominators and simplify.

$$\frac{(x + 2)(x - 2)}{(x + 3)(x - 2)} \cdot \frac{-2x}{(x + 2)(x + 3)}$$

$$= \frac{-2x(x + 2)(x - 2)}{(x + 3)(x - 2)(x + 2)(x + 3)}$$

$$= \frac{-2x}{(x + 3)^2} = -\frac{2x}{(x + 3)^2} \qquad \text{Lowest terms}$$

In the last step, we used the fact that the quotient of a negative number and a positive number is negative to write the negative sign in front of the fraction.

◀◀ **WORK PROBLEM 7 AT THE SIDE.**

E X A M P L E 8   Dividing Rational Expressions

Divide: $\dfrac{m^2 - 4}{m^2 - 1} \div \dfrac{2m^2 + 4m}{1 - m}$.

$$\frac{m^2 - 4}{m^2 - 1} \div \frac{2m^2 + 4m}{1 - m} = \frac{m^2 - 4}{m^2 - 1} \cdot \frac{1 - m}{2m^2 + 4m} \qquad \text{Definition of division}$$

$$= \frac{(m + 2)(m - 2)}{(m + 1)(m - 1)} \cdot \frac{1 - m}{2m(m + 2)} \qquad \text{Factor.}$$

As shown in Section 8.1, $\dfrac{1 - m}{m - 1} = -1$, so

$$\frac{(m + 2)(m - 2)}{(m + 1)(m - 1)} \cdot \frac{1 - m}{2m(m + 2)} = \frac{-1(m - 2)}{2m(m + 1)}$$

$$= \frac{2 - m}{2m(m + 1)}. \qquad -1(m - 2) = 2 - m$$

◀◀ **WORK PROBLEM 8 AT THE SIDE.**

8.2 Exercises

Multiply. Write answers in lowest terms. See Examples 1 and 2.

1. $\dfrac{10m^2}{7} \cdot \dfrac{14}{15m}$

2. $\dfrac{36z^3}{6z} \cdot \dfrac{28}{z^2}$

3. $\dfrac{16y^4}{18y^5} \cdot \dfrac{15y^5}{y^2}$

4. $\dfrac{20x^5}{-2x^2} \cdot \dfrac{8x^4}{35x^3}$

5. $\dfrac{2(c + d)}{3} \cdot \dfrac{18}{6(c + d)^2}$

6. $\dfrac{4(y - 2)}{x} \cdot \dfrac{3x}{6(y - 2)^2}$

Find the reciprocal. See Example 4.

7. $\dfrac{3p^3}{16q}$

8. $\dfrac{6x^4}{9y^2}$

9. $\dfrac{r^2 + rp}{7}$

10. $\dfrac{16}{9a^2 + 36a}$

11. $\dfrac{z^2 + 7z + 12}{z^2 - 9}$

12. $\dfrac{p^2 - 4p + 3}{p^2 - 3p}$

Divide. Write answers in lowest terms. See Examples 5 and 6.

13. $\dfrac{9z^4}{3z^5} \div \dfrac{3z^2}{5z^3}$

14. $\dfrac{35q^8}{9q^5} \div \dfrac{25q^6}{10q^5}$

15. $\dfrac{4t^4}{2t^5} \div \dfrac{(2t)^3}{-6}$

16. $\dfrac{-12a^6}{3a^2} \div \dfrac{(2a)^3}{27a}$

17. $\dfrac{3}{2y - 6} \div \dfrac{6}{y - 3}$

18. $\dfrac{4m + 16}{10} \div \dfrac{3m + 12}{18}$

19. Explain in your own words how to multiply rational expressions.

20. Explain in your own words how to divide rational expressions.

Multiply or divide. Write answers in lowest terms. See Examples 3, 7, and 8.

21. $\dfrac{5x - 15}{3x + 9} \cdot \dfrac{4x + 12}{6x - 18}$

22. $\dfrac{8r + 16}{24r - 24} \cdot \dfrac{6r - 6}{3r + 6}$

23. $\dfrac{2 - t}{8} \div \dfrac{t - 2}{6}$

24. $\dfrac{4}{m-2} \div \dfrac{16}{2-m}$

25. $\dfrac{27-3z}{4} \cdot \dfrac{12}{2z-18}$

26. $\dfrac{5-x}{5+x} \cdot \dfrac{x+5}{x-5}$

27. $\dfrac{6(m-2)^2}{5(m+4)^2} \cdot \dfrac{15(m+4)}{2(2-m)}$

28. $\dfrac{7(q-1)}{3(q+1)^2} \cdot \dfrac{6(q+1)}{3(1-q)^2}$

29. $\dfrac{p^2+4p-5}{p^2+7p+10} \div \dfrac{p-1}{p+4}$

30. $\dfrac{z^2-3z+2}{z^2+4z+3} \div \dfrac{z-1}{z+1}$

31. $\dfrac{2k^2-k-1}{2k^2+5k+3} \div \dfrac{4k^2-1}{2k^2+k-3}$

32. $\dfrac{2m^2-5m-12}{m^2+m-20} \div \dfrac{4m^2-9}{m^2+4m-5}$

33. $\dfrac{2k^2+3k-2}{6k^2-7k+2} \cdot \dfrac{4k^2-5k+1}{k^2+k-2}$

34. $\dfrac{2m^2-5m-12}{m^2-10m+24} \div \dfrac{4m^2-9}{m^2-9m+18}$

35. $\dfrac{m^2+2mp-3p^2}{m^2-3mp+2p^2} \div \dfrac{m^2+4mp+3p^2}{m^2+2mp-8p^2}$

36. $\dfrac{r^2+rs-12s^2}{r^2-rs-20s^2} \div \dfrac{r^2-2rs-3s^2}{r^2+rs-30s^2}$

37. $\left(\dfrac{x^2+10x+25}{x^2+10x} \cdot \dfrac{10x}{x^2+15x+50}\right) \div \dfrac{x+5}{x+10}$

38. $\left(\dfrac{m^2-12m+32}{8m} \cdot \dfrac{m^2-8m}{m^2-8m+16}\right) \div \dfrac{m-8}{m-4}$

39. Consider the division problem $\dfrac{x-6}{x+4} \div \dfrac{x+7}{x+5}$. We know that division by 0 is undefined, so the restrictions on x are $x \neq -4$, $x \neq -5$, and $x \neq -7$. Why is the last restriction needed?

40. In the problem shown in Exercise 39, why is 6 allowed as a replacement for x, even though 6 causes a numerator to be 0?

8.3 Least Common Denominators

OBJECTIVE 1 In this section, we demonstrate a preliminary step needed to add or subtract rational expressions with different denominators. Just as with rational numbers, adding or subtracting rational expressions (to be discussed in the next section) often requires a **least common denominator,** the simplest expression that is divisible by all denominators. For example, the least common denominator for $\frac{2}{9}$ and $\frac{5}{12}$ is 36 because 36 is the smallest number divisible by both 9 and 12.

Least common denominators often can be found by inspection. For example, the least common denominator for $\frac{1}{6}$ and $\frac{2}{3m}$ is $6m$. In other cases, a least common denominator can be found by a procedure similar to that used in Chapter 7 for finding the greatest common factor.

Finding a Least Common Denominator

1. Factor each denominator into prime factors.
2. List each different denominator factor the *greatest* number of times it appears in any denominator.
3. Multiply the denominators from Step 2 to get the least common denominator.

When each denominator is factored into prime factors, every prime factor must be a factor of the least common denominator. The least common denominator is often abbreviated LCD.

In Example 1, the least common denominator is found for both numerical denominators and algebraic denominators.

EXAMPLE 1　Finding the Least Common Denominator

Find the least common denominator for each pair of fractions.

(a) $\dfrac{1}{24}, \dfrac{7}{15}$ 　　　　**(b)** $\dfrac{1}{8x}, \dfrac{3}{10x}$

Write each denominator in factored form, with numerical coefficients in prime factored form.

$$24 = 2 \cdot 2 \cdot 2 \cdot 3 = 2^3 \cdot 3 \qquad 8x = 2 \cdot 2 \cdot 2 \cdot x = 2^3 \cdot x$$
$$15 = 3 \cdot 5 \qquad\qquad\qquad 10x = 2 \cdot 5 \cdot x$$

The LCD is found by taking each different factor the greatest number of times it appears as a factor in any denominator. That is, each factor must be in the LCD and raised to its highest power.

$$\text{LCD} = 2 \cdot 2 \cdot 2 \cdot 3 \cdot 5 \qquad \text{LCD} = 2 \cdot 2 \cdot 2 \cdot 5 \cdot x$$
$$= 2^3 \cdot 3 \cdot 5 = 120 \qquad\quad = 2^3 \cdot 5 \cdot x = 40x$$

■

WORK PROBLEM 1 AT THE SIDE. ▶▶

OBJECTIVES

1 ▶ Find least common denominators.
2 ▶ Rewrite rational expressions with the least common denominator.

FOR EXTRA HELP

Tutorial　　　　Tape 16　　　SSM, Sec. 8.3

1. Find the least common denominator.

(a) $\dfrac{7}{20p}, \dfrac{11}{30p}$

(b) $\dfrac{9}{8m^4}, \dfrac{11}{12m^6}$

ANSWERS

1. (a) $60p$ (b) $24m^6$

485

2. Find the LCD.

(a) $\dfrac{4}{16m^3 n}, \dfrac{5}{9m^5}$

(b) $\dfrac{3}{25a^2}, \dfrac{2}{10a^3 b}$

3. Find the LCD.

(a) $\dfrac{7}{3a}, \dfrac{5}{3a - 10}$

(b) $\dfrac{1}{12a}, \dfrac{5}{a^2 - 4a}$

(c)

$\dfrac{2m}{m^2 - 3m + 2}, \dfrac{5m - 3}{m^2 + 3m - 10}$

(d) $\dfrac{6}{x - 4}, \dfrac{3x - 1}{4 - x}$

E X A M P L E 2   Finding the LCD

Find the LCD for $\dfrac{5}{6r^2}$ and $\dfrac{3}{4r^3}$.

Factor each denominator.

$$6r^2 = 2 \cdot 3 \cdot r^2$$
$$4r^3 = 2^2 \cdot r^3$$

The highest power of 2 is 2, the highest power of 3 is 1, and the highest power of r is 3; therefore,

$$\text{LCD} = 2^2 \cdot 3^1 \cdot r^3 = 12r^3.$$

■

◀◀ **WORK PROBLEM 2 AT THE SIDE.**

E X A M P L E 3   Finding the LCD

Find the LCD.

(a) $\dfrac{6}{5m}, \dfrac{4}{m^2 - 3m}$

Factor each denominator.

$$5m = 5 \cdot m$$
$$m^2 - 3m = m(m - 3)$$
$$\text{LCD} = 5 \cdot m \cdot (m - 3) = 5m(m - 3)$$

Because m is not a *factor* of $m - 3$, both factors, m and $m - 3$, must appear in the least common denominator.

(b) $\dfrac{1}{r^2 - 4r - 5}, \dfrac{3}{r^2 - r - 20}$

Factor each denominator.

$$r^2 - 4r - 5 = (r - 5)(r + 1)$$
$$r^2 - r - 20 = (r - 5)(r + 4)$$

The LCD is $(r - 5)(r + 1)(r + 4)$.

(c) $\dfrac{1}{q - 5}, \dfrac{3}{5 - q}$

The expressions $q - 5$ and $5 - q$ are negatives of each other because

$$-(q - 5) = -q + 5 = 5 - q.$$

Therefore, either $q - 5$ or $5 - q$ can be used as the LCD.

■

◀◀ **WORK PROBLEM 3 AT THE SIDE.**

OBJECTIVE ▶2▶ Once the least common denominator has been found, the next step in preparing two fractions for addition or subtraction is to use the fundamental property to rewrite each fraction with the least common denominator. The next example shows how to do this with both numerical and algebraic fractions.

ANSWERS

2. (a) $144m^5 n$ **(b)** $50a^3 b$
3. (a) $3a(3a - 10)$
  (b) $12a(a - 4)$
  (c) $(m - 1)(m - 2)(m + 5)$
  (d) either $x - 4$ or $4 - x$

EXAMPLE 4 Writing a Fraction with a Given Denominator

Rewrite each rational expression with the indicated denominator.

(a) $\dfrac{3}{8} = \dfrac{}{40}$

(b) $\dfrac{9k}{25} = \dfrac{}{50k}$

For each example, first factor the denominator on the right. Then compare the denominator on the left with the one on the right to decide what factors are missing. (It may be necessary to factor both denominators.)

$\dfrac{3}{8} = \dfrac{}{5 \cdot 8}$

$\dfrac{9k}{25} = \dfrac{}{25 \cdot 2k}$

A factor of 5 is missing. Using the property of 1, multiply $\frac{3}{8}$ by $\frac{5}{5}$.

Factors of 2 and k are missing. Multiply by $\frac{2k}{2k}$.

$\dfrac{3}{8} = \dfrac{3}{8} \cdot \dfrac{5}{5} = \dfrac{15}{40}$

$\dfrac{9k}{25} = \dfrac{9k}{25} \cdot \dfrac{2k}{2k} = \dfrac{18k^2}{50k}$

$\dfrac{5}{5} = 1$ ⬑

$\dfrac{2k}{2k} = 1$ ⬑

EXAMPLE 5 Writing a Fraction with a Given Denominator

Rewrite the following rational expression with the indicated denominator.

$$\dfrac{12p}{p^2 + 8p} = \dfrac{}{p(p+8)(p-4)}$$

Factor $p^2 + 8p$ as $p(p + 8)$. Compare with the denominator on the right. The factor $p - 4$ is missing, so multiply $\dfrac{12p}{p(p+8)}$ by $\dfrac{p-4}{p-4}$.

$$\dfrac{12p}{p^2 + 8p} = \dfrac{12p}{p(p+8)} \cdot \dfrac{p-4}{p-4} = \dfrac{12p(p-4)}{p(p+8)(p-4)}$$

WORK PROBLEM 4 AT THE SIDE. ▶▶

4. Rewrite each rational expression with the indicated denominator.

(a) $\dfrac{7k}{5} = \dfrac{}{30p}$

(b) $\dfrac{9}{2a + 5} = \dfrac{}{6a + 15}$

(c)

$\dfrac{5k + 1}{k^2 + 2k} = \dfrac{}{k(k + 2)(k - 1)}$

ANSWERS

4. (a) $\dfrac{42kp}{30p}$ **(b)** $\dfrac{27}{6a + 15}$

(c) $\dfrac{(5k + 1)(k - 1)}{k(k + 2)(k - 1)}$

NUMBERS IN THE
Real World *collaborative investigations*

Statistics Require Careful Interpretation

In baseball statistics, a player's **batting average** gives the average number of hits per time at bat. For example, a player who has gotten 84 hits in 250 times at bat has a batting average of 84/250 = .336. This average can be interpreted as the empirical probability of that player's getting a hit the next time at bat.

The following are actual comparisons of hits and at-bats for two major league players in the 1989 and 1990 seasons. The numbers illustrate a puzzling statistical occurrence known as **Simpson's paradox.** (This information was reported by Richard J. Friedlander on page 845 of the November 1992 issue of the *MAA Journal*.)

	Dave Justice			Andy Van Slyke		
	Hits	At-bats	Batting Average	Hits	At-bats	Batting Average
1989	12	51	_____	113	476	_____
1990	124	439	_____	140	493	_____
Combined (1989–90)	_____	_____	_____	_____	_____	_____

FOR GROUP DISCUSSION

1. Fill in the ten blanks in the table, giving batting averages to three decimal places.

2. Which player had a better average in 1989?

3. Which player had a better average in 1990?

4. Which player had a better average in 1989 and 1990 combined?

5. Did the results above surprise you? How can it be that one player's batting average leads another's for each of two years, and yet trails the other's for the combined years?

8.3 Exercises

Decide whether the statement is true or false.

1. The LCD for the two fractions $\frac{1}{a}$ and $\frac{1}{b}$ is ab if the greatest common factor of a and b is 1.

2. If a is a factor of b, then the LCD for $\frac{1}{a}$ and $\frac{1}{b}$ is b.

3. If x^a and x^b are denominators of two fractions and $a < b$, then x^a is the LCD.

4. If a fraction with denominator $x - 4$ must be written as an equivalent fraction with denominator $(x - 4)^3$, then the original fraction must be multiplied by $x - 4$ in both the numerator and denominator.

Find the least common denominator for each list. See Examples 1–3.

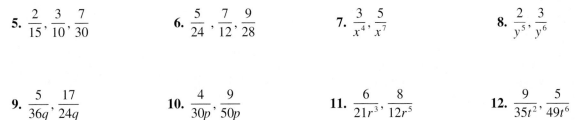

5. $\dfrac{2}{15}, \dfrac{3}{10}, \dfrac{7}{30}$

6. $\dfrac{5}{24}, \dfrac{7}{12}, \dfrac{9}{28}$

7. $\dfrac{3}{x^4}, \dfrac{5}{x^7}$

8. $\dfrac{2}{y^5}, \dfrac{3}{y^6}$

9. $\dfrac{5}{36q}, \dfrac{17}{24q}$

10. $\dfrac{4}{30p}, \dfrac{9}{50p}$

11. $\dfrac{6}{21r^3}, \dfrac{8}{12r^5}$

12. $\dfrac{9}{35t^2}, \dfrac{5}{49t^6}$

13. If the denominators of two fractions in prime factored form are $2^3 \cdot 3$ and $2^2 \cdot 5$, what is the factored form of their LCD?

14. Suppose two algebraic fractions have denominators $(t + 4)^3(t - 3)$ and $(t + 4)^2(t + 8)$. Find the factored form of their LCD. What is the similarity between the answers for this problem and for Exercise 13?

15. If two denominators have greatest common factor equal to 1, how can you easily find their least common denominator?

16. Suppose two fractions have denominators a^k and a^r, where k and r are natural numbers, with $k > r$. What is their least common denominator?

Find the least common denominator for each pair. See Examples 1–3.

17. $\dfrac{9}{28m^2}, \dfrac{3}{12m - 20}$

18. $\dfrac{15}{27a^3}, \dfrac{8}{9a - 45}$

19. $\dfrac{7}{5b - 10}, \dfrac{11}{6b - 12}$

20. $\dfrac{3}{7x^2 + 21x}, \dfrac{1}{5x^2 + 15x}$

21. $\dfrac{5}{c - d}, \dfrac{8}{d - c}$

22. $\dfrac{4}{y - x}, \dfrac{7}{x - y}$

23. $\dfrac{3}{k^2 + 5k}, \dfrac{2}{k^2 + 3k - 10}$

24. $\dfrac{1}{z^2 - 4z}, \dfrac{4}{z^2 - 3z - 4}$

25. $\dfrac{5}{p^2 + 8p + 15}, \dfrac{3}{p^2 - 3p - 18}$

26. $\dfrac{10}{y^2 - 10y + 21}, \dfrac{2}{y^2 - 2y - 3}$

Rewrite each rational expression with the given denominator. See Examples 4 and 5.

27. $\dfrac{4}{11} = \dfrac{}{55}$

28. $\dfrac{6}{7} = \dfrac{}{42}$

29. $\dfrac{-5}{k} = \dfrac{}{9k}$

30. $\dfrac{-3}{q} = \dfrac{}{6q}$

31. $\dfrac{13}{40y} = \dfrac{}{80y^3}$

32. $\dfrac{5}{27p} = \dfrac{}{108p^4}$

33. $\dfrac{5t^2}{6r} = \dfrac{}{42r^4}$

34. $\dfrac{8y^2}{3x} = \dfrac{}{30x^3}$

35. $\dfrac{5}{2(m + 3)} = \dfrac{}{8(m + 3)}$

36. $\dfrac{7}{4(y - 1)} = \dfrac{}{16(y - 1)}$

37. $\dfrac{-4t}{3t - 6} = \dfrac{}{6t - 12}$

38. $\dfrac{-7k}{5k + 20} = \dfrac{}{15k + 60}$

39. $\dfrac{14}{z^2 - 3z} = \dfrac{}{z(z - 3)(z - 2)}$

40. $\dfrac{12}{x(x + 4)} = \dfrac{}{x(x + 4)(x - 9)}$

41. $\dfrac{2(b - 1)}{b^2 + b} = \dfrac{}{b^3 + 3b^2 + 2b}$

42. $\dfrac{3(c + 2)}{c(c - 1)} = \dfrac{}{c^3 - 5c^2 + 4c}$

8.4 *Addition and Subtraction of Rational Expressions*

We are now ready to add and subtract rational expressions. We will need the skills developed in the previous section to find least common denominators and to write fractions with the least common denominator.

OBJECTIVE 1 We find the sum of two rational expressions with a procedure similar to the one that we use for adding two fractions in arithmetic.

Adding Rational Expressions

If P/Q and R/Q are rational expressions, then

$$\frac{P}{Q} + \frac{R}{Q} = \frac{P + R}{Q}.$$

Again, the first example shows how the addition of rational expressions compares with that of rational numbers.

E X A M P L E 1   Adding Rational Expressions with the Same Denominator

Add.

(a) $\dfrac{4}{7} + \dfrac{2}{7}$

(b) $\dfrac{3x}{x + 1} + \dfrac{2x}{x + 1}$

The denominators are the same, so the sum is found by adding the two numerators and keeping the same (common) denominator.

$$\frac{4}{7} + \frac{2}{7} = \frac{4 + 2}{7}$$

$$= \frac{6}{7}$$

$$\frac{3x}{x + 1} + \frac{2x}{x + 1} = \frac{3x + 2x}{x + 1}$$

$$= \frac{5x}{x + 1}$$

WORK PROBLEM 1 AT THE SIDE. ▶▶

OBJECTIVE 2 We use the steps given below to add two rational expressions with different denominators. These are the same steps that we use to add fractions with different denominators in arithmetic.

Adding Rational Expressions with Different Denominators

Step 1 Find the least common denominator (LCD).

Step 2 Rewrite each rational expression as an equivalent fraction with the least common denominator as the denominator.

Step 3 Add the numerators to get the numerator of the sum. The least common denominator is the denominator of the sum.

Step 4 Write the answer in lowest terms.

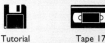
1. Find each sum.

(a) $\dfrac{3}{y + 4} + \dfrac{2}{y + 4}$

(b) $\dfrac{x}{x + y} + \dfrac{1}{x + y}$

(c) $\dfrac{a}{a + b} + \dfrac{b}{a + b}$

ANSWERS

1. (a) $\dfrac{5}{y + 4}$ **(b)** $\dfrac{x + 1}{x + y}$ **(c)** 1

491

2. Find each sum.

(a) $\dfrac{6}{5x} + \dfrac{9}{2x}$

(b) $\dfrac{m}{3n} + \dfrac{2}{7n}$

3. Find the sums.

(a) $\dfrac{2p}{3p+3} + \dfrac{5p}{2p+2}$

(b) $\dfrac{4}{y^2-1} + \dfrac{6}{y+1}$

(c) $\dfrac{-2}{p+1} + \dfrac{4p}{p^2-1}$

ANSWERS

2. (a) $\dfrac{57}{10x}$ (b) $\dfrac{7m+6}{21n}$

3. (a) $\dfrac{19p}{6(p+1)}$ (b) $\dfrac{2(3y-1)}{(y+1)(y-1)}$

(c) $\dfrac{2}{p-1}$

E X A M P L E   2   Adding Rational Expressions with Different Denominators

Add.

(a) $\dfrac{1}{12} + \dfrac{7}{15}$

First find the LCD.

$$\text{LCD} = 2^2 \cdot 3 \cdot 5 = 60$$

Now rewrite each rational expression as a fraction with the LCD, either 60 or $12y$, as the denominator.

$$\dfrac{1}{12} + \dfrac{7}{15} = \dfrac{1}{12} \cdot \dfrac{5}{5} + \dfrac{7}{15} \cdot \dfrac{4}{4}$$

$$= \dfrac{5}{60} + \dfrac{28}{60}$$

Since the fractions now have common denominators, add the numerators, and use the LCD as the denominator of the sum. Write in lowest terms.

$$\dfrac{5}{60} + \dfrac{28}{60} = \dfrac{5+28}{60}$$

$$= \dfrac{33}{60} = \dfrac{11}{20}$$

(b) $\dfrac{2}{3y} + \dfrac{1}{4y}$

$$\text{LCD} = 2^2 \cdot 3 \cdot y = 12y$$

$$\dfrac{2}{3y} + \dfrac{1}{4y} = \dfrac{2}{3y} \cdot \dfrac{4}{4} + \dfrac{1}{4y} \cdot \dfrac{3}{3}$$

$$= \dfrac{8}{12y} + \dfrac{3}{12y}$$

$$\dfrac{8}{12y} + \dfrac{3}{12y} = \dfrac{8+3}{12y}$$

$$= \dfrac{11}{12y}$$

◀◀ **WORK PROBLEM 2 AT THE SIDE.**

E X A M P L E   3   Adding Rational Expressions

Add $\dfrac{2x}{x^2-1}$ and $\dfrac{-1}{x+1}$.

Find the least common denominator by factoring both denominators.

$$x^2 - 1 = (x+1)(x-1); \qquad x+1 \text{ cannot be factored.}$$

Write the sum with denominators in factored form as

$$\dfrac{2x}{(x+1)(x-1)} + \dfrac{-1}{x+1}.$$

The LCD is $(x+1)(x-1)$. Here only the second fraction must be rewritten. Multiply the numerator and denominator of the second fraction by $x-1$.

$$\dfrac{2x}{(x+1)(x-1)} + \dfrac{-1(x-1)}{(x+1)(x-1)} \qquad \text{Multiply by } \dfrac{x-1}{x-1}.$$

With both denominators now the same, add the numerators and use the LCD as the denominator of the sum.

$$\dfrac{2x - 1(x-1)}{(x+1)(x-1)} = \dfrac{2x - x + 1}{(x+1)(x-1)} \qquad \text{Add numerators.}$$

$$= \dfrac{x+1}{(x+1)(x-1)} \qquad \text{Combine terms.}$$

$$= \dfrac{1}{x-1} \qquad \text{Lowest terms}$$

◀◀ **WORK PROBLEM 3 AT THE SIDE.**

EXAMPLE 4 Adding Rational Expressions

Add $\dfrac{2x}{x^2 + 5x + 6}$ and $\dfrac{x + 1}{x^2 + 2x - 3}$.

$$\frac{2x}{(x + 2)(x + 3)} + \frac{x + 1}{(x + 3)(x - 1)} \qquad \text{Factor denominators.}$$

The LCD is $(x + 2)(x + 3)(x - 1)$. By the fundamental property,

$$\frac{2x}{(x + 2)(x + 3)} + \frac{x + 1}{(x + 3)(x - 1)}$$

$$= \frac{2x\,(x - 1)}{(x + 2)(x + 3)(x - 1)} + \frac{(x + 1)(x + 2)}{(x + 3)(x - 1)(x + 2)}.$$

Since the two rational expressions now have the same denominator, add their numerators. The LCD is the denominator of the sum.

$$\frac{2x(x - 1)}{(x + 2)(x + 3)(x - 1)} + \frac{(x + 1)(x + 2)}{(x + 3)(x - 1)(x + 2)}$$

$$= \frac{2x(x - 1) + (x + 1)(x + 2)}{(x + 2)(x + 3)(x - 1)} \qquad \text{Add numerators.}$$

$$= \frac{2x^2 - 2x + x^2 + 3x + 2}{(x + 2)(x + 3)(x - 1)} \qquad \text{Distributive property}$$

$$= \frac{3x^2 + x + 2}{(x + 2)(x + 3)(x - 1)} \qquad \text{Combine terms.}$$

It is usually more convenient to leave the denominator in factored form. The numerator cannot be factored here, so the expression is in lowest terms. ∎

WORK PROBLEM 4 AT THE SIDE. ▶▶

OBJECTIVE **3**▶ We subtract rational expressions as follows.

> **Subtracting Rational Expressions**
>
> If P/Q and R/Q are rational expressions, then
>
> $$\frac{P}{Q} - \frac{R}{Q} = \frac{P - R}{Q}.$$

We will not show a parallel subtraction problem from arithmetic because the steps for subtraction are essentially the same as for addition.

EXAMPLE 5 Subtracting Rational Expressions

Subtract: $\dfrac{2m}{m - 1} - \dfrac{m + 3}{m - 1}$.

$$\frac{2m}{m - 1} - \frac{m + 3}{m - 1} = \frac{2m - (m + 3)}{m - 1} \qquad \text{Subtract numerators.}$$

$$= \frac{2m - m - 3}{m - 1} \qquad \text{Distributive property}$$

$$= \frac{m - 3}{m - 1} \qquad \text{Combine terms.}$$

4. Add.

(a) $\dfrac{2k}{k^2 - 5k + 4} + \dfrac{3}{k^2 - 1}$

(b)

$$\frac{4m}{m^2 + 3m + 2} + \frac{2m - 1}{m^2 + 6m + 5}$$

ANSWERS

4. (a) $\dfrac{(2k - 3)(k + 4)}{(k - 4)(k - 1)(k + 1)}$

(b) $\dfrac{6m^2 + 23m - 2}{(m + 2)(m + 1)(m + 5)}$

5. Find each difference. Write answers in lowest terms.

(a) $\dfrac{3}{m^2} - \dfrac{2}{m^2}$

(b) $\dfrac{x}{2x + 3} - \dfrac{3x + 4}{2x + 3}$

6. Subtract.

(a) $\dfrac{1}{k + 4} - \dfrac{2}{k}$

(b) $\dfrac{6}{a + 2} - \dfrac{1}{a - 3}$

7. Subtract.

(a) $\dfrac{5}{x - 1} - \dfrac{3x}{1 - x}$

(b) $\dfrac{2y}{y - 2} - \dfrac{1 + y}{2 - y}$

◀◀ WORK PROBLEM 5 AT THE SIDE.

E X A M P L E 6 Subtracting Rational Expressions

Subtract: $\dfrac{9}{x - 2} - \dfrac{3}{x}$.

The LCD is $x(x - 2)$.

$$\dfrac{9}{x - 2} - \dfrac{3}{x} = \dfrac{9x}{x(x - 2)} - \dfrac{3(x - 2)}{x(x - 2)} \quad \text{Get least common denominator.}$$

$$= \dfrac{9x - 3(x - 2)}{x(x - 2)} \quad \text{Subtract numerators; keep same denominator.}$$

$$= \dfrac{9x - 3x + 6}{x(x - 2)} \quad \text{Distributive property}$$

$$= \dfrac{6x + 6}{x(x - 2)} \quad \text{Combine terms in the numerator.}$$

$$= \dfrac{6(x + 1)}{x(x - 2)} \quad \text{Factor.}$$

◀◀ WORK PROBLEM 6 AT THE SIDE.

E X A M P L E 7 Subtracting Rational Expressions

Subtract: $\dfrac{3x}{x - 5} - \dfrac{2x - 25}{5 - x}$.

The denominators are negatives (or opposites), so either may be used as the common denominator. Let us choose $x - 5$.

$$\dfrac{3x}{x - 5} - \dfrac{2x - 25}{5 - x} = \dfrac{3x}{x - 5} - \dfrac{2x - 25}{5 - x} \cdot \dfrac{-1}{-1} \quad \text{Fundamental property}$$

$$= \dfrac{3x}{x - 5} - \dfrac{-2x + 25}{x - 5} \quad \text{Multiply.}$$

$$= \dfrac{3x - (-2x + 25)}{x - 5} \quad \text{Subtract numerators.}$$

$$= \dfrac{3x + 2x - 25}{x - 5} \quad \text{Distributive property}$$

$$= \dfrac{5x - 25}{x - 5} \quad \text{Combine terms.}$$

$$= \dfrac{5(x - 5)}{x - 5} \quad \text{Factor.}$$

$$= 5 \quad \text{Lowest terms}$$

◀◀ WORK PROBLEM 7 AT THE SIDE.

ANSWERS

5. (a) $\dfrac{1}{m^2}$ (b) $\dfrac{-2(x + 2)}{2x + 3}$

6. (a) $\dfrac{-k - 8}{k(k + 4)}$ (b) $\dfrac{5(a - 4)}{(a + 2)(a - 3)}$

7. (a) $\dfrac{5 + 3x}{x - 1}$ (b) $\dfrac{3y + 1}{y - 2}$

┌─ **E X A M P L E 8**   **Subtracting Rational Expressions**

Find $\dfrac{6x}{x^2 - 2x + 1} - \dfrac{1}{x^2 - 1}$.

Begin by factoring the denominators.

$$x^2 - 2x + 1 = (x - 1)(x - 1) \quad \text{and} \quad x^2 - 1 = (x - 1)(x + 1)$$

From the factored denominators, identify the LCD, $(x - 1)(x - 1)(x + 1)$. Use the factor $x - 1$ twice because it appears twice in the first denominator.

$$\frac{6x}{(x - 1)(x - 1)} - \frac{1}{(x - 1)(x + 1)}$$

$$= \frac{6x(x + 1)}{(x - 1)(x - 1)(x + 1)} - \frac{1(x - 1)}{(x - 1)(x - 1)(x + 1)} \qquad \text{Fundamental property}$$

$$= \frac{6x(x + 1) - 1(x - 1)}{(x - 1)(x - 1)(x + 1)} \qquad \text{Subtract numerators.}$$

$$= \frac{6x^2 + 6x - x + 1}{(x - 1)(x - 1)(x + 1)} \qquad \text{Distributive property}$$

$$= \frac{6x^2 + 5x + 1}{(x - 1)(x - 1)(x + 1)} \qquad \text{Combine terms.}$$

$$= \frac{(3x + 1)(2x + 1)}{(x - 1)^2(x + 1)} \qquad \text{Factor.}$$

WORK PROBLEM 8 AT THE SIDE. ▶▶

┌─ **E X A M P L E 9**   **Subtracting Rational Expressions**

Find $\dfrac{q}{q^2 - 4q - 5} - \dfrac{3}{2q^2 - 13q + 15}$.

To find the LCD, factor each denominator.

$$q^2 - 4q - 5 = (q + 1)(q - 5)$$
$$2q^2 - 13q + 15 = (q - 5)(2q - 3)$$

The LCD is $(q + 1)(q - 5)(2q - 3)$. Rewrite each of the two rational expressions with the LCD, using the fundamental property.

$$\frac{q}{(q + 1)(q - 5)} - \frac{3}{(q - 5)(2q - 3)}$$

$$= \frac{q(2q - 3)}{(q + 1)(q - 5)(2q - 3)} - \frac{3(q + 1)}{(q + 1)(q - 5)(2q - 3)}$$

$$= \frac{q(2q - 3) - 3(q + 1)}{(q + 1)(q - 5)(2q - 3)} \qquad \text{Subtract numerators.}$$

$$= \frac{2q^2 - 3q - 3q - 3}{(q + 1)(q - 5)(2q - 3)} \qquad \text{Distributive property}$$

$$= \frac{2q^2 - 6q - 3}{(q + 1)(q - 5)(2q - 3)} \qquad \text{Combine terms.}$$

WORK PROBLEM 9 AT THE SIDE. ▶▶

---

8. Subtract.

(a) $\dfrac{4y}{y^2 - 1} - \dfrac{5}{y^2 + 2y + 1}$

(b) $\dfrac{3r}{r^2 - 5r} - \dfrac{4}{r^2 - 10r + 25}$

9. Subtract.

(a) $\dfrac{2}{p^2 - 5p + 4} - \dfrac{3}{p^2 - 1}$

(b)

$\dfrac{q}{2q^2 + 5q - 3} - \dfrac{3q + 4}{3q^2 + 10q + 3}$

ANSWERS

8. (a) $\dfrac{4y^2 - y + 5}{(y + 1)^2(y - 1)}$ **(b)** $\dfrac{3r - 19}{(r - 5)^2}$

9. (a) $\dfrac{14 - p}{(p - 4)(p - 1)(p + 1)}$

(b) $\dfrac{(-3q + 2)(q + 2)}{(2q - 1)(q + 3)(3q + 1)}$

Number Magic

Performers like David Copperfield, Doug Henning, the late Harry Blackstone, Jr., and street magician David Blaine have at one time or another used the properties of numbers to mystify their audiences. Numerous card tricks require no sleight of hand, as they are "self-working" due to number properties. Several years ago, the Kellogg's Company included on the back of Rice Krispies boxes a variation of an old number trick, billed as the Age Detector Magic Trick. The following rectangular arrays of numbers are found on the back of the box, with directions to cut out the six cards.

32	37	42	47	52	57
33	38	43	48	53	58
34	39	44	49	54	59
35	40	45	50	55	60
36	41	46	51	56	★

1	11	21	31	41	51
3	13	23	33	43	53
5	15	25	35	45	55
7	17	27	37	47	57
9	19	29	39	49	59

2	11	22	31	42	51
3	14	23	34	43	54
6	15	26	35	46	55
7	18	27	38	47	58
10	19	30	39	50	59

16	21	26	31	52	57
17	22	27	48	53	58
18	23	28	49	54	59
19	24	29	50	55	60
20	25	30	51	56	★

4	13	22	31	44	53
5	14	23	36	45	54
6	15	28	37	46	55
7	20	29	38	47	60
12	21	30	39	52	★

8	13	26	31	44	57
9	14	27	40	45	58
10	15	28	41	46	59
11	24	29	42	47	60
12	25	30	43	56	★

Cut along dashed lines ✂

You, the performer, are to spread the six cards on a table face up, and someone in the audience is chosen to think of their age (from one to sixty). The person is then to hand you the cards with his or her age on them. Immediately you can guess the person's age. For example, if you are handed the top left and bottom right cards, you can immediately tell the person his or her age: 40. How is it done? Simply add the numbers in the upper left-hand corners of the cards you are handed. In the example given, you would add 32 + 8 to get 40. This trick works because of a property of the binary number system—a system that uses two as its base (rather than ten, as in our decimal system).

FOR GROUP DISCUSSION

Investigate the following number tricks and try to determine why they work.

1. Choose any number. Multiply it by 2. Add 8. Divide by 2. Subtract the number you started with. The answer is 4.

2. Repeat the trick in Exercise 1, but add 10 instead of 8. The answer is 5.
   (Note: In the trick above, the answer will always be $\frac{1}{2}$ the number that was added.)

3. Choose any three-digit number with all different digits. Now reverse the digits, and subtract the smaller from the larger. Repeat this process several times, choosing different digits each time. You will notice that the middle digit of the answer is always 0, and the sum of the first and last digits is always 9. Discuss how you can use this fact to perform a number trick.

Data: "Age Detector Magic Trick," as appeared on Kellogg's Rice Krispies.

8.4 Exercises

Fill in the blanks.

1. To add rational expressions with the same denominator, we add the _____ and keep the _____ .

2. To subtract two rational expressions with the same denominator, we subtract the _____ and keep the _____ .

Add or subtract as indicated. Write each answer in lowest terms. See Examples 1 and 5.

3. $\dfrac{4}{m} + \dfrac{7}{m}$

4. $\dfrac{5}{p} + \dfrac{11}{p}$

5. $\dfrac{a+b}{2} - \dfrac{a-b}{2}$

6. $\dfrac{x-y}{2} - \dfrac{x+y}{2}$

7. $\dfrac{x^2}{x+5} + \dfrac{5x}{x+5}$

8. $\dfrac{t^2}{t-3} + \dfrac{-3t}{t-3}$

9. $\dfrac{y^2 - 3y}{y+3} + \dfrac{-18}{y+3}$

10. $\dfrac{r^2 - 8r}{r-5} + \dfrac{15}{r-5}$

11. Explain how to add rational expressions with different denominators.

12. Explain how to subtract rational expressions with different denominators.

Add or subtract as indicated. Write each answer in lowest terms. See Examples 2–3 and 5–7.

13. $\dfrac{z}{5} + \dfrac{1}{3}$

14. $\dfrac{p}{8} + \dfrac{3}{5}$

15. $\dfrac{5}{7} - \dfrac{r}{2}$

16. $\dfrac{10}{9} - \dfrac{z}{3}$

17. $-\dfrac{3}{4} - \dfrac{1}{2x}$

18. $-\dfrac{5}{8} - \dfrac{3}{2a}$

19. $\dfrac{5 + 5k}{4} + \dfrac{1 + k}{8}$

20. $\dfrac{6 - 5r}{9} - \dfrac{2 - 3r}{6}$

21. $\dfrac{b + 3}{b} + \dfrac{b + 7}{3b}$

22. $\dfrac{3q - 1}{q} + \dfrac{q + 2}{4q}$

23. $\dfrac{7}{3p^2} - \dfrac{2}{p}$

24. $\dfrac{12}{5m^2} - \dfrac{2}{m}$

25. $\dfrac{1}{p - 2} - \dfrac{3}{p}$

26. $\dfrac{2}{k - 4} - \dfrac{1}{k}$

27. $\dfrac{4}{x - 5} + \dfrac{6}{5 - x}$

28. $\dfrac{10}{m - 2} + \dfrac{5}{2 - m}$

29. $\dfrac{-1}{1 - y} - \dfrac{3}{y - 1}$

30. $\dfrac{-4}{p - 3} - \dfrac{7}{3 - p}$

31. $\dfrac{6}{c - 2} - \dfrac{8}{c + 2}$

32. $\dfrac{5}{r - 3} - \dfrac{2}{r + 3}$

33. $\dfrac{2m}{m - n} - \dfrac{5m + n}{2m - 2n}$

34. $\dfrac{5p}{p - q} - \dfrac{3p + 1}{4p - 4q}$

35. $\dfrac{-2}{x^2 - 4} + \dfrac{7}{4x + 8}$

36. $\dfrac{-4}{z^2 - 16} + \dfrac{3}{2z + 8}$

37. What are the two possible LCDs that could be used for the sum $\dfrac{10}{m - 2} + \dfrac{5}{2 - m}$?

38. If one form of the correct answer to a sum or difference of rational expressions is $\dfrac{4}{k - 3}$, what would an alternate form of the answer be if the denominator is $3 - k$?

Add or subtract as indicated. Write each answer in lowest terms. See Examples 4, 8, and 9.

39. $\dfrac{1}{a^2 - 1} - \dfrac{a - 1}{a^2 + 3a - 4}$

40. $\dfrac{5}{x^2 - 9} - \dfrac{x + 2}{x^2 + 4x + 3}$

41. $\dfrac{8}{m - 2} + \dfrac{3}{5m} + \dfrac{7}{5m(m - 2)}$

42. $\dfrac{-1}{7z} + \dfrac{3}{z + 2} + \dfrac{4}{7z(z + 2)}$

43. $\dfrac{4y - 1}{2y^2 + 5y - 3} - \dfrac{y + 3}{6y^2 + y - 2}$

44. $\dfrac{2q + 1}{3q^2 + 10q - 8} - \dfrac{3q + 5}{2q^2 + 5q - 12}$

8.5 Complex Fractions

A rational expression with fractions in the numerator, denominator, or both, is called a **complex (kahm-PLEKS) fraction.** Examples of complex fractions include

$$\frac{3 + \dfrac{4}{x}}{5}, \quad \frac{\dfrac{3x^2 - 5x}{6x^2}}{2x - \dfrac{1}{x}}, \quad \text{and} \quad \frac{3x + x}{5 - \dfrac{2}{x}}.$$

The parts of a complex fraction are named as follows.

$$\frac{\dfrac{2}{p} - \dfrac{1}{q}}{\dfrac{3}{p} + \dfrac{5}{q}}$$

$\leftarrow$ Numerator of complex fraction
$\leftarrow$ Main fraction bar (indicates division)
$\leftarrow$ Denominator of complex fraction

Complex fractions can always be simplified to rational expressions without fractions in the numerator and denominator. Two methods are commonly used to do this. We show both methods in this section.

OBJECTIVE 1 **SIMPLIFYING COMPLEX FRACTIONS: METHOD 1** Because a fraction represents a quotient, one method of simplifying complex fractions is to rewrite both the numerator and denominator as single fractions, and then to perform the indicated division.

E X A M P L E 1 **Simplifying Complex Fractions by Method 1**

Simplify each complex fraction.

(a) $\dfrac{\dfrac{2}{3} + \dfrac{5}{9}}{\dfrac{1}{4} + \dfrac{1}{12}}$

(b) $\dfrac{6 + \dfrac{3}{x}}{\dfrac{x}{4} + \dfrac{1}{8}}$

First, write each numerator as a single fraction.

$$\frac{2}{3} + \frac{5}{9} = \frac{2(3)}{3(3)} + \frac{5}{9}$$

$$= \frac{6}{9} + \frac{5}{9} = \frac{11}{9}$$

$$6 + \frac{3}{x} = \frac{6}{1} + \frac{3}{x}$$

$$= \frac{6x}{x} + \frac{3}{x} = \frac{6x + 3}{x}$$

Do the same thing with each denominator.

$$\frac{1}{4} + \frac{1}{12} = \frac{1(3)}{4(3)} + \frac{1}{12}$$

$$= \frac{3}{12} + \frac{1}{12} = \frac{4}{12} = \frac{1}{3}$$

$$\frac{x}{4} + \frac{1}{8} = \frac{x(2)}{4(2)} + \frac{1}{8}$$

$$= \frac{2x}{8} + \frac{1}{8} = \frac{2x + 1}{8}$$

The original complex fraction can now be rewritten as follows.

$$\frac{\dfrac{11}{9}}{\dfrac{1}{3}}$$

$$\frac{\dfrac{6x + 3}{x}}{\dfrac{2x + 1}{8}}$$

CONTINUED ON NEXT PAGE

OBJECTIVES

1. Simplify complex fractions by simplifying numerator and denominator (Method 1).

2. Simplify complex fractions by multiplying by the least common denominator (Method 2).

FOR EXTRA HELP

Tutorial   Tape 17   SSM, Sec. 8.5

1. Simplify the complex fractions.

(a) $\dfrac{6 + \dfrac{1}{x}}{5 - \dfrac{2}{x}}$

(b) $\dfrac{9 - \dfrac{4}{p}}{\dfrac{2}{p} + 1}$

2. Simplify the complex fractions.

(a) $\dfrac{\dfrac{rs^2}{t}}{\dfrac{r^2 s}{t^2}}$

(b) $\dfrac{\dfrac{m^2 n^3}{p}}{\dfrac{m^4 n}{p^2}}$

Now use the rule for division and the fundamental property.

$$\frac{11}{9} \div \frac{1}{3} = \frac{11}{9} \cdot \frac{3}{1}$$

$$= \frac{11 \cdot 3}{3 \cdot 3 \cdot 1}$$

$$= \frac{11}{3}$$

$$\frac{6x + 3}{x} \div \frac{2x + 1}{8}$$

$$= \frac{6x + 3}{x} \cdot \frac{8}{2x + 1}$$

$$= \frac{3(2x + 1)}{x} \cdot \frac{8}{2x + 1} = \frac{24}{x}$$

◀◀ **WORK PROBLEM 1 AT THE SIDE.**

E X A M P L E 2   Simplifying a Complex Fraction by Method 1

Simplify the complex fraction $\dfrac{\dfrac{xp}{q^3}}{\dfrac{p^2}{qx^2}}$.

Here the numerator and denominator are already single fractions, so use the rule for division and then the fundamental property.

$$\frac{xp}{q^3} \div \frac{p^2}{qx^2} = \frac{xp}{q^3} \cdot \frac{qx^2}{p^2} = \frac{x^3}{q^2 p}$$

◀◀ **WORK PROBLEM 2 AT THE SIDE.**

E X A M P L E 3   Simplifying a Complex Fraction by Method 1

Simplify $\dfrac{\dfrac{3}{x + 2} - 4}{\dfrac{2}{x + 2} + 1}$.

$$\frac{\dfrac{3}{x + 2} - 4}{\dfrac{2}{x + 2} + 1} = \frac{\dfrac{3}{x + 2} - \dfrac{4(x + 2)}{x + 2}}{\dfrac{2}{x + 2} + \dfrac{1(x + 2)}{x + 2}}$$ Write both second terms with a denominator of $x + 2$.

$$= \frac{\dfrac{3 - 4(x + 2)}{x + 2}}{\dfrac{2 + 1(x + 2)}{x + 2}}$$ Subtract in the numerator.

Add in the denominator.

$$= \frac{\dfrac{3 - 4x - 8}{x + 2}}{\dfrac{2 + x + 2}{x + 2}}$$ Distributive property

CONTINUED ON NEXT PAGE

ANSWERS

1. (a) $\dfrac{6x + 1}{5x - 2}$ (b) $\dfrac{9p - 4}{2 + p}$

2. (a) $\dfrac{st}{r}$ (b) $\dfrac{n^2 p}{m^2}$

$$= \frac{\dfrac{-5 - 4x}{x + 2}}{\dfrac{4 + x}{x + 2}} \qquad \text{Combine terms.}$$

$$= \frac{-5 - 4x}{x + 2} \cdot \frac{x + 2}{4 + x} \qquad \begin{array}{l}\text{Multiply by the}\\ \text{reciprocal.}\end{array}$$

$$= \frac{-5 - 4x}{4 + x} \qquad \text{Lowest terms}$$

WORK PROBLEM 3 AT THE SIDE. ▶▶

3. Simplify the complex fraction

$$\frac{\dfrac{2}{x - 1} + \dfrac{1}{x + 1}}{\dfrac{3}{x - 1} - \dfrac{4}{x + 1}}.$$

OBJECTIVE ②▶ **SIMPLIFYING COMPLEX FRACTIONS: METHOD 2** As an alternative method, complex fractions may be simplified by multiplying both numerator and denominator by the least common denominator of all the denominators appearing in the complex fraction. In the next example, this second method is used to simplify the same complex fractions as in Example 1.

┌ **E X A M P L E   4   Simplifying Complex Fractions by Method 2**

Simplify each complex fraction.

(a) $\dfrac{\dfrac{2}{3} + \dfrac{5}{9}}{\dfrac{1}{4} + \dfrac{1}{12}}$ **(b)** $\dfrac{6 + \dfrac{3}{x}}{\dfrac{x}{4} + \dfrac{1}{8}}$

Find the least common denominator for all the denominators in the complex fraction.

The LCD for 3, 9, 4, and 12 is 36. | The LCD for x, 4, and 8 is $8x$.

Multiply the numerator and denominator of the complex fraction by the LCD. Then use the distributive property to simplify and combine terms. Write the answers in lowest terms.

$$\frac{\dfrac{2}{3} + \dfrac{5}{9}}{\dfrac{1}{4} + \dfrac{1}{12}} = \frac{36\left(\dfrac{2}{3} + \dfrac{5}{9}\right)}{36\left(\dfrac{1}{4} + \dfrac{1}{12}\right)} \qquad\qquad \frac{6 + \dfrac{3}{x}}{\dfrac{x}{4} + \dfrac{1}{8}} = \frac{8x\left(6 + \dfrac{3}{x}\right)}{8x\left(\dfrac{x}{4} + \dfrac{1}{8}\right)}$$

$$= \frac{36\left(\dfrac{2}{3}\right) + 36\left(\dfrac{5}{9}\right)}{36\left(\dfrac{1}{4}\right) + 36\left(\dfrac{1}{12}\right)} \qquad\qquad = \frac{8x(6) + 8x\left(\dfrac{3}{x}\right)}{8x\left(\dfrac{x}{4}\right) + 8x\left(\dfrac{1}{8}\right)}$$

$$= \frac{24 + 20}{9 + 3} \qquad\qquad\qquad = \frac{48x + 24}{2x^2 + x}$$

$$= \frac{44}{12} = \frac{11}{3} \qquad\qquad\qquad = \frac{24(2x + 1)}{x(2x + 1)} = \frac{24}{x}$$

3. $\dfrac{3x + 1}{-x + 7}$

4. Simplify by the second method.

(a) $\dfrac{2 - \dfrac{6}{a}}{3 + \dfrac{4}{a}}$

(b) $\dfrac{\dfrac{p}{5 - p}}{\dfrac{4p}{2p + 1}}$

5. Simplify. Use either method.

$$\dfrac{\dfrac{1}{x} + \dfrac{2}{x - 1}}{\dfrac{4}{x - 1}}$$

E X A M P L E   5   **Simplifying a Complex Fraction by Method 2**

Simplify $\dfrac{\dfrac{3}{m + 5}}{\dfrac{9}{m + 2}}$.

The LCD is $(m + 5)(m + 2)$. Multiply the numerator and denominator of the complex fraction by the LCD.

$$\dfrac{\dfrac{3}{m + 5}}{\dfrac{9}{m + 2}} \cdot \dfrac{(m + 5)(m + 2)}{(m + 5)(m + 2)} = \dfrac{3(m + 2)}{9(m + 5)} \qquad \text{Fundamental property}$$

$$= \dfrac{m + 2}{3(m + 5)} \qquad \text{Lowest terms}$$

◀◀ **WORK PROBLEM 4 AT THE SIDE.**

You may want to choose one method of simplifying complex fractions and stick to it. Although either method can be used correctly to simplify any complex fraction, some students prefer to use Method 1 for problems like Example 2 and Example 5, which are the quotient of two fractions. They find Method 2 works best for problems like Example 1 (or Example 4), which has a sum or difference in the numerator or denominator or both.

E X A M P L E   6   **Simplifying a Complex Fraction by Method 2**

Simplify $\dfrac{\dfrac{1}{y} + \dfrac{2}{y + 2}}{\dfrac{4}{y} - \dfrac{3}{y + 2}}$.

The LCD is $y(y + 2)$. Multiply the numerator and denominator by the LCD.

$$\dfrac{\dfrac{1}{y} + \dfrac{2}{y + 2}}{\dfrac{4}{y} - \dfrac{3}{y + 2}} \cdot \dfrac{y(y + 2)}{y(y + 2)} = \dfrac{1(y + 2) + 2y}{4(y + 2) - 3y} \qquad \text{Fundamental property}$$

$$= \dfrac{y + 2 + 2y}{4y + 8 - 3y} \qquad \text{Distributive property}$$

$$= \dfrac{3y + 2}{y + 8} \qquad \text{Combine terms.}$$

◀◀ **WORK PROBLEM 5 AT THE SIDE.**

ANSWERS

4. (a) $\dfrac{2a - 6}{3a + 4}$ **(b)** $\dfrac{2p + 1}{4(5 - p)}$

5. $\dfrac{3x - 1}{4x}$

8.5 Exercises

1. In a fraction, what operation does the fraction bar represent?

2. What property of real numbers justifies Method 2 for simplifying complex fractions?

Simplify each complex fraction. Use either method. See Examples 1–6.

3. $\dfrac{-\dfrac{4}{3}}{\dfrac{2}{9}}$

4. $\dfrac{-\dfrac{5}{6}}{\dfrac{5}{4}}$

5. $\dfrac{\dfrac{p}{q^2}}{\dfrac{p^2}{q}}$

6. $\dfrac{\dfrac{a}{x}}{\dfrac{a^2}{2x}}$

7. $\dfrac{\dfrac{x}{y^2}}{\dfrac{x^2}{y}}$

8. $\dfrac{\dfrac{p^4}{r}}{\dfrac{p^2}{r^2}}$

9. $\dfrac{\dfrac{4a^4b^3}{3a}}{\dfrac{2ab^4}{b^2}}$

10. $\dfrac{\dfrac{2r^4t^2}{3t}}{\dfrac{5r^2t^5}{3r}}$

11. $\dfrac{\dfrac{m+2}{3}}{\dfrac{m-4}{m}}$

12. $\dfrac{\dfrac{q-5}{q}}{\dfrac{q+5}{3}}$

13. $\dfrac{\dfrac{2}{x}-3}{\dfrac{2-3x}{2}}$

14. $\dfrac{6+\dfrac{2}{r}}{\dfrac{3r+1}{4}}$

15. $\dfrac{\dfrac{1}{x}+x}{\dfrac{x^2+1}{8}}$

16. $\dfrac{\dfrac{3}{m}-m}{\dfrac{3-m^2}{4}}$

17. $\dfrac{a-\dfrac{5}{a}}{a+\dfrac{1}{a}}$

18. $\dfrac{q+\dfrac{1}{q}}{q+\dfrac{4}{q}}$

19. $\dfrac{\dfrac{1}{2}+\dfrac{1}{p}}{\dfrac{2}{3}+\dfrac{1}{p}}$

20. $\dfrac{\dfrac{3}{4}-\dfrac{1}{r}}{\dfrac{1}{5}+\dfrac{1}{r}}$

21. $\dfrac{\dfrac{t}{t+2}}{\dfrac{4}{t^2-4}}$

22. $\dfrac{\dfrac{m}{m+1}}{\dfrac{3}{m^2-1}}$

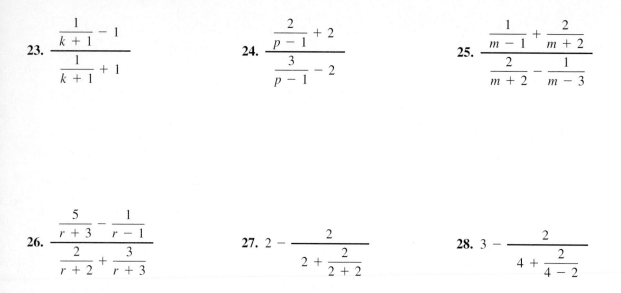

23. $\dfrac{\dfrac{1}{k+1}-1}{\dfrac{1}{k+1}+1}$

24. $\dfrac{\dfrac{2}{p-1}+2}{\dfrac{3}{p-1}-2}$

25. $\dfrac{\dfrac{1}{m-1}+\dfrac{2}{m+2}}{\dfrac{2}{m+2}-\dfrac{1}{m-3}}$

26. $\dfrac{\dfrac{5}{r+3}-\dfrac{1}{r-1}}{\dfrac{2}{r+2}+\dfrac{3}{r+3}}$

27. $2-\dfrac{2}{2+\dfrac{2}{2+2}}$

28. $3-\dfrac{2}{4+\dfrac{2}{4-2}}$

--------------------- **MATHEMATICAL CONNECTIONS (EXERCISES 29–32)** ---------------------

In order to find the average of two numbers, we add them and divide by 2. Suppose that we wish to find the average of $\dfrac{3}{8}$ and $\dfrac{5}{6}$. Work Exercises 29–32, in order, to see how a complex fraction occurs in a problem like this.

29. Write in symbols: the sum of $\dfrac{3}{8}$ and $\dfrac{5}{6}$, divided by 2. Your result should be a complex fraction.

30. Simplify the complex fraction from Exercise 29 using Method 1.

31. Simplify the complex fraction from Exercise 29 using Method 2.

32. Your answers in Exercises 30 and 31 should be the same. Which method did you prefer? Why?

8.6 Equations Involving Rational Expressions

OBJECTIVES

1. Solve equations involving rational expressions.
2. Solve a formula for a specified variable.

FOR EXTRA HELP

Tutorial   Tape 17   SSM, Sec. 8.6

OBJECTIVE 1 When an equation involves fractions, the multiplication property of equality can be used first to clear it of fractions. The equation can then be solved in the usual way. The goal is to get an equivalent equation (one with the same solutions) that does not have fractions. Choose as multiplier the least common denominator of all denominators in the fractions of the equation.

EXAMPLE 1 Solving an Equation Involving Rational Expressions

Solve $\dfrac{x}{3} + \dfrac{x}{4} = 10 + x$.

Because the least common denominator of the two fractions is 12, we begin by multiplying each side of the equation by 12.

$$12\left(\frac{x}{3} + \frac{x}{4}\right) = 12(10 + x)$$

$$12\left(\frac{x}{3}\right) + 12\left(\frac{x}{4}\right) = 12(10) + 12x \qquad \text{Distributive property}$$

$$\frac{12x}{3} + \frac{12x}{4} = 120 + 12x$$

$$4x + 3x = 120 + 12x$$

This equation has no fractions. Solve it using the methods given earlier for solving linear equations.

$$7x = 120 + 12x \qquad \text{Combine terms.}$$

$$-5x = 120 \qquad \text{Subtract } 12x.$$

$$x = -24 \qquad \text{Divide by } -5.$$

Check: $\quad \dfrac{-24}{3} + \dfrac{-24}{4} = 10 - 24 \qquad ? \qquad \text{Let } x = -24.$

$$-8 - 6 = -14 \qquad \text{True}$$

The solution set is $\{-24\}$.

WORK PROBLEM 1 AT THE SIDE. ▶▶

Caution

Note that use of the LCD here is different from its use in the previous section. Here, we use the Multiplication Property of Equality to multiply each side of an *equation* by the LCD. Earlier, we used the property of 1 to multiply a *fraction* by another fraction that had the LCD as both its numerator and denominator. Be careful not to confuse these two methods.

EXAMPLE 2 Solving an Equation Involving Rational Expressions

Solve $\dfrac{p}{2} - \dfrac{p-1}{3} = 1$.

$$6\left(\frac{p}{2} - \frac{p-1}{3}\right) = 6 \cdot 1 \qquad \text{Multiply by the LCD, 6.}$$

$$6\left(\frac{p}{2}\right) - 6\left(\frac{p-1}{3}\right) = 6 \qquad \text{Distributive property}$$

$$3p - 2(p - 1) = 6$$

— **CONTINUED ON NEXT PAGE**

1. Solve each equation and check your answer.

(a) $\dfrac{x}{5} + 3 = \dfrac{3}{5}$

(b) $\dfrac{x}{2} - \dfrac{x}{3} = \dfrac{5}{6}$

ANSWERS

1. (a) $\{-12\}$ (b) $\{5\}$

505

2. Solve each equation, and check your answer.

(a) $\dfrac{k}{6} - \dfrac{k+1}{4} = -\dfrac{1}{2}$

Be very careful to put parentheses around $p - 1$; otherwise you may find an incorrect solution. Continue simplifying and solve.

$$3p - 2p + 2 = 6 \qquad \text{Distributive property}$$
$$p + 2 = 6 \qquad \text{Combine terms.}$$
$$p = 4 \qquad \text{Subtract 2.}$$

Check to see that the solution set is {4} by replacing p with 4 in the original equation.

◀◀ **WORK PROBLEM 2 AT THE SIDE.**

In solving equations that have a variable in the denominator, remember that the number 0 cannot be used as a denominator. Therefore, the solution cannot be a number that will make the denominator equal 0.

(b) $\dfrac{2m-3}{5} - \dfrac{m}{3} = -\dfrac{6}{5}$

E X A M P L E 3 **Solving an Equation Involving Rational Expressions**

Solve $\dfrac{x}{x-2} = \dfrac{2}{x-2} + 2$.

The common denominator is $x - 2$. Because $x = 2$ makes a denominator 0, x cannot equal 2. Solve the equation by multiplying each side of the equation by $x - 2$.

$$(x-2)\left(\frac{x}{x-2}\right) = (x-2)\left(\frac{2}{x-2} + 2\right)$$
$$(x-2)\left(\frac{x}{x-2}\right) = (x-2)\left(\frac{2}{x-2}\right) + (x-2)(2)$$
$$x = 2 + 2x - 4$$
$$x = -2 + 2x \qquad \text{Combine terms.}$$
$$-x = -2 \qquad \text{Subtract } 2x.$$
$$x = 2 \qquad \text{Divide by } -1.$$

3. Solve $1 - \dfrac{2}{x+1} = \dfrac{2x}{x+1}$ and check your answer.

The proposed solution is 2. However, as shown above, 2 cannot be a solution because 2 makes a denominator equal 0. *The equation has no solution and the solution set is ∅.* (Equations with no solutions are one of the main reasons that it is important to always check proposed solutions.)

◀◀ **WORK PROBLEM 3 AT THE SIDE.**

In summary, solve equations with rational expressions as follows.

Solving Equations with Rational Expressions

Step 1 Multiply each side of the equation by the least common denominator. (This clears the equation of fractions.)

Step 2 Solve the resulting equation.

Step 3 Check each proposed solution by substituting it in the original equation.

ANSWERS

2. (a) {3} **(b)** {−9}

3. When the equation is solved, −1 is found. However, because $x = -1$ leads to a 0 denominator in the original equation, there is no solution. The solution set is ∅.

---E X A M P L E 4   **Solving an Equation Involving Rational Expressions**

Solve $\dfrac{2m}{m^2 - 4} + \dfrac{1}{m - 2} = \dfrac{2}{m + 2}$.

Multiply by the LCD, $(m + 2)(m - 2)$.

$$(m + 2)(m - 2)\left(\dfrac{2m}{m^2 - 4} + \dfrac{1}{m - 2}\right)$$
$$= (m + 2)(m - 2)\dfrac{2}{m + 2}$$

$$(m + 2)(m - 2)\dfrac{2m}{m^2 - 4} + (m + 2)(m - 2)\dfrac{1}{m - 2}$$
$$= (m + 2)(m - 2)\dfrac{2}{m + 2}$$

$$2m + m + 2 = 2(m - 2)$$

$$3m + 2 = 2m - 4 \qquad \text{Distributive property; combine terms.}$$

$$m + 2 = -4 \qquad \text{Subtract } 2m.$$

$$m = -6 \qquad \text{Subtract } 2.$$

Check to see that $\{-6\}$ is indeed the solution set for the given equation.

WORK PROBLEMS 4 AND 5 AT THE SIDE. ▶▶

---E X A M P L E 5   **Solving an Equation Involving Rational Expressions**

Solve $\dfrac{2}{x^2 - x} = \dfrac{1}{x^2 - 1}$.

Begin by finding a least common denominator. Since $x^2 - x = x(x - 1)$, and $x^2 - 1 = (x + 1)(x - 1)$, the LCD is $x(x + 1)(x - 1)$. Multiply each side of the equation by the LCD.

$$x(x + 1)(x - 1)\dfrac{2}{x(x - 1)} = x(x + 1)(x - 1)\dfrac{1}{(x + 1)(x - 1)}$$

$$2(x + 1) = x$$
$$2x + 2 = x \qquad \text{Distributive property}$$
$$x + 2 = 0 \qquad \text{Subtract } x.$$
$$x = -2 \qquad \text{Subtract } 2.$$

To be sure that $x = -2$ is a solution, substitute -2 for x in the original equation. Since -2 satisfies the equation, the solution set is $\{-2\}$.

WORK PROBLEM 6 AT THE SIDE. ▶▶

4. Check -6 as a solution to Example 4. Is the solution correct?

5. Solve each equation, and check your answer.

(a) $\dfrac{2p}{p^2 - 1} = \dfrac{2}{p + 1} - \dfrac{1}{p - 1}$

(b)
$$\dfrac{8r}{4r^2 - 1} = \dfrac{3}{2r + 1} + \dfrac{3}{2r - 1}$$

6. Solve each equation, and check your answer.

(a) $\dfrac{4}{3m + 3} = \dfrac{m + 1}{m^2 + m}$

(b) $\dfrac{2}{p^2 - 2p} = \dfrac{3}{p^2 - p}$

EXAMPLE 6 Solving an Equation Involving Rational Expressions

Solve $\dfrac{1}{x-1} + \dfrac{1}{2} = \dfrac{2}{x^2-1}$.

The least common denominator is $2(x+1)(x-1)$. Multiply each side of the equation by the LCD, $2(x+1)(x-1)$.

$$2(x+1)(x-1)\left(\frac{1}{x-1} + \frac{1}{2}\right) = 2(x+1)(x-1)\frac{2}{(x+1)(x-1)}$$

$$2(x+1)(x-1)\frac{1}{x-1} + 2(x+1)(x-1)\frac{1}{2}$$

$$= 2(x+1)(x-1)\frac{2}{(x+1)(x-1)}$$

$$2(x+1) + (x+1)(x-1) = 4$$

$$2x + 2 + x^2 - 1 = 4 \qquad \text{Distributive property}$$

$$x^2 + 2x + 1 = 4 \qquad \text{Combine terms.}$$

$$x^2 + 2x - 3 = 0 \qquad \text{Subtract 4.}$$

$$(x+3)(x-1) = 0 \qquad \text{Factor.}$$

Solving this equation suggests that $x = -3$ or $x = 1$. But 1 makes a denominator of the original equation equal 0, so 1 is not a solution. However, -3 is a solution, as shown by substituting -3 for x in the original equation.

$$\frac{1}{x-1} + \frac{1}{2} = \frac{2}{x^2-1}$$

$$\frac{1}{-3-1} + \frac{1}{2} = \frac{2}{(-3)^2-1} \qquad ? \qquad \text{Let } x = -3.$$

$$\frac{1}{-4} + \frac{1}{2} = \frac{2}{9-1} \qquad ? \qquad \text{Simplify.}$$

$$\frac{1}{4} = \frac{1}{4} \qquad \text{True}$$

The check shows that $\{-3\}$ is the solution set.

EXAMPLE 7 Solving an Equation Involving Rational Expressions

Solve $\dfrac{1}{k^2+4k+3} + \dfrac{1}{2k+2} = \dfrac{3}{4k+12}$.

Factor the three denominators to get the common denominator, $4(k+1)(k+3)$. Multiply each side by this product.

CONTINUED ON NEXT PAGE

$$4(k+1)(k+3)\left(\frac{1}{(k+1)(k+3)} + \frac{1}{2(k+1)}\right)$$

$$= 4(k+1)(k+3)\frac{3}{4(k+3)}$$

$$4(k+1)(k+3)\frac{1}{(k+1)(k+3)} + 2 \cdot 2(k+1)(k+3)\frac{1}{2(k+1)}$$

$$= 4(k+1)(k+3)\frac{3}{4(k+3)}$$

$4 + 2(k+3) = 3(k+1)$	Simplify.
$4 + 2k + 6 = 3k + 3$	Distributive property
$2k + 10 = 3k + 3$	Combine terms.
$7 = k$	Subtract $2k$ and 3.

Check to see that $\{7\}$ is the solution set.

WORK PROBLEM 7 AT THE SIDE. ▶▶

OBJECTIVE 2 Solving a formula for a specified variable was discussed in Chapter 2. In the next example this process is applied to a formula with fractions.

E X A M P L E 8   Solving for a Specified Variable

Solve the formula $S = \dfrac{a(r^n - 1)}{r - 1}$ for a.

We need to isolate a on one side of the equation.

$$S = \frac{a(r^n - 1)}{r - 1}$$

$$(r-1)S = (r-1)\frac{a(r^n - 1)}{r - 1} \qquad \text{Multiply each side by the LCD, } r - 1.$$

$$(r-1)S = a(r^n - 1)$$

$$\frac{(r-1)S}{r^n - 1} = a \qquad \text{Divide each side by } r^n - 1.$$

The equation is now solved for a.

WORK PROBLEM 8 AT THE SIDE. ▶▶

7. Solve each equation and check your answer.

(a) $\dfrac{1}{x-2} + \dfrac{1}{5} = \dfrac{2}{5(x^2 - 4)}$

(b) $\dfrac{6}{5a+10} - \dfrac{1}{a-5} = \dfrac{4}{a^2 - 3a - 10}$

8. Solve each equation for the specified variable.

(a) $z = \dfrac{x}{x+y}$ for y

(b) $a = \dfrac{v-w}{t}$ for v

Real World *collaborative investigations*

Geometry in Our World

An interesting occurrence of geometry in the world around us involves three-dimensional figures known as **regular polyhedra** (singular: **polyhedron**). Because they were studied by Plato and his followers, they are also known as the platonic solids.

Tetrahedron   Hexahedron (Cube)   Octahedron   Dodecahedron   Icosahedron

Each regular polyhedron is formed by regular polygons (two-dimensional figures having equal sides and equal angles). The polygons are called faces. They meet at edges, and the corners formed are called vertices. There are only five regular polyhedra, pictured above.

Many crystals and some viruses are constructed in the shapes of regular polyhedra. Many viruses that infect animals were once thought to have spherical shapes, until X-ray analysis revealed that their shells are icosahedral in nature. *Radiolaria* is a group of microorganisms. The skeletons of these single-celled animals take on beautiful geometric shapes. The one pictured is based on the tetrahedron.

The Swiss mathematician Leonhard Euler (1707–1783) investigated a remarkable relationship among the numbers of faces, vertices, and edges. The group discussion exercise below will allow you to discover it as well.

FOR GROUP DISCUSSION

As a class, use the figures in the text, or models brought to class by your instructor or class members, to complete the following table.

Polyhedron	Faces (F)	Vertices (V)	Edges (E)	Value of $F + V - E$
Tetrahedron				
Hexahedron (Cube)				
Octahedron				
Dodecahedron				
Icosahedron				

Now, state your conjecture. Verify it by looking up **Euler's formula** in a more advanced book on geometry or the history of mathematics.

8.6 Exercises

Identify as an expression or an equation. If it is an expression, simplify it. If it is an equation, solve it. See Example 1.

1. $\dfrac{7}{8}x + \dfrac{1}{5}x$

2. $\dfrac{4}{7}x + \dfrac{3}{5}x$

3. $\dfrac{7}{8}x + \dfrac{1}{5}x = 1$

4. $\dfrac{4}{7}x + \dfrac{3}{5}x = 1$

5. $\dfrac{3}{5}y - \dfrac{7}{10}y$

6. $\dfrac{3}{5}y - \dfrac{7}{10}y = 1$

7. Explain how the LCD is used in a different way when adding and subtracting rational expressions compared to solving equations with rational expressions.

8. If we multiply both sides of the equation $\dfrac{6}{x+5} = \dfrac{6}{x+5}$ by $x + 5$, we get $6 = 6$. Is {all real numbers} the solution set of this equation? Explain.

Solve each equation, and check your answers. See Examples 1–3.

9. $\dfrac{5}{y} + 4 = \dfrac{2}{y}$

10. $\dfrac{11}{q} = 3 - \dfrac{1}{q}$

11. $\dfrac{p}{3} - \dfrac{p}{6} = 4$

12. $\dfrac{x}{15} + \dfrac{x}{5} = 4$

13. $\dfrac{3x}{5} - 6 = x$

14. $\dfrac{5t}{4} + t = 9$

15. $\dfrac{4m}{7} + m = 11$

16. $a - \dfrac{3a}{2} = 1$

17. $\dfrac{z-1}{4} = \dfrac{z+3}{3}$

18. $\dfrac{r-5}{2} = \dfrac{r+2}{3}$

19. $\dfrac{3p+6}{8} = \dfrac{3p-3}{16}$

20. $\dfrac{2z+1}{5} = \dfrac{7z+5}{15}$

21. $\dfrac{2x+3}{x} = \dfrac{3}{2}$

22. $\dfrac{5-2y}{y} = \dfrac{1}{4}$

23. $\dfrac{k}{k-4} - 5 = \dfrac{4}{k-4}$

24. $\dfrac{-5}{a+5} = \dfrac{a}{a+5} + 2$

25. $\dfrac{q+2}{3} + \dfrac{q-5}{5} = \dfrac{7}{3}$

26. $\dfrac{b+7}{8} - \dfrac{b-2}{3} = \dfrac{4}{3}$

27. $\dfrac{t}{6} + \dfrac{4}{3} = \dfrac{t-2}{3}$

28. $\dfrac{x}{2} = \dfrac{5}{4} + \dfrac{x-1}{4}$

29. $\dfrac{3m}{5} - \dfrac{3m-2}{4} = \dfrac{1}{5}$

30. $\dfrac{8p}{5} = \dfrac{3p-4}{2} + \dfrac{5}{2}$

31. What values of x cannot be solutions of the equation $\dfrac{1}{x-4} = \dfrac{3}{2x}$?

32. What is wrong with the following problem? "Solve $\dfrac{2}{3x} + \dfrac{1}{5x}$."

Solve each equation, and check your answers. See Examples 3–7.

33. $\dfrac{3}{x-1} + \dfrac{2}{4x-4} = \dfrac{7}{4}$

34. $\dfrac{2}{p+3} + \dfrac{3}{8} = \dfrac{5}{4p+12}$

35. $\dfrac{y}{3y+3} = \dfrac{2y-3}{y+1} - \dfrac{2y}{3y+3}$

36. $\dfrac{2k+3}{k+1} - \dfrac{3k}{2k+2} = \dfrac{-2k}{2k+2}$

37. $\dfrac{2}{m} = \dfrac{m}{5m+12}$

38. $\dfrac{x}{4-x} = \dfrac{2}{x}$

39. $\dfrac{-2}{z+5} + \dfrac{3}{z-5} = \dfrac{20}{z^2-25}$

40. $\dfrac{3}{r+3} - \dfrac{2}{r-3} = \dfrac{-12}{r^2-9}$

41. $\dfrac{3y}{y^2+5y+6} = \dfrac{5y}{y^2+2y-3} - \dfrac{2}{y^2+y-2}$

42. $\dfrac{x+4}{x^2-3x+2} - \dfrac{5}{x^2-4x+3} = \dfrac{x-4}{x^2-5x+6}$

43. $\dfrac{5x}{14x+3} = \dfrac{1}{x}$

44. $\dfrac{m}{8m+3} = \dfrac{1}{3m}$

45. $\dfrac{2}{z-1} - \dfrac{5}{4} = \dfrac{-1}{z+1}$

46. $\dfrac{5}{p-2} = 7 - \dfrac{10}{p+2}$

Solve each formula for the specified variable. See Example 8.

47. $m = \dfrac{kF}{a}$ for F

48. $I = \dfrac{kE}{R}$ for E

49. $m = \dfrac{kF}{a}$ for a

50. $I = \dfrac{kE}{R}$ for R

51. $I = \dfrac{E}{R+r}$ for R

52. $I = \dfrac{E}{R+r}$ for r

53. $h = \dfrac{2A}{B+b}$ for A

54. $d = \dfrac{2S}{n(a+L)}$ for S

55. $d = \dfrac{2S}{n(a+L)}$ for a

56. $h = \dfrac{2A}{B+b}$ for B

SUMMARY EXERCISES ON RATIONAL EXPRESSIONS

A common error when working with rational expressions is to confuse *operations* on rational expressions with the *solution of equations* with rational expressions. For example, the four possible operations on the rational expressions

$$\frac{1}{x} \quad \text{and} \quad \frac{1}{x - 2}$$

can be performed as follows.

Add. $\quad \dfrac{1}{x} + \dfrac{1}{x - 2} = \dfrac{x - 2}{x(x - 2)} + \dfrac{x}{x(x - 2)} = \dfrac{x - 2 + x}{x(x - 2)} = \dfrac{2x - 2}{x(x - 2)}$

Subtract. $\quad \dfrac{1}{x} - \dfrac{1}{x - 2} = \dfrac{x - 2}{x(x - 2)} - \dfrac{x}{x(x - 2)} = \dfrac{x - 2 - x}{x(x - 2)} = \dfrac{-2}{x(x - 2)}$

Multiply. $\quad \dfrac{1}{x} \cdot \dfrac{1}{x - 2} = \dfrac{1}{x(x - 2)}$

Divide. $\quad \dfrac{1}{x} \div \dfrac{1}{x - 2} = \dfrac{1}{x} \cdot \dfrac{x - 2}{1} = \dfrac{x - 2}{x}$

On the other hand, the equation

$$\frac{1}{x} + \frac{1}{x - 2} = \frac{3}{4}$$

is solved by multiplying each side by the least common denominator, $4x(x - 2)$, giving an equation with no denominators.

$$4x(x - 2)\frac{1}{x} + 4x(x - 2)\frac{1}{x - 2} = 4x(x - 2)\frac{3}{4}$$
$$4x - 8 + 4x = 3x^2 - 6x$$
$$0 = 3x^2 - 14x + 8$$
$$0 = (3x - 2)(x - 4)$$
$$x = \frac{2}{3} \quad \text{or} \quad x = 4$$

Solution set: $\left\{\dfrac{2}{3}, 4\right\}$

In each of the following exercises, first decide whether it is a rational expression to be added, subtracted, multiplied, or divided or an equation to be solved. Then perform the operation, or solve the given equation.

1. $\dfrac{4}{p} + \dfrac{6}{p}$

2. $\dfrac{x^3 y^2}{x^2 y^4} \cdot \dfrac{y^5}{x^4}$

3. $\dfrac{1}{x^2 + x - 2} \div \dfrac{4x^2}{2x - 2}$

4. $\dfrac{8}{m - 5} = 2$

5. $\dfrac{2y^2 + y - 6}{2y^2 - 9y + 9} \cdot \dfrac{y^2 - 2y - 3}{y^2 - 1}$

6. $\dfrac{2}{k^2 - 4k} + \dfrac{3}{k^2 - 16}$

7. $\dfrac{x - 4}{5} = \dfrac{x + 3}{6}$

8. $\dfrac{3t^2 - t}{6t^2 + 15t} \div \dfrac{6t^2 + t - 1}{2t^2 - 5t - 25}$

9. $\dfrac{4}{p + 2} + \dfrac{1}{3p + 6}$

10. $\dfrac{1}{y} + \dfrac{1}{y - 3} = -\dfrac{5}{4}$

11. $\dfrac{3}{t - 1} + \dfrac{1}{t} = \dfrac{7}{2}$

12. $\dfrac{6}{y} - \dfrac{2}{3y}$

13. $\dfrac{5}{4z} - \dfrac{2}{3z}$

14. $\dfrac{k + 2}{3} = \dfrac{2k - 1}{5}$

15. $\dfrac{1}{m^2 + 5m + 6} + \dfrac{2}{m^2 + 4m + 3}$

16. $\dfrac{2k^2 - 3k}{20k^2 - 5k} \div \dfrac{2k^2 - 5k + 3}{4k^2 + 11k - 3}$

8.7 *Applications of Rational Expressions*

Each time we learn to solve a new type of equation, we are able to apply the knowledge to new applications. We can now solve applications involving rational expressions. The problem-solving techniques of earlier chapters still apply. The main difference between the problems in this section and those of earlier sections is that the equations for these problems involve rational expressions.

OBJECTIVE ▶ In order to prepare for more meaningful applications, we begin with an example about an unknown number.

┌─
│ **E X A M P L E 1   Solving a Problem About Numbers**
│
│ If the same number is added to both the numerator and denominator of the fraction $\frac{3}{4}$, the result is $\frac{5}{6}$. Find the number.
│   Let x = the number added to numerator and denominator. Then
│
│ $$\frac{3 + x}{4 + x} \qquad \begin{array}{l}\text{Same number added to}\\ \text{numerator and denominator}\end{array}$$
│
│ represents the result of adding the same number to both the numerator and denominator. Since this result is to equal $\frac{5}{6}$,
│
│ $$\frac{3 + x}{4 + x} = \frac{5}{6}.$$
│
│ Solve this equation by multiplying each side by the common denominator, $6(4 + x)$.
│
│ $$6(4 + x)\frac{3 + x}{4 + x} = 6(4 + x)\frac{5}{6}$$
│ $$6(3 + x) = 5(4 + x)$$
│ $$18 + 6x = 20 + 5x \qquad \text{Distributive property}$$
│ $$x = 2 \qquad \text{Subtract 18 and } 5x.$$
│
│ Solution set: {2}
│ Check this solution in the words of the original problem: If 2 is added to both the numerator and denominator of $\frac{3}{4}$, the result is $\frac{5}{6}$, as needed.
└─

WORK PROBLEM 1 AT THE SIDE. ▶▶

OBJECTIVE ▶ If an automobile travels at an average rate of 50 miles per hour for two hours, then it travels $50 \times 2 = 100$ miles. This is an example of the basic relationship between distance, rate, and time:

$$\text{distance} = \text{rate} \times \text{time}.$$

This relationship is given by the formula $d = rt$. By solving, in turn, for r and t in the formula, we obtain two other equivalent forms of the formula. The three forms are given below.

Distance, Rate, and Time Relationship

$$d = rt \qquad r = \frac{d}{t} \qquad t = \frac{d}{r}$$

The next example illustrates the uses of these formulas.

OBJECTIVES

1 ▶ Solve problems about numbers using rational expressions.

2 ▶ Solve applied problems about distance, rate, and time using rational expressions.

3 ▶ Solve applied problems about work using rational expressions.

4 ▶ Solve applied problems about variation using rational expressions.

FOR EXTRA HELP

Tutorial   Tape 18   SSM, Sec. 8.7

1. Solve each problem.

(a) A certain number is added to the numerator and subtracted from the denominator of $\frac{5}{8}$. The new fraction equals the reciprocal of $\frac{5}{8}$. Find the number.

(b) The denominator of a fraction is 1 more than the numerator. If 6 is added to the numerator and subtracted from the denominator, the result is $\frac{15}{4}$. Find the original fraction.

ANSWERS

1. (a) 3 (b) $\frac{9}{10}$

515

2. Solve each problem.

(a) In 1994, Leroy Burrell of the United States won the world record in the 100 meter dash in 9.85 seconds. What was his average speed?

(b) The winner of the 2000 meter women's track event in 1994 was Sonia O'Sullivan of Ireland, whose average speed was 369 meters per minute. What was her winning time?

(c) A small plane flew from Warsaw to Rome averaging 164 miles per hour. The trip took 2 hours. What is the distance between Warsaw and Rome?

EXAMPLE 2 Finding Distance, Rate, or Time

(a) The speed of sound is 1088 feet per second at sea level at 32°F. In 5 seconds under these conditions, sound travels

$$1088 \times 5 = 5440 \text{ feet.}$$
$$\underset{\text{Rate}}{\uparrow} \times \underset{\text{Time}}{\uparrow} = \underset{\text{Distance}}{\uparrow}$$

Here, we found distance given rate and time, using $d = rt$.

(b) Over a short distance, an elephant can travel at a rate of 25 miles per hour. In order to travel $\frac{1}{4}$ mile, it would take an elephant

$$\underset{\text{Rate}}{\overset{\text{Distance} \rightarrow}{}} \frac{\frac{1}{4}}{25} = \frac{1}{4} \times \frac{1}{25} = \frac{1}{100} \text{ hour.} \quad \leftarrow \text{Time}$$

Here, we find time given rate and distance, using $t = \frac{d}{r}$. To convert $\frac{1}{100}$ hour to minutes, multiply $\frac{1}{100}$ by 60 to get $\frac{60}{100}$ or $\frac{3}{5}$ minute. To convert $\frac{3}{5}$ minute to seconds, multiply $\frac{3}{5}$ by 60 to get 36 seconds.

(c) In the 1994 Winter Olympics, Dan Jansen of the U.S. won the 1000 meter event with a time of 72.6 seconds. His rate was

$$\underset{\text{Time} \rightarrow}{\overset{\text{Distance} \rightarrow}{}} \frac{1000}{72.6} = 13.77 \text{ (rounded) meters per second.} \quad \leftarrow \text{Rate}$$

This answer was obtained using a calculator. Here, we found rate given distance and time, using $r = \frac{d}{t}$.

◀◀ **WORK PROBLEM 2 AT THE SIDE.**

Many applied problems use the formulas just discussed. The next example shows how to solve a typical application of the formula $d = rt$. A strategy for solving such problems involves two major steps:

Solving Motion Problems

Step 1 Set up a sketch showing what is happening in the problem.

Step 2 Make a chart using the information given in the problem, along with the unknown quantities.

The chart will help you organize the information, and the sketch will help you set up the equation.

EXAMPLE 3 Solving a Motion Problem

Two cars leave Baton Rouge, Louisiana, at the same time and travel east on Interstate 12. One travels at a constant speed of 55 miles per hour and the other travels at a constant speed of 63 miles per hour. In how many hours will the distance between them be 24 miles?

Since we are looking for time, let $t =$ the number of hours until the distance between them is 24 miles. The sketch in Figure 1 shows what is happening in the problem. Now, construct a chart like the one that follows. Fill in the information given in the problem, and use t for the time traveled by each car. Multiply rate by time to get the expressions for distances traveled.

CONTINUED ON NEXT PAGE

ANSWERS
2. (a) 10.15 meters per second
(b) 5.42 minutes **(c)** 328 miles

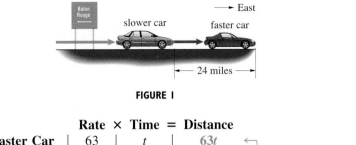

FIGURE 1

	Rate	× Time	= Distance
Faster Car	63	t	$63t$
Slower Car	55	t	$55t$

Difference is 24 miles.

The quantities $63t$ and $55t$ represent the different distances. Refer to Figure 1, and notice that the *difference* between the larger distance and the smaller distance is 24 miles. Now write the equation, and solve it.

$$63t - 55t = 24$$
$$8t = 24 \qquad \text{Combine terms.}$$
$$t = 3 \qquad \text{Divide by 8.}$$

Solution set: {3}

After 3 hours the faster car will have traveled $63 \times 3 = 189$ miles, and the slower car will have traveled $55 \times 3 = 165$ miles. Since $189 - 165 = 24$, the conditions of the problem are satisfied. It will take 3 hours for the distance between them to be 24 miles.

Note

In motion problems like the one in Example 3, once you have filled in two pieces of information in each row of the chart, you should automatically fill in the third piece of information, using the appropriate form of the formula relating distance, rate, and time. Set up the equation based upon your sketch and the information in the chart.

WORK PROBLEM 3 AT THE SIDE. ▶▶

E X A M P L E 4   Solving a Problem About Distance, Rate, and Time

The Big Muddy River has a current of 3 miles per hour. A motorboat takes as long to go 12 miles downstream as to go 8 miles upstream. What is the speed of the boat in still water?

Let x = the speed of the boat in still water. Because the current pushes the boat when the boat is going downstream, the speed of the boat downstream will be the sum of the speed of the boat and the speed of the current, $x + 3$ miles per hour. Also, the boat's speed going upstream is given by $x - 3$ miles per hour. This information is summarized in the following chart.

	d	r	t
Downstream	12	$x + 3$	
Upstream	8	$x - 3$	

CONTINUED ON NEXT PAGE

3. Work each problem.

(a) From a point on a straight road, Lupe and Maria ride bicycles in opposite directions. Lupe rides 10 miles per hour and Maria rides 12 miles per hour. In how many hours will they be 55 miles apart?

(b) At a given hour, two steamboats leave a city in the same direction on a straight canal. One travels at 18 miles per hour, and the other travels at 25 miles per hour. In how many hours will the boats be 35 miles apart?

4. Work each problem.

(a) A boat can go 20 miles against the current in the same time it can go 60 miles with the current. The current is flowing at 4 miles per hour. Find the speed of the boat with no current.

Fill in the column representing time by using the formula $t = \frac{d}{r}$. Then the time upstream is the distance divided by the rate, or

$$t = \frac{d}{r} = \frac{8}{x - 3},$$

and the time downstream is also the distance divided by the rate, or

$$t = \frac{d}{r} = \frac{12}{x + 3}.$$

Now complete the chart.

	d	r	t
Downstream	12	$x + 3$	$\dfrac{12}{x + 3}$
Upstream	8	$x - 3$	$\dfrac{8}{x - 3}$

Times are equal.

According to the original problem, the time upstream equals the time downstream. The two times from the chart must therefore be equal, giving the equation

$$\frac{12}{x + 3} = \frac{8}{x - 3}.$$

Solve this equation by multiplying each side by $(x + 3)(x - 3)$.

$$(x + 3)(x - 3)\frac{12}{x + 3} = (x + 3)(x - 3)\frac{8}{x - 3}$$
$$12(x - 3) = 8(x + 3)$$
$$12x - 36 = 8x + 24 \qquad \text{Distributive property}$$
$$4x = 60 \qquad \text{Subtract } 8x; \text{ add } 36.$$
$$x = 15 \qquad \text{Divide by 4.}$$

(b) An airplane, maintaining a constant airspeed, takes as long to go 450 miles with the wind as it does to go 375 miles against the wind. If the wind is blowing at 15 miles per hour, what is the speed of the plane?

Solution set: $\{15\}$

The speed of the boat in still water is 15 miles per hour.

Check the solution by first finding the speed of the boat downstream, which is $15 + 3 = 18$ miles per hour. Traveling 12 miles would take

$$d = rt$$
$$12 = 18t$$
$$t = \frac{2}{3} \text{ hour.}$$

On the other hand, the speed of the boat upstream is $15 - 3 = 12$ miles per hour, and traveling 8 miles would take

$$d = rt$$
$$8 = 12t$$
$$t = \frac{2}{3} \text{ hour.}$$

The time upstream equals the time downstream, as required.

◀◀ **WORK PROBLEM 4 AT THE SIDE.**

ANSWERS

4. (a) 8 miles per hour
   (b) 165 miles per hour

OBJECTIVE 3 Suppose that you can mow your lawn in 4 hours. Then after 1 hour, you will have mowed $\frac{1}{4}$ of the lawn. After 2 hours, you will have mowed $\frac{2}{4}$ or $\frac{1}{2}$ of the lawn, and so on. This idea is generalized as follows.

Rate of Work

If a job can be completed in t units of time, then the rate of work is

$$\frac{1}{t} \text{ job per unit of time.}$$

The relationship between problems involving work and problems involving distance is a very close one. Recall that the formula $d = rt$ says that distance traveled is equal to rate of travel multiplied by time traveled. Similarly, the fractional part of a job accomplished is equal to the rate of the work multiplied by the time worked. In the lawn mowing example, after 3 hours, the fractional part of the job done is

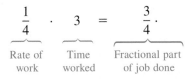

$$\underbrace{\frac{1}{4}}_{\substack{\text{Rate of}\\\text{work}}} \cdot \underbrace{3}_{\substack{\text{Time}\\\text{worked}}} = \underbrace{\frac{3}{4}}_{\substack{\text{Fractional part}\\\text{of job done}}}.$$

After 4 hours, $(\frac{1}{4})(4) = 1$ whole job has been done.

These ideas are used in solving problems about the length of time needed to do a job. These problems are often called work problems.

EXAMPLE 5 Solving a Problem About Work Rates

With spraying equipment, Mateo can paint the woodwork in a small house in 8 hours. His assistant, Chet, needs 14 hours to complete the same job painting by hand. If both Mateo and Chet work together, how long will it take them to paint the woodwork?

Let $x =$ the number of hours it will take for Mateo and Chet to paint the woodwork, working together.

Certainly, x will be less than 8, since Mateo alone can complete the job in 8 hours. Begin by making a chart as shown. Remember that based on the previous discussion, Mateo's rate alone is $\frac{1}{8}$ job per hour, and Chet's rate is $\frac{1}{14}$ job per hour.

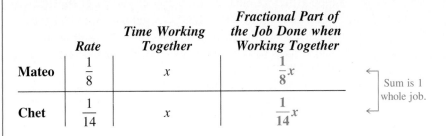

	Rate	Time Working Together	Fractional Part of the Job Done when Working Together	
Mateo	$\frac{1}{8}$	x	$\frac{1}{8}x$	← Sum is 1
Chet	$\frac{1}{14}$	x	$\frac{1}{14}x$	← whole job.

Since together Mateo and Chet complete 1 whole job, we must add their individual fractional parts and set the sum equal to 1.

CONTINUED ON NEXT PAGE

5. Work each problem.

(a) Michael can paint a room, working alone, in 8 hours. Lindsay can paint the same room, working alone, in 6 hours. How long will it take them if they work together?

$$\underbrace{\frac{1}{8}x}_{\text{Fractional part done by Mateo}} + \underbrace{\frac{1}{14}x}_{\text{Fractional part done by Chet}} = \underbrace{1}_{\text{1 whole job}}$$

$$56\left(\frac{1}{8}x + \frac{1}{14}x\right) = 56(1) \quad \text{Multiply by LCD, 56.}$$

$$56\left(\frac{1}{8}x\right) + 56\left(\frac{1}{14}x\right) = 56(1) \quad \text{Distributive property}$$

$$7x + 4x = 56$$

$$11x = 56 \quad \text{Combine like terms.}$$

$$x = \frac{56}{11} \quad \text{Divide by 11.}$$

Solution set: $\left\{\dfrac{56}{11}\right\}$

Working together, Mateo and Chet can paint the woodwork in $\frac{56}{11}$ hours, or $5\frac{1}{11}$ hours. Check to be sure the answer is reasonable.

An alternative approach to work problems is to consider the part of the job can that can be done in 1 hour. For instance, in Example 5 Mateo can do the entire job in 8 hours, and Chet can do it in 14. Thus, their work rates, as we saw in Example 5, are $\frac{1}{8}$ and $\frac{1}{14}$, respectively. Since it takes them x hours to complete the job when working together, in one hour they can paint $\frac{1}{x}$ of the woodwork. The amount painted by Mateo in one hour plus the amount painted by Chet in one hour must equal the amount they can do together. This leads to the equation

(b) Roberto can tune up his Bronco in 2 hours working alone. His brother Marco can do the job in 3 hours working alone. How long would it take them if they worked together?

$$\text{Amount by Mateo} \rightarrow \frac{1}{8} + \overset{\overset{\textstyle\text{Amount by Chet}}{\downarrow}}{\frac{1}{14}} = \frac{1}{x} \cdot \leftarrow \text{Amount together}$$

Compare this with the equation in Example 5. Both lead to the same solution.

◄◄ **WORK PROBLEM 5 AT THE SIDE.**

OBJECTIVE ▶4 Suppose that gasoline costs $1.50 per gallon. Then 1 gallon costs $1.50, 2 gallons cost 2($1.50) = $3.00, 3 gallons cost 3($1.50) = $4.50, and so on. Each time, the total cost is obtained by multiplying the number of gallons by the price per gallon. In general, if k equals the price per gallon and x equals the number of gallons, then the total cost y is equal to kx. Notice that as the number of gallons increases, the total cost increases.

The preceding discussion is an example of variation (vair-ee-AY-shun). Equations with fractions often result when discussing variation. As in the gasoline example, two variables **vary directly** if one is a constant multiple of the other.

Direct Variation

y varies directly as x if there exists a constant k such that

$$y = kx.$$

ANSWERS

5. (a) $3\frac{3}{7}$ hours **(b)** $1\frac{1}{5}$ hours

┌─
EXAMPLE 6   Using Direct Variation

Suppose y varies directly as x, and $y = 20$ when $x = 4$. Find y when $x = 9$.

Since y varies directly as x, there is a constant k such that $y = kx$. We know that $y = 20$ when $x = 4$. Substituting these values into $y = kx$ gives

$$y = kx$$
$$20 = k \cdot 4,$$

from which $\qquad\qquad k = 5.$

Since $y = kx$ and $k = 5$,

$$y = 5x. \qquad\qquad \text{Let } k = 5.$$

When $x = 9$, $\qquad y = 5x = 5 \cdot 9 = 45.$ $\quad$ Let $x = 9$.

Thus, $y = 45$ when $x = 9$.

─────────────────────────────────────

WORK PROBLEM 6 AT THE SIDE. ▶▶

In another common type of variation, the value of one variable increases while the value of another decreases. For example, an increase in the supply of an item causes a decrease in the price of the item.

Inverse Variation

y varies inversely (IN-vers-lee) **as x** if there exists a constant k such that

$$y = \frac{k}{x}.$$

┌─
EXAMPLE 7   Using Inverse Variation

In a certain manufacturing process, the cost of producing an item varies inversely as the number of items produced. If 10,000 items are produced, the cost is $2 per item. Find the cost per item to produce 25,000 items.

Let x = the number items produced and

c = the cost per item.

Since c varies inversely as x, there is a constant k such that

$$c = \frac{k}{x}.$$

First find k by replacing c with 2 and x with 10,000. Then c can be found for $x = 25,000$.

$$2 = \frac{k}{10{,}000}$$

$$20{,}000 = k \qquad\qquad \text{Multiply by } 10{,}000.$$

$$c = \frac{20{,}000}{25{,}000} = .80 \qquad \begin{array}{l}\text{Let } k = 20{,}000 \text{ and} \\ x = 25{,}000.\end{array}$$

The cost per item to make 25,000 items is $.80.

─────────────────────────────────────

WORK PROBLEM 7 AT THE SIDE. ▶▶

6. Work each problem.

(a) If z varies directly as t, and $z = 11$ when $t = 4$, find z when $t = 32$.

(b) The circumference of a circle varies directly as the radius. A circle with a radius of 7 centimeters has a circumference of 43.96 centimeters. Find the circumference if the radius is 11 centimeters.

7. Work each problem.

(a) Suppose z varies inversely as t, and $z = 8$ when $t = 2$. Find z when $t = 32$.

(b) The current in a simple electrical circuit varies inversely as the resistance. If the current is 80 amperes when the resistance is 10 ohms, find the current if the resistance is 16 ohms.

ANSWERS

6. (a) 88 **(b)** 69.08 centimeters

7. (a) $\frac{1}{2}$ **(b)** 50 amperes

Numbers in the

Real World *collaborative investigations*

Big Numbers in Our World (and Others) ... And Just How Big is a Googol?

The term **googol**, meaning 10^{100}, was coined by Professor Edward Kasner of Columbia University. A googol is made up of a 1 with 100 zeros following it. This number exceeds the estimated number of electrons in the universe, which is 10^{79}.

From *Mathematical Circles Revisited* by Howard Eves comes this list of estimates of large numbers:

$$10^{100} =$$

10,000,000,000,
000,000,000,00
0,000,000,000,
000,000,000,00
0,000,000,000,
000,000,000,00
0,000,000,000,
000,000,000,00
0,000,000,000,
000,000

1. the boiling point of iron is 5.4×10^3 degrees Fahrenheit;

2. the temperature at the center of an atomic bomb explosion is 2×10^8 degrees Fahrenheit;

3. the total number of possible bridge hands is 6.35×10^{11};

4. the total number of words spoken since the beginning of the world is about 10^{16};

5. the total number of printed words since the Gutenberg Bible appeared is somewhat larger than 10^{16};

6. the age of the earth is set at about 3350 million years, or about 10^{17} seconds;

7. the half-life of uranium 238 is 1.42×10^{17} seconds;

8. the number of grains of sand on the beach at Coney Island, New York, is about 10^{20};

9. the total age of the expanding universe is probably less than 10^{22} seconds;

10. the mass of the earth is about 1.2×10^{25} pounds;

11. the number of atoms of oxygen in an average thimble is perhaps about 10^{27};

12. the diameter of the universe, as assigned by relativity theory, is about 10^{29} centimeters;

13. the number of snow crystals necessary to form the ice age would be about 10^{30};

14. the total number of ways of arranging 52 cards is of the order 8×10^{67};

15. the total number of electrons in the universe is, by an estimate made by Sir Arthur Eddington, about 10^{79}.

FOR GROUP DISCUSSION

Suppose that you have just applied for and been offered a new job with "salary negotiable." You and your new employer meet to discuss the salary, and you present her with this offer: For 30 days, you will work and be paid 1¢ on Day one, 2¢ on Day two, 4¢ on Day three, 8¢ on Day four, and so on, doubling your pay each day. Your employer then accepts your offer.

1. When will you have received half of your month's wages?

2. When will you have received one-fourth of your total month's wages?

3. What do you think your approximate monthly salary will be? Use your calculator to find out.

4. How many days would it take for you to become a "googolaire?"

8.7 Exercises

Set up the equation you would use to solve the problem. Use x as the variable. In Exercises 1 and 2, work parts (a) and (b) first. Do not actually solve the equation.

1. One-third of a number is two more than one-sixth of the same number. What is the number?
   (a) Write an expression for "one-third of a number."
   (b) Write an expression for "two more than one-sixth of the same number."
   (c) Set up an equation to solve the problem.

2. The numerator of the fraction $\frac{13}{15}$ is increased by an amount so that the value of the resulting fraction is $\frac{7}{5}$. By what amount was the numerator increased?
   (a) What is the numerator of the fraction $\frac{13}{15}$?
   (b) Write an expression for "the numerator of the fraction $\frac{13}{15}$ increased by an amount."
   (c) Set up an equation to solve the problem.

Solve each problem. See Example 1.

3. In a certain fraction, the denominator is 4 less than the numerator. If 3 is added to both the numerator and the denominator, the resulting fraction is equal to $\frac{3}{2}$. Find the original fraction.

4. The denominator of a certain fraction is 3 times the numerator. If 2 is added to the numerator and subtracted from the denominator, the resulting fraction is equal to 1. Find the original fraction.

5. Calgene, a company that sells genetically engineered tomatoes, found that these tomatoes provided about $\frac{1}{2}$ of its total revenue in 1995. Production of oils contributed about $\frac{1}{4}$ of its total revenue in 1995. If these two products produced $74.8 million that year, what was the total revenue?

6. The profits from a carnival are to be given to two scholarships so that one scholarship receives $\frac{3}{2}$ as much money as the other. If the total amount given to the two scholarships is $780, how much goes to the scholarship that receives the lesser amount?

7. A recent news article stated that Germany invests $\frac{1}{10}$ as much as the United States in Mexico. If the total amount invested in the Mexican economy by these countries is $2 billion, how much does each country invest?

8. Japan invests about $\frac{3}{4}$ as much as Germany in the Mexican economy, according to a recent news article. If Japan and Germany invest a total of $.32 billion, how much does each of these countries invest?

Solve each problem. See Examples 2–4.

9. In the 1994 Olympic Nordic Skiing, Vladimir Smimov of Kazakhstan won the 50 kilometer event in 2.07 hours. What was his average speed?

10. Bonnie Blair won the women's 500 meter speed skating in the 1994 Olympics. Her average speed was 12.74 meters per second. What was her time?

11. Suppose Amanda walks D miles at R miles per hour in the same time that Kenneth walks d miles at r miles per hour. Give an equation relating D, R, d, and r.

12. If a migrating hawk travels m miles per hour in still air, what is its rate when it flies into a steady headwind of 5 miles per hour? What is its rate with a tailwind of 5 miles per hour?

13. Sandi Goldstein flew from Dallas to Indianapolis at 180 miles per hour and then flew back at 150 miles per hour. The trip at the slower speed took 1 hour longer than the trip at the higher speed. Find the distance between the two cities.

14. The distance from Seattle, Washington, to Victoria, British Columbia, is about 148 miles by ferry. It takes about 4 hours less to travel by the same ferry to Vancouver, British Columbia, a distance of about 74 miles. What is the average speed of the ferry?

15. Cosmas N'Deti of Kenya won the 1995 Boston marathon (a 26 mile race) in (about) .8 of an hour less time than the winner of the first Boston marathon in 1897, John H. McDermott of New York. Find each runner's speed, if McDermott's speed was (about) .73 times N'Deti's speed. (*Hint:* Don't round until the end of the calculations.)

	d	r	t
N'Deti			
McDermott			

16. Women first ran in the Boston marathon in 1972, when Nina Kuscsik of New York won the race. In 1995, the winner was Uta Pippig of Germany, whose time was .8 of an hour less than Kuscsik's in 1972. If Pippig ran $\frac{4}{3}$ as fast as Kuscsik, find each runner's speed. (See Exercise 15 for distance.)

17. A boat can go 20 miles against a current in the same time that it can go 60 miles with the current. The current is 4 miles per hour. Find the speed of the boat in still water.

	d	r	t
Against Current			
With Current			

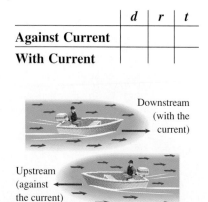

Downstream (with the current)

Upstream (against the current)

18. A plane flies 350 miles with the wind in the same time that it can fly 310 miles against the wind. The plane has a still-air speed of 165 miles per hour. Find the speed of the wind.

Solve each problem. See Example 5.

19. If it takes John Young 10 hours to do a job, what is his work rate?

20. If it takes Shelley McCarthy 12 hours to do a job, what part of the job does she do in 8 hours?

21. Geraldo and Luisa Hernandez operate a small cleaners. Luisa, working alone, can clean a day's laundry in 9 hours. Geraldo can clean a day's laundry in 8 hours. How long would it take them if they work together?

22. Lea can groom the horses in her boarding stable in 5 hours, while Tran needs 4 hours. How long will it take them to groom the horses if they work together?

23. A pump can pump the water out of a flooded basement in 10 hours. A smaller pump takes 12 hours. How long would it take to pump the water from the basement using both pumps?

24. A copier can do a printing job in 7 hours. A smaller copier can do the same job in 12 hours. How long would it take to do the job using both copiers?

25. An experienced employee can enter tax data into a computer twice as fast as a new employee. Working together, it takes the employees 2 hours. How long would it take the experienced employee working alone?

26. One roofer can put a new roof on a house three times faster than another. Working together they can roof a house in 4 days. How long would it take the faster roofer working alone?

27. If y varies directly as x, then y increases as x _____?_____ .
(decreases/increases)

28. If y varies inversely as x, then y increases as x _____?_____ .
(decreases/increases)

Solve the following variation problems. See Examples 6 and 7.

29. The interest on an investment varies directly as the rate of interest. If the interest is $48 when the interest rate is 5%, find the interest when the rate is 4.2%.

30. For a given base, the area of a triangle varies directly as its height. Find the area of a triangle with a height of 6 inches, if the area is 10 square inches when the height is 4 inches.

31. Over a specified distance, speed varies inversely with time. If a Dodge Viper on a test track goes a certain distance in one-half minute at 160 miles per hour, what speed is needed to go the same distance in three-fourths of a minute?

32. For a constant area, the length of a rectangle varies inversely as the width. The length of a rectangle is 27 feet when the width is 10 feet. Find the length of a rectangle with the same area if the width is 18 feet.

33. The pressure exerted by water at a given point varies directly with the depth of the point beneath the surface of the water. Water exerts 4.34 pounds per square inch for every 10 feet travelled below the water's surface. What is the pressure exerted on a scuba diver at 20 feet?

34. If the volume is constant, the pressure of a gas in a container varies directly as the temperature. If the pressure is 5 pounds per square inch at a temperature of 200 Kelvin, what is the pressure at a temperature of 300 Kelvin?

35. If the temperature is constant, the pressure of a gas in a container varies inversely as the volume of the container. If the pressure is 10 pounds per square foot in a container with 3 cubic feet, what is the pressure in a container with 1.5 cubic feet?

36. The force required to compress a spring varies directly as the change in the length of the spring. If a force of 12 pounds is required to compress a certain spring 3 inches, how much force is required to compress the spring 5 inches?

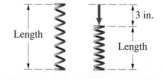

KEY TERMS

8.1 **rational expression** — The quotient of two polynomials with denominator not zero is called a rational expression.

  lowest terms — A rational expression is written in lowest terms if there are no common factors in the numerator and denominator (except 1).

8.3 **least common denominator** — (LCD) The smallest expression that is divisible by all denominators is called the least common denominator.

8.5 **complex fraction** — A rational expression with one or more fractions in the numerator, denominator, or both is a complex fraction.

8.7 **direct variation** — y varies directly as x if there is a constant k such that $y = kx$.

  inverse variation — y varies inversely as x if there is a constant k such that $y = \frac{k}{x}$.

QUICK REVIEW

Concepts	*Examples*

8.1 The Fundamental Property of Rational Expressions

To find the values for which a rational expression is not defined, set the denominator equal to zero and solve the equation.

Given: $\dfrac{x - 4}{x^2 - 16}$.

$$x^2 - 16 = 0$$
$$(x - 4)(x + 4) = 0$$
$$x - 4 = 0 \quad \text{or} \quad x + 4 = 0$$
$$x = 4 \quad \text{or} \quad x = -4$$

The rational expression is not defined for 4 or -4.

To write a rational expression in lowest terms, (1) factor; and (2) use the fundamental property to replace the quotient of common factors with 1.

Given: $\dfrac{x^2 - 1}{(x - 1)^2}$.

$$\frac{(x - 1)(x + 1)}{(x - 1)(x - 1)} = 1\left(\frac{x + 1}{x - 1}\right) = \frac{x + 1}{x - 1}$$

8.2 Multiplication and Division of Rational Expressions

Multiplication

$$\frac{3x + 9}{x - 5} \cdot \frac{x^2 - 3x - 10}{x^2 - 9}$$

1. Multiply numerators and multiply denominators.

$$= \frac{(3x + 9)(x^2 - 3x - 10)}{(x - 5)(x^2 - 9)}$$

2. Factor.

$$= \frac{3(x + 3)(x - 5)(x + 2)}{(x - 5)(x + 3)(x - 3)}$$

3. Write in lowest terms.

$$= \frac{3(x + 2)}{x - 3}$$

Division

$$\frac{2x + 1}{x + 5} \div \frac{6x^2 - x - 2}{x^2 - 25}$$

1. Multiply the first rational expression by the reciprocal of the second rational expression.

$$= \frac{2x + 1}{x + 5} \cdot \frac{x^2 - 25}{6x^2 - x - 2}$$

2. Factor.

$$= \frac{2x + 1}{x + 5} \cdot \frac{(x + 5)(x - 5)}{(2x + 1)(3x - 2)}$$

3. Write in lowest terms.

$$= \frac{x - 5}{3x - 2}$$

Concepts	*Examples*
8.3 Least Common Denominators **Finding the LCD**	Find the LCD for $\dfrac{3}{k^2 - 16}$ and $\dfrac{1}{4k^2 - 16k}$.
1. Factor each denominator into prime factors.	$k^2 - 16 = (k + 4)(k - 4)$ $4k^2 - 16k = 4k(k - 4)$
2. Choose each different denominator factor the greatest number of times it appears in any denominator.	
3. Multiply the factors from Step 2 to get the LCD.	$\text{LCD} = 4k(k + 4)(k - 4)$
To write a rational expression with the LCD as denominator:	$\dfrac{5}{2z^2 - 6z} = \dfrac{}{4z^3 - 12z^2}$
1. Factor both denominators.	$\dfrac{5}{2z(z - 3)} = \dfrac{}{4z^2(z - 3)}$
2. Decide what factors the denominator must be multiplied by to equal the LCD.	$2z(z - 3)$ must be multiplied by $2z$.
3. Multiply the rational expression by that factor over itself (multiply by 1).	$\dfrac{5}{2z(z - 3)} \cdot \dfrac{2z}{2z} = \dfrac{10z}{4z^2(z - 3)}$
8.4 Addition and Subtraction of Rational Expressions **Adding Rational Expressions**	Add $\dfrac{2}{3m + 6} + \dfrac{m}{m^2 - 4}$.
1. Find the LCD.	$\dfrac{2}{3(m + 2)} + \dfrac{m}{(m + 2)(m - 2)}$ *Factor.* The LCD is $3(m + 2)(m - 2)$.
2. Rewrite each rational expression as a fraction with the LCD as denominator.	$= \dfrac{2(m - 2)}{3(m + 2)(m - 2)} + \dfrac{3m}{3(m + 2)(m - 2)}$
3. Add the numerators to get the numerator of the sum. The LCD is the denominator of the sum.	$= \dfrac{2m - 4 + 3m}{3(m + 2)(m - 2)}$
4. Write in lowest terms.	$= \dfrac{5m - 4}{3(m + 2)(m - 2)}$
Subtracting Rational Expressions	Subtract $\dfrac{6}{k + 4} - \dfrac{2}{k}$.
Follow the same steps as for addition, but subtract in Step 3.	The LCD is $k(k + 4)$. $\dfrac{6k}{(k + 4)k} - \dfrac{2(k + 4)}{k(k + 4)} = \dfrac{6k - 2(k + 4)}{k(k + 4)}$ $= \dfrac{6k - 2k - 8}{k(k + 4)}$ $= \dfrac{4k - 8}{k(k + 4)}$

Concepts	Examples

8.5 Complex Fractions

Simplifying Complex Fractions

Method 1 Simplify the numerator and denominator separately. Then divide the simplified numerator by the simplified denominator.

$$\frac{\dfrac{1}{a} - a}{1 - a} = \frac{\dfrac{1}{a} - \dfrac{a^2}{a}}{1 - a} = \frac{\dfrac{1 - a^2}{a}}{1 - a}$$

$$= \frac{1 - a^2}{a} \cdot \frac{1}{1 - a}$$

$$= \frac{(1 - a)(1 + a)}{a(1 - a)} = \frac{1 + a}{a}$$

Method 2 Multiply numerator and denominator of the complex fraction by the LCD of all the denominators in the complex fraction.

$$\frac{\dfrac{1}{a} - a}{1 - a} = \frac{\dfrac{1}{a} - a}{1 - a} \cdot \frac{a}{a} = \frac{\dfrac{a}{a} - a^2}{(1 - a)a}$$

$$= \frac{1 - a^2}{(1 - a)a} = \frac{(1 + a)(1 - a)}{(1 - a)a}$$

$$= \frac{1 + a}{a}$$

8.6 Equations Involving Rational Expressions

1. Find the LCD of all denominators in the equation.

2. Using the multiplication property of equality, multiply each side of the equation by the LCD.

3. Solve the resulting equation, which should have no fractions.

4. Check each proposed solution.

Solve $\dfrac{2}{x - 1} + \dfrac{3}{4} = \dfrac{5}{4}$.

The LCD is $4(x - 1)$.

$$4(x - 1)\left(\frac{2}{x - 1} + \frac{3}{4}\right) = 4(x - 1)\left(\frac{5}{4}\right)$$

$$4(x - 1)\left(\frac{2}{x - 1}\right) + 4(x - 1)\left(\frac{3}{4}\right) = 4(x - 1)\left(\frac{5}{4}\right)$$

$$8 + 3(x - 1) = (x - 1)5$$
$$8 + 3x - 3 = 5x - 5$$
$$-2x = -10$$
$$x = 5$$

The proposed solution, 5, checks. The solution set is {5}.

8.7 Applications of Rational Expressions

Solving Motion Problems

1. Draw a sketch showing what is happening in the problem.

2. Make a chart showing the information in the problem and assigning the variables.

3. Use a form of the equation $d = rt$ to complete the chart.

On a trip from Sacramento to Monterey, Marge traveled at an average speed of 60 miles per hour. The return trip, at an average speed of 64 miles per hour, took $\frac{1}{4}$ hour less. How far did she travel between the two cities?

$$\text{Sacramento} \xrightarrow[\;\;64 \text{ mph}\;\;]{\;\;60 \text{ mph}\;\;} \text{Monterey}$$

	d	r	$t = \dfrac{d}{r}$
Going	x	60	$\dfrac{x}{60}$
Returning	x	64	$\dfrac{x}{64}$

Concepts	Examples

8.7 Applications of Rational Expressions (Continued)

4. Use the sketch and the chart to write an equation.

Since the time for the return trip was $\frac{1}{4}$ hour less, the time going equals the time returning plus $\frac{1}{4}$.

$$\frac{x}{60} = \frac{x}{64} + \frac{1}{4}$$

5. Solve the equation.

The solution set for this equation is $\{240\}$. She traveled 240 miles.

6. Check the answer.

Traveling 240 miles at 60 miles per hour takes 4 hours. Traveling 240 miles at 64 miles per hour takes $3\frac{3}{4}$ hours, which is $\frac{1}{4}$ hour less, as required.

Solving Problems About Work Rates

It takes the regular mail carrier 6 hours to cover the route. A substitute took 8 hours to cover the route. How long would it take them together?

1. If a job can be completed in t units of time, then the rate of work is $\frac{1}{t}$ job per unit of time.

Let t be the number of hours to cover the route together.

2. Make a chart giving the work rate, time, and part of the job done for each person.

3. Use either of the following approaches to get an equation.
  (a) Add the fractional parts of the job done by each person and set the sum equal to 1 (the whole job).
  (b) Add the fractional parts of the job that can be done in 1 hour by each person and set the sum equal to the part that can be done working together in 1 hour.

	Rate	Time Working Together	Part of Job Done
Regular	$\frac{1}{6}$	t	$\left(\frac{1}{6}\right)t$
Sub	$\frac{1}{8}$	t	$\left(\frac{1}{8}\right)t$

(a) $\frac{1}{6}t + \frac{1}{8}t = 1$ or **(b)** $\frac{1}{6} + \frac{1}{8} = \frac{1}{t}$

4. Solve the equation.

5. Check the solution.

The solution set of either equation is $\{3\frac{3}{7}\}$. It would take them $3\frac{3}{7}$ hours to cover the route together.

Solving Variation Problems

1. Write the variation equation using $y = kx$ or $y = \frac{k}{x}$.

If a varies inversely as b, and $a = 5$ when $b = 4$, find a when $b = 6$.

The equation is $a = \frac{k}{b}$.

2. Find k by substituting the given values of x and y into the equation.

Substitute $a = 5$ and $b = 4$.

$$5 = \frac{k}{4}$$

3. Write the equation with the value of k from Step 2 and the given value of x or y. Solve for the remaining variable.

The solution is $k = 20$. Let $k = 20$ and $b = 6$ in the variation equation.

$$a = \frac{20}{6} = \frac{10}{3}$$

CHAPTER 8 REVIEW EXERCISES

[8.1] *Find any values of the variables for which the following rational expressions are undefined.*

1. $\dfrac{4}{x-3}$

2. $\dfrac{y+3}{2y}$

3. $\dfrac{m-2}{m^2-2m-3}$

4. $\dfrac{2k+1}{3k^2+17k+10}$

Find the numerical value of each rational expression when (**a**) $x = -2$ *and* (**b**) $x = 4$.

5. $\dfrac{x^2}{x-5}$

6. $\dfrac{4x-3}{5x+2}$

7. $\dfrac{3x}{x^2-4}$

8. $\dfrac{x-1}{x+2}$

Write each rational expression in lowest terms.

9. $\dfrac{5a^3b^3}{15a^4b^2}$

10. $\dfrac{m-4}{4-m}$

11. $\dfrac{4x^2-9}{6-4x}$

12. $\dfrac{4p^2+8pq-5q^2}{10p^2-3pq-q^2}$

[8.2] *Find each product or quotient. Write each answer in lowest terms.*

13. $\dfrac{18p^3}{6} \cdot \dfrac{24}{p^4}$

14. $\dfrac{8x^2}{12x^5} \cdot \dfrac{6x^4}{2x}$

15. $\dfrac{9m^2}{(3m)^4} \div \dfrac{6m^5}{36m}$

16. $\dfrac{x-3}{4} \cdot \dfrac{5}{2x-6}$

17. $\dfrac{3q+3}{5-6q} \div \dfrac{4q+4}{2(5-6q)}$

18. $\dfrac{2r+3}{r-4} \cdot \dfrac{r^2-16}{6r+9}$

19. $\dfrac{6a^2+7a-3}{2a^2-a-6} \div \dfrac{a+5}{a-2}$

20. $\dfrac{y^2-6y+8}{y^2+3y-18} \div \dfrac{y-4}{y+6}$

21. $\dfrac{2p^2+13p+20}{p^2+p-12} \cdot \dfrac{p^2+2p-15}{2p^2+7p+5}$

22. $\dfrac{3z^2+5z-2}{9z^2-1} \cdot \dfrac{9z^2+6z+1}{z^2+5z+6}$

[8.3] *Find the least common denominator for each list of fractions.*

23. $\dfrac{1}{8}, \dfrac{5}{12}, \dfrac{7}{32}$

24. $\dfrac{4}{9y}, \dfrac{7}{12y^2}, \dfrac{5}{27y^4}$

25. $\dfrac{1}{m^2+2m}, \dfrac{4}{m^2+7m+10}$

26. $\dfrac{3}{x^2+4x+3}, \dfrac{5}{x^2+5x+4}$

Rewrite each rational expression with the given denominator.

27. $\dfrac{5}{8} = \dfrac{}{56}$

28. $\dfrac{10}{k} = \dfrac{}{4k}$

29. $\dfrac{3}{2a^3} = \dfrac{}{10a^4}$

30. $\dfrac{9}{x-3} = \dfrac{}{18-6x}$

31. $\dfrac{-3y}{2y-10} = \dfrac{}{50-10y}$

32. $\dfrac{4b}{b^2+2b-3} = \dfrac{}{(b+3)(b-1)(b+2)}$

[8.4] *Add or subtract as indicated. Write each answer in lowest terms.*

33. $\dfrac{10}{x} + \dfrac{5}{x}$

34. $\dfrac{6}{3p} - \dfrac{12}{3p}$

35. $\dfrac{9}{k} - \dfrac{5}{k-5}$

36. $\dfrac{4}{y} + \dfrac{7}{7+y}$

37. $\dfrac{m}{3} - \dfrac{2+5m}{6}$

38. $\dfrac{12}{x^2} - \dfrac{3}{4x}$

39. $\dfrac{5}{a-2b} + \dfrac{2}{a+2b}$

40. $\dfrac{4}{k^2-9} - \dfrac{k+3}{3k-9}$

41. $\dfrac{8}{z^2+6z} - \dfrac{3}{z^2+4z-12}$

42. $\dfrac{11}{2p-p^2} - \dfrac{2}{p^2-5p+6}$

[8.5] *Simplify each complex fraction.*

43. $\dfrac{\dfrac{a^4}{b^2}}{\dfrac{a^3}{b}}$

44. $\dfrac{\dfrac{y-3}{y}}{\dfrac{y+3}{4y}}$

45. $\dfrac{\dfrac{3m+2}{m}}{\dfrac{2m-5}{6m}}$

46. $\dfrac{\dfrac{1}{p} - \dfrac{1}{q}}{\dfrac{1}{q-p}}$

47. $\dfrac{x + \dfrac{1}{w}}{x - \dfrac{1}{w}}$

48. $\dfrac{\dfrac{1}{r+t} - 1}{\dfrac{1}{r+t} + 1}$

[8.6] *Solve each equation. Check your answer.*

49. $\dfrac{k}{5} - \dfrac{2}{3} = \dfrac{1}{2}$

50. $\dfrac{4-z}{z} + \dfrac{3}{2} = \dfrac{-4}{z}$

51. $\dfrac{x}{2} - \dfrac{x-3}{7} = -1$

52. $\dfrac{3y - 1}{y - 2} = \dfrac{5}{y - 2} + 1$

53. $\dfrac{3}{m - 2} + \dfrac{1}{m - 1} = \dfrac{7}{m^2 - 3m + 2}$

Solve for the specified variable.

54. $m = \dfrac{Ry}{t}$ for t

55. $x = \dfrac{3y - 5}{4}$ for y

56. $p^2 = \dfrac{4}{3m - q}$ for m

[8.7] *Solve each problem.*

57. When half a number is subtracted from two-thirds of the number, the answer is 2. Find the number.

58. The commission received by a salesperson for selling a small car is $\frac{2}{3}$ that received for selling a large car. On a recent day, Linda sold one of each, earning a commission of $300. Find the commission for each type of car.

59. In 1911, at the first Indianapolis 500 (mile) race, Ray Harroun won with an average speed of 74.59 miles per hour. What was his time?

60. A man can plant his garden in 5 hours, working alone. His daughter can do the same job in 8 hours. How long would it take them if they worked together?

61. The head gardener can mow the lawns in the city park twice as fast as his assistant. Working together, they can complete the job in $1\frac{1}{3}$ hours. How long would it take the head gardener working alone?

62. The area of a circle varies directly as the square of its radius. The world's largest pecan pie was baked in Oklahoma and it measured 40 feet in diameter. Its area was approximately 1256.64 square feet. Find the radius of the world's largest cherry pie, whose area was 314.16 square feet.

63. If a parallelogram has a fixed area, the height varies inversely as the base. A parallelogram has a height of 8 centimeters and a base of 12 centimeters. Find the height if the base is changed to 24 centimeters.

MIXED REVIEW EXERCISES

Perform the indicated operations.

64. $\dfrac{\dfrac{5}{x-y}+2}{3-\dfrac{2}{x+y}}$

65. $\dfrac{4}{m-1}-\dfrac{3}{m+1}$

66. $\dfrac{8p^5}{5}\div\dfrac{2p^3}{10}$

67. $\dfrac{r-3}{8}\div\dfrac{3r-9}{4}$

68. $\dfrac{\dfrac{5}{x}-1}{\dfrac{5-x}{3x}}$

69. $\dfrac{4}{z^2-2z+1}-\dfrac{3}{z^2-1}$

Solve.

70. $F=\dfrac{k}{d-D}$ for d

71. $\dfrac{2}{z}-\dfrac{z}{z+3}=\dfrac{1}{z+3}$

72. About $\frac{1}{10}$ as many people in the United States speak French at home as speak Spanish. A total of 19.1 million U.S. residents speak one of these two languages at home. How many speak Spanish?

73. One pipe can fill a swimming pool in 6 hours, and another pipe can do it in 9 hours. How long will it take the two pipes working together to fill the pool $\frac{3}{4}$ full?

74. Anne Kelly flew her plane 400 kilometers with the wind in the same time it took her to go 200 kilometers against the wind. The speed of the wind is 50 kilometers per hour. Find the speed of the plane in still air.

75. Phillip Morris and Ford are among the top ten U.S. companies in sales. In a recent year, Phillip Morris's sales were close to $\frac{1}{2}$ Ford's sales. If total sales for the two companies were \$151 billion, what were Ford's sales that year?

1. Find any values for which $\dfrac{3x - 1}{x^2 - 2x - 8}$ is undefined.

1. _____

Write each rational expression in lowest terms.

2. $\dfrac{4m^2 - 2m}{6m - 3}$

2. _____

3. $\dfrac{6a^2 + a - 2}{2a^2 - 3a + 1}$

3. _____

Multiply or divide. Write all answers in lowest terms.

4. $\dfrac{x^6 y}{x^3} \cdot \dfrac{y^2}{x^2 y^3}$

4. _____

5. $\dfrac{5(d - 2)}{9} \div \dfrac{3(d - 2)}{5}$

5. _____

6. $\dfrac{6k^2 - k - 2}{8k^2 + 10k + 3} \cdot \dfrac{4k^2 + 7k + 3}{3k^2 + 5k + 2}$

6. _____

7. $\dfrac{4a^2 + 9a + 2}{3a^2 + 11a + 10} \div \dfrac{4a^2 + 17a + 4}{3a^2 + 2a - 5}$

7. _____

Rewrite each rational expression with the given denominator.

8. $\dfrac{15}{4p} = \dfrac{}{64p^3}$

8. _____

9. $\dfrac{3}{6m - 12} = \dfrac{}{42m - 84}$

9. _____

Add or subtract as indicated. Write each answer in lowest terms.

10. $\dfrac{8}{c} - \dfrac{5}{c}$

10. _____

11. $\dfrac{-4}{y + 2} + \dfrac{6}{5y + 10}$

11. _____

12. _____

12. $\dfrac{3}{2m^2 - 9m - 5} - \dfrac{m + 1}{2m^2 - m - 1}$

Simplify each complex fraction.

13. _____

13. $\dfrac{\dfrac{2p}{k^2}}{\dfrac{3p^2}{k^3}}$

14. _____

14. $\dfrac{\dfrac{1}{x + 3} - 1}{1 + \dfrac{1}{x + 3}}$

15. _____

15. What values of x could not be solutions of the equation

$$\dfrac{2}{x + 1} - \dfrac{3}{x - 4} = 6?$$

Solve each equation.

16. _____

16. $\dfrac{4}{3y} + \dfrac{3}{5y} = -\dfrac{11}{30}$

17. _____

17. $\dfrac{2}{p^2 - 2p - 3} = \dfrac{3}{p - 3} + \dfrac{2}{p + 1}$

For each problem, write an equation, and solve it.

18. _____

18. If four times a number is added to the reciprocal of twice the number, the result is 3. Find the number.

19. _____

19. A boat goes 7 miles per hour in still water. It takes as long to go 20 miles upstream as 50 miles downstream. Find the speed of the current.

20. _____

20. The current in a simple electrical circuit varies inversely as the resistance. If the current is 50 amperes (an *ampere* is a unit for measuring current) when the resistance is 10 ohms (an *ohm* is a unit for measuring resistance), find the current if the resistance is 5 ohms.

1. Solve for x: $3x + 2(x - 4) = 5x - 8$.

2. Solve the inequality, and graph its solution set.

$$-3 \leq \frac{2}{3}x - 1 \leq 1$$

⟶

3. Otis Taylor invested some money at 4% interest and twice as much at 3% interest. His interest for the first year was $400. How much did he invest at each rate?

4. A student must have an average grade of at least 80 on the four tests in a course to earn a grade of B. David Hingle had grades of 79, 75, and 88 on the first three tests. What possible scores can he make on the fourth test so that he can get a B in the course?

5. Solve for t: $|6t - 4| + 8 = 3$.

6. Evaluate $-4^2 + (-4)^2$.

7. Solve for x: $3x - 5 \geq 1$ or $2x + 7 \leq 9$.

8. Solve for t: $.04t + .06(t - 1) = 1.04$.

9. Solve and graph the solution set.

$$4 - 7(q + 4) < -3q$$

⟶

10. Solve the compound inequality and graph the solution set.

$$3x - 2 < 10 \text{ and } -2x < 10$$

⟶

Find the intercepts of each line.

11. $-4x + 5y = 20$ **12.** $y = -4$ **13.** $x = 3y$

14. Solve: $5x + 2y = 16$
$3x - 3y = 18.$

15. If $f(x) = \dfrac{x + 1}{3 - x}$, find $f(2)$.

Write in scientific notation.

16. .000076

Write without scientific notation.

17. 5.6×10^9

Simplify. Write answers with only positive exponents. Assume that all variables represent nonzero real numbers.

18. $\dfrac{3x^{-4}y^3}{7^{-1}x^5y^{-4}}$ **19.** $\left(\dfrac{a^{-3}b^4}{a^2b^{-1}}\right)^{-2}$ **20.** $\left(\dfrac{m^{-4}n^2}{m^2n^{-3}}\right) \cdot \left(\dfrac{m^5n^{-1}}{m^{-2}n^5}\right)$

Perform the indicated operations.

21. $9(-3x^3 - 4x + 12) + 2(x^3 - x^2 + 3)$ **22.** $(4f + 3)(3f - 1)$

23. $(x + y)(x^2 - xy + y^2)$ **24.** $(7t^3 + 8)(7t^3 - 8)$

25. $\left(\dfrac{1}{4}x + 5\right)^2$

26. If $P(x) = -4x^3 + 2x - 8$, find $P(-2)$.

27. Divide $(2x^4 + 3x^3 - 8x^2 + x + 2)$ by $(x - 1)$.

Factor each of the following polynomials completely.

28. $2x^2 - 13x - 45$

29. $100t^4 - 25$

30. Solve the equation $3x^2 + 4x = 7$.

31. For what values of x is

$$\dfrac{x + 8}{x^2 - 3x - 4}$$

undefined?

32. Write in lowest terms: $\dfrac{8x^2 - 18}{8x^2 + 4x - 12}$.

Perform the indicated operations. Express answers in lowest terms.

33. $\dfrac{x + 4}{x - 2} + \dfrac{2x - 10}{x - 2}$

34. $\dfrac{2}{a + b} - \dfrac{3}{a - b}$

35. $\dfrac{2x}{2x - 1} + \dfrac{4}{2x + 1} + \dfrac{8}{4x^2 - 1}$

36. $\dfrac{3}{x^2 - y^2} - \dfrac{2}{x - y}$

37. Solve the equation $\dfrac{-3x}{x+1} + \dfrac{4x+1}{x} = \dfrac{-3}{x^2+x}$.

38. Solve the formula for q: $\dfrac{1}{f} = \dfrac{1}{p} + \dfrac{1}{q}$.

Solve each problem.

39. Lucinda can fly her plane 200 miles against the wind in the same time it takes her to fly 300 miles with the wind. The wind blows at 30 miles per hour. Find the speed of her plane in still air.

40. Machine A can complete a certain job in 2 hours. To speed up the work, Machine B, which could complete the job alone in 3 hours, is brought in to help. How long will it take the two machines to complete the job working together?

Roots and Radicals

9.1 Finding Roots

In Section 1.1, we discussed the idea of the *square* of a number. Recall that squaring a number means multiplying the number by itself.

$$\text{If } a = 8, \quad \text{then} \quad a^2 = \mathbf{8 \cdot 8} = 64.$$
$$\text{If } a = -4, \quad \text{then} \quad a^2 = (-4)(-4) = 16.$$
$$\text{If } a = -\frac{1}{2}, \quad \text{then} \quad a^2 = \left(-\frac{1}{2}\right)\left(-\frac{1}{2}\right) = \frac{1}{4}.$$

In this chapter, the opposite problem is considered.

$$\text{If } a^2 = 49, \quad \text{then} \quad a = ?$$
$$\text{If } a^2 = 100, \quad \text{then} \quad a = ?$$
$$\text{If } a^2 = 25, \quad \text{then} \quad a = ?$$

OBJECTIVE 1 ▶ Finding a in the three statements above requires finding a number that can be multiplied by itself to result in the given number. The number a is called a **square root** of the number a^2.

E X A M P L E 1   Finding the Square Roots of a Number

Find all square roots of 49.

Find a square root of 49 by thinking of a number that multiplied by itself gives 49. One square root is 7 because $7 \cdot 7 = 49$. Another square root of 49 is -7 because $(-7)(-7) = 49$. The number 49 has two square roots 7 and -7. One is positive, and one is negative.

WORK PROBLEM 1 AT THE SIDE. ▶▶

All numbers that have rational number square roots (such as 121 and $\frac{9}{4}$) are called **perfect squares.** The positive square root of a number is written with the symbol $\sqrt{}$. For example, the positive square root of 121 is 11 and that of $\frac{9}{4}$ is $\frac{3}{2}$. These are written

$$\sqrt{121} = 11 \quad \text{and} \quad \sqrt{\frac{9}{4}} = \frac{3}{2}.$$

The symbol $-\sqrt{}$ is used for the negative square root of a number. For example, the negative square roots of 121 and $\frac{9}{4}$ are written

$$-\sqrt{121} = -11 \quad \text{and} \quad -\sqrt{\frac{9}{4}} = -\frac{3}{2}.$$

www.mathnotes.com

OBJECTIVES

1 ▶ Find square roots.

2 ▶ Decide whether a given root is rational, irrational, or not a real number.

3 ▶ Find decimal approximations for irrational square roots.

4 ▶ Use the Pythagorean formula.

5 ▶ Find higher roots.

FOR EXTRA HELP

Tutorial   Tape 18   SSM, Sec. 9.1

1. Find all square roots.

  (a) 100

  (b) 25

  (c) 36

  (d) $\dfrac{25}{36}$

ANSWERS

1. (a) 10, -10 **(b)** 5, -5
  (c) 6, -6 **(d)** $\dfrac{5}{6}, -\dfrac{5}{6}$

2. Find the *square* of each radical expression.

(a) $\sqrt{41}$

Most calculators have a square root key, usually labeled $\boxed{\sqrt{x}}$, that will allow us to find the square root of a number. For example, if we enter 121 and press the square root key, the display will show 11.

The symbol $\sqrt{}$ is called a **radical sign** and, used alone, always represents the positive square root (except that $\sqrt{0} = 0$). The number inside the radical sign is called the **radicand** (RAD-ih-kand), and the entire expression, radical sign and radicand, is called a **radical** (RAD-ih-kul). An algebraic expression containing a radical is called a **radical expression.**

> If a is a positive real number,
>
> $$\sqrt{a} \text{ is the positive square root of } a,$$
> $$-\sqrt{a} \text{ is the negative square root of } a.$$
>
> Also, for nonnegative a,
>
> $$\sqrt{a} \cdot \sqrt{a} = (\sqrt{a})^2 = a \quad \text{and} \quad -\sqrt{a} \cdot -\sqrt{a} = (-\sqrt{a})^2 = a.$$
>
> Also, $\sqrt{0} = 0$.

(b) $-\sqrt{39}$

When the square root of a positive real number is squared, the result is that positive real number. (Also, $(\sqrt{0})^2 = 0$.) This is illustrated in the next example.

E X A M P L E   2   Squaring Radical Expressions

Find the *square* of each radical expression.

(a) $\sqrt{13}$
   $(\sqrt{13})^2 = 13$, by the definition of square root.

(b) $-\sqrt{29}$
   $(-\sqrt{29})^2 = 29$ The square of a *negative* number is positive.

(c) $\sqrt{p^2 + 1}$
   $(\sqrt{p^2 + 1})^2 = p^2 + 1$ Remember that $p^2 + 1$ is positive.

◀◀ **WORK PROBLEM 2 AT THE SIDE.**

(c) $\sqrt{2x^2 + 3}$

E X A M P L E   3   Finding Square Roots

Find each square root.

(a) $\sqrt{36}$

We find square roots by factoring to prime factors and looking for pairs of factors.

Because the root is 2, circle like numbers in *pairs*. Each pair gives a factor of the square root: $\sqrt{36} = 2 \cdot 3 = 6$.

(b) $-\sqrt{144}$

This symbol represents the negative square root of 144. By finding pairs of prime factors of 144, you should get

$$144 = \underbrace{2 \cdot 2} \cdot \underbrace{2 \cdot 2} \cdot \underbrace{3 \cdot 3}$$
$$-\sqrt{144} = -(2 \quad \cdot \quad 2 \quad \cdot \quad 3) = -12.$$

ANSWERS

2. (a) 41 **(b)** 39 **(c)** $2x^2 + 3$

CONTINUED ON NEXT PAGE

(c) $-\sqrt{\dfrac{16}{49}}$

Because $\dfrac{16}{49} = \dfrac{4}{7} \cdot \dfrac{4}{7}$, $-\sqrt{\dfrac{16}{49}} = -\dfrac{4}{7}$.

WORK PROBLEM 3 AT THE SIDE. ▶▶

OBJECTIVE 2▶ A number that is not a perfect square has a square root that is not a rational number. For example, $\sqrt{5}$ is not a rational number because it cannot be written as the ratio of two integers. However, $\sqrt{5}$ is a real number and corresponds to a point on the number line. As mentioned in Chapter 1, a real number that is not rational is called an *irrational number*. The number $\sqrt{5}$ is irrational. Many square roots of integers are irrational.

Not every number has a *real number* square root. For example, there is no real number that can be squared to get -36. (The square of a real number can never be negative.) Because of this, $\sqrt{-36}$ is not a real number.

If a is a negative real number, $\sqrt{a}$ is not a real number.

E X A M P L E   4   Identifying Types of Square Roots

Tell whether each square root is rational, irrational, or not a real number.

(a) $\sqrt{17}$

Because 17 is not a perfect square, $\sqrt{17}$ is irrational.

(b) $\sqrt{64}$

The number 64 is a perfect square, 8^2, so $\sqrt{64} = 8$, a rational number.

(c) $\sqrt{-25}$

There is no real number whose square is -25. Therefore, $\sqrt{-25}$ is not a real number.

WORK PROBLEM 4 AT THE SIDE. ▶▶

Note
Not all irrational numbers are square roots of integers. For example, π (approximately 3.14159) is an irrational number that is not a square root of any integer.

OBJECTIVE 3▶ Even if a number is irrational, a decimal that approximates the number can be found by using a calculator.

For example, if we use a calculator to find $\sqrt{10}$, the display will show 3.16227766, which is only an approximation of $\sqrt{10}$, and not an exact rational value.

---

3. Find each square root.

(a) $\sqrt{16}$

(b) $-\sqrt{169}$

(c) $-\sqrt{225}$

(d) $\sqrt{729}$

(e) $\sqrt{\dfrac{36}{25}}$

4. Tell whether each square root is *rational, irrational,* or *not a real number.*

(a) $\sqrt{9}$

(b) $\sqrt{7}$

(c) $\sqrt{\dfrac{4}{9}}$

(d) $\sqrt{72}$

(e) $\sqrt{-43}$

ANSWERS

3. (a) 4 **(b)** -13 **(c)** -15 **(d)** 27
(e) $\dfrac{6}{5}$

4. (a) rational **(b)** irrational
(c) rational **(d)** irrational
(e) not a real number

5. Find a decimal approximation for each square root.

(a) $\sqrt{28}$

(b) $\sqrt{63}$

(c) $\sqrt{190}$

(d) $\sqrt{1000}$

E X A M P L E 5 **Approximating Irrational Square Roots**

Use a calculator to find a decimal approximation for each square root. Round answers to the nearest thousandth.

(a) $\sqrt{11}$

Using the square root key of a calculator gives $3.31662479 \approx 3.317$, where $\approx$ means "is approximately equal to."

(b) $\sqrt{39} \approx 6.245$ Use a calculator.

(c) $\sqrt{740} \approx 27.203$

◀◀ **WORK PROBLEM 5 AT THE SIDE.**

OBJECTIVE ▶ 4 One application of square roots comes from the Pythagorean formula. By this formula, if c is the length of the hypotenuse of a right triangle, and a and b are the lengths of the two legs, then

$$a^2 + b^2 = c^2.$$

See Figure 1.

FIGURE I

E X A M P L E 6 **Using the Pythagorean Formula**

Find the third side of each right triangle with sides a, b, and c, where c is the hypotenuse.

(a) $a = 3, b = 4$

Use the formula to find c^2 first.

$$c^2 = a^2 + b^2$$
$$c^2 = 3^2 + 4^2 \qquad \text{Let } a = 3 \text{ and } b = 4.$$
$$c^2 = 9 + 16 = 25 \qquad \text{Square and add.}$$

Now find the positive square root of 25 to get c.

$$c = \sqrt{25} = 5$$

(Although -5 is also a square root of 25, the length of a side of a triangle must be a positive number. Therefore we discard the negative square root.)

(b) $c = 9, b = 5$

Substitute the given values in the formula, $c^2 = a^2 + b^2$. Then solve for a^2.

$$9^2 = a^2 + 5^2 \qquad \text{Let } c = 9 \text{ and } b = 5.$$
$$81 = a^2 + 25 \qquad \text{Square.}$$
$$56 = a^2$$

Use a calculator to find $a = \sqrt{56} \approx 7.483$.

ANSWERS

5. **(a)** 5.292 **(b)** 7.937 **(c)** 13.784
    (d) 31.623

Caution
Be careful not to make the common mistake of thinking that $\sqrt{a^2 + b^2}$ equals $a + b$. As Example 6(a) shows, $\sqrt{9 + 16} = \sqrt{25} = 5$. However, $\sqrt{9} + \sqrt{16} = 3 + 4 = 7$. Since $5 \neq 7$, in general,

$$\sqrt{a^2 + b^2} \neq a + b.$$

WORK PROBLEM 6 AT THE SIDE. ▶▶

The Pythagorean formula can be used to solve applied problems that involve right triangles.

E X A M P L E 7 Using the Pythagorean Formula

A ladder 10 feet long leans against a wall. The foot of the ladder is 6 feet from the base of the wall. How high up the wall does the top of the ladder rest?

As shown in Figure 2, a right triangle is formed with the ladder as the hypotenuse.

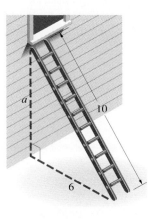

FIGURE 2

Let a represent the height of the top of the ladder when measured straight down to the ground. By the Pythagorean formula,

$$c^2 = a^2 + b^2$$
$$10^2 = a^2 + 6^2 \qquad \text{Let } c = 10 \text{ and } b = 6.$$
$$100 = a^2 + 36 \qquad \text{Square.}$$
$$64 = a^2 \qquad \text{Subtract 36.}$$
$$\sqrt{64} = a$$
$$a = 8. \qquad \sqrt{64} = 8$$

Choose the positive square root of 64 because a represents a length. The top of the ladder rests 8 feet up the wall.

■

WORK PROBLEM 7 AT THE SIDE. ▶▶

OBJECTIVE 5▶ Finding the square root of a number is the inverse of squaring a number. In a similar way, there are inverses to finding the cube of a number, or finding the fourth or higher power of a number. These inverses are called finding the **cube root**, written $\sqrt[3]{a}$, the **fourth root**, written $\sqrt[4]{a}$, and so on. In $\sqrt[n]{a}$, the number n is the **index** or **order** of the radical. It would be possible

6. Find the unknown side in each right triangle.

(a) $a = 7, b = 24$

(b) $c = 15, b = 13$

(c) $c = 11, a = 8$

7. A rectangle has dimensions 5 feet by 12 feet. Find the length of its diagonal.

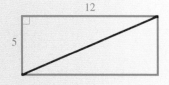

ANSWERS
6. (a) 25 **(b)** $\sqrt{56} \approx 7.483$
(c) $\sqrt{57} \approx 7.550$
7. 13 feet

8. Find each cube root.

(a) $\sqrt[3]{27}$

(b) $\sqrt[3]{64}$

(c) $\sqrt[3]{-125}$

9. Find each root.

(a) $\sqrt[4]{81}$

(b) $\sqrt[4]{-81}$

(c) $-\sqrt[4]{625}$

(d) $\sqrt[5]{243}$

(e) $\sqrt[5]{-32}$

to write $\sqrt[2]{a}$, instead of $\sqrt{a}$, but the simpler symbol $\sqrt{a}$ is customary because the square root is the most commonly used root. A scientific calculator can be used to find these roots. When working with cube roots or fourth roots, it is helpful to memorize the first few *perfect cubes* ($2^3 = 8$, $3^3 = 27$, and so on), and the first few perfect fourth powers.

EXAMPLE 8 Finding Cube Roots

Find each cube root.

(a) $\sqrt[3]{8}$

Look for a number that can be cubed to give 8. Because $2^3 = 8$, $\sqrt[3]{8} = 2$.

(b) $\sqrt[3]{-8}$

$\sqrt[3]{-8} = -2$ because $(-2)^3 = -8$.

(c) $\sqrt[3]{216}$

Factor prime factors as we did with square roots.

216
2 ⟍ 108
   2 ⟍ 54
      2 ⟍ 27
         3 ⟍ 9
            3   3

> Because the root is 3, circle groups of 3 like factors. Each group gives a factor of the cube root.

Therefore, $\sqrt[3]{216} = 2 \cdot 3 = 6$.

As these examples suggest, the cube root of a positive number is positive, and the cube root of a negative number is negative. *There is only one real number cube root for each real number.*

◀◀ WORK PROBLEM 8 AT THE SIDE.

When the index of the radical is even (square root, fourth root, and so on), the radicand must be nonnegative to get a real number root. Also, for even indexes the symbols $\sqrt{}$, $\sqrt[4]{}$, $\sqrt[6]{}$, and so on are used for the *nonnegative* roots, which are called **principal** (PRIN-sih-pul) **roots.**

EXAMPLE 9 Finding Higher Roots

Find each root.

(a) $\sqrt[4]{16}$

$\sqrt[4]{16} = 2$ because 2 is positive and $2^4 = 16$.

(b) $-\sqrt[4]{16} = -2$

(c) $\sqrt[4]{-16}$

There is no real number that equals $\sqrt[4]{-16}$ because a fourth power of a real number must be nonnegative.

(d) $-\sqrt[5]{32}$

First find $\sqrt[5]{32}$. Because 2 is the number whose fifth power is 32, $\sqrt[5]{32} = 2$. If $\sqrt[5]{32} = 2$, then $-\sqrt[5]{32} = -2$.

◀◀ WORK PROBLEM 9 AT THE SIDE.

ANSWERS

8. (a) 3 **(b)** 4 **(c)** −5

9. (a) 3
  (b) not a real number
  (c) −5 **(d)** 3 **(e)** −2

9.1 Exercises

Decide whether the statement is true or false. If false tell why.

1. Every nonnegative number has two square roots.

2. A negative number has negative square roots.

3. Every positive number has two square roots.

4. Every positive number has three real cube roots.

5. The cube root of every real number has the same sign as the number itself.

6. The positive square root of a positive number is its principal square root.

Find all square roots of each number. See Example 1.

7. 16

8. 9

9. 144

10. 225

11. $\dfrac{25}{196}$

12. $\dfrac{81}{400}$

13. 900

14. 1600

Find the square of each radical expression. See Example 2.

15. $\sqrt{100}$

16. $\sqrt{36}$

17. $-\sqrt{19}$

18. $-\sqrt{99}$

19. $\sqrt{3x^2 + 4}$

20. $\sqrt{9y^2 + 3}$

Find each square root that is a real number. See Examples 3 and 4(c).

21. $\sqrt{49}$

22. $\sqrt{81}$

23. $-\sqrt{121}$

24. $\sqrt{196}$

25. $-\sqrt{\dfrac{144}{121}}$

26. $-\sqrt{\dfrac{49}{36}}$

27. $\sqrt{-121}$

28. $\sqrt{-25}$

What must be true about a for each statement in Exercises 29–32 to be true?

29. $\sqrt{a}$ represents a positive number.

30. $-\sqrt{a}$ represents a negative number.

31. $\sqrt{a}$ is not a real number.

32. $-\sqrt{a}$ is not a real number.

Write rational, irrational, *or* not a real number *for each number. If a number is rational, give its exact value. If a number is irrational, give a decimal approximation to the nearest thousandth. Use a calculator as necessary. See Examples 4 and 5.*

33. $\sqrt{25}$

34. $\sqrt{169}$

35. $\sqrt{29}$

36. $\sqrt{33}$

37. $-\sqrt{64}$

38. $-\sqrt{900}$

39. $-\sqrt{300}$

40. $-\sqrt{500}$

41. $\sqrt{-29}$

42. $\sqrt{-47}$

43. Explain why the answers to Exercises 23 and 27 are different.

44. Explain why $\sqrt[3]{-8}$ and $-\sqrt[3]{8}$ represent the same number.

Find the length of the unknown side of each right triangle with sides a, b, and c, where c is the hypotenuse. See Figure 1 and Example 6.

45. $a = 8, b = 15$

46. $a = 24, b = 10$

47. $a = 6, c = 10$

48. $b = 12, c = 13$

49. $a = 11, b = 4$

50. $a = 13, b = 9$

Use the Pythagorean formula to solve each problem. In Exercises 57 and 58, round the answer to the nearest thousandth. See Example 7.

51. The diagonal of a rectangle measures 25 centimeters. The width of the rectangle is 7 centimeters. Find the length of the rectangle.

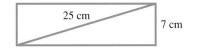

52. The length of a rectangle is 40 meters, and the width is 9 meters. Find the measure of the diagonal of the rectangle.

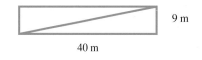

53. Megan Wacaser is flying a kite on 100 feet of string. How high is it above her hand (vertically) if the horizontal distance between Megan and the kite is 60 feet?

54. A guy wire is attached to the mast of a shortwave transmitting antenna. It is attached 96 feet above ground level. If the wire is staked to the ground 72 feet from the base of the mast, how long is the wire?

55. Two cars leave Tomball, Texas, at the same time. One travels north at 25 miles per hour, and the other travels west at 60 miles per hour. How far apart are they after 3 hours?

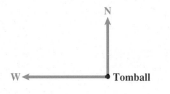

56. A boat is being pulled toward a dock with a rope attached at water level. When the boat is 24 feet from the dock, 30 feet of rope is extended. What is the height of the dock above the water?

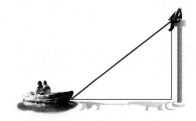

57. What is the value of *x* (to the nearest thousandth) in the figure?

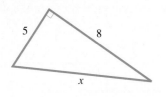

58. What is the value of *y* (to the nearest thousandth) in the figure?

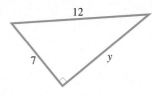

59. Use specific values for *a* and *b* different from those given in the "Caution" following Example 6 to show that $\sqrt{a^2 + b^2} \neq a + b$.

60. Why would the values *a* = 0 and *b* = 1 *not* be satisfactory in Exercise 59?

Find each of the following roots that are real numbers. See Examples 8 and 9.

61. $\sqrt[3]{1000}$

62. $\sqrt[3]{8}$

63. $\sqrt[3]{125}$

64. $\sqrt[3]{216}$

65. $\sqrt[4]{625}$

66. $\sqrt[4]{10,000}$

67. $\sqrt[4]{-1}$

68. $\sqrt[4]{-625}$

INTERPRETING TECHNOLOGY (EXERCISES 69–72)

Exercises 69–72 illustrate the way one graphing calculator shows nth roots. Give each indicated root. Exercises 71 and 72 indicate 5th roots.

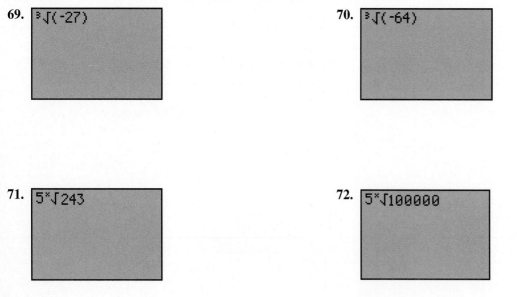

69. ³√(-27)

70. ³√(-64)

71. 5ˣ√243

72. 5ˣ√100000

9.2 *Multiplication and Division of Radicals*

OBJECTIVE 1 We develop several useful rules for finding products and quotients of radicals in this section. To illustrate the rule for products, notice that

$$\sqrt{4} \cdot \sqrt{9} = 2 \cdot 3 = 6 \quad \text{and} \quad \sqrt{4 \cdot 9} = \sqrt{36} = 6,$$

showing that

$$\sqrt{4} \cdot \sqrt{9} = \sqrt{4 \cdot 9}.$$

This result is a particular case of the more general *product rule for radicals*.

Product Rule for Radicals

For nonnegative real numbers a and b,

$$\sqrt{a} \cdot \sqrt{b} = \sqrt{a \cdot b} \quad \text{and} \quad \sqrt{a \cdot b} = \sqrt{a} \cdot \sqrt{b}.$$

The product of two radicals is the radical of the product, and the radical of a product is the product of the radicals.

EXAMPLE 1 Using the Product Rule to Multiply Radicals

Use the product rule for radicals to find each product.

(a) $\sqrt{2} \cdot \sqrt{3} = \sqrt{2 \cdot 3} = \sqrt{6}$

(b) $\sqrt{7} \cdot \sqrt{5} = \sqrt{35}$

(c) $\sqrt{11} \cdot \sqrt{a} = \sqrt{11a}$ Assume $a \geq 0$.

WORK PROBLEM 1 AT THE SIDE. ▶▶

OBJECTIVE 2 A very important use of the product rule is in simplifying radical expressions. As a first step, a radical expression is *simplified* when no perfect square factor remains under the radical sign. This is accomplished by using the product rule in the form $\sqrt{a \cdot b} = \sqrt{a} \cdot \sqrt{b}$. Example 2 shows how a radical may be simplified using the product rule.

EXAMPLE 2 Using the Product Rule to Simplify Radicals

Simplify each radical.

(a) $\sqrt{20}$

Because 20 has a perfect square factor of 4, we can write

$$\sqrt{20} = \sqrt{4 \cdot 5} \qquad \text{4 is a perfect square.}$$
$$= \sqrt{4} \cdot \sqrt{5} \qquad \text{Product rule}$$
$$= 2\sqrt{5}. \qquad \sqrt{4} = 2$$

Thus, $\sqrt{20} = 2\sqrt{5}$. Because 5 has no perfect square factor (other than 1), $2\sqrt{5}$ is called the *simplified form* of $\sqrt{20}$.

We could also factor 20 to prime factors and look for pairs of like factors.

```
    20
   ╱  ╲
  2    10        Each pair of like factors produces a rational factor
      ╱  ╲       of the simplified form.
     2    5
```

Therefore, $\sqrt{20} = \sqrt{2 \cdot 2 \cdot 5} = 2\sqrt{5}$.

— **CONTINUED ON NEXT PAGE**

OBJECTIVES

1 ▶ Multiply radicals.

2 ▶ Simplify radicals using the product rule.

3 ▶ Simplify radical quotients.

4 ▶ Simplify higher roots.

FOR EXTRA HELP

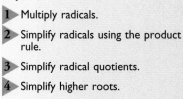

Tutorial   Tape 18   SSM, Sec. 9.2

1. Use the product rule for radicals to find each product.

(a) $\sqrt{6} \cdot \sqrt{11}$

(b) $\sqrt{2} \cdot \sqrt{5}$

(c) $\sqrt{10} \cdot \sqrt{r}, \; r \geq 0$

ANSWERS

1. (a) $\sqrt{66}$ (b) $\sqrt{10}$ (c) $\sqrt{10r}$

551

2. Simplify each radical.

(a) $\sqrt{8}$

(b) $\sqrt{27}$

(c) $\sqrt{50}$

(d) $\sqrt{60}$

3. Find each product and simplify.

(a) $\sqrt{3} \cdot \sqrt{15}$

(b) $\sqrt{10} \cdot \sqrt{50}$

(c) $\sqrt{12} \cdot \sqrt{2}$

(d) $\sqrt{7} \cdot \sqrt{14}$

(b) $\sqrt{72}$

Notice that 9 is a perfect square factor of 72. We could begin by factoring 72 as $9 \cdot 8$, to get

$$\sqrt{72} = \sqrt{9 \cdot 8} = 3\sqrt{8},$$

but then we would have to factor 8 as $4 \cdot 2$ in order to complete the simplification.

$$\sqrt{72} = 3\sqrt{8} = 3\sqrt{4 \cdot 2} = 3\sqrt{4} \cdot \sqrt{2} = 3 \cdot 2\sqrt{2} = 6\sqrt{2}$$

We could also factor 72 by finding pairs of like factors (not necessarily prime).

 Because $36 = 6 \cdot 6$, we can shortcut the prime factor process.

Thus, $\sqrt{72} = \sqrt{6 \cdot 6 \cdot 2} = 6\sqrt{2}$. In either case, we obtain $6\sqrt{2}$ as the simplified form of $\sqrt{72}$; however, our work is simpler if we begin with the largest perfect square factor.

(c) $\sqrt{300} = \sqrt{100 \cdot 3}$ 100 is a perfect square.

$\quad\quad = \sqrt{100} \cdot \sqrt{3}$ Product rule

$\quad\quad = 10\sqrt{3}$ $\sqrt{100} = 10$

(d) $\sqrt{15}$

The number 15 has no perfect square factors (except 1), so $\sqrt{15}$ cannot be simplified further.

◀◀ **WORK PROBLEM 2 AT THE SIDE.**

Sometimes the product rule can be used to simplify an answer, as Example 3 shows.

E X A M P L E   3   Multiplying and Simplifying Radicals

Find each product and simplify.

(a) $\sqrt{9} \cdot \sqrt{75} = 3\sqrt{75}$ $\sqrt{9} = 3$

$\quad\quad\quad\quad\quad = 3\sqrt{25 \cdot 3}$ 25 is a perfect square.

$\quad\quad\quad\quad\quad = 3\sqrt{25} \cdot \sqrt{3}$ Product rule

$\quad\quad\quad\quad\quad = 3 \cdot 5\sqrt{3}$ $\sqrt{25} = 5$

$\quad\quad\quad\quad\quad = 15\sqrt{3}$ Multiply.

Notice that we could have used the product rule to get $\sqrt{9} \cdot \sqrt{75} = \sqrt{675}$, and then simplified. However, the product rule as used above allows us to obtain the final answer without using a number as large as 675.

(b) $\sqrt{8} \cdot \sqrt{12} = \sqrt{8 \cdot 12}$ Product rule

$\quad\quad\quad\quad\quad = \sqrt{8 \cdot 2 \cdot 6}$ Factor 12.

$\quad\quad\quad\quad\quad = \sqrt{16 \cdot 6}$ 16 is a perfect square.

$\quad\quad\quad\quad\quad = \sqrt{16} \cdot \sqrt{6}$ Product rule

$\quad\quad\quad\quad\quad = 4\sqrt{6}$ $\sqrt{16} = 4$

◀◀ **WORK PROBLEM 3 AT THE SIDE.**

OBJECTIVE 3 The quotient rule for radicals is very similar to the product rule. It, too, can be used either way.

Quotient Rule for Radicals

If a and b are nonnegative real numbers and b is not 0,

$$\sqrt{\frac{a}{b}} = \frac{\sqrt{a}}{\sqrt{b}} \quad \text{and} \quad \frac{\sqrt{a}}{\sqrt{b}} = \sqrt{\frac{a}{b}}.$$

The radical of a quotient is the quotient of the radicals, and the quotient of two radicals is the radical of the quotient.

E X A M P L E 4 **Using the Quotient Rule to Simplify Radicals**

Simplify each radical.

(a) $\sqrt{\frac{25}{9}} = \frac{\sqrt{25}}{\sqrt{9}} = \frac{5}{3}$ Quotient rule

(b) $\frac{\sqrt{288}}{\sqrt{2}} = \sqrt{\frac{288}{2}} = \sqrt{144} = 12$ Quotient rule

(c) $\sqrt{\frac{3}{4}} = \frac{\sqrt{3}}{\sqrt{4}} = \frac{\sqrt{3}}{2}$ Quotient rule

E X A M P L E 5 **Using the Quotient Rule**

Simplify $\frac{27\sqrt{15}}{9\sqrt{3}}$.

Use multiplication of fractions and the quotient rule as follows.

$$\frac{27\sqrt{15}}{9\sqrt{3}} = \frac{27}{9} \cdot \frac{\sqrt{15}}{\sqrt{3}} = \frac{27}{9} \cdot \sqrt{\frac{15}{3}} = 3\sqrt{5}$$

WORK PROBLEM 4 AT THE SIDE. ▶▶

Some problems require both the product and the quotient rules, as Example 6 shows.

E X A M P L E 6 **Using Both the Product and the Quotient Rules**

Simplify $\sqrt{\frac{3}{5}} \cdot \sqrt{\frac{1}{5}}$.

Use the product and quotient rules.

$$\sqrt{\frac{3}{5}} \cdot \sqrt{\frac{1}{5}} = \sqrt{\frac{3}{5} \cdot \frac{1}{5}} \quad \text{Product rule}$$

$$= \sqrt{\frac{3}{25}} \quad \text{Multiply fractions.}$$

$$= \frac{\sqrt{3}}{\sqrt{25}} \quad \text{Quotient rule}$$

$$= \frac{\sqrt{3}}{5} \quad \sqrt{25} = 5$$

WORK PROBLEM 5 AT THE SIDE. ▶▶

---

4. Use the quotient rule to simplify each radical.

(a) $\sqrt{\frac{81}{16}}$

(b) $\frac{\sqrt{192}}{\sqrt{3}}$

(c) $\sqrt{\frac{10}{49}}$

(d) $\frac{8\sqrt{50}}{4\sqrt{5}}$

5. Multiply and then simplify each product.

(a) $\sqrt{\frac{5}{6}} \cdot \sqrt{120}$

(b) $\sqrt{\frac{3}{8}} \cdot \sqrt{\frac{7}{2}}$

ANSWERS

4. (a) $\frac{9}{4}$ **(b)** 8 **(c)** $\frac{\sqrt{10}}{7}$ **(d)** $2\sqrt{10}$

5. (a) 10 **(b)** $\frac{\sqrt{21}}{4}$

6. Simplify each radical. Assume all variables represent nonnegative real numbers.

(a) $\sqrt{36y^6}$

(b) $\sqrt{100p^8}$

(c) $\sqrt{a^5}$

Finally, the properties of this section are also valid when variables appear under the radical sign, as long as all the variables represent only nonnegative real numbers. For example, $\sqrt{5^2} = 5$, but $\sqrt{(-5)^2} = \sqrt{25} \neq -5$. This means that the square root of a squared number is always nonnegative. We can use absolute value to express this.

For any real number a,
$$\sqrt{a^2} = |a|.$$

In examples and exercises where variables are assumed to be nonnegative, absolute value bars are not necessary, because for $x \geq 0$, $|x| = x$.

EXAMPLE 7 Simplifying Radicals Involving Variables

Simplify each radical. Assume all variables represent nonnegative real numbers.

(a) $\sqrt{25m^4} = \sqrt{25} \cdot \sqrt{m^4}$ Product rule

           $= 5m^2$

(b) $\sqrt{64p^{10}} = 8p^5$ Product rule

(c) $\sqrt{r^9} = \sqrt{r^8 \cdot r}$

       $= \sqrt{r^8} \cdot \sqrt{r} = r^4\sqrt{r}$ Product rule

◀◀ WORK PROBLEM 6 AT THE SIDE.

7. Simplify each root.

(a) $\sqrt[3]{108}$

OBJECTIVE 4 The product rule and the quotient rule for radicals also work for other roots. To simplify cube roots, look for factors that are *perfect cubes*. A **perfect cube** is a number with a rational cube root. For example, $\sqrt[3]{64} = 4$, and because 4 is a rational number, 64 is a perfect cube. Higher roots are handled in a similar manner.

Properties of Radicals

For all real numbers where the indicated roots exist,
$$\sqrt[n]{x} \cdot \sqrt[n]{y} = \sqrt[n]{xy} \quad \text{and} \quad \frac{\sqrt[n]{x}}{\sqrt[n]{y}} = \sqrt[n]{\frac{x}{y}} \quad (y \neq 0).$$

(b) $\sqrt[4]{160}$

EXAMPLE 8 Simplifying Higher Roots

Simplify each radical.

(a) $\sqrt[3]{32} = \sqrt[3]{8 \cdot 4}$ 8 is a perfect cube.

       $= \sqrt[3]{8} \cdot \sqrt[3]{4} = 2\sqrt[3]{4}$

(c) $\sqrt[4]{\dfrac{16}{625}}$

(b) $\sqrt[4]{32} = \sqrt[4]{16} \cdot \sqrt[4]{2} = 2\sqrt[4]{2}$ 16 is a perfect fourth power.

(c) $\sqrt[3]{\dfrac{8}{125}} = \dfrac{\sqrt[3]{8}}{\sqrt[3]{125}} = \dfrac{2}{5}$

◀◀ WORK PROBLEM 7 AT THE SIDE.

9.2 Exercises

Decide whether each statement is true or false.

1. $\sqrt{(-6)^2} = -6$

2. $\sqrt[3]{(-6)^3} = -6$

Use the product rule for radicals to find each product. See Example 1.

3. $\sqrt{3} \cdot \sqrt{27}$

4. $\sqrt{2} \cdot \sqrt{8}$

5. $\sqrt{6} \cdot \sqrt{15}$

6. $\sqrt{10} \cdot \sqrt{15}$

7. $\sqrt{13} \cdot \sqrt{13}$

8. $\sqrt{17} \cdot \sqrt{17}$

9. $\sqrt{13} \cdot \sqrt{r}, \, r \geq 0$

10. $\sqrt{19} \cdot \sqrt{k}, \, k \geq 0$

11. Which one of the following radicals is simplified according to the guidelines of Objective 2?

  (a) $\sqrt{47}$ **(b)** $\sqrt{45}$ **(c)** $\sqrt{48}$ **(d)** $\sqrt{44}$

12. If p is a prime number, is $\sqrt{p}$ in simplified form? Explain your answer.

Simplify each radical. See Example 2.

13. $\sqrt{45}$

14. $\sqrt{27}$

15. $\sqrt{90}$

16. $\sqrt{56}$

17. $\sqrt{75}$

18. $\sqrt{18}$

19. $\sqrt{125}$

20. $\sqrt{80}$

21. $-\sqrt{700}$

22. $-\sqrt{600}$

23. $3\sqrt{27}$

24. $9\sqrt{8}$

Find each product and simplify. See Example 3.

25. $\sqrt{3} \cdot \sqrt{18}$

26. $\sqrt{3} \cdot \sqrt{21}$

27. $\sqrt{12} \cdot \sqrt{48}$

28. $\sqrt{50} \cdot \sqrt{72}$

29. $\sqrt{12} \cdot \sqrt{30}$

30. $\sqrt{30} \cdot \sqrt{24}$

31. Simplify the product $\sqrt{8} \cdot \sqrt{32}$ in two different ways. First, multiply 8 by 32 and simplify the square root of this product. Second, simplify $\sqrt{8}$ and simplify $\sqrt{32}$ and then multiply. Do you get the same answer? Make a conjecture (an educated guess) about whether the correct answer can always be obtained using either method in a simplification such as this.

32. Simplify the radical $\sqrt{288}$ in two ways. First, factor 288 as $144 \cdot 2$ and then simplify. Second, factor 288 as $48 \cdot 6$ and then simplify, performing any additional steps as needed. Do you get the same answer? Make a conjecture concerning the quickest way to simplify such a radical.

Use the quotient rule and the product rule, as necessary, to simplify each of the following. See Examples 4–6.

33. $\sqrt{\dfrac{16}{225}}$

34. $\sqrt{\dfrac{9}{100}}$

35. $\sqrt{\dfrac{7}{16}}$

36. $\sqrt{\dfrac{13}{25}}$

37. $\sqrt{\dfrac{5}{7}} \cdot \sqrt{35}$

38. $\sqrt{\dfrac{10}{13}} \cdot \sqrt{130}$

39. $\sqrt{\dfrac{5}{2}} \cdot \sqrt{\dfrac{125}{8}}$

40. $\sqrt{\dfrac{8}{3}} \cdot \sqrt{\dfrac{512}{27}}$

41. $\dfrac{30\sqrt{10}}{5\sqrt{2}}$

42. $\dfrac{50\sqrt{20}}{2\sqrt{10}}$

Simplify each radical. Assume that all variables represent nonnegative real numbers. See Example 7.

43. $\sqrt{m^2}$

44. $\sqrt{k^2}$

45. $\sqrt{y^4}$

46. $\sqrt{s^4}$

47. $\sqrt{36z^2}$

48. $\sqrt{49n^2}$

49. $\sqrt{400x^6}$

50. $\sqrt{900y^8}$

51. $\sqrt{z^5}$

52. $\sqrt{a^{13}}$

53. $\sqrt{x^6 y^{12}}$

54. $\sqrt{a^8 b^{10}}$

Simplify each radical. See Example 8.

55. $\sqrt[3]{40}$

56. $\sqrt[3]{48}$

57. $\sqrt[3]{54}$

58. $\sqrt[3]{135}$

59. $\sqrt[3]{128}$

60. $\sqrt[3]{192}$

61. $\sqrt[4]{80}$

62. $\sqrt[4]{243}$

63. $\sqrt[3]{\dfrac{8}{27}}$

64. $\sqrt[3]{\dfrac{64}{125}}$

65. $\sqrt[3]{-\dfrac{216}{125}}$

66. $\sqrt[3]{-\dfrac{1}{64}}$

67. In Example 2(a) we showed *algebraically* that $\sqrt{20}$ is equal to $2\sqrt{5}$. To give *numerical support* to this result, use a calculator to do the following:

(a) Find a decimal approximation for $\sqrt{20}$ using your calculator. Record as many digits as the calculator shows.

(b) Find a decimal approximation for $\sqrt{5}$ using your calculator, and then multiply the result by 2. Record as many digits as the calculator shows.

(c) Your results in parts (a) and (b) should be the same. A mathematician would not accept this numerical exercise as *proof* that $\sqrt{20}$ is equal to $2\sqrt{5}$. Can you explain why?

68. On your calculator, multiply the approximations for $\sqrt{3}$ and $\sqrt{5}$. Now, predict what your calculator will show when you find an approximation for $\sqrt{15}$. What rule stated in this section justifies your answer?

The volume of a cube is found with the formula $V = s^3$, where s is the length of an edge of the cube. Use this information in Exercises 69 and 70.

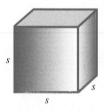

69. A container in the shape of a cube has a volume of 216 cubic centimeters. What is the depth of the container?

70. A cube-shaped box must be constructed to contain 128 cubic feet. What should the dimensions (height, width, and length) of the box be?

The volume of a sphere is found with the formula $V = (4/3)\pi r^3$, where r is the length of the radius of the sphere. Use this information in Exercises 71 and 72.

71. A ball in the shape of a sphere has a volume of 288π cubic inches. What is the radius of the ball?

72. Suppose that the volume of the ball described in Exercise 71 is multiplied by 8. How is the radius affected?

73. When we multiply two radicals with variables under the radical sign, such as $\sqrt{a} \cdot \sqrt{b} = \sqrt{ab}$, why is it important to know that both a and b represent nonnegative numbers?

74. Is it necessary to restrict k to a nonnegative number to say that $\sqrt[3]{k} \cdot \sqrt[3]{k} \cdot \sqrt[3]{k} = k$? Why?

9.3 Addition and Subtraction of Radicals

OBJECTIVES

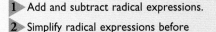

1. Add and subtract radical expressions.
2. Simplify radical expressions before adding or subtracting.
3. Simplify radical expressions involving multiplication.

FOR EXTRA HELP

Tutorial   Tape 19   SSM, Sec. 9.3

OBJECTIVE 1 We add or subtract radical expressions by using the distributive property. For example,

$$8\sqrt{3} + 6\sqrt{3} = (8 + 6)\sqrt{3} = 14\sqrt{3}.$$

Also,

$$2\sqrt{11} - 7\sqrt{11} = -5\sqrt{11}.$$

Only **like radicals,** those that are multiples of the same root of the same number, can be combined in this way.

E X A M P L E 1   Adding and Subtracting Like Radicals

Add or subtract, as indicated.

(a) $3\sqrt{6} + 5\sqrt{6} = (3 + 5)\sqrt{6} = 8\sqrt{6}$ Distributive property

(b) $5\sqrt{10} - 7\sqrt{10} = (5 - 7)\sqrt{10} = -2\sqrt{10}$

(c) $\sqrt{7} + 2\sqrt{7} = 1\sqrt{7} + 2\sqrt{7} = 3\sqrt{7}$

(d) $\sqrt{5} + \sqrt{5} = 1\sqrt{5} + 1\sqrt{5} = (1 + 1)\sqrt{5} = 2\sqrt{5}$

(e) $\sqrt{3} + \sqrt{7}$ Cannot be simplified further

WORK PROBLEM 1 AT THE SIDE. ▶▶

OBJECTIVE 2 Sometimes each radical expression in a sum or difference must be simplified first. Doing this might cause like radicals to appear, which then can be added or subtracted.

E X A M P L E 2   Adding and Subtracting Radicals That Require Simplification

Add or subtract, as indicated.

(a) $3\sqrt{2} + \sqrt{8} = 3\sqrt{2} + \sqrt{4 \cdot 2}$ Simplify $\sqrt{8}$.

$\qquad\qquad\quad = 3\sqrt{2} + \sqrt{4} \cdot \sqrt{2}$ Product rule

$\qquad\qquad\quad = 3\sqrt{2} + 2\sqrt{2}$ $\sqrt{4} = 2$

$\qquad\qquad\quad = 5\sqrt{2}$ Add like radicals.

(b) $\sqrt{18} - \sqrt{27} = \sqrt{9 \cdot 2} - \sqrt{9 \cdot 3}$ Simplify $\sqrt{18}$ and $\sqrt{27}$.

$\qquad\qquad\quad = \sqrt{9} \cdot \sqrt{2} - \sqrt{9} \cdot \sqrt{3}$ Product rule

$\qquad\qquad\quad = 3\sqrt{2} - 3\sqrt{3}$ $\sqrt{9} = 3$

Because $\sqrt{2}$ and $\sqrt{3}$ are unlike radicals, this difference cannot be simplified further.

(c) $2\sqrt{12} + 3\sqrt{75} = 2(\sqrt{4} \cdot \sqrt{3}) + 3(\sqrt{25} \cdot \sqrt{3})$ Product rule

$\qquad\qquad\quad\quad = 2(2\sqrt{3}) + 3(5\sqrt{3})$ $\sqrt{4} = 2$ and $\sqrt{25} = 5$

$\qquad\qquad\quad\quad = 4\sqrt{3} + 15\sqrt{3}$ Multiply.

$\qquad\qquad\quad\quad = 19\sqrt{3}$ Add like radicals.

WORK PROBLEM 2 AT THE SIDE. ▶▶

OBJECTIVE 3 Some radical expressions require both multiplication and addition (or subtraction). The order of operations presented earlier still applies.

---

1. Add or subtract, as indicated.

   (a) $8\sqrt{5} + 2\sqrt{5}$

   (b) $-4\sqrt{3} + 9\sqrt{3}$

   (c) $12\sqrt{11} - 3\sqrt{11}$

   (d) $\sqrt{15} + \sqrt{15}$

   (e) $2\sqrt{7} + 2\sqrt{10}$

2. Add or subtract, as indicated.

   (a) $\sqrt{8} + 4\sqrt{2}$

   (b) $\sqrt{27} + \sqrt{12}$

   (c) $5\sqrt{200} - 6\sqrt{18}$

ANSWERS

1. **(a)** $10\sqrt{5}$ **(b)** $5\sqrt{3}$ **(c)** $9\sqrt{11}$
   (d) $2\sqrt{15}$
   (e) cannot be simplified further
2. **(a)** $6\sqrt{2}$ **(b)** $5\sqrt{3}$ **(c)** $32\sqrt{2}$

3. Multiply and combine terms. Assume all variables represent nonnegative real numbers.

(a) $\sqrt{7} \cdot \sqrt{21} + 2\sqrt{27}$

(b) $\sqrt{3r} \cdot \sqrt{6} + \sqrt{8r}$

(c) $\sqrt[3]{81x^4} + 5\sqrt[3]{24x^4}$

E X A M P L E 3 **Multiplying and Combining Terms in Radical Expressions**

Multiply and combine terms. Assume all variables represent nonnegative real numbers.

(a)
$$
\begin{aligned}
\sqrt{5} \cdot \sqrt{15} + 4\sqrt{3} &= \sqrt{5 \cdot 15} + 4\sqrt{3} && \text{Product rule} \\
&= \sqrt{75} + 4\sqrt{3} && \text{Multiply.} \\
&= \sqrt{25 \cdot 3} + 4\sqrt{3} && \text{25 is a perfect square.} \\
&= \sqrt{25} \cdot \sqrt{3} + 4\sqrt{3} && \text{Product rule} \\
&= 5\sqrt{3} + 4\sqrt{3} && \sqrt{25} = 5 \\
&= 9\sqrt{3} && \text{Add like radicals.}
\end{aligned}
$$

(b)
$$
\begin{aligned}
\sqrt{2} \cdot \sqrt{6k} + \sqrt{27k} &= \sqrt{12k} + \sqrt{27k} && \text{Product rule} \\
&= \sqrt{4 \cdot 3k} + \sqrt{9 \cdot 3k} && \text{Factor.} \\
&= \sqrt{4} \cdot \sqrt{3k} + \sqrt{9} \cdot \sqrt{3k} && \text{Product rule} \\
&= 2\sqrt{3k} + 3\sqrt{3k} && \sqrt{4} = 2 \text{ and} \\
& && \sqrt{9} = 3 \\
&= 5\sqrt{3k} && \text{Add like radicals.}
\end{aligned}
$$

(c)
$$
\begin{aligned}
2\sqrt[3]{32m^3} - \sqrt[3]{108m^3} &= 2\sqrt[3]{(8m^3)4} - \sqrt[3]{(27m^3)4} && \text{Product rule} \\
&= 2(2m)\sqrt[3]{4} - 3m\sqrt[3]{4} && \sqrt[3]{8m^3} = 2m; \\
& && \sqrt[3]{27m^3} = 3m \\
&= 4m\sqrt[3]{4} - 3m\sqrt[3]{4} && \text{Multiply.} \\
&= m\sqrt[3]{4} && \text{Subtract like radicals.}
\end{aligned}
$$

Caution

A sum or difference of radicals can be simplified only if the radicals are **like radicals.** For example, $\sqrt{5} + 3\sqrt{5} = 4\sqrt{5}$, but $\sqrt{5} + 5\sqrt{3}$ cannot be simplified further. Also, $2\sqrt{3} + 5\sqrt[3]{3}$ cannot be simplified further.

◀◀ **WORK PROBLEM 3 AT THE SIDE.**

9.3 Exercises

Simplify and add or subtract wherever possible. See Examples 1 and 2.

1. $14\sqrt{7} - 19\sqrt{7}$

2. $16\sqrt{2} - 18\sqrt{2}$

3. $\sqrt{17} + 4\sqrt{17}$

4. $5\sqrt{19} + \sqrt{19}$

5. $6\sqrt{7} - \sqrt{7}$

6. $11\sqrt{14} - \sqrt{14}$

7. $\sqrt{45} + 4\sqrt{20}$

8. $\sqrt{24} + 6\sqrt{54}$

9. $5\sqrt{72} - 3\sqrt{50}$

10. $6\sqrt{18} - 5\sqrt{32}$

11. $-5\sqrt{32} + 2\sqrt{98}$

12. $-4\sqrt{75} + 3\sqrt{12}$

13. $5\sqrt{7} - 3\sqrt{28} + 6\sqrt{63}$

14. $3\sqrt{11} + 5\sqrt{44} - 8\sqrt{99}$

15. $2\sqrt{8} - 5\sqrt{32} - 2\sqrt{48}$

16. $5\sqrt{72} - 3\sqrt{48} + 4\sqrt{128}$

17. $4\sqrt{50} + 3\sqrt{12} - 5\sqrt{45}$

18. $6\sqrt{18} + 2\sqrt{48} + 6\sqrt{28}$

19. $\frac{1}{4}\sqrt{288} + \frac{1}{6}\sqrt{72}$

20. $\frac{2}{3}\sqrt{27} + \frac{3}{4}\sqrt{48}$

21. The distributive property, which says $a(b + c) = ab + ac$ and $ba + ca = (b + c)a$, provides the justification for adding and subtracting like radicals. While we usually skip the step that indicates this property, we could not make the statement $2\sqrt{3} + 4\sqrt{3} = 6\sqrt{3}$ without it. Write an equation showing how the distributive property is actually used in this statement.

22. In Example 1(e), we state that $\sqrt{3} + \sqrt{7}$ cannot be simplified further. Why is this so? Show, by using calculator approximations, that $\sqrt{3} + \sqrt{7}$ is *not* equal to $\sqrt{10}$.

Perform the indicated operations. Assume that all variables represent nonnegative real numbers. See Example 3.

23. $\sqrt{6} \cdot \sqrt{2} + 9\sqrt{3}$

24. $4\sqrt{15} \cdot \sqrt{3} + 4\sqrt{5}$

25. $\sqrt{9x} + \sqrt{49x} - \sqrt{25x}$

26. $\sqrt{4a} - \sqrt{16a} + \sqrt{100a}$

27. $\sqrt{6x^2} + x\sqrt{24}$

28. $\sqrt{75x^2} + x\sqrt{108}$

29. $3\sqrt{8x^2} - 4x\sqrt{2} - x\sqrt{8}$ **30.** $\sqrt{2b^2} + 3b\sqrt{18} - b\sqrt{200}$ **31.** $-8\sqrt{32k} + 6\sqrt{8k}$

32. $4\sqrt{12x} + 2\sqrt{27x}$ **33.** $2\sqrt{125x^2z} + 8x\sqrt{80z}$ **34.** $\sqrt{48x^2y} + 5x\sqrt{27y}$

35. $4\sqrt[3]{16} - 3\sqrt[3]{54}$ **36.** $5\sqrt[3]{128} + 3\sqrt[3]{250}$ **37.** $6\sqrt[3]{8p^2} - 2\sqrt[3]{27p^2}$

38. $8k\sqrt[3]{54k} + 6\sqrt[3]{16k^4}$ **39.** $5\sqrt[4]{m^3} + 8\sqrt[4]{16m^3}$ **40.** $5\sqrt[4]{m^5} + 3\sqrt[4]{81m^5}$

41. Despite the fact that $\sqrt{25}$ and $\sqrt[3]{8}$ are radicals that have different root indexes, they can be added to obtain a single term: $\sqrt{25} + \sqrt[3]{8} = 5 + 2 = 7$. Make up a similar sum of radicals that leads to an answer of 10.

42. In the directions for Exercises 23–40, we made the assumption that all variables represent nonnegative real numbers. However, in Exercise 38, variables actually *may* represent negative numbers. Explain why this is so.

_____ **MATHEMATICAL CONNECTIONS (EXERCISES 43–46)** _____

Addition and subtraction of like radicals is no different than addition and subtraction of like terms. Work Exercises 43–46 in order.

43. Combine like terms: $5x^2y + 3x^2y - 14x^2y$.

44. Combine like terms: $5(p - 2q)^2(a + b) + 3(p - 2q)^2(a + b) - 14(p - 2q)^2(a + b)$.

45. Combine like radicals: $5a^2\sqrt{xy} + 3a^2\sqrt{xy} - 14a^2\sqrt{xy}$.

46. Compare your answers in Exercises 43–45. How are they alike? How are they different?

9.4 *Rationalizing the Denominator*

OBJECTIVE **1** We found decimal approximations for radicals in the first section of this chapter. For more complicated radical expressions, it is easier to find these decimals if the denominators do not contain any radicals. For example, the radical in the denominator of

$$\frac{\sqrt{3}}{\sqrt{2}}$$

can be eliminated by multiplying the numerator and the denominator by $\sqrt{2}$.

$$\frac{\sqrt{3}}{\sqrt{2}} = \frac{\sqrt{3} \cdot \sqrt{2}}{\sqrt{2} \cdot \sqrt{2}} = \frac{\sqrt{6}}{2} \qquad \sqrt{2} \cdot \sqrt{2} = \sqrt{4} = 2$$

This process of changing the denominator from a radical (irrational number) to a rational number is called **rationalizing** (RA-shun-ul-eye-zing) **the denominator.** The value of the radical expression is not changed; only the form is changed, because the expression has been multiplied by 1 in the form of $\sqrt{2}/\sqrt{2}$.

E X A M P L E 1 **Rationalizing Denominators**

Rationalize each denominator.

(a) $\dfrac{9}{\sqrt{6}}$

$$\frac{9}{\sqrt{6}} = \frac{9 \cdot \sqrt{6}}{\sqrt{6} \cdot \sqrt{6}} \qquad \text{Multiply numerator and denominator by } \sqrt{6}.$$

$$= \frac{9\sqrt{6}}{6} \qquad \sqrt{6} \cdot \sqrt{6} = 6$$

$$= \frac{3\sqrt{6}}{2} \qquad \text{Lowest terms}$$

(b) $\dfrac{12}{\sqrt{8}}$

The denominator could be rationalized here by multiplying by $\sqrt{8}$. However, the result can be found more directly by multiplying numerator and denominator by $\sqrt{2}$. This is because $\sqrt{8} \cdot \sqrt{2} = \sqrt{16} = 4$, a rational number.

$$\frac{12}{\sqrt{8}} = \frac{12 \cdot \sqrt{2}}{\sqrt{8} \cdot \sqrt{2}} \qquad \text{Multiply by } \sqrt{2} \text{ in numerator and denominator.}$$

$$= \frac{12\sqrt{2}}{\sqrt{16}} \qquad \text{Product rule}$$

$$= \frac{12\sqrt{2}}{4} \qquad \sqrt{16} = 4$$

$$= 3\sqrt{2} \qquad \text{Lowest terms}$$

WORK PROBLEM 1 AT THE SIDE. ▶▶

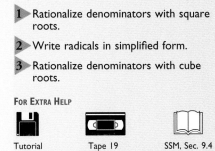
1. Rationalize each denominator.

(a) $\dfrac{3}{\sqrt{5}}$

(b) $\dfrac{-6}{\sqrt{11}}$

(c) $-\dfrac{\sqrt{7}}{\sqrt{2}}$

(d) $\dfrac{20}{\sqrt{18}}$

ANSWERS

1. (a) $\dfrac{3\sqrt{5}}{5}$ **(b)** $\dfrac{-6\sqrt{11}}{11}$

(c) $-\dfrac{\sqrt{14}}{2}$ **(d)** $\dfrac{10\sqrt{2}}{3}$

2. Simplify by rationalizing each denominator.

(a) $\sqrt{\dfrac{16}{11}}$

(b) $\sqrt{\dfrac{5}{18}}$

(c) $\sqrt{\dfrac{8}{32}}$

OBJECTIVE A radical is considered to be in simplified form if the following three conditions are met.

> **Simplified Form of a Radical**
>
> **1.** The radicand contains no factor (except 1) that is a perfect square.
> **2.** The radicand has no fractions.
> **3.** No denominator contains a radical.

In the following examples, radicals are simplified according to these conditions.

E X A M P L E   2   Simplifying a Radical

Simplify $\sqrt{\frac{27}{5}}$.

    This violates condition 2. To begin, use the quotient rule for radicals.

$$\sqrt{\frac{27}{5}} = \frac{\sqrt{27}}{\sqrt{5}} \qquad \text{Quotient rule}$$

$$\frac{\sqrt{27}}{\sqrt{5}} = \frac{\sqrt{27} \cdot \sqrt{5}}{\sqrt{5} \cdot \sqrt{5}} \qquad \begin{array}{l}\text{Multiply both numerator and} \\ \text{denominator by } \sqrt{5}.\end{array}$$

$$= \frac{\sqrt{9 \cdot 3} \cdot \sqrt{5}}{5} \qquad \text{Product rule; } \sqrt{5} \cdot \sqrt{5} = 5$$

$$= \frac{\sqrt{9} \cdot \sqrt{3} \cdot \sqrt{5}}{5} \qquad \text{Product rule}$$

$$= \frac{3 \cdot \sqrt{3} \cdot \sqrt{5}}{5} \qquad \sqrt{9} = 3$$

$$= \frac{3\sqrt{15}}{5} \qquad \text{Product rule}$$

▬◀ **WORK PROBLEM 2 AT THE SIDE.**

E X A M P L E   3   Simplifying a Product of Radicals

Simplify $\sqrt{\frac{5}{8}} \cdot \sqrt{\frac{1}{6}}$.

    Use both the quotient rule and the product rule.

$$\sqrt{\frac{5}{8}} \cdot \sqrt{\frac{1}{6}} = \sqrt{\frac{5}{8} \cdot \frac{1}{6}} \qquad \text{Product rule}$$

$$= \sqrt{\frac{5}{48}} \qquad \text{Multiply fractions.}$$

$$= \frac{\sqrt{5}}{\sqrt{48}} \qquad \text{Quotient rule}$$

Now rationalize the denominator by multiplying both the numerator and the denominator by $\sqrt{3}$ (because $\sqrt{48} \cdot \sqrt{3} = \sqrt{48 \cdot 3} = \sqrt{144} = 12$). One way to find the smallest radical to multiply by is to first factor the denominator: $\sqrt{48} = \sqrt{3} \cdot \sqrt{16} = \sqrt{3} \cdot 4$. Since $\sqrt{3} \cdot \sqrt{3} = 3$, multiplying by $\sqrt{3}$ will produce a rational denominator, as required.

CONTINUED ON NEXT PAGE

$$\frac{\sqrt{5}}{\sqrt{48}} = \frac{\sqrt{5} \cdot \sqrt{3}}{\sqrt{48} \cdot \sqrt{3}}$$

$$= \frac{\sqrt{15}}{\sqrt{144}} \qquad \text{Product rule}$$

$$= \frac{\sqrt{15}}{12} \qquad \sqrt{144} = 12$$

WORK PROBLEM 3 AT THE SIDE. ▶▶

E X A M P L E 4 Simplifying a Quotient of Radicals

Simplify $\dfrac{\sqrt{4x}}{\sqrt{y}}$. Assume that x and y are positive real numbers.

Multiply numerator and denominator by $\sqrt{y}$.

$$\frac{\sqrt{4x}}{\sqrt{y}} = \frac{\sqrt{4x} \cdot \sqrt{y}}{\sqrt{y} \cdot \sqrt{y}} = \frac{\sqrt{4xy}}{y} = \frac{2\sqrt{xy}}{y}$$

WORK PROBLEM 4 AT THE SIDE. ▶▶

E X A M P L E 5 Simplifying a Radical Quotient

Simplify the expression $\sqrt{\dfrac{2x^2y}{3}}$. Assume that x and y are nonnegative real numbers.

First use the quotient rule.

$$\sqrt{\frac{2x^2y}{3}} = \frac{\sqrt{2x^2y}}{\sqrt{3}} \qquad \text{Quotient rule}$$

Next, multiply both the numerator and denominator by $\sqrt{3}$ to rationalize the denominator.

$$\frac{\sqrt{2x^2y}}{\sqrt{3}} = \frac{\sqrt{2x^2y} \cdot \sqrt{3}}{\sqrt{3} \cdot \sqrt{3}}$$

$$= \frac{\sqrt{6x^2y}}{3} \qquad \text{Product rule}$$

$$= \frac{\sqrt{x^2}\,\sqrt{6y}}{3} \qquad \text{Product rule}$$

$$= \frac{x\sqrt{6y}}{3} \qquad \sqrt{x^2} = x, \text{ since } x \geq 0$$

WORK PROBLEM 5 AT THE SIDE. ▶▶

OBJECTIVE 3 ▶ A denominator with a cube root is rationalized by changing the radicand in the denominator to a perfect cube, as shown in the next example.

---

3. Simplify.

(a) $\sqrt{\dfrac{1}{2}} \cdot \sqrt{\dfrac{5}{6}}$

(b) $\sqrt{\dfrac{1}{10}} \cdot \sqrt{20}$

(c) $\sqrt{\dfrac{5}{8}} \cdot \sqrt{\dfrac{24}{10}}$

4. Simplify $\dfrac{\sqrt{5p}}{\sqrt{q}}$. Assume that p and q are positive real numbers.

5. Simplify $\sqrt{\dfrac{5r^2t^2}{7}}$. Assume that r and t represent nonnegative real numbers.

ANSWERS

3. (a) $\dfrac{\sqrt{15}}{6}$ (b) $\sqrt{2}$ (c) $\dfrac{\sqrt{6}}{2}$

4. $\dfrac{\sqrt{5pq}}{q}$

5. $\dfrac{rt\sqrt{35}}{7}$

6. Simplify. Rationalize each denominator.

(a) $\sqrt[3]{\dfrac{5}{7}}$

(b) $\dfrac{\sqrt[3]{5}}{\sqrt[3]{9}}$

E X A M P L E   6   Rationalizing a Cube Root Denominator

Simplify. Rationalize each denominator.

(a) $\sqrt[3]{\dfrac{3}{2}}$

Multiply the numerator and the denominator by enough factors of 2 to make the denominator a perfect cube. This will eliminate the radical in the denominator. Here, multiply by $\sqrt[3]{2^2}$.

$$\sqrt[3]{\dfrac{3}{2}} = \dfrac{\sqrt[3]{3}}{\sqrt[3]{2}} = \dfrac{\sqrt[3]{3} \cdot \sqrt[3]{2^2}}{\sqrt[3]{2} \cdot \sqrt[3]{2^2}} = \dfrac{\sqrt[3]{3 \cdot 2^2}}{\sqrt[3]{2^3}} = \dfrac{\sqrt[3]{12}}{2} \qquad \text{Since } \sqrt[3]{2^3} = \sqrt[3]{8} = 2$$

(b) $\dfrac{\sqrt[3]{3}}{\sqrt[3]{4}}$

Since $4 \cdot 2 = 2^2 \cdot 2 = 2^3$, multiply numerator and denominator by $\sqrt[3]{2}$.

$$\dfrac{\sqrt[3]{3}}{\sqrt[3]{4}} = \dfrac{\sqrt[3]{3} \cdot \sqrt[3]{2}}{\sqrt[3]{4} \cdot \sqrt[3]{2}} = \dfrac{\sqrt[3]{6}}{\sqrt[3]{8}} = \dfrac{\sqrt[3]{6}}{2}$$

Caution
A common error in Example 6(a) is to multiply by $\sqrt[3]{2}$ instead of $\sqrt[3]{2^2}$ or $\sqrt[3]{4}$. Notice that this would give a denominator of $\sqrt[3]{2} \cdot \sqrt[3]{2} = \sqrt[3]{4}$. Because 4 is not a perfect cube, the denominator is still not rationalized.

◀◀ **WORK PROBLEM 6 AT THE SIDE.**

9.4 Exercises

Rationalize each denominator. See Examples 1 and 2.

1. $\dfrac{8}{\sqrt{2}}$

2. $\dfrac{12}{\sqrt{3}}$

3. $\dfrac{-\sqrt{11}}{\sqrt{3}}$

4. $\dfrac{-\sqrt{13}}{\sqrt{5}}$

5. $\dfrac{7\sqrt{3}}{\sqrt{5}}$

6. $\dfrac{4\sqrt{6}}{\sqrt{5}}$

7. $\dfrac{24\sqrt{10}}{16\sqrt{3}}$

8. $\dfrac{18\sqrt{15}}{12\sqrt{2}}$

9. $\dfrac{16}{\sqrt{27}}$

10. $\dfrac{24}{\sqrt{18}}$

11. $\dfrac{-3}{\sqrt{50}}$

12. $\dfrac{-5}{\sqrt{75}}$

13. $\dfrac{63}{\sqrt{45}}$

14. $\dfrac{27}{\sqrt{32}}$

15. $\dfrac{\sqrt{24}}{\sqrt{8}}$

16. $\dfrac{\sqrt{36}}{\sqrt{18}}$

17. $\sqrt{\dfrac{1}{2}}$

18. $\sqrt{\dfrac{1}{3}}$

19. $\sqrt{\dfrac{13}{5}}$

20. $\sqrt{\dfrac{17}{11}}$

21. When we rationalize the denominator of an expression such as $\dfrac{4}{\sqrt{3}}$, we multiply both the numerator and the denominator by $\sqrt{3}$. By what number are we actually multiplying the given expression, and what property of real numbers justifies the fact that our result is equal to the given expression?

22. In Example 1(a), we show algebraically that $\dfrac{9}{\sqrt{6}}$ is equal to $\dfrac{3\sqrt{6}}{2}$. Give numerical support to this result by finding the decimal approximation of $\dfrac{9}{\sqrt{6}}$ on your calculator, and then finding the decimal approximation of $\dfrac{3\sqrt{6}}{2}$. Are they the same?

Rationalize each denominator. See Example 3.

23. $\sqrt{\dfrac{7}{13}} \cdot \sqrt{\dfrac{13}{3}}$

24. $\sqrt{\dfrac{19}{20}} \cdot \sqrt{\dfrac{20}{3}}$

25. $\sqrt{\dfrac{21}{7}} \cdot \sqrt{\dfrac{21}{8}}$

26. $\sqrt{\dfrac{5}{8}} \cdot \sqrt{\dfrac{5}{6}}$

27. $\sqrt{\dfrac{1}{12}} \cdot \sqrt{\dfrac{1}{3}}$

28. $\sqrt{\dfrac{1}{8}} \cdot \sqrt{\dfrac{1}{2}}$

29. $\sqrt{\dfrac{2}{9}} \cdot \sqrt{\dfrac{9}{2}}$

30. $\sqrt{\dfrac{4}{3}} \cdot \sqrt{\dfrac{3}{4}}$

Simplify each radical. Assume that all variables represent positive real numbers. See Examples 4 and 5.

31. $\sqrt{\dfrac{7}{x}}$

32. $\sqrt{\dfrac{19}{y}}$

33. $\sqrt{\dfrac{4x^3}{y}}$

34. $\sqrt{\dfrac{9t^3}{s}}$

35. $\sqrt{\dfrac{18x^3}{6y}}$

36. $\sqrt{\dfrac{24t^3}{8p}}$

37. $\sqrt{\dfrac{9a^2r^5}{7t}}$

38. $\sqrt{\dfrac{16x^3y^2}{13z}}$

*In Exercises 39 and 40, (**a**) give the answer as a simplified radical and (**b**) use a calculator to give the answer correct to the nearest thousandth.*

39. The period p of a pendulum is the time it takes for it to swing from one extreme to the other and back again. The value of p in seconds is given by

$$p = k \cdot \sqrt{\dfrac{L}{g}},$$

where L is the length of the pendulum, g is the acceleration due to gravity, and k is a constant. Find the period when $k = 6$, $L = 9$ feet, and $g = 32$ feet per second per second.

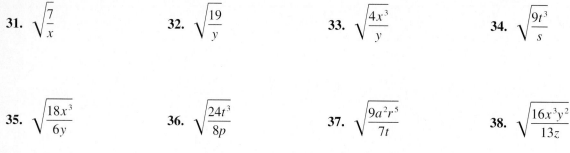

40. The velocity v of a meteorite approaching Earth is given by

$$v = \dfrac{k}{\sqrt{d}}$$

kilometers per second, where d is its distance from the center of Earth and k is a constant. What is the velocity of a meteorite that is 6000 kilometers away from the center of Earth, if $k = 450$?

41. Which one of the following would be an appropriate choice for multiplying the numerator and the denominator of $\dfrac{\sqrt[3]{2}}{\sqrt[3]{5}}$ in order to rationalize the denominator?
(**a**) $\sqrt[3]{5}$ (**b**) $\sqrt[3]{25}$ (**c**) $\sqrt[3]{2}$
(**d**) $\sqrt[3]{3}$

42. In Example 6(b), we multiply numerator and denominator of $\dfrac{\sqrt[3]{3}}{\sqrt[3]{4}}$ by $\sqrt[3]{2}$ to rationalize the denominator. Suppose we had chosen to multiply by $\sqrt[3]{16}$ instead. Would we have obtained the correct answer after all simplifications were done?

Simplify. Rationalize each denominator. Assume that variables in the denominator are nonzero. See Example 6.

43. $\sqrt[3]{\dfrac{3}{2}}$

44. $\sqrt[3]{\dfrac{2}{5}}$

45. $\dfrac{\sqrt[3]{4}}{\sqrt[3]{7}}$

46. $\dfrac{\sqrt[3]{5}}{\sqrt[3]{10}}$

47. $\sqrt[3]{\dfrac{3}{4y^2}}$

48. $\sqrt[3]{\dfrac{3}{25x^2}}$

49. $\dfrac{\sqrt[3]{7m}}{\sqrt[3]{36n}}$

50. $\dfrac{\sqrt[3]{11p}}{\sqrt[3]{49q}}$

9.5 *Simplifying Radical Expressions*

It can be difficult to decide on the "simplest" form of a radical. The conditions for which a square root radical is in simplest form were listed in the previous section. Below is a set of guidelines you should follow when you are simplifying radical expressions.

Simplifying Radical Expressions

1. If a radical represents a rational number, then that rational number should be used in place of the radical.

    For example, $\sqrt{49}$ is simplified by writing 7; $\sqrt{64}$ by writing 8; $\sqrt{\frac{169}{9}}$ by writing $\frac{13}{3}$.

2. If a radical expression contains products of radicals, the product rule for radicals, $\sqrt{x} \cdot \sqrt{y} = \sqrt{xy}$, should be used to get a single radical.

    For example, $\sqrt{3} \cdot \sqrt{2}$ is simplified to $\sqrt{6}$; $\sqrt{5} \cdot \sqrt{x}$ to $\sqrt{5x}$.

3. If a radicand has a factor that is a perfect square, the radical should be expressed as the product of the positive square root of the perfect square and the remaining radical factor. A similar statement applies to higher roots.

    For example, $\sqrt{20}$ is simplified to $\sqrt{20} = 2\sqrt{5}$; $\sqrt[3]{16} = 2\sqrt[3]{2}$.

4. If a radical expression contains sums or differences of radicals, the distributive property should be used to combine like radicals.

    For example, $3\sqrt{2} + 4\sqrt{2}$ is combined as $7\sqrt{2}$, but $3\sqrt{2} + 4\sqrt{3}$ cannot be further combined.

5. Any denominator containing a radical should be rationalized.

    For example, $\dfrac{5}{\sqrt{3}}$ is rationalized as $\dfrac{5}{\sqrt{3}} = \dfrac{5\sqrt{3}}{\sqrt{3} \cdot \sqrt{3}} = \dfrac{5\sqrt{3}}{3}$, and $\sqrt{\dfrac{2}{5}} = \dfrac{\sqrt{2}}{\sqrt{5}} = \dfrac{\sqrt{2} \cdot \sqrt{5}}{\sqrt{5} \cdot \sqrt{5}} = \dfrac{\sqrt{10}}{5}$.

OBJECTIVE 1   The first example involves sums of radical expressions.

EXAMPLE 1   Adding Radical Expressions

Perform the indicated operations.

(a) $\sqrt{16} + \sqrt{9}$

Here $\sqrt{16} + \sqrt{9} = 4 + 3 = 7$.

(b) $5\sqrt{2} + 2\sqrt{18}$

First simplify $\sqrt{18}$.

$$
\begin{aligned}
5\sqrt{2} + 2\sqrt{18} &= 5\sqrt{2} + 2(\sqrt{9} \cdot \sqrt{2}) &&\text{9 is a perfect square.}\\
&= 5\sqrt{2} + 2(3\sqrt{2}) &&\sqrt{9} = 3\\
&= 5\sqrt{2} + 6\sqrt{2} &&\text{Multiply.}\\
&= 11\sqrt{2} &&\text{Add like radicals.}
\end{aligned}
$$

WORK PROBLEM 1 AT THE SIDE. ▶▶

OBJECTIVE 2   The next examples show how to simplify radical expressions with products.

OBJECTIVES

1. ▶ Simplify radical expressions with sums.
2. ▶ Simplify radical expressions with products.
3. ▶ Simplify radical expressions with quotients.
4. ▶ Write radical expressions with quotients in lowest terms.

FOR EXTRA HELP

Tutorial   Tape 19   SSM, Sec. 9.5

1. Perform the indicated operations.

    (a) $\sqrt{36} + \sqrt{25}$

    (b) $3\sqrt{3} + 2\sqrt{27}$

    (c) $4\sqrt{8} - 2\sqrt{32}$

    (d) $2\sqrt{12} - 5\sqrt{48}$

ANSWERS

1. **(a)** 11 **(b)** $9\sqrt{3}$
   (c) 0 **(d)** $-16\sqrt{3}$

2. Find each product. Simplify the answers.

(a) $\sqrt{7}(\sqrt{2} + \sqrt{5})$

(b) $\sqrt{2}(\sqrt{8} + \sqrt{20})$

(c) $(\sqrt{2} + 5\sqrt{3})(\sqrt{3} - 2\sqrt{2})$

(d) $(\sqrt{2} - \sqrt{5})(\sqrt{10} + \sqrt{2})$

E X A M P L E   2   Multiplying Radical Expressions

Multiply $\sqrt{5}(\sqrt{8} - \sqrt{32})$ and simplify the product.

Start by simplifying $\sqrt{8}$ and $\sqrt{32}$.

$$\sqrt{8} = 2\sqrt{2} \quad \text{and} \quad \sqrt{32} = 4\sqrt{2}$$

Now simplify inside the parentheses.

$$\sqrt{5}(\sqrt{8} - \sqrt{32}) = \sqrt{5}(2\sqrt{2} - 4\sqrt{2})$$
$$= \sqrt{5}(-2\sqrt{2}) \qquad \text{Subtract like radicals.}$$
$$= -2\sqrt{5 \cdot 2} \qquad \text{Product rule}$$
$$= -2\sqrt{10} \qquad \text{Multiply.}$$

E X A M P L E   3   Multiplying Radical Expressions

Find each product and simplify the answers.

(a) $(\sqrt{3} + 2\sqrt{5})(\sqrt{3} - 4\sqrt{5})$

We can find the products of these sums of radicals in the same way that we found the product of binomials in Chapter 6. The pattern of multiplication is the same, using the FOIL method.

$$(\sqrt{3} + 2\sqrt{5})(\sqrt{3} - 4\sqrt{5})$$
$$= \underbrace{\sqrt{3} \cdot \sqrt{3}}_{\text{First}} + \underbrace{\sqrt{3}(-4\sqrt{5})}_{\text{Outside}} + \underbrace{2\sqrt{5}(\sqrt{3})}_{\text{Inside}} + \underbrace{2\sqrt{5}(-4\sqrt{5})}_{\text{Last}}$$
$$= 3 - 4\sqrt{15} + 2\sqrt{15} - 8 \cdot 5 \qquad \text{Product rule}$$
$$= 3 - 2\sqrt{15} - 40 \qquad \text{Add like radicals.}$$
$$= -37 - 2\sqrt{15} \qquad \text{Combine terms.}$$

(b) $(\sqrt{3} + \sqrt{21})(\sqrt{3} - \sqrt{7})$
$$= \sqrt{3}(\sqrt{3}) + \sqrt{3}(-\sqrt{7}) + \sqrt{21}(\sqrt{3})$$
$$\quad + \sqrt{21}(-\sqrt{7}) \qquad \text{FOIL}$$
$$= 3 - \sqrt{21} + \sqrt{63} - \sqrt{147} \qquad \text{Product rule}$$
$$= 3 - \sqrt{21} + \sqrt{9} \cdot \sqrt{7} - \sqrt{49} \cdot \sqrt{3} \qquad \text{9 and 49 are perfect squares.}$$
$$= 3 - \sqrt{21} + 3\sqrt{7} - 7\sqrt{3} \qquad \sqrt{9} = 3 \text{ and } \sqrt{49} = 7$$

Since there are no like radicals, no terms can be combined.

◄◄ **WORK PROBLEM 2 AT THE SIDE.**

Since radicals represent real numbers, the special products of binomials discussed in Chapter 6 can be used to find products of radicals. Example 4 uses the rule for the product of the sum and difference of two terms,

$$(a + b)(a - b) = a^2 - b^2.$$

E X A M P L E   4   Using a Special Product with Radicals

Find each product.

(a) $(4 + \sqrt{3})(4 - \sqrt{3})$

Follow the pattern given above. Let $a = 4$ and $b = \sqrt{3}$.

$$(4 + \sqrt{3})(4 - \sqrt{3}) = 4^2 - (\sqrt{3})^2$$
$$= 16 - 3 \qquad 4^2 = 16 \text{ and } (\sqrt{3})^2 = 3$$
$$= 13$$

CONTINUED ON NEXT PAGE

(b) $(\sqrt{12} - \sqrt{6})(\sqrt{12} + \sqrt{6}) = (\sqrt{12})^2 - (\sqrt{6})^2$

$$= 12 - 6 \quad \begin{array}{l}(\sqrt{12})^2 = 12 \text{ and} \\ (\sqrt{6})^2 = 6\end{array}$$

$$= 6$$

WORK PROBLEM 3 AT THE SIDE. ▶▶

Both products in Example 4 resulted in rational numbers. The pairs of expressions in those products, such as $4 + \sqrt{3}$ and $4 - \sqrt{3}$, and $\sqrt{12} - \sqrt{6}$ and $\sqrt{12} + \sqrt{6}$, are called **conjugates** (KAHN-juh-guts) of each other.

OBJECTIVE 3 Products of conjugates similar to those in Example 4 can be used to rationalize the denominators in more complicated quotients, such as

$$\frac{2}{4 - \sqrt{3}}.$$

By Example 4(a), if this denominator, $4 - \sqrt{3}$, is multiplied by $4 + \sqrt{3}$, then the product $(4 - \sqrt{3})(4 + \sqrt{3})$ is the rational number 13. Multiplying numerator and denominator of the quotient by $4 + \sqrt{3}$ gives

$$\frac{2}{4 - \sqrt{3}} = \frac{2(4 + \sqrt{3})}{(4 - \sqrt{3})(4 + \sqrt{3})}$$

$$= \frac{2(4 + \sqrt{3})}{13}.$$

The denominator now has been rationalized; it contains no radical signs.

Using Conjugates to Simplify a Radical Expression

To simplify a radical expression with two terms in the denominator, where at least one of those terms is a radical, multiply both the numerator and the denominator by the conjugate of the denominator.

> **E X A M P L E 5   Using Conjugates to Rationalize a Denominator**
>
> Rationalize the denominator in the quotient
>
> $$\frac{5}{3 + \sqrt{5}}.$$
>
> The radical in the denominator can be eliminated by multiplying both numerator and denominator by $3 - \sqrt{5}$.
>
> $$\frac{5}{3 + \sqrt{5}} = \frac{5(3 - \sqrt{5})}{(3 + \sqrt{5})(3 - \sqrt{5})}$$
>
> $$= \frac{5(3 - \sqrt{5})}{3^2 - (\sqrt{5})^2} \qquad (a + b)(a - b) = a^2 - b^2$$
>
> $$= \frac{5(3 - \sqrt{5})}{9 - 5} \qquad 3^2 = 9 \text{ and } (\sqrt{5})^2 = 5$$
>
> $$= \frac{5(3 - \sqrt{5})}{4}$$

3. Find each product. Simplify the answers.

(a) $(3 + \sqrt{5})(3 - \sqrt{5})$

(b) $(\sqrt{3} - 2)(\sqrt{3} + 2)$

(c) $(\sqrt{5} + \sqrt{3})(\sqrt{5} - \sqrt{3})$

ANSWERS

3. (a) 4 **(b)** −1 **(c)** 2

4. Rationalize each denominator.

(a) $\dfrac{5}{4 + \sqrt{2}}$

(b) $\dfrac{\sqrt{5} + 3}{2 - \sqrt{5}}$

(c) $\dfrac{1}{\sqrt{6} + \sqrt{3}}$

5. Write each quotient in lowest terms.

(a) $\dfrac{5\sqrt{3} - 15}{10}$

(b) $\dfrac{8\sqrt{5} + 12}{16}$

E X A M P L E   6   Using Conjugates to Rationalize a Denominator

Simplify $\dfrac{6 + \sqrt{2}}{\sqrt{2} - 5}$.

Multiply numerator and denominator by $\sqrt{2} + 5$.

$$\frac{6 + \sqrt{2}}{\sqrt{2} - 5} = \frac{(6 + \sqrt{2})(\sqrt{2} + 5)}{(\sqrt{2} - 5)(\sqrt{2} + 5)}$$

$$= \frac{6\sqrt{2} + 30 + 2 + 5\sqrt{2}}{2 - 25} \qquad \text{FOIL}$$

$$= \frac{11\sqrt{2} + 32}{-23} \qquad \text{Combine terms.}$$

$$= \frac{-11\sqrt{2} - 32}{23} \qquad \frac{a}{-b} = \frac{-a}{b}$$

◀◀ **WORK PROBLEM 4 AT THE SIDE.**

OBJECTIVE ▶ 4 The final example shows how to write certain quotients in lowest terms.

E X A M P L E   7   Writing a Radical Quotient in Lowest Terms

Write $\dfrac{3\sqrt{3} + 9}{12}$ in lowest terms.

Factor the numerator and denominator, and then use the fundamental property from Section 8.1 to replace common factors with 1.

$$\frac{3\sqrt{3} + 9}{12} = \frac{3(\sqrt{3} + 3)}{3(4)} = 1 \cdot \frac{\sqrt{3} + 3}{4} = \frac{\sqrt{3} + 3}{4}$$

Caution

A common error is to try to reduce an expression like the one in Example 7 to lowest terms before factoring. For example,

$$\frac{4 + 8\sqrt{5}}{4} \neq 1 + 8\sqrt{5}.$$

The correct simplification is $\dfrac{4 + 8\sqrt{5}}{4} = \dfrac{4(1 + 2\sqrt{5})}{4} = 1 + 2\sqrt{5}.$

◀◀ **WORK PROBLEM 5 AT THE SIDE.**

ANSWERS

4. (a) $\dfrac{5(4 - \sqrt{2})}{14}$ **(b)** $-11 - 5\sqrt{5}$

(c) $\dfrac{\sqrt{6} - \sqrt{3}}{3}$

5. (a) $\dfrac{\sqrt{3} - 3}{2}$ **(b)** $\dfrac{2\sqrt{5} + 3}{4}$

9.5 *Exercises*

Based on the work so far in this chapter, many simple operations involving radicals should now be performed mentally. In Exercises 1–4, perform the operations mentally, and write the answers without doing intermediate steps.

1. $\sqrt{49} + \sqrt{36}$ **2.** $\sqrt{100} - \sqrt{81}$ **3.** $\sqrt{2} \cdot \sqrt{8}$ **4.** $\sqrt{8} \cdot \sqrt{8}$

Simplify each expression. Use the five guidelines given in the text. See Examples 1–4.

5. $3\sqrt{5} + 2\sqrt{45}$ **6.** $2\sqrt{2} + 4\sqrt{18}$ **7.** $8\sqrt{50} - 4\sqrt{72}$

8. $4\sqrt{80} - 5\sqrt{45}$ **9.** $\sqrt{5}(\sqrt{3} - \sqrt{7})$ **10.** $\sqrt{7}(\sqrt{10} + \sqrt{3})$

11. $2\sqrt{5}(\sqrt{2} + 3\sqrt{5})$ **12.** $3\sqrt{7}(2\sqrt{7} + 4\sqrt{5})$ **13.** $3\sqrt{14} \cdot \sqrt{2} - \sqrt{28}$

14. $7\sqrt{6} \cdot \sqrt{3} - 2\sqrt{18}$ **15.** $(2\sqrt{6} + 3)(3\sqrt{6} + 7)$ **16.** $(4\sqrt{5} - 2)(2\sqrt{5} - 4)$

17. $(5\sqrt{7} - 2\sqrt{3})(3\sqrt{7} + 4\sqrt{3})$ **18.** $(2\sqrt{10} + 5\sqrt{2})(3\sqrt{10} - 3\sqrt{2})$ **19.** $(2\sqrt{7} + 3)^2$

20. $(4\sqrt{5} + 5)^2$

21. $(5 - \sqrt{2})(5 + \sqrt{2})$

22. $(3 - \sqrt{5})(3 + \sqrt{5})$

23. $(\sqrt{8} - \sqrt{7})(\sqrt{8} + \sqrt{7})$

24. $(\sqrt{12} - \sqrt{11})(\sqrt{12} + \sqrt{11})$

25. $(\sqrt{2} + \sqrt{3})(\sqrt{6} - \sqrt{2})$

26. $(\sqrt{3} + \sqrt{5})(\sqrt{15} - \sqrt{5})$

27. $(\sqrt{10} - \sqrt{5})(\sqrt{5} + \sqrt{20})$

28. $(\sqrt{6} - \sqrt{3})(\sqrt{3} + \sqrt{18})$

29. $(\sqrt{5} + \sqrt{30})(\sqrt{6} + \sqrt{3})$

30. $(\sqrt{10} - \sqrt{20})(\sqrt{2} - \sqrt{5})$

31. In Example 3(a), the original expression simplifies to $-37 - 2\sqrt{15}$. Students often try to simplify expressions like this by combining the -37 and the -2 to get $-39\sqrt{15}$, which is incorrect. Explain why this is incorrect.

32. If you were to attempt to rationalize the denominator of $\dfrac{2}{4 + \sqrt{3}}$ by multiplying the numerator and the denominator by $4 + \sqrt{3}$, what problem would arise? What should you multiply by?

Rationalize the denominators. See Examples 5 and 6.

33. $\dfrac{1}{3 + \sqrt{2}}$

34. $\dfrac{1}{4 - \sqrt{3}}$

35. $\dfrac{14}{2 - \sqrt{11}}$

36. $\dfrac{19}{5 - \sqrt{6}}$

37. $\dfrac{\sqrt{2}}{2 - \sqrt{2}}$

38. $\dfrac{\sqrt{7}}{7 - \sqrt{7}}$

39. $\dfrac{\sqrt{5}}{\sqrt{2} + \sqrt{3}}$

40. $\dfrac{\sqrt{3}}{\sqrt{2} + \sqrt{3}}$

41. $\dfrac{\sqrt{12}}{\sqrt{3} + 1}$

42. $\dfrac{\sqrt{18}}{\sqrt{2} - 1}$

43. $\dfrac{\sqrt{5} + 2}{2 - \sqrt{3}}$

44. $\dfrac{\sqrt{7} + 3}{4 - \sqrt{5}}$

Write each quotient in lowest terms. See Example 7.

45. $\dfrac{6\sqrt{11} - 12}{6}$

46. $\dfrac{12\sqrt{5} - 24}{12}$

47. $\dfrac{2\sqrt{3} + 10}{16}$

48. $\dfrac{4\sqrt{6} + 24}{20}$

49. $\dfrac{12 - \sqrt{40}}{4}$

50. $\dfrac{9 - \sqrt{72}}{12}$

───────────── **MATHEMATICAL CONNECTIONS (EXERCISES 51–56)** ─────────────

Work Exercises 51–56 in order. They are designed to help you see why a common student error is indeed an error.

51. Use the distributive property to write $6(5 + 3x)$ as a sum.

52. Your answer in Exercise 51 should be $30 + 18x$. Why can we not combine these two terms to get $48x$?

53. Repeat Exercise 18 from earlier in this exercise set.

54. Your answer in Exercise 53 should be $30 + 18\sqrt{5}$. Many students will, in error, try to combine these terms to get $48\sqrt{5}$. Why is this wrong?

55. Write the expression similar to $30 + 18x$ that simplifies to $48x$. Then write the expression similar to $30 + 18\sqrt{5}$ that simplifies to $48\sqrt{5}$.

56. Write a short paragraph explaining the similarities between combining like terms and combining like radicals.

───

Solve the problem.

57. The radius of the circular top or bottom of a tin can with a surface area S and a height h is given by

$$r = \frac{-h + \sqrt{h^2 + .64S}}{2}.$$

What radius should be used to make a can with a height of 12 inches and a surface area of 400 square inches?

9.6 Equations with Radicals

The addition and multiplication properties of equality are not enough to solve an equation with radicals such as

$$\sqrt{x + 1} = 3.$$

OBJECTIVE 1 Solving equations that have radicals requires a new property, the *squaring property*.

Squaring Property of Equality

If each side of a given equation is squared, all solutions of the original equation are *among* the solutions of the squared equation.

Caution

Be very careful with the squaring property: Using this property can give a new equation with *more* solutions than the original equation. For example, starting with the equation $y = 4$ and squaring each side gives

$$y^2 = 4^2, \quad \text{or} \quad y^2 = 16.$$

This last equation, $y^2 = 16$, has *two* solutions, 4 or -4, while the original equation, $y = 4$, has only *one* solution, 4. Because of this possibility, checking is more than just a guard against algebraic errors when solving an equation with radicals. It is an essential part of the solution process. *All potential solutions from the squared equation must be checked in the original equation.*

E X A M P L E I   Using the Squaring Property of Equality

Solve the equation $\sqrt{p + 1} = 3$.

Use the squaring property of equality to square each side of the equation.

$$(\sqrt{p + 1})^2 = 3^2$$
$$p + 1 = 9 \qquad (\sqrt{p + 1})^2 = p + 1$$
$$p = 8 \qquad \text{Subtract 1.}$$

Now check this answer in the original equation.

Check:
$$\sqrt{p + 1} = 3$$
$$\sqrt{8 + 1} = 3 \quad ? \qquad \text{Let } p = 8.$$
$$\sqrt{9} = 3 \quad ?$$
$$3 = 3 \qquad \text{True}$$

Because this statement is true, {8} is the solution set of $\sqrt{p + 1} = 3$. In this case the squared equation had just one solution, which also satisfied the original equation.

WORK PROBLEM I AT THE SIDE. ▶▶

OBJECTIVES

1. Solve equations with radicals.
2. Identify equations with no solutions.
3. Solve equations by squaring a binomial.

FOR EXTRA HELP

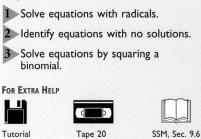

Tutorial   Tape 20   SSM, Sec. 9.6

1. Solve each equation.

   (a) $\sqrt{k} = 3$

   (b) $\sqrt{m - 2} = 4$

   (c) $\sqrt{9 - y} = 4$

ANSWERS

1. (a) {9} (b) {18} (c) {−7}

577

2. Solve each equation.

(a) $\sqrt{3x + 9} = 2\sqrt{x}$

E X A M P L E 2   **Using the Squaring Property of Equality**

Solve $3\sqrt{x} = \sqrt{x + 8}$.

Squaring each side gives

$$(3\sqrt{x})^2 = (\sqrt{x + 8})^2$$

$$3^2(\sqrt{x})^2 = (\sqrt{x + 8})^2 \qquad (ab)^2 = a^2b^2$$

$$9x = x + 8 \qquad (\sqrt{x})^2 = x; (\sqrt{x + 8})^2 = x + 8$$

$$8x = 8 \qquad \text{Subtract } x.$$

$$x = 1. \qquad \text{Divide by 8.}$$

Check:
$$3\sqrt{x} = \sqrt{x + 8}$$
$$3\sqrt{1} = \sqrt{1 + 8} \qquad ? \qquad \text{Let } x = 1.$$
$$3(1) = \sqrt{9} \qquad ? \qquad \sqrt{1} = 1$$
$$3 = 3 \qquad \text{True}$$

The solution set of $3\sqrt{x} = \sqrt{x + 8}$ is $\{1\}$.

Caution
Avoid the common error of writing the solution as the final result obtained in the check. In Example 2, the solution is 1, *not* 3.

◀◀ **WORK PROBLEM 2 AT THE SIDE.**

OBJECTIVE 2   Not all equations with radicals have a solution, as shown by the equations in Examples 3 and 4.

(b) $5\sqrt{a} = \sqrt{20a + 5}$

E X A M P L E 3   **Using the Squaring Property of Equality**

Solve the equation $\sqrt{y} = -3$.

Square each side.

$$(\sqrt{y})^2 = (-3)^2$$

$$y = 9$$

Check this proposed answer in the original equation.

Check:
$$\sqrt{y} = -3$$
$$\sqrt{9} = -3 \qquad ? \qquad \text{Let } y = 9.$$
$$3 = -3 \qquad \text{False}$$

Because the statement $3 = -3$ is false, the number 9 is not a solution of the given equation and is said to be *extraneous*. In fact, $\sqrt{y} = -3$ has $\emptyset$ as its solution set. Because $\sqrt{y}$ represents the *nonnegative* square root of y, we might have seen immediately that there is no solution.

ANSWERS

2. (a) $\{9\}$ **(b)** $\{1\}$

─ **E X A M P L E   4**   **Using the Squaring Property of Equality**

Solve $a = \sqrt{a^2 + 5a + 10}$.

Square each side.

$$a^2 = (\sqrt{a^2 + 5a + 10})^2$$

$$a^2 = a^2 + 5a + 10 \qquad (\sqrt{a^2 + 5a + 10})^2 = a^2 + 5a + 10$$

$$0 = 5a + 10 \qquad \text{Subtract } a^2.$$

$$-10 = 5a \qquad \text{Subtract } 10.$$

$$a = -2 \qquad \text{Divide by 5.}$$

Check this potential solution in the original equation.

Check: $a = \sqrt{a^2 + 5a + 10}$

$$-2 = \sqrt{(-2)^2 + 5(-2) + 10} \quad ? \qquad \text{Let } a = -2.$$

$$-2 = \sqrt{4 - 10 + 10} \qquad\quad ? \qquad \text{Multiply.}$$

$$-2 = 2 \qquad\qquad\qquad\qquad \text{False}$$

Because $a = -2$ leads to a false result, the solution set is $\emptyset$.

■

> **WORK PROBLEM 3 AT THE SIDE.** ▶▶

The steps to use in solving an equation with radicals are summarized below.

> **Solving an Equation with Radicals**
>
> *Step 1* Arrange the terms so that a radical is alone on one side of the equation.
>
> *Step 2* Square each side.
>
> *Step 3* Combine like terms. Repeat Steps 1–3 if a radical is still present.
>
> *Step 4* Solve the equation for potential solutions.
>
> *Step 5* Check all solutions from Step 4 in the *original* equation.

OBJECTIVE ▶3 The next examples use the following facts from Chapter 6:

$$(a + b)^2 = a^2 + 2ab + b^2$$

and

$$(a - b)^2 = a^2 - 2ab + b^2.$$

By these patterns, for example,

$$(y - 3)^2 = y^2 - 2(y)(3) + (3)^2$$
$$= y^2 - 6y + 9.$$

> **WORK PROBLEM 4 AT THE SIDE.** ▶▶

---

3. Solve each equation that has a solution. (*Hint:* In (a) subtract 4 from each side.)

(a) $\sqrt{y} + 4 = 0$

(b) $m = \sqrt{m^2 - 4m - 16}$

4. Square each expression.

(a) $m - 5$

(b) $2k - 5$

(c) $3m - 2p$

5. Solve each equation.

(a) $\sqrt{6w + 6} = w + 1$

EXAMPLE 5 **Using the Squaring Property of Equality**

Solve the equation $\sqrt{2y - 3} = y - 3$.

Square each side, using the result above on the right side of the equation.

$$(\sqrt{2y - 3})^2 = (y - 3)^2$$
$$2y - 3 = y^2 - 6y + 9$$

This equation is quadratic because of the y^2 term. As shown in Section 7.5, solving this equation requires that one side be equal to 0. Subtract $2y$ and add 3, getting

$$0 = y^2 - 8y + 12.$$

Solve this equation by factoring.

$$0 = (y - 6)(y - 2)$$
$$y - 6 = 0 \quad \text{or} \quad y - 2 = 0 \qquad \text{Set each factor equal to 0.}$$
$$y = 6 \quad \text{or} \qquad y = 2 \qquad \text{Solve.}$$

Check both of these potential solutions in the original equation.
Check:

If $y = 6$,
$$\sqrt{2y - 3} = y - 3$$
$$\sqrt{2(6) - 3} = 6 - 3 \quad ? \quad \text{Let } y = 6.$$
$$\sqrt{12 - 3} = 3 \quad ? \quad \text{Multiply.}$$
$$\sqrt{9} = 3 \quad ?$$
$$3 = 3. \qquad \text{True}$$

If $y = 2$,
$$\sqrt{2y - 3} = y - 3$$
$$\sqrt{2(2) - 3} = 2 - 3 \quad ? \quad \text{Let } y = 2.$$
$$\sqrt{4 - 3} = -1 \quad ? \quad \text{Multiply.}$$
$$\sqrt{1} = -1 \quad ?$$
$$1 = -1. \qquad \text{False}$$

Only 6 is a valid solution of the equation, so the solution set is $\{6\}$.

(b) $2u - 1 = \sqrt{10u + 9}$

◀◀ **WORK PROBLEM 5 AT THE SIDE.**

Sometimes it is necessary to write an equation in a different form before squaring each side. The next example shows why.

EXAMPLE 6 **Using the Squaring Property of Equality**

Solve the equation $\sqrt{x} + 1 = 2x$.

Squaring each side gives

$$(\sqrt{x} + 1)^2 = (2x)^2$$
$$x + 2\sqrt{x} + 1 = 4x^2,$$

an equation that is more complicated, and still contains a radical. It would be better instead to rewrite the original equation so that the radical is alone on one side of the equal sign. Isolate the radical by subtracting 1 from each side.

CONTINUED ON NEXT PAGE

ANSWERS

5. **(a)** $\{5, -1\}$ **(b)** $\{4\}$

$$\sqrt{x} = 2x - 1$$

$$(\sqrt{x})^2 = (2x - 1)^2 \qquad \text{Square each side.}$$

$$x = 4x^2 - 4x + 1$$

$$0 = 4x^2 - 5x + 1 \qquad \text{Subtract } x.$$

$$0 = (4x - 1)(x - 1) \qquad \text{Factor.}$$

$$4x - 1 = 0 \quad \text{or} \quad x - 1 = 0 \qquad \text{Set each factor equal to 0.}$$

$$x = \frac{1}{4} \quad \text{or} \qquad x = 1 \qquad \text{Solve.}$$

Check:

If $x = \dfrac{1}{4}$,

$$\sqrt{x} + 1 = 2x$$

$$\sqrt{\frac{1}{4}} + 1 = 2\left(\frac{1}{4}\right) \quad ? \quad \text{Let } x = \frac{1}{4}.$$

$$\frac{3}{2} = \frac{1}{2}. \qquad \text{False}$$

If $x = 1$,

$$\sqrt{x} + 1 = 2x$$

$$\sqrt{1} + 1 = 2(1) \quad ? \quad \text{Let } x = 1.$$

$$2 = 2. \qquad \text{True}$$

The only solution to the original equation is 1, so the solution set is $\{1\}$.

Caution

Errors often occur when each side of an equation is squared. For example, in Example 6 after the equation is rewritten as

$$\sqrt{x} = 2x - 1,$$

it would be *incorrect* to write the next step as

$$x = 4x^2 + 1.$$

Don't forget that the binomial $2x - 1$ must be squared to get $4x^2 - 4x + 1$.

WORK PROBLEM 6 AT THE SIDE. ▶▶

6. Solve each equation.

(a) $\sqrt{x - 3} = x - 15$

(b) $\sqrt{z + 5} + 2 = z + 5$

Numbers in the Real World

Real World *collaborative investigations*

Are You a Good "Guesstimator"?

While calculators can make life easier when it comes to computations, keep in mind that many times a good guess, or estimate (hence, the term **guesstimate**) is sufficient. In such cases, a calculator may not be necessary, or even appropriate. For example, consider the following problem:

> A birdhouse for purple martins can accommodate up to 6 nests. How many birdhouses would be necessary to accommodate 92 nests?

If we divide 92 by 6 either by hand or with a calculator, we get an answer of 15.333333 (rounded). Can this possibly be the desired number? Of course not, because we cannot consider fractions of birdhouses. Now we must ask ourselves should we decide on 15 or 16 birdhouses? Because we need to provide nesting space for the nests left over after the 15 birdhouses (as indicated by the decimal fraction), we should plan to use 16 birdhouses. Notice that, in this problem, we must round our answer *up* to the next counting number.

FOR GROUP DISCUSSION

Use estimation techniques to answer the following. Once you answer correctly, share the technique you used with a class member. (Remember, when it comes to guesstimating, not all people are created equal!)

1. A certain type of carrying case will hold a maximum of 48 audiocassettes. If you need to store 490 audiocassettes, how many carrying cases will you need?

2. A plastic page designed to store baseball cards will hold up to 16 cards. If you must store your collection of 484 cards, how many pages will you need?

3. Each room available for administering a placement test will hold up to 40 students. Two hundred fifty students have signed up for the test. How many rooms will be used?

4. A gardener wants to fertilize 2000 tomato plants. Each bag of fertilizer will supply up to 150 plants. How many bags does she need to do the job?

5. The planet Mercury takes 88.0 Earth days to revolve around the sun. Pluto takes 90,824.2 days to do the same. When Pluto has revolved around the sun once, about how many times will Mercury have revolved around the sun?
   (a) 100   **(b)** 1,000   **(c)** 10,000   **(d)** 100,000

6. Hale County, Texas, has a population of 34,671 and covers 1005 square miles. About how many inhabitants per square mile does the county have?
   (a) 35   **(b)** 350   **(c)** 3500   **(d)** 35,000

7. The 1990 United States census showed that the total population of the country was 248,709,873. Of these, 25,223,086 were in the 45–50-year age bracket. On the average, about one in every _____ citizens is in that age bracket.
   (a) 5   **(b)** 10   **(c)** 15   **(d)** 20

8. In voting for which age Elvis was to be in his portrayal on a U.S. postage stamp, voters cast 851,000 votes for the younger Elvis against 277,723 votes for the more mature Elvis. A newspaper article read "Younger Elvis is Winner by _____ to 1 Margin." What number belongs in the blank?
   (a) 5   **(b)** 4   **(c)** 3   **(d)** 2

9.6 Exercises

1. How can you tell that the equation $\sqrt{x} = -8$ has no real number solution without performing any algebraic steps?

2. Explain why the equation $x^2 = 36$ has two real number solutions, while the equation $\sqrt{x} = 6$ has only one real number solution.

Find the solution set for each of the following equations. See Examples 1–4.

3. $\sqrt{x} = 7$

4. $\sqrt{k} = 10$

5. $\sqrt{y + 2} = 3$

6. $\sqrt{x + 7} = 5$

7. $\sqrt{r - 4} = 9$

8. $\sqrt{k - 12} = 3$

9. $\sqrt{4 - t} = 7$

10. $\sqrt{9 - s} = 5$

11. $\sqrt{2t + 3} = 0$

12. $\sqrt{5x - 4} = 0$

13. $\sqrt{3x - 8} = -2$

14. $\sqrt{6y + 4} = -3$

15. $\sqrt{m} - 4 = 7$

16. $\sqrt{t} + 3 = 10$

17. $\sqrt{10x - 8} = 3\sqrt{x}$

18. $\sqrt{17t - 4} = 4\sqrt{t}$

19. $5\sqrt{x} = \sqrt{10x + 15}$

20. $4\sqrt{y} = \sqrt{20y - 16}$

21. $\sqrt{3x - 5} = \sqrt{2x + 1}$

22. $\sqrt{5y + 2} = \sqrt{3y + 8}$

23. $k = \sqrt{k^2 - 5k - 15}$

24. $s = \sqrt{s^2 - 2s - 6}$

25. $7x = \sqrt{49x^2 + 2x - 10}$

26. $6m = \sqrt{36m^2 + 5m - 5}$

27. The first step in solving the equation $\sqrt{2x + 1} = x - 7$ is to square both sides of the equation. Errors often occur in solving equations such as this one when the right side of the equation is squared incorrectly. Why is the square of the right side *not* equal to $x^2 + 49$? What is the correct answer for the square of the right side?

28. Explain why the equation $x = 3\sqrt{x + 13}$ cannot have a negative solution.

Find the solution set for each equation. See Examples 5 and 6.

29. $\sqrt{2x + 1} = x - 7$

30. $\sqrt{3x + 3} = x - 5$

31. $\sqrt{3k + 10} + 5 = 2k$

32. $\sqrt{4t + 13} + 1 = 2t$

33. $\sqrt{5x + 1} - 1 = x$

34. $\sqrt{x + 1} - x = 1$

35. $\sqrt{6t + 7} + 3 = t + 5$

36. $\sqrt{10x + 24} = x + 4$

37. $x - 4 - \sqrt{2x} = 0$

38. $x - 3 - \sqrt{4x} = 0$

39. $\sqrt{x + 6} = 2x$

40. $\sqrt{k + 12} = k$

Solve each problem.

41. Police sometimes use the following procedure to estimate the speed at which a car was traveling at the time of an accident. A police officer drives the car involved in the accident under conditions similar to those during which the accident took place and then skids to a stop. If the car is driven at 30 miles per hour, then the speed at the time of the accident is given by

$$s = 30\sqrt{\dfrac{a}{p}}$$

where a is the length of the skid marks left at the time of the accident and p is the length of the skid marks in the police test. Find s for the following values of a and p. Round to the nearest tenth.

(a) $a = 862$ feet; $p = 156$ feet

(b) $a = 382$ feet; $p = 96$ feet

(c) $a = 84$ feet; $p = 26$ feet

42. A function for calculating the distance one can see from an airplane to the horizon on a clear day is given by

$$d(x) = 1.22\sqrt{x}$$

where x is the altitude of the plane in feet and $d(x)$ is in miles. How far can one see to the horizon in a plane flying at the following altitudes? Round to the nearest tenth.

(a) 15,000 feet

(b) 18,000 feet

(c) 24,000 feet

Solve each problem. Give answers to the nearest tenth.

43. A surveyor wants to find the height of a building. At a point 110.0 feet from the base of the building he sights to the top of the building and finds the distance to be 193.0 feet. See the figure. How high is the building?

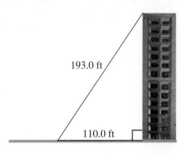

193.0 ft

110.0 ft

44. Two towns are separated by dense woods. To go from Town B to Town A, it is necessary to travel due west for 19.0 miles, then turn due north and travel for 14.0 miles. See the figure. How far apart are the towns?

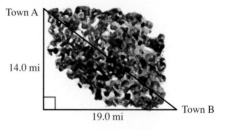

Town A

14.0 mi

19.0 mi

Town B

9.7 Complex Numbers

As we saw in Chapter 1, the set of real numbers includes many other number sets (the rational numbers, integers, and natural numbers, for example). In this section a new set of numbers is introduced that includes the set of real numbers, as well as numbers that are even roots of negative numbers, like $\sqrt{-2}$.

OBJECTIVE 1 The equation $x^2 + 1 = 0$ has no real number solutions, since any solution must be a number whose square is -1. In the set of real numbers all squares are nonnegative numbers, because multiplication is defined in such a way that the product of two positive numbers or two negative numbers is always positive. To provide a solution for the equation $x^2 + 1 = 0$, a new number i, called the **imaginary** (ih-MAJ-uh-nair-ee) **unit,** is defined so that

$$i^2 = -1.$$

That is, i is a number whose square is -1. This definition of i makes it possible to define any square root of a negative number as follows.

For any positive number b,
$$\sqrt{-b} = i\sqrt{b}.$$

E X A M P L E   1   Simplifying Square Roots of Negative Numbers

Write each number as a product of a real number and i.

(a) $\sqrt{-100} = i\sqrt{100} = 10i$

(b) $\sqrt{-2} = i\sqrt{2}$

Caution
It is easy to mistake $\sqrt{2}i$ for $\sqrt{2i}$, with the i under the radical. For this reason, we often write $\sqrt{2}i$ as $i\sqrt{2}$.

WORK PROBLEM 1 AT THE SIDE. ▶▶

When finding a product such as $\sqrt{-4} \cdot \sqrt{-9}$, we cannot use the product rule for radicals because that rule applies only when no more than one radicand is negative. For this reason, we change $\sqrt{-b}$ to the form $i\sqrt{b}$ before performing any multiplications or divisions. For example,

$$\sqrt{-4} \cdot \sqrt{-9} = i\sqrt{4} \cdot i\sqrt{9}$$
$$= i \cdot 2 \cdot i \cdot 3$$
$$= 6i^2$$
$$= 6(-1) \qquad \text{Let } i^2 = -1.$$
$$= -6.$$

An *incorrect* use of the product rule for radicals would give a *wrong* answer.

$$\sqrt{-4} \cdot \sqrt{-9} = \sqrt{(-4)(-9)}$$
$$= \sqrt{36}$$
$$= 6 \qquad \qquad \text{INCORRECT}$$

OBJECTIVES

1 ▶ Simplify numbers of the form $\sqrt{-b}$, where $b > 0$.

2 ▶ Recognize imaginary and complex numbers.

3 ▶ Add and subtract complex numbers.

4 ▶ Find products of complex numbers.

5 ▶ Find quotients of complex numbers.

6 ▶ Find powers of i.

FOR EXTRA HELP

Tutorial   Tape 20   SSM, Sec. 9.7

1. Write each number as a product of a real number and i.

(a) $\sqrt{-16}$

(b) $-\sqrt{-81}$

(c) $\sqrt{-7}$

ANSWERS

1. (a) $4i$ **(b)** $-9i$ **(c)** $i\sqrt{7}$

2. Multiply.

(a) $\sqrt{-7} \cdot \sqrt{-7}$

(b) $\sqrt{-5} \cdot \sqrt{-10}$

(c) $\sqrt{-15} \cdot \sqrt{2}$

3. Divide.

(a) $\dfrac{\sqrt{-32}}{\sqrt{-2}}$

(b) $\dfrac{\sqrt{-27}}{\sqrt{-3}}$

(c) $\dfrac{\sqrt{-40}}{\sqrt{10}}$

EXAMPLE 2   Multiplying Square Roots of Negative Numbers

Multiply.

(a)
$$\sqrt{-3} \cdot \sqrt{-7} = i\sqrt{3} \cdot i\sqrt{7}$$
$$= i^2\sqrt{3 \cdot 7}$$
$$= (-1)\sqrt{21}$$
$$= -\sqrt{21}$$

(b)
$$\sqrt{-2} \cdot \sqrt{-8} = i\sqrt{2} \cdot i\sqrt{8}$$
$$= i^2\sqrt{2 \cdot 8}$$
$$= (-1)\sqrt{16}$$
$$= (-1)4 = -4$$

(c) $\sqrt{-5} \cdot \sqrt{6} = i\sqrt{5} \cdot \sqrt{6} = i\sqrt{30}$

◄◄ **WORK PROBLEM 2 AT THE SIDE.**

The methods used to find the products in Example 2 also apply to quotients, as the next example shows.

EXAMPLE 3   Dividing Square Roots of Negative Numbers

Divide.

(a) $\dfrac{\sqrt{-75}}{\sqrt{-3}} = \dfrac{i\sqrt{75}}{i\sqrt{3}} = \sqrt{\dfrac{75}{3}} = \sqrt{25} = 5$

(b) $\dfrac{\sqrt{-32}}{\sqrt{8}} = \dfrac{i\sqrt{32}}{\sqrt{8}} = i\sqrt{\dfrac{32}{8}} = i\sqrt{4} = 2i$

◄◄ **WORK PROBLEM 3 AT THE SIDE.**

OBJECTIVE 2 With the new number i and the real numbers, a new set of numbers can be formed that includes the real numbers as a subset. The *complex numbers* are defined as follows.

If a and b are real numbers, then any number of the form $a + bi$ is called a **complex number.**

In the complex number $a + bi$, the number a is called the **real part** and b is called the **imaginary part.*** When $b = 0$, $a + bi$ is a real number, so the real numbers are a subset of the complex numbers. Complex numbers with $b \neq 0$ are called **imaginary numbers.**** In spite of their name, imaginary numbers are very useful in applications, particularly in work with electricity.

* In some texts, bi is called the imaginary part.
** Imaginary numbers are sometimes defined as complex numbers with $a = 0$ and $b \neq 0$.

ANSWERS

2. (a) -7 (b) $-5\sqrt{2}$ (c) $i\sqrt{30}$
3. (a) 4 (b) 3 (c) $2i$

The relationships among the various sets of numbers discussed in this book are shown in Figure 3.

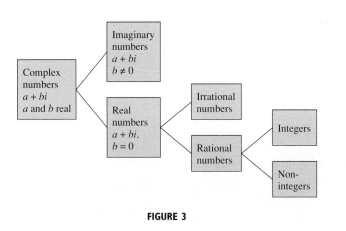

FIGURE 3

OBJECTIVE ▶3▶ The commutative, associative, and distributive properties for real numbers are also valid for complex numbers. Thus, to add complex numbers, we add their real parts and add their imaginary parts.

E X A M P L E   4   Adding Complex Numbers

Add.

(a) $(2 + 3i) + (6 + 4i)$

$\qquad = (2 + 6) + (3 + 4)i$ Commutative and associative properties

$\qquad = 8 + 7i$

(b) $5 + (9 - 3i) = (5 + 9) - 3i$

$\qquad\qquad\qquad = 14 - 3i$

WORK PROBLEM 4 AT THE SIDE. ▶▶

We subtract complex numbers by subtracting their real parts and subtracting their imaginary parts.

E X A M P L E   5   Subtracting Complex Numbers

Subtract.

(a) $(6 + 5i) - (3 + 2i) = (6 - 3) + (5 - 2)i$

$\qquad\qquad\qquad\qquad = 3 + 3i$

(b) $(7 - 3i) - (8 - 6i) = (7 - 8) + [-3 - (-6)]i$

$\qquad\qquad\qquad\qquad = -1 + 3i$

(c) $(-9 + 4i) - (-9 + 8i) = (-9 + 9) + (4 - 8)i$

$\qquad\qquad\qquad\qquad = 0 - 4i = -4i$

WORK PROBLEM 5 AT THE SIDE. ▶▶

4. Add.

(a) $(4 + 6i) + (-3 + 5i)$

(b) $(-1 + 8i) + (9 - 3i)$

5. Subtract.

(a) $(7 + 3i) - (4 + 2i)$

(b) $(-6 - i) - (-5 - 4i)$

(c) $8 - (3 - 2i)$

6. Multiply.

(a) $6i(4 + 3i)$

OBJECTIVE 4 Complex numbers of the form $a + bi$ have the same form as binomials, so we multiply two complex numbers by using the FOIL method for multiplying binomials. (Recall that FOIL stands for *First-Outside-Inside-Last*.)

EXAMPLE 6   Multiplying Complex Numbers

(a) $4i(2 + 3i) = (4i)(2) + (4i)(3i) = 8i + 12i^2 = 8i + 12(-1)$
$$= -12 + 8i$$

(b) $(3 + 5i)(4 - 2i) = \underbrace{3(4)}_{\text{First}} + \underbrace{3(-2i)}_{\text{Outside}} + \underbrace{5i(4)}_{\text{Inside}} + \underbrace{5i(-2i)}_{\text{Last}}$

Now simplify. (Remember that $i^2 = -1$.)

$$= 12 - 6i + 20i - 10i^2$$
$$= 12 + 14i - 10(-1) \qquad \text{Let } i^2 = -1.$$
$$= 12 + 14i + 10$$
$$= 22 + 14i$$

(c) $(2 + 3i)(1 - 5i) = 2(1) + 2(-5i) + 3i(1) + 3i(-5i)$
$$= 2 - 10i + 3i - 15i^2$$
$$= 2 - 7i - 15(-1)$$
$$= 2 - 7i + 15$$
$$= 17 - 7i$$

(b) $(6 - 4i)(2 + 4i)$

◀◀ WORK PROBLEM 6 AT THE SIDE.

The two complex numbers $a + bi$ and $a - bi$ are called **conjugates** of each other. The product of a complex number and its conjugate is always a real number, as shown here.

$$(a + bi)(a - bi) = a \cdot a - abi + abi - b^2i^2$$
$$= a^2 - b^2(-1)$$
$$(a + bi)(a - bi) = a^2 + b^2$$

For example, $(3 + 7i)(3 - 7i) = 3^2 + 7^2 = 9 + 49 = 58$.

(c) $(3 - 2i)(3 + 2i)$

OBJECTIVE 5 The quotient of two complex numbers should be a complex number. To write the quotient as a complex number, we need to eliminate i in the denominator. We use conjugates to do this.

EXAMPLE 7   Dividing Complex Numbers

Find the quotients.

(a) $\dfrac{4 - 3i}{5 + 2i}$

Using the property of 1, we multiply the numerator and denominator by the conjugate of the denominator. The conjugate of $5 + 2i$ is $5 - 2i$.

CONTINUED ON NEXT PAGE

$$\frac{4 - 3i}{5 + 2i} = \frac{(4 - 3i)(5 - 2i)}{(5 + 2i)(5 - 2i)}$$

$$= \frac{20 - 8i - 15i + 6i^2}{5^2 + 2^2} \qquad \text{FOIL method}$$

$$= \frac{14 - 23i}{29} \quad \text{or} \quad \frac{14}{29} - \frac{23}{29}i \qquad \text{Write as } a + bi.$$

Notice that this is just like rationalizing the denominator.

(b) $\dfrac{1 + i}{i}$

The conjugate of i is $-i$. Multiply the numerator and denominator by $-i$.

$$\frac{1 + i}{i} = \frac{(1 + i)(-i)}{i(-i)}$$

$$= \frac{-i - i^2}{-i^2}$$

$$= \frac{-i - (-1)}{-(-1)}$$

$$= \frac{-i + 1}{1} = 1 - i$$

$\blacksquare$

WORK PROBLEM 7 AT THE SIDE. ▶▶

Objective 6 ▶ The fact that i^2 is equal to -1 can be used to find higher powers of i, as shown below.

$$i^3 = i \cdot i^2 = i(-1) = -i \qquad i^6 = i^2 \cdot i^4 = (-1) \cdot 1 = -1$$

$$i^4 = i^2 \cdot i^2 = (-1)(-1) = 1 \qquad i^7 = i^3 \cdot i^4 = (-i) \cdot 1 = -i$$

$$i^5 = i \cdot i^4 = i \cdot 1 = i \qquad i^8 = i^4 \cdot i^4 = 1 \cdot 1 = 1$$

As these examples show, the powers of i rotate through four numbers: i, -1, $-i$, and 1. Larger powers of i can be simplified by using the fact that $i^4 = 1$. For example, $i^{75} = (i^4)^{18} \cdot i^3 = 1^{18} \cdot i^3 = 1 \cdot i^3 = i^3 = -i$. This example suggests a quick method for simplifying large powers of i.

E X A M P L E   8   Simplifying Powers of i

Find each power of i.

(a) $i^{12} = (i^4)^3 = 1^3 = 1$

(b) $i^{39} = i^{36} \cdot i^3$

$$= (i^4)^9 \cdot i^3$$

$$= 1^9 \cdot (-i)$$

$$= -i$$

(c) $i^{-2} = \dfrac{1}{i^2} = \dfrac{1}{-1} = -1$

CONTINUED ON NEXT PAGE

7. Find the quotients.

(a) $\dfrac{2 + i}{3 - i}$

(b) $\dfrac{6 + 2i}{4 - 3i}$

(c) $\dfrac{5}{3 - 2i}$

(d) $\dfrac{5 - i}{i}$

Answers

7. (a) $\dfrac{1}{2} + \dfrac{1}{2}i$ **(b)** $\dfrac{18}{25} + \dfrac{26}{25}i$

(c) $\dfrac{15}{13} + \dfrac{10}{13}i$ **(d)** $-1 - 5i$

8. Find each power of i.

(a) i^{21}

(d) $i^{-1} = \dfrac{1}{i}$

    To simplify this quotient, multiply numerator and denominator by $-i$, the conjugate of i.

$$\frac{1}{i} = \frac{1(-i)}{i(-i)}$$

$$= \frac{-i}{-i^2}$$

$$= \frac{-i}{-(-1)}$$

$$= \frac{-i}{1}$$

$$= -i$$

(b) i^{36}

◀◀ **WORK PROBLEM 8 AT THE SIDE.**

(c) i^{50}

(d) i^{-9}

ANSWERS

8. (a) i **(b)** 1 **(c)** -1 **(d)** $-i$

9.7 Exercises

Complete each statement.

1. _____ is the imaginary unit.

2. $i^2 = $ _____ .

3. $\sqrt{-b} = i$ _____ , for $b > 0$.

4. The real part of the complex number $a + bi$ is _____ .

5. Every real number is a complex number. Explain why this is so.

6. Not every complex number is a real number. Give an example of this, and explain why this statement is true.

Write each as a product of a real number and i. Simplify all radical expressions. See Example 1.

7. $\sqrt{-169}$ **8.** $\sqrt{-225}$ **9.** $-\sqrt{-144}$ **10.** $-\sqrt{-196}$

11. $\sqrt{-5}$ **12.** $\sqrt{-21}$ **13.** $\sqrt{-48}$ **14.** $\sqrt{-96}$

Multiply or divide as indicated. See Examples 2 and 3.

15. $\sqrt{-15} \cdot \sqrt{-15}$ **16.** $\sqrt{-19} \cdot \sqrt{-19}$ **17.** $\sqrt{-4} \cdot \sqrt{-25}$ **18.** $\sqrt{-9} \cdot \sqrt{-81}$

19. $\dfrac{\sqrt{-300}}{\sqrt{-100}}$ **20.** $\dfrac{\sqrt{-40}}{\sqrt{-10}}$ **21.** $\dfrac{\sqrt{-75}}{\sqrt{3}}$ **22.** $\dfrac{\sqrt{-160}}{\sqrt{10}}$

Add or subtract as indicated. Write your answers in the form a + bi. See Examples 4 and 5.

23. $(3 + 2i) + (-4 + 5i)$ **24.** $(7 + 15i) + (-11 + 14i)$ **25.** $(5 - i) + (-5 + i)$

26. $(-2 + 6i) + (2 - 6i)$ **27.** $(4 + i) - (-3 - 2i)$ **28.** $(9 + i) - (3 + 2i)$

29. $(-3 - 4i) - (-1 - 4i)$

30. $(-2 - 3i) - (-5 - 3i)$

31. $(-4 + 11i) + (-2 - 4i) + (7 + 6i)$

32. $(-1 + i) + (2 + 5i) + (3 + 2i)$

33. $[(7 + 3i) - (4 - 2i)] + (3 + i)$

34. $[(7 + 2i) + (-4 - i)] - (2 + 5i)$

35. Fill in the blank with the correct response: Because $(4 + 2i) - (3 + i) = 1 + i$, using the definition of subtraction we can check this to find that $(1 + i) + (3 + i) = $ _____ .

36. Fill in the blank with the correct response: Because $\frac{-5}{2 - i} = -2 - i$, using the definition of division we can check this to find that $(-2 - i)(2 - i) = $ _____ .

Multiply. See Example 6.

37. $(3i)(27i)$

38. $(5i)(125i)$

39. $(-8i)(-2i)$

40. $(-32i)(-2i)$

41. $5i(-6 + 2i)$

42. $3i(4 + 9i)$

43. $(4 + 3i)(1 - 2i)$

44. $(7 - 2i)(3 + i)$

45. $(4 + 5i)^2$

46. $(3 + 2i)^2$

47. $(12 + 3i)(12 - 3i)$

48. $(6 + 7i)(6 - 7i)$

49. **(a)** What is the conjugate of $a + bi$?
(b) If we multiply $a + bi$ by its conjugate, we get _____ + _____ , which is always a real number.

50. Explain the procedure you would use to find the quotient
$$\frac{-1 + 5i}{3 + 2i}.$$

Write in the form $a + bi$. See Example 7.

51. $\dfrac{2}{1 - i}$

52. $\dfrac{29}{5 + 2i}$

53. $\dfrac{-7 + 4i}{3 + 2i}$

54. $\dfrac{-38 - 8i}{7 + 3i}$

55. $\dfrac{8i}{2 + 2i}$ **56.** $\dfrac{-8i}{1 + i}$ **57.** $\dfrac{2 - 3i}{2 + 3i}$ **58.** $\dfrac{-1 + 5i}{3 + 2i}$

MATHEMATICAL CONNECTIONS (EXERCISES 59–64)

Consider the following expressions:

Binomials	**Complex Numbers**
$x + 2, \quad 3x - 1$	$1 + 2i, \quad 3 - i.$

When we add, subtract, or multiply complex numbers in standard form, the rules are the same as those for the corresponding operations on binomials. That is, we add or subtract like terms, and we use FOIL *to multiply. Division, however, is comparable to division by the sum or difference of radicals, where we multiply by the conjugate to get a rational denominator. To express the quotient of two complex numbers in standard form, we also multiply by the conjugate of the denominator.*

The following exercises illustrate these ideas. Work them in order.

59. (a) Add the two binomials.   **(b)** Add the two complex numbers.

60. (a) Subtract the second binomial from the first.
  (b) Subtract the second complex number from the first.

61. (a) Multiply the two binomials.   **(b)** Multiply the two complex numbers.

62. (a) Rationalize the denominator: $\dfrac{\sqrt{3} - 1}{1 + \sqrt{2}}$.   **(b)** Write in standard form: $\dfrac{3 - i}{1 + 2i}$.

63. Explain why the answers for parts (a) and (b) in Exercise 61 do not correspond as the answers in Exercises 59 and 60 do.

64. Explain why the answers for parts (a) and (b) in Exercise 62 do not correspond as the answers in Exercises 59 and 60 do.

65. Recall that if $a \neq 0$, $\frac{1}{a}$ is called the reciprocal of a. Use this definition to express the reciprocal of $5 - 4i$ in the form $a + bi$.

66. Recall that if $a \neq 0$, a^{-1} is defined to be $\frac{1}{a}$. Use this definition to express $(4 - 3i)^{-1}$ in the form $a + bi$.

Find each power of i. See Example 8.

67. i^{18} **68.** i^{26} **69.** i^{89} **70.** i^{45}

71. i^{96} **72.** i^{48} **73.** i^{-5} **74.** i^{-17}

75. A student simplified i^{-18} as follows:

$$i^{-18} = i^{-18} \cdot i^{20} = i^{-18+20} = i^2 = -1.$$

Explain the mathematical justification for this correct work.

76. Add: $3(2 - i)^{-1} + 5(1 + i)^{-1}$.

Ohm's law for the current I in a circuit with voltage E, resistance R, capacitance reactance X_c, and inductive reactance X_L is

$$I = \frac{E}{R + (X_L - X_c)i}.$$

77. Find I if $E = 2 + 3i$, $R = 5$, $X_L = 4$, and $X_c = 3$.

78. Using the law given for Exercise 77, find E if $I = 1 - i$, $R = 2$, $X_L = 3$, and $X_c = 1$.

79. Show that $1 + 5i$ is a solution of
$$x^2 - 2x + 26 = 0.$$

80. Show that $3 + 2i$ is a solution of
$$x^2 - 6x + 13 = 0.$$

9.1	**square root**	The square roots of a^2 are a and $-a$ (a is nonnegative).
	perfect square	A number with a rational square root is called a perfect square.
	radicand	The number or expression under a radical sign is called the radicand.
	radical	A radical sign with a radicand is called a radical.
	radical expression	An algebraic expression containing a radical is called a radical expression.
9.3	**like radicals**	Like radicals are multiples of the same radical.
9.4	**rationalizing the denominator**	The process of changing the denominator of a fraction from a radical (irrational number) to a rational number is called rationalizing the denominator.
	simplified form	A radical is in simplified form if the radicand contains no factor (except 1) that is a perfect square, the radicand contains no fractions, and no denominator contains a radical.
9.5	**conjugates**	The conjugate of $a + b$ is $a - b$.
9.7	**imaginary unit**	The number i is called the imaginary unit.
	complex number	A complex number is a number that can be written in the form $a + bi$, where a and b are real numbers.
	real part	The real part of $a + bi$ is a.
	imaginary part	The imaginary part of $a + bi$ is b.
	imaginary number	A complex number $a + bi$ with $b \neq 0$ is called an imaginary number.

$\sqrt{}$	radical sign
$\sqrt[n]{a}$	radical; principal nth root of a
i	a number whose square is -1

QUICK REVIEW

Concepts	Examples

9.1 Finding Roots

If a is a positive real number,
$\sqrt{a}$ is the positive square root of a;
$-\sqrt{a}$ is the negative square root of a; $\sqrt{0} = 0$.

If a is a negative real number, $\sqrt{a}$ is not a real number.

If a is a positive rational number, $\sqrt{a}$ is rational if a is a perfect square. $\sqrt{a}$ is irrational if a is not a perfect square.

Each real number has exactly one real cube root.

$\sqrt{49} = 7$

$-\sqrt{81} = -9$

$\sqrt{-25}$ is not a real number.

$\sqrt{\dfrac{4}{9}}, \sqrt{16}$ are rational. $\sqrt{\dfrac{2}{3}}, \sqrt{21}$ are irrational.

$\sqrt[3]{27} = 3$; $\sqrt[3]{-8} = -2$

9.2 Multiplication and Division of Radicals

Product Rule for Radicals

For nonnegative real numbers a and b,

$$\sqrt{a} \cdot \sqrt{b} = \sqrt{ab} \quad \text{and} \quad \sqrt{ab} = \sqrt{a} \cdot \sqrt{b}.$$

Quotient Rule for Radicals

If a and b are nonnegative real numbers and b is not 0,

$$\frac{\sqrt{a}}{\sqrt{b}} = \sqrt{\frac{a}{b}} \quad \text{and} \quad \sqrt{\frac{a}{b}} = \frac{\sqrt{a}}{\sqrt{b}}.$$

$\sqrt{5} \cdot \sqrt{7} = \sqrt{5 \cdot 7} = \sqrt{35}$

$\sqrt{8} \cdot \sqrt{2} = \sqrt{16} = 4$

$\sqrt{48} = \sqrt{16} \cdot \sqrt{3} = 4\sqrt{3}$

$\sqrt{\dfrac{25}{64}} = \dfrac{\sqrt{25}}{\sqrt{64}} = \dfrac{5}{8}$

$\dfrac{\sqrt{8}}{\sqrt{2}} = \sqrt{\dfrac{8}{2}} = \sqrt{4} = 2$

9.3 Addition and Subtraction of Radicals

Add and subtract like radicals by using the distributive property. Only like radicals can be combined in this way.

$2\sqrt{5} + 4\sqrt{5} = (2 + 4)\sqrt{5}$
$\qquad\qquad = 6\sqrt{5}$

$\sqrt{8} + \sqrt{32} = 2\sqrt{2} + 4\sqrt{2}$
$\qquad\qquad = 6\sqrt{2}$

9.4 Rationalizing the Denominator

The denominator of a radical is rationalized by multiplying both the numerator and denominator by the same number.

$\dfrac{2}{\sqrt{3}} = \dfrac{2 \cdot \sqrt{3}}{\sqrt{3} \cdot \sqrt{3}} = \dfrac{2\sqrt{3}}{3}$

$\sqrt[3]{\dfrac{5}{121}} = \dfrac{\sqrt[3]{5} \cdot \sqrt[3]{11}}{\sqrt[3]{11^2} \cdot \sqrt[3]{11}} = \dfrac{\sqrt[3]{55}}{11}$

Concepts	Examples
9.5 Simplifying Radical Expressions When appropriate, use the rules for adding and multiplying polynomials to simplify radical expressions. If a radical expression contains two terms in the denominator and at least one of those terms is a square root radical, multiply both the numerator and the denominator by the conjugate of the denominator.	$$\sqrt{6}(\sqrt{5} - \sqrt{7}) = \sqrt{30} - \sqrt{42}$$ $$(\sqrt{5} - \sqrt{3})(\sqrt{5} + \sqrt{3}) = 5 - 3 = 2$$ $$\frac{6}{\sqrt{7} - \sqrt{2}} = \frac{6}{\sqrt{7} - \sqrt{2}} \cdot \frac{\sqrt{7} + \sqrt{2}}{\sqrt{7} + \sqrt{2}}$$ $$= \frac{6(\sqrt{7} + \sqrt{2})}{7 - 2} \quad \text{Multiply fractions.}$$ $$= \frac{6(\sqrt{7} + \sqrt{2})}{5} \quad \text{Simplify.}$$
9.6 Equations with Radicals **Solving an Equation with Radicals** *Step 1* Arrange the terms so that a radical is alone on one side of the equation. *Step 2* Square each side. (By the squaring property of equality, all solutions of the original equation are *among* the solutions of the squared equation.) *Step 3* Combine like terms. Repeat Steps 1–3 if necessary. *Step 4* Solve the equation for potential solutions. *Step 5* Check all potential solutions from Step 4 in the original equation.	Solve: $\sqrt{2x - 3} + x = 3$. $$\sqrt{2x - 3} = 3 - x \qquad \text{Isolate the radical.}$$ $$(\sqrt{2x - 3})^2 = (3 - x)^2 \qquad \text{Square.}$$ $$2x - 3 = 9 - 6x + x^2$$ $$0 = x^2 - 8x + 12 \qquad \text{Put in standard form.}$$ $$0 = (x - 2)(x - 6) \qquad \text{Factor.}$$ $$x - 2 = 0 \quad \text{or} \quad x - 6 = 0 \qquad \text{Set each factor} = 0.$$ $$x = 2 \quad \text{or} \qquad x = 6 \qquad \text{Solve.}$$ A check is essential here. Verify that $\{2\}$ is the solution set (6 is extraneous).
9.7 Complex Numbers $$i^2 = -1$$ For any positive number b, $\sqrt{-b} = i\sqrt{b}$. To multiply $\sqrt{-3} \cdot \sqrt{-27}$, first change each factor to the form $i\sqrt{b}$, then multiply. Quotients such as $$\frac{\sqrt{-18}}{\sqrt{-2}}$$ are found similarly.	$$\sqrt{-3} \cdot \sqrt{-27} = i\sqrt{3} \cdot i\sqrt{27}$$ $$= i^2\sqrt{81}$$ $$= -1 \cdot 9 = -9$$ $$\frac{\sqrt{-18}}{\sqrt{-2}} = \frac{i\sqrt{18}}{i\sqrt{2}} = \frac{\sqrt{18}}{\sqrt{2}} = \sqrt{9} = 3$$

Concepts	*Examples*
9.7 Complex Numbers (Continued)	

Adding and Subtracting Complex Numbers
Add (or subtract) the real parts and add (or subtract) the imaginary parts.

$$(5 + 3i) + (8 - 7i) = 13 - 4i$$
$$(5 + 3i) - (8 - 7i)$$
$$= (5 - 8) + [3 - (-7)]i$$
$$= -3 + 10i$$

Multiplying and Dividing Complex Numbers
Multiply complex numbers by using the FOIL method. Divide complex numbers by multiplying the numerator and the denominator by the conjugate of the denominator.

$$(2 + i)(5 - 3i) = 10 - 6i + 5i - 3i^2$$
$$= 10 - i - 3(-1)$$
$$= 10 - i + 3$$
$$= 13 - i$$

$$\frac{2}{3 + i} = \frac{2}{3 + i} \cdot \frac{3 - i}{3 - i}$$
$$= \frac{2(3 - i)}{9 - i^2}$$
$$= \frac{2(3 - i)}{10}$$
$$= \frac{3 - i}{5}$$

[9.1] *Find all square roots of each number.*

1. 49   **2.** 81   **3.** 196   **4.** 121

5. 225   **6.** 729

Find each root that is a real number.

7. $\sqrt{16}$ **8.** $-\sqrt{36}$ **9.** $\sqrt[3]{1000}$ **10.** $\sqrt[4]{81}$

11. $\sqrt{-8100}$ **12.** $-\sqrt{4225}$ **13.** $\sqrt{\dfrac{49}{36}}$ **14.** $\sqrt{\dfrac{100}{81}}$

15. If $\sqrt{a}$ is not a real number, then what kind of number must a be?

16. Find the value of x.

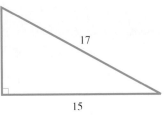

Write rational, irrational, *or* not a real number *for each number. If a number is rational, give its exact value. If a number is irrational, give a decimal approximation for the number. Round approximations to the nearest thousandth.*

17. $\sqrt{23}$ **18.** $\sqrt{169}$ **19.** $-\sqrt{25}$ **20.** $\sqrt{-4}$

[9.2] *Use the product rule to simplify each expression.*

21. $\sqrt{2} \cdot \sqrt{7}$

22. $\sqrt{12} \cdot \sqrt{3}$

23. $\sqrt{5} \cdot \sqrt{15}$

24. $\sqrt{12} \cdot \sqrt{12}$

25. $-\sqrt{27}$

26. $\sqrt{48}$

27. $\sqrt{160}$

28. $\sqrt[3]{-125}$

29. $\sqrt[3]{1728}$

30. $\sqrt{12} \cdot \sqrt{27}$

31. $\sqrt{32} \cdot \sqrt{48}$

32. $\sqrt{50} \cdot \sqrt{125}$

Use the product rule and the quotient rule, as necessary, to simplify each expression.

33. $\sqrt{\dfrac{9}{4}}$

34. $-\sqrt{\dfrac{121}{400}}$

35. $\sqrt{\dfrac{3}{49}}$

36. $\sqrt{\dfrac{7}{169}}$

37. $\sqrt{\dfrac{1}{6}} \cdot \sqrt{\dfrac{5}{6}}$

38. $\sqrt{\dfrac{2}{5}} \cdot \sqrt{\dfrac{2}{45}}$

39. $\dfrac{3\sqrt{10}}{\sqrt{5}}$

40. $\dfrac{24\sqrt{12}}{6\sqrt{3}}$

41. $\dfrac{8\sqrt{150}}{4\sqrt{75}}$

Simplify each expression. Assume that all variables represent nonnegative real numbers.

42. $\sqrt{p} \cdot \sqrt{p}$

43. $\sqrt{k} \cdot \sqrt{m}$

44. $\sqrt{r^{18}}$

45. $\sqrt{x^{10}y^{16}}$

46. $\sqrt{x^9}$

47. $\sqrt{\dfrac{36}{p^2}}, \ p \neq 0$

48. $\sqrt{a^{15}b^{21}}$

49. $\sqrt{121x^6y^{10}}$

50. Use a calculator to find approximations for $\sqrt{.5}$ and $\dfrac{\sqrt{2}}{2}$. Based on your results, do you

think that these two expressions represent the same number? If so, verify it *algebraically*.

[9.3] *Simplify and combine terms where possible.*

51. $\sqrt{11} + \sqrt{11}$

52. $3\sqrt{2} + 6\sqrt{2}$

53. $3\sqrt{75} + 2\sqrt{27}$

54. $4\sqrt{12} + \sqrt{48}$

55. $4\sqrt{24} - 3\sqrt{54} + \sqrt{6}$

56. $2\sqrt{7} - 4\sqrt{28} + 3\sqrt{63}$

57. $\dfrac{2}{5}\sqrt{75} + \dfrac{3}{4}\sqrt{160}$

58. $\dfrac{1}{3}\sqrt{18} + \dfrac{1}{4}\sqrt{32}$

59. $\sqrt{15} \cdot \sqrt{2} + 5\sqrt{30}$

Simplify each expression. Assume that all variables represent nonnegative real numbers.

60. $\sqrt{4x} + \sqrt{36x} - \sqrt{9x}$

61. $\sqrt{16p} + 3\sqrt{p} - \sqrt{49p}$

62. $\sqrt{20m^2} - m\sqrt{45}$

63. $3k\sqrt{8k^2 n} + 5k^2\sqrt{2n}$

[9.4] *Perform the indicated operations, and write all answers in simplest form. Rationalize all denominators. Assume that all variables represent nonnegative real numbers.*

64. $\dfrac{10}{\sqrt{3}}$

65. $\dfrac{15}{\sqrt{2}}$

66. $\dfrac{8\sqrt{2}}{\sqrt{5}}$

67. $\dfrac{5}{\sqrt{5}}$

68. $\dfrac{12}{\sqrt{24}}$

69. $\dfrac{\sqrt{2}}{\sqrt{15}}$

70. $\sqrt{\dfrac{2}{5}}$

71. $\sqrt{\dfrac{5}{14}} \cdot \sqrt{28}$

72. $\sqrt{\dfrac{2}{7}} \cdot \sqrt{\dfrac{1}{3}}$

73. $\sqrt{\dfrac{r^2}{16x}}, \; x \neq 0$

74. $\sqrt[3]{\dfrac{1}{3}}$

75. $\sqrt[3]{\dfrac{2}{7}}$

[9.5]

76. Explain how you would show, without using a calculator, that $\dfrac{\sqrt{6}}{4}$ and $\sqrt{\dfrac{48}{128}}$ represent the exact same number. Then actually perform the necessary steps.

Simplify each expression.

77. $-\sqrt{3}(\sqrt{5} + \sqrt{27})$

78. $3\sqrt{2}(\sqrt{3} + 2\sqrt{2})$

79. $(2\sqrt{3} - 4)(5\sqrt{3} + 2)$

80. $(5\sqrt{7} + 2)^2$

81. $(\sqrt{5} - \sqrt{7})(\sqrt{5} + \sqrt{7})$

82. $(2\sqrt{3} + 5)(2\sqrt{3} - 5)$

83. $(\sqrt{7} + 2\sqrt{6})(\sqrt{12} - \sqrt{2})$

Rationalize the denominators.

84. $\dfrac{1}{2 + \sqrt{5}}$

85. $\dfrac{2}{\sqrt{2} - 3}$

86. $\dfrac{\sqrt{8}}{\sqrt{2} + 6}$

87. $\dfrac{\sqrt{3}}{1 + \sqrt{3}}$ **88.** $\dfrac{\sqrt{5} - 1}{\sqrt{2} + 3}$ **89.** $\dfrac{2 + \sqrt{6}}{\sqrt{3} - 1}$

Write each quotient in lowest terms.

90. $\dfrac{15 + 10\sqrt{6}}{15}$ **91.** $\dfrac{3 + 9\sqrt{7}}{12}$ **92.** $\dfrac{6 + \sqrt{192}}{2}$

[9.6] *Find the solution set for each equation.*

93. $\sqrt{m} + 5 = 0$ **94.** $\sqrt{p} + 4 = 0$ **95.** $\sqrt{k + 1} = 7$

96. $\sqrt{5m + 4} = 3\sqrt{m}$ **97.** $\sqrt{2p + 3} = \sqrt{5p - 3}$ **98.** $\sqrt{4y + 1} = y - 1$

99. $\sqrt{-2k - 4} = k + 2$ **100.** $\sqrt{2 - x} + 3 = x + 7$ **101.** $\sqrt{x} - x + 2 = 0$

102. $\sqrt{2 - x} + x = 0$ **103.** $\sqrt{4y - 2} = \sqrt{3y + 1}$ **104.** $\sqrt{2x + 3} = x + 2$

[9.7] *Write as a product of a real number and i.*

105. $\sqrt{-25}$ **106.** $\sqrt{-200}$

107. If a is a positive real number, is $-\sqrt{-a}$ a real number?

Perform the indicated operations. Write each imaginary number answer in the form $a + bi$.

108. $(-2 + 5i) + (-8 - 7i)$ **109.** $(5 + 4i) - (-9 - 3i)$ **110.** $\sqrt{-5} \cdot \sqrt{-7}$

111. $\sqrt{-25} \cdot \sqrt{-81}$ **112.** $\dfrac{\sqrt{-72}}{\sqrt{-8}}$ **113.** $(2 + 3i)(1 - i)$

114. $(6 - 2i)^2$

115. $\dfrac{3 - i}{2 + i}$

116. $\dfrac{5 + 14i}{2 + 3i}$

Find each power of i.

117. i^{11}

118. i^{52}

119. i^{-13}

MIXED REVIEW EXERCISES

Simplify each expression if possible. Assume all variables represent nonnegative real numbers.

120. $\sqrt{3} \cdot \sqrt{27}$

121. $2\sqrt{27} + 3\sqrt{75} - \sqrt{300}$

122. $\sqrt{\dfrac{121}{t^2}}, \, t \neq 0$

123. $\dfrac{1}{5 + \sqrt{2}}$

124. $\sqrt{\dfrac{1}{3}} \cdot \sqrt{\dfrac{24}{5}}$

125. $\sqrt{50y^2}$

126. $\sqrt[3]{-125}$

127. $-\sqrt{5}(\sqrt{2} + \sqrt{75})$

128. $\sqrt{\dfrac{16r^3}{3s}}, \, s \neq 0$

129. $\dfrac{12 + 6\sqrt{13}}{12}$

130. $-\sqrt{162} + \sqrt{8}$

131. $(\sqrt{5} - \sqrt{2})^2$

132. i^{24}

133. $(5 - i)(5 + i)$

134. Express $\sqrt{-98}$ as a product of i and a real number.

Solve.

135. $\sqrt{x + 2} = x - 4$

136. $\sqrt{k} + 3 = 0$

137. $\sqrt{1 + 3t} - t = -3$

CHAPTER 9 TEST

On this test, assume that all variables represent nonnegative real numbers.

1. Find all square roots of 196.

1. _____

2. Consider $\sqrt{142}$.
  (a) Determine whether it is rational or irrational.
  (b) Find a decimal approximation to the nearest thousandth.

2. (a) _____

  (b) _____

3. Simplify $\sqrt[3]{216}$.

3. _____

Simplify where possible.

4. $-\sqrt{27}$

4. _____

5. $\sqrt{\dfrac{128}{25}}$

5. _____

6. $\sqrt[3]{32}$

6. _____

7. $\dfrac{20\sqrt{18}}{5\sqrt{3}}$

7. _____

8. $3\sqrt{28} + \sqrt{63}$

8. _____

9. $3\sqrt{27x} - 4\sqrt{48x} + 2\sqrt{3x}$

9. _____

10. $\sqrt[3]{32x^2y^3}$

10. _____

11. $(6 - \sqrt{5})(6 + \sqrt{5})$

11. _____

12. $(2 - \sqrt{7})(3\sqrt{2} + 1)$

12. _____

13. $(\sqrt{5} + \sqrt{6})^2$

13. _____

14. (a) _____

(b) _____

Solve the following problem.

14. The hypotenuse of a right triangle measures 9 inches, and one leg measures 3 inches. Find the measure of the other leg.
(a) Give its length in simplified radical form.
(b) Round the answer to the nearest thousandth.

15. _____

Rationalize each denominator.

15. $\dfrac{5\sqrt{2}}{\sqrt{7}}$

16. $\sqrt{\dfrac{2}{3x}}$ $(x > 0)$

16. _____

17. _____

17. $\dfrac{-2}{\sqrt[3]{4}}$

18. $\dfrac{-3}{4 - \sqrt{3}}$

18. _____

19. _____

Solve each equation.

19. $\sqrt{x + 1} = 5 - x$

20. _____

20. $3\sqrt{x} - 1 = 2x$

21. (a) _____

21. Perform the indicated operations. Express answers in the form $a + bi$.

(a) $(-2 + 5i) - (3 + 6i) - 7i$ (b) $\dfrac{7 + i}{1 - i}$

(b) _____

22. _____

22. Simplify i^{35}.

Solve the following equations.

1. $7 - (4 + 3t) + 2t = -6(t - 2) - 5$

2. $|6x - 9| = |-4x + 2|$

Solve the following inequalities.

3. $-5 - 3(m - 2) < 11 - 2(m + 2)$ **4.** $1 + 4x > 5$ and $-2x > -6$ **5.** $-2 < 1 - 3y < 7$

6. Write an equation of the line through the points $(-4, 6)$ and $(7, -6)$.

7. The lines with equations $2x + 3y = 8$ and $6y = 4x + 16$ are **(a)** parallel **(b)** perpendicular **(c)** neither.

8. For the graph of $f(x) = -3x + 6$,
(a) what is the y-intercept?
(b) what is the x-intercept?

9. For many items, the cost per item to manufacture the item varies inversely as the number made. It costs \$200 each to manufacture 1500 widgets. How much will it cost per item to make 2500 widgets, if widgets are that type of item?

10. Graph the inequality $-2x + y < -6$.

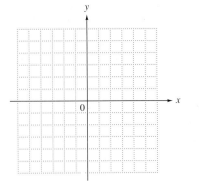

11. Find the measures of the marked angles.

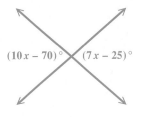

$(10x - 70)°$ $(7x - 25)°$

Solve each system.

12. $5x + 2y = 7$
    $10x + 4y = 12$

13. $2x + y - z = 5$
    $3x + 2y + z = 8$
    $4x + 2y - 2z = 10$

Evaluate each determinant.

14. $\begin{vmatrix} 1 & 5 & 2 \\ 2 & 7 & 4 \\ 3 & -3 & 6 \end{vmatrix}$

15. $\begin{vmatrix} 5 & 6 \\ 7 & 3 \end{vmatrix}$

Perform the indicated operations.

16. $(3k^3 - 5k^2 + 8k - 2) - (4k^3 + 11k + 7) + (2k^2 - 5k)$

17. $(8x - 7)(x + 3)$

18. $\dfrac{8z^3 - 16z^2 + 24z}{8z^2}$

19. $\dfrac{6y^4 - 3y^3 + 5y^2 + 6y - 9}{2y + 1}$

Factor each polynomial completely.

20. $2p^2 - 5pq + 3q^2$

21. $18k^4 + 9k^2 - 20$

22. $25x^2 - 49y^2$

Perform each operation and express answers in lowest terms.

23. $\dfrac{y^2 + y - 12}{y^3 + 9y^2 + 20y} \div \dfrac{y^2 - 9}{y^3 + 3y^2}$

24. $\dfrac{1}{x + y} + \dfrac{3}{x - y}$

Simplify each complex fraction.

25. $\dfrac{\dfrac{-6}{x - 2}}{\dfrac{8}{3x - 6}}$

26. $\dfrac{\dfrac{1}{a} - \dfrac{1}{b}}{\dfrac{a}{b} - \dfrac{b}{a}}$

Solve by factoring.

27. $2x^2 + 11x + 15 = 0$

28. $5t(t - 1) = 2(1 - t)$

Simplify.

29. $-\sqrt{144}$

30. $\sqrt[3]{-64}$

31. $8\sqrt{20} + 3\sqrt{80} - 2\sqrt{500}$

32. $\dfrac{-9}{\sqrt{80}}$

33. $\dfrac{4}{\sqrt{6} - \sqrt{5}}$

34. $\dfrac{12}{\sqrt[3]{2}}$

35. Write $\sqrt{-29}$ as a product of i and a real number.

36. Solve $\sqrt{8x - 4} - \sqrt{7x + 2} = 0$.

Solve each problem.

37. The current of a river runs at 3 miles per hour. Brent's boat can go 36 miles downstream in the same time that it takes to go 24 miles upstream. Find the speed of the boat in still water.

38. Brenda rides her bike 4 miles per hour faster than her husband, Chuck. If Brenda can ride 48 miles in the same time that Chuck can ride 24 miles, what are their speeds?

39. A jar containing only dimes and quarters has 29 coins with a face value of $4.70. How many of each denomination are there?

40. How many liters of pure alcohol must be mixed with 40 liters of 18% alcohol to obtain a 22% alcohol solution?

Quadratic Equations and Inequalities

10.1 Solving Quadratic Equations by the Square Root Property

Recall that a *quadratic equation* is an equation that can be written in the standard form $ax^2 + bx + c = 0$ for real numbers a, b, and c, with $a \neq 0$. Quadratic equations where $ax^2 + bx + c$ is factorable were solved earlier using the zero-factor property. For example, to solve $x^2 + 4x + 3 = 0$, we begin by factoring on the left side and then set each factor equal to zero.

$$x^2 + 4x + 3 = 0$$

$$(x + 3)(x + 1) = 0 \qquad \text{Factor.}$$

$$x + 3 = 0 \quad \text{or} \quad x + 1 = 0 \qquad \text{Zero-factor property}$$

$$x = -3 \quad \text{or} \qquad x = -1$$

The solution set is $\{-3, -1\}$.

OBJECTIVE ▶ We can solve equations such as $x^2 = 9$ by factoring as follows.

$$x^2 = 9$$

$$x^2 - 9 = 0 \qquad \text{Subtract 9.}$$

$$(x + 3)(x - 3) = 0 \qquad \text{Factor.}$$

$$x + 3 = 0 \quad \text{or} \quad x - 3 = 0 \qquad \text{Zero-factor property}$$

$$x = -3 \quad \text{or} \qquad x = 3$$

This result is generalized as the **square root property of equations.**

Square Root Property of Equations

If x and b are complex numbers and if $x^2 = b$, then $x = \sqrt{b}$ or $x = -\sqrt{b}$.

Note
If b is a negative number, then $x^2 = b$ has two imaginary solutions.

www.mathnotes.com

OBJECTIVES

1 ▶ Solve equations of the form $x^2 = $ a number.

2 ▶ Solve equations of the form $(ax + b)^2 = $ a number.

FOR EXTRA HELP

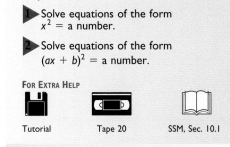

Tutorial   Tape 20   SSM, Sec. 10.1

1. Solve each equation. Write radicals in simplified form.

(a) $k^2 = 49$

(b) $b^2 = 11$

(c) $c^2 = 12$

(d) $x^2 = -9$

E X A M P L E 1 **Solving Quadratic Equations by the Square Root Property**

Solve each equation. Write radicals in simplified form.

(a) $x^2 = 16$

By the square root property, if $x^2 = 16$, then

$$x = \sqrt{16} = 4 \quad \text{or} \quad x = -\sqrt{16} = -4.$$

An abbreviation for $x = 4$ or $x = -4$ is written $x = \pm 4$ (read "plus or minus 4"). Check each solution by substituting it for x in the original equation. The solution set is $\{-4, 4\}$.

(b) $z^2 = 5$

The solutions are $z = \sqrt{5}$ or $z = -\sqrt{5}$, which may be written $\pm\sqrt{5}$. The solution set is $\{\pm\sqrt{5}\}$.

(c) $m^2 = 8$

Use the square root property to get $m = \sqrt{8}$ or $m = -\sqrt{8}$. Simplify $\sqrt{8}$ as $\sqrt{8} = 2\sqrt{2}$, so the solution set is $\{\pm 2\sqrt{2}\}$.

(d) $y^2 = -4$

$$y = \sqrt{-4} \quad \text{or} \quad y = -\sqrt{-4}$$
$$y = i\sqrt{4} \quad \text{or} \quad y = -i\sqrt{4}$$
$$y = 2i \quad \text{or} \quad y = -2i$$

The solution set is $\{-2i, 2i\}$.

◄◄ **WORK PROBLEM 1 AT THE SIDE.**

OBJECTIVE 2 In each of the equations in Example 1, the exponent 2 appeared with a single variable as its base. The square root property of equations can be extended to solve equations where the base is a binomial, as shown in the next example.

2. Solve each equation.

(a) $(m + 2)^2 = 36$

(b) $(p - 4)^2 = 3$

E X A M P L E 2 **Solving a Quadratic Equation by the Square Root Property**

Solve $(x - 1)^2 = 6$.

$$x - 1 = \sqrt{6} \qquad \text{or} \quad x - 1 = -\sqrt{6} \qquad \text{Square root property}$$
$$x = 1 + \sqrt{6} \quad \text{or} \qquad x = 1 - \sqrt{6} \qquad \text{Add 1.}$$

Check: $$(1 + \sqrt{6} - 1)^2 = (\sqrt{6})^2 = 6;$$
$$(1 - \sqrt{6} - 1)^2 = (-\sqrt{6})^2 = 6.$$

The solution set is $\{1 + \sqrt{6}, 1 - \sqrt{6}\}$.

◄◄ **WORK PROBLEM 2 AT THE SIDE.**

ANSWERS

1. (a) $\{-7, 7\}$ (b) $\{-\sqrt{11}, \sqrt{11}\}$
(c) $\{-2\sqrt{3}, 2\sqrt{3}\}$
(d) $\{-3i, 3i\}$
2. (a) $\{-8, 4\}$ (b) $\{4 + \sqrt{3}, 4 - \sqrt{3}\}$

E X A M P L E 3 **Solving a Quadratic Equation by the Square Root Property**

Solve the equation $(3r - 2)^2 = 27$.

$$3r - 2 = \sqrt{27} \quad \text{or} \quad 3r - 2 = -\sqrt{27} \qquad \text{Square root property}$$

Now simplify the radical: $\sqrt{27} = \sqrt{9 \cdot 3} = \sqrt{9} \cdot \sqrt{3} = 3\sqrt{3}$.

$$3r - 2 = 3\sqrt{3} \qquad \text{or} \quad 3r - 2 = -3\sqrt{3}$$
$$3r = 2 + 3\sqrt{3} \quad \text{or} \qquad 3r = 2 - 3\sqrt{3} \qquad \text{Add 2.}$$
$$r = \frac{2 + 3\sqrt{3}}{3} \quad \text{or} \qquad r = \frac{2 - 3\sqrt{3}}{3} \qquad \text{Divide by 3.}$$

Solution set: $\left\{ \dfrac{2 + \sqrt{3}}{2}, \dfrac{2 - \sqrt{3}}{2} \right\}$

E X A M P L E 4 **Solving a Quadratic Equation with Imaginary Solutions**

Solve $(x + 3)^2 = -9$.

$$(x + 3)^2 = -9$$
$$x + 3 = \pm\sqrt{9}$$
$$x + 3 = \pm 3i$$
$$x = -3 \pm 3i$$

The solution set is $\{-3 + 3i, -3 - 3i\}$.

WORK PROBLEM 3 AT THE SIDE. ▶▶

---

3. Solve each equation.

(a) $(2x - 5)^2 = 18$

(b) $(x + 5)^2 = -36$

A Mathematical Model of the Spread of AIDS

In 1981 the first case of AIDS (acquired immune deficiency syndrome) was reported in the United States. According to the Centers for Disease Control and Prevention, over 360,000 individuals have been diagnosed with AIDS and of them over 220,000 have died. AIDS is one of the most devastating diseases of our time. The first two columns of the table list the total (cumulative) number of AIDS cases diagnosed in the United States through 1993.

Year	AIDS Cases Reported	AIDS Cases Predicted
1982	1,563	1,563
1983	4,647	2,876
1984	10,845	10,589
1985	22,620	24,702
1986	41,662	45,215
1987	70,222	72,128
1988	105,489	105,441
1989	147,170	145,154
1990	193,245	191,267
1991	248,023	243,780
1992	315,329	302,693
1993	361,509	368,006

Mathematics can be used to make predictions about the number of future AIDS cases. A relatively simple algebraic equation that models this data is

$$N = 3200(Y - 1982)^2 - 1887(Y - 1982) + 1563,$$

where N is the number of AIDS cases predicted in year Y. The number of AIDS cases predicted by the model is given in the third column of the table. Although this equation does not model the data exactly, it is capable of giving an estimate of future AIDS cases. To predict the number of cases in the year 2000, let $Y = 2000$ in this equation.

FOR GROUP DISCUSSION

1. Use the model equation given to approximate the number of predicted AIDS cases in 1998.

2. Repeat Exercise 1 for 1999.

3. Repeat Exercise 1 for 2000.

4. Discuss the pitfalls of using a model equation such as this to predict too far into the future.

Source: Wright, J. (editor), *The Universal Almanac*, Universal Press Syndicate Company, 1994.

10.1 Exercises

1. If k is a positive perfect square, then $x^2 = k$ has two _____ solutions.
   rational/irrational

2. If k is a prime number, then $x^2 = k$ has two _____ solutions.
   rational/irrational

3. Which one of these equations has exactly one real number solution?
   (a) $x^2 = 4$ **(b)** $y^2 = -4$ **(c)** $(x - 4)^2 = 1$ **(d)** $t^2 = 0$

4. Which one of the equations in Exercise 3 has two imaginary solutions?

5. Why can't the square of a real number be negative?

6. Why does $x^2 = -6$ have no real solutions?

7. Does the equation $-x^2 = 16$ have real solutions? Explain.

8. When a student was asked to solve $x^2 = 81$, she wrote 9 as her answer. Her teacher did not give her full credit, and the student argued that because $9^2 = 81$, her answer had to be correct. Why was her answer not completely correct?

Solve each equation by using the square root property. Express all radicals in simplest form. See Example 1.

9. $x^2 = 81$ **10.** $y^2 = 121$ **11.** $k^2 = 14$ **12.** $m^2 = 22$

13. $t^2 = 48$ **14.** $x^2 = 54$ **15.** $y^2 = \dfrac{25}{4}$ **16.** $m^2 = \dfrac{36}{121}$

17. $z^2 = 2.25$ **18.** $w^2 = 56.25$ **19.** $r^2 - 3 = 0$ **20.** $x^2 - 13 = 0$

Solve each equation by using the square root property. Express all radicals in simplest form. See Examples 2–4.

21. $(x - 3)^2 = 25$ **22.** $(y - 7)^2 = 16$ ***23.** $(z + 5)^2 = -16$

*24. $(m + 2)^2 = -81$ **25.** $(x - 8)^2 = 27$ **26.** $(y - 5)^2 = 40$

27. $(3k + 2)^2 = 49$ **28.** $(5t + 3)^2 = 36$ **29.** $(4x - 3)^2 = 9$

---

*Exercises identified with asterisks have imaginary number solutions.

30. $(7y - 5)^2 = 25$

31. $(5 - 2x)^2 = 30$

32. $(3 - 2a)^2 = 70$

33. $(3k + 1)^2 = 18$

34. $(5z + 6)^2 = 75$

35. $\left(\dfrac{1}{2}x + 5\right)^2 = 12$

36. $\left(\dfrac{1}{3}y + 4\right)^2 = 27$

37. $(4k - 1)^2 - 48 = 0$

38. $(2s - 5)^2 - 180 = 0$

39. Johnny solved the equation in Exercise 31 and wrote his answer as $\left\{\dfrac{5 + \sqrt{30}}{2}, \dfrac{5 - \sqrt{30}}{2}\right\}$. Linda solved the same equation and wrote her answer as $\left\{\dfrac{-5 + \sqrt{30}}{-2}, \dfrac{-5 - \sqrt{30}}{-2}\right\}$. The teacher gave them both full credit. Explain why both students were correct, although their answers seem to differ.

40. In the solutions found in Example 3 of this section, why is it not valid to reduce the answers by dividing out the threes in the numerator and denominator?

If one side of a quadratic equation is a perfect square trinomial, then we can factor the trinomial and use the method of Examples 2–4 to solve it. Solve each of the following equations in this way.

41. $x^2 + 4x + 4 = 25$

42. $x^2 + 6x + 9 = 100$

Solve each problem.

43. One expert at marksmanship can hold a silver dollar at forehead level, drop it, draw his gun, and shoot the coin as it passes waist level. The distance traveled by a falling object is given by

$$d = 16t^2,$$

where d is the distance (in feet) the object falls in t seconds. If the coin falls about 4 feet, use the formula to estimate the time that elapses between the dropping of the coin and the shot.

44. The illumination produced by a light source depends on the distance from the source. For a particular light source, this relationship can be expressed as

$$d^2 = \dfrac{4050}{I},$$

where d is the distance from the source (in feet) and I is the amount of illumination in foot-candles. How far from the source is the illumination equal to 50 foot-candles?

45. Becky and Brad are the owners of Cole's Baseball Cards. They have found that the price p, in dollars, of a particular Kirby Puckett baseball card depends on the demand d, in hundreds, for the card, according to the formula $p = (d - 2)^2$. What demand produces a price of $5 for the card?

46. The amount A that P dollars invested at a rate of interest r will grow to in 2 years is

$$A = P(1 + r)^2.$$

At what interest rate will $100 grow to $110.25 in two years?

10.2 Solving Quadratic Equations by Completing the Square

OBJECTIVES

1. Solve quadratic equations by completing the square when the coefficient of the squared term is 1.

2. Solve quadratic equations by completing the square when the coefficient of the squared term is not 1.

3. Simplify an equation before solving.

4. Solve applied problems that require quadratic equations.

FOR EXTRA HELP

Tutorial   Tape 21   SSM, Sec. 10.2

OBJECTIVE ▶ The properties studied so far are not enough to solve the equation

$$x^2 + 6x + 7 = 0.$$

If we could write the equation in a form like $(x + 3)^2 = 2$, we could solve it with the square root property discussed in the previous section. To do that, we need to have a perfect square trinomial on one side. The next example shows how to rewrite the equation $x^2 + 6x + 7 = 0$ so it can be solved by that method.

WORK PROBLEM 1 AT THE SIDE. ▶▶

E X A M P L E   1   Rewriting an Equation to Use the Square Root Property

Solve $x^2 + 6x + 7 = 0$.

Start by subtracting 7 from each side of the equation.

$$x^2 + 6x = -7$$

The quantity on the left-hand side of $x^2 + 6x = -7$ must be made into a perfect square trinomial. The expression $x^2 + 6x + 9$ is a perfect square, since

$$x^2 + 6x + 9 = (x + 3)^2.$$

Therefore, if 9 is added to each side, the equation will have a perfect square trinomial on the left-hand side, as needed.

$$x^2 + 6x + 9 = -7 + 9 \qquad \text{Add 9.}$$
$$(x + 3)^2 = 2 \qquad\qquad \text{Factor.}$$

Now use the square root property to complete the solution.

$$x + 3 = \sqrt{2} \qquad \text{or} \quad x + 3 = -\sqrt{2}$$
$$x = -3 + \sqrt{2} \quad \text{or} \qquad x = -3 - \sqrt{2}$$

The solution set is $\{-3 + \sqrt{2}, -3 - \sqrt{2}\}$. Check by substituting $-3 + \sqrt{2}$ and $-3 - \sqrt{2}$ for x in the original equation. ∎

The process of changing the form of the equation in Example 1 from

$$x^2 + 6x + 7 = 0 \quad \text{to} \quad (x + 3)^2 = 2$$

is called **completing the square.** When completing the square, only the *form* of the equation is changed. To see this, simplify $(x + 3)^2 = 2$; the result will be $x^2 + 6x + 7 = 0$.

E X A M P L E   2   Solving a Quadratic Equation by Completing the Square

Solve $m^2 - 8m = 5$.

A suitable number must be added to each side to make the left side a perfect square. This number can be found as follows: Recall from Chapter 6 that

$$(m + a)^2 = m^2 + 2am + a^2.$$

CONTINUED ON NEXT PAGE

---

1. As a review, factor each of these perfect square trinomials.

(a) $x^2 + 6x + 9$

(b) $q^2 - 20q + 100$

ANSWERS

1. (a) $(x + 3)^2$ **(b)** $(q - 10)^2$

617

2. Solve by completing the square.

(a) $a^2 + 4a = 1$

We want to find the value of a^2, the number to be added to each side. First we must find a. Here, the middle term of the trinomial $m^2 - 8m + a^2$ is $-8m$, so

$$2am = -8m$$
$$2a = -8 \quad \text{Divide each side by } m.$$
$$a = -4. \quad \text{Divide each side by 2.}$$

Then $a^2 = (-4)^2 = 16$, and 16 should be added to each side of the given equation.

$$m^2 - 8m = 5$$
$$m^2 - 8m + 16 = 5 + 16 \quad \text{Add 16.} \qquad \textbf{(1)}$$

The trinomial $m^2 - 8m + 16$ is a perfect square trinomial. Factor this trinomial to get

$$m^2 - 8m + 16 = (m - 4)^2.$$

Equation (1) becomes

$$(m - 4)^2 = 21.$$

Now use the square root property.

$$m - 4 = \sqrt{21} \quad \text{or} \quad m - 4 = -\sqrt{21}$$
$$m = 4 + \sqrt{21} \quad \text{or} \quad m = 4 - \sqrt{21}$$

The solution set is

$$\{4 + \sqrt{21},\ 4 - \sqrt{21}\}.$$

(b) $z^2 + 6z - 3 = 0$

Let us summarize what we did to find $a = -4$ above. The coefficient of m in the middle term was -8.

1. We multiplied -8 by $\frac{1}{2}$ (took half of -8) to get -4.
2. We squared -4 to get 16.
3. We added 16 to each side of the given equation.

Thus, to find the number to add to both sides, we take half the coefficient of the first degree term, square it, and add it to both sides.

◀◀ **WORK PROBLEM 2 AT THE SIDE.**

OBJECTIVE ▶ The process of completing the square discussed above requires the coefficient of the squared term to be 1. With an equation of the form $ax^2 + bx + c = 0$, to get 1 as a coefficient of x^2, first divide each side of the equation by a. The next examples illustrate this approach.

E X A M P L E   3   Solving a Quadratic Equation by Completing the Square

Solve $4y^2 + 16y = 9$.

Before completing the square, the coefficient of y^2 must be 1. Here the coefficient of y^2 is 4. Make the coefficient 1 by dividing each side of the equation by 4.

$$y^2 + 4y = \frac{9}{4} \quad \text{Divide by 4.}$$

ANSWERS
2. (a) $\{-2 + \sqrt{5},\ -2 - \sqrt{5}\}$
(b) $\{-3 + 2\sqrt{3},\ -3 - 2\sqrt{3}\}$

CONTINUED ON NEXT PAGE

Next, complete the square by taking half the coefficient of y, or $(\frac{1}{2})(4) = 2$, and squaring the result: $2^2 = 4$. Add 4 to each side of the equation, and perform the addition on the right-hand side.

$$y^2 + 4y + 4 = \frac{9}{4} + 4 \qquad \text{Add 4.}$$

$$y^2 + 4y + 4 = \frac{25}{4} \qquad \text{Combine terms.}$$

$$(y + 2)^2 = \frac{25}{4} \qquad \text{Factor.}$$

Use the square root property of equations and solve for y.

$$y + 2 = \frac{5}{2} \qquad \text{or} \quad y + 2 = -\frac{5}{2} \qquad \text{Square root property}$$

$$y = -2 + \frac{5}{2} \quad \text{or} \qquad y = -2 - \frac{5}{2} \qquad \text{Subtract 2.}$$

$$y = \frac{1}{2} \qquad \text{or} \qquad y = -\frac{9}{2} \qquad \text{Combine terms.}$$

Check:

$$4y^2 + 16y = 9 \qquad\qquad\qquad 4y^2 + 16y = 9$$

$$4\left(\frac{1}{2}\right)^2 + 16\left(\frac{1}{2}\right) = 9 \quad ? \qquad 4\left(-\frac{9}{2}\right)^2 + 16\left(-\frac{9}{2}\right) = 9 \quad ?$$

$$4\left(\frac{1}{4}\right) + 8 = 9 \quad ? \qquad\qquad 4\left(\frac{81}{4}\right) - 72 = 9 \quad ?$$

$$1 + 8 = 9 \qquad \text{True} \qquad\qquad 81 - 72 = 9 \qquad \text{True}$$

The solution set is $\{\frac{1}{2}, -\frac{9}{2}\}$.

The steps in solving a quadratic equation by completing the square are summarized below.

Completing the Square

Use *completing the square* to solve the quadratic equation $ax^2 + bx + c = 0$ as follows.

Step 1 If the coefficient of the squared term is 1, proceed to Step 2. If the coefficient of the squared term is not 1 but some other nonzero number, divide each side of the equation by this coefficient. This gives an equation that has 1 as the coefficient of x^2.

Step 2 Make sure that all terms with variables are on one side of the equals sign and that all constants are on the other side.

Step 3 Take half the coefficient of x and square the result. Add the square to each side of the equation. By factoring, the side containing the variables can now be written as a perfect square.

Step 4 Apply the square root property of equations.

3. Solve by completing the square.

(a) $9m^2 + 18m + 5 = 0$

(b) $4k^2 - 24k + 11 = 0$

Answers

3. (a) $\left\{-\frac{1}{3}, -\frac{5}{3}\right\}$ **(b)** $\left\{\frac{11}{2}, \frac{1}{2}\right\}$

4. Solve by completing the square.

(a) $3x^2 + 5x - 2 = 0$

(b) $2y^2 - 4y = 1$

E X A M P L E   4   Solving a Quadratic Equation by Completing the Square

Solve $2x^2 - 7x = 9$.

Divide each side of the equation by 2 to get a coefficient of 1 for the x^2 term.

$$x^2 - \frac{7}{2}x = \frac{9}{2} \qquad \text{Divide by 2.}$$

Now take half the coefficient of x and square it. Half of $-\frac{7}{2}$ is $-\frac{7}{4}$, and $(-\frac{7}{4})^2 = \frac{49}{16}$. Add $\frac{49}{16}$ to each side of the equation, and write the left side as a perfect square.

$$x^2 - \frac{7}{2}x + \frac{49}{16} = \frac{9}{2} + \frac{49}{16} \qquad \text{Add } \tfrac{49}{16}.$$

$$\left(x - \frac{7}{4}\right)^2 = \frac{121}{16} \qquad \text{Factor.}$$

Use the square root property.

$$x - \frac{7}{4} = \sqrt{\frac{121}{16}} \quad \text{or} \quad x - \frac{7}{4} = -\sqrt{\frac{121}{16}}$$

Because $\sqrt{\dfrac{121}{16}} = \dfrac{11}{4}$,

$$x - \frac{7}{4} = \frac{11}{4} \quad \text{or} \quad x - \frac{7}{4} = -\frac{11}{4}$$

$$x = \frac{18}{4} \quad \text{or} \quad x = -\frac{4}{4} \qquad \text{Add } \tfrac{7}{4}.$$

$$x = \frac{9}{2} \quad \text{or} \quad x = -1.$$

The solution set is $\{\frac{9}{2}, -1\}$.

◀◀ **WORK PROBLEM 4 AT THE SIDE.**

E X A M P L E   5   Solving a Quadratic Equation by Completing the Square

Solve $4p^2 + 8p + 5 = 0$.

$$p^2 + 2p + \frac{5}{4} = 0 \qquad \text{Divide each side by 4.}$$

$$p^2 + 2p = -\frac{5}{4} \qquad \text{Subtract } \tfrac{5}{4} \text{ from each side.}$$

The coefficient of p is 2. Take half of 2, square the result, and add this square to each side. The left-hand side can then be written as a perfect square.

$$p^2 + 2p + 1 = -\frac{5}{4} + 1 \qquad \text{Add 1 on each side.}$$

$$(p + 1)^2 = -\frac{1}{4} \qquad \text{Factor.}$$

$$p + 1 = \pm\sqrt{-\frac{1}{4}} \qquad \text{Square root property.}$$

$$p = -1 \pm \frac{1}{2}i \qquad \sqrt{-\tfrac{1}{4}} = \tfrac{1}{2}i$$

The solution set is $\{-1 + \frac{1}{2}i, -1 - \frac{1}{2}i\}$.

ANSWERS

4. (a) $\left\{-2, \dfrac{1}{3}\right\}$ **(b)** $\left\{\dfrac{2 + \sqrt{6}}{2}, \dfrac{2 - \sqrt{6}}{2}\right\}$

WORK PROBLEM 5 AT THE SIDE. ▶▶

5. Solve $9x^2 + 36x + 37 = 0$ by completing the square.

OBJECTIVE ▶ The next example shows how to simplify an equation before solving it.

E X A M P L E 6 Solving a Quadratic Equation by Completing the Square

Solve $(m - 2)(m + 1) = 5$.

Before we can use the method of completing the square, the equation must be in the form $ax^2 + bx + c = 0$. Start by multiplying on the left.

$$(m - 2)(m + 1) = 5$$
$$m^2 - m - 2 = 5 \qquad \text{Use FOIL.}$$
$$m^2 - m = 7 \qquad \text{Add 2.}$$

Now complete the square. Half of -1 is $-\frac{1}{2}$, and $(-\frac{1}{2})^2 = \frac{1}{4}$. Add $\frac{1}{4}$ to each side.

$$m^2 - m + \frac{1}{4} = 7 + \frac{1}{4} \qquad \text{Add } \tfrac{1}{4}.$$

$$\left(m - \frac{1}{2}\right)^2 = \frac{29}{4} \qquad \text{Factor; combine terms.}$$

$$m - \frac{1}{2} = \sqrt{\frac{29}{4}} \qquad \text{or} \quad m - \frac{1}{2} = -\sqrt{\frac{29}{4}} \qquad \text{Square root property}$$

$$m - \frac{1}{2} = \frac{\sqrt{29}}{2} \qquad \text{or} \quad m - \frac{1}{2} = -\frac{\sqrt{29}}{2}$$

$$m = \frac{1}{2} + \frac{\sqrt{29}}{2} \qquad \text{or} \qquad m = \frac{1}{2} - \frac{\sqrt{29}}{2} \qquad \text{Add } \tfrac{1}{2}.$$

$$m = \frac{1 + \sqrt{29}}{2} \qquad \text{or} \qquad m = \frac{1 - \sqrt{29}}{2}$$

We check each of these solutions by substituting it in the original equation. To do this for one of these values, let $m = \dfrac{1 + \sqrt{29}}{2}$.

Check:
$$(m - 2)(m + 1) = 5$$

$$\left(\frac{1 + \sqrt{29}}{2} - 2\right)\left(\frac{1 + \sqrt{29}}{2} + 1\right) = 5 \qquad ?$$

$$\left(\frac{1 + \sqrt{29}}{2} - \frac{4}{2}\right)\left(\frac{1 + \sqrt{29}}{2} + \frac{2}{2}\right) = 5 \qquad ?$$

$$\left(\frac{-3 + \sqrt{29}}{2}\right)\left(\frac{3 + \sqrt{29}}{2}\right) = 5 \qquad ?$$

$$\frac{(-3 + \sqrt{29})(3 + \sqrt{29})}{2 \cdot 2} = 5 \qquad ?$$

$$\frac{-9 - 3\sqrt{29} + 3\sqrt{29} + 29}{4} = 5 \qquad ?$$

$$\frac{20}{4} = 5 \qquad \text{True}$$

6. Solve each equation.

(a) $r^2 + 1 = 3r$

(b) $(x + 2)(x + 1) = 5$

◀◀ **WORK PROBLEM 6 AT THE SIDE.**

OBJECTIVE ▶   There are many practical applications of quadratic equations. The next example illustrates an application from physics.

E X A M P L E   7   Solving a Velocity Problem

If a ball is thrown into the air from ground level with an initial velocity of 64 feet per second, its height s (in feet) in t seconds is given by the function $s(t) = -16t^2 + 64t$. How long will it take the ball to reach a height of 48 feet?

Since s represents the height, let $s(t) = 48$ in the formula to get

$$48 = -16t^2 + 64t.$$

To solve this equation for the time, t, by completing the square, we should divide both sides by -16. Let us also reverse the sides of the equation.

$-3 = t^2 - 4t$	Divide by -16.
$t^2 - 4t = -3$	Reverse the sides.
$t^2 - 4t + 4 = -3 + 4$	Add $[(\frac{1}{2})4]^2 = 4$.
$(t - 2)^2 = 1$	Factor.
$t - 2 = 1$ or $t - 2 = -1$	Square root property
$t = 3$ or $t = 1$	Add 2.

You may wonder how we can get two correct answers for the time required for the ball to reach a height of 48 feet. The ball reaches that height twice, once on the way up and again on the way down. So it takes 1 second to reach 48 feet on the way up, and then after 3 seconds, the ball reaches 48 feet again on the way down.

7. Suppose a ball is propelled upward with an initial velocity of 128 feet per second. Its height in seconds at time t is given by $s(t) = -16t^2 + 128t$, where $s(t)$ is in feet. At what times will it be 48 feet above the ground? Give answers to the nearest tenth.

◀◀ **WORK PROBLEM 7 AT THE SIDE.**

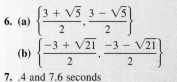

10.2 Exercises

1. What is the first step that you would perform in order to solve the equation $4x^2 + 8x = 3$ by completing the square?

2. Why is it not possible to solve the equation $2x^3 - x - 1 = 0$ by completing the square?

Solve each equation by completing the square. See Examples 1 and 2.

3. $x^2 - 4x = -3$

4. $y^2 - 2y = 8$

5. $x^2 + 2x - 5 = 0$

6. $r^2 + 4r + 1 = 0$

7. $z^2 + 6z + 9 = 0$

8. $k^2 - 8k + 16 = 0$

Find the number that should be added to each expression to make it a perfect square. See Example 2.

9. $y^2 + 14y$

10. $z^2 + 18z$

11. $k^2 - 5k$

12. $m^2 - 9m$

13. $r^2 + \frac{1}{2}r$

14. $s^2 - \frac{1}{3}s$

15. Which one of the following steps is an appropriate way to begin solving the quadratic equation

$$2x^2 - 4x = 9$$

by completing the square?
(a) Add 4 to both sides of the equation.
(b) Factor the left side as $2x(x - 2)$.
(c) Factor the left side as $x(2x - 4)$.
(d) Divide both sides by 2.

16. In an earlier chapter, we solved the quadratic equation

$$4p^2 - 26p + 40 = 0$$

by the factoring method. If we were to solve this by the method of completing the square, would we get the same solutions, $\frac{5}{2}$ and 4?

Solve each equation by completing the square. See Examples 3–6.

17. $2x^2 - 4x - 5 = 0$

18. $2x^2 - 6x - 3 = 0$

19. $4y^2 + 4y = 3$

20. $9x^2 + 3x = 2$

*21. $16x^2 + 32x + 17 = 0$

*22. $4t^2 + 16t + 17 = 0$

23. $3k^2 + 7k = 4$

24. $2k^2 + 5k = 1$

25. $(x + 3)(x - 1) = 5$

26. $(y - 8)(y + 2) = 24$

27. $-x^2 + 2x = -5$

28. $-r^2 + 3r = -2$

Solve each problem. See Example 7.

29. If an object is thrown upward from ground level with an initial velocity of 96 feet per second, its height, *s*, (in feet) in *t* seconds is given by the function $s(t) = -16t^2 + 96t$. In how many seconds will the object be at a height of 80 feet?

30. How much time will it take the object described in Exercise 29 to be at a height of 100 feet? Round your answers to the nearest tenth.

INTERPRETING TECHNOLOGY (EXERCISES 31–32)

A quadratic equation in standard form can be solved graphically by using a graphing calculator. The figures for Exercises 31 and 32 show two views of the graph of $y = x^2 + bx + c$. The displays at the bottom of the screens indicate the roots (or solutions) of the corresponding equation $x^2 + bx + c = 0$ (that is, the x-values where $y = 0$). What is the solution set of each equation? Verify your answers by solving each equation algebraically.

31. $x^2 - 5x - 6 = 0$

32. $x^2 + 6x + 5 = 0$

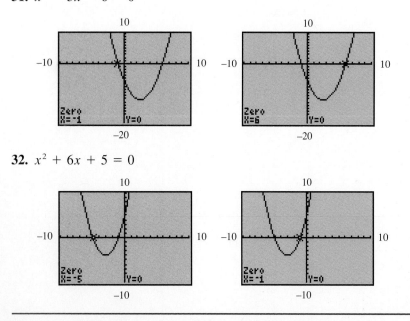

---

*Exercises marked with asterisks have imaginary number solutions.

10.3 Solving Quadratic Equations by the Quadratic Formula

The examples in the previous section showed that we can solve any quadratic equation by completing the square; however, completing the square is often tedious and time consuming. Later in this section, we will complete the square on the general quadratic equation to develop a formula that can be used to solve any quadratic equation. For now, we state the formula and show how it is used.

> **Quadratic Formula**
>
> The solutions of $ax^2 + bx + c = 0$ $(a \neq 0)$ are
>
> $$x = \frac{-b \pm \sqrt{b^2 - 4ac}}{2a}.$$

OBJECTIVE ▶ To use the quadratic formula, first write the given equation in standard form $ax^2 + bx + c = 0$; then identify the values of a, b, and c and substitute them into the quadratic formula, as shown in the next examples.

WORK PROBLEM I AT THE SIDE. ▶▶

EXAMPLE 1 Using the Quadratic Formula

Solve $6x^2 - 5x - 4 = 0$.

First, identify the letters a, b, and c of the general quadratic equation, $ax^2 + bx + c = 0$. Here a, the coefficient of the second-degree term, is 6, while b, the coefficient of the first-degree term, is -5, and the constant c is -4. Substitute these values into the quadratic formula.

$$x = \frac{-b \pm \sqrt{b^2 - 4ac}}{2a}$$

$$x = \frac{-(-5) \pm \sqrt{(-5)^2 - 4(6)(-4)}}{2(6)} \qquad a = 6, b = -5, c = -4$$

$$= \frac{5 \pm \sqrt{25 + 96}}{12}$$

$$= \frac{5 \pm \sqrt{121}}{12}$$

$$x = \frac{5 \pm 11}{12}$$

This last statement leads to two solutions, one from $+$ and one from $-$, giving

$$x = \frac{5 + 11}{12} = \frac{16}{12} = \frac{4}{3} \quad \text{or} \quad x = \frac{5 - 11}{12} = \frac{-6}{12} = -\frac{1}{2}.$$

Check these solutions by substituting each one in the original equation. The solution set is $\{-\frac{1}{2}, \frac{4}{3}\}$.

> **Caution**
> Notice in the quadratic formula that the square root is added to or subtracted from the value of $-b$ *before* dividing by $2a$.

1. Identify the letters a, b, and c. (*Hint:* If necessary, write the equation in standard form first.) *Do not solve.*

(a) $-3q^2 + 9q - 4 = 0$

(b) $3y^2 = 6y + 2$

ANSWERS

1. (a) -3; 9; -4 **(b)** 3; -6; -2

625

2. Solve $6x^2 + 4x - 1 = 0$ using the quadratic formula.

◄◄ **WORK PROBLEM 2 AT THE SIDE.**

EXAMPLE 2 Using the Quadratic Formula

Solve $4r^2 = 8r - 1$.

Rewrite the equation as $4r^2 - 8r + 1 = 0$, and identify $a = 4$, $b = -8$, and $c = 1$. Now use the quadratic formula.

$$r = \frac{-b \pm \sqrt{b^2 - 4ac}}{2a}$$

$$r = \frac{-(-8) \pm \sqrt{(-8)^2 - 4(4)(1)}}{2(4)} \qquad a = 4, b = -8, c = 1$$

$$= \frac{8 \pm \sqrt{64 - 16}}{8}$$

$$= \frac{8 \pm \sqrt{48}}{8} = \frac{8 \pm 4\sqrt{3}}{8}$$

$$r = \frac{4(2 \pm \sqrt{3})}{8} = \frac{2 \pm \sqrt{3}}{2}$$

The solution set for $4r^2 = 8r - 1$ is $\left\{ \dfrac{2 + \sqrt{3}}{2}, \dfrac{2 - \sqrt{3}}{2} \right\}$.

The solutions to the equation in the next example are imaginary numbers.

EXAMPLE 3 Using the Quadratic Formula

Solve $(9q + 3)(q - 1) = -8$.

Every quadratic equation must be in standard form before we begin to solve it, whether we are factoring or using the quadratic formula. To put this equation in standard form, we first multiply on the left, then collect all nonzero terms on the left.

$$(9q + 3)(q - 1) = -8$$
$$9q^2 - 6q - 3 = -8$$
$$9q^2 - 6q + 5 = 0$$

After writing the equation in the required form, we can identify $a = 9$, $b = -6$, and $c = 5$. Then we use the quadratic formula.

$$q = \frac{-(-6) \pm \sqrt{(-6)^2 - 4(9)(5)}}{2(9)}$$

$$= \frac{6 \pm \sqrt{-144}}{18} = \frac{6 \pm 12i}{18} \qquad i = \sqrt{-1}$$

$$= \frac{6(1 \pm 2i)}{18} = \frac{1 \pm 2i}{3}$$

The solution set is $\left\{ \dfrac{1 + 2i}{3}, \dfrac{1 - 2i}{3} \right\}$.

We could have written the solutions in Example 3 as $a + bi$, the standard form for complex numbers, as follows:

$$\frac{1 \pm 2i}{3} = \frac{1}{3} \pm \frac{2}{3}i.$$

WORK PROBLEM 3 AT THE SIDE. ▶▶

OBJECTIVE ▶ The next example shows how the quadratic formula is used to solve applied problems that cannot be solved by factoring.

E X A M P L E 4 **Solving an Applied Problem Using the Quadratic Formula**

If a rock is thrown upward from the top of a 144-foot building, its position (in feet above the ground) is given by $s(t) = -16t^2 + 112t + 144$, where t is time in seconds after it was dropped. When does it hit the ground?

When the rock hits the ground, its distance above the ground is zero. Find t when s is zero by solving the equation

$$0 = -16t^2 + 112t + 144. \quad \text{Let } s(t) = 0.$$
$$0 = t^2 - 7t - 9 \quad \text{Divide both sides by } -16.$$
$$t = \frac{7 \pm \sqrt{49 + 36}}{2} \quad \text{Quadratic formula}$$
$$t = \frac{7 \pm \sqrt{85}}{2}$$

In applied problems, we often prefer approximations to exact values. Using a calculator, we find $\sqrt{85} \approx 9.2$, so

$$t \approx \frac{7 \pm 9.2}{2},$$

giving the solutions $t \approx 8.1$ or $t \approx -1.1$. Discard the negative solution. The rock will hit the ground about 8.1 seconds after it is thrown.

WORK PROBLEM 4 AT THE SIDE. ▶▶

As mentioned earlier, the quadratic formula is developed by completing the square to solve the general quadratic equation $ax^2 + bx + c = 0$ ($a \neq 0$). We show the steps here for the case where $a > 0$. First divide both sides by a (using the fact that $a > 0$).

$$x^2 + \frac{b}{a}x + \frac{c}{a} = 0 \quad \text{Step 1}$$

Now subtract $\frac{c}{a}$ from each side.

$$x^2 + \frac{b}{a}x = -\frac{c}{a} \quad \text{Step 2}$$

Take half of $\frac{b}{a}$ and square it.

$$\frac{1}{2}\left(\frac{b}{a}\right) = \frac{b}{2a} \quad \text{Step 3}$$
$$\left(\frac{b}{2a}\right)^2 = \frac{b^2}{4a^2}$$

3. Solve each equation using the quadratic formula.

(a) $2k^2 + 19 = 14k$

(b) $z^2 = 4z - 5$

4. A ball is thrown vertically upward from the ground. Its distance in feet from the ground at t seconds is $s(t) = -16t^2 + 64t$. At what times will the ball be 32 feet from the ground? Use a calculator and round answers to the nearest tenth. (*Hint:* There are two answers.)

ANSWERS

3. (a) $\left\{\frac{7 + \sqrt{11}}{2}, \frac{7 - \sqrt{11}}{2}\right\}$
   (b) $\{2 + i, 2 - i\}$
4. at .6 second and at 3.4 seconds

Add the square to both sides.

$$x^2 + \frac{b}{a}x + \frac{b^2}{4a^2} = -\frac{c}{a} + \frac{b^2}{4a^2} \qquad \text{Step 4}$$

Write the left side as a perfect square and rearrange the right side.

$$\left(x + \frac{b}{2a}\right)^2 = \frac{b^2}{4a^2} + \frac{-c}{a} \qquad \text{Step 5}$$

$$\left(x + \frac{b}{2a}\right)^2 = \frac{b^2}{4a^2} + \frac{-4ac}{4a^2} \qquad \text{Write with a common denominator.}$$

$$\left(x + \frac{b}{2a}\right)^2 = \frac{b^2 - 4ac}{4a^2} \qquad \text{Add fractions.}$$

By the square root property,

$$x + \frac{b}{2a} = \sqrt{\frac{b^2 - 4ac}{4a^2}} \quad \text{or} \quad x + \frac{b}{2a} = -\sqrt{\frac{b^2 - 4ac}{4a^2}}. \qquad \text{Step 6}$$

Because

$$\sqrt{\frac{b^2 - 4ac}{4a^2}} = \frac{\sqrt{b^2 - 4ac}}{\sqrt{4a^2}} = \frac{\sqrt{b^2 - 4ac}}{2a},$$

the result above can be expressed as

$$x + \frac{b}{2a} = \frac{\sqrt{b^2 - 4ac}}{2a} \qquad \text{or} \quad x + \frac{b}{2a} = \frac{-\sqrt{b^2 - 4ac}}{2a}$$

$$x = -\frac{b}{2a} + \frac{\sqrt{b^2 - 4ac}}{2a} \qquad \text{or} \qquad x = -\frac{b}{2a} - \frac{\sqrt{b^2 - 4ac}}{2a}$$

$$x = \frac{-b + \sqrt{b^2 - 4ac}}{2a} \qquad \text{or} \qquad x = \frac{-b - \sqrt{b^2 - 4ac}}{2a}.$$

This result agrees with the solution in the quadratic formula, $x = \frac{-b \pm \sqrt{b^2 - 4ac}}{2a}$. If $a < 0$, going through the same steps gives the same two solutions.

OBJECTIVE ▶ The quadratic formula gives the solutions of the quadratic equation $ax^2 + bx + c = 0$ as

$$x = \frac{-b \pm \sqrt{b^2 - 4ac}}{2a}.$$

If a, b, and c are integers, the type of solutions of a quadratic equation (that is, rational, irrational, or imaginary) is determined by the quantity under the square root sign, $b^2 - 4ac$. Because it distinguishes among the three types of solutions, the quantity $b^2 - 4ac$ is called the **discriminant** (dis-KRIM-ih-nunt). By calculating the discriminant before solving a quadratic equation, we can predict whether the solutions will be rational numbers, irrational numbers, or imaginary numbers. This can be useful in an applied problem, for example, where imaginary number solutions are not acceptable. If the discriminant is a perfect square (including 0), we can solve the equation by factoring. Otherwise, the quadratic formula should be used.

Discriminant

The discriminant of $ax^2 + bx + c = 0$ is given by $b^2 - 4ac$. If a, b, and c are integers, then the type of solution is determined as follows.

Discriminant	Solutions
Positive, and the square of an integer	Two different rational solutions
Positive, but not the square of an integer	Two different irrational solutions
Zero	One rational solution
Negative	Two different imaginary solutions

E X A M P L E 5   Using the Discriminant

Given the equation $6x^2 - x - 15 = 0$, find the discriminant and determine whether the solutions of the equation will be rational, irrational, or imaginary.

We find the discriminant by evaluating $b^2 - 4ac$. In this example, $a = 6$, $b = -1$, $c = -15$, and the discriminant is

$$b^2 - 4ac = (-1)^2 - 4(6)(-15) = 1 + 360 = 361.$$

A calculator shows that $361 = 19^2$. Because the discriminant is a perfect square and a, b, and c are integers, the solutions to the given equation will be two different rational numbers, and the equation can be solved by factoring.

WORK PROBLEM 5 AT THE SIDE. ▶▶

E X A M P L E 6   Using the Discriminant

Predict the number and type of solutions for each of the following equations.

(a) $4y^2 + 9 = 12y$

Rewrite the equation as $4y^2 - 12y + 9 = 0$ to find $a = 4$, $b = -12$, and $c = 9$. The discriminant is

$$b^2 - 4ac = (-12)^2 - 4(4)(9) = 144 - 144 = 0.$$

Because the discriminant is 0, the quantity under the radical in the quadratic formula is 0, and there is only one rational solution. Again, we can solve the equation by factoring.

(b) $3m^2 - 4m = 5$

Rewrite the equation as $3m^2 - 4m - 5 = 0$. Then $a = 3$, $b = -4$, $c = -5$, and the discriminant is

$$b^2 - 4ac = (-4)^2 - 4(3)(-5) = 16 + 60 = 76.$$

Because 76 is not a perfect square, $\sqrt{76}$ is irrational, and since a, b, and c are integers, the given equation will have two different irrational solutions, one from using $\sqrt{76}$ and one from using $-\sqrt{76}$.

— **CONTINUED ON NEXT PAGE**

5. Find the discriminant and decide whether it is a perfect square.

(a) $6m^2 - 13m - 28 = 0$

(b) $4y^2 + 2y + 1 = 0$

(c) $15k^2 + 11k = 14$

6. Predict the number and type of solutions for each equation.

(a) $2x^2 + 3x = 4$

(b) $2x^2 + 3x + 4 = 0$

(c) $x^2 + 20x + 100 = 0$

(d) $3x^2 + 7x = 0$

7. Use the discriminant to decide whether each trinomial can be factored; then factor it, if possible.

(a) $2y^2 + 13y - 7$

(b) $6z^2 - 11z + 18$

(c) $4x^2 + x + 1 = 0$

Here $a = 4$, $b = 1$, $c = 1$, and the discriminant is

$$b^2 - 4ac = 1^2 - 4(4)(1) = 1 - 16 = -15.$$

Because the discriminant is negative, the equation $4x^2 + x + 1 = 0$ will have two imaginary number solutions.

◀◀ **WORK PROBLEM 6 AT THE SIDE.**

OBJECTIVE ▶ As mentioned earlier, a quadratic trinomial can be factored with rational coefficients only if the corresponding quadratic equation has rational solutions. Thus, we can use the discriminant to decide whether a given trinomial is factorable.

E X A M P L E 7 **Deciding Whether a Trinomial Is Factorable**

Decide whether the following trinomials can be factored.

(a) $24x^2 + 7x - 5$

To decide whether the solutions of $24x^2 + 7x - 5 = 0$ are rational numbers, evaluate the discriminant.

$$b^2 - 4ac = 7^2 - 4(24)(-5) = 49 + 480 = 529 = 23^2$$

Because 529 is a perfect square, the solutions are rational and the trinomial can be factored. It factors as

$$24x^2 + 7x - 5 = (3x - 1)(8x + 5).$$

(b) $11m^2 - 9m + 12$

The discriminant is $b^2 - 4ac = (-9)^2 - 4(11)(12) = -447$. This number is negative, so the corresponding quadratic equation has imaginary number solutions and therefore the trinomial cannot be factored.

◀◀ **WORK PROBLEM 7 AT THE SIDE.**

10.3 Exercises

Decide whether the statement is true or false based on the discussion in Sections 10.1–10.3.

1. The equation $(x + 4)^2 = -8$ has two imaginary solutions.

2. The discriminant of $x^2 + x + 1 = 0$ is -3, and so the equation has two imaginary solutions.

3. The equations $x^2 + 3x - 4 = 0$ and $-x^2 - 3x + 4 = 0$ have the same solution set.

4. An equation of the form $x^2 + kx = 0$ will have two real solutions, one of which is 0.

5. If p is a prime number, $x^2 = p$ will have two irrational solutions.

6. The equation $x^2 - 5 = 0$ cannot be solved using the quadratic formula, since there is no value for b.

Use the quadratic formula to solve each equation. (All solutions for these equations are real numbers.) See Examples 1 and 2.

7. $m^2 - 8m + 15 = 0$ **8.** $x^2 + 3x - 28 = 0$ **9.** $2k^2 + 4k + 1 = 0$

10. $2y^2 + 3y - 1 = 0$ **11.** $2x^2 - 2x = 1$ **12.** $9t^2 + 6t = 1$

13. $x^2 + 18 = 10x$ **14.** $x^2 - 4 = 2x$ **15.** $-2t(t + 2) = -3$

16. $-3x(x + 2) = -4$ **17.** $(r - 3)(r + 5) = 2$ **18.** $(k + 1)(k - 7) = 1$

Use the quadratic formula to solve each equation. (All solutions for these equations are imaginary numbers.) See Example 3.

19. $k^2 + 47 = 0$

20. $x^2 + 19 = 0$

21. $r^2 - 6r + 14 = 0$

22. $t^2 + 4t + 11 = 0$

23. $4x^2 - 4x = -7$

24. $9x^2 - 6x = -7$

25. $x(3x + 4) = -2$

26. $y(2y + 3) = -2$

27. $\dfrac{x + 5}{2x - 1} = \dfrac{x - 4}{x - 6}$

28. $\dfrac{3x - 4}{2x - 5} = \dfrac{x + 5}{x + 2}$

29. $\dfrac{1}{x^2} + 1 = -\dfrac{1}{x}$

30. $\dfrac{4}{r^2} + 3 = \dfrac{1}{r}$

Solve each problem. See Example 4.

31. A ball is thrown vertically upward from the ground. Its distance in feet from the ground in t seconds is $s(t) = -16t^2 + 128t$. At what times will the ball be 240 feet from the ground?

32. A toy rocket is launched from ground level. Its distance in feet from the ground in t seconds is $s(t) = -16t^2 + 208t$. At what times will the rocket be 640 feet from the ground?

A rock is thrown upward from ground level, and its distance from the ground in t seconds is $s(t) = -16t^2 + 160t$. Use algebra and a short explanation to answer Exercises 33 and 34.

33. After how many seconds does it reach a height of 400 feet? How would you describe in words its position at this height?

34. After how many seconds does it reach a height of 425 feet? How would you interpret the mathematical result here?

MATHEMATICAL CONNECTIONS (EXERCISES 35–40)

In earlier chapters we saw how linear functions can be used to model certain data. In some cases, quadratic functions provide better models, as the data may not be linear in nature. For example, during the years 1988–1990, the quadratic function $f(x) = -6.5x^2 + 132.5x + 2117$ modeled the number of prisoners in the United States under the sentence of death, according to statistics from the U.S. Justice Department. In this model, $x = 0$ corresponds to 1988, $x = 1$ to 1989, and $x = 2$ to 1990. The accompanying graph models these data in another way.

PRISONERS ON DEATH ROW

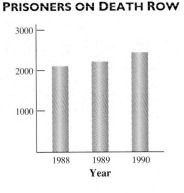

Year

Now work Exercises 35–40 in order, using the quadratic function f above.

35. How many such prisoners were there in 1988?

36. How many such prisoners were there in 1989?

37. How many such prisoners were there in 1990?

38. The equation given was based only on information for the three years named. If we wanted to make a prediction for the number of prisoners in 1991, how would we do it based on this quadratic function?

39. Use this function to predict the year in which the number of prisoners was 2543. (*Hint:* Replace $f(x)$ by 2543 and solve using the quadratic formula.)

40. Suppose that the number of people in a small town grew according to the quadratic function $f(x) = 5x^2 + 45x + 350$, where $x = 0$ corresponds to 1994, $x = 1$ to 1995, and so on. In what year would the population be 700?

Use the discriminant to determine whether the solutions to each of the following equations are
(**a**) *two distinct rational numbers,* (**b**) *exactly one rational number,*
(**c**) *two distinct irrational numbers,* (**d**) *two distinct imaginary numbers.*

Do not solve. See Examples 5 and 6.

41. $25x^2 + 70x + 49 = 0$

42. $4k^2 - 28k + 49 = 0$

43. $x^2 + 4x + 2 = 0$

44. $9x^2 - 12x - 1 = 0$

45. $3x^2 = 5x + 2$

46. $4x^2 = 4x + 3$

47. $3y^2 - 10y + 15 = 0$

48. $18x^2 + 60x + 82 = 0$

Use the discriminant to determine whether each polynomial is factorable. If it can be factored, do so. See Example 7.

49. $24x^2 - 34x - 45$

50. $36y^2 + 69y + 28$

51. $36x^2 + 21x - 24$

52. $18k^2 + 13k - 12$

53. $12x^2 - 83x - 7$

54. $16y^2 - 61y - 12$

55. Find all values of k such that $2x^2 + kx + 2 = 0$ has exactly one rational number solution.

56. Find all values of k such that $3x^2 - 2x + k = 0$ has two distinct imaginary number solutions.

SUMMARY EXERCISES ON QUADRATIC EQUATIONS

Four methods have now been introduced for solving quadratic equations written in the form $ax^2 + bx + c = 0$. The chart below shows some advantages and some disadvantages of each method.

Method	Advantages	Disadvantages
1. Factoring	Usually the fastest method	Not all equations can be solved by factoring. Some factorable polynomials are hard to factor.
2. Square root property	Simplest method for solving equations of the form $(ax + b)^2 =$ a number	Few equations are given in this form.
3. Completing the square	Can always be used (also, the procedure is useful in other areas of mathematics)	Requires more steps than other methods
4. Quadratic formula	Can always be used	More difficult than factoring because of the $\sqrt{b^2 - 4ac}$ expression

Solve each quadratic equation by the method of your choice.

1. $s^2 = 36$

2. $x^2 + 3x = -1$

3. $y^2 - \dfrac{100}{81} = 0$

4. $81t^2 = 49$

5. $z^2 - 4z + 3 = 0$

6. $w^2 + 3w + 2 = 0$

7. $z(z - 9) = -20$

8. $x^2 + 3x - 2 = 0$

9. $(3k - 2)^2 = 9$

10. $(2s - 1)^2 = 10$

11. $(x + 6)^2 = 121$

12. $(5k + 1)^2 = 36$

13. $(3r - 7)^2 = 24$

14. $(7p - 1)^2 = 32$

*15. $x^2 + x + 1 = 0$

_____

*Exercises identified with asterisks have imaginary number solutions.

*16. $2t^2 + 1 = t$

17. $-2x^2 = -3x - 2$

18. $-2x^2 + x = -1$

19. $8z^2 = 15 + 2z$

20. $3k^2 = 3 - 8k$

21. $0 = -x^2 + 2x + 1$

22. $3x^2 + 5x = -1$

23. $5y^2 - 22y = -8$

24. $y(y + 6) + 4 = 0$

25. $(x + 2)(x + 1) = 10$

26. $16x^2 + 40x + 25 = 0$

27. $4x^2 = -1 + 5x$

28. $2p^2 = 2p + 1$

29. $3m(3m + 4) = 7$

30. $5x - 1 + 4x^2 = 0$

31. $\dfrac{r^2}{2} + \dfrac{7r}{4} + \dfrac{11}{8} = 0$

32. $t(15t + 58) = -48$

33. $9k^2 = 16(3k + 4)$

34. $\dfrac{1}{5}x^2 + x + 1 = 0$

*35. $y^2 - y + 3 = 0$

*36. $4m^2 - 11m + 8 = -2$

37. $-3x^2 + 4x = -4$

38. $z^2 - \dfrac{5}{12}z = \dfrac{1}{6}$

39. $5k^2 + 19k = 2k + 12$

40. $\dfrac{1}{2}n^2 - n = \dfrac{15}{2}$

41. $k^2 - \dfrac{4}{15} = -\dfrac{4}{15}k$

42. If $D > 0$ and $\dfrac{5 + \sqrt{D}}{3}$ is a solution of $ax^2 + bx + c = 0$, what must be another solution of the equation?

10.4 Equations Quadratic in Form

OBJECTIVE ▶ A variety of nonquadratic equations can be written in the form of a quadratic equation and solved by using these methods. For example, some equations with fractions lead to quadratic equations. As you solve the equations in this section, try to decide which is the best method for each equation.

┌───

E X A M P L E   1   **Writing an Equation in Quadratic Form**

Solve $\dfrac{1}{x} + \dfrac{1}{x-1} = \dfrac{7}{12}$.

To clear the equation of fractions, multiply each term by the common denominator, $12x(x-1)$.

$$12x(x-1)\frac{1}{x} + 12x(x-1)\frac{1}{x-1} = 12x(x-1)\frac{7}{12}$$

$$12(x-1) + 12x = 7x(x-1)$$

$$12x - 12 + 12x = 7x^2 - 7x \qquad \text{Distributive property}$$

$$24x - 12 = 7x^2 - 7x \qquad \text{Combine terms.}$$

A quadratic equation must be in the form $ax^2 + bx + c = 0$ before we can solve it. Combine terms and arrange them so that one side of the equation is zero.

$$0 = 7x^2 - 31x + 12 \qquad \text{Subtract } 24x \text{ and add 12.}$$

$$0 = (7x - 3)(x - 4) \qquad \text{Factor.}$$

Setting each factor equal to 0 and solving the two linear equations gives the solutions $\frac{3}{7}$ and 4. Check by substituting these solutions in the original equation. The solution set is $\{\frac{3}{7}, 4\}$.

└───

> **WORK PROBLEM 1 AT THE SIDE.** ▶▶

OBJECTIVE ▶ Earlier we solved distance-rate-time (or motion) problems that led to linear equations or rational equations. Now we can extend that work to motion problems that lead to quadratic equations. We can write an equation to solve the problem just as we did earlier. Distance-rate-time applications often lead to equations with fractions, as in the next example.

1. Solve.

(a) $\dfrac{5}{m} + \dfrac{12}{m^2} = 2$

(b) $\dfrac{2}{x} + \dfrac{1}{x-2} = \dfrac{5}{3}$

(c) $\dfrac{4}{m-1} + 9 = -\dfrac{7}{m}$

ANSWERS

1. (a) $\left\{ -\dfrac{3}{2}, 4 \right\}$ (b) $\left\{ \dfrac{4}{5}, 3 \right\}$ (c) $\left\{ \dfrac{7}{9}, -1 \right\}$

E X A M P L E 2   Solving a Motion Problem

A riverboat for tourists averages 12 miles per hour in still water. It takes the boat 1 hour, 4 minutes to go 6 miles upstream and return. Find the speed of the current. See Figure 1.

FIGURE 1

For a problem about rate (or speed), we use the distance formula, $d = rt$.

Let $\quad\quad x =$ the speed of the current;

$\quad 12 - x =$ the rate upstream;

$\quad 12 + x =$ the rate downstream.

The rate upstream is the difference of the speed of the boat in still water and the speed of the current, or $12 - x$. The speed downstream is, in the same way, $12 + x$. To find the time, rewrite the formula $d = rt$ as

$$t = \frac{d}{r}.$$

This information was used to complete the following chart.

	d	r	t
Upstream	6	$12 - x$	$\dfrac{6}{12 - x}$
Downstream	6	$12 + x$	$\dfrac{6}{12 + x}$

Times in hours

The total time, 1 hour and 4 minutes, can be written as

$$1 + \frac{4}{60} = 1 + \frac{1}{15} = \frac{16}{15} \text{ hours.}$$

Because the time upstream plus the time downstream equals $\frac{16}{15}$ hours,

$$\frac{6}{12 - x} + \frac{6}{12 + x} = \frac{16}{15}.$$

CONTINUED ON NEXT PAGE

Now multiply both sides of the equation by the common denominator $15(12 - x)(12 + x)$ and solve the resulting quadratic equation.

$$15(12 + x)6 + 15(12 - x)6 = 16(12 - x)(12 + x)$$

$$90(12 + x) + 90(12 - x) = 16(144 - x^2)$$

$$1080 + 90x + 1080 - 90x = 2304 - 16x^2 \quad \text{Distributive property}$$

$$2160 = 2304 - 16x^2 \quad \text{Combine terms.}$$

$$16x^2 = 144$$

$$x^2 = 9$$

Solve $x^2 = 9$ by using the square root property to get the two solutions

$$x = 3 \quad \text{or} \quad x = -3.$$

The speed of the current cannot be -3, so the solution is $x = 3$ miles per hour.

Caution

As shown in Example 2, when a quadratic equation is used to solve an applied problem, sometimes only *one* answer satisfies the application. It is *always necessary* to check each answer in the words of the stated problem.

WORK PROBLEM 2 AT THE SIDE. ▶▶

In an earlier chapter we solved problems about work rates, using the formula $r = 1/t$ for the rate at which a person completes 1 job in t time units. Now we extend this idea to problems which produce quadratic equations. The method of solution is the same.

EXAMPLE 3   Solving a Work Problem

It takes two carpet layers 4 hours to carpet a room. If each worked alone, one of them could do the job in one hour less time than the other. How long would it take the slower one to complete the job alone?

Let x represent the number of hours for the slower carpet layer to complete the job alone. Then the faster carpet layer could do the entire job in $x - 1$ hours. Together, they do the job in 4 hours. The slower person's rate is $\frac{1}{x}$, and the faster person's rate is $\frac{1}{x-1}$. Fill in a chart as shown.

Worker	Rate	Time Working Together	Fractional Part of the Job Done
Slower	$\dfrac{1}{x}$	4	$\dfrac{1}{x}(4)$
Faster	$\dfrac{1}{x-1}$	4	$\dfrac{1}{x-1}(4)$

Sum is 1 whole job.

Together they complete one whole job, so the sum of the two fractional parts is 1.

$$\frac{1}{x}(4) + \frac{1}{x-1}(4) = 1$$

CONTINUED ON NEXT PAGE

---

2. (a) In 4 hours Kerrie can go 15 miles upriver and come back. The speed of the current is 5 miles per hour. Complete this chart.

	d	r	t
Up			
Down			

(b) Find the speed of the boat from part (a) in still water.

(c) In $1\frac{3}{4}$ hours Ken rows his boat 5 miles upriver and comes back. The speed of the current is 3 miles per hour. How fast does Ken row?

3. Carlos can complete a certain lab test in 2 hours less time than Jaime can. If they can finish the job together in 2 hours, how long would it take each of them working alone? Round answers to the nearest tenth.

Multiply both sides by the common denominator, $x(x - 1)$.

$$4(x - 1) + 4x = x(x - 1)$$
$$4x - 4 + 4x = x^2 - x \qquad \text{Distributive property}$$
$$0 = x^2 - 9x + 4 \qquad \text{Standard form}$$

Now use the quadratic formula.

$$x = \frac{9 \pm \sqrt{81 - 16}}{2} = \frac{9 \pm \sqrt{65}}{2} \qquad a = 1, b = -9, c = 4$$

From a calculator, $\sqrt{65} \approx 8.062$, so

$$x \approx \frac{9 \pm 8.062}{2}.$$

Using the $+$ sign gives $x \approx 8.5$, while the $-$ sign leads to $x \approx .5$. (Here we rounded to the nearest tenth.) Only the solution 8.5 makes sense in the original problem. (Why?) Thus, the slower carpet layer can do the job in about 8.5 hours and the faster in about $8.5 - 1 = 7.5$ hours.

◄◄ **WORK PROBLEM 3 AT THE SIDE.**

OBJECTIVE ▶ In the previous chapter we saw that some equations with radicals lead to quadratic equations.

EXAMPLE 4 Writing an Equation in Quadratic Form

Solve each equation.

(a) $k = \sqrt{6k - 8}$

This equation is not quadratic. However, squaring both sides of the equation gives $k^2 = 6k - 8$, which is a quadratic equation that we can solve by factoring.

$$k^2 = 6k - 8$$
$$k^2 - 6k + 8 = 0$$
$$(k - 4)(k - 2) = 0$$
$$k = 4 \quad \text{or} \quad k = 2 \qquad \text{Potential solutions}$$

Check both of these numbers in the original (and *not* the squared) equation to be sure they are solutions. (This check is an essential step.)

Check: If $k = 4$,   If $k = 2$,

$$4 = \sqrt{6(4) - 8} \quad ? \qquad\qquad 2 = \sqrt{6(2) - 8} \quad ?$$
$$4 = \sqrt{16} \quad ? \qquad\qquad\qquad 2 = \sqrt{4} \quad ?$$
$$4 = 4. \qquad \text{True} \qquad\qquad 2 = 2. \qquad \text{True}$$

Both numbers check, so the solution set is {2, 4}.

(b) $x + \sqrt{x} = 6$

$$\sqrt{x} = 6 - x \qquad \text{Isolate the radical on one side.}$$
$$x = 36 - 12x + x^2 \qquad \text{Square both sides.}$$
$$0 = x^2 - 13x + 36 \qquad \text{Write in standard form.}$$
$$0 = (x - 4)(x - 9) \qquad \text{Factor.}$$
$$x - 4 = 0 \quad \text{or} \quad x - 9 = 0 \qquad \text{Set each factor equal to 0.}$$
$$x = 4 \quad \text{or} \qquad x = 9$$

ANSWERS

3. Jaime: 5.2 hours; Carlos: 3.2 hours

CONTINUED ON NEXT PAGE ──

Check both potential solutions.

Check: If $x = 4$, If $x = 9$,

$4 + \sqrt{4} = 6.$ True $9 + \sqrt{9} = 6.$ False

The solution set is {4}.

WORK PROBLEM 4 AT THE SIDE.

OBJECTIVE ▶ An equation that can be written in the form $au^2 + bu + c = 0$, for $a \neq 0$ and an algebraic expression u, is called **quadratic in form.**

E X A M P L E   5 **Solving an Equation That Is Quadratic in Form**

Solve each of the following.

(a) $x^4 - 13x^2 + 36 = 0$

Because $x^4 = (x^2)^2$, we can write this equation as $(x^2)^2 - 13(x^2) + 36 = 0$, so it is in quadratic form with $u = x^2$ and can be solved by factoring. Since $u = x^2$, $u^2 = x^4$, and the equation becomes

$$u^2 - 13u + 36 = 0 \quad \text{Let } u = x^2.$$
$$(u - 4)(u - 9) = 0 \quad \text{Factor.}$$
$$u - 4 = 0 \quad \text{or} \quad u - 9 = 0 \quad \text{Set each factor equal to 0.}$$
$$u = 4 \qquad\qquad u = 9. \quad \text{Solve.}$$

To find x, substitute x^2 for u.

$$x^2 = 4 \quad \text{or} \quad x^2 = 9$$
$$x = \pm 2 \qquad x = \pm 3 \quad \text{Square root property}$$

The equation $x^4 - 13x^2 + 36 = 0$, a fourth-degree equation, has four solutions.* The solution set is $\{-3, -2, 2, 3\}$, as can be verified by substituting into the equation.

(b) $4x^4 + 1 = 5x^2$

Use the fact that $x^4 = (x^2)^2$ again, and let $u = x^2$ and $u^2 = x^4$.

$$4u^2 + 1 = 5u \quad \text{Let } u = x^2.$$
$$4u^2 - 5u + 1 = 0 \quad \text{Write in standard form.}$$
$$(4u - 1)(u - 1) = 0 \quad \text{Factor.}$$
$$4u - 1 = 0 \quad \text{or} \quad u - 1 = 0 \quad \text{Set each factor equal to 0.}$$
$$u = \frac{1}{4} \quad \text{or} \qquad u = 1 \quad \text{Solve.}$$
$$x^2 = \frac{1}{4} \quad \text{or} \qquad x^2 = 1 \quad u = x^2$$
$$x = \pm\frac{1}{2} \quad \text{or} \qquad x = \pm 1 \quad \text{Square root property}$$

The solution set is $\{-1, -\frac{1}{2}, \frac{1}{2}, 1\}$.

WORK PROBLEM 5 AT THE SIDE.

---

* In general, an equation in which an nth-degree polynomial equals 0 has n solutions, although some of them may be repeated, and thus are the same.

4. Solve. Check each answer.

(a) $x = \sqrt{7x - 10}$

(b) $2x = \sqrt{x} + 1$

5. Solve.

(a) $m^4 - 10m^2 + 9 = 0$

(b) $9k^4 - 37k^2 + 4 = 0$

ANSWERS

4. (a) {2, 5} **(b)** {1}

5. (a) $\{-3, -1, 1, 3\}$

(b) $\left\{-2, -\frac{1}{3}, \frac{1}{3}, 2\right\}$

6. Solve.

(a) $5(r + 3)^2 + 9(r + 3) = 2$

E X A M P L E 6 Solving an Equation That Is Quadratic in Form

Solve each equation.

(a) $2(4m - 3)^2 + 7(4m - 3) + 5 = 0$

Because of the repeated quantity $4m - 3$, this equation is quadratic in form with $u = 4m - 3$. (Any letter except m could be used instead of u.)

$$2(4m - 3)^2 + 7(4m - 3) + 5 = 0$$
$$2u^2 + 7u + 5 = 0 \qquad \text{Let } 4m - 3 = u.$$
$$(2u + 5)(u + 1) = 0 \qquad \text{Factor.}$$
$$u = -\frac{5}{2} \quad \text{or} \quad u = -1 \qquad \text{Zero-factor property}$$

To find m, substitute $4m - 3$ for u.

$$4m - 3 = -\frac{5}{2} \quad \text{or} \quad 4m - 3 = -1$$
$$4m = \frac{1}{2} \quad \text{or} \quad 4m = 2$$
$$m = \frac{1}{8} \quad \text{or} \quad m = \frac{1}{2}$$

The solution set of the original equation is $\{\frac{1}{8}, \frac{1}{2}\}$.

(b) $4m^{2/3} = 3m^{1/3} + 1$

(b) $2a^{2/3} - 11a^{1/3} + 12 = 0$

Let $a^{1/3} = u$; then $a^{2/3} = u^2$. Substitute into the given equation.

$$2u^2 - 11u + 12 = 0 \qquad \text{Let } a^{1/3} = u, a^{2/3} = u^2.$$
$$(2u - 3)(u - 4) = 0 \qquad \text{Factor.}$$
$$2u - 3 = 0 \quad \text{or} \quad u - 4 = 0$$
$$u = \frac{3}{2} \quad \text{or} \quad u = 4$$
$$a^{1/3} = \frac{3}{2} \quad \text{or} \quad a^{1/3} = 4 \qquad u = a^{1/3}$$
$$a = \left(\frac{3}{2}\right)^3 = \frac{27}{8} \quad \text{or} \quad a = 4^3 = 64 \qquad \text{Cube both sides.}$$

(Recall that in the previous chapter we solved equations with radicals by raising both sides to the same power.) Check that the solution set is $\{\frac{27}{8}, 64\}$.

◀◀ **WORK PROBLEM 6 AT THE SIDE.**

ANSWERS

6. (a) $\left\{-5, -\frac{14}{5}\right\}$ **(b)** $\left\{-\frac{1}{64}, 1\right\}$

10.4 Exercises

Based on the discussion and examples of this section, write a sentence describing the first step you would take in solving each of the following equations. Do not actually solve the equation.

1. $\dfrac{14}{x} = x - 5$

2. $\sqrt{1 + x} + x = 5$

3. $(r^2 + r)^2 - 8(r^2 + r) + 12 = 0$

4. $3t = \sqrt{16 - 10t}$

Solve by first clearing each equation of fractions. Check your answers. See Example 1.

5. $1 - \dfrac{3}{x} - \dfrac{28}{x^2} = 0$

6. $4 - \dfrac{7}{r} - \dfrac{2}{r^2} = 0$

7. $3 - \dfrac{1}{t} = \dfrac{2}{t^2}$

8. $1 + \dfrac{2}{k} = \dfrac{3}{k^2}$

9. $\dfrac{1}{x} + \dfrac{2}{x + 2} = \dfrac{17}{35}$

10. $\dfrac{2}{m} + \dfrac{3}{m + 9} = \dfrac{11}{4}$

11. $\dfrac{2}{x + 1} + \dfrac{3}{x + 2} = \dfrac{7}{2}$

12. $\dfrac{4}{3 - y} + \dfrac{2}{5 - y} = \dfrac{26}{15}$

13. $\dfrac{3}{2x} - \dfrac{1}{2(x + 2)} = 1$

14. $\dfrac{4}{3x} - \dfrac{1}{2(x + 1)} = 1$

15. If it takes m hours to grade a set of papers, what is the grader's rate (in job per hour)?

16. A boat goes 20 miles per hour in still water, and the rate of the current is t miles per hour.
   (a) What is the rate of the boat when it travels upstream?
   (b) What is the rate of the boat when it travels downstream?

Solve each problem by writing an equation with fractions and solving it. See Examples 2 and 3.

17. On a windy day Yoshiaki found that he could go 16 miles downstream and then 4 miles back upstream at top speed in a total of 48 minutes. What was the top speed of Yoshiaki's boat if the current was 15 miles per hour?

18. Lekesha flew her plane for 6 hours at a constant speed. She traveled 810 miles with the wind, then turned around and traveled 720 miles against the wind. The wind speed was a constant 15 miles per hour. Find the speed of the plane.

19. Albuquerque and Amarillo are 300 miles apart. Steve rides his Honda 20 miles per hour faster than Paula rides her Yamaha. Find Steve's average speed if he travels from Albuquerque to Amarillo in $1\frac{1}{4}$ hours less time than Paula.

20. The distance from Jackson to Lodi is about 40 miles, as is the distance from Lodi to Manteca. Rico drove from Jackson to Lodi during the rush hour, stopped in Lodi for a root beer, and then drove on to Manteca at 10 miles per hour faster. Driving time for the entire trip was 88 minutes. Find his speed from Jackson to Lodi.

21. A washing machine can be filled in 6 minutes if both the hot and cold water taps are fully opened. Filling the washer with hot water alone takes 9 minutes longer than filling it with cold water alone. How long does it take to fill the washer with cold water?

22. Two pipes together can fill a large tank in 2 hours. One of the pipes, used alone, takes 3 hours longer than the other to fill the tank. How long would each pipe take to fill the tank alone?

Find all solutions by first squaring. Check your answers. See Example 4.

23. $2x = \sqrt{11x + 3}$

24. $4x = \sqrt{6x + 1}$

25. $3y = \sqrt{16 - 10y}$

26. $4t = \sqrt{8t + 3}$

27. $p - 2\sqrt{p} = 8$

28. $k + \sqrt{k} = 12$

29. $m = \sqrt{\dfrac{6 - 13m}{5}}$

30. $r = \sqrt{\dfrac{20 - 19r}{6}}$

Find all solutions to the following equations. Check your answers. See Examples 5 and 6.

31. $t^4 - 18t^2 + 81 = 0$

32. $y^4 - 8y^2 + 16 = 0$

33. $4k^4 - 13k^2 + 9 = 0$

34. $9x^4 - 25x^2 + 16 = 0$

35. $(x + 3)^2 + 5(x + 3) + 6 = 0$

36. $(k - 4)^2 + (k - 4) - 20 = 0$

37. $(t + 5)^2 + 6 = 7(t + 5)$

38. $3(m + 4)^2 - 8 = 2(m + 4)$

39. $2 + \dfrac{5}{3k - 1} = \dfrac{-2}{(3k - 1)^2}$

40. $3 - \dfrac{7}{2p + 2} = \dfrac{6}{(2p + 2)^2}$

41. $2 - 6(m - 1)^{-2} = (m - 1)^{-1}$

42. $3 - 2(x - 1)^{-1} = (x - 1)^{-2}$

Use substitution to solve the following equations. Check your answers. See Example 6.

43. $x^{2/3} + x^{1/3} - 2 = 0$

44. $3x^{2/3} - x^{1/3} - 24 = 0$

45. $2(1 + \sqrt{y})^2 = 13(1 + \sqrt{y}) - 6$

46. $(k^2 + k)^2 + 12 = 8(k^2 + k)$

MATHEMATICAL CONNECTIONS (EXERCISES 47–52)

Consider the following equation, which contains variable expressions in the denominators.

$$\frac{x^2}{(x-3)^2} + \frac{3x}{x-3} - 4 = 0$$

Work Exercises 47–52 in order. They all pertain to this equation.

47. Why can 3 not possibly be a solution for this equation?

48. Multiply both sides of the equation by the LCD, $(x-3)^2$, and solve. There is only one solution—what is it?

49. Write the equation in a different manner so that it is quadratic in form using the expression $\dfrac{x}{x-3}$.

50. In your own words, explain why the expression $\dfrac{x}{x-3}$ cannot equal 1.

51. Solve the equation from Exercise 49 by making the substitution $t = \dfrac{x}{x-3}$. You should get two values for t. Why is one of them impossible for this equation?

52. Solve the equation $x^2(x-3)^{-2} + 3x(x-3)^{-1} - 4 = 0$ by letting $s = (x-3)^{-1}$. You should get two values for s. Why is this impossible for this equation?

10.5 *Formulas and Applications Involving Quadratic Equations*

OBJECTIVE Many useful formulas have a second-degree term. We can use the methods presented earlier in this chapter to solve a formula for a variable that is squared.

The formula in Example 1 is the Pythagorean theorem of geometry.

Pythagorean Theorem

If c is the length of the longest side of a right triangle and a and b are the lengths of the shorter sides, then

$$c^2 = a^2 + b^2.$$

See the figure.

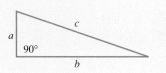

The longest side is the **hypotenuse** and the two shorter sides are the **legs** of the triangle.

EXAMPLE 1 Solving for a Squared Variable

Solve the Pythagorean theorem $c^2 = a^2 + b^2$ for b.

Think of c^2 and a^2 as constants. We solve for b by first getting b^2 alone on one side of the equation. Begin by subtracting a^2 from both sides.

$$c^2 = a^2 + b^2$$
$$c^2 - a^2 = b^2$$

Now use the square root property.

$$b = \sqrt{c^2 - a^2} \quad \text{or} \quad b = -\sqrt{c^2 - a^2}$$

Because b represents the side of a triangle, b must be positive. Because of this,

$$b = \sqrt{c^2 - a^2}$$

is the only solution. The solution cannot be simplified further.

WORK PROBLEM 1 AT THE SIDE. ▶▶

EXAMPLE 2 Solving for a Squared Variable

Solve $s = 2t^2 + kt$ for t.

Because the equation has terms with t^2 and t, first put it in the quadratic form $ax^2 + bx + c = 0$ with t as the variable instead of x.

$$s = 2t^2 + kt$$
$$0 = 2t^2 + kt - s$$

Now use the quadratic formula with $a = 2$, $b = k$, and $c = -s$.

$$t = \frac{-k \pm \sqrt{k^2 - 4(2)(-s)}}{2(2)}$$
$$= \frac{-k \pm \sqrt{k^2 + 8s}}{4}$$

The solutions are $t = \dfrac{-k + \sqrt{k^2 + 8s}}{4}$ and $t = \dfrac{-k - \sqrt{k^2 + 8s}}{4}$.

OBJECTIVES

1. ▶ Solve second-degree formulas for a specified variable.

2. ▶ Solve applied problems using the Pythagorean theorem.

3. ▶ Solve applied problems using formulas for area.

4. ▶ Solve applied problems about work.

5. ▶ Solve applied problems using quadratic functions as models.

FOR EXTRA HELP

Tutorial   Tape 22   SSM, Sec. 10.5

1. Solve $5y^2 = z^2 + 9x^2$ for x. Assume the variables represent the lengths of the sides of a right triangle.

ANSWERS

1. $x = \dfrac{\sqrt{5y^2 - z^2}}{3}$

647

2. Solve $2t^2 - 5t + k = 0$ for t.

◀◀ WORK PROBLEM 2 AT THE SIDE.

Caution

The following examples show that it is important to check all proposed solutions of applied problems against the information given in the original problem. Numbers that are valid solutions of the equation may not satisfy the physical conditions of the problem.

OBJECTIVE ▶ The Pythagorean theorem is used again in the solution of the next example.

EXAMPLE 3 Using the Pythagorean Theorem

Two cars left an intersection at the same time, one heading due north, the other due west. Some time later, they were exactly 100 miles apart. The car headed north had gone 20 miles farther than the car headed west. How far had each car traveled?

Let x be the distance traveled by the car headed west. Then $x + 20$ is the distance traveled by the car headed north. These distances are shown in Figure 2. The cars are 100 miles apart, so the hypotenuse of the right triangle equals 100 and the two legs are equal to x and $x + 20$.

3. A 13-foot ladder is leaning against a house. The distance from the bottom of the ladder to the house is 7 feet less than the distance from the top of the ladder to the ground. How far is the bottom of the ladder from the house?

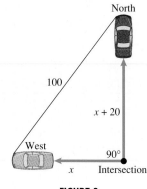

FIGURE 2

By the Pythagorean theorem,
$$c^2 = a^2 + b^2$$
$$100^2 = x^2 + (x + 20)^2$$
$$10{,}000 = x^2 + x^2 + 40x + 400$$
$$0 = 2x^2 + 40x - 9600$$
$$0 = 2(x^2 + 20x - 4800). \qquad \text{Factor out the common factor.}$$

Divide both sides by 2 to get $x^2 + 20x - 4800 = 0$. Although this equation can be solved by factoring, we use the quadratic formula to find x. Here, $a = 1$, $b = 20$, and $c = -4800$.

$$x = \frac{-20 \pm \sqrt{400 - 4(-4800)}}{2} = \frac{-20 \pm \sqrt{19{,}600}}{2}$$

From a calculator, $\sqrt{19{,}600} = 140$, so $x = \dfrac{-20 \pm 140}{2}$.

The solutions are $x = 60$ or $x = -80$. Since distance cannot be negative, we reject -80. Thus 60 and $60 + 20 = 80$ are the required distances in miles.

◀◀ WORK PROBLEM 3 AT THE SIDE.

OBJECTIVE ▶ Formulas for area may also result in quadratic equations, as the next example shows.

EXAMPLE 4   Solving an Area Problem

A rectangular reflecting pool in a park is 20 feet wide and 30 feet long. The park gardener wants to plant a strip of grass of uniform width around the edge of the pool. She has enough seed to cover 336 square feet. How wide will the strip be?

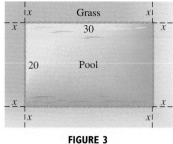

FIGURE 3

The pool is shown in Figure 3. If x represents the unknown width of the grass strip, the width of the large rectangle is given by $20 + 2x$ (the width of the pool plus two grass strips), and the length is given by $30 + 2x$. The area of the large rectangle is given by the product of its length and width, $(20 + 2x)(30 + 2x)$. The area of the pool is $20 \cdot 30 = 600$ square feet. The area of the large rectangle, minus the area of the pool, should equal the area of the grass strip. Since the area of the grass strip is to be 336 square feet, the equation is

$$\underset{\substack{\uparrow \\ \text{Area of rectangle}}}{(20 + 2x)(30 + 2x)} - \underset{\substack{\uparrow \\ \text{Area of pool}}}{600} = \underset{\substack{\uparrow \\ \text{Area of strip}}}{336.}$$

$$600 + 100x + 4x^2 - 600 = 336 \qquad \text{Multiply.}$$

$$4x^2 + 100x - 336 = 0 \qquad \text{Collect terms.}$$

$$x^2 + 25x - 84 = 0 \qquad \text{Divide by 4.}$$

$$(x + 28)(x - 3) = 0 \qquad \text{Factor.}$$

$$x = -28 \quad \text{or} \quad x = 3$$

The width of a grass strip cannot be -28 feet, so 3 feet is the desired width of the strip.

WORK PROBLEM 4 AT THE SIDE. ▶▶

OBJECTIVE ▶   As shown earlier, applied problems about work may result in quadratic equations.

EXAMPLE 5   Solving a Work Problem

A janitorial service provides two people to clean an office building. Working together, the two can clean the building in 5 hours. One person is new to the job and would take 2 hours longer than the other person to clean the building working alone. How long would it take the experienced worker to clean the building working alone?

Let $x =$ time in hours for the experienced worker to clean the building;

$x + 2 =$ time in hours for the new worker to clean the building;

$5 =$ time in hours for the two workers to clean the building together.

Then $\dfrac{1}{x} =$ the rate for the experienced worker;

$\dfrac{1}{x + 2} =$ the rate for the new worker.

CONTINUED ON NEXT PAGE

---

4. Suppose the pool in Example 4 is 20 feet by 40 feet and there is enough seed to cover 700 square feet. How wide should the grass strip be?

5. If the new worker in Example 5 takes 1 hour longer than the experienced worker does to clean the building, and together they complete the job in 5 hours, how long would the experienced worker require to complete the job working alone? Round the answer to the nearest tenth of an hour.

The parts of the job done by each are $\frac{1}{x}(5)$ and $\frac{1}{x+2}(5)$, so

$$\frac{1}{x}(5) + \frac{1}{x+2}(5) = 1.$$

$$5(x+2) + 5x = x(x+2) \qquad \text{Multiply by the common denominator.}$$

$$5x + 10 + 5x = x^2 + 2x \qquad \text{Distributive property}$$

$$0 = x^2 - 8x - 10 \qquad \text{Collect terms.}$$

$$x = \frac{8 \pm \sqrt{64 + 40}}{2} \qquad \text{Quadratic formula}$$

$$= \frac{8 \pm \sqrt{104}}{2}$$

$$\approx \frac{8 \pm 10.20}{2} \qquad \text{From a calculator}$$

$$x = 9.1 \qquad \text{Discard negative solution.}$$

To the nearest tenth, the experienced worker requires 9.1 hours to clean the building working alone.

◀◀ **WORK PROBLEM 5 AT THE SIDE.**

OBJECTIVE 5 Quadratic functions can be used to model data, as shown in the following example.

6. Evaluate

$$\frac{1.22 \pm \sqrt{(-1.22)^2 - 4(.011)(23.625)}}{2(.011)}$$

for both solutions. Which one is the smaller positive solution?

E X A M P L E   6   Using a Quadratic Function as a Model

Union membership in the United States rose from 1930 to 1960 and then declined. The quadratic function

$$f(x) = -.011x^2 + 1.22x - 8.5$$

approximates the number of union members, in millions, in the years 1950 through 1990, where x is the number of years since 1930 (when unions were in their formative stages). (*Source:* U.S. Union Membership, 1930–1990 Table. Bureau of Labor Statistics, U.S. Department of Labor)

(a) Approximate the union membership for the year 1950.

The year 1950 is 20 years from 1930, so we let $x = 20$ and find $f(20)$:

$$f(20) = -.011(20)^2 + 1.22(20) - 8.5 = 11.5 \text{ (million)}.$$

(b) In what year did union membership reach 15.125 million?

Here we must find the value of x that makes $f(x) = 15.125$.

$$f(x) = -.011x^2 + 1.22x - 8.5$$

$$15.125 = -.011x^2 + 1.22x - 8.5 \qquad \text{Let } f(x) = 15.125.$$

$$.011x^2 - 1.22x + 23.625 = 0 \qquad \text{Standard form}$$

Now use $a = .011$, $b = -1.22$, and $c = 23.625$ in the quadratic formula to find the smaller positive solution.

◀◀ **WORK PROBLEM 6 AT THE SIDE.**

The smaller positive solution is $x = 25$, so the year is 1930 + 25 = 1955. (We reject the larger positive solution since it leads to a later year than 1990.)

10.5 Exercises

In Exercises 1 and 2, solve for m in terms of the other variables (m > 0).

1.

2.

Solve each equation for the indicated variable. (While in practice we would often reject a negative value due to the physical nature of the quantity represented by the variable, leave ± in your answers here.) See Examples 1 and 2.

3. $d = kt^2$ for t

4. $s = kwd^2$ for d

5. $I = \dfrac{ks}{d^2}$ for d

6. $R = \dfrac{k}{d^2}$ for d

7. $F = \dfrac{kA}{v^2}$ for v

8. $L = \dfrac{kd^4}{h^2}$ for h

9. $V = \dfrac{1}{3}\pi r^2 h$ for r

10. $V = \pi(r^2 + R^2)h$ for r

11. $At^2 + Bt = -C$ for t

12. $S = 2\pi rh + \pi r^2$ for r

13. $D = \sqrt{kh}$ for h

14. $F = \dfrac{k}{\sqrt{d}}$ for d

15. $p = \sqrt{\dfrac{k\ell}{g}}$ for ℓ

16. $p = \sqrt{\dfrac{k\ell}{g}}$ for g

17. In the formula of Exercise 15, if g is a positive number, explain why k and ℓ cannot have different signs if the equation is to have a real value for p.

18. In Example 2 of this section, suppose that k and s are both positive numbers. Which one of the two solutions given is positive? Which one is negative?

Solve the following problems. See Example 3.

19. Two ships leave port at the same time, one heading due south and the other heading due east. Several hours later, they are 170 miles apart. If the ship traveling south traveled 70 miles farther than the other, how many miles did they each travel?

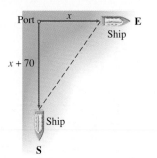

20. Rita is flying a kite that is 30 feet farther above her hand than its horizontal distance from her. The string from her hand to the kite is 150 feet long. How high is the kite?

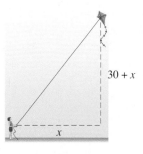

21. The longer leg of a right triangle is 1 meter longer than the shorter leg, while the hypotenuse is 8 meters longer than the longer leg. Find the lengths of the three sides.

22. The hypotenuse of a right triangle is 1 meter longer than twice the length of the shorter leg, and the longer leg is 9 meters shorter than 3 times the length of the shorter leg. Find the lengths of the three sides of the triangle.

Solve the following problems. See Example 4.

23. A square has an area of 256 square centimeters. If the same amount is removed from one dimension and added to the other, the resulting rectangle has an area 16 square centimeters less. Find the dimensions of the rectangle.

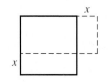

24. A rectangular piece of sheet metal is 2 inches longer than it is wide. A square piece 3 inches on a side is cut from each corner. The sides are then turned up to form an uncovered box of volume 765 cubic inches. Find the dimensions of the original piece of metal.

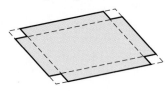

25. A couple wants to buy a rug for a room that is 20 feet long and 15 feet wide. They want to leave an even strip of flooring uncovered around the edges of the room. How wide a strip will they have if they buy a rug with an area of 234 square feet?

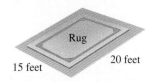

26. A club swimming pool is 30 feet wide and 40 feet long. The club members want an exposed aggregate border in a strip of uniform width around the pool. They have enough material for 296 square feet. How wide can the strip be?

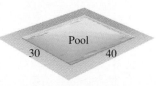

Use the quadratic formula to solve the following problems. Round answers to the nearest tenth. See Example 5.

27. Carmen and Paul can clean the house together in 2 hours. Working alone, it takes Carmen $\frac{1}{2}$ hour longer than it takes Paul to do the job. How long would it take Carmen alone?

28. Mashari and Jamal are distributing brochures for a fund-raising campaign. Together they can complete the job in 3 hours. If Mashari could do the job alone in one hour more than Jamal, how long would it take Jamal working alone?

29. Two pipes can fill a tank in 4 hours when used together. Alone, one can fill the tank in .5 hour more than the other. How long will it take each pipe to fill the tank alone?

30. Charlie Dawkins can process a stack of invoices 1 hour faster than Arnold Parker can. Working together, they take 1.5 hours. How long would it take each person working alone?

For Exercises 31–40, refer to Example 6.

The adjusted poverty threshold for a single person from the year 1984 to the year 1990 is approximated by the quadratic model $f(x) = 18.7x^2 + 105.3x + 4814.1$, where $x = 0$ corresponds to 1984, and $f(x)$ is in dollars. (Source: Congressional Budget Office.) Use this model to answer the questions in Exercises 31–34.

POVERTY THRESHOLD FOR A SINGLE PERSON

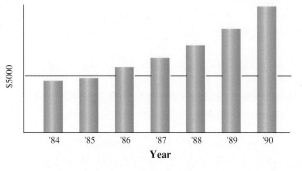

31. What was the threshold in 1984?

32. What was the threshold in 1986?

33. In what year during this period was the threshold $f(x)$ approximately $5300?

34. In what year during this period was the threshold $f(x)$ approximately $6119?

35. The function $D(t) = 13t^2 - 100t$ gives the distance a car going approximately 68 miles per hour will skid in t seconds. Find the time it would take for the car to skid 180 feet.

36. The function in Exercise 35 becomes $D(t) = 13t^2 - 73t$ for a car going 50 miles per hour. Find the time for this car to skid 218 feet.

37. An object is projected directly upward from the ground. After t seconds its distance above the ground is given by the function $f(t) = 144t - 16t^2$ feet. After how many seconds will it be 128 feet above the ground?

38. Refer to Exercise 37. When does the object strike the ground?

39. The formula $A = P(1 + r)^2$ gives the amount A in dollars that P dollars will grow to in 2 years at interest rate r (where r is given as a decimal), using compound interest. What interest rate will cause \$2000 to grow to \$2142.25 in 2 years?

40. If a square piece of cardboard has 3-inch squares cut from its corners and then has the flaps folded up to form an open-top box, the volume of the box is given by the function $V(x) = 3(x - 6)^2$, where x is the length of each side of the original piece of cardboard in inches. What original length would yield a box with a volume of 432 cubic inches?

William Froude was a 19th century naval architect who used the expression

$$\frac{v^2}{g\ell}$$

in shipbuilding. This expression, known as the Froude number, was also used by R. McNeill Alexander in his research on dinosaurs. (See "How Dinosaurs Ran" in Scientific American, *April, 1991, pp. 130–136.) For each of the following, ℓ is given, as well as the value of the Froude number. Find the value of v (in meters per second). It is known that $g = 9.8$ meters per second squared.*

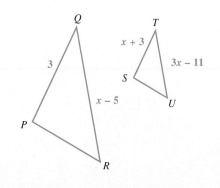

41. rhinoceros: $\ell = 1.2$; Froude number $= 2.57$

42. triceratops: $\ell = 2.8$; Froude number $= .16$

The corresponding sides of similar triangles are proportional. Use this fact to find the lengths of the indicated sides of the pair of similar triangles. Check all possible solutions in both triangles. Sides of a triangle cannot be negative.

43. side AC

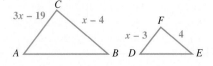

44. side RQ

10.6 Nonlinear and Fractional Inequalities

OBJECTIVES

1. ▶ Recognize a quadratic inequality.
2. ▶ Solve quadratic inequalities.
3. ▶ Solve polynomial inequalities of degree 3 or more.
4. ▶ Solve fractional inequalities.
5. ▶ Solve fractional inequalities using an alternative method.
6. ▶ Understand special cases of quadratic inequalities.

FOR EXTRA HELP

Tutorial   Tape 22   SSM, Sec. 10.6

OBJECTIVE ▶ We have discussed methods of solving linear inequalities (earlier) and methods of solving quadratic equations (in this chapter). Now this work can be extended to include solving *quadratic inequalities*.

> **Quadratic Inequality**
>
> A **quadratic inequality** can be written in the form
>
> $$ax^2 + bx + c < 0 \quad \text{or} \quad ax^2 + bx + c > 0$$
>
> where a, b, and c are real numbers, with $a \neq 0$.

As before, $<$ and $>$ can be replaced with $\leq$ and $\geq$ as necessary.

OBJECTIVE ▶ A method for solving quadratic inequalities is shown in the next example.

EXAMPLE 1   Solving a Quadratic Inequality

Solve $x^2 - x - 12 > 0$.

First solve the quadratic equation $x^2 - x - 12 = 0$ by factoring.

$$(x - 4)(x + 3) = 0$$
$$x - 4 = 0 \quad \text{or} \quad x + 3 = 0$$
$$x = 4 \quad \text{or} \quad x = -3$$

The numbers 4 and -3 divide the number line into the three regions shown in Figure 4. (Be careful to put the smaller number on the left.)

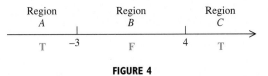

FIGURE 4

The numbers 4 and -3 are the only numbers that make the expression $x^2 - x - 12$ equal to zero. All other numbers make the expression either positive or negative. The sign of the expression can change from positive to negative or from negative to positive only at a number that makes it zero. Therefore, if one number in a region satisfies the inequality, then all the numbers in that region will satisfy the inequality. Choose any number from Region *A* in Figure 4 (any number less than -3). Substitute this number for x in the inequality $x^2 - x - 12 > 0$. If the result is *true,* then all numbers in Region *A* satisfy the original inequality.

Let us choose -5 from Region *A*. Substitute -5 into $x^2 - x - 12 > 0$, getting

$$(-5)^2 - (-5) - 12 > 0 \quad ?$$
$$25 + 5 - 12 > 0 \quad ?$$
$$18 > 0. \quad \text{True}$$

Because -5 from Region *A* satisfies the inequality, all numbers from Region *A* are solutions.

Try 0 from Region *B*. If $x = 0$, then

$$0^2 - 0 - 12 > 0 \quad ?$$
$$-12 > 0. \quad \text{False}$$

The numbers in Region *B* are *not* solutions.

CONTINUED ON NEXT PAGE

1. Does the number 5 from Region C satisfy $x^2 - x - 12 > 0$?

◀◀ **WORK PROBLEM 1 AT THE SIDE.**

In Problem 1 at the side, the number 5 satisfies the inequality, so the numbers in Region C are also solutions to the inequality.

Based on these results (shown by the colored letters in Figure 4), the solution set includes the numbers in Regions A and C, as shown on the graph in Figure 5. The solution set is written in interval notation as

$$(-\infty, -3) \cup (4, \infty).$$

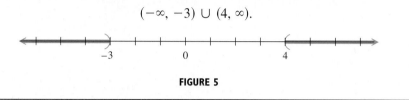

FIGURE 5

2. Solve. Graph each solution.

(a) $x^2 + x - 6 > 0$

In summary, a quadratic inequality is solved by following these steps.

Solving a Quadratic Inequality

Step 1 Replace the inequality symbol with = and solve the equation.

Step 2 Place the numbers found in Step 1 on a number line. These numbers divide the number line into regions.

Step 3 Substitute a number from each region into the inequality to determine the intervals that make the inequality true. If one number in a region satisfies the inequality, then all the numbers in that region will satisfy the inequality. The numbers in those intervals that make the inequality true are in the solution set.

Step 4 The numbers found in Step 1 are included in the solution set if the symbol is ≤ or ≥; they are not included if it is < or >.

(b) $3m^2 - 13m - 10 \leq 0$

◀◀ **WORK PROBLEM 2 AT THE SIDE.**

OBJECTIVE ▶ Higher-degree polynomial inequalities that are factorable can be solved in the same way as quadratic inequalities.

E X A M P L E   2   Solving a Third-Degree Polynomial Inequality

Solve $(x - 1)(x + 2)(x - 4) \leq 0$.

This is a *cubic* (third-degree) inequality rather than a quadratic inequality, but it can be solved by the method shown above and by extending the zero-factor property to more than two factors. Begin by setting the factored polynomial *equal* to 0 and solving the equation.

$$(x - 1)(x + 2)(x - 4) = 0$$

$x - 1 = 0$ or $x + 2 = 0$ or $x - 4 = 0$

$x = 1$ or $x = -2$ or $x = 4$

Locate the numbers -2, 1, and 4 on a number line as in Figure 6 to determine the regions A, B, C, and D. The numbers -2, 1, and 4 are in the solution set because of the "or equal to" part of the inequality symbol.

ANSWERS

1. yes

2. (a) $(-\infty, -3) \cup (2, \infty)$

(b) $\left[-\dfrac{2}{3}, 5\right]$

CONTINUED ON NEXT PAGE ─

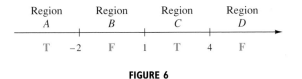

Region A | Region B | Region C | Region D

T   -2   F   1   T   4   F

FIGURE 6

Substitute a number from each region into the original inequality to determine which regions satisfy the inequality. These results are shown below the number line in Figure 6. For example, in Region A, using $x = -3$ gives

$$(-3 - 1)(-3 + 2)(-3 - 4) \leq 0$$
$$(-4)(-1)(-7) \leq 0$$
$$-28 \leq 0. \quad \text{True}$$

The numbers in Region A are in the solution set. Verify that the numbers in Region C are also in the solution set, which is written

$$(-\infty, -2] \cup [1, 4].$$

The solution set is graphed in Figure 7.

FIGURE 7

WORK PROBLEM 3 AT THE SIDE. ▶▶

OBJECTIVE ▶ Inequalities involving fractions are solved in a similar manner, by going through the following steps.

Solving Inequalities with Fractions

Step 1 Rewrite the inequality as an equation and solve the equation.

Step 2 Set the denominator equal to zero and solve that equation.

Step 3 Use the solutions from Steps 1 and 2 to divide a number line into regions.

Step 4 Test a number from each region by substitution in the inequality to determine the intervals that satisfy the inequality.

Step 5 Be sure to exclude any values that make the denominator equal to zero.

Caution
Don't forget Step 2. Any number that makes the denominator zero *must* separate two regions on the number line because there will be no point on the number line for that value of the variable.

3. Solve. Graph each solution.

(a) $(y - 3)(y + 2)(y + 1) > 0$

(b) $(k - 5)(k + 1)(k - 3) \leq 0$

ANSWERS

3. (a) $(-2, -1) \cup (3, \infty)$

-2 -1 0 1 2 3 4

(b) $(-\infty, -1] \cup [3, 5]$

-1 0 1   3   5

4. Solve. Graph each solution.

(a) $\dfrac{2}{x - 4} < 3$

——————————————→

(b) $\dfrac{5}{y + 1} > 4$

——————————————→

4. (a) $(-\infty, 4) \cup \left(\dfrac{14}{3}, \infty\right)$

$$\dfrac{14}{3}$$

←——————————→
0 1 2 3 4 5 6

(b) $\left(-1, \dfrac{1}{4}\right)$

$$\dfrac{1}{4}$$

←——————————→
-2 -1 0 1 2

E X A M P L E 3 Solving an Inequality with a Fraction

Solve the inequality $\dfrac{-1}{p - 3} > 1$.

Step 1 Write the corresponding equation and solve it.

$$\dfrac{-1}{p - 3} = 1$$

$$-1 = p - 3 \qquad \text{Multiply by the common denominator.}$$

$$2 = p$$

Step 2 Find the number that makes the denominator 0.

$$p - 3 = 0$$

$$p = 3$$

Step 3 These two numbers, 2 and 3, divide a number line into three regions. (See Figure 8.)

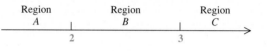

FIGURE 8

Step 4 Testing one number from each region in the given inequality shows that the solution set is the interval (2, 3).

Step 5 This interval does not include any value that might make the denominator of the original inequality equal to zero. A graph of this solution set is given in Figure 9.

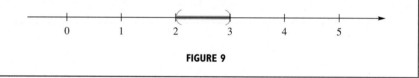

FIGURE 9

◀◀ **WORK PROBLEM 4 AT THE SIDE.**

E X A M P L E 4 Solving an Inequality with a Fraction

Solve $\dfrac{m - 2}{m + 2} \leq 2$.

Write the corresponding equation and solve it. (Step 1)

$$\dfrac{m - 2}{m + 2} = 2$$

$$m - 2 = 2(m + 2) \qquad \text{Multiply by the common denominator.}$$

$$m - 2 = 2m + 4 \qquad \text{Distributive property}$$

$$-6 = m$$

Set the denominator equal to zero and solve the equation. (Step 2)

$$m + 2 = 0$$

$$m = -2$$

CONTINUED ON NEXT PAGE

The numbers -6 and -2 determine three regions (Step 3). Test one number from each region (Step 4) to see that the solution set is the interval

$$(-\infty, -6] \cup (-2, \infty).$$

The number -6 satisfies the equality in $\leq$, but -2 cannot be used as a solution since it makes the denominator equal to zero (Step 5). The graph of the solution set is shown in Figure 10.

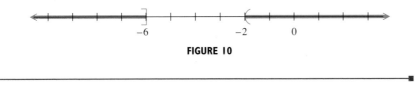

FIGURE 10

WORK PROBLEM 5 AT THE SIDE. ▶▶

OBJECTIVE ▶ There is an alternative method for solving fractional inequalities like those in Examples 3 and 4. It requires writing the inequality with 0 on one side. The other side is then expressed as a single fraction. The values that cause the numerator or the denominator to equal 0 are then used to divide a number line into regions, and then a test value is used from each region to determine the solution set.

EXAMPLE 5 Using an Alternative Method to Solve a Fractional Inequality

Use the alternative method described above to solve the inequality from Example 3, $\dfrac{-1}{p - 3} > 1$.

$$\frac{-1}{p - 3} > 1$$

$$\frac{-1}{p - 3} - 1 > 0 \qquad \text{Subtract 1 so that one side is equal to 0.}$$

$$\frac{-1}{p - 3} - \frac{p - 3}{p - 3} > 0 \qquad \text{Use } p - 3 \text{ as the common denominator.}$$

$$\frac{-1 - p + 3}{p - 3} > 0 \qquad \text{Write as a single fraction.}$$

$$\frac{-p + 2}{p - 3} > 0 \qquad \text{Combine terms.}$$

The number 2 makes the numerator 0, and 3 makes the denominator 0. Notice that these determine the same regions as indicated in Figure 8. Testing one number from each region shows that only Region B, $(2, 3)$, makes the final inequality true. The solution set is graphed in Figure 9.

WORK PROBLEM 6 AT THE SIDE. ▶▶

OBJECTIVE ▶ Special cases of quadratic inequalities may occur, such as those discussed in the next example.

---

5. Solve $\dfrac{k + 2}{k - 1} \leq 5$, and graph the solution.

6. Use the alternative method to solve the inequality in Example 4,

$$\frac{m - 2}{m + 2} \leq 2.$$

ANSWERS

5. $(-\infty, 1) \cup \left[\dfrac{7}{4}, \infty \right)$

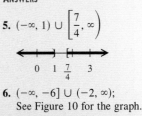

6. $(-\infty, -6] \cup (-2, \infty)$; See Figure 10 for the graph.

7. Solve.

(a) $(3k - 2)^2 > -2$

(b) $(5z + 3)^2 < -3$

E X A M P L E   6   Solving Special Cases

Solve $(2y - 3)^2 > -1$.

Because $(2y - 3)^2$ is never negative, it is always greater than -1. Thus, the solution is the set of all real numbers, $(-\infty, \infty)$. In the same way, there is no solution for $(2y - 3)^2 < -1$ and the solution set is $\emptyset$.

◀◀ **WORK PROBLEM 7 AT THE SIDE.**

10.6 Exercises

1. Explain how you determine whether to include or exclude endpoints when solving a quadratic or higher-degree inequality.

2. Explain why the number 7 cannot possibly be a solution of a fractional inequality that has $x - 7$ as the denominator of a fraction.

3. The solution set of the inequality $x^2 + x - 12 < 0$ is the interval $(-4, 3)$. Without actually performing any work, give the solution set of the inequality $x^2 + x - 12 \geq 0$.

4. Without actually performing any work, give the solution set of the fractional inequality $\frac{3}{x^2 + 1} > 0$. (*Hint:* Determine the sign of the numerator. Determine what the sign of the denominator *must* be. Then consider the inequality symbol.)

Solve the following inequalities, and graph each solution set. See Example 1. (Hint: In Exercises 17 and 18, use the quadratic formula.)

5. $(x + 1)(x - 5) > 0$

6. $(m + 6)(m - 2) > 0$

7. $(r + 4)(r - 6) < 0$

8. $(y + 4)(y - 8) < 0$

9. $x^2 - 4x + 3 \geq 0$

10. $m^2 - 3m - 10 \geq 0$

11. $10a^2 + 9a \geq 9$

12. $3r^2 + 10r \geq 8$

13. $9p^2 + 3p < 2$

14. $2y^2 + y < 15$

15. $6x^2 + x \geq 1$

16. $4y^2 + 7y \geq -3$

17. $y^2 - 6y + 6 \geq 0$

18. $3k^2 - 6k + 2 \leq 0$

Solve the following inequalities and graph each solution set. See Example 2.

19. $(p - 1)(p - 2)(p - 4) < 0$

20. $(2r + 1)(3r - 2)(4r + 7) < 0$

21. $(a - 4)(2a + 3)(3a - 1) \geq 0$

22. $(z + 2)(4z - 3)(2z + 7) \geq 0$

Solve the following inequalities and graph each solution set. See Examples 3–5.

23. $\dfrac{x - 1}{x - 4} > 0$

24. $\dfrac{x + 1}{x - 5} > 0$

25. $\dfrac{2y + 3}{y - 5} \leq 0$

26. $\dfrac{3t + 7}{t - 3} \leq 0$

⟶

27. $\dfrac{8}{x - 2} \geq 2$

⟶

28. $\dfrac{20}{y - 1} \geq 1$

⟶

29. $\dfrac{3}{2t - 1} < 2$

⟶

30. $\dfrac{6}{m - 1} < 1$

⟶

31. $\dfrac{a}{a + 2} \geq 2$

⟶

32. $\dfrac{m}{m + 5} \geq 2$

⟶

33. $\dfrac{4k}{2k - 1} < k$

⟶

34. $\dfrac{r}{r + 2} < 2r$

⟶

Solve the following inequalities. See Example 6.

35. $(4 - 3x)^2 \geq -2$

36. $(6y + 7)^2 \geq -1$

37. $(3x + 5)^2 \leq -4$

38. $(8t + 5)^2 \leq -5$

MATHEMATICAL CONNECTIONS (EXERCISES 39–42)

A rock is projected vertically upward from the ground. Its distance s in feet above the ground after t seconds is given by the quadratic function $s(t) = -16t^2 + 256t$. Work Exercises 39–42 in order to see how quadratic equations and inequalities are connected.

39. At what times will the rock be 624 feet above the ground? (*Hint:* Set $s(t) = 624$ and solve the quadratic *equation.*)

40. At what times will the rock be more than 624 feet above the ground? (*Hint:* Set $s(t) > 624$ and solve the quadratic *inequality.*)

41. At what times will the rock be at ground level? (*Hint:* Set $s(t) = 0$ and solve the quadratic *equation.*)

42. At what times will the rock be less than 624 feet above the ground? (*Hint:* Set $s(t) < 624$, solve the quadratic *inequality,* and observe the solutions in Exercises 40 and 41 to determine the smallest and largest possible values of *t.*)

10.3	quadratic formula	The quadratic formula is a formula for solving quadratic equations.
	discriminant	The discriminant is the quantity under the radical in the quadratic formula.
10.4	quadratic in form	An equation that can be written as a quadratic equation is called quadratic in form.
10.5	Pythagorean theorem	The Pythagorean theorem states that in a right triangle the square of the length of the hypotenuse equals the sum of the squares of the lengths of the legs.
	hypotenuse	The hypotenuse is the longest side in a right triangle.
	leg	The two shorter sides of a right triangle are called the legs.
10.6	quadratic inequality	A quadratic inequality is an inequality that can be written in the form $ax^2 + bx + c < 0$ or $ax^2 + bx + c > 0$, or with $\leq$ or $\geq$.

NEW SYMBOLS

$\pm$ plus or minus

QUICK REVIEW

Concepts	*Examples*
10.1 Solving Quadratic Equations by the Square Root Property	
Square Root Property of Equations	Solve $(2x + 1)^2 = 5$.
If x and b are complex numbers and if $x^2 = b$, then $x = \sqrt{b}$ or $x = -\sqrt{b}$.	$$2x + 1 = \pm\sqrt{5}$$ $$2x = -1 \pm \sqrt{5}$$ $$x = \frac{-1 \pm \sqrt{5}}{2}$$ Solution set: $\left\{ \dfrac{-1 + \sqrt{5}}{2}, \dfrac{-1 - \sqrt{5}}{2} \right\}$
10.2 Solving Quadratic Equations by Completing the Square	
Completing the Square	Solve $2x^2 + 4x - 1 = 0$.
1. If the coefficient of the squared term is 1, go to Step 2. If it is not 1, divide each side of the equation by this coefficient.	1. $x^2 + 2x - \dfrac{1}{2} = 0$
2. Make sure that all variable terms are on one side of the equation, and all constant terms are on the other.	2. $x^2 + 2x = \dfrac{1}{2}$

Concepts	Examples
10.2 Solving Quadratic Equations by Completing the Square (Continued)	
Completing the Square	
3. Take half the coefficient of x, square it, and add the square to each side of the equation.	3. $x^2 + 2x + 1 = \dfrac{1}{2} + 1$
4. Factor the variable side.	4. $(x + 1)^2 = \dfrac{3}{2}$
5. Use the square root property to solve the equation.	5. $x + 1 = \pm\sqrt{\dfrac{3}{2}} = \pm\dfrac{\sqrt{6}}{2}$ $x = -1 \pm \dfrac{\sqrt{6}}{2}$ $x = \dfrac{-2 \pm \sqrt{6}}{2}$ Solution set: $\left\{ \dfrac{-2 + \sqrt{6}}{2}, \dfrac{-2 - \sqrt{6}}{2} \right\}$
10.3 Solving Quadratic Equations by the Quadratic Formula	
Quadratic Formula The solutions of $ax^2 + bx + c = 0$ $(a \neq 0)$ are $x = \dfrac{-b \pm \sqrt{b^2 - 4ac}}{2a}.$	Solve $3x^2 + 5x + 2 = 0.$ $x = \dfrac{-5 \pm \sqrt{5^2 - 4(3)(2)}}{2(3)}$ $x = -1 \quad \text{or} \quad x = -\dfrac{2}{3}$ Solution set: $\left\{ -1, -\dfrac{2}{3} \right\}$
The Discriminant If a, b, and c are integers, then the discriminant, $b^2 - 4ac$, of $ax^2 + bx + c = 0$ determines the type of solutions as follows. | Discriminant	Solutions
Positive square of an integer	2 rational solutions
Positive, not square of an integer	2 irrational solutions
Zero	1 rational solution
Negative	2 imaginary solutions		For $x^2 + 3x - 10 = 0$, the discriminant is $3^2 - 4(1)(-10) = 49.$ There are **2 rational** solutions. For $2x^2 + 5x + 1 = 0$, the discriminant is $5^2 - 4(2)(1) = 17.$ There are **2 irrational** solutions. For $9x^2 - 6x + 1 = 0$, the discriminant is $(-6)^2 - 4(9)(1) = 0.$ There is **1 rational** solution. For $4x^2 + x + 1 = 0$, the discriminant is $1^2 - 4(4)(1) = -15.$ There are **2 imaginary** solutions.
If the discriminant of $ax^2 + bx + c$ is a perfect square, the trinomial is factorable.	The discriminant of $30x^2 - 13x - 10$ is $b^2 - 4ac = (-13)^2 - 4(30)(-10) = 1369 = 37^2$, so it is factorable. It factors as $(5x + 2)(6x - 5).$

Concepts	Examples
10.4 Equations Quadratic in Form An equation that can be written in the form $au^2 + bu + c = 0$, for $a \neq 0$ and an algebraic expression u, is called quadratic in form. Substitute u for the expression, solve for u, and then solve for the variable in the expression.	Solve $3(x + 5)^2 + 7(x + 5) + 2 = 0$. Here $u = x + 5$, so the equation can be written as $3u^2 + 7u + 2 = 0$. Solve for u by factoring. $$(3u + 1)(u + 2) = 0$$ $$u = -\frac{1}{3} \quad \text{or} \quad u = -2$$ Now solve for x. $u = x + 5 \qquad\qquad u = x + 5$ $-\dfrac{1}{3} = x + 5 \qquad -2 = x + 5$ $x = -\dfrac{16}{3} \qquad\qquad x = -7$ Solution set: $\left\{ -7, -\dfrac{16}{3} \right\}$
10.5 Formulas and Applications Involving Quadratic Equations **Formulas** **(a)** If the variable appears only to the second degree, isolate the squared variable on one side of the equation. Then use the square root property.	Solve $A = \dfrac{2mp}{r^2}$ for r. $r^2 A = 2mp \qquad$ Multiply by r^2. $r^2 = \dfrac{2mp}{A} \qquad$ Divide by A. $r = \pm \sqrt{\dfrac{2mp}{A}} \qquad$ Take square roots. $r = \pm \dfrac{\sqrt{2mpA}}{A} \qquad$ Rationalize.
(b) If the variable appears to the first and second degree, write the equation in standard quadratic form. Then use the quadratic formula to solve.	Solve $m^2 + rm = t$ for m. $m^2 + rm - t = 0 \qquad$ Standard form $m = \dfrac{-r \pm \sqrt{r^2 - 4(1)(-t)}}{2(1)} \quad$ $a = 1, b = r,$ $c = -t$ $m = \dfrac{-r \pm \sqrt{r^2 + 4t}}{2}$

Applications
Quadratic functions can often be used to model data that are not linear.
	The quadratic function $f(x) = .0234x^2 - .5029x + 12.5$ can be used to model the number of infant deaths per 1000 live births between 1980 and 1989, where $x = 0$ corresponds to 1980, $x = 1$ to 1981, and so on.
	
	    To find the number of deaths in 1985, let $x = 5$ and find $f(5)$. Since $f(5) \approx 10.6$, there were about 10.6 deaths per 1000 live births in 1985.
	
	    To find the year in which the number of deaths was 11.6, solve $11.6 = .0234x^2 - .5029x + 12.5$ to find the approximate solutions 2 and 19.5. Only 2 applies here, and $x = 2$ corresponds to 1982.

Concepts	Examples
10.6 Nonlinear and Fractional Inequalities	Solve $2x^2 + 5x + 2 < 0$.
Solving a Quadratic Inequality	
Step 1 Rewrite the inequality as an equation and solve.	$$2x^2 + 5x + 2 = 0$$ $$x = -\frac{1}{2}, x = -2$$
Step 2 Place the numbers found in Step 1 on a number line. These numbers divide the line into regions.	
Step 3 Substitute a number from each region into the inequality to determine the intervals that belong in the solution set—those intervals containing numbers that make the inequality true.	$x = -3$ makes it false; $x = -1$ makes it true; $x = 0$ makes it false. Solution set: $\left(-2, -\frac{1}{2}\right)$
This method can be extended to inequalities having more than two factors.	
Solving Inequalities with Fractions	Solve $\dfrac{x}{x + 2} > 4$.
Step 1 Rewrite the inequality as an equation and solve the equation.	Solving $\dfrac{x}{x + 2} = 4$ yields $x = -\dfrac{8}{3}$.
Step 2 Set the denominator equal to zero and solve that equation.	$$x + 2 = 0$$ $$x = -2$$
Step 3 Use the solutions from Steps 1 and 2 to divide a number line into regions.	
Step 4 Test a number from each region in the inequality to determine the regions that satisfy the inequality.	-4 makes it false; $-\frac{7}{3}$ makes it true; 0 makes it false.
Step 5 Exclude any values that make the denominator zero.	The solution set is $\left(-\frac{8}{3}, -2\right)$, since -2 makes the denominator 0 and $-\frac{8}{3}$ gives a false sentence.
An alternative method of solving inequalities with fractions is discussed in Section 10.6.	

[10.1] *In Exercises 1–7, solve each equation by using the square root property. Express all radicals in simplest form.*

1. $y^2 = 144$

2. $x^2 = 37$

3. $m^2 = 128$

4. $(k + 2)^2 = 25$

5. $(r - 3)^2 = 10$

6. $(2p + 1)^2 = 14$

*7. $(3k + 2)^2 = -9$

8. The square root property can be applied only to equations of the form _____ .

[10.2] *Solve each equation by completing the square.*

9. $m^2 + 6m + 5 = 0$

10. $p^2 + 4p = 7$

11. $-x^2 + 5 = 2x$

12. $2y^2 - 3 = -8y$

13. $5k^2 - 3k - 2 = 0$

*14. $(4a + 1)(a - 1) = -7$

_____

*Exercises identified with asterisks have imaginary number solutions.

Solve each problem.

15. If an object is thrown upward from a height of 50 feet, with an initial velocity of 32 feet per second, then its height after t seconds is given by the function $h(t) = -16t^2 + 32t + 50$, where h is in feet. After how many seconds will it reach a height of 30 feet?

16. A certain projectile is located $d(t) = 2t^2 - 5t + 2$ feet from the ground after t seconds have elapsed. How many seconds will it take the projectile to be 14 feet from the ground?

[10.3] *Solve each equation using the quadratic formula.*

17. $2y^2 + y - 21 = 0$

18. $k^2 + 5k = 7$

19. $(t + 3)(t - 4) = -2$

20. $9p^2 = 42p - 49$

*21. $3p^2 = 2(2p - 1)$

22. $m(2m - 7) = 3m^2 + 3$

*23. $2x^2 + 3x + 4 = 0$

24. A student wrote the following as the quadratic formula for solving $ax^2 + bx + c = 0$, $a \neq 0$: $x = -b \pm \dfrac{\sqrt{b^2 - 4ac}}{2a}$. Was this correct? If not, what is wrong with it?

25. A rock is projected vertically upward from the ground. Its distance in feet from the ground in t seconds is $s(t) = -16t^2 + 256t$. At what times will the rock be 768 feet from the ground?

26. Explain why the problem in Exercise 25 has two answers.

Use the quadratic equation $x^2 + 2x + k = 0$ for Exercises 27 and 28.

27. What value(s) of k will give only one real solution?

28. What value(s) of k will give two real solutions?

Use the discriminant to predict whether the solutions to the following equations are
(a) *two distinct rational numbers;*   **(b)** *exactly one rational number;*
(c) *two distinct irrational numbers;*   **(d)** *two distinct imaginary numbers.*

29. $a^2 + 5a + 2 = 0$ **30.** $4c^2 = 3 - 4c$ **31.** $4x^2 = 6x - 8$ **32.** $9z^2 + 30z + 25 = 0$

Use the discriminant to tell which polynomials can be factored. If a polynomial can be factored, factor it.

33. $24x^2 - 74x + 45$ **34.** $36x^2 + 69x - 34$

[10.4] *Solve each equation.*

35. $\dfrac{15}{x} = 2x - 1$ **36.** $\dfrac{1}{y} + \dfrac{2}{y + 1} = 2$ **37.** $8(3x + 5)^2 + 2(3x + 5) - 1 = 0$

38. $-2r = \sqrt{\dfrac{48 - 20r}{2}}$ **39.** $2x^{2/3} - x^{1/3} - 28 = 0$ **40.** $(x^2 + x)^2 = 8(x^2 + x) - 12$

41. Lisa Wunderle drove 8 miles to pick up her friend Laurie, and then drove 11 miles to a mall at a speed 15 miles per hour faster. If Lisa's total travel time was 24 minutes, what was her speed on the trip to pick up Laurie?

42. It takes Linda Youngman 2 hours longer to write a report to her boss than it takes Ed Moura. Working together, it would take them 3 hours. How long would it take each one to do the job alone? Round your answer to the nearest tenth of an hour.

43. Why can't the equation $x = \sqrt{2x + 4}$ have a negative solution?

44. If you were to use the quadratic formula to solve $x^4 - 5x^2 + 6 = 0$, what would you have to remember after you applied the formula?

[10.5] *Solve each formula for the indicated variable. (Give answers with ±.)*

45. $S = \dfrac{Id^2}{k}$ for d

46. $k = \dfrac{rF}{wv^2}$ for v

47. $S = 2\pi rh + 2\pi r^2$ for r

48. $mt^2 = 3mt + 6$ for t

Solve each of the following problems.

49. The Mart Hotel in Dallas, Texas, is 400 feet high. Suppose that a ball is projected upward from the top of the Mart, and its position in feet above the ground is given by the quadratic function $f(t) = -16t^2 + 45t + 400$, where t is the number of seconds elapsed. How long will it take for the ball to reach a height of 200 feet above the ground?

50. The Toronto Dominion Center in Winnipeg, Manitoba, is 407 feet high. Suppose that a ball is projected upward from the top of the Center, and its position in feet above the ground is given by the quadratic function $s(t) = -16t^2 + 75t + 407$, where t is the number of seconds elapsed. How long will it take for the ball to reach a height of 450 feet above the ground?

51. A rectangle has a length 2 meters more than its width. If one meter is cut from the length and one meter is added to the width, the resulting figure is a square with an area of 121 square meters. Find the dimensions of the original rectangle.

52. The hypotenuse of a right triangle is 9 feet shorter than twice the length of the longer leg. The shorter leg is 3 feet shorter than the longer leg. Find the lengths of the three sides of the triangle.

53. The product of the page numbers of two pages facing each other in this book is 4692. What are the page numbers?

54. Nancy Boyle wants to buy a mat for a photograph that measures 14 inches by 20 inches. She wants to have an even border around the picture when it is mounted on the mat. If the area of the mat she chooses is 352 square inches, how wide will the border be?

55. A search light moves horizontally back and forth along a wall with the distance of the light from a starting point at t minutes given by the quadratic function $f(t) = 100t^2 - 300t$. How long will it take before the light returns to the starting point?

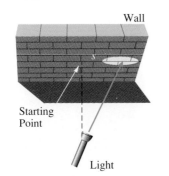

Wall

Starting Point

Light

56. The manager of a fast-food outlet has determined that the demand for frozen yogurt is $\frac{25}{p}$ units per day, where p is the price (in dollars) per unit. The supply is $70p + 15$ units per day. Find the price at which supply and demand are equal.

57. Use the formula $A = P(1 + r)^2$ to find the interest rate r at which a principal P of $10,000 will increase to $10,920.25 in 2 years.

58. Jim can complete a job alone in one hour less than his brother, Jake. Together they can complete the job in 4 hours. How long does the job take each brother working alone?

[10.6] *Solve the following inequalities and graph the solution sets.*

59. $(x - 4)(2x + 3) > 0$

60. $x^2 + x \leq 12$

61. $2k^2 > 5k + 3$

62. $\dfrac{3y + 4}{y - 2} \leq 1$

63. $(x + 2)(x - 3)(x + 5) \leq 0$

64. $(4m + 3)^2 \leq -4$

MIXED REVIEW EXERCISES

Solve.

65. $V = r^2 + R^2 h$ for R

*66. $3t^2 - 6t = -4$

67. $(b^2 - 2b)^2 = 11(b^2 - 2b) - 24$

68. $(r - 1)(2r + 3)(r + 6) < 0$

69. $(3k + 11)^2 = 7$

70. $p = \sqrt{\dfrac{yz}{6}}$ for y

71. $-5x^2 = -8x + 3$

72. $6 + \dfrac{15}{s^2} = -\dfrac{19}{s}$

73. $\dfrac{-2}{x + 5} \leq -5$

74. Two pipes together can fill a large tank in 2 hours. One of the pipes, used alone, takes 3 hours longer than the other does to fill the tank. How long would each pipe take to fill the tank alone?

75. Phong paddled his canoe 20 miles upstream, then paddled back. If the speed of the current was 3 miles per hour and the total trip took 7 hours, what was Phong's speed?

76. $\dfrac{2}{x - 4} + \dfrac{1}{x} = \dfrac{11}{5}$

*77. $y^2 = -242$

78. $(8k - 7)^2 \geq -1$

CHAPTER 10 TEST

*Items marked * require knowledge of imaginary numbers.*

Solve using the square root property.

1. $t^2 = 54$

1. _____

2. $(7x + 3)^2 = 25$

2. _____

Solve by completing the square.

3. $x^2 + 2x = 1$

3. _____

Solve using the quadratic formula.

4. $2x^2 - 3x - 1 = 0$

4. _____

*5. $3t^2 - 4t = -5$

5. _____

6. $3x = \sqrt{\dfrac{9x + 2}{2}}$

6. _____

7. Maretha and Lillaana are typesetters. For a certain report, Lillaana can set the type 2 hours faster than Maretha can. If they work together, they can do the entire report in 5 hours. How long will it take each of them working alone to prepare the report? Round your answers to the nearest tenth of an hour.

7. _____

*8. If k is a negative number, then which one of the following equations will have two imaginary solutions?
   (a) $x^2 = 4k$ **(b)** $x^2 = -4k$
   (c) $(x + 2)^2 = -k$ **(d)** $x^2 + k = 0$

8. _____

9. What is the discriminant for $2x^2 - 8x - 3 = 0$? How many and what type of solutions does this equation have? (Do not actually solve.)

9. _____

Solve by any method.

10. $3 - \dfrac{16}{x} - \dfrac{12}{x^2} = 0$

10. _____

11. $4x^2 + 7x - 3 = 0$

11. _____

12. _____

12. $9x^4 + 4 = 37x^2$

13. _____

13. $12 = (2d + 1)^2 + (2d + 1)$

14. _____

14. Solve for r: $S = 4\pi r^2$ (Leave $\pm$ in your answer.)

Solve each problem.

15. _____

15. The quadratic function $f(x) = 3.23x^2 - 1.89x + 1.06$ approximates the number of AIDS cases diagnosed in the United States between 1982 and 1993. Here the number of cases is in thousands, and $x = 0$ corresponds to 1982, $x = 1$ to 1983, and so on. In what year did the number of cases diagnosed reach 24.64 thousand?

16. _____

16. Sandi Goldstein paddled her canoe 10 miles upstream, and then paddled back to her starting point. If the rate of the current was 3 miles per hour and the entire trip took $3\frac{1}{2}$ hours, what was Sandi's rate?

17. _____

17. Adam Bryer has a pool 24 feet long and 10 feet wide. He wants to construct a concrete walk around the pool. If he plans for the walk to be of uniform width and cover 152 square feet, what will the width of the walk be?

18. _____

18. At a point 30 meters from the base of a tower, the distance to the top of the tower is 2 meters more than twice the height of the tower. Find the height of the tower.

Solve. Graph each solution set.

19. _____

19. $2x^2 + 7x > 15$

20. _____

20. $\dfrac{5}{t - 4} \le 1$

Let $S = \{-\frac{7}{3}, -2, -\sqrt{3}, 0, .7, \sqrt{12}, \sqrt{-8}, 7, \frac{32}{3}\}$. List the elements of S that are elements of the following sets.

1. Integers   **2.** Rational numbers   **3.** Real numbers   **4.** Complex numbers

Simplify each of the following.

5. $|-3| + 8 - |-9| - (-7 + 3)$ **6.** $2(-3)^2 + (-8)(-5) + (-17)$

Solve the following.

7. $-2x + 4 = 5(x - 4) + 17$ **8.** $|3y - 7| \leq 1$ **9.** $|4z + 2| > 7$

10. Find the slope and y-intercept of the line with equation $2x - 4y = 7$.

11. Write an equation in standard form of the line through $(2, -1)$ and perpendicular to $-3x + y = 5$.

Write with positive exponents only. Assume variables represent positive real numbers.

12. $\left(\dfrac{x^{-3}y^2}{x^5y^{-2}}\right)^{-1}$ **13.** $\dfrac{(4x^{-2})^2(2y^3)}{8x^{-3}y^5}$

14. If $f(x) = \sqrt{2x - 5}$, find $f(10)$. What is the domain of this function?

15. Solve the system $5x - 3y = 17$
$$2y = 4x - 12.$$

16. For a certain system of equations, $D_x = 4$, $D_y = 3$, $D_z = -6$, and $D = 12$. Find the solution set of the system.

Perform the indicated operations.

17. $\left(\dfrac{2}{3}t + 9\right)^2$

18. $(3t^3 + 5t^2 - 8t + 7) - (6t^3 + 4t - 8)$

19. Divide $4x^3 + 2x^2 - x + 26$ by $x + 2$.

Factor completely.

20. $16x - x^3$

21. $24m^2 + 2m - 15$

22. $9x^2 - 30xy + 25y^2$

23. $2ax - 10a + bx - 5b$

Perform the operations, and express answers in lowest terms. Assume denominators are nonzero.

24. $\dfrac{x^2 - 3x - 10}{x^2 + 3x + 2} \cdot \dfrac{x^2 - 2x - 3}{x^2 + 2x - 15}$

25. $\dfrac{5t + 2}{-6} \div \dfrac{15t + 6}{5}$

26. $\dfrac{3}{2 - k} - \dfrac{5}{k} + \dfrac{6}{k^2 - 2k}$

27. $\dfrac{\dfrac{r}{s} - \dfrac{s}{r}}{\dfrac{r}{s} + 1}$

Simplify the following radical expressions.

28. $\sqrt[3]{\dfrac{27}{16}}$

29. $\dfrac{2}{\sqrt{7} - \sqrt{5}}$

Solve the following equations.

30. $2x = \sqrt{\dfrac{5x + 2}{3}}$

31. $\dfrac{3}{x - 3} - \dfrac{2}{x - 2} = \dfrac{3}{x^2 - 5x + 6}$

32. $(r - 5)(2r + 3) = 1$

33. $b^4 - 5b^2 + 4 = 0$

Solve the following problems.

34. The perimeter of a rectangle is 20 inches, and its area is 21 square inches. What are the dimensions of the rectangle?

35. Two cars left an intersection at the same time, one heading due south and the other due east. Later they were exactly 95 miles apart. The car heading east had gone 38 miles less than twice as far as the car heading south. How far had each car traveled?

36. A knitting shop orders yarn from three suppliers, I, II, and III. One month the shop ordered a total of 100 units of yarn from these suppliers. The delivery costs were $80, $50, and $65 per unit for the orders from suppliers I, II, and III, respectively, with total delivery costs of $5990. The shop ordered the same amount from suppliers I and III. How many units were ordered from each supplier?

37. An object is propelled upward. Its height in feet is given by the function

$$f(t) = -16t^2 + 80t + 50.$$

At what times t is its height 70 feet? Round to the nearest tenth of a second.

Solve each equation by the method stated.

38. $z^2 - 2z = 15$ (completing the square)

39. $2x^2 - 4x - 3 = 0$ (quadratic formula)

40. Solve the fractional inequality $\dfrac{x + 3}{x - 5} > 2$.

Graphs of Nonlinear Functions and Conic Sections

11

INTERNET

www.mathnotes.com

11.1 Graphs of Quadratic Functions; Vertical Parabolas

In Chapter 4, we saw that the graphs of first-degree equations are straight lines. In this chapter we discuss the graphs of nonlinear equations. In particular, we will investigate graphs of second-degree equations, which are equations with one or more second-degree terms. These graphs are the intersections of an infinite cone and a plane, as shown in Figure 1. Because of this, the graphs are called **conic sections.**

OBJECTIVES

1. Graph parabolas that are examples of quadratic functions.

2. Graph parabolas with horizontal and vertical shifts.

3. Predict the shape and direction of the graph of a parabola from the coefficient of x^2.

FOR EXTRA HELP

Tutorial

Tape 22

SSM, Sec. 11.1

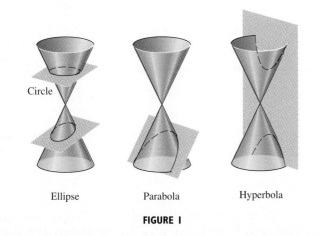

Circle

Ellipse   Parabola   Hyperbola

FIGURE 1

OBJECTIVE 1 Let us begin by graphing the equation $y = x^2$. First, we make a table of ordered pairs satisfying the equation.

x	-2	$-\frac{3}{2}$	-1	$-\frac{1}{2}$	0	$\frac{1}{2}$	1	$\frac{3}{2}$	2
y	4	$\frac{9}{4}$	1	$\frac{1}{4}$	0	$\frac{1}{4}$	1	$\frac{9}{4}$	4

We then plot these points and draw a smooth curve through them to get the graph shown in Figure 2. This graph is called a **parabola** (puh-RAB-uh-luh). The point (0, 0), with the smallest y-value of any point on the curve, is the **vertex** of this parabola. The vertical line through the vertex is the **axis** of this parabola. The parabola is **symmetric with respect to its axis;** that is, if the graph were folded along the axis, the two portions of the curve would coincide.

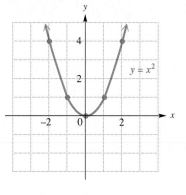

FIGURE 2

Because the graph of $y = x^2$ satisfies the conditions of the graph of a function (Section 4.5), we may write its equation as $f(x) = x^2$. This is the simplest example of a quadratic function.

> **Quadratic Function**
>
> A function defined by an equation that can be written in the form
> $$f(x) = ax^2 + bx + c$$
> for real numbers, a, b, and c, with $a \neq 0$, is a **quadratic function.**

The graph of any quadratic function is a parabola with a vertical axis.

Parabolas have many applications. If an object is thrown upward, then (disregarding air resistance) the path it follows is a parabola. The large disks seen on the sidelines of televised football games, which are used by television crews to pick up the shouted signals of the players on the field, have cross sections that are parabolas. Cross sections of radar dishes and automobile headlights also form parabolas. Additional applications of parabolas are given in the next section.

For the rest of this section, we shall use the function notation $f(x)$ in discussing parabolas.

OBJECTIVE 2 Parabolas need not have their vertices at the origin, as does $f(x) = x^2$. For example, to graph a parabola of the form $f(x) = x^2 + k$, we start by selecting the sample values of x that were used to graph $f(x) = x^2$. The corresponding values of $f(x)$ in $f(x) = x^2 + k$ differ by k from those of $f(x) = x^2$. For this reason, the graph of $f(x) = x^2 + k$ is shifted, or translated, k units vertically compared with that of $f(x) = x^2$.

EXAMPLE 1   Graphing a Parabola with a Vertical Shift

Graph $f(x) = x^2 - 2$.

As we mentioned before, this graph has the same shape as $f(x) = x^2$, but since k here is -2, the graph is shifted 2 units downward, with vertex

CONTINUED ON NEXT PAGE

at $(0, -2)$. Every function value is 2 less than the corresponding function value of $f(x) = x^2$. Plotting points gives the graph in Figure 3. The graph of $f(x) = x^2$ is also shown for comparison.

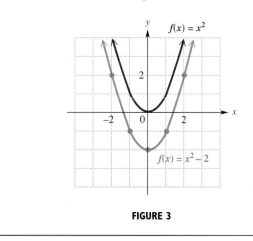

FIGURE 3

Vertical Shift

The graph of $f(x) = x^2 + k$ is a parabola with the same shape as the graph of $f(x) = x^2$. The parabola is shifted k units upward if $k > 0$, and $|k|$ units downward if $k < 0$. The vertex is $(0, k)$.

WORK PROBLEM 1 AT THE SIDE. ▶▶

The graph of $f(x) = (x - h)^2$ is also a parabola with the same shape as $f(x) = x^2$. Because $(x - h)^2 \geq 0$ for all x, the vertex of the parabola $f(x) = (x - h)^2$ should be the lowest point on the parabola. The lowest point occurs here when $f(x)$ is 0. To get $f(x)$ equal to 0, let $x = h$ so the vertex of $f(x) = (x - h)^2$ is at $(h, 0)$. Based on this, the graph of $f(x) = (x - h)^2$ is shifted h units horizontally compared with that of $f(x) = x^2$.

EXAMPLE 2 Graphing a Parabola with a Horizontal Shift

Graph $f(x) = (x - 2)^2$.

When $x = 2$, then $f(x) = 0$, giving the vertex $(2, 0)$. The parabola $f(x) = (x - 2)^2$ has the same shape as $f(x) = x^2$ but is shifted 2 units to the right, as shown in Figure 4. Again, we show the graph of $f(x) = x^2$ for comparison.

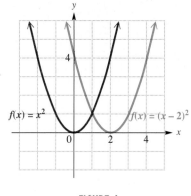

FIGURE 4

1. Graph each parabola.

(a) $f(x) = x^2 + 3$

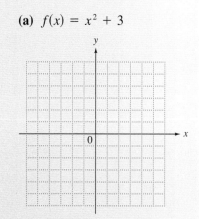

(b) $f(x) = x^2 - 1$

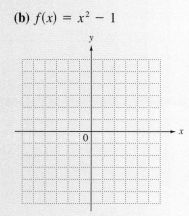

ANSWERS

1. (a)

(b)

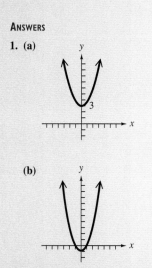

2. Graph each parabola.

(a) $f(x) = (x - 3)^2$

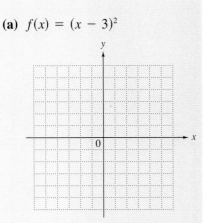

(b) $f(x) = (x + 2)^2$

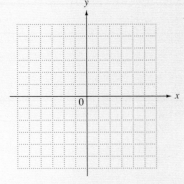

Horizontal Shift

The graph of $f(x) = (x - h)^2$ is a parabola with the same shape as the graph of $f(x) = x^2$. The parabola is shifted h units horizontally: h units to the right if $h > 0$, and $|h|$ units to the left if $h < 0$. The vertex is $(h, 0)$.

Caution

Errors frequently occur when horizontal shifts are involved. In order to determine the direction and magnitude of horizontal shifts, find the value that would cause the expression $x - h$ to equal 0. For example, the graph of $f(x) = (x - 5)^2$ would be shifted 5 units to the *right*, because $+5$ would cause $x - 5$ to equal 0. On the other hand, the graph of $f(x) = (x + 4)^2$ would be shifted 4 units to the *left*, because -4 would cause $x + 4$ to equal 0.

◀◀ **WORK PROBLEM 2 AT THE SIDE.**

A parabola can have both a horizontal and a vertical shift, as in Example 3.

E X A M P L E   3   Graphing a Parabola with Horizontal and Vertical Shifts

Graph $f(x) = (x + 3)^2 - 2$.

This graph has the same shape as $f(x) = x^2$, but is shifted 3 units to the left (since $x + 3 = 0$ if $x = -3$) and 2 units downward (because of the -2). As shown in Figure 5, the vertex is at $(-3, -2)$.

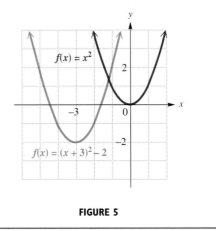

FIGURE 5

E X A M P L E   4   Graphing a Parabola with Horizontal and Vertical Shifts

Graph $f(x) = (x - 1)^2 + 3$.

The graph is shifted one unit to the right and three units up, so the vertex is at $(1, 3)$. See Figure 6.

CONTINUED ON NEXT PAGE

ANSWERS

2. (a)

(b)

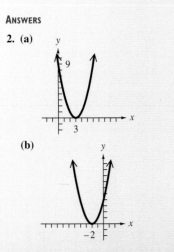

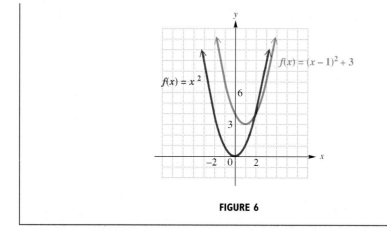

FIGURE 6

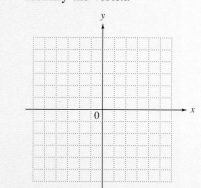

3. (a) Graph $f(x) = (x + 2)^2 - 1$.
Identify the vertex.

The characteristics of the graph of a function of the form $f(x) = (x - h)^2 + k$ are summarized as follows.

> The graph of $f(x) = (x - h)^2 + k$ is a parabola with the same shape as $f(x) = x^2$ and with vertex at (h, k). The axis is the vertical line $x = h$.

WORK PROBLEM 3 AT THE SIDE. ▶▶

OBJECTIVE 3 ▶ Not all parabolas open upward, and not all parabolas have the same shape as $f(x) = x^2$. In the next example we show how to identify parabolas opening downward and having a different shape from that of $f(x) = x^2$.

(b) Identify the vertex
of the graph of
$f(x) = (x - 2)^2 + 5$.

E X A M P L E 5   **Graphing a Parabola
That Opens Downward**

Graph $f(x) = -\frac{1}{2}x^2$.

This parabola is shown in Figure 7. Some ordered pairs that satisfy the equation are $(0, 0)$, $(1, -\frac{1}{2})$, $(2, -2)$, $(-1, -\frac{1}{2})$, and $(-2, -2)$. A table with these ordered pairs is given next to the graph. The coefficient $-\frac{1}{2}$ affects the shape of the graph; the $\frac{1}{2}$ makes the parabola wider (since the values of $f(x)$ grow more slowly than they would for $f(x) = x^2$), and the negative sign makes the parabola open downward. The graph is not shifted in any direction; the vertex is still at $(0, 0)$. Here, the vertex has the *largest* function value (y-value) of any point on the graph.

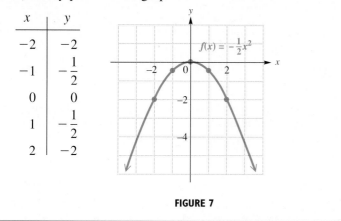

x	y
-2	-2
-1	$-\frac{1}{2}$
0	0
1	$-\frac{1}{2}$
2	-2

FIGURE 7

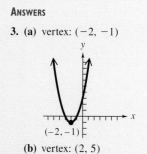

4. Decide whether each parabola opens upward or downward.

(a) $f(x) = -\dfrac{2}{3}x^2$

(b) $f(x) = \dfrac{3}{4}x^2 + 1$

(c) $f(x) = -2x^2 - 3$

(d) $f(x) = 3x^2 + 2$

5. Decide whether each parabola in Problem 4 is wider or narrower than $f(x) = x^2$.

6. Graph $f(x) = \dfrac{1}{2}(x - 2)^2 + 1$.

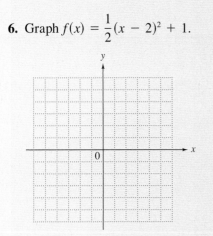

Some general principles concerning the graph of $f(x) = a(x - h)^2 + k$ are summarized as follows.

> **1.** The graph of the quadratic function defined by
>
> $$f(x) = a(x - h)^2 + k, \qquad a \ne 0$$
>
> is a parabola with vertex at (h, k) and the vertical line $x = h$ as axis.
>
> **2.** The graph opens upward if a is positive and downward if a is negative.
>
> **3.** The graph is wider than $f(x) = x^2$ if $0 < |a| < 1$. The graph is narrower than $f(x) = x^2$ if $|a| > 1$.

◀◀ **WORK PROBLEMS 4 AND 5 AT THE SIDE.**

E X A M P L E   6   Using the General Principles to Graph a Parabola

Graph $f(x) = -2(x + 3)^2 + 4$.

The parabola opens downward (because $a < 0$), and is narrower than the graph of $f(x) = x^2$, since $|-2| = 2 > 1$, causing the values of $f(x)$ to grow more quickly than they would for $f(x) = x^2$. This parabola has its vertex at $(-3, 4)$, as shown in Figure 8. To complete the graph, we plotted the ordered pairs $(-4, 2)$ and $(-2, 2)$, which are shown in the table next to the graph. Notice that these two points are symmetric with respect to the axis of the parabola. This symmetry is very useful for finding additional ordered pairs that satisfy the equation.

x	y
-4	2
-3	4
-2	2

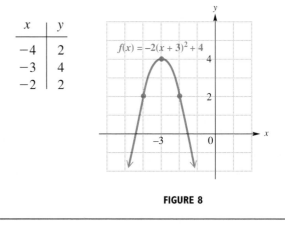

FIGURE 8

◀◀ **WORK PROBLEM 6 AT THE SIDE.**

Note
In this section we have seen how the graph of $f(x) = x^2$ can be shifted horizontally or vertically, and how the shape of the graph can be altered. The principles discussed here can be generalized to graphs of other kinds of functions, as we shall see later in this chapter.

ANSWERS
4. (a) downward **(b)** upward
    (c) downward **(d)** upward
5. (a) wider **(b)** wider **(c)** narrower
    (d) narrower
6.

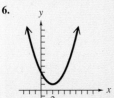

11.1 Exercises

1. Match each quadratic function with its graph from choices A–D.

_____ **(a)** $f(x) = (x + 2)^2 - 1$

_____ **(b)** $f(x) = (x + 2)^2 + 1$

_____ **(c)** $f(x) = (x - 2)^2 - 1$

_____ **(d)** $f(x) = (x - 2)^2 + 1$

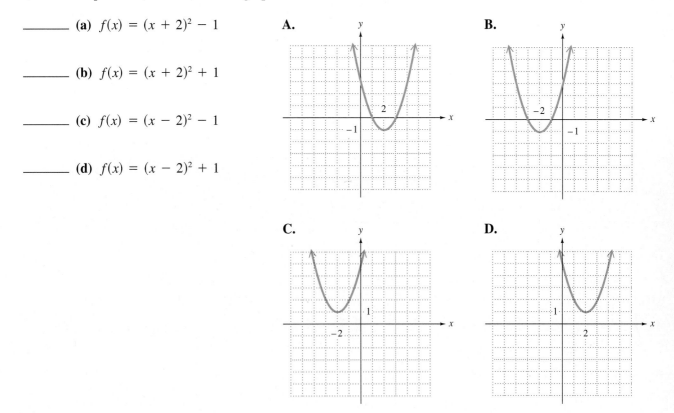

A.

B.

C.

D.

2. Match each quadratic function with its graph from choices A–D.

_____ **(a)** $f(x) = -x^2 + 2$

_____ **(b)** $f(x) = -x^2 - 2$

_____ **(c)** $f(x) = -(x + 2)^2$

_____ **(d)** $f(x) = -(x - 2)^2$

A.

B.

C.

D.

Identify the vertex of each parabola. See Examples 1–4.

3. $f(x) = -3x^2$

4. $f(x) = .5x^2$

5. $f(x) = x^2 + 4$

6. $f(x) = x^2 - 4$

7. $f(x) = (x - 1)^2$

8. $f(x) = (x + 3)^2$

9. $f(x) = (x + 3)^2 - 4$

10. $f(x) = (x - 5)^2 - 8$

For each quadratic function, tell whether the graph opens upward or downward and whether the graph is wider, narrower, or the same as $f(x) = x^2$. See Examples 5 and 6.

11. $f(x) = -.4x^2$

12. $f(x) = -2x^2$

13. $f(x) = 3x^2 + 1$

14. $f(x) = \frac{2}{3}x^2 - 4$

15. Describe how each of the parabolas in Exercises 9 and 10 is shifted compared to the graph of $y = x^2$.

16. What does the value of a in $y = a(x - h)^2 + k$ tell you about the graph of the equation compared to the graph of $y = x^2$?

17. For $f(x) = a(x - h)^2 + k$, in what quadrant is the vertex if
(a) $h > 0, k > 0$;   **(b)** $h > 0, k < 0$;   **(c)** $h < 0, k > 0$;   **(d)** $h < 0, k < 0$?

18. Think of how the graph of the linear function $f(x) = x + 5$ compares to the graph of $g(x) = x$, and fill in the blank: To graph $f(x) = x + 5$, shift the graph of $g(x) = x$ _____ units in the _____ direction.

Sketch the graph of each parabola. Plot at least two points in addition to the vertex.
See Examples 1–6.

19. $f(x) = -2x^2$

20. $f(x) = \dfrac{1}{3}x^2$

21. $f(x) = x^2 - 1$

22. $f(x) = x^2 + 3$

23. $f(x) = -x^2 + 2$

24. $f(x) = 2x^2 - 2$

25. $f(x) = .5(x - 4)^2$

26. $f(x) = -2(x + 1)^2$

27. $f(x) = (x + 2)^2 - 1$

28. $f(x) = -2(x + 3)^2 + 4$

29. $f(x) = -.5(x + 1)^2 + 2$

30. $f(x) = -\dfrac{2}{3}(x + 2)^2 + 1$

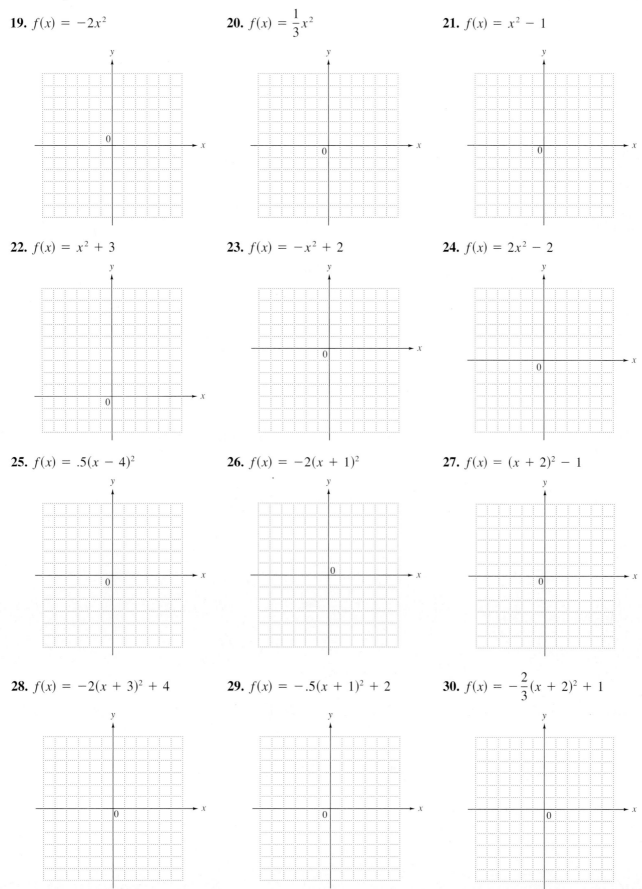

MATHEMATICAL CONNECTIONS (EXERCISES 31–36)

*The procedures described in this section that allow the
graph of $f(x) = x^2$ to be shifted vertically and
horizontally are applicable to other types of functions as
well. A linear function is one of the form $f(x) = ax + b$.
Consider the graph of the simplest linear function,
$f(x) = x$, shown here, and then work through Exercises
31–36 in order.*

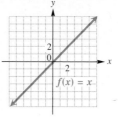

31. Based on the concepts of this section, how does the graph of $f(x) = x^2 + 6$ compare to
the graph of $g(x) = x^2$ if a *vertical* shift is considered?

32. Graph the linear function $f(x) = x + 6$.

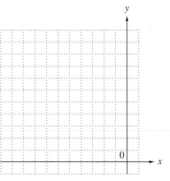

33. How does the graph of $f(x) = x + 6$ compare to the graph of $g(x) = x$ if a vertical shift
is considered? (*Hint:* Look at the y-intercept.)

34. Based on the concepts of this section, how does the graph of $f(x) = (x - 6)^2$ compare to the
graph of $g(x) = x^2$ if a *horizontal* shift is considered?

35. Graph the linear function $f(x) = x - 6$.

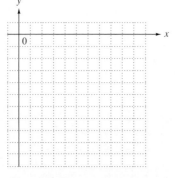

36. How does the graph of $f(x) = x - 6$ compare to the graph of $g(x) = x$ if a horizontal
shift is considered? (*Hint:* Look at the x-intercept.)

11.2 *More about Quadratic Functions; Horizontal Parabolas*

OBJECTIVE 1 When the equation of a parabola is given in the form $f(x) = ax^2 + bx + c$, we need to locate the vertex in order to sketch an accurate graph. This can be done in two ways. The first is by completing the square, as shown in Examples 1 and 2. The second is by using a formula that may be derived by completing the square.

EXAMPLE 1 Completing the Square to Find the Vertex

Find the vertex of the graph of $f(x) = x^2 - 4x + 5$.

To find the vertex, we need to express $x^2 - 4x + 5$ in the form $(x - h)^2 + k$. This is done by completing the square on $x^2 - 4x$, as in Section 10.2. The process is a little different here because we want to keep $f(x)$ alone on one side of the equation. Instead of adding the appropriate number to both sides, *add and subtract* it on the right. This is equivalent to adding 0.

$$f(x) = x^2 - 4x + 5$$
$$= (x^2 - 4x \quad) + 5$$

Half of -4 is -2; $(-2)^2 = 4$.

$$= (x^2 - 4x + 4 - 4) + 5 \qquad \text{Add and subtract 4.}$$
$$= (x^2 - 4x + 4) - 4 + 5 \qquad \text{Bring } -4 \text{ outside the parentheses.}$$
$$f(x) = (x - 2)^2 + 1 \qquad \text{Factor; combine terms.}$$

As we saw in the previous section, the vertex of this parabola is $(2, 1)$.

WORK PROBLEM 1 AT THE SIDE. ▶▶

EXAMPLE 2 Completing the Square to Find the Vertex When $a \neq 1$

Find the vertex of the graph of $f(x) = -3x^2 + 6x - 1$.

We must complete the square on $-3x^2 + 6x$. Because the x^2 term has a coefficient other than 1, factor that coefficient out of the first two terms, and then proceed as in Example 1.

$$f(x) = -3x^2 + 6x - 1$$
$$= -3(x^2 - 2x) - 1$$

Half of -2 is -1, and $(-1)^2 = 1$.

$$= -3(x^2 - 2x + 1 - 1) - 1 \qquad \text{Add and subtract 1.}$$
$$= -3(x^2 - 2x + 1) + (-3)(-1) - 1 \qquad \text{Distributive property}$$
$$= -3(x^2 - 2x + 1) + 3 - 1$$
$$f(x) = -3(x - 1)^2 + 2 \qquad \text{Factor; combine terms.}$$

The vertex is at $(1, 2)$.

WORK PROBLEM 2 AT THE SIDE. ▶▶

OBJECTIVES

1 Find the vertex of a vertical parabola.

2 Graph a quadratic function.

3 Use the discriminant to find the number of x-intercepts of a vertical parabola.

4 Use quadratic functions to solve problems involving maximum or minimum value.

5 Graph horizontal parabolas.

FOR EXTRA HELP

Tutorial   Tape 23   SSM, Sec. 11.2

1. Find the vertex of each parabola by completing the square.

(a) $f(x) = x^2 - 6x + 7$

(b) $f(x) = x^2 + 4x - 9$

2. Find the vertex of each parabola.

(a) $f(x) = 2x^2 - 4x + 1$

(b) $f(x) = -\dfrac{1}{2}x^2 + 2x - 3$

ANSWERS

1. (a) $(3, -2)$ (b) $(-2, -13)$
2. (a) $(1, -1)$ (b) $(2, -1)$

691

3. Use the formula to find the vertex of the graph of each quadratic function.

(a) $f(x) = -2x^2 + 3x - 1$

A formula for the vertex of the graph of the quadratic function defined by $f(x) = ax^2 + bx + c$ can be found by completing the square for the standard form of the equation.

$$f(x) = ax^2 + bx + c \quad (a \neq 0) \qquad \text{Standard form}$$

$$f(x) = a\left(x^2 + \frac{b}{a}x\right) + c \qquad \text{Factor } a \text{ from the first two terms.}$$

$$f(x) = a\left(x^2 + \frac{b}{a}x + \frac{b^2}{4a^2} - \frac{b^2}{4a^2}\right) + c \qquad \text{Add and subtract } \frac{b^2}{4a^2} \text{ within the parentheses.}$$

$$f(x) = a\left(x^2 + \frac{b}{a}x + \frac{b^2}{4a^2}\right) - \frac{b^2}{4a} + c \qquad \text{Distributive property}$$

$$f(x) = a\left(x + \frac{b}{2a}\right)^2 + \frac{4ac - b^2}{4a} \qquad \text{Factor and combine terms.}$$

$$f(x) = a\underbrace{\left[x - \left(\frac{-b}{2a}\right)\right]^2}_{h} + \underbrace{\frac{4ac - b^2}{4a}}_{k}$$

The final equation shows that the vertex (h, k) can be expressed in terms of a, b, and c. However, it is not necessary to memorize k, since $k = f(h)$.

(b) $f(x) = 4x^2 - x + 5$

Vertex Formula

The graph of the quadratic function defined by $f(x) = ax^2 + bx + c$ has its vertex at

$$\left(\frac{-b}{2a}, f\left(\frac{-b}{2a}\right)\right),$$

and the axis of the parabola is the line $x = \frac{-b}{2a}$.

E X A M P L E 3   Using the Formula to Find the Vertex

Use the vertex formula to find the vertex of the graph of

$$f(x) = x^2 - x - 6.$$

For this function, $a = 1$, $b = -1$, and $c = -6$. The x-coordinate of the vertex of the parabola is given by

$$\frac{-b}{2a} = \frac{-(-1)}{2(1)} = \frac{1}{2}.$$

The y-coordinate is $f\left(\frac{-b}{2a}\right) = f\left(\frac{1}{2}\right)$.

$$f\left(\frac{1}{2}\right) = \left(\frac{1}{2}\right)^2 - \frac{1}{2} - 6 = \frac{1}{4} - \frac{1}{2} - 6 = -\frac{25}{4}$$

Finally, the vertex is $\left(\frac{1}{2}, -\frac{25}{4}\right)$.

◄◄ WORK PROBLEM 3 AT THE SIDE.

OBJECTIVE 2 ▶ Parabolas were graphed in the previous section. A more general approach involving finding intercepts and the vertex is given here.

ANSWERS

3. (a) $\left(\frac{3}{4}, \frac{1}{8}\right)$ **(b)** $\left(\frac{1}{8}, \frac{79}{16}\right)$.

Graphing a Quadratic Function f

Step 1 Find the y-intercept by evaluating $f(0)$.

Step 2 Find any x-intercepts by solving $f(x) = 0$.

Step 3 Find the vertex either by using the formula or by completing the square.

Step 4 Find and plot additional points as needed, using the symmetry about the axis.

Step 5 Verify that the graph opens upward (if $a > 0$) or opens downward (if $a < 0$).

The domain of a quadratic function is $(-\infty, \infty)$, unless otherwise specified. The range is determined by the y-value of the vertex, and whether the parabola opens upward or downward.

E X A M P L E   4   Using the Steps for Graphing a Quadratic Function

Graph the quadratic function with $f(x) = x^2 - x - 6$. Give the domain and the range.

Begin by finding the y-intercept.

$$f(x) = x^2 - x - 6$$
$$f(0) = 0^2 - 0 - 6 \qquad \text{Find } f(0).$$
$$f(0) = -6$$

The y-intercept is $(0, -6)$. Now find any x-intercepts.

$$f(x) = x^2 - x - 6$$
$$0 = x^2 - x - 6 \qquad \text{Let } f(x) = 0.$$
$$0 = (x - 3)(x + 2) \qquad \text{Factor.}$$
$$x - 3 = 0 \quad \text{or} \quad x + 2 = 0 \qquad \text{Set each factor equal to 0 and solve.}$$
$$x = 3 \quad \text{or} \qquad x = -2$$

The x-intercepts are $(3, 0)$ and $(-2, 0)$. The vertex, found in Example 3, is $\left(\frac{1}{2}, -\frac{25}{4}\right)$. Plot the points found so far, and plot any additional points as needed. The symmetry of the graph is helpful here. The graph is shown in Figure 9. The domain is $(-\infty, \infty)$ and the range is $\left[-\frac{25}{4}, \infty\right)$.

x	y
-2	0
-1	-4
0	-6
$\frac{1}{2}$	$-\frac{25}{4}$
2	-4
3	0

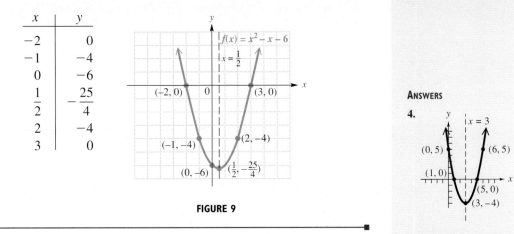

FIGURE 9

WORK PROBLEM 4 AT THE SIDE. ▶▶

4. Graph the quadratic function

$$f(x) = x^2 - 6x + 5.$$

Give the domain and the range.

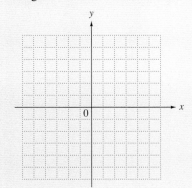

ANSWERS

4.

domain: $(-\infty, \infty)$;
range: $[-4, \infty)$

5. Use the discriminant to determine the number of x-intercepts for each graph.

(a) $f(x) = 4x^2 - 20x + 25$

(b) $f(x) = 2x^2 + 3x + 5$

(c) $f(x) = -3x^2 - x + 2$

OBJECTIVE 3 ▶ The graph of a quadratic function may have two x-intercepts, one x-intercept, or no x-intercepts, as shown in Figure 10.

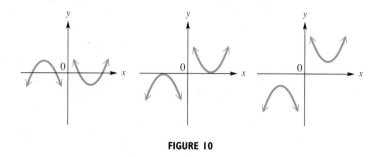

FIGURE 10

Recall from Section 10.3 that the value of $b^2 - 4ac$ is called the *discriminant* of the quadratic equation $ax^2 + bx + c = 0$. It can be used to determine the number of real solutions of a quadratic equation. In a similar way, the discriminant of a quadratic *function* can be used to determine the number of x-intercepts of its graph. If the discriminant is positive, the parabola will have two x-intercepts. If the discriminant is 0, there will be only one x-intercept, and it will be the vertex of the parabola. If the discriminant is negative, the graph will have no x-intercepts.

EXAMPLE 5 Using the Discriminant to Determine the Number of x-Intercepts

Determine the number of x-intercepts of the graph of each quadratic function. Use the discriminant.

(a) $f(x) = 2x^2 + 3x - 5$

The discriminant is $b^2 - 4ac$. Here $a = 2$, $b = 3$, and $c = -5$, so

$$b^2 - 4ac = 9 - 4(2)(-5) = 49.$$

Since the discriminant is positive, the parabola has two x-intercepts.

(b) $f(x) = -3x^2 - 1$

In this equation, $a = -3$, $b = 0$, and $c = -1$. The discriminant is

$$b^2 - 4ac = 0 - 4(-3)(-1) = -12.$$

The discriminant is negative, so the graph has no x-intercepts.

(c) $f(x) = 9x^2 + 6x + 1$

Here, $a = 9$, $b = 6$, and $c = 1$. The discriminant is

$$b^2 - 4ac = 36 - 4(9)(1) = 0.$$

The parabola has only one x-intercept (its vertex) because the value of the discriminant is 0.

■

◀◀ WORK PROBLEM 5 AT THE SIDE.

OBJECTIVE 4 ▶ As we have seen, the vertex of a parabola is either the highest or the lowest point on the parabola. The y-value of the vertex gives the maximum or minimum value of y, while the x-value tells where that maximum or minimum occurs. In many practical problems we want to know the largest or smallest value of some quantity. When that quantity can be expressed as a quadratic function with $y = ax^2 + bx + c$, as in the next example, the vertex can be used to find the desired value.

ANSWERS
5. (a) discriminant is 0; one x-intercept
(b) discriminant is -31;
no x-intercepts
(c) discriminant is 25;
two x-intercepts

E X A M P L E   6   Finding the Maximum Area of a Rectangular Region

6. Solve Example 6 if the farmer has only 100 feet of fencing.

A farmer has 120 feet of fencing. He wants to put a fence around a rectangular field next to a river. Find the maximum area he can enclose.

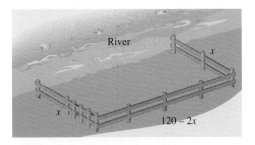

FIGURE 11

Figure 11 shows the field. Let x represent the width of the field. Then, since there are 120 feet of fencing,

$$x + x + \text{length} = 120 \qquad \text{Sum of the sides is 120 feet.}$$
$$2x + \text{length} = 120 \qquad \text{Combine terms.}$$
$$\text{length} = 120 - 2x. \qquad \text{Subtract } 2x.$$

The area is given by the product of the length and width, or

$$A = x(120 - 2x) = 120x - 2x^2.$$

To make the area (and thus $120x - 2x^2$) as large as possible, first find the vertex of the parabola $A = 120x - 2x^2$. Writing the equation in standard form as $A = -2x^2 + 120x$ shows that $a = -2$, $b = 120$, and $c = 0$, so

$$h = -\frac{b}{2a} = -\frac{120}{2(-2)} = -\frac{120}{-4} = 30$$
$$k = f(30) = -2(30)^2 + 120(30) = -2(900) + 3600 = 1800.$$

The graph is a parabola that opens downward, and its vertex is $(30, 1800)$. The vertex of the graph shows that the maximum area will be 1800 square feet. This area will occur if x, the width of the field, is 30 feet.

WORK PROBLEM 6 AT THE SIDE. ▶▶

OBJECTIVE 5   If x and y are exchanged in the equation $y = ax^2 + bx + c$, the equation becomes $x = ay^2 + by + c$. Because of the interchange of the roles of x and y, these parabolas are horizontal (with horizontal lines as axes), compared with the vertical ones graphed previously.

E X A M P L E   7   Graphing a Horizontal Parabola

Graph $x = (y - 2)^2 - 3$.

This graph has its vertex at $(-3, 2)$, because the roles of x and y are reversed. It opens to the right, the positive x-direction, and has the same shape as $y = x^2$. Plotting a few additional points gives the graph shown in Figure 12.

— CONTINUED ON NEXT PAGE

ANSWERS

6. The field should be 25 feet by 50 feet with a maximum area of 1250 square feet.

7. Graph $x = (y + 1)^2 - 4$.

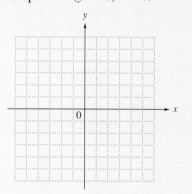

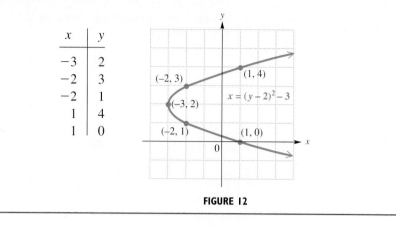

FIGURE 12

Horizontal Parabola

The graph of $x = a(y - k)^2 + h$ is a horizontal parabola with vertex at (h, k) and the horizontal line $y = k$ as axis. The graph opens to the right if a is positive and to the left if a is negative.

◄◄ WORK PROBLEM 7 AT THE SIDE.

When a quadratic equation is given in the form $x = ay^2 + by + c$, completing the square on y will allow us to find the vertex.

EXAMPLE 8   Completing the Square to Graph a Horizontal Parabola

Graph $x = -2y^2 + 4y - 3$. Give the domain and the range of the relation.

$$
\begin{aligned}
x &= -2y^2 + 4y - 3 \\
&= -2(y^2 - 2y) - 3 && \text{Factor out } -2. \\
&= -2(y^2 - 2y + 1 - 1) - 3 && \text{Add } 0\ (1 - 1 = 0). \\
&= -2(y^2 - 2y + 1) + 2 - 3 && \text{Distributive property} \\
x &= -2(y - 1)^2 - 1 && \text{Factor.}
\end{aligned}
$$

Because of the negative coefficient (-2), the graph opens to the left (the negative x direction) and is narrower than $y = x^2$. As shown in Figure 13, the vertex is $(-1, 1)$. The domain is $(-\infty, -1]$ and the range is $(-\infty, \infty)$.

8. Find the vertex of each parabola. Tell whether the graph opens to the right or to the left. Give the domain and the range.

(a) $x = 2y^2 - 6y + 5$

(b) $x = -y^2 + 2y + 5$

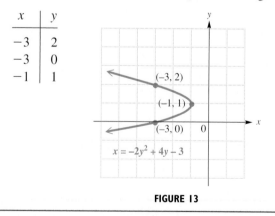

FIGURE 13

◄◄ WORK PROBLEM 8 AT THE SIDE.

Caution

Only quadratic equations that are solved for y define functions. The graphs of the equations in Examples 7 and 8 are not graphs of functions. They do not satisfy the conditions of the vertical line test.

ANSWERS

7.

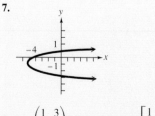

8. (a) $\left(\frac{1}{2}, \frac{3}{2}\right)$; right; domain: $\left[\frac{1}{2}, \infty\right)$; range: $(-\infty, \infty)$ **(b)** $(6, 1)$; left; domain: $(-\infty, 6]$; range: $(-\infty, \infty)$

11.2 Exercises

1. How can you determine just by looking at the equation of a parabola whether it has a vertical or a horizontal axis?

2. Why can't the graph of a quadratic function be a horizontal parabola?

3. How can you determine the number of x-intercepts of the graph of a quadratic function without graphing the function?

4. If the vertex of the graph of a quadratic function is $(1, -3)$, and the graph opens downward, how many x-intercepts does the graph have?

Find the vertex of each parabola. For each equation, decide whether the graph opens upward, downward, to the left, or to the right; and whether it is wider, narrower, or the same shape as the graph of $y = x^2$. If it is a vertical parabola, use the discriminant to determine the number of x-intercepts. See Examples 1–3, 5, 7, and 8.

5. $y = 2x^2 + 4x + 5$

6. $y = 3x^2 - 6x + 4$

7. $y = -x^2 + 5x + 3$

8. $x = -y^2 + 7y - 2$

9. $x = \dfrac{1}{3}y^2 + 6y + 24$

10. $x = .5y^2 + 10y - 5$

Graph each parabola using the techniques described in this section. In Exercises 11–14 give the domain and range. See Examples 4, 7, and 8.

11. $f(x) = x^2 + 4x + 3$

12. $f(x) = x^2 + 2x - 2$

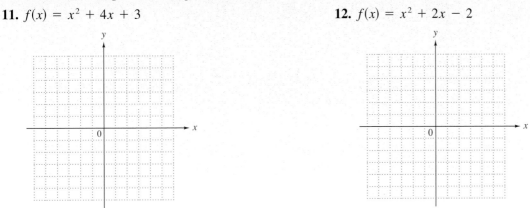

13. $f(x) = -2x^2 + 4x - 5$

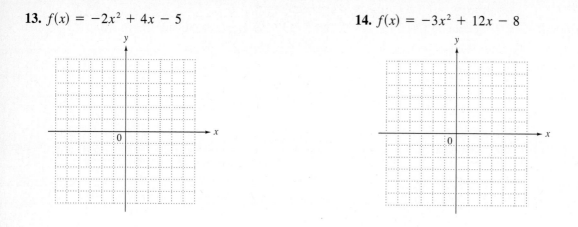

14. $f(x) = -3x^2 + 12x - 8$

15. $x = -\dfrac{1}{5}y^2 + 2y - 4$

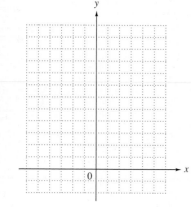

16. $x = -.5y^2 - 4y - 6$

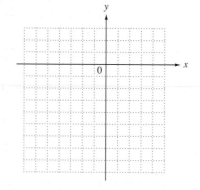

17. $x = 3y^2 + 12y + 5$

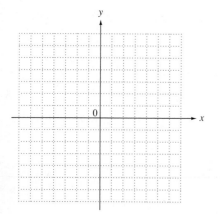

18. $x = 4y^2 + 16y + 11$

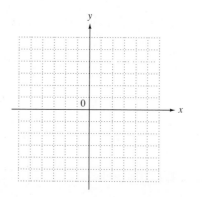

INTERPRETING TECHNOLOGY (EXERCISES 19–22)

Graphing calculators are capable of determining the coordinates of "peaks" and "valleys" of graphs. In the case of quadratic functions, these peaks and valleys are the vertices, and are called maximum and minimum points. For example, the vertex of the graph of
$f(x) = -x^2 - 6x - 13$ *is* $(-3, -4)$, *as indicated in the display at the bottom of the screen. In this case, the vertex is a maximum point.*

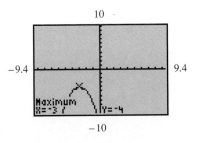

In Exercises 19–22, match the function with its calculator-generated graph by determining the vertex and using the display at the bottom of the screen.

19. $f(x) = x^2 - 8x + 18$

20. $f(x) = x^2 + 8x + 18$

21. $f(x) = x^2 - 8x + 14$

22. $f(x) = x^2 + 8x + 14$

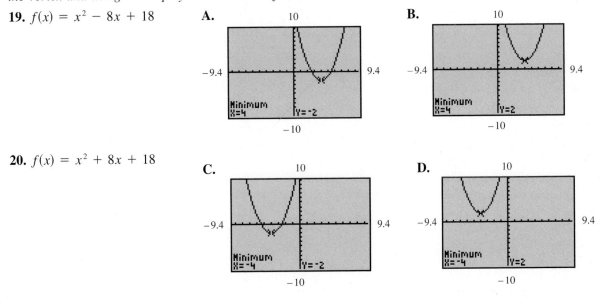

Solve each of the following applied problems. See Example 6.

23. Palo Alto College is planning to construct a rectangular parking lot on land bordered on one side by a highway. The plan is to use 640 feet of fencing to fence off the other three sides. What should the dimensions of the lot be if the enclosed area is to be a maximum?

24. Martha Wacaser has 100 meters of fencing material to enclose a rectangular exercise run for her dog. What width will give the enclosure the maximum area?

25. If an object is thrown upward with an initial velocity of 32 feet per second, then its height (in feet) after t seconds is given by

$$h(t) = 32t - 16t^2.$$

Find the maximum height attained by the object and the number of seconds it takes to hit the ground.

26. A projectile is fired straight upward so that its distance (in feet) above the ground t seconds after firing is

$$s(t) = -16t^2 + 400t.$$

Find the maximum height it reaches and the number of seconds it takes to reach that height.

27. Find the pair of numbers with a sum of 60 whose product is the maximum. (*Hint:* Let x and $60 - x$ represent the two numbers.)

28. For a trip to a resort, a charter bus company charges a fare of $48 per person, plus $2 per person for each unsold seat on the bus. If the bus has 42 seats and x represents the number of unsold seats, find the following:

(a) a function defined by $R(x)$ that describes the total revenue from the trip (*Hint:* Multiply the total number riding, $42 - x$, by the price per ticket, $48 + 2x$);

(b) the number of unsold seats that produces the maximum revenue;

(c) the maximum revenue.

The accompanying bar graph shows the annual average number of nonfarm payroll jobs in California for the years 1988 through 1992. If the tops of the bars were joined by a smooth curve, the curve would resemble the graph of a quadratic function (that is, a parabola). Using a technique from statistics it can be determined that this function can be described approximately as

$$f(x) = -.10x^2 + .42x + 11.90,$$

where x = 0 corresponds to 1988, x = 1 corresponds to 1989, and so on, and f(x) represents the number of payroll jobs in millions. (Source: California Employment Development Department)

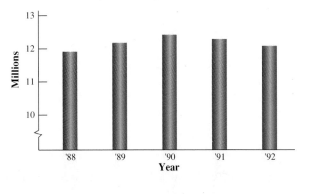

NONFARM PAYROLL JOBS IN CALIFORNIA

29. Explain why the coefficient of x^2 in the function is negative, based on the graph formed by joining the tops of the bars.

30. Determine the coordinates of the vertex of the graph using algebraic methods.

31. How does the x-coordinate of the vertex of the parabola indicate that during the time period under consideration, the maximum number of payroll jobs was in 1990?

32. What does the y-coordinate of the vertex of the parabola indicate?

MATHEMATICAL CONNECTIONS (EXERCISES 33–36)

In Example 1 of Section 10.6, we determined the solution set of the quadratic inequality $x^2 - x - 12 > 0$ by using regions on a number line and testing values in the inequality. If we graph $f(x) = x^2 - x - 12$, the x-intercepts will determine the solutions of the quadratic equation $x^2 - x - 12 = 0$. The solution set is $\{-3, 4\}$. The x-values of the points on the graph that are above *the x-axis form the solution set of $x^2 - x - 12 > 0$. As seen in the figure, this solution set is $(-\infty, -3) \cup (4, \infty)$, which supports the result found in Section 10.6. Similarly, the solution set of the quadratic inequality $x^2 - x - 12 < 0$ is found by locating the points on the graph that lie* below *the x-axis. Those x-values belong to the open interval $(-3, 4)$.*

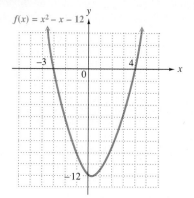

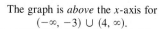

The graph is *above* the x-axis for
$(-\infty, -3) \cup (4, \infty)$.

In Exercises 33–36, the graph of a quadratic function f is given. Use only the graph to find the solution set of the equation or inequality. Work through parts (a)–(c) in order each time.

33. $f(x) = x^2 - 4x + 3$

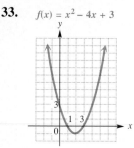

(a) $x^2 - 4x + 3 = 0$
(b) $x^2 - 4x + 3 > 0$
(c) $x^2 - 4x + 3 < 0$

34. $f(x) = 3x^2 + 10x - 8$

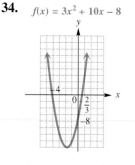

(a) $3x^2 + 10x - 8 = 0$
(b) $3x^2 + 10x - 8 \geq 0$
(c) $3x^2 + 10x - 8 < 0$

35. $f(x) = -x^2 + 3x + 10$

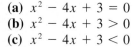

(a) $-x^2 + 3x + 10 = 0$
(b) $-x^2 + 3x + 10 \geq 0$
(c) $-x^2 + 3x + 10 \leq 0$

36. $f(x) = -2x^2 - x + 15$

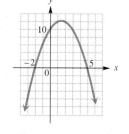

(a) $-2x^2 - x + 15 = 0$
(b) $-2x^2 - x + 15 \geq 0$
(c) $-2x^2 - x + 15 \leq 0$

11.3 *Graphs of Elementary Functions and Circles*

In the first two sections of this chapter, we introduced an important function, $f(x) = x^2$. This quadratic function is sometimes called the **squaring function,** and it is one of the most important elementary functions in algebra.

OBJECTIVE 1 Three other elementary functions are those defined by $|x|$, $\frac{1}{x}$, and $\sqrt{x}$. The first of these, with $f(x) = |x|$, is called the **absolute value function.** Its graph, along with a table of selected ordered pairs, is shown in Figure 14. Its domain is $(-\infty, \infty)$ and its range is $[0, \infty)$.

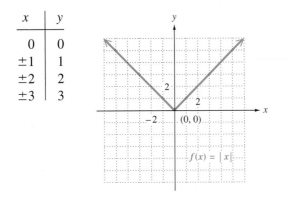

x	y
0	0
±1	1
±2	2
±3	3

FIGURE 14

Another important elementary function is the **reciprocal function,** defined by $f(x) = \frac{1}{x}$. Its graph is shown in Figure 15, along with a table of selected ordered pairs. Notice that x can never equal zero for this function, and as a result, as x gets closer and closer to 0, the graph either approaches ∞ or $-\infty$. Also, $\frac{1}{x}$ can never equal 0, and as x approaches ∞ or $-\infty$, $\frac{1}{x}$ approaches 0. The axes are called **asymptotes** (ASS-im-tohts) for the function. (Asymptotes are studied in more detail in college algebra courses.) For the reciprocal function, the domain and the range are both $(-\infty, 0) \cup (0, \infty)$.

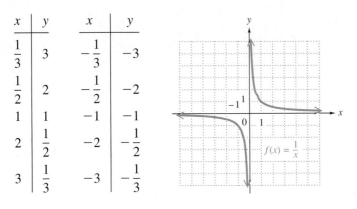

x	y	x	y
$\frac{1}{3}$	3	$-\frac{1}{3}$	-3
$\frac{1}{2}$	2	$-\frac{1}{2}$	-2
1	1	-1	-1
2	$\frac{1}{2}$	-2	$-\frac{1}{2}$
3	$\frac{1}{3}$	-3	$-\frac{1}{3}$

FIGURE 15

The function with $f(x) = \sqrt{x}$ is called the **square root function.** Its graph is shown in Figure 16. Notice that since we restrict function values to be real numbers, x cannot take on negative values. Thus, the domain of the square root function is $[0, \infty)$. Because the principal square root is always nonnegative, the range is also $[0, \infty)$. A partial table of values is shown along with the graph.

x	y
0	0
1	1
4	2

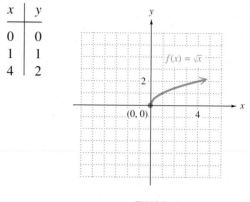

FIGURE 16

Just as the graph of $f(x) = x^2$ can be translated, as seen in Section 11.1, so can the graphs of these other elementary functions. Example 1 shows how this is done.

EXAMPLE 1 Graphing Translations of Other Elementary Functions

Sketch the graph of each function, using translations as discussed in Section 11.1.

(a) $f(x) = |x - 2|$

The graph of $y = (x - 2)^2$ is obtained by shifting the graph of $y = x^2$ two units to the right. In a similar manner, the graph of $f(x) = |x - 2|$ is found by shifting the graph of $y = |x|$ two units to the right, as shown in Figure 17. The table of ordered pairs accompanying the graph supports this.

x	y
0	2
1	1
2	0
3	1
4	2

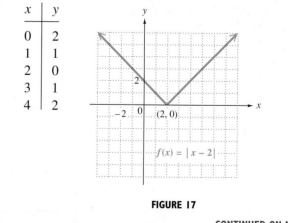

FIGURE 17

CONTINUED ON NEXT PAGE

(b) $f(x) = \dfrac{1}{x} + 3$

The graph of this function is found by translating the graph of $y = \dfrac{1}{x}$ three units upward. See Figure 18.

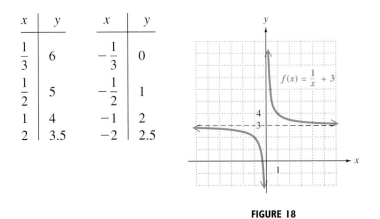

x	y	x	y
$\frac{1}{3}$	6	$-\frac{1}{3}$	0
$\frac{1}{2}$	5	$-\frac{1}{2}$	1
1	4	-1	2
2	3.5	-2	2.5

FIGURE 18

(c) $f(x) = \sqrt{x + 1} - 4$

The graph of $y = (x + 1)^2 - 4$ is obtained by shifting the graph of $y = x^2$ one unit to the left and four units downward. Following this pattern, we shift the graph of $y = \sqrt{x}$ one unit to the left and four units downward to get the graph of $f(x) = \sqrt{x + 1} - 4$. See Figure 19.

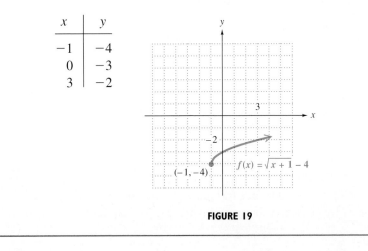

x	y
-1	-4
0	-3
3	-2

FIGURE 19

WORK PROBLEM 1 AT THE SIDE. ▶▶

OBJECTIVE 2 Using the Pythagorean formula, it can be shown that the distance d between the points (x_1, y_1) and (x_2, y_2) is given by the formula

$$d = \sqrt{(x_2 - x_1)^2 + (y_2 - y_1)^2}.$$

We can use this formula to determine the equation of a circle. Notice that a circle is not the graph of a function; it does not pass the vertical line test.

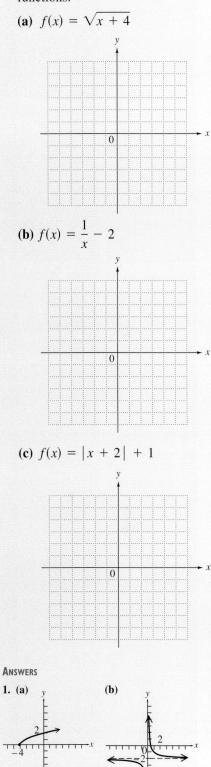

1. Graph each of the following functions.

(a) $f(x) = \sqrt{x + 4}$

(b) $f(x) = \dfrac{1}{x} - 2$

(c) $f(x) = |x + 2| + 1$

ANSWERS

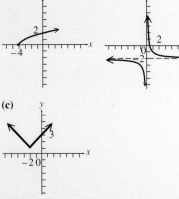

1. (a)   **(b)**

(c)

2. Find the equation of the circle with radius 4 and center $(0, 0)$. Sketch its graph.

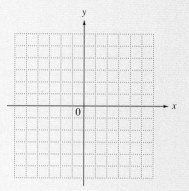

A **circle** is the set of all points in a plane that lie a fixed distance from a fixed point. The fixed point is called the **center** and the fixed distance is called the **radius.**

EXAMPLE 2 Finding the Equation of a Circle and Graphing It

Find the equation of the circle with radius 3 and center at $(0, 0)$. Draw the graph.

If the point (x, y) is on the circle, the distance from (x, y) to the center $(0, 0)$ is 3. By the distance formula,

$$\sqrt{(x_2 - x_1)^2 + (y_2 - y_1)^2} = d$$
$$\sqrt{(x - 0)^2 + (y - 0)^2} = 3$$
$$x^2 + y^2 = 9. \quad \text{Square both sides.}$$

The equation of this circle is $x^2 + y^2 = 9$. Its graph is given in Figure 20.

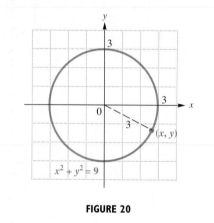

FIGURE 20

◀◀ **WORK PROBLEM 2 AT THE SIDE.**

EXAMPLE 3 Graphing a Circle Given Its Center and Radius

Find an equation for the circle that has its center at $(4, -3)$ and radius 5, and draw the graph.

Again, use the distance formula with the points (x, y) and $(4, -3)$. The graph of this circle is shown in Figure 21. The equation is

$$\sqrt{(x - 4)^2 + (y + 3)^2} = 5$$
$$(x - 4)^2 + (y + 3)^2 = 25.$$

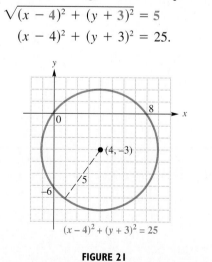

FIGURE 21

ANSWERS

2. $x^2 + y^2 = 16$

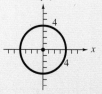

The results of Examples 2 and 3 can be generalized to derive an equation of a circle with radius r and center at (h, k). If (x, y) is a point on the circle, the distance from the center (h, k) to the point (x, y) is r. Then by the distance formula with $d = r$, $x_1 = x$, $x_2 = h$, $y_1 = y$, and $y_2 = k$,

$$\sqrt{(x - h)^2 + (y - k)^2} = r.$$

Squaring both sides of the equation gives the following result.

Equation of a Circle

The circle with radius r and center at (h, k) has an equation of the form

$$(x - h)^2 + (y - k)^2 = r^2.$$

WORK PROBLEMS 3 AND 4 AT THE SIDE. ▶▶

OBJECTIVE ▸**3**▸ In the equation found in Example 3, multiplying out $(x - 4)^2$ and $(y + 3)^2$ and then combining like terms gives

$$(x - 4)^2 + (y + 3)^2 = 25$$
$$x^2 - 8x + 16 + y^2 + 6y + 9 = 25$$
$$x^2 + y^2 - 8x + 6y = 0.$$

This result suggests that an equation that has both x^2 and y^2 terms with the same coefficient may represent a circle. In many cases it does, and the next example shows how to determine the center and radius of the circle.

E X A M P L E 4   Finding the Center and Radius of a Circle

Find the center and radius of the circle whose equation is

$$x^2 + y^2 + 2x + 4y - 4 = 0.$$

Since the equation has x^2 and y^2 terms with equal coefficients, its graph might be that of a circle. To find the center and radius, we complete the square on x and the square on y as follows. Keep only the terms with the variables on the left side and group the x terms and the y terms.

$$(x^2 + 2x \quad) + (y^2 + 4y \quad) = 4$$

Add the appropriate constants to complete both squares on the left.

$$(x^2 + 2x + 1 - 1) + (y^2 + 4y + 4 - 4) = 4 \qquad \text{Add 0 twice.}$$
$$(x^2 + 2x + 1) - 1 + (y^2 + 4y + 4) - 4 = 4 \qquad \text{Associative property}$$
$$(x^2 + 2x + 1) + (y^2 + 4y + 4) = 4 + 5 \qquad \text{Add 5 on both sides.}$$
$$(x + 1)^2 + (y + 2)^2 = 9 \qquad \text{Factor.}$$

The last equation shows that the center of the circle is $(-1, -2)$ and the radius is 3.

Caution

If the procedure of Example 4 leads to an equation of the form $(x - h)^2 + (y - k)^2 = 0$, the graph is the single point (h, k). If the constant on the right side is negative, the equation has no graph.

---

3. Find the equation of the circle with center at $(3, -2)$ and radius 4. Graph the circle.

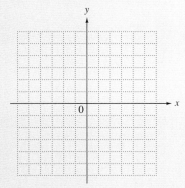

4. Determine the center and radius of $(x - 5)^2 + (y + 2)^2 = 9$ and graph the circle.

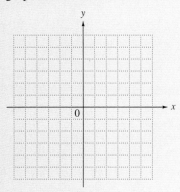

ANSWERS

3. $(x - 3)^2 + (y + 2)^2 = 16$

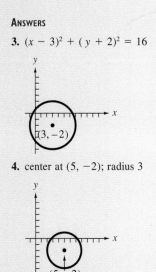

4. center at $(5, -2)$; radius 3

5. Find the center and radius of the circle with equation

$$x^2 + y^2 - 6x + 8y - 11 = 0.$$

◀◀ **WORK PROBLEM 5 AT THE SIDE.**

OBJECTIVE 4 Earlier in this section we examined the graph of the square root function with $f(x) = \sqrt{x}$. By replacing x with more complicated expressions, other types of functions can be obtained. In particular, if we let the radicand be of the form $k - x^2$, where k is a positive constant, the graph is a semicircle.

E X A M P L E   5   Graphing a Semicircle

Graph $f(x) = \sqrt{25 - x^2}$.

Replace $f(x)$ with y and square both sides to get the equation

$$y^2 = 25 - x^2, \quad \text{or} \quad x^2 + y^2 = 25.$$

This is the graph of a circle with center at $(0, 0)$ and radius 5. Since $f(x)$, or y, represents a principal square root in the original equation, $f(x)$ must be nonnegative. This restricts the graph to the upper half of the circle, as shown in Figure 22. Use the graph and the vertical line test to verify that it is indeed a function.

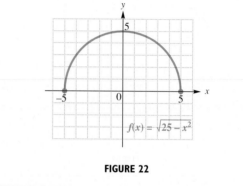

FIGURE 22

6. Graph the function with $f(x) = \sqrt{36 - x^2}$.

◀◀ **WORK PROBLEM 6 AT THE SIDE.**

ANSWERS

5. center at $(3, -4)$; radius 6

6.

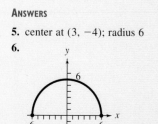

11.3 Exercises

Fill in the blank with the correct response.

1. For the reciprocal function $f(x) = \dfrac{1}{x}$,

   _____ is the only real number not in the domain.

2. The range of the square root function, $f(x) = \sqrt{x}$, is _____.

3. The lowest point on the graph of $f(x) = |x|$ has coordinates (_____ , _____).

4. The vertical line with equation $x =$ _____ is the axis of symmetry of the graph of $f(x) = |x|$, since the left and right halves of the graph are mirror images of each other across this line.

Without actually plotting points, match the function defined by the absolute value expression with its graph. See Example 1.

5. $f(x) = |x - 2| + 2$

6. $f(x) = |x + 2| + 2$

7. $f(x) = |x - 2| - 2$

8. $f(x) = |x + 2| - 2$

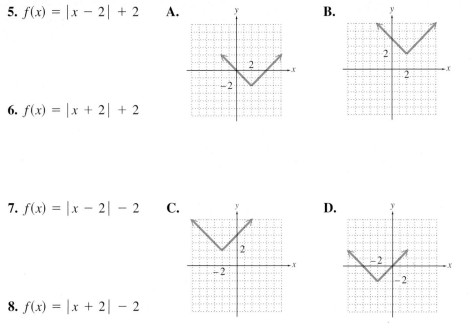

Graph each of the following functions. See Example 1.

9. $f(x) = |x + 1|$

10. $f(x) = |x - 1|$

11. $f(x) = \dfrac{1}{x} + 1$

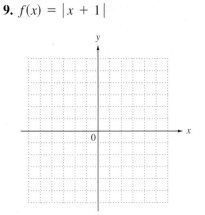

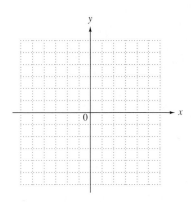

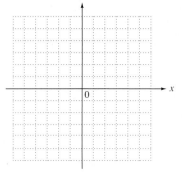

12. $f(x) = \dfrac{1}{x} - 1$

13. $f(x) = \sqrt{x - 2}$

14. $f(x) = \sqrt{x + 5}$

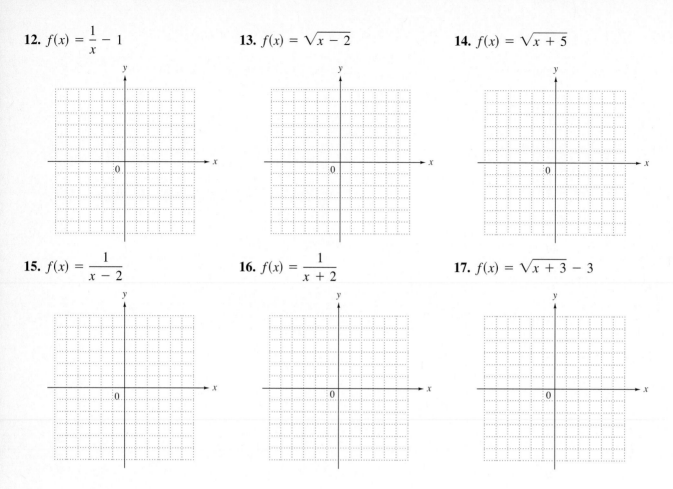

15. $f(x) = \dfrac{1}{x - 2}$

16. $f(x) = \dfrac{1}{x + 2}$

17. $f(x) = \sqrt{x + 3} - 3$

18. What is the center of a circle with equation $x^2 + y^2 = r^2$ $(r > 0)$?

Match the equation with the correct graph. See Example 3.

19. $(x - 3)^2 + (y - 2)^2 = 25$

A.

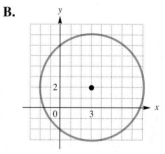

B.

20. $(x - 3)^2 + (y + 2)^2 = 25$

21. $(x + 3)^2 + (y - 2)^2 = 25$

C.

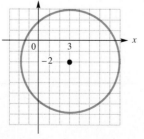

D.

22. $(x + 3)^2 + (y + 2)^2 = 25$

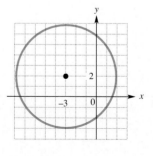

23. Use the distance formula to find the distance between the given pair of points.
   (a) $(2, 3)$ and $(5, 7)$ **(b)** $(-1, 4)$ and $(5, 6)$
   (c) $(0, 5)$ and $(5, 0)$ **(d)** $(0, -8)$ and $(-8, 0)$

Find the center and radius of each circle. (Hint: In Exercises 28 and 29, divide both sides by a common factor.) See Example 4.

24. $x^2 + y^2 + 4x + 6y + 9 = 0$

25. $x^2 + y^2 - 8x - 12y + 3 = 0$

26. $x^2 + y^2 + 10x - 14y - 7 = 0$

27. $x^2 + y^2 - 2x + 4y - 4 = 0$

28. $3x^2 + 3y^2 - 12x - 24y + 12 = 0$

29. $2x^2 + 2y^2 + 20x + 16y + 10 = 0$

30. A circle can be drawn on a piece of poster-board by fastening one end of a string with a thumbtack, pulling the string taut with a pencil, and tracing a curve as shown in the figure. Explain why this method works.

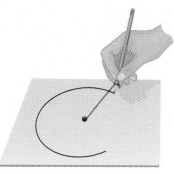

Graph the following. See Examples 2–4.

31. $x^2 + y^2 = 9$

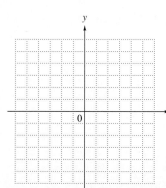

32. $x^2 + y^2 = 4$

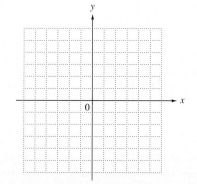

33. $2y^2 = 10 - 2x^2$

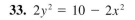

34. $3x^2 = 48 - 3y^2$

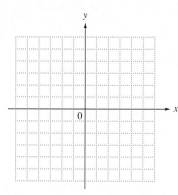

35. $(x + 3)^2 + (y - 2)^2 = 9$

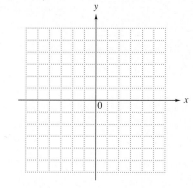

36. $(x - 1)^2 + (y + 3)^2 = 16$

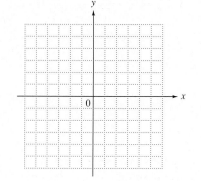

37. $x^2 + y^2 - 4x - 6y + 9 = 0$

38. $x^2 + y^2 + 8x + 2y - 8 = 0$

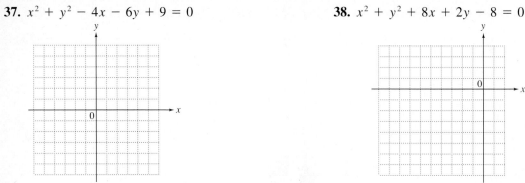

Graph each function involving a square root. See Example 5.

39. $f(x) = \sqrt{16 - x^2}$

40. $f(x) = \sqrt{9 - x^2}$

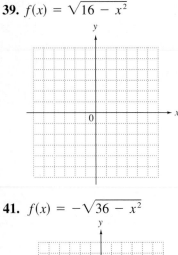

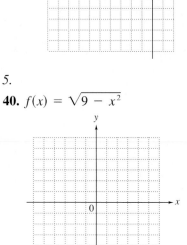

41. $f(x) = -\sqrt{36 - x^2}$

42. $f(x) = -\sqrt{25 - x^2}$

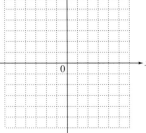

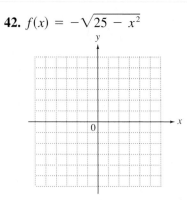

43. A parabola can be defined as the set of all points in a plane equally distant from a given point and a given line not containing the point. See the graph.

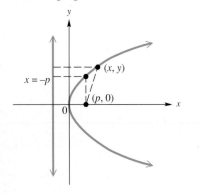

(a) Suppose (x, y) is to be on the parabola. Suppose the line mentioned in the definition is given by $x = -p$. Find the distance between (x, y) and the line. (The distance from a point to a line is the length of the perpendicular from the point to the line.)

(b) If $x = -p$ is the line mentioned in the definition, why should the given point have coordinates $(p, 0)$? (*Hint:* See the graph.)

(c) Find an expression for the distance from (x, y) to $(p, 0)$.

(d) Find an equation for the parabola in the graph. (*Hint:* Use the results of parts (a) and (c) and the fact that (x, y) is equally distant from the point and the line.)

11.4 Ellipses and Hyperbolas

OBJECTIVE 1 An **ellipse** (ee-LIPS) is the set of all points in a plane the sum of whose distances from two fixed points is constant. These fixed points are called **foci** (singular: *focus*). Figure 23 shows an ellipse centered at the origin, with foci at $(c, 0)$ and $(-c, 0)$, x-intercepts $(a, 0)$ and $(-a, 0)$, and y-intercepts $(0, b)$ and $(0, -b)$. From the definition above, it can be shown by the distance formula that an ellipse has the following equation.

Equation of an Ellipse

The ellipse whose x-intercepts are $(a, 0)$ and $(-a, 0)$ and whose y-intercepts are $(0, b)$ and $(0, -b)$ has an equation of the form

$$\frac{x^2}{a^2} + \frac{y^2}{b^2} = 1.$$

Note that a circle is a special case of an ellipse, where $a^2 = b^2$.

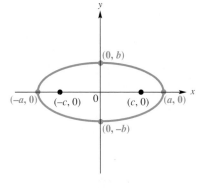

FIGURE 23

The paths of the earth and other planets around the sun are approximately ellipses; the sun is at one focus and a point in space is at the other. The orbits of communication satellites and other space vehicles are elliptical. Elliptical bicycle gears are designed to respond to the legs' natural strengths and weaknesses. At the top and bottom of the powerstroke, where the legs have the least leverage, the gear offers little resistance, but as the gear rotates, the resistance increases. This allows the legs to apply more power where it is most naturally available. See Figure 24.

FIGURE 24

OBJECTIVE 2 To graph an ellipse, plot the four intercepts $(a, 0)$, $(-a, 0)$, $(0, b)$, and $(0, -b)$, and sketch an ellipse through the intercepts.

EXAMPLE 1 Graphing an Ellipse

Graph $\dfrac{x^2}{49} + \dfrac{y^2}{36} = 1$.

The x-intercepts of this ellipse are $(7, 0)$ and $(-7, 0)$. The y-intercepts are $(0, 6)$ and $(0, -6)$. Additional points can be found by choosing a value

CONTINUED ON NEXT PAGE

OBJECTIVES

1. Recognize the equation of an ellipse.
2. Graph ellipses.
3. Recognize the equation of a hyperbola.
4. Graph hyperbolas by using the asymptotes.
5. Identify conic sections by name from their equations.
6. Graph portions of conic sections defined by functions involving square roots.

FOR EXTRA HELP

Tutorial   Tape 23   SSM, Sec. 11.4

for x (or y) and substituting into the equation to find the corresponding value for y (or x). For example, let's choose $y = 3$.

$$\frac{x^2}{49} + \frac{y^2}{36} = 1$$

$$\frac{x^2}{49} + \frac{9}{36} = 1 \qquad \text{If } y = 3, \text{ then } y^2 = 9.$$

$$\frac{x^2}{49} + \frac{1}{4} = 1 \qquad \text{Reduce.}$$

$$196\left(\frac{x^2}{49}\right) + 196\left(\frac{1}{4}\right) = 196 \qquad \text{Multiply by } 49 \cdot 4 = 196.$$

$$4x^2 + 49 = 196$$

$$4x^2 = 147 \qquad \text{Subtract 49.}$$

$$x^2 = 36.75 \qquad \text{Divide by 4.}$$

$$x \approx \pm 6.1$$

Thus, the points $(6.1, 3)$ and $(-6.1, 3)$ are on the graph. Choosing $y = -3$ leads to the same x-values so $(6.1, -3)$ and $(-6.1, -3)$ are also on the graph. Plotting these points and the intercepts and sketching the ellipse through them gives the graph in Figure 25.

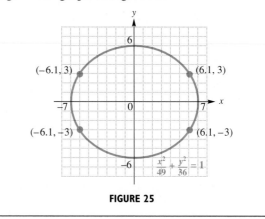

FIGURE 25

EXAMPLE 2 Graphing an Ellipse

Graph $\dfrac{x^2}{36} + \dfrac{y^2}{121} = 1$.

The x-intercepts for this ellipse are $(6, 0)$ and $(-6, 0)$, and the y-intercepts are $(0, 11)$ and $(0, -11)$. Additional points could be found, as shown in Example 1. The graph has been sketched in Figure 26.

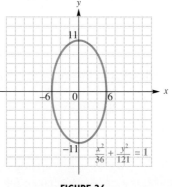

FIGURE 26

WORK PROBLEM I AT THE SIDE. ▶▶

1. Graph.

(a) $\dfrac{x^2}{4} + \dfrac{y^2}{25} = 1$

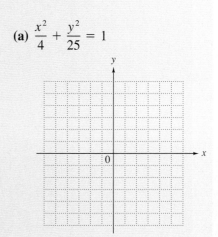

OBJECTIVE 3 A **hyperbola** (hy-PUR-buh-luh) is the set of all points in a plane the *difference* of whose distances from two foci is constant. Figure 27 shows a hyperbola; it can be shown, using this definition and the distance formula, that this hyperbola has equation

$$\frac{x^2}{16} - \frac{y^2}{12} = 1.$$

The x-intercepts are $(4, 0)$ and $(-4, 0)$. When $x = 0$ the equation becomes

$$-\frac{y^2}{12} = 1 \quad \text{or} \quad y^2 = -12.$$

This equation has no real number solutions, so there is no real number value for y corresponding to $x = 0$; that is, there are no y-intercepts.

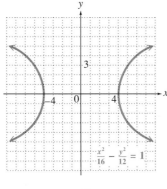

FIGURE 27

Figure 28 shows the graph of the hyperbola

$$\frac{y^2}{25} - \frac{x^2}{9} = 1.$$

Here the y-intercepts are $(0, 5)$ and $(0, -5)$, and there are no x-intercepts.

(b) $\dfrac{x^2}{64} + \dfrac{y^2}{49} = 1$

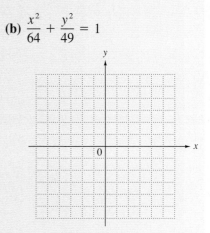

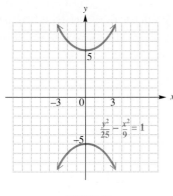

FIGURE 28

ANSWERS

1. (a)

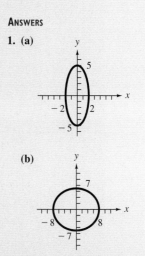

(b)

This discussion about hyperbolas is summarized as follows.

Equations of Hyperbolas

A hyperbola with x-intercepts $(a, 0)$ and $(-a, 0)$ has an equation of the form

$$\frac{x^2}{a^2} - \frac{y^2}{b^2} = 1,$$

and a hyperbola with y-intercepts $(0, b)$ and $(0, -b)$ has an equation of the form

$$\frac{y^2}{b^2} - \frac{x^2}{a^2} = 1.$$

A cross section of a large microwave antenna system consists of a parabola and a hyperbola, with the focus of the parabola coinciding with one focus of the hyperbola. See Figure 29. The incoming microwaves that are parallel to the axis of the parabola are reflected from the parabola up toward the hyperbola and back to the other focus of the hyperbola, where the cone of the antenna is located to capture the signal.

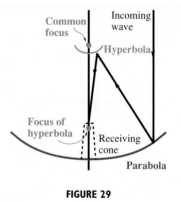

FIGURE 29

OBJECTIVE 4 The two branches of the graph of a hyperbola approach a pair of intersecting straight lines. As mentioned in the previous section when the graph of $f(x) = \dfrac{1}{x}$ was studied, these lines are called *asymptotes*. (See Figure 30.) These lines are useful for sketching the graph of the hyperbola. We find the asymptotes as follows.

Asymptotes of Hyperbolas

The extended diagonals of the rectangle with corners at the points (a, b), $(-a, b)$, $(-a, -b)$, and $(a, -b)$ are the asymptotes of either of the hyperbolas

$$\frac{x^2}{a^2} - \frac{y^2}{b^2} = 1 \quad \text{or} \quad \frac{y^2}{b^2} - \frac{x^2}{a^2} = 1.$$

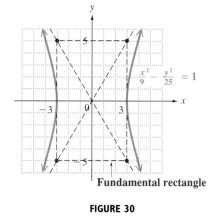

FIGURE 30

This rectangle is called the **fundamental rectangle.**

E X A M P L E 3 Graphing a Hyperbola

Graph $\dfrac{x^2}{9} - \dfrac{y^2}{25} = 1$.

Here $a = 3$ and $b = 5$. The x-intercepts are $(3, 0)$ and $(-3, 0)$. The four points $(3, 5)$, $(3, -5)$, $(-3, 5)$, and $(-3, -5)$ are the corners of the fundamental rectangle that determines the asymptotes shown in Figure 30. The hyperbola approaches these lines as x and y get larger and larger in absolute value.

In summary, to graph either of the two forms of hyperbolas,

$$\frac{x^2}{a^2} - \frac{y^2}{b^2} = 1 \quad \text{or} \quad \frac{y^2}{b^2} - \frac{x^2}{a^2} = 1,$$

follow these steps.

Graphing a Hyperbola

Step 1 Locate the intercepts. They are at $(a, 0)$ and $(-a, 0)$ if the x^2 term has a positive coefficient, or at $(0, b)$ and $(0, -b)$ if the y^2 term has a positive coefficient.

Step 2 Locate the corners of a rectangle at (a, b), $(a, -b)$, $(-a, -b)$, and $(-a, b)$.

Step 3 Sketch the asymptotes (the extended diagonals of the rectangle).

Step 4 Sketch each branch of the hyperbola through an intercept and approaching the asymptotes.

WORK PROBLEM 2 AT THE SIDE. ▶▶

OBJECTIVE 5▶ By rewriting a second-degree equation in one of the forms given for ellipses, hyperbolas, circles, or parabolas, we can determine when the graph is one of these figures. A summary of the equations and graphs of the conic sections is given here.

---

2. Use asymptotes to graph the following.

(a) $\dfrac{x^2}{4} - \dfrac{y^2}{25} = 1$

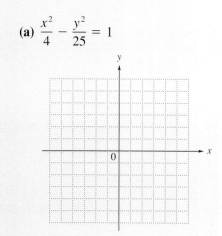

(b) $\dfrac{y^2}{81} - \dfrac{x^2}{64} = 1$

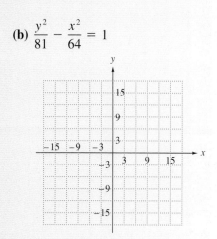

ANSWERS

2. (a)

(b)

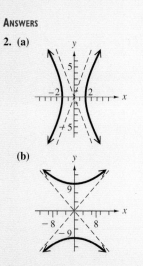

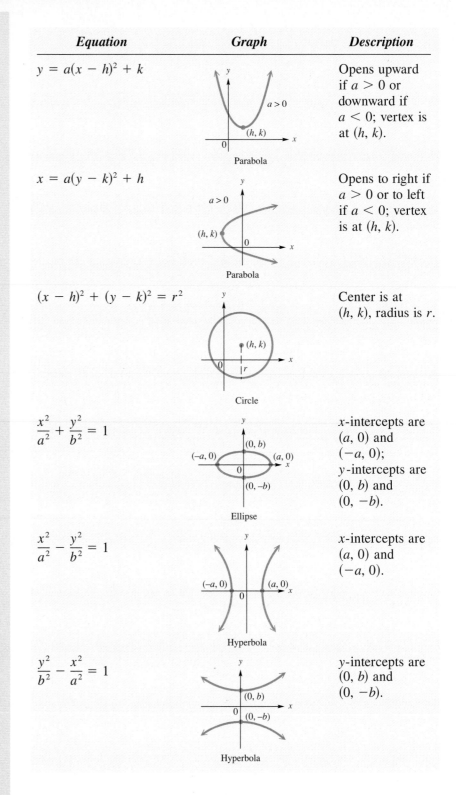

Equation	Graph	Description
$y = a(x - h)^2 + k$	Parabola	Opens upward if $a > 0$ or downward if $a < 0$; vertex is at (h, k).
$x = a(y - k)^2 + h$	Parabola	Opens to right if $a > 0$ or to left if $a < 0$; vertex is at (h, k).
$(x - h)^2 + (y - k)^2 = r^2$	Circle	Center is at (h, k), radius is r.
$\dfrac{x^2}{a^2} + \dfrac{y^2}{b^2} = 1$	Ellipse	x-intercepts are $(a, 0)$ and $(-a, 0)$; y-intercepts are $(0, b)$ and $(0, -b)$.
$\dfrac{x^2}{a^2} - \dfrac{y^2}{b^2} = 1$	Hyperbola	x-intercepts are $(a, 0)$ and $(-a, 0)$.
$\dfrac{y^2}{b^2} - \dfrac{x^2}{a^2} = 1$	Hyperbola	y-intercepts are $(0, b)$ and $(0, -b)$.

E X A M P L E   4   Identifying the Graph of a Given Equation

Identify the graph of each equation.

(a) $9x^2 = 108 + 12y^2$

Both variables are squared, so the graph is either an ellipse or a hyperbola. (This situation also occurs for a circle, which may be considered a

CONTINUED ON NEXT PAGE

special case of the ellipse.) To see which one it is, rewrite the equation so that the x and y terms are on one side of the equation and 1 is on the other.

$$9x^2 = 108 + 12y^2$$

$$9x^2 - 12y^2 = 108 \qquad \text{Subtract } 12y^2.$$

$$\frac{x^2}{12} - \frac{y^2}{9} = 1 \qquad \text{Divide by 108.}$$

Because of the minus sign, the graph of this equation is a hyperbola.

(b) $x^2 = y - 3$

Only one of the two variables is squared, x, so this is the vertical parabola $y = x^2 + 3$.

(c) $x^2 = 9 - y^2$

Get the variable terms on the same side of the equation.

$$x^2 = 9 - y^2$$

$$x^2 + y^2 = 9 \qquad \text{Add } y^2.$$

This equation represents a circle with center at the origin and radius 3.

WORK PROBLEM 3 AT THE SIDE. ▶▶

OBJECTIVE 6 In the previous section we saw that semicircles can be defined by functions involving square root radicals. Similarly, portions of parabolas, ellipses, and hyperbolas can be defined by functions involving square root radicals. The final example illustrates such a function.

┌ **E X A M P L E   5   Graphing a Portion of an Ellipse**

Graph $\dfrac{y}{6} = \sqrt{1 - \dfrac{x^2}{16}}$.

Square both sides to get an equation whose form is known.

$$\frac{y^2}{36} = 1 - \frac{x^2}{16}$$

$$\frac{x^2}{16} + \frac{y^2}{36} = 1 \qquad \text{Add } \tfrac{x^2}{16}.$$

This is the equation of an ellipse with x-intercepts $(4, 0)$ and $(-4, 0)$ and y-intercepts $(0, 6)$ and $(0, -6)$. Since $\tfrac{y}{6}$ equals a principal square root in the original equation, y must be nonnegative, restricting the graph to the upper half of the ellipse, as shown in Figure 31. Verify that this is the graph of a function, using the vertical line test.

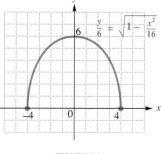

FIGURE 31

3. Identify the graph of each equation.

(a) $3x^2 = 27 - 4y^2$

(b) $6x^2 = 100 + 2y^2$

(c) $3x^2 = 27 - 4y$

(d) $3x^2 = 27 - 3y^2$

4. Graph

$$\frac{y}{3} = -\sqrt{1 - \frac{x^2}{4}}.$$

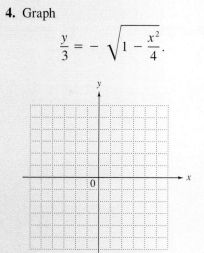

ANSWERS

3. (a) ellipse **(b)** hyperbola **(c)** parabola **(d)** circle

4.

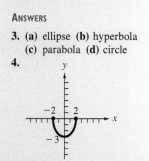

WORK PROBLEM 4 AT THE SIDE. ▶▶

Numbers in the

Real World *a graphing calculator minicourse*

Lesson 5: Matrices

Graphing calculators are capable of working with matrices and associated topics such as their determinants. At the beginning of Section 5.6, we briefly introduced matrices, and then showed how determinants of square matrices can be used to solve systems of equations with Cramer's rule. This can be done efficiently with a graphing calculator.

Suppose that we wish to solve the system $5x + 7y = -1$
$6x + 8y = 1.$

Let $D = \begin{bmatrix} 5 & 7 \\ 6 & 8 \end{bmatrix}$ be the matrix of coefficients, $A = D_x = \begin{bmatrix} -1 & 7 \\ 1 & 8 \end{bmatrix}$ be the matrix obtained by substituting the constants for the coefficients of x, and $B = D_y = \begin{bmatrix} 5 & -1 \\ 6 & 1 \end{bmatrix}$ be the matrix obtained by substituting the constants for the coefficients of y. These matrices can be entered into a graphing calculator. (See the owner's manual for instructions.) The calculator can calculate the determinants and the correct quotients. As seen in the first three figures at the left, the quotient for x is 7.5 and the quotient for y is –5.5. Thus the solution set is $\{(7.5, -5.5)\}$.

In more advanced courses such as college algebra, matrices are studied in a more theoretical way. A method of solving systems called **row reduction** employing properties of matrices is very efficient and is used by computers to solve large systems. Arithmetic operations with matrices are defined, and identities and inverses for multiplication are studied. For example, two matrices with the same dimensions are added or subtracted by adding or subtracting the corresponding entries. The next-to-last screen at the left shows how the matrices A and B defined above are added and subtracted. Multiplication of matrices is defined in what might seem to be a rather strange way. (Consult a college algebra text for the definition.) A very interesting property of matrix multiplication is that it is not commutative. The bottom screen shows that the matrix products AB and BA are not equal.

Graphing Calculator Explorations

1. Use a graphing calculator to apply Cramer's rule to find the solution of each system.

   (a) $x + y = 4$
   $2x - y = 2$

   (b) $x + y - z = -2$
   $2x - y + z = -5$
   $x - 2y + 3z = 4$

2. In Section 5.6 we saw that if the determinant of the coefficient matrix is 0, the system cannot be solved by Cramer's rule. Use a graphing calculator to show that this is the case for the following systems.

   (a) $3x + 2y = 9$
   $6x + 4y = 3$

   (b) $2x - 3y + 4z = 10$
   $6x - 9y + 12z = 24$
   $x + 2y - 3z = 5$

[D]
[[5 7]
[6 8]]

[A]
[[-1 7]
[1 8]]

[B]
[[5 -1]
[6 1]]

det([A])/det([D]
)
         7.5
det([B])/det([D]
)
         -5.5

[A]+[B]
[[4 6]
[7 9]]
[A]-[B]
[[-6 8]
[-5 7]]

[A]*[B]
[[37 8]
[53 7]]
[B]*[A]
[[-6 27]
[-5 50]]

11.4 Exercises

Match the equation with the correct graph. See Examples 1–3.

1. $\dfrac{x^2}{25} + \dfrac{y^2}{9} = 1$

2. $\dfrac{x^2}{9} + \dfrac{y^2}{25} = 1$

3. $\dfrac{x^2}{9} - \dfrac{y^2}{25} = 1$

4. $\dfrac{x^2}{25} - \dfrac{y^2}{9} = 1$

A.   **B.**

C.   **D.**

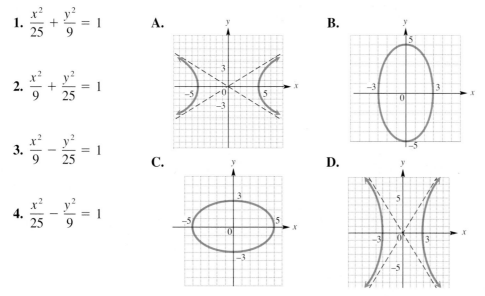

5. Write an explanation of how you can tell from the equation whether the branches of a hyperbola open up and down or left and right.

6. Explain why a set of points that form an ellipse does not satisfy the definition of a function.

Graph each ellipse. See Examples 1 and 2.

7. $\dfrac{x^2}{9} + \dfrac{y^2}{25} = 1$

8. $\dfrac{x^2}{9} + \dfrac{y^2}{16} = 1$

9. $\dfrac{x^2}{36} + \dfrac{y^2}{16} = 1$

10. $\dfrac{x^2}{9} + \dfrac{y^2}{4} = 1$

11. $\dfrac{x^2}{49} + \dfrac{y^2}{25} = 1$

12. $\dfrac{x^2}{16} + \dfrac{y^2}{9} = 1$

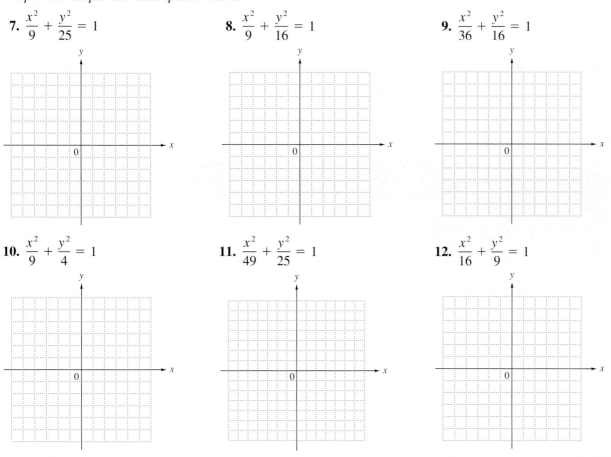

13. Describe how the fundamental rectangle is used to sketch a hyperbola.

14. Explain why the graph of a hyperbola does not satisfy the conditions for the graph of a function.

Graph each hyperbola. See Example 3.

15. $\dfrac{x^2}{16} - \dfrac{y^2}{9} = 1$

16. $\dfrac{y^2}{4} - \dfrac{x^2}{25} = 1$

17. $\dfrac{y^2}{9} - \dfrac{x^2}{9} = 1$

18. $\dfrac{x^2}{49} - \dfrac{y^2}{16} = 1$

19. $\dfrac{x^2}{25} - \dfrac{y^2}{36} = 1$

20. $\dfrac{y^2}{9} - \dfrac{x^2}{4} = 1$

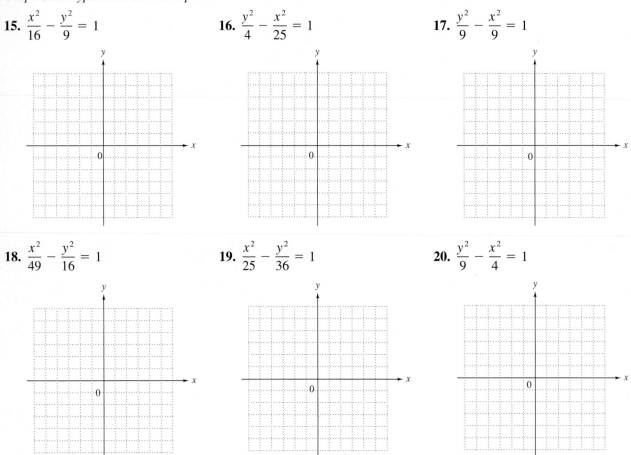

Identify each of the following as a parabola, circle, ellipse, or hyperbola. Sketch the graph.
See Example 4.

21. $x^2 - y^2 = 16$

22. $x^2 + y^2 = 16$

23. $4x^2 + y^2 = 16$

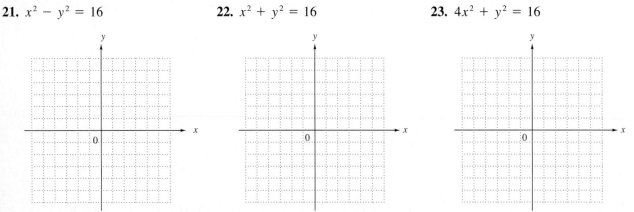

24. $x^2 - 2y = 0$

25. $y^2 = 36 - x^2$

26. $9x^2 + 25y^2 = 225$

27. $9x^2 = 144 + 16y^2$

28. $x^2 + 9y^2 = 9$

29. $y^2 = 4 + x^2$

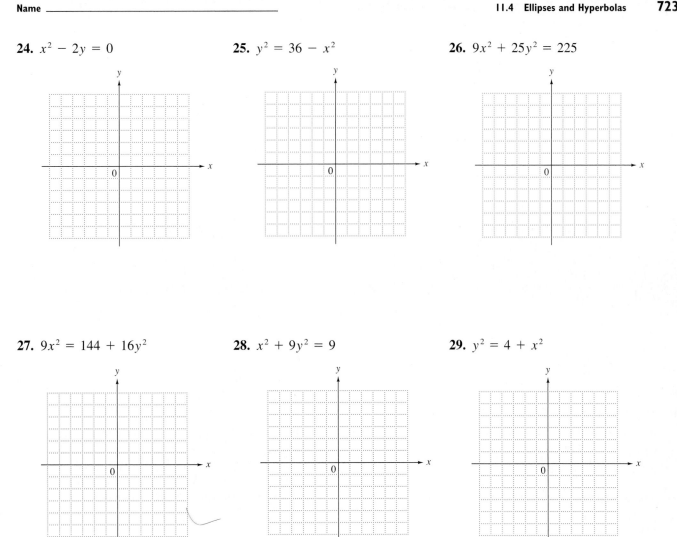

Each of the following is a portion of a conic section. Sketch each graph. See Example 5.

30. $\dfrac{y}{3} = \sqrt{1 + \dfrac{x^2}{9}}$

31. $y = \sqrt{\dfrac{x + 4}{2}}$

32. $y = -2\sqrt{\dfrac{9 - x^2}{9}}$

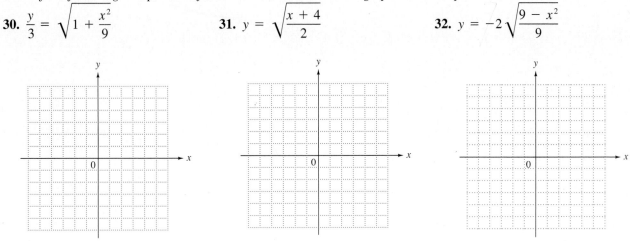

33. It is possible to sketch an ellipse on a piece of posterboard by fastening two ends of a length of string with thumbtacks, pulling the string taut with a pencil, and tracing a curve, as shown in the drawing. Explain why this method works.

34. A pair of buildings in a sports complex are shaped and positioned like a portion of the branches of the hyperbola $400x^2 - 625y^2 = 250{,}000$ where x and y are in meters.
(a) How far apart are the buildings at their closest point?
(b) Find the distance d in the figure.

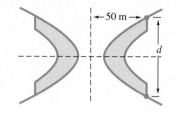

35. The orbit of Venus around the sun (one of the foci) is an ellipse with equation

$$\frac{x^2}{5013} + \frac{y^2}{4970} = 1$$

where x and y are measured in millions of miles.
(a) Find the farthest distance between Venus and the sun.
(b) Find the smallest distance between Venus and the sun. (*Hint:* See Figure 23 and use the fact that $c^2 = a^2 - b^2$.)

36. The graph shown here resembles a portion of a hyperbola. It shows the number of gun-related deaths per 100,000 for young African-American males starting in 1985. Here, 1985 is represented by $x = 0$, 1986 by $x = 1$, and so on. It can be shown that the function

$$f(x) = 23.8\sqrt{4.41 + x^2}$$

provides a good model for these data.

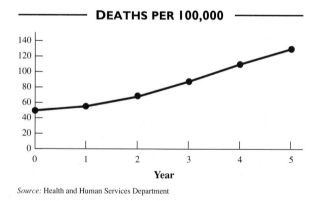

DEATHS PER 100,000

Source: Health and Human Services Department

(a) According to the graph, what is the number of deaths for the year 1988?
(b) According to the function, what is the number of deaths for the year 1988?

11.5 *Nonlinear Systems of Equations*

An equation in which some terms have more than one variable, or a variable of degree two or higher, is called a **nonlinear** (non-LIN-ee-er) **equation.** A **nonlinear system of equations** includes at least one nonlinear equation. Nonlinear systems can be solved by the elimination method, the substitution method, or a combination of the two. The following examples illustrate the use of these methods for solving nonlinear systems.

OBJECTIVE 1 The substitution method usually is most useful when one of the equations is linear. The first two examples illustrate this kind of system.

E X A M P L E I **Using Substitution When One Equation Is Linear**

Solve the system

$$x^2 + y^2 = 9 \qquad (1)$$
$$2x - y = 3. \qquad (2)$$

Solve the linear equation for one of the two variables, then substitute the resulting expression into the nonlinear equation to obtain an equation in one variable. Let us solve equation (2) for y.

$$2x - y = 3 \qquad (2)$$
$$y = 2x - 3 \qquad (3)$$

Substituting $2x - 3$ for y in equation (1) gives

$$x^2 + (2x - 3)^2 = 9 \qquad \text{Replace } y \text{ with } 2x - 3.$$
$$x^2 + 4x^2 - 12x + 9 = 9 \qquad \text{Square the binomial.}$$
$$5x^2 - 12x = 0 \qquad \text{Standard form}$$
$$x(5x - 12) = 0 \qquad \text{Factor.}$$
$$x = 0 \quad \text{or} \quad x = \frac{12}{5}. \qquad \begin{array}{l}\text{Set each factor equal}\\\text{to 0; solve.}\end{array}$$

Let $x = 0$ in the equation $y = 2x - 3$ to get $y = -3$. If $x = \frac{12}{5}$, then $y = \frac{9}{5}$. The solution set of the system is $\{(0, -3), (\frac{12}{5}, \frac{9}{5})\}$. The graph of the system, shown in Figure 32, confirms the solution.

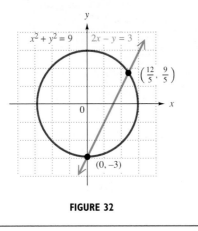

FIGURE 32

WORK PROBLEM I AT THE SIDE. ▶▶

OBJECTIVES

1 ▶ Solve a nonlinear system by substitution.

2 ▶ Use the elimination method to solve a system with two second-degree equations.

3 ▶ Solve a system that requires a combination of methods.

FOR EXTRA HELP

Tutorial Tape 24 SSM, Sec. 11.5

1. Solve each system.

   (a) $x^2 + y^2 = 10$
   $\qquad x = y + 2$

   (b) $x^2 - 2y^2 = 8$
   $\qquad y + x = 6$

ANSWERS

1. (a) $\{(3, 1), (-1, -3)\}$
   (b) $\{(4, 2), (20, -14)\}$

2. Solve each system.

(a) $xy = 8$
$x + y = 6$

EXAMPLE 2 Using Substitution When One Equation Is Linear

Solve the system

$$6x - y = 5 \qquad \text{(4)}$$
$$xy = 4. \qquad \text{(5)}$$

Although we could solve the linear equation for either x or y and then substitute into the other equation, let us instead solve $xy = 4$ for x, to get $x = \frac{4}{y}$. Substituting $\frac{4}{y}$ for x in equation (4) gives

$$6\left(\frac{4}{y}\right) - y = 5. \qquad \text{Replace } x \text{ with } \frac{4}{y}.$$

$$\frac{24}{y} - y = 5$$

$$24 - y^2 = 5y \qquad \text{Multiply by } y. \text{ (Assume } y \neq 0.\text{)}$$

$$0 = y^2 + 5y - 24 \qquad \text{Standard form}$$

$$0 = (y - 3)(y + 8) \qquad \text{Factor.}$$

$$y = 3 \quad \text{or} \quad y = -8 \qquad \begin{array}{l}\text{Set each factor equal} \\ \text{to 0; solve.}\end{array}$$

Substitute these results into $x = \frac{4}{y}$ to obtain the corresponding values for x.

$$\text{If } y = 3, \text{ then } x = \frac{4}{3}.$$

$$\text{If } y = -8, \text{ then } x = -\frac{1}{2}.$$

(b) $xy + 10 = 0$
$4x + 9y = -2$

The solution set is $\{(\frac{4}{3}, 3), (-\frac{1}{2}, -8)\}$.

◀◀ **WORK PROBLEM 2 AT THE SIDE.**

OBJECTIVE 2▶ The elimination method is often useful when both equations are second-degree equations. This method is used in the following example.

EXAMPLE 3 Solving a Nonlinear System by Elimination

Solve the system

$$x^2 + y^2 = 9 \qquad \text{(6)}$$
$$2x^2 - y^2 = -6. \qquad \text{(7)}$$

Adding the two equations will eliminate y, leaving an equation that can be solved for x.

$$\begin{array}{rcll} x^2 + y^2 = & 9 & \quad \text{(6)} \\ \underline{2x^2 - y^2 = -6} & & \quad \text{(7)} \\ 3x^2 = & 3 & \end{array}$$

$$x^2 = 1$$

$$x = 1 \quad \text{or} \quad x = -1$$

CONTINUED ON NEXT PAGE ──

ANSWERS

2. (a) $\{(4, 2), (2, 4)\}$

(b) $\left\{(-5, 2), \left(\frac{9}{2}, -\frac{20}{9}\right)\right\}$

Each value of x gives corresponding values for y when substituted into one of the original equations. Using equation (6) gives the following results.

If $x = 1$,

$(1)^2 + y^2 = 9$

$y^2 = 8$

$y = \sqrt{8}$ or $-\sqrt{8}$

$y = 2\sqrt{2}$ or $-2\sqrt{2}$.

If $x = -1$,

$(-1)^2 + y^2 = 9$

$y^2 = 8$

$y = 2\sqrt{2}$ or $-2\sqrt{2}$.

The solution set is $\{(1, 2\sqrt{2}), (1, -2\sqrt{2}), (-1, 2\sqrt{2}), (-1, -2\sqrt{2})\}$. The graph in Figure 33 shows the four points of intersection.

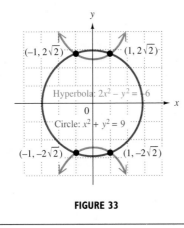

FIGURE 33

WORK PROBLEM 3 AT THE SIDE. ▶▶

OBJECTIVE 3 The next example shows a system of second-degree equations that can be solved only by a combination of methods.

┌ **E X A M P L E   4**   **Solving a Nonlinear System by a Combination of Methods**

Solve the system

$$x^2 + 2xy - y^2 = 7 \tag{8}$$

$$x^2 - y^2 = 3. \tag{9}$$

The elimination method is used here in combination with the substitution method. To begin, we eliminate the squared terms by multiplying both sides of equation (9) by -1 and then adding the result to (8).

$$\begin{array}{rl} x^2 + 2xy - y^2 = & 7 \\ \underline{-x^2 \qquad\quad + y^2 = } & \underline{-3} \\ 2xy \qquad\quad = & 4 \end{array}$$

Next, we solve $2xy = 4$ for either variable. Let us solve for y.

$$2xy = 4$$

$$y = \frac{2}{x} \tag{10}$$

└ CONTINUED ON NEXT PAGE

3. Solve each system.

(a) $x^2 + y^2 = 41$
    $x^2 - y^2 = 9$

(b) $x^2 + 3y^2 = 40$
    $4x^2 - y^2 = 4$

4. Solve each system.

(a) $x^2 + xy + y^2 = 3$
$ x^2 + y^2 = 5$

Now substitute $y = \frac{2}{x}$ into one of the original equations. It is easier to do this with (9).

$$x^2 - y^2 = 3 \tag{9}$$

$$x^2 - \left(\frac{2}{x}\right)^2 = 3$$

$$x^2 - \frac{4}{x^2} = 3$$

To clear the equation of fractions, we multiply both sides by x^2.

$$x^4 - 4 = 3x^2$$

$$x^4 - 3x^2 - 4 = 0$$

$$(x^2 - 4)(x^2 + 1) = 0$$

$$x^2 - 4 = 0 \quad \text{or} \quad x^2 + 1 = 0$$

$$x^2 = 4 \qquad\qquad x^2 = -1$$

$$x = 2 \ \text{ or } \ x = -2 \qquad x = i \ \text{ or } \ x = -i$$

By substituting the four values of x from above into equation (10), we get the corresponding values of y.

If $x = 2$, then $y = 1$.   If $x = i$, then $y = -2i$.

If $x = -2$, then $y = -1$.   If $x = -i$, then $y = 2i$.

There are four solutions, two ordered pairs of real numbers and two ordered pairs of imaginary numbers, in the solution set: $\{(2, 1), (-2, -1), (i, -2i), (-i, 2i)\}$.

Note that if you substitute the x-values found above into equation (8) or (9) instead of into equation (10), you get extraneous solutions. It is always wise to check all solutions in both of the given equations. The graph of the system, shown in Figure 34, shows only the two real intersection points because the graph is in the real number plane. The two ordered pairs with imaginary components are solutions of the system but do not show up on the graph.

(b) $x^2 + 7xy - 2y^2 = -8$
$-2x^2 + 4y^2 = 16$

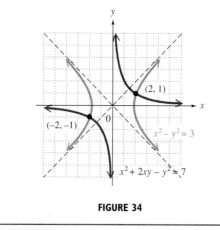

FIGURE 34

◀◀ **WORK PROBLEM 4 AT THE SIDE.**

ANSWERS

4. (a) $\{(1, -2), (-1, 2), (2, -1),$
$(-2, 1)\}$

(b) $\{(0, 2), (0, -2), (2i\sqrt{2}, 0),$
$(-2i\sqrt{2}, 0)\}$

11.5 Exercises

1. Write an explanation of the steps you would use to solve the system

$$x^2 + y^2 = 25$$
$$y = x - 1$$

by the substitution method.

2. Write an explanation of the steps you would use to solve the system

$$x^2 + y^2 = 12$$
$$x^2 - y^2 = 13$$

by the elimination method.

3. Is it possible for a nonlinear system consisting of a line and a circle to have three ordered pairs in its solution set? Explain.

4. Suppose that a nonlinear system consists of two ellipses whose graphs look like this:

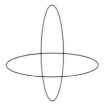

How many ordered pairs are there in the solution set of this system? Explain.

Solve the following systems by the substitution method. See Examples 1 and 2.

5. $y = 4x^2 - x$
$\quad y = x$

6. $y = x^2 + 6x$
$\quad 3y = 12x$

7. $\quad y = x^2 + 6x + 9$
$\quad x + y = 3$

8. $\quad y = x^2 + 8x + 16$
$\quad x - y = -4$

9. $x^2 + y^2 = 2$
$\quad 2x + y = 1$

10. $2x^2 + 4y^2 = 4$
$\quad x = 4y$

11. $xy = 4$
       $3x + 2y = -10$

12. $xy = -5$
       $2x + y = 3$

13. $xy = -3$
       $x + y = -2$

14. $xy = 12$
       $x + y = 8$

15. $y = 3x^2 + 6x$
       $y = x^2 - x - 6$

16. $y = 2x^2 + 1$
       $y = 5x^2 + 2x - 7$

17. $2x^2 - y^2 = 6$
       $y = x^2 - 3$

18. $x^2 + y^2 = 4$
       $y = x^2 - 2$

Solve the following systems using the elimination method or a combination of the elimination and substitution methods. See Examples 3 and 4.

19. $3x^2 + 2y^2 = 12$
       $x^2 + 2y^2 = 4$

20. $2x^2 + y^2 = 28$
       $4x^2 - 5y^2 = 28$

21. $xy = 6$
       $3x^2 - y^2 = 12$

22. $xy = 5$
       $2y^2 - x^2 = 5$

23. $2x^2 + 2y^2 = 8$
$3x^2 + 4y^2 = 24$

24. $5x^2 + 5y^2 = 20$
$x^2 + 2y^2 = 2$

25. $x^2 + xy + y^2 = 15$
$x^2 + y^2 = 10$

26. $2x^2 + 3xy + 2y^2 = 21$
$x^2 + y^2 = 6$

27. $3x^2 + 2xy - 3y^2 = 5$
$-x^2 - 3xy + y^2 = 3$

28. $-2x^2 + 7xy - 3y^2 = 4$
$2x^2 - 3xy + 3y^2 = 4$

INTERPRETING TECHNOLOGY (EXERCISES 29–32)

If the two equations making up a nonlinear system are graphed with a graphing calculator, the calculator can be used to solve the system by finding the coordinates of the points of intersection of the two graphs. For example, the nonlinear system

$$y = x^2 - 3$$
$$2x + y = 0$$

can be solved by substitution to find that the solution set is $\{(-3, 6), (1, -2)\}$. *The graphs of* $y_1 = x^2 - 3$ *and* $y_2 = -2x$ *shown in the two screens here indicate that these are indeed the points of intersection.*

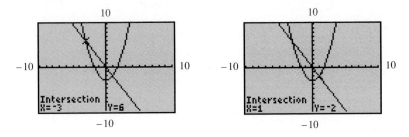

In Exercises 29–32, nonlinear systems are given, along with a screen showing the coordinates of one of the points of intersection of the two graphs. Solve the system using substitution or elimination to find the coordinates of the other *point of intersection.*

29. $y = x^2 + 1$
    $x + y = 1$

30. $y = -x^2$
    $x + y = 0$

31. $y = \dfrac{1}{2}x^2$
    $x + y = 4$

32. $y = -\dfrac{1}{3}x^2$
    $2x - y = 9$

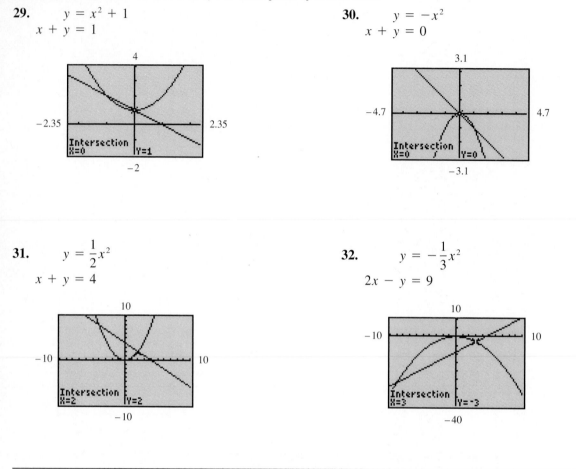

Write a nonlinear system of equations and solve.

33. The area of a rectangular rug is 84 square feet, and its perimeter is 38 feet. Find the length and width of the rug.

34. Find the length and width of a rectangular room whose perimeter is 50 meters and whose area is 100 square meters.

35. A company has found that the price p (in dollars) of its scientific calculator is related to the supply x (in thousands) by the equation $px = 16$. The price is related to the demand x (in thousands) for the calculator by the equation $p = 10x + 12$. The *equilibrium price* is the value of p where demand equals supply. Find the equilibrium price and the supply/demand at that price by solving a system of equations.

36. The calculator company in Exercise 35 has also determined that the cost y to make x (thousand) calculators is $y = 4x^2 + 36x + 20$, while the revenue y from the sale of x (thousand) calculators is $36x^2 - 3y = 0$. Find the *break-even point*, where cost just equals revenue, by solving a system of equations.

11.6 Second-Degree Inequalities; Systems of Inequalities

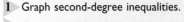

OBJECTIVE 1 ▸ Recall from Section 4.4 that a linear inequality such as $3x + 2y \leq 5$ is graphed by first graphing the boundary line $3x + 2y = 5$. **Second-degree inequalities** such as $x^2 + y^2 \leq 36$ are graphed in much the same way. The boundary of $x^2 + y^2 \leq 36$ is the graph of the equation $x^2 + y^2 = 36$, a circle with radius 6 and center at the origin, as shown in Figure 35. As with linear inequalities, the inequality $x^2 + y^2 \leq 36$ will include either the points outside the circle or the points inside the circle. Decide which region to shade by substituting any point not on the circle, such as $(0, 0)$, into the inequality. Since $0^2 + 0^2 < 36$ is a true statement, the inequality includes the points inside the circle, the shaded region in Figure 35.

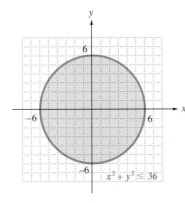

1. Graph $y \geq (x + 1)^2 - 5$.

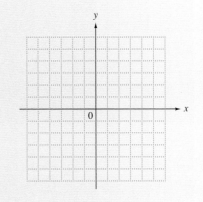

FIGURE 35

E X A M P L E I   Graphing a Second-Degree Inequality

Graph $y < -2(x - 4)^2 - 3$.

The boundary, $y = -2(x - 4)^2 - 3$, is a parabola opening downward with vertex at $(4, -3)$. Using the point $(0, 0)$ as a test point gives

$$0 < -2(0 - 4)^2 - 3$$
$$0 < -32 - 3$$
$$0 < -35. \qquad \text{False}$$

This is a false statement, so the points in the region containing $(0, 0)$ do not satisfy the inequality. Figure 36 shows the final graph; the parabola is drawn with a dashed line since the points of the parabola itself do not satisfy the inequality. The solution includes all points inside (below) the parabola.

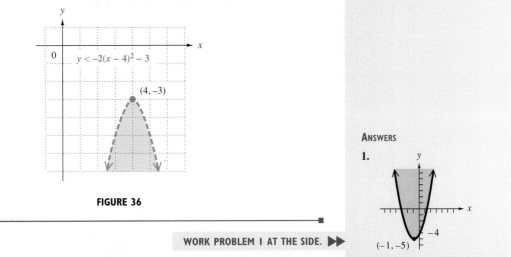

FIGURE 36

WORK PROBLEM I AT THE SIDE. ▸▸

2. Graph $x^2 + 4y^2 > 36$.

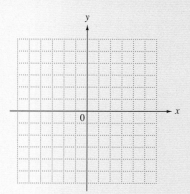

EXAMPLE 2 Graphing a Second-Degree Inequality

Graph $16y^2 \leq 144 + 9x^2$.

First rewrite the inequality as follows.

$$16y^2 - 9x^2 \leq 144$$

$$\frac{y^2}{9} - \frac{x^2}{16} \leq 1 \qquad \text{Divide both sides by 144.}$$

This form of the inequality shows that the boundary is the hyperbola

$$\frac{y^2}{9} - \frac{x^2}{16} = 1.$$

Since the test point $(0, 0)$ satisfies the inequality $16y^2 \leq 144 + 9x^2$, the region containing $(0, 0)$, is shaded, as shown in Figure 37.

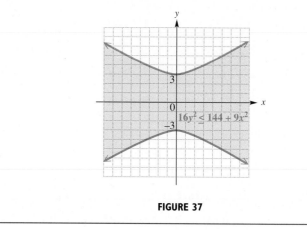

FIGURE 37

◀◀ WORK PROBLEM 2 AT THE SIDE.

OBJECTIVE 2 The solution set of a **system of inequalities,** such as

$$2x + 3y > 6$$
$$x^2 + y^2 < 16,$$

is the intersection of the solution sets of the individual inequalities.

The graph of the solution set of this system is the set of all points on the plane that belong to the graphs of both inequalities in the system. Graph this system by graphing both inequalities on the same coordinate axes, as shown in Figure 38. The heavily shaded region containing those points that belong to both graphs is the graph of the system. In this case the points of the boundary lines are not included.

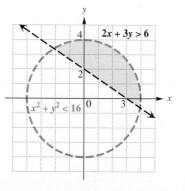

FIGURE 38

ANSWERS

2.

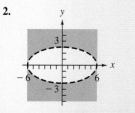

E X A M P L E 3   **Graphing a System of Inequalities**

Graph the solution set of the system

$$x + y < 1$$
$$y \leq 2x + 3$$
$$y \geq -2.$$

Graph each inequality separately, on the same axes. The graph of the system is the triangular region enclosed by the three lines in Figure 39. It contains all points that satisfy all three inequalities. One boundary line $(x + y = 1)$ is dashed, while the other two are solid.

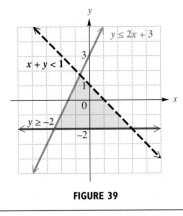

FIGURE 39

WORK PROBLEM 3 AT THE SIDE. ▶▶

E X A M P L E 4   **Graphing a System of Inequalities**

Graph the solution set of the system

$$y \geq x^2 - 2x + 1$$
$$2x^2 + y^2 > 4$$
$$y < 4.$$

The graph of $y = x^2 - 2x + 1$ is a parabola with vertex at $(1, 0)$. Those points that are above the parabola satisfy the condition $y > x^2 - 2x + 1$. Thus points on the parabola or above it are in the solution set of $y \geq x^2 - 2x + 1$. The graph of $2x^2 + y^2 = 4$ is an ellipse. To satisfy the inequality $2x^2 + y^2 > 4$, a point must lie outside the ellipse. The graph of $y < 4$ includes all points below the line $y = 4$. Finally, the graph of the system is the shaded region in Figure 40 that lies outside the ellipse, above or on the boundary of the parabola, and below the line $y = 4$.

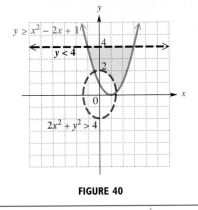

FIGURE 40

WORK PROBLEM 4 AT THE SIDE. ▶▶

3. Graph the solution set of the system

$$3x - 4y \geq 12$$
$$x + 3y \geq 6$$
$$y \leq 2.$$

4. Graph the solution set of the system

$$y \geq x^2 + 1$$
$$\frac{x^2}{9} + \frac{y^2}{4} \geq 1$$
$$y \leq 5.$$

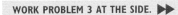

ANSWERS

3.

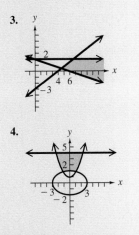

4.

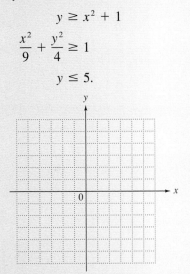

NUMBERS IN THE

Real World *a graphing calculator*
minicourse

Lesson 6: Solution of Quadratic Equations and Inequalities

Suppose that we want to use a graphing calculator to solve the quadratic equation $2x^2 + 3x - 2 = 0$. The graph of $y = 2x^2 + 3x - 2$ is a parabola, as discussed in Section 11.1. Recall that the x-intercept of the line $y = mx + b$ is a solution of the equation $mx + b = 0$. Extending this idea to quadratic equations, we may say that the x-intercept(s), if any, of the graph of $y = ax^2 + bx + c$, are the real solutions of the equation $ax^2 + bx + c = 0$. Therefore, based on the graph and displays shown below, the solutions of $2x^2 + 3x - 2 = 0$ are .5 (or $\frac{1}{2}$) and -2.

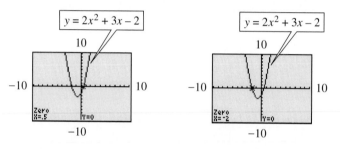

In Section 10.6 we examined algebraic methods for solving quadratic inequalities. We can use graphs of quadratic functions to solve such inequalities graphically. Based on the figures above, we see that the graph of $y = 2x^2 + 3x - 2$ lies *below* the x-axis for x-values between -2 and .5. Thus the solution set of $2x^2 + 3x - 2 < 0$ is the interval $(-2, .5)$. Because the graph of $y = 2x^2 + 3x - 2$ lies *above* the x-axis to the left of -2 and to the right of .5, the solution set of $2x^2 + 3x - 2 > 0$ is $(-\infty, -2) \cup (.5, \infty)$.

GRAPHING CALCULATOR EXPLORATIONS

1. Graph the function $y = x^2 - 4x - 5$ in a window that shows both x-intercepts.

   (a) Solve the equations $x^2 - 4x - 5 = 0$ algebraically. How do the solutions compare to the x-intercepts of the graph?

   (b) Use the graph to give the solution set of $x^2 - 4x - 5 < 0$.

   (c) Use the graph to give the solution set of $x^2 - 4x - 5 > 0$.

2. The figures below show the graph of $y = -x^2 + x + 12$. Use the graph to solve each of the following:

   (a) $-x^2 + x + 12 = 0$ (b) $-x^2 + x + 12 \leq 0$ (c) $-x^2 + x + 12 \geq 0$

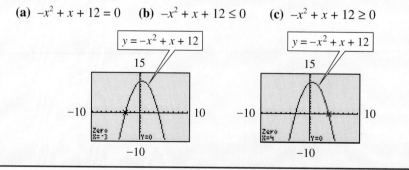

11.6 Exercises

1. Match the nonlinear inequality with its graph.

    (a) $y \geq x^2 + 4$

    (b) $y \leq x^2 + 4$

    (c) $y < x^2 + 4$

    (d) $y > x^2 + 4$

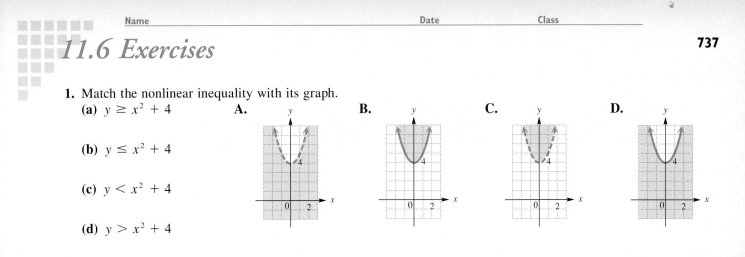

Graph each of the following inequalities. See Examples 1 and 2.

  2. $y > x^2 - 1$ **3.** $y^2 > 4 + x^2$ **4.** $y^2 \leq 4 - 2x^2$

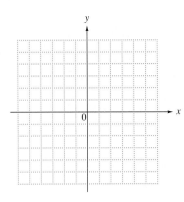 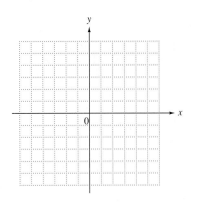 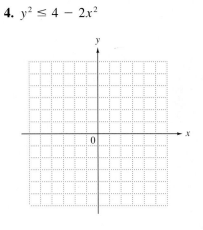

  5. $y + 2 \geq x^2$ **6.** $x^2 \leq 16 - y^2$ **7.** $2y^2 \geq 8 - x^2$

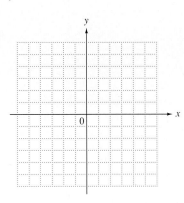 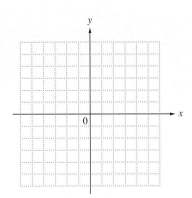 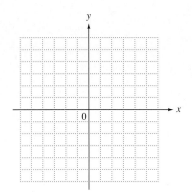

8. $x^2 \leq 16 + 4y^2$

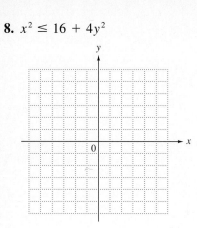

9. $y \leq x^2 + 4x + 2$

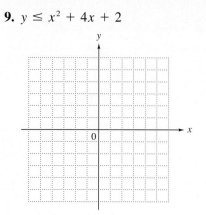

10. $9x^2 < 16y^2 - 144$

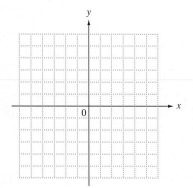

11. $9x^2 > 16y^2 + 144$

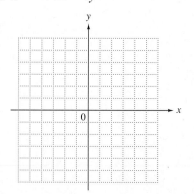

12. $4y^2 \leq 36 - 9x^2$

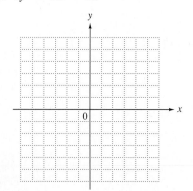

13. $x^2 - 4 \geq -4y^2$

14. $x \geq y^2 - 8y + 14$

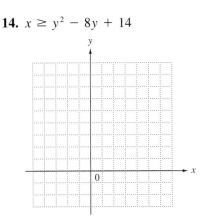

15. $x \leq -y^2 + 6y - 7$

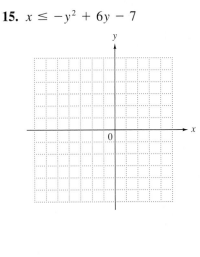

16. $y^2 - 16x^2 \leq 16$

17. $25x^2 \leq 9y^2 + 225$

18. Explain how to graph the solution set of a system of nonlinear inequalities.

19. Which one of the following is a description of the graph of the solution set of the system below?

$$x^2 + y^2 < 25$$
$$y > -2$$

(a) all points outside the circle $x^2 + y^2 = 25$ and above the line $y = -2$

(b) all points outside the circle $x^2 + y^2 = 25$ and below the line $y = -2$

(c) all points inside the circle $x^2 + y^2 = 25$ and above the line $y = -2$

(d) all points inside the circle $x^2 + y^2 = 25$ and below the line $y = -2$

20. Fill in the blank with the appropriate response: The graph of the system

$$y > x^2 + 1$$
$$\frac{x^2}{9} + \frac{y^2}{4} > 1$$
$$y < 5$$

consists of all points _____ the
                        (above/below)

parabola $y = x^2 + 1$, _____ the
                         (inside/outside)

ellipse $x^2/9 + y^2/4 = 1$, and _____
                                  (above/below)

the line $y = 5$.

Graph the following systems of inequalities. See Examples 3 and 4.

21. $2x + 5y < 10$
　　$x - 2y < 4$

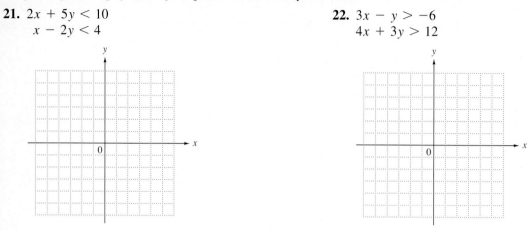

22. $3x - y > -6$
　　$4x + 3y > 12$

23. $5x - 3y \leq 15$
　　$4x + y \geq 4$

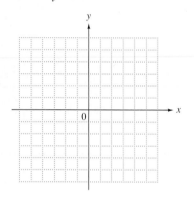

24. $4x - 3y \leq 0$
　　$x + y \leq 5$

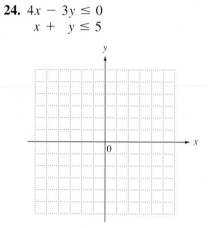

25. $x \leq 5$
　　$y \leq 4$

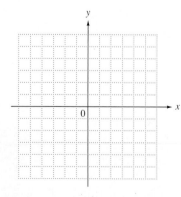

26. $x \geq -2$
　　$y \leq 4$

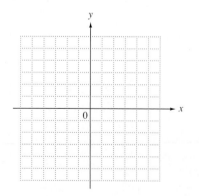

27. $y > x^2 - 4$
  $y < -x^2 + 3$

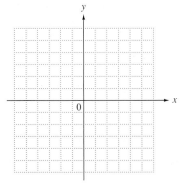

28. $x^2 - y^2 \geq 9$
  $\dfrac{x^2}{16} + \dfrac{y^2}{9} \leq 1$

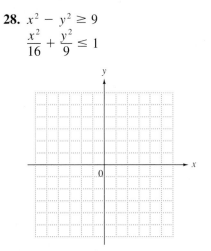

29. $y^2 - x^2 \geq 4$
  $-5 \leq y \leq 5$

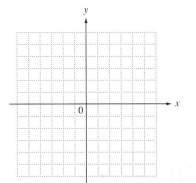

30. $x \geq 0$
       $y \geq 0$
  $x^2 + y^2 \geq 4$
   $x + y \leq 5$

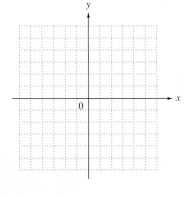

31. $y \leq -x^2$
  $y \geq x - 3$
  $y \leq -1$
  $x < 1$

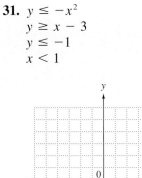

32. $y < x^2$
     $y > -2$
   $x + y < 3$
  $3x - 2y > -6$

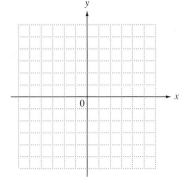

INTERPRETING TECHNOLOGY (EXERCISES 33–36)

Graphing calculators have the capability of shading above or below graphs, thus allowing them to illustrate nonlinear inequalities and systems of nonlinear inequalities. For example, the inequality discussed in Example 1 and graphed in Figure 36,

$$y < -2(x - 4)^2 - 3$$

is shown in the accompanying screen.

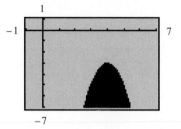

Match the nonlinear inequality or system of nonlinear inequalities with its calculator-generated graph.

33. $y > x^2 + 2$

34. $y < x^2 + 2$

35. $y > x^2 + 2$
       $y < 5$

36. $y < x^2 + 2$
       $y > -5$

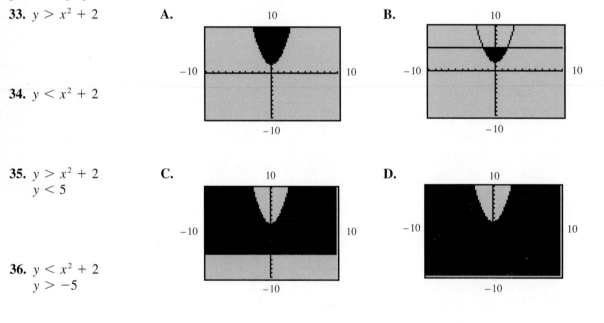

KEY TERMS

11.1	**conic sections**	Graphs that result from cutting an infinite cone with a plane are called conic sections.
	parabola	The graph of a second-degree function is a parabola.
	vertex	The point on a parabola that has the smallest y-value (if the parabola opens upward) or the largest y-value (if the parabola opens downward) is called the vertex of the parabola.
	axis	The vertical or horizontal line through the vertex of a parabola is its axis.
	quadratic function	A function defined by $f(x) = ax^2 + bx + c$, for real numbers a, b, and c, with $a \neq 0$, is a quadratic function.
11.3	**squaring function**	The function defined by $f(x) = x^2$ is called the squaring function.
	absolute value function	The function defined by $f(x) = |x|$ is called the absolute value function.
	reciprocal function	The function defined by $f(x) = \dfrac{1}{x}$ is called the reciprocal function.
	asymptotes	Lines that a graph approaches, such as the x- and y-axes for the reciprocal function, are called asymptotes of the graph.
	square root function	The function defined by $f(x) = \sqrt{x}$ is called the square root function.
	circle	A circle is the set of all points in a plane that lie a fixed distance from a fixed point.
	center	The fixed point discussed in the definition of a circle is the center of the circle.
	radius	The radius of a circle is the fixed distance between the center and any point on the circle.
11.4	**ellipse**	An ellipse is the set of all points in a plane the sum of whose distances from two fixed points is constant.
	hyperbola	A hyperbola is the set of all points in a plane the difference of whose distances from two fixed points is constant.
	asymptotes of a hyperbola	The two intersecting lines that the branches of a hyperbola approach are called asymptotes of the hyperbola.
	fundamental rectangle	The asymptotes of a hyperbola are the extended diagonals of its fundamental rectangle.
11.5	**nonlinear equation**	An equation that cannot be written in the form $Ax + By = C$, for real numbers A, B, and C, is a nonlinear equation.
	nonlinear system of equations	A nonlinear system of equations is a system with at least one nonlinear equation.
11.6	**second-degree inequality**	A second-degree inequality is an inequality with at least one variable of degree two and no variable with degree greater than two.
	system of inequalities	A system of inequalities consists of two or more inequalities to be solved at the same time.

QUICK REVIEW

Concepts	Examples

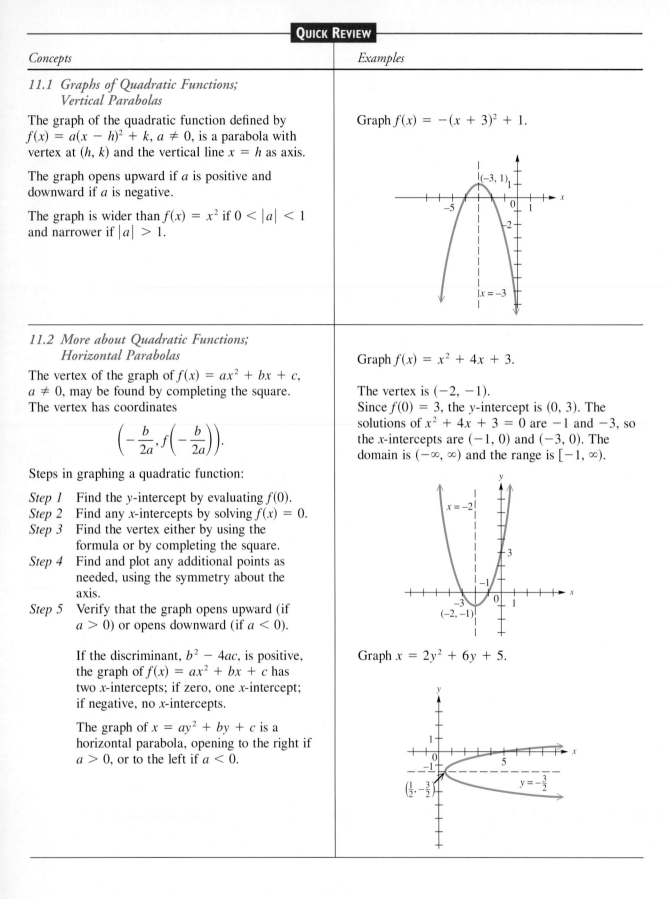

11.1 Graphs of Quadratic Functions; Vertical Parabolas

The graph of the quadratic function defined by $f(x) = a(x - h)^2 + k$, $a \neq 0$, is a parabola with vertex at (h, k) and the vertical line $x = h$ as axis.

The graph opens upward if a is positive and downward if a is negative.

The graph is wider than $f(x) = x^2$ if $0 < |a| < 1$ and narrower if $|a| > 1$.

Graph $f(x) = -(x + 3)^2 + 1$.

11.2 More about Quadratic Functions; Horizontal Parabolas

The vertex of the graph of $f(x) = ax^2 + bx + c$, $a \neq 0$, may be found by completing the square. The vertex has coordinates

$$\left(-\frac{b}{2a}, f\left(-\frac{b}{2a} \right) \right).$$

Steps in graphing a quadratic function:

Step 1   Find the y-intercept by evaluating $f(0)$.
Step 2   Find any x-intercepts by solving $f(x) = 0$.
Step 3   Find the vertex either by using the formula or by completing the square.
Step 4   Find and plot any additional points as needed, using the symmetry about the axis.
Step 5   Verify that the graph opens upward (if $a > 0$) or opens downward (if $a < 0$).

If the discriminant, $b^2 - 4ac$, is positive, the graph of $f(x) = ax^2 + bx + c$ has two x-intercepts; if zero, one x-intercept; if negative, no x-intercepts.

The graph of $x = ay^2 + by + c$ is a horizontal parabola, opening to the right if $a > 0$, or to the left if $a < 0$.

Graph $f(x) = x^2 + 4x + 3$.

The vertex is $(-2, -1)$.
Since $f(0) = 3$, the y-intercept is $(0, 3)$. The solutions of $x^2 + 4x + 3 = 0$ are -1 and -3, so the x-intercepts are $(-1, 0)$ and $(-3, 0)$. The domain is $(-\infty, \infty)$ and the range is $[-1, \infty)$.

Graph $x = 2y^2 + 6y + 5$.

Concepts	Examples

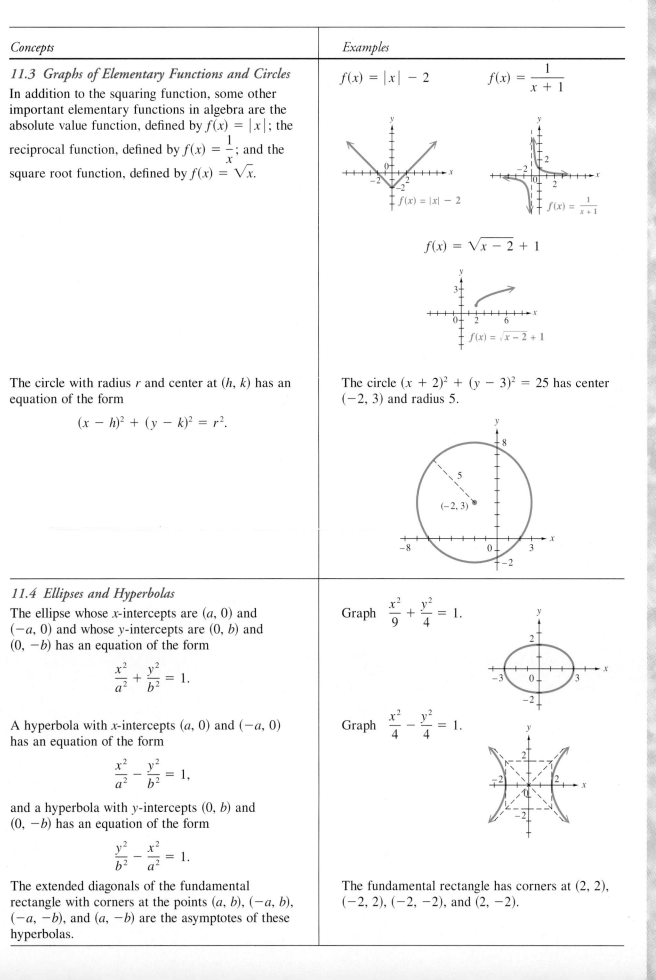

Concepts

11.3 Graphs of Elementary Functions and Circles

In addition to the squaring function, some other important elementary functions in algebra are the absolute value function, defined by $f(x) = |x|$; the reciprocal function, defined by $f(x) = \dfrac{1}{x}$; and the square root function, defined by $f(x) = \sqrt{x}$.

The circle with radius r and center at (h, k) has an equation of the form

$$(x - h)^2 + (y - k)^2 = r^2.$$

11.4 Ellipses and Hyperbolas

The ellipse whose x-intercepts are $(a, 0)$ and $(-a, 0)$ and whose y-intercepts are $(0, b)$ and $(0, -b)$ has an equation of the form

$$\frac{x^2}{a^2} + \frac{y^2}{b^2} = 1.$$

A hyperbola with x-intercepts $(a, 0)$ and $(-a, 0)$ has an equation of the form

$$\frac{x^2}{a^2} - \frac{y^2}{b^2} = 1,$$

and a hyperbola with y-intercepts $(0, b)$ and $(0, -b)$ has an equation of the form

$$\frac{y^2}{b^2} - \frac{x^2}{a^2} = 1.$$

The extended diagonals of the fundamental rectangle with corners at the points (a, b), $(-a, b)$, $(-a, -b)$, and $(a, -b)$ are the asymptotes of these hyperbolas.

Examples

$f(x) = |x| - 2 \qquad f(x) = \dfrac{1}{x + 1}$

$f(x) = \sqrt{x - 2} + 1$

The circle $(x + 2)^2 + (y - 3)^2 = 25$ has center $(-2, 3)$ and radius 5.

Graph $\dfrac{x^2}{9} + \dfrac{y^2}{4} = 1.$

Graph $\dfrac{x^2}{4} - \dfrac{y^2}{4} = 1.$

The fundamental rectangle has corners at $(2, 2)$, $(-2, 2)$, $(-2, -2)$, and $(2, -2)$.

Concepts	Examples
11.5 Nonlinear Systems of Equations Nonlinear systems can be solved by the substitution method, the elimination method, or a combination of the two.	Solve the system $$x^2 + 2xy - y^2 = 14$$ $$x^2 \qquad - y^2 = -16. \qquad (*)$$ Multiply equation (*) by -1 and use elimination. $$\begin{aligned} x^2 + 2xy - y^2 &= 14 \\ -x^2 \qquad + y^2 &= 16 \\ \hline 2xy \qquad\qquad &= 30 \\ xy &= 15 \end{aligned}$$ Solve for y to obtain $y = \frac{15}{x}$, and substitute into equation (*). $$x^2 - \left(\frac{15}{x}\right)^2 = -16$$ This simplifies to $$x^2 - \frac{225}{x^2} = -16.$$ Multiply by x^2 and get one side equal to 0. $$x^4 + 16x^2 - 225 = 0$$ Factor and solve. $$(x^2 - 9)(x^2 + 25) = 0$$ $$x = \pm 3 \qquad x = \pm 5i$$ Find corresponding y values to get the solution set $\{(3, 5), (-3, -5), (5i, -3i), (-5i, 3i)\}$.
11.6 Second-Degree Inequalities; **Systems of Inequalities** To graph a second-degree inequality, graph the corresponding equation as a boundary and use test points to determine which region(s) form the solution. Shade the appropriate region(s).	Graph $y \geq x^2 - 2x + 3$.
The solution set of a system of inequalities is the intersection of the solution sets of the individual inequalities.	Graph the solution set of the system $$3x - 5y > -15$$ $$x^2 + y^2 \leq 25.$$

[11.1–11.2] *Identify the vertex of each parabola.*

1. $f(x) = 3x^2 - 2$ **2.** $f(x) = 6 - 2x^2$ **3.** $f(x) = (x + 2)^2$ **4.** $f(x) = -(x - 1)^2$

5. $f(x) = (x - 3)^2 + 7$ **6.** $f(x) = \frac{4}{3}(x - 2)^2 + 1$ **7.** $x = (y + 2)^2 + 3$ **8.** $x = (y - 3)^2 - 4$

Graph each of the following. In Exercises 9 and 10 give the domain and range.

9. $f(x) = 4x^2 + 4x - 2$ **10.** $f(x) = -2x^2 + 8x - 5$

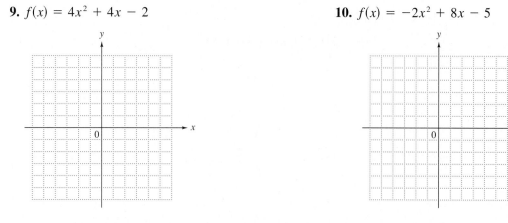

11. $x = -\frac{1}{2}y^2 + 6y - 14$ **12.** $x = 2y^2 + 8y + 3$

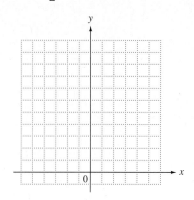   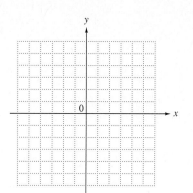

13. Explain how the discriminant can be used to determine the number of x-intercepts of the graph of a quadratic function.

14. Which of the following would most closely resemble the graph of $f(x) = a(x - h)^2 + k$, if $a < 0$, $h > 0$, and $k < 0$?

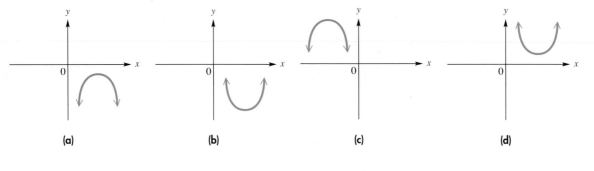

(a)   (b)   (c)   (d)

15. From 1982 to 1990, sales (in billions of dollars) of video games were approximated by the function

$$f(x) = .2x^2 - 1.6x + 3.3,$$

where x is the number of years since 1982. In what year during that period were sales a minimum? What were the minimum sales?

16. The height (in feet) of a projectile t seconds after being fired into the air is given by $f(t) = -16t^2 + 160t$. Find the number of seconds required for the projectile to reach maximum height. What is the maximum height?

17. Find the length and width of a rectangle having a perimeter of 200 meters if the area is to be a maximum.

18. Find the two numbers whose sum is 10 and whose product is a maximum.

[11.3] *Graph each of the following.*

19. $f(x) = |x + 4|$

20. $f(x) = \dfrac{1}{x - 4}$

21. $f(x) = \sqrt{x} + 3$

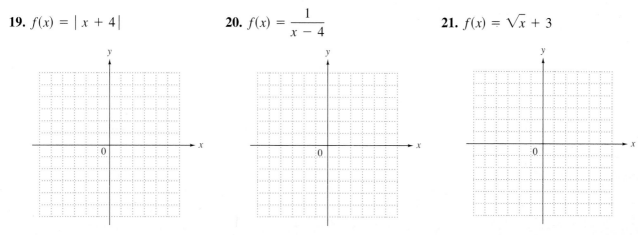

22. Find the distance between the points $(-4, 3)$ and $(5, 4)$.

Write an equation for each of the following circles.

23. Center $(-1, -3)$, $r = 5$

24. Center $(4, 2)$, $r = 6$

Find the center and radius of each circle.

25. $x^2 + y^2 + 6x - 4y - 3 = 0$

26. $x^2 + y^2 - 8x - 2y + 13 = 0$

27. $2x^2 + 2y^2 + 4x + 20y = -34$

28. $4x^2 + 4y^2 - 24x + 16y = 48$

[11.4] *Graph each of the following.*

29. $\dfrac{x^2}{16} + \dfrac{y^2}{9} = 1$

30. $\dfrac{x^2}{49} + \dfrac{y^2}{25} = 1$

31. $\dfrac{x^2}{16} - \dfrac{y^2}{25} = 1$

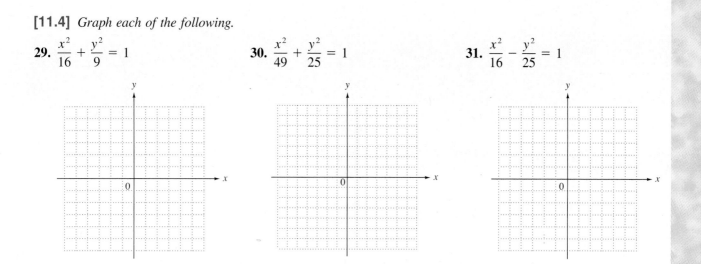

32. $\dfrac{y^2}{25} - \dfrac{x^2}{4} = 1$

33. $x^2 + 9y^2 = 9$

34. $f(x) = -\sqrt{16 - x^2}$

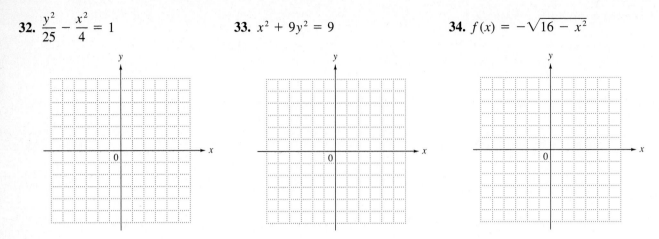

Identify each of the following as a parabola, circle, ellipse, or hyperbola.

35. $x^2 + y^2 = 64$

36. $y = 2x^2 - 3$

37. $y^2 = 2x^2 - 8$

38. $y^2 = 8 - 2x^2$

39. $x = y^2 + 4$

40. $x^2 - y^2 = 64$

[11.5] *Solve each system.*

41. $\quad 2y = 3x - x^2$
$\quad\;\; x + 2y = -12$

42. $y + 1 \;\; = x^2 + 2x$
$\quad\; y + 2x = 4$

43. $x^2 + 3y^2 = 28$
$\quad\;\; y - x \;\; = -2$

44. $\quad\;\; xy = 8$
$\quad x - 2y = 6$

45. $x^2 + \;\; y^2 = 6$
$\quad x^2 - 2y^2 = -6$

46. $3x^2 - 2y^2 = 12$
$\quad\;\; x^2 + 4y^2 = 18$

47. How many solutions are possible for a system of two equations whose graphs are a circle and a line?

48. How many solutions are possible for a system of two equations whose graphs are a parabola and a hyperbola?

[11.6] *Graph each inequality.*

49. $9x^2 \geq 16y^2 + 144$

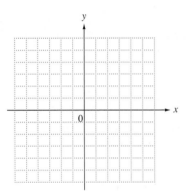

50. $4x^2 + y^2 \geq 16$

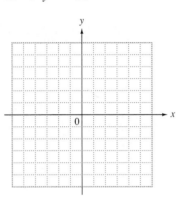

51. $y < -(x + 2)^2 + 1$

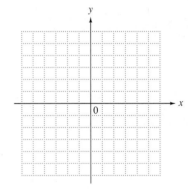

Graph each system of inequalities.

52. $2x + 5y \leq 10$
$3x - y \leq 6$

53. $|x| \leq 2$
$|y| > 1$
$4x^2 + 9y^2 \leq 36$

54. $9x^2 \leq 4y^2 + 36$
$x^2 + y^2 \leq 16$

Graph.

55. $\dfrac{x^2}{64} + \dfrac{y^2}{25} = 1$

56. $\dfrac{y^2}{4} - 1 = \dfrac{x^2}{9}$

57. $x^2 + y^2 = 25$

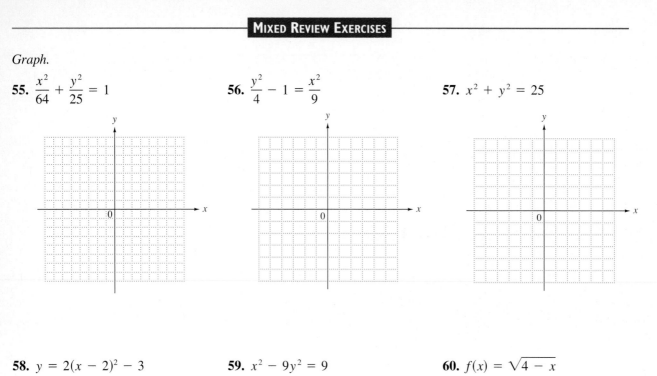

58. $y = 2(x - 2)^2 - 3$

59. $x^2 - 9y^2 = 9$

60. $f(x) = \sqrt{4 - x}$

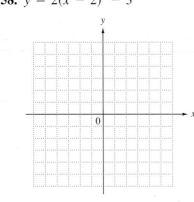
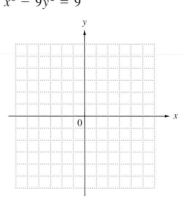
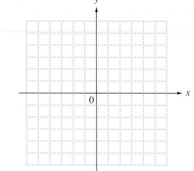

61. $3x + 2y \geq 0$
     $y \leq 4$
     $x \leq 4$

62. $4y > 3x - 12$
     $x^2 < 16 - y^2$

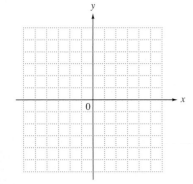
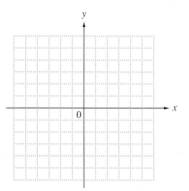

1. Graph the quadratic function defined by $f(x) = \dfrac{1}{2}x^2 - 2$. Identify the vertex.

2. Identify the vertex of the graph of $f(x) = -x^2 + 4x - 1$. Sketch the graph. Give the domain and the range.

3. The number of chinook salmon that returned each year through the Lower Granite Dam in the state of Washington during the period from 1982 to 1987 can be approximated by the quadratic function with

$$f(x) = -2.6x^2 + 11.7x + 22.5,$$

where $x = 0$ corresponds to 1982, $x = 1$ corresponds to 1983, and so on, and $f(x)$ is in thousands. (*Source:* Idaho Department of Fish and Game)

(a) Based on this model, how many returned in 1983?

(b) In what year did the return reach a maximum? To the nearest thousand, how many salmon returned that year?

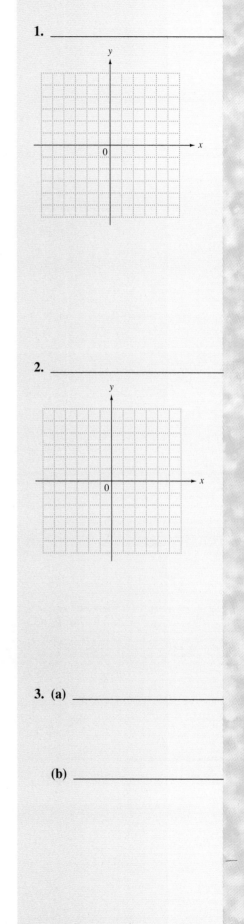

1. _____

2. _____

3. (a) _____

(b) _____

4. _____

4. Which one of these graphs most closely resembles the graph of
$f(x) = (x - 2)^2 + 2$?

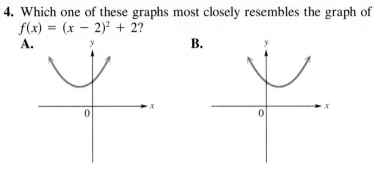

A.   **B.**

5.

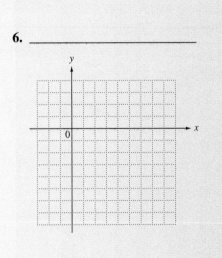

5. Sketch the graph of $f(x) = |x - 3| + 4$.

6. _____

6. Find the center and radius of the circle whose equation is
$(x - 2)^2 + (y + 3)^2 = 16$. Sketch the graph.

7. _____

7. Find the center and radius of the circle whose equation is
$x^2 + y^2 + 8x - 2y = 8$.

8. _____

8. If the discriminant of $f(x) = ax^2 + bx + c$ is negative, what can
you conclude about its graph?

Graph each of the following.

9. $f(x) = \sqrt{9 - x^2}$

10. $4x^2 + 9y^2 = 36$

11. $16y^2 - 4x^2 = 64$

12. $x = -(y - 2)^2 + 2$

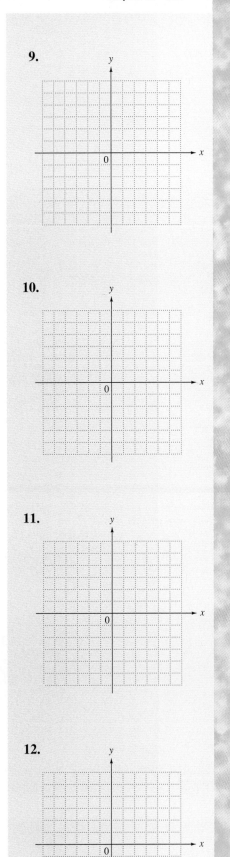

9.

10.

11.

12.

Identify each of the following as the equation of a parabola, hyperbola, ellipse, or circle.

13. _____

13. $6x^2 + 4y^2 = 12$

14. _____

14. $16x^2 = 144 + 9y^2$

15. _____

15. $4y^2 + 4x = 9$

Solve each nonlinear system.

16. _____

16. $2x - y = 9$
     $xy = 5$

17. _____

17. $x - 4 = 3y$
     $x^2 + y^2 = 8$

18. _____

18. $x^2 + \ y^2 = 25$
     $x^2 - 2y^2 = 16$

19.

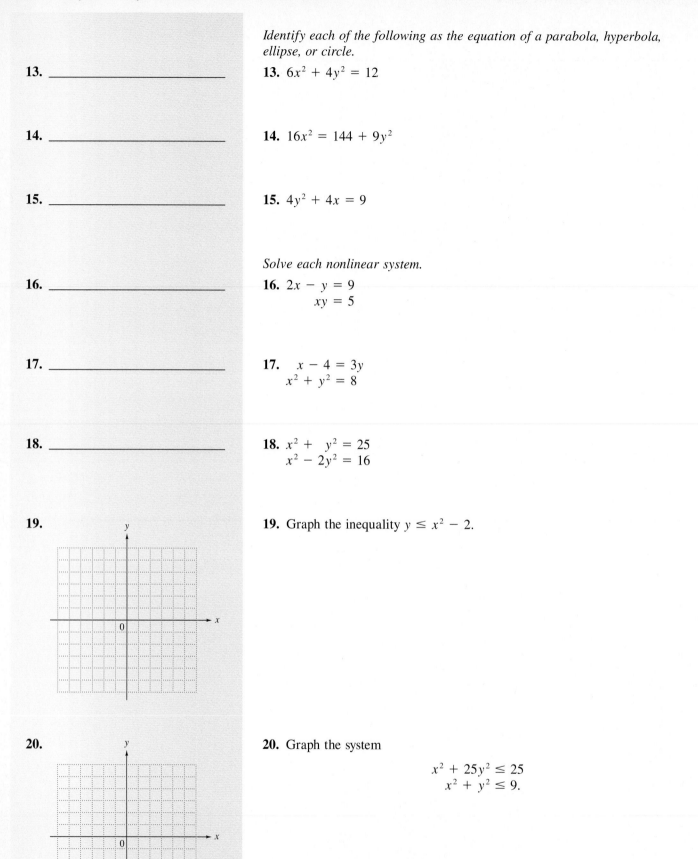

19. Graph the inequality $y \le x^2 - 2$.

20.

20. Graph the system

$$x^2 + 25y^2 \le 25$$
$$x^2 + y^2 \le 9.$$

1. Simplify $-10 + |-5| - |3| + 4$.

Solve.

2. $4 - (2x + 3) + x = 5x - 3$

3. $-4k + 7 \geq 6k + 1$

4. $|5m| - 6 = 14$

5. $|2p - 5| > 15$

6. Find the slope of the line through $(2, 5)$ and $(-4, 1)$.

7. Find the equation of the line through $(-3, -2)$ and perpendicular to $2x - 3y = 7$.

Perform the indicated operations.

8. $(5y - 3)^2$

9. $(2r + 7)(6r - 1)$

10. $(8x^4 - 4x^3 + 2x^2 + 13x + 8) \div (2x + 1)$

Factor.

11. $12x^2 - 7x - 10$ **12.** $2y^4 + 5y^2 - 3$ **13.** $z^4 - 1$ **14.** $100x^4 - 81z^2$

Simplify.

15. $\dfrac{5x - 15}{24} \cdot \dfrac{64}{3x - 9}$

16. $\dfrac{y^2 - 4}{y^2 - y - 6} \div \dfrac{y^2 - 2y}{y - 1}$

17. $\dfrac{5}{c + 5} - \dfrac{2}{c + 3}$

18. $\dfrac{p}{p^2 + p} + \dfrac{1}{p^2 + p}$

19. Kareem and Jamal want to clean their office. Kareem can do the job alone in 3 hours, while Jamal can do it alone in 2 hours. How long will it take them if they work together?

Simplify. Assume all variables represent positive real numbers.

20. $\left(\dfrac{4}{3}\right)^{-1}$

21. $\dfrac{(2a)^{-2}a^4}{a^{-3}}$

22. $4\sqrt[3]{16} - 2\sqrt[3]{54}$

23. $\dfrac{3\sqrt{5x}}{\sqrt{2x}}$

24. $\dfrac{5 + 3i}{2 - i}$

Solve.

25. $2\sqrt{k} = \sqrt{5k + 3}$

26. $10q^2 + 13q = 3$

27. $(4x - 1)^2 = 8$

28. $3k^2 - 3k - 2 = 0$

29. $2(x^2 - 3)^2 - 5(x^2 - 3) = 12$

30. $F = \dfrac{kwv^2}{r}$ for v

Graph.

31. $f(x) = -3x + 5$

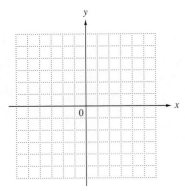

32. $f(x) = -2(x - 1)^2 + 3$

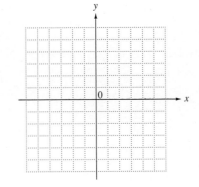

33. $f(x) = \sqrt{x - 2}$

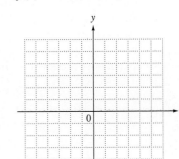

34. $\dfrac{x^2}{4} - \dfrac{y^2}{16} = 1$

35. $\dfrac{x^2}{25} + \dfrac{y^2}{16} \le 1$

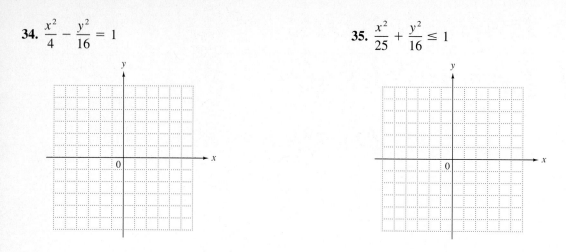

Solve each system.

36. $3x - y = 12$
$2x + 3y = -3$

37. $x + y - 2z = 9$
$2x + y + z = 7$
$3x - y - z = 13$

38. $xy = -5$
$2x + y = 3$

Solve each problem.

39. Al and Bev traveled from their apartment to a picnic 20 miles away. Al traveled on his bike; Bev, who left later, took her car. Al's average speed was half of Bev's average speed. The trip took Al 1/2 hour longer than Bev. What was Bev's average speed?

40. A cash drawer contains only fives and twenties. There are eight more fives than twenties. The total value of the money is $215. How many of each type of bill are in the drawer?

Exponential and Logarithmic Functions

12

12.1 Inverse Functions

In this chapter we will study two important types of functions, *exponential* and *logarithmic*. These functions are related in a special way. They are *inverses* of one another. We begin by discussing inverse functions in general.

> **Note**
> A calculator with the following keys will be very helpful in this chapter.
>
> $\boxed{y^x}$, $\boxed{10^x}$ or $\boxed{\log x}$, $\boxed{e^x}$ or $\boxed{\ln x}$
>
> We will explain how these keys are used at appropriate places in the chapter.

OBJECTIVE 1 Suppose G is the function $\{(-2, 2), (-1, 1), (0, 0), (1, 3), (2, 5)\}$. We can form another set of ordered pairs from G by exchanging the x- and y-values of each pair in G. Call this set F, with

$$F = \{(2, -2), (1, -1), (0, 0), (3, 1), (5, 2)\}.$$

To show that these two sets are related, F is called the *inverse* of G. For a function to have an inverse that is also a function, the given function must be *one-to-one*. In a **one-to-one function** each x-value corresponds to only one y-value and each y-value corresponds to only one x-value.

The function shown in Figure 1(a) is not one-to-one because the y-value 7 corresponds to *two* x-values, 2 and 3. That is, the ordered pairs $(2, 7)$ and $(3, 7)$ both appear in the function. The function in Figure 1(b) is one-to-one.

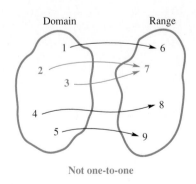
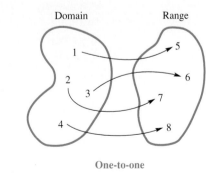

(a)   **(b)**

FIGURE 1

OBJECTIVES

1. Decide whether a function is one-to-one and, if it is, find its inverse.

2. Use the horizontal line test to determine whether a function is one-to-one.

3. Find the equation of the inverse of a function.

4. Graph the inverse f^{-1} from the graph of f.

FOR EXTRA HELP

Tutorial   Tape 24   SSM, Sec. 12.1

1. Decide whether or not each function is one-to-one. If it is, find the inverse.

(a) {(1, 2), (2, 4), (3, 3), (4, 5)}

(b) {(0, 3), (−1, 2), (1, 3)}

(c) How Far Can You See from a Plane?

Height	Miles
5,000	87
15,000	149
20,000	172
35,000	228

The *inverse* (IN-vers) of any one-to-one function f is found by exchanging the components of the ordered pairs of f. The inverse of f is written f^{-1}. Read f^{-1} as "the inverse of f" or "f-inverse." The definition of the inverse of a function follows.

The **inverse** of a one-to-one function f, written f^{-1}, is the set of all ordered pairs of the form (y, x), where (x, y) belongs to f.

Caution

The symbol $f^{-1}(x)$ does not represent $\dfrac{1}{f(x)}$.

Because we form the inverse by interchanging x and y, the domain of f becomes the range of f^{-1}, and the range of f becomes the domain of f^{-1}.

E X A M P L E 1   Deciding Whether a Function Is One-to-One

Decide whether each function is one-to-one. If it is, find its inverse.

(a) $F = \{(-2, 1), (-1, 0), (0, 1), (1, 2), (2, 2)\}$

   Each x-value in F corresponds to just one y-value. However, the y-value 2 corresponds to two x-values, 1 and 2. Also, the y-value 1 corresponds to both -2 and 0. Because some y-values correspond to more than one x-value, F is not one-to-one and does not have an inverse.

(b) $G = \{(3, 1), (0, 2), (2, 3), (4, 0)\}$

   Every x-value in G corresponds to only one y-value, and every y-value corresponds to only one x-value, so G is a one-to-one function. The inverse function is found by exchanging the numbers in each ordered pair.

$$G^{-1} = \{(1, 3), (2, 0), (3, 2), (0, 4)\}$$

Notice how the domain and range of G become the range and domain, respectively, of G^{-1}.

(c) Let f be the function defined by the correspondence in the table. Then f is not one-to-one because 16 minutes corresponds to vacuuming and bicycling. Also, 13 minutes corresponds to jogging and skiing.

Minutes Needed to Burn 100 Calories by Exercising	
vacuuming	16
walking	27
jogging	13
running	6
skiing	13
bicycling	16

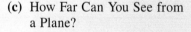 **WORK PROBLEM 1 AT THE SIDE.**

ANSWERS
1. **(a)** {(2, 1), (4, 2), (3, 3), (5, 4)}
(b) not a one-to-one function
(c)

Miles	Height
87	5,000
149	15,000
172	20,000
228	35,000

OBJECTIVE 2 It may be difficult to decide whether a function is one-to-one just by looking at the equation that defines the function. However, by graphing the function and observing the graph, we can use the following *horizontal line test* to tell whether the function is one-to-one.

Horizontal Line Test

A function is one-to-one if every horizontal line intersects the graph of the function at most once.

The horizontal line test follows from the definition of a one-to-one function. Any two points that lie on the same horizontal line have the same y-coordinate. No two ordered pairs that belong to a one-to-one function may have the same y-coordinate, and therefore no horizontal line will intersect the graph of a one-to-one function more than once.

┌─ **E X A M P L E   2   Using the Horizontal Line Test**

Use the horizontal line test to determine whether the graphs in Figures 2 and 3 are graphs of one-to-one functions.

(a)

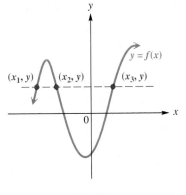

FIGURE 2

Because the horizontal line shown in Figure 2 intersects the graph in more than one point (actually three points in this case), the function is not one-to-one.

(b)

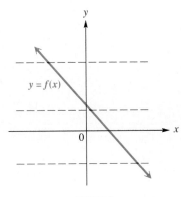

FIGURE 3

Every horizontal line will intersect the graph in Figure 3 in exactly one point. This function is one-to-one.

└─

WORK PROBLEM 2 AT THE SIDE. ▶▶

OBJECTIVE **3**▶ By definition, the inverse of a function is found by exchanging the x- and y-values of the ordered pairs of the function, which reverses the correspondence. The equation of the inverse of a function defined by $y = f(x)$ is found in the same way.

---

2. Use the horizontal line test to determine whether each graph is the graph of a one-to-one function.

(a)

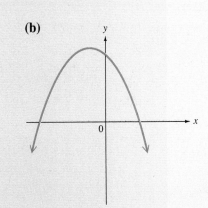

(b)

ANSWERS

2. (a) one-to-one
   (b) not one-to-one

3. Decide whether each equation defines a one-to-one function. If so, find the equation of the inverse.

(a) $f(x) = 3x - 4$

(b) $f(x) = x^3 + 1$

(c) $f(x) = (x - 3)^2$

Inverse of $y = f(x)$

For a one-to-one function f defined by an equation $y = f(x)$, find the defining equation of the inverse as follows.

Step 1 Exchange x and y.

Step 2 Solve for y.

Step 3 Replace y with $f^{-1}(x)$.

This procedure is illustrated in the following example.

E X A M P L E   3   Finding the Equation of the Inverse

Decide whether each of the following defines a one-to-one function. If so, find the equation of the inverse.

(a) $f(x) = 2x + 5$

This linear function has a straight line graph, so by the horizontal line test it is a one-to-one function. Use the steps given above to find the inverse. Let $y = f(x)$ so that

$$y = 2x + 5.$$

Step 1 $\qquad x = 2y + 5$ $\qquad$ Exchange x and y.

Step 2 $\qquad 2y = x - 5$

$$y = \frac{x - 5}{2} \qquad \text{Solve for } y.$$

Step 3 $\qquad f^{-1}(x) = \frac{x - 5}{2}.$ $\qquad$ Replace y with $f^{-1}(x)$.

In the function defined by $y = 2x + 5$, we find the value of y by starting with a value for x, multiplying by 2, and adding 5. The inverse function has us *subtract* 5 and then *divide* by 2. This shows how an inverse "undoes" what a function does to the variable x.

(b) $f(x) = (x - 2)^3$

Because of the cube, each value of x produces just one value of y, so this is a one-to-one function. We find the inverse by replacing $f(x)$ with y and then exchanging x and y.

$$y = (x - 2)^3$$
$$x = (y - 2)^3 \qquad \text{Exchange } x \text{ and } y.$$
$$\sqrt[3]{x} = \sqrt[3]{(y - 2)^3} \qquad \text{Take cube roots.}$$
$$\sqrt[3]{x} = y - 2$$
$$\sqrt[3]{x} + 2 = y \qquad \text{Solve for } y.$$
$$f^{-1}(x) = \sqrt[3]{x} + 2 \qquad \text{Replace } y \text{ with } f^{-1}(x).$$

(c) $f(x) = x^2 + 2$

Both $x = 3$ and $x = -3$ correspond to $y = 11$. Because of the x^2 term, there are many pairs of x-values that correspond to the same y-value. This means that the function defined by $f(x) = x^2 + 2$ is not one-to-one.

◀◀ WORK PROBLEM 3 AT THE SIDE.

ANSWERS

3. (a) $f^{-1}(x) = \dfrac{x + 4}{3}$

(b) $f^{-1}(x) = \sqrt[3]{x - 1}$

(c) not a one-to-one function

─ **E X A M P L E   4**   **Finding $f(a)$ and $f^{-1}(a)$ Where a Is a Constant**

Let $f(x) = x^3$. Find the following.

(a) $f(-2) = (-2)^3 = -8$

(b) $f(3) = 3^3 = 27$

(c) $f^{-1}(-8)$

From part (a), $(-2, -8)$ belongs to the function f. Because f is one-to-one, $(-8, -2)$ belongs to f^{-1}, with $f^{-1}(-8) = -2$.

(d) $f^{-1}(27)$

Because $(3, 27)$ belongs to f, it follows that $(27, 3)$ belongs to f^{-1} and $f^{-1}(27) = 3$.

> **WORK PROBLEM 4 AT THE SIDE.** ▶▶

OBJECTIVE 4 Suppose the point (a, b) shown in Figure 4 belongs to a one-to-one function f. Then the point (b, a) belongs to f^{-1}. The line segment connecting (a, b) and (b, a) is perpendicular to, and cut in half by, the line $y = x$. The points (a, b) and (b, a) are "mirror images" of each other with respect to $y = x$. For this reason we can find the graph of f^{-1} from the graph of f by locating the mirror image of each point of f with respect to the line $y = x$.

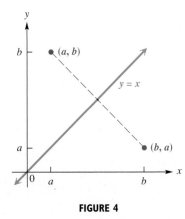

FIGURE 4

4. Find the following for
$$f(x) = \sqrt[3]{1 - x}.$$

(a) $f(2)$

(b) $f(9)$

(c) $f^{-1}(-1)$

(d) $f^{-1}(-2)$

5. Use the given graphs to graph each inverse.

(a)

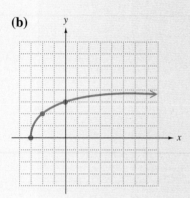

(b)

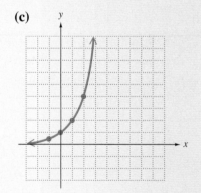

(c)

EXAMPLE 5   Finding the Equation of the Inverse

Graph the inverses of the functions shown in Figure 5.

In Figure 5 the graphs of two functions are shown in **blue**. Their inverses are shown in **red**.

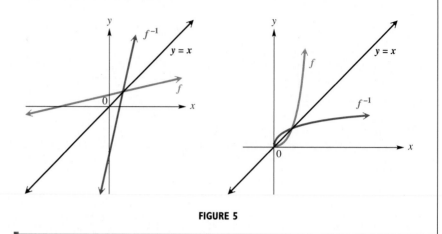

FIGURE 5

◀◀ **WORK PROBLEM 5 AT THE SIDE.**

ANSWERS

5. (a)

(b)

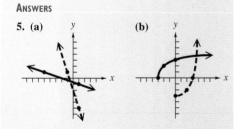

(c)

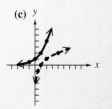

12.1 Exercises

1. Is the function defined by $f(x) = x^2$ a one-to-one function?

2. Does the function defined by $f(x) = x^3$ have an inverse? If so, is it defined by $f^{-1}(x) = \sqrt[3]{x}$?

3. If a function f is one-to-one and the point (a, b) lies on the graph of f, which one of the following *must* lie on the graph of f^{-1}?
  (a) (b, a) **(b)** $(-a, b)$ **(c)** $(-b, -a)$ **(d)** $(a, -b)$

4. Suppose that f is a one-to-one function.
  (a) If $f(3) = 5$, then $f^{-1}(5) =$ _____ .
                (b) If $f(-2) = 4$, then $f^{-1}(4) =$ _____ .
  (c) If $f(-19) = 3$, then $f^{-1}(3) =$ _____ .
         (d) If $f(a) = b$, then $f^{-1}(b) =$ _____ .

5. Suppose you consider the set of ordered pairs (x, y) such that x represents a person in your mathematics class and y represents that person's mother. Explain how this function might not be a one-to-one function.

6. The road mileage between Denver, Colorado, and several selected U.S. cities is shown in the table below.

City	Distance to Denver in Miles
Atlanta	1398
Dallas	781
Indianapolis	1058
Kansas City, MO	600
Los Angeles	1059
San Francisco	1235

If we consider this as a function that pairs a city with a distance, is it a one-to-one function? How could we change the answer to this question by adding 1 mile to one of the distances shown?

If the function is one-to-one, find the inverse. See Examples 1–3.

7. $\{(3, 6), (2, 10), (5, 12)\}$

8. $\{(-1, 3), (0, 5), (5, 0), (7, -\frac{1}{2})\}$

9. $\{(-1, 3), (2, 7), (4, 3), (5, 8)\}$

10. $\{(-8, 6), (-4, 3), (0, 6), (5, 10)\}$

11. $f(x) = 2x + 4$

12. $f(x) = 3x + 1$

13. $g(x) = \sqrt{x - 3}, x \geq 3$

14. $g(x) = \sqrt{x + 2}, x \geq -2$ **15.** $f(x) = 3x^2 + 2$ **16.** $f(x) = -4x^2 - 1$

17. $f(x) = x^3 - 4$ **18.** $f(x) = x^3 - 3$

19. If $m \neq 0$, does the linear function defined by $f(x) = mx + b$ have an inverse? Explain.

20. Does the function defined by $f(x) = x^2$ have an inverse? Explain.

*Let $f(x) = 2^x$. We will see in the next section that this function is one-to-one. Find each of the following, always working part **(a)** before part **(b)**. See Example 4.*

21. (a) $f(3)$
    (b) $f^{-1}(8)$

22. (a) $f(4)$
    (b) $f^{-1}(16)$

23. (a) $f(0)$
    (b) $f^{-1}(1)$

24. (a) $f(-2)$
    (b) $f^{-1}(\frac{1}{4})$

*The graphs of some functions are given in Exercises 25–30. **(a)** Use the horizontal line test to determine whether the function is one-to-one. **(b)** If the function is one-to-one, graph the inverse of the function with a dashed line (or curve) on the same set of axes. (Remember that if f is one-to-one and $f(a) = b$, then $f^{-1}(b) = a$.) See Examples 2 and 5.*

25.   **26.**

27.   **28.**

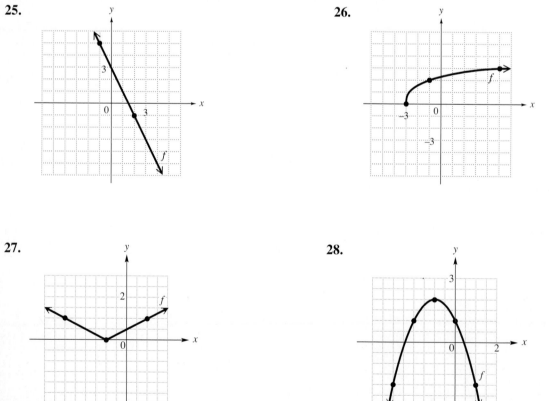

29.

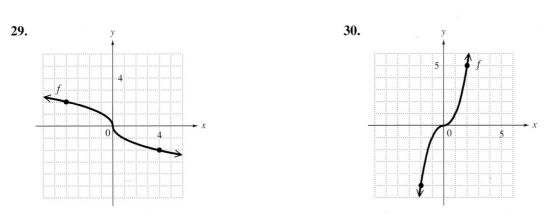

30.

Each function in Exercises 31–38 is a one-to-one function. Graph the function as a solid line (or curve); then graph its inverse on the same set of axes as a dashed line (or curve). In Exercises 35–38 you are given a table to complete so that graphing the function will be a bit easier. See Example 5.

31. $f(x) = 2x - 1$

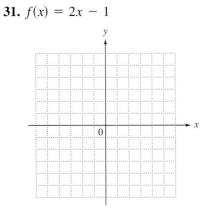

32. $f(x) = 2x + 3$

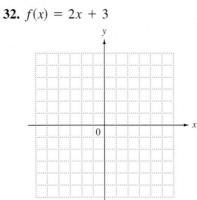

33. $g(x) = -4x$

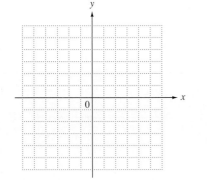

34. $g(x) = -2x$

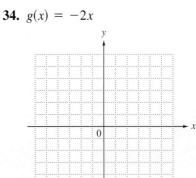

35. $f(x) = \sqrt{x}, x \geq 0$

x	$f(x)$
0	
1	
4	

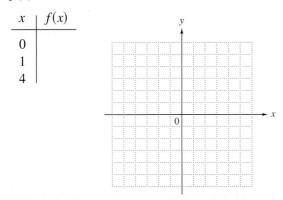

36. $f(x) = -\sqrt{x}, x \geq 0$

x	$f(x)$
0	
1	
4	

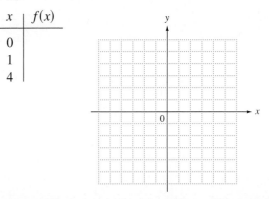

37. $y = x^3 - 2$

38. $y = x^3 + 3$

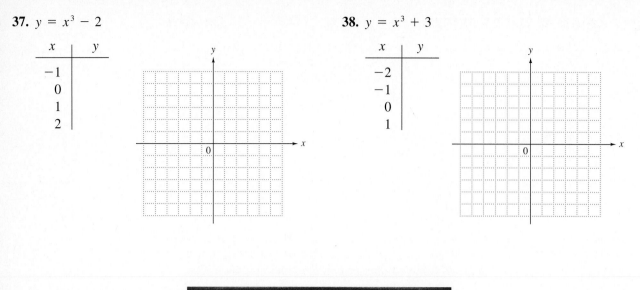

x	y
-1	
0	
1	
2	

x	y
-2	
-1	
0	
1	

INTERPRETING TECHNOLOGY (EXERCISES 39–41)

The screens show the graphs of a pair of functions. In each case, decide whether the relations are inverse functions, and if they are, determine the expression that defines y_2.

39.

40.

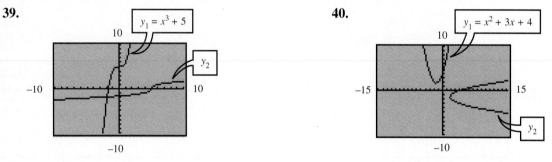

41. Explain why the "inverse" of the function in Exercise 40 does not actually satisfy the definition of inverse as given in this section.

---

42. Which one of the following graphs does not have an inverse as defined in this section?

(a)   **(b)**

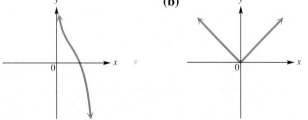

12.2 *Exponential Functions*

OBJECTIVES

1. Identify exponential functions.
2. Graph exponential functions.
3. Solve exponential equations of the form $a^x = a^k$ for x.
4. Use exponential functions in applications.

FOR EXTRA HELP

Tutorial   Tape 25   SSM, Sec. 12.2

OBJECTIVE 1 Earlier, expressions such as 2^x were evaluated for integer values. To define a *rational* power of a positive number, such as $2^{1/2}$, consider that $(\sqrt{2})^2 = 2$ and applying the rule for raising a power to a power leads to $(2^{1/2})^2 = 2^1 = 2$. It seems reasonable to define $2^{1/2}$ as $\sqrt{2}$. In general

$$a^{1/n} = \sqrt[n]{a} \text{ if } a > 0 \text{ and } n > 1$$

and

$$a^{m/n} = (\sqrt[n]{a})^m = \sqrt[n]{a^m} \text{ where these roots exist.}$$

In more advanced courses it is shown that 2^x exists for all real-number values of x, both rational and irrational. (Later in the chapter, we will see how to approximate the value of 2^x for irrational x.) With this assumption, we can now define an exponential function.

Exponential Function

For $a > 0$, $a \neq 1$, and all real numbers x,

$$f(x) = a^x$$

defines an **exponential function.**

Note

The two restrictions on a in the definition of an exponential function are important. The restriction that a must be positive is necessary so that the function can be defined for all real numbers x. For example, letting a be negative ($a = -2$, for instance) and letting $x = 1/2$ would give the expression $(-2)^{1/2} = \sqrt{-2}$, which is not real. The other restriction, $a \neq 1$, is necessary because 1 raised to any power is equal to 1, and the function would then be the linear function $f(x) = 1$.

OBJECTIVE 2 We can graph an exponential function as we do other functions, by finding several ordered pairs that belong to the function. Plotting these points and connecting them with a smooth curve gives the graph.

EXAMPLE 1 Graphing an Exponential Function

Graph the exponential function $f(x) = 2^x$.

We choose values of x and find the corresponding values of y.

x	-3	-2	-1	0	1	2	3	4
$y = 2^x$	$\dfrac{1}{8}$	$\dfrac{1}{4}$	$\dfrac{1}{2}$	1	2	4	8	16

Plotting these points and drawing a smooth curve through them gives the graph shown in Figure 6. This graph is typical of the graphs of exponential functions of the form $f(x) = a^x$, where $a > 1$. The larger the value of a, the faster the graph rises. To see this, compare the graph of $F(x) = 5^x$ with the graph of $f(x) = 2^x$ in Figure 6.

CONTINUED ON NEXT PAGE

Chapter 12 Exponential and Logarithmic Functions

1. Graph.

(a) $f(x) = 10^x$

(b) $g(x) = \left(\dfrac{1}{4}\right)^x$

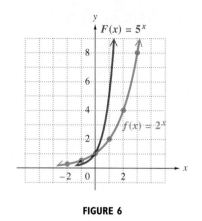

FIGURE 6

By the vertical line test, the graphs in Figure 6 represent functions. As these graphs suggest, the domain of an exponential function includes all real numbers. Because y is always positive, the range is $(0, \infty)$. Figure 6 also shows an important characteristic of exponential functions where $a > 1$; as x gets larger, y increases at a faster and faster rate.

Caution
Be sure to plot enough points to see how rapidly the graph rises.

E X A M P L E 2   Graphing an Exponential Function

Graph $g(x) = \left(\dfrac{1}{2}\right)^x$.

Again, find some points on the graph.

x	-3	-2	-1	0	1	2	3
$y = \left(\dfrac{1}{2}\right)^x$	8	4	2	1	$\dfrac{1}{2}$	$\dfrac{1}{4}$	$\dfrac{1}{8}$

The graph, shown in Figure 7, is the graph of a function. The graph is very similar to that of $f(x) = 2^x$, shown in Figure 6, with the same domain and range, except that here as x gets larger, y *decreases*. This graph is typical of the graph of a function of the form $f(x) = a^x$, where $0 < a < 1$.

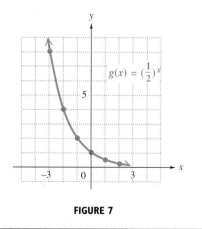

FIGURE 7

1. (a)   **(b)**

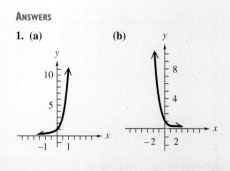

◀◀ **WORK PROBLEM I AT THE SIDE.**

Characteristics of the Graph of $f(x) = a^x$

1. The graph contains the point $(0, 1)$.
2. When $a > 1$, the graph will *rise* from left to right. When $0 < a < 1$, the graph will *fall* from left to right. In both cases, the graph goes from the second quadrant to the first.
3. The graph will approach the x-axis, but never touch it. (It is an asymptote.)
4. The domain is $(-\infty, \infty)$, and the range is $(0, \infty)$.

To graph a more complicated exponential function, we plot carefully selected points, as shown in the next example.

E X A M P L E 3   Graphing an Exponential Function

Graph $f(x) = 3^{2x-4}$.

Find some ordered pairs.

$$\text{If } x = 0, \quad y = 3^{2(0)-4} = 3^{-4} = \frac{1}{81}.$$

$$\text{If } x = 2, \quad y = 3^{2(2)-4} = 3^0 = 1.$$

These ordered pairs, $(0, \frac{1}{81})$ and $(2, 1)$, and other ordered pairs, are shown in the table next to Figure 8. The graph is similar to the graph of $f(x) = 2^x$ except that it is shifted to the right and rises more rapidly.

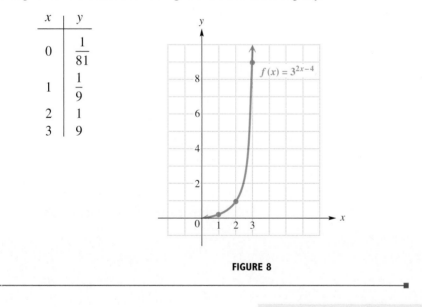

x	y
0	$\frac{1}{81}$
1	$\frac{1}{9}$
2	1
3	9

FIGURE 8

WORK PROBLEM 2 AT THE SIDE. ▶▶

OBJECTIVE 3 ▶ Until now, we have solved only equations that had the variable as a base; all exponents have been constants. An **exponential equation** is an equation that has a variable in an exponent, such as

$$9^x = 27.$$

Because the exponential function $f(x) = a^x$ is a one-to-one function, we can use the following property to solve many exponential equations.

For $a > 0$ and $a \neq 1$, if $a^x = a^y$ then $x = y$.

This property would not necessarily be true if $a = 1$.

2. Graph $y = 2^{4x-3}$.

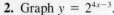

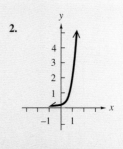

3. Solve. Check your answers.

(a) $25^x = 125$

To solve an exponential equation using this property, go through the following steps.

> **Solving Exponential Equations**
>
> *Step 1* Write each side as a power of the same base.
>
> *Step 2* If necessary, use the rules of exponents to simplify the exponents.
>
> *Step 3* Set the exponents equal.
>
> *Step 4* Solve the equation from Step 3.

> **Note**
> The steps above cannot be applied to an exponential equation like
> $$3^x = 12$$
> because Step 1 cannot easily be done. A method for solving such equations is given in Section 12.6.

(b) $4^x = 32$

E X A M P L E   4   Solving an Exponential Equation

Solve $5^x = 125$.

$$5^x = 125$$

Step 1	$5^x = 5^3$	Get the same base.
Step 3	$x = 3$	Set the exponents equal.

Steps 2 and 4 were not needed here. Check by substituting in the given equation:

$$5^x = 125$$
$$5^3 = 125 \quad ?$$
$$125 = 125 \quad \text{True}$$

The solution set is $\{3\}$.

(c) $81^p = 27$

E X A M P L E   5   Solving an Exponential Equation

Solve $9^x = 27$.

Step 1	$(3^2)^x = 3^3$	Get the same base.
Step 2	$3^{2x} = 3^3$	Simplify exponents.
Step 3	$2x = 3$	Set exponents equal.
Step 4	$x = \dfrac{3}{2}$	Solve.

Check that the solution set is $\{\frac{3}{2}\}$ by substituting $\frac{3}{2}$ for x in the given equation.

◀◀ **WORK PROBLEM 3 AT THE SIDE.**

OBJECTIVE ▶ Exponential equations frequently occur in applications describing growth or decay of some quantity. In particular, they are used to describe the growth and decay of populations.

ANSWERS

3. (a) $\left\{\dfrac{3}{2}\right\}$ **(b)** $\left\{\dfrac{5}{2}\right\}$ **(c)** $\left\{\dfrac{3}{4}\right\}$

EXAMPLE 6   Solving a Growth and Decay Problem

The air pollution, y, in appropriate units, in a large industrial city has been growing according to the equation

$$y = 1000(2)^{.3x}$$

where x is time in years from 1985. That is, $x = 0$ represents 1985, $x = 2$ represents 1987, and so on.

(a) Find the amount of pollution in 1985.
Let $x = 0$, and solve for y.

$$
\begin{aligned}
y &= 1000(2)^{.3x} \\
&= 1000(2)^{(.3)(0)} \qquad \text{Let } x = 0. \\
&= 1000(2)^0 \\
&= 1000(1) \\
&= 1000
\end{aligned}
$$

The pollution in 1985 was 1000 units.

(b) Assuming that growth continued at the same rate, estimate the pollution in 1995.
Here, $x = 10$ represents 1995.

$$
\begin{aligned}
y &= 1000(2)^{.3x} \\
&= 1000(2)^{(.3)(10)} \qquad \text{Let } x = 10. \\
&= 1000(2)^3 \\
&= 1000(8) \\
&= 8000
\end{aligned}
$$

In 1995 the pollution was about 8000 units.

(c) Graph $y = 1000(2)^{.3x}$.
The scale on the y-axis must be quite large to allow for the very large y-values. A calculator can be used to find a few more ordered pairs. The y^x (or x^y) key on the calculator is used to find values of numbers to a variable power. For example, to find y if $x = 15$ the equation gives $y = 1000(2)^{4.5}$. To evaluate $2^{4.5}$ with a calculator, touch 2, then the $\boxed{y^x}$ key, then touch $\boxed{4.5}$, then the $\boxed{\text{Enter}}$ or $\boxed{=}$ key. You should get $2^{4.5} \approx 22.6$, so $y \approx 1000(22.6) = 22{,}600$. The graph is shown in Figure 9.

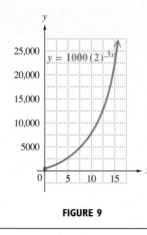

FIGURE 9

WORK PROBLEM 4 AT THE SIDE. ▶▶

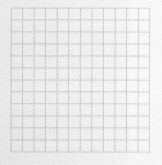
ANSWERS

4. **(a)** 100 grams **(b)** $33\dfrac{1}{3}$ grams

(c) .41 gram
(d)

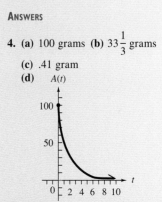

NUMBERS IN THE
Real World *a graphing calculator minicourse*

Lesson 7: Solving a Nonlinear System of Equations

In Section 11.5 we saw how nonlinear systems of equations may be solved by elimination, substitution, or a combination of the methods. We can support our solutions graphically by graphing the equations that make up the system, and then finding the point(s) of intersection of the graphs. Suppose that we wish to solve the system found in Exercise 7 of Section 11.5 using graphical methods. The system is $y = x^2 + 6x + 9$, $x + y = 3$. The first equation has a parabola as its graph, and the second has a line as its graph. If we graph $y_1 = x^2 + 6x + 9$ and $y_2 = -x + 3$ on the same set of axes, we see that one point of intersection is $(-6, 9)$. The other is $(-1, 4)$. See the first figure. These are the same solutions that we obtain if we use an algebraic method.

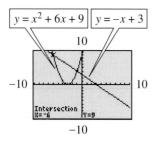

A nonlinear system such as the one found in Example 1 of Section 11.5 is a bit more difficult to solve graphically, since the first equation of the system has a circle as its graph. A circle does not represent the graph of a function, so it will be necessary to solve $x^2 + y^2 = 9$ for y as follows: $x^2 + y^2 = 9$, $y^2 = 9 - x^2$, $y = \pm\sqrt{9 - x^2}$.

If we graph $y_1 = \sqrt{9 - x^2}$, $y_2 = -\sqrt{9 - x^2}$, and $y_3 = 2x - 3$, we can use the intersection-of-graphs capability of the calculator to find the solution set of the system $x^2 + y^2 = 9$, $2x - y = 3$. As seen in the second figure, one point of intersection is $(2.4, 1.8)$. The other is $(0, -3)$. This supports the algebraic results found earlier.

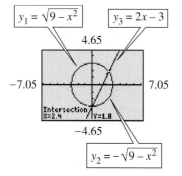

(In order to obtain an accurate depiction of a circle on a graphing calculator screen, we must use a "square" window. Consult your owner's manual for details.)

GRAPHING CALCULATOR EXPLORATIONS

1. The nonlinear system $y = x^2 + 6x$, $3y = 12x$ appears in Exercise 6 of Section 11.5.

   (a) Solve the system algebraically.

   (b) Graph the first equation as $y_1 = x^2 + 6x$ and the second as $y_2 = 4x$.
   Use the intersection-of-graphs capability of your calculator to support your algebraic solution from part (a).

2. The nonlinear system $x^2 + y^2 = 4$, $y = x^2 - 2$ appears in Exercise 18 of Section 11.5.

   (a) Solve the system algebraically. (*Hint:* There are three solutions, and two of them have irrational x-coordinates.)

   (b) Graph the circle $x^2 + y^2 = 4$ by graphing $y_1 = \sqrt{4 - x^2}$ and $y_2 = -\sqrt{4 - x^2}$. Use a square viewing window. Then graph the parabola $y_3 = x^2 - 2$. Support your results of part (a) using intersection of graphs.

   (c) Show that the irrational x-coordinates have decimal approximations equal to the values you find when you use the square root key of your calculator.

12.2 Exercises

Fill in the blanks with the correct responses.

1. For an exponential function $f(x) = a^x$, if $a > 1$, the graph _____ from left to right.
 (rises/falls)

If $0 < a < 1$, the graph _____ from left to right.
 (rises/falls)

2. The y-intercept of the graph of $y = a^x$ is _____ .

3. The graph of the exponential function $y = a^x$ _____ have an x-intercept.
 (does/does not)

4. The point $(2, \underline{\hspace{1cm}})$ is on the graph of $f(x) = 3^{4x-3}$.

Graph the following exponential functions. See Examples 1–3.

5. $f(x) = 3^x$

6. $f(x) = 5^x$

7. $g(x) = \left(\dfrac{1}{3}\right)^x$

8. $g(x) = \left(\dfrac{1}{5}\right)^x$

9. $y = 2^{2x-2}$

10. $y = 2^{2x+1}$

11. Based on your answer to Exercise 1, make a conjecture (an educated guess) about whether an exponential function $f(x) = a^x$ is one-to-one. Then decide whether it has an inverse based on the concepts of Section 12.1.

12. What is the domain of an exponential function $f(x) = a^x$? What is its range?

Solve the following equations. See Examples 4 and 5.

13. $6^x = 36$

14. $8^x = 64$

15. $100^x = 1000$

16. $8^x = 4$

17. $16^{2x+1} = 64^{x+3}$

18. $9^{2x-8} = 27^{x-4}$

19. $5^x = \dfrac{1}{125}$

20. $3^x = \dfrac{1}{81}$

21. $5^x = .2$

22. $10^x = .1$

23. $\left(\dfrac{3}{2}\right)^x = \dfrac{8}{27}$

24. $\left(\dfrac{4}{3}\right)^x = \dfrac{27}{64}$

Solve each of the following applications of exponential functions. Use a calculator with a y^x key. See Example 6.

25. The amount of radioactive material in an ore sample is given by the function

$$A(t) = 100(3.2)^{-.5t},$$

where $A(t)$ is the amount present, in grams, in the sample t months after the initial measurement.

(a) How much was present at the initial measurement? *(Hint: $t = 0$.)*

(b) How much was present 2 months later?

(c) How much was present 10 months later?

26. The population of Brazil, in millions, is approximated by the function

$$f(x) = 155.3(2)^{.025x},$$

where $x = 0$ corresponds to 1994, $x = 1$ corresponds to 1995, and so on.

(a) What will be the population of Brazil in the year 2000 according to this model?

(b) What will be the population in 2034?

(c) How will the population in 2034 compare to the population in 1994?

The bar graph shows the average annual major league baseball player's salary for each year since free agency began. Using a technique from statistics, it was determined that the function

$$S(x) = 74{,}741(1.17)^x$$

approximates the salary, where $x = 0$ corresponds to 1976, and so on, up to $x = 18$ representing 1994. (Salary is in dollars.)

27. Based on this model, what was the average salary in 1986?

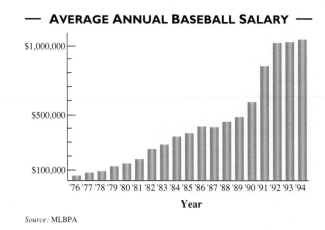

— **AVERAGE ANNUAL BASEBALL SALARY** —

Source: MLBPA

28. Based on the graph, in what year did the average salary first exceed \$1,000,000?

29. The accompanying graphing calculator screen shows the function $S(x)$ graphed from $x = 0$ to $x = 20$. Interpret the display at the bottom of the screen.

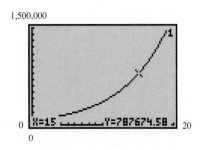

30. In the bar graph, we see that the tops of the bars rise from left to right, and in the calculator-generated graph, the curve rises from left to right. What part of the equation $S(x) = 74{,}741(1.17)^x$ indicates that during this time period, baseball salaries were *rising*?

12.3 *Logarithmic Functions*

OBJECTIVES

1. ▶ Define a logarithm.

2. ▶ Write exponential statements in logarithmic form and logarithmic statements in exponential form.

3. ▶ Solve logarithmic equations of the form $\log_a b = k$ for a, b, or k.

4. ▶ Graph logarithmic functions.

5. ▶ Use logarithmic functions in applications.

FOR EXTRA HELP

Tutorial   Tape 25   SSM, Sec. 12.3

The graph of $y = 2^x$ is the curve shown in **blue** in Figure 10. Because $y = 2^x$ is a one-to-one function, it has an inverse. By interchanging x and y, we get $x = 2^y$, the inverse of $y = 2^x$. As we saw in Section 12.1, the graph of the inverse is found by reflecting the graph of $y = 2^x$ about the line $y = x$. The graph of $x = 2^y$ is shown as a **red** curve in Figure 10.

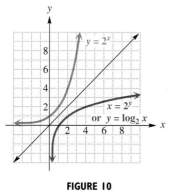

FIGURE 10

OBJECTIVE 1 ▶ We cannot solve the equation $x = 2^y$ for the dependent variable y with the methods presented up to now. The following definition is used to solve $x = 2^y$ for y.

Definition of Logarithm

For all positive numbers a, $a \neq 1$, and all positive numbers x,

$$y = \log_a x \text{ means the same as } x = a^y.$$

The abbreviation **log** is used for **logarithm** (LOG-uh-rith-im). Read $\log_a x$ as "the logarithm of x to the base a." **This key statement should be memorized.** To remember the location of the base and exponent in each form, refer to the diagram below.

$$\text{Logarithmic form: } y = \log_a x$$

$$\text{Exponent} \qquad \text{Base}$$

$$\text{Exponential form: } x = a^y$$

$$\text{Exponent} \qquad \text{Base}$$

When working with logarithmic (log-uh-RITH-mik) form and exponential form, remember the following.

A logarithm is an exponent; $\log_a x$ is the exponent on the base a that yields the number x.

OBJECTIVE 2 ▶ We can use the definition of logarithm to write exponential statements in logarithmic form and logarithmic statements in exponential form.

EXAMPLE 1   **Converting between Exponential and Logarithmic Form**

The following list shows several pairs of equivalent statements. The same statement is written in both exponential and logarithmic form.

—— **CONTINUED ON NEXT PAGE**

1. Complete the chart.

Exponential Form	Logarithmic Form
$2^5 = 32$	——
$100^{1/2} = 10$	——
——	$\log_8 4 = \dfrac{2}{3}$
——	$\log_6 \dfrac{1}{1296} = -4$

Exponential Form	Logarithmic Form
$3^2 = 9$	$\log_3 9 = 2$
$\left(\dfrac{1}{5}\right)^{-2} = 25$	$\log_{1/5} 25 = -2$
$10^5 = 100{,}000$	$\log_{10} 100{,}000 = 5$
$4^{-3} = \dfrac{1}{64}$	$\log_4 \dfrac{1}{64} = -3$

◀◀ **WORK PROBLEM 1 AT THE SIDE.**

OBJECTIVE 3 A **logarithmic equation** is an equation with a logarithm in at least one term. We can solve logarithmic equations of the form $\log_a b = k$ for any of the three variables by first writing the equation in exponential form.

EXAMPLE 2 Solving Logarithmic Equations

Solve the following equations.

2. Solve each equation.

(a) $\log_4 x = -2$

$$\log_4 x = -2$$
$$x = 4^{-2} \qquad \text{Convert to exponential form.}$$
$$x = \frac{1}{16}$$

The solution set is $\left\{\frac{1}{16}\right\}$.

(a) $\log_3 27 = x$

(b) $\log_{1/2} 16 = y$

$$\log_{1/2} 16 = y$$
$$\left(\frac{1}{2}\right)^y = 16 \qquad \text{Convert to exponential form.}$$
$$(2^{-1})^y = 2^4 \qquad \text{Write with same base.}$$
$$2^{-y} = 2^4 \qquad \text{Multiply exponents.}$$
$$-y = 4 \qquad \text{Set exponents equal.}$$
$$y = -4 \qquad \text{Solve.}$$

The solution set is $\{-4\}$.

(b) $\log_5 p = 2$

(c) $\log_m \dfrac{1}{16} = -4$

◀◀ **WORK PROBLEM 2 AT THE SIDE.**

For any positive real number b, we know that $b^1 = b$ and $b^0 = 1$. Writing these two statements in logarithmic form gives the following two properties of logarithms.

For any positive real number b, $b \neq 1$,
$$\log_b b = 1 \quad \text{and} \quad \log_b 1 = 0.$$

EXAMPLE 3 Using Properties of Logarithms

(a) $\log_7 7 = 1$ **(b)** $\log_{\sqrt{2}} \sqrt{2} = 1$

(c) $\log_9 1 = 0$ **(d)** $\log_{.2} 1 = 0$

WORK PROBLEM 3 AT THE SIDE. ▶▶

Now we can define the logarithmic function with base a as follows.

Logarithmic Function

If a and x are positive numbers, with $a \neq 1$, then

$$f(x) = \log_a x$$

defines the **logarithmic function with base a.**

OBJECTIVE 4 To graph a logarithmic function, it is helpful to write it in exponential form first. Then we can plot selected ordered pairs to determine the graph.

┌─ **E X A M P L E 4   Graphing a Logarithmic Function**

Graph $y = \log_{1/2} x$.

Writing $y = \log_{1/2} x$ in its exponential form as $x = (\frac{1}{2})^y$ helps us identify ordered pairs that satisfy the equation. Here it is easier to choose values for y and find the corresponding values of x. Doing this gives the following pairs.

x	$\frac{1}{4}$	$\frac{1}{2}$	1	2	4	8
y	2	1	0	-1	-2	-3

Be careful to get these in the right order.

Plotting these points and connecting them with a smooth curve, we get the graph in Figure 11. This graph is typical of logarithmic functions with base $0 < a < 1$. The graph of $x = 2^y$ in Figure 10, which is equivalent to $y = \log_2 x$, is typical of graphs of logarithmic functions with base $a > 1$.

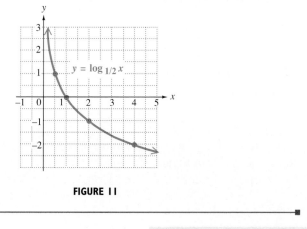

FIGURE 11

WORK PROBLEM 4 AT THE SIDE. ▶▶

Based on the graphs of the functions $y = \log_2 x$ in Figure 10 and $y = \log_{1/2} x$ in Figure 11, we can make the following generalizations about the graphs of logarithmic functions of the form $g(x) = \log_a x$.

Characteristics of the Graph of $g(x) = \log_a x$

1. The graph contains the point $(1, 0)$.
2. When $a > 1$, the graph will *rise* from left to right, from the fourth quadrant to the first. When $0 < a < 1$, the graph will *fall* from left to right, from the first quadrant to the fourth.
3. The graph will approach the y-axis, but never touch it. (It is an asymptote.)
4. The domain is $(0, \infty)$, and the range is $(-\infty, \infty)$.

3. Find the value of each of the following.

(a) $\log_{2/5} \dfrac{2}{5}$

(b) $\log_{.4} 1$

4. Graph.

(a) $y = \log_3 x$

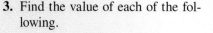

(b) $y = \log_{1/10} x$

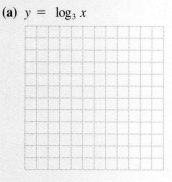

ANSWERS

3. (a) 1 **(b)** 0

4. (a)

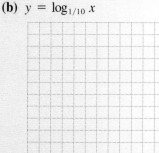

(b)

5. A population of mites in a laboratory is growing so that the population is

$$P(t) = 80 \log_{10}(t + 10),$$

where t is the number of days after a study began. Find $P(t)$ for the following values of t.

(a) $t = 0$

(b) $t = 90$

(c) $t = 990$

(d) Graph the function.

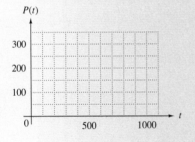

Compare these generalizations to the similar ones for exponential functions in Section 12.2.

OBJECTIVE 5 Logarithmic functions, like exponential functions, are used in applications to describe growth and decay.

E X A M P L E   5   Solving a Logarithmic Growth Problem

Sales (in thousands of units) of a new product are approximated by

$$S(t) = 100 + 30 \log_3(2t + 1),$$

where t is the number of years after the product is introduced.

(a) What were the sales after 1 year?
Let $t = 1$, and find $S(1)$.

$$S(t) = 100 + 30 \log_3(2t + 1)$$
$$S(1) = 100 + 30 \log_3(2 \cdot 1 + 1) \qquad \text{Let } t = 1.$$
$$= 100 + 30 \log_3 3$$
$$= 100 + 30(1) \qquad\qquad \log_3 3 = 1$$
$$= 130$$

Sales were 130 thousand units after 1 year.

(b) Find the sales after 13 years.
Find $S(13)$.

$$S(t) = 100 + 30 \log_3(2t + 1)$$
$$S(13) = 100 + 30 \log_3(2 \cdot 13 + 1) \qquad \text{Let } t = 13.$$
$$= 100 + 30 \log_3 27$$
$$= 100 + 30(3) \qquad\qquad \log_3 27 = 3$$
$$= 190$$

After 13 years, sales had increased to 190 thousand units.

(c) Graph $y = S(t)$.
Use the two ordered pairs $(1, 130)$ and $(13, 190)$ found above. Check that $(0, 100)$ and $(40, 220)$ also satisfy the equation. Use these ordered pairs and a knowledge of the general shape of the graph of a logarithmic function to get the graph in Figure 12.

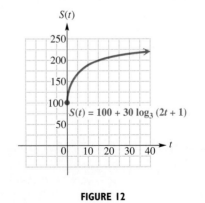

FIGURE 12

◄◄ **WORK PROBLEM 5 AT THE SIDE.**

ANSWERS

5. (a) 80 **(b)** 160 **(c)** 240
 (d)

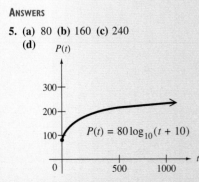

12.3 Exercises

1. By definition $\log_a x$ is the exponent to which the base a must be raised in order to obtain x. Use this to simplify each of the following, without doing any written work. (Example: $\log_3 9$ is 2, because 2 is the exponent to which 3 must be raised in order to obtain 9.)
   (a) $\log_4 16$ (b) $\log_3 81$ (c) $\log_3(\frac{1}{3})$
   (d) $\log_{10} .01$ (e) $\log_5 \sqrt{5}$ (f) $\log_{12} 1$

2. Compare the summary of facts about the graph of $f(x) = a^x$ in Section 12.2 with the similar summary of facts about the graph of $g(x) = \log_a x$ in this section. Make a list of the facts that reinforce the concept that f and g are inverse functions.

Write in logarithmic form. See Example 1.

3. $4^5 = 1024$

4. $3^6 = 729$

5. $\left(\dfrac{1}{2}\right)^{-3} = 8$

6. $\left(\dfrac{1}{6}\right)^{-3} = 216$

7. $10^{-3} = .001$

8. $36^{1/2} = 6$

Write in exponential form. See Example 1.

9. $\log_4 64 = 3$

10. $\log_2 512 = 9$

11. $\log_{10} \dfrac{1}{10,000} = -4$

12. $\log_{100} 100 = 1$

13. $\log_6 1 = 0$

14. $\log_\pi 1 = 0$

15. When a student asked his teacher to explain to him how to evaluate $\log_9 3$ without showing any work, his teacher told him, "Think radically." Explain what the teacher meant by this hint.

16. A student told her teacher "I know that $\log_2 1$ is the exponent to which 2 must be raised in order to obtain 1, but I can't think of any such number." How would you explain to the student that the value of $\log_2 1$ is 0?

Solve each equation for x. See Examples 2 and 3.

17. $x = \log_{27} 3$

18. $x = \log_{125} 5$

19. $\log_x 9 = \dfrac{1}{2}$

20. $\log_x 5 = \dfrac{1}{2}$

21. $\log_x 125 = -3$

22. $\log_x 64 = -6$

23. $\log_{12} x = 0$

24. $\log_4 x = 0$

25. $\log_x x = 1$

26. $\log_x 1 = 0$

27. $\log_x \dfrac{1}{25} = -2$

28. $\log_x \dfrac{1}{10} = -1$

29. $\log_8 32 = x$

30. $\log_{81} 27 = x$

31. $\log_\pi \pi^4 = x$

32. $\log_{\sqrt{2}} \sqrt{2}^9 = x$

33. $\log_6 \sqrt{216} = x$

34. $\log_4 \sqrt{64} = x$

If the point (p, q) is on the graph of $f(x) = a^x$ (for $a > 0$ and $a \neq 1$), then the point (q, p) is on the graph of $f^{-1}(x) = \log_a x$. Use this fact, and refer to the graphs required in Exercises 5–8 in Section 12.2 to graph the following logarithmic functions. See Example 4.

35. $y = \log_3 x$

36. $y = \log_5 x$

37. $y = \log_{1/3} x$

38. $y = \log_{1/5} x$

39. Graph the function $y = \log_{2.718} x$ using the exponential key of a calculator, and rewriting it as $2.718^y = x$. Choose -1, 0, and 1 as y values, and approximate x to the nearest tenth.

40. Use the exponential key of your calculator to find approximations for the expression $(1 + \frac{1}{x})^x$, using x values of 1, 10, 100, 1000, and 10,000. Explain what seems to be happening as x gets larger and larger. (*Hint*: Look at the base in Exercise 39.)

Solve each application of logarithmic functions. See Example 5.

41. A study showed that the number of mice in an old abandoned house was approximated by the function

$$M(t) = 6 \log_4(2t + 4),$$

where t is measured in months and $t = 0$ corresponds to January, 1993. Find the number of mice in the house in
(a) January, 1993
(b) July, 1993
(c) July, 1995.
(d) Graph the function.

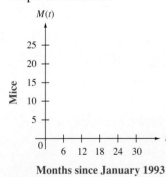

Months since January 1993

42. A supply of hybrid striped bass were introduced into a lake in January, 1980. Biologists researching the bass population over the next decade found that the number of bass in the lake was approximated by the function

$$B(t) = 500 \log_3(2t + 3),$$

where $t = 0$ corresponds to January, 1980, $t = 1$ to January, 1981, $t = 2$ to January, 1982, and so on. Use this function to find the bass population in
(a) January, 1980
(b) January, 1983
(c) January, 1992.
(d) Graph the function for $0 \le t \le 12$.

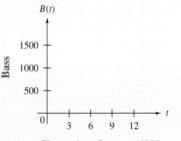

Years since January 1980

Use the graph below to predict the value of $f(t)$ for the given values of t.

43. $t = 0$

44. $t = 10$

45. $t = 60$

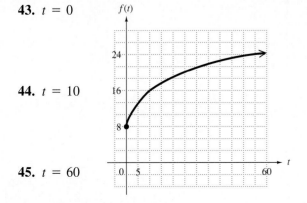

46. Show that the points determined in Exercises 43–45 lie on the graph of $f(t) = 8 \log_5(2t + 5)$.

47. Explain why 1 is not allowed as a base for a logarithmic function.

48. Explain why $\log_a 1$ is 0 for any value of a that is allowed as the base of a logarithm. Use a rule of exponents introduced earlier in your explanation.

49. The domain of $f(x) = a^x$ is $(-\infty, \infty)$, while the range is $(0, \infty)$. Therefore, because $g(x) = \log_a x$ is the inverse of f, the domain of g is _____, while the range of g is _____.

50. The graphs of both $f(x) = 3^x$ and $g(x) = \log_3 x$ rise from left to right. Which one rises at a faster rate?

In the United States, the intensity of an earthquake is rated using the Richter scale. *The Richter scale rating of an earthquake of intensity x is given by*

$$R = \log_{10} \frac{x}{x_0},$$

where x_0 is the intensity of an earthquake of a certain (small) size. The figure shows Richter scale ratings for major Southern California earthquakes since 1920. As the figure indicates, earthquakes "come in bunches" and the 1990s have been an especially busy time.

51. The 1994 Northridge earthquake had a Richter scale rating of 6.7; the Landers earthquake had a rating of 7.3. How much more powerful was the Landers earthquake than the Northridge earthquake?

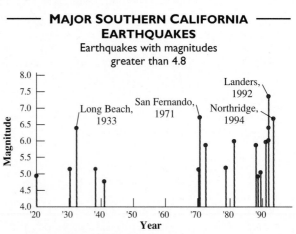

MAJOR SOUTHERN CALIFORNIA EARTHQUAKES

Earthquakes with magnitudes greater than 4.8

Sources: Caltech; U.S. Geological Survey

52. Compare the smallest rated earthquake in the figure (at 4.8) with the Landers quake. How much more powerful was the Landers quake?

12.4 *Properties of Logarithms*

Logarithms are very important in applications and in further work in mathematics. The properties that make logarithms so useful are given in this section.

OBJECTIVE 1 One way in which logarithms simplify problems is by changing a problem of multiplication into one of addition. This is done with the following property of logarithms.

OBJECTIVES
1 Use the multiplication property for logarithms.
2 Use the division property for logarithms.
3 Use the power property for logarithms.
4 Use the properties of logarithms to write logarithmic expressions in alternative forms.

FOR EXTRA HELP

Tutorial   Tape 25   SSM, Sec. 12.4

Multiplication Property for Logarithms

If x, y, and b are positive real numbers, where $b \neq 1$, then

$$\log_b xy = \log_b x + \log_b y.$$

The logarithm of a product is the sum of the logarithms.

Note

The word statement of the product rule can be restated by replacing "logarithm" with "exponent." The rule then becomes the familiar rule for multiplying exponential expressions: The *exponent* of a product is equal to the sum of the *exponents* of the factors.

To prove this rule, let $m = \log_b x$ and $n = \log_b y$, and recall that

$$\log_b x = m \qquad \text{means} \qquad b^m = x,$$
$$\log_b y = n \qquad \text{means} \qquad b^n = y.$$

Now consider the product xy.

$xy = b^m \cdot b^n$	Substitution
$xy = b^{m+n}$	Product rule for exponents
$\log_b xy = m + n$	Convert to logarithmic form.
$\log_b xy = \log_b x + \log_b y$	Substitution

The last statement is the result we wish to prove.

EXAMPLE 1 Using the Multiplication Property

Use the multiplication property to rewrite the following (assume $x > 0$).

(a) $\log_5(6 \cdot 9)$

By the multiplication property,

$$\log_5(6 \cdot 9) = \log_5 6 + \log_5 9.$$

(b) $\log_7 8 + \log_7 12$

$$\log_7 8 + \log_7 12 = \log_7(8 \cdot 12) = \log_7 96$$

(c) $\log_3(3x)$

$$\log_3(3x) = \log_3 3 + \log_3 x$$

Because $\log_3 3 = 1$,

$$\log_3(3x) = 1 + \log_3 x.$$

(d) $\log_4 x^3$

$$\log_4 x^3 = \log_4(x \cdot x \cdot x) \qquad x^3 = x \cdot x \cdot x$$
$$= \log_4 x + \log_4 x + \log_4 x$$
$$= 3 \log_4 x$$

1. Use the multiplication property to rewrite the following.

(a) $\log_6(5 \cdot 8)$

(b) $\log_4 3 + \log_4 7$

(c) $\log_8 8k \quad (k > 0)$

(d) $\log_5 m^2$

2. Use the division property to rewrite the following.

(a) $\log_7 \dfrac{9}{4}$

(b) $\log_3 p - \log_3 q$
$(p > 0, q > 0)$

(c) $\log_4 \dfrac{3}{16}$

◀◀ WORK PROBLEM I AT THE SIDE.

OBJECTIVE 2 The rule for division is similar to the rule for multiplication.

Division Property for Logarithms

If x, y, and b are positive real numbers, where $b \neq 1$, then

$$\log_b \frac{x}{y} = \log_b x - \log_b y.$$

The logarithm of the quotient is the difference of the logarithms.

The proof of this rule is very similar to the proof of the multiplication property.

E X A M P L E 2   Using the Division Property

Use the division property to rewrite the following.

(a) $\log_4 \dfrac{7}{9} = \log_4 7 - \log_4 9$

(b) If $x > 0$, then $\log_5 6 - \log_5 x = \log_5 \dfrac{6}{x}$.

(c) $\log_3 \dfrac{27}{5} = \log_3 27 - \log_3 5$

$$= 3 - \log_3 5 \qquad \log_3 27 = 3$$

◀◀ WORK PROBLEM 2 AT THE SIDE.

OBJECTIVE 3 The next rule gives a method for evaluating powers and roots such as

$$2^{\sqrt{2}}, \quad (\sqrt{2})^{3/4}, \quad (.032)^{5/8}, \quad \text{and} \quad \sqrt[5]{12}.$$

This rule makes it possible to find approximations for numbers that could not be evaluated before. By the multiplication property for logarithms,

$$\log_5 2^3 = \log_5(2 \cdot 2 \cdot 2)$$
$$= \log_5 2 + \log_5 2 + \log_5 2$$
$$= 3 \log_5 2.$$

Also,

$$\log_2 7^4 = \log_2(7 \cdot 7 \cdot 7 \cdot 7)$$
$$= \log_2 7 + \log_2 7 + \log_2 7 + \log_2 7$$
$$= 4 \log_2 7.$$

Furthermore, we saw in Example 1(d) that $\log_4 x^3 = 3 \log_4 x$. These examples suggest the following generalization.

Power Property for Logarithms

If x and b are positive real numbers, where $b \neq 1$, and r is any real number, then

$$\log_b x^r = r(\log_b x).$$

The logarithm of a number to a power equals the exponent times the logarithm of the number.

ANSWERS

1. (a) $\log_6 5 + \log_6 8$ **(b)** $\log_4 21$
(c) $1 + \log_8 k$ **(d)** $2 \log_5 m \quad (m > 0)$

2. (a) $\log_7 9 - \log_7 4$ **(b)** $\log_3 \dfrac{p}{q}$

(c) $\log_4 3 - 2$

As examples of this result,

$$\log_b m^5 = 5 \log_b m \quad \text{and} \quad \log_3 5^{3/4} = \frac{3}{4} \log_3 5.$$

To prove the power property, let

$$\log_b x = m.$$

$b^m = x$	Convert to exponential form.
$(b^m)^r = x^r$	Raise to the power r.
$b^{mr} = x^r$	Power rule for exponents
$\log_b x^r = mr$	Convert to logarithmic form.
$\log_b x^r = rm$	
$\log_b x^r = r \log_b x$	$m = \log_b x$

This is the statement to be proved.

As a special case of this rule, let $r = \frac{1}{p}$, so that $\log_b \sqrt[p]{x} = \log_b x^{1/p} = \frac{1}{p} \log_b x$. For example, using this result with $x > 0$,

$$\log_b \sqrt[5]{x} = \frac{1}{5} \log_b x \quad \text{and} \quad \log_b \sqrt[3]{x^4} = \frac{4}{3} \log_b x.$$

EXAMPLE 3 Using the Power Rule

Use the power rule to rewrite each of the following. Assume $b > 0$, $x > 0$, and $b \neq 1$.

(a) $\log_5 4^2 = 2 \log_5 4$

(b) $\log_b x^5 = 5 \log_b x$

(c) $\log_b \sqrt{7}$

When using the power rule with logarithms of expressions involving radicals, begin by rewriting the radical expression with a rational exponent, as shown earlier.

$$\log_b \sqrt{7} = \log_b 7^{1/2} \qquad \sqrt{x} = x^{1/2}$$

$$= \frac{1}{2} \log_b 7 \qquad \text{Power rule}$$

(d) $\log_2 \sqrt[5]{x^2} = \log_2 x^{2/5} = \frac{2}{5} \log_2 x$

WORK PROBLEM 3 AT THE SIDE. ▶▶

Two special properties involving both exponential and logarithmic expressions come directly from the fact that logarithmic and exponential functions are inverses of each other.

If $b > 0$ and $b \neq 1$, then

$$b^{\log_b x} = x \quad (x > 0) \qquad \text{and} \qquad \log_b b^x = x.$$

To prove the first statement, let

$$y = \log_b x.$$

$b^y = x$	Convert to exponential form.
$b^{\log_b x} = x$	Replace y with $\log_b x$.

The proof of the second statement is similar.

3. Use the power rule to rewrite the following. Assume $a > 0$, $b > 0$, $a \neq 1$, and $b \neq 1$.

(a) $\log_3 5^2$

(b) $\log_a x^4 \quad (x > 0)$

(c) $\log_b \sqrt{8}$

(d) $\log_2 \sqrt[3]{2}$

4. Find the value of each expression.

   (a) $\log_{10} 10^3$

   (b) $\log_2 8$

   (c) $5^{\log_5 3}$

E X A M P L E   4   Using the Special Properties

Find the value of the following logarithmic expressions.

(a) $\log_5 5^4 = 4$, by the second property.

(b) $\log_3 9$

$$\log_3 9 = \log_3 3^2 = 2$$

The second property was used in the last step.

(c) $4^{\log_4 10} = 10$

◀◀ **WORK PROBLEM 4 AT THE SIDE.**

OBJECTIVE 4▶ The properties of logarithms are useful for writing expressions in an alternative form. This use of logarithms is important in calculus.

E X A M P L E   5   Writing Logarithms in Alternative Forms

Use the properties of logarithms to rewrite each expression. Assume all variables represent positive real numbers.

(a) $\log_4 4x^3 = \log_4 4 + \log_4 x^3$ Multiplication property

$\qquad\qquad = 1 + 3 \log_4 x$ $\log_4 4 = 1$; Power property

(b) $\log_7 \sqrt{\dfrac{m}{n}} = \log_7 \left(\dfrac{m}{n}\right)^{1/2}$

$\qquad\qquad = \dfrac{1}{2} \log_7 \dfrac{m}{n}$ Power property

$\qquad\qquad = \dfrac{1}{2}(\log_7 m - \log_7 n)$ Division property

(c) $\log_5 \dfrac{a}{bc} = \log_5 a - \log_5 bc$ Division property

$\qquad\qquad = \log_5 a - (\log_5 b + \log_5 c)$ Multiplication property

$\qquad\qquad = \log_5 a - \log_5 b - \log_5 c$

Notice the careful use of parentheses in the second step. Because we are subtracting the logarithm of a product and rewriting it as a sum of two terms, we must place parentheses around the sum.

(d) $\log_8(2p + 3r)$ cannot be rewritten by the properties of logarithms.

◀◀ **WORK PROBLEM 5 AT THE SIDE.**

5. Write as a sum or difference of logarithms. Assume all variable expressions are positive.

   (a) $\log_6 36m^5$

   (b) $\log_2 \sqrt{9z}$

   (c) $\log_q \dfrac{8r}{m-1}$ $(m \neq 1, q \neq 1)$

   (d) $\log_4(3x + y)$

> **Caution**
> Remember that there is no property of logarithms to rewrite the logarithm of a *sum*. That is,
>
> $$\log_b(x + y) \neq \log_b x + \log_b y.$$

ANSWERS

4. (a) 3 (b) 3 (c) 3

5. (a) $2 + 5 \log_6 m$ (b) $\log_2 3 + \dfrac{1}{2} \log_2 z$

   (c) $\log_q 8 + \log_q r - \log_q(m - 1)$
   (d) cannot be rewritten

12.4 Exercises

Fill in the blanks with the correct responses in Exercises 1–4.

1. The logarithm of the product of two numbers is equal to the _____ of the logarithms of the numbers.

2. The logarithm of the quotient of two numbers is equal to the _____ of the logarithms of the numbers.

3. $\log_a b^k =$ _____ $\log_a b$ $(a > 0, a \neq 1, b > 0)$

4. The logarithm of the square root of a number is equal to _____ times the logarithm of the number.

Use the properties of logarithms introduced in this section to express each of the following as a sum or difference of logarithms, or as a single number if possible. Assume that all variables represent positive real numbers. See Examples 1–5.

5. $\log_7 \dfrac{4}{5}$

6. $\log_8 \dfrac{9}{11}$

7. $\log_2 8^{1/4}$

8. $\log_3 9^{3/4}$

9. $\log_4 \dfrac{3\sqrt{x}}{y}$

10. $\log_5 \dfrac{6\sqrt{z}}{w}$

11. $\log_3 \dfrac{\sqrt[3]{4}}{x^2 y}$

12. $\log_7 \dfrac{\sqrt[3]{13}}{pq^2}$

13. $\log_3 \sqrt{\dfrac{xy}{5}}$

14. $\log_6 \sqrt{\dfrac{pq}{7}}$

15. $\log_2 \dfrac{\sqrt[3]{x} \cdot \sqrt[5]{y}}{r^2}$

16. $\log_4 \dfrac{\sqrt[4]{z} \cdot \sqrt[5]{w}}{s^2}$

17. A student erroneously wrote $\log_a(x + y) = \log_a x + \log_a y$. When his teacher explained that this was indeed wrong, the student claimed he had used the distributive property. Write a few sentences explaining why the distributive property does not apply in this case.

18. Write a few sentences explaining how the rules for multiplying and dividing powers of the same base are similar to the rules for finding logarithms of products and quotients.

Use the properties of logarithms introduced in this section to express each of the following as a single logarithm. Assume all variables are defined in such a way that the variable expressions are positive, and bases are positive numbers not equal to 1. See Examples 1–5.

19. $\log_b x + \log_b y$

20. $\log_b 2 + \log_b z$

21. $3 \log_a m - \log_a n$

22. $5 \log_b x - \log_b y$

23. $(\log_a r - \log_a s) + 3 \log_a t$

24. $(\log_a p - \log_a q) + 2 \log_a r$

25. $3 \log_a 5 - 4 \log_a 3$

26. $3 \log_a 5 + \dfrac{1}{2} \log_a 9$

27. $\log_{10}(x + 3) + \log_{10}(x - 3)$

28. $\log_{10}(y + 4) + \log_{10}(y - 4)$

29. $3 \log_p x + \dfrac{1}{2} \log_p y - \dfrac{3}{2} \log_p z - 3 \log_p a$

30. $\dfrac{1}{3} \log_b x + \dfrac{2}{3} \log_b y - \dfrac{3}{4} \log_b s - \dfrac{2}{3} \log_b t$

31. Explain why the statement for the power rule for logarithms requires that x be a positive real number.

32. What is wrong with the following "proof" that $\log_2 16$ does not exist?

$$\log_2 16 = \log_2 (-4)(-4)$$
$$= \log_2(-4) + \log_2(-4)$$

Since the logarithm of a negative number is not defined, the final step cannot be evaluated, and so $\log_2 16$ does not exist.

MATHEMATICAL CONNECTIONS (EXERCISES 33–38)

Work Exercises 33–38 in order.

33. Evaluate $\log_3 81$.

34. Write the *meaning* of the expression $\log_3 81$.

35. Evaluate $3^{\log_3 81}$.

36. Write the *meaning* of the expression $\log_2 19$.

37. Evaluate $2^{\log_2 19}$.

38. Keeping in mind that a logarithm is an exponent, and using the results from Exercises 33–37, what is the simplest form of the expression $k^{\log_k m}$?

12.5 *Evaluating Logarithms*

As mentioned earlier, logarithms are important in many applications of mathematics to everyday problems, particularly in biology, engineering, economics, and social science. In this section we show how to find numerical approximations for logarithms. Traditionally base 10 logarithms have been used most extensively, because our number system is base 10. Logarithms to base 10 are called **common logarithms,** and $\log_{10} x$ is abbreviated as simply $\log x$, where the base is understood to be 10.

OBJECTIVE 1 We use calculators to evaluate common logarithms. In the next example we give the results of evaluating some common logarithms using a calculator with a log key. (This may be a second function key on some calculators.) For simple scientific calculators, just enter the number, then touch the log key. For graphing calculators, these steps are reversed. We will give all logarithms to four decimal places.

┌─ **E X A M P L E I**   **Evaluating Common Logarithms**

Evaluate each logarithm.

(a) $\log 327.1 \approx 2.5147$

(b) $\log 437{,}000 \approx 5.6405$

(c) $\log .0615 \approx -1.2111$

In Example 1(c), $\log .0615 \approx -1.2111$, a negative result. The common logarithm of a number between 0 and 1 is always negative because the logarithm is the exponent on 10 that produces the number. For example,

$$10^{-1.2111} \approx .0615.$$

If the exponent (the logarithm) were positive, the result would be greater than 1 because $10^0 = 1$.

└───

> **WORK PROBLEM I AT THE SIDE.** ▶▶

OBJECTIVE 2 In chemistry, the **pH** (pronounced "p" "h") of a solution is defined as follows.

$$\text{pH} = -\log[\text{H}_3\text{O}^+],$$

where $[\text{H}_3\text{O}^+]$ is the hydronium ion concentration in moles per liter.

The pH is a measure of the acidity (uh-SID-uh-tee) or alkalinity (al-kuh-LIN-uh-tee) of a solution; water, for example, has a pH of 7. In general, acids have pH numbers less than 7, and alkaline solutions have pH values greater than 7.

1. Find the following.

(a) log 41,600

(b) log 43.5

(c) log .442

2. Find the pH of solutions with the following hydronium ion concentrations.

(a) 3.7×10^{-8}

(b) 1.2×10^{-3}

E X A M P L E   2   **Finding pH**

Find the pH of grapefruit with a hydronium (hy-DROH-nee-um) ion (EYE-on) concentration of 6.3×10^{-4}.

Use the definition of pH.

$$\begin{aligned}
\text{pH} &= -\log(6.3 \times 10^{-4}) \\
&= -(\log 6.3 + \log 10^{-4}) \qquad \text{Multiplication property} \\
&\approx -[.7993 - 4] \\
&= -.7993 + 4 \approx 3.2
\end{aligned}$$

It is customary to round pH values to the nearest tenth.

◀◀ **WORK PROBLEM 2 AT THE SIDE.**

E X A M P L E   3   **Finding Hydronium Ion Concentration**

Find the hydronium ion concentration of drinking water with a pH of 6.5.

$$\text{pH} = 6.5 = -\log[\text{H}_3\text{O}^+]$$
$$\log[\text{H}_3\text{O}^+] = -6.5$$

This last line indicates that $10^{-6.5} = [\text{H}_3\text{O}^+]$. Using a calculator, we find

$$[\text{H}_3\text{O}^+] = 3.2 \times 10^{-7}.$$

---

3. Find the hydronium ion concentrations of solutions with the following pHs.

(a) 4.6

(b) 7.5

◀◀ **WORK PROBLEM 3 AT THE SIDE.**

OBJECTIVE ▶ **3** The most important logarithms used in applications are **natural logarithms,** which have as base the number e. The number e is irrational, like π: $e \approx 2.7182818$. Logarithms to base e are called natural logarithms because they occur in biology and the social sciences in natural situations that involve growth or decay. The base e logarithm of x is written $\ln x$ (read "el en x"). A graph of $y = \ln x$, the natural logarithm function, is given in Figure 13.

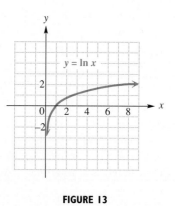

FIGURE 13

A calculator key labeled $\boxed{\ln x}$ is used to evaluate natural logarithms. If your calculator has an $\boxed{e^x}$ key, but not a key labeled $\boxed{\ln x}$, find natural logarithms by entering the number, touching the $\boxed{\text{INV}}$ key, and then touching the $\boxed{e^x}$ key. This works because $y = e^x$ is the inverse function of $y = \ln x$ (or $y = \log_e x$).

EXAMPLE 4 Finding Natural Logarithms

Find each of the following logarithms to the nearest ten-thousandth.

(a) ln .5841

$$\ln .5841 \approx -.5377$$

As with common logarithms, a number between 0 and 1 has a negative natural logarithm.

(b) ln 192.7 ≈ 5.2611

(c) ln 10.84 ≈ 2.3832

WORK PROBLEM 4 AT THE SIDE.

OBJECTIVE 4 One of the most common applications of exponential functions depends on the fact that in many situations involving growth or decay of a population, the amount or number of some quantity present at time t can be closely approximated by

$$y = y_0 e^{kt}$$

where y_0 is the amount or number present at time $t = 0$, k is a constant, and e is the base of natural logarithms mentioned earlier.

EXAMPLE 5 Applying Natural Logarithms

The population of Canada, in millions, is approximated by the exponential function with

$$f(t) = 29.5e^{.052t},$$

where t represents time in years from 1995. Thus, 1995 is represented by $t = 0$, 1996 is represented by $t = 1$, and so on. To find the population in 1995, we substitute 0 for t to get

$$f(0) = 29.5e^{(.052)(0)} = 29.5(1) = 29.5$$

million. Similarly, the population in 5 years (the year 2000) will be about

$$f(5) = 29.5e^{(.052)(5)} = 38.3$$

million.

WORK PROBLEM 5 AT THE SIDE.

EXAMPLE 6 Applying Natural Logarithms

The number of years, $N(r)$, since two independently evolving languages split off from a common ancestral language is approximated by

$$N(r) = -5000 \ln r,$$

where r is the percent of words from the ancestral language common to both languages now. Find N if $r = 70\%$.

Write 70% as .7, and find $N(.7)$.

$$N(.7) = -5000 \ln .7 \approx 1783$$

Approximately 1800 years have passed since the two languages separated.

---

4. Find each logarithm to the nearest ten-thousandth.

(a) ln .01

(b) ln 27

(c) ln 529

5. In Example 5, find the population in 1998.

6. Find $N(.2)$ in Example 6.

ANSWERS
4. (a) −4.6052 **(b)** 3.2958 **(c)** 6.2710
5. about 34.5 million
6. about 8000 years

WORK PROBLEM 6 AT THE SIDE.

7. Find each logarithm.

(a) $\log_3 17$
(Use common logarithms.)

In Examples 5 and 6, the final answers were obtained *without* rounding off the intermediate values. In general, it is best to wait until the final step to round off the answer; otherwise, a build-up of round-off error may cause the final answer to have an incorrect final decimal place digit.

Objective 5 A calculator can be used to approximate the values of common logarithms (base 10) or natural logarithms (base *e*). However, sometimes we need to use logarithms to other bases. The following rule is used to convert logarithms from one base to another.

Change-of-Base Rule

If $a > 0$, $a \neq 1$, $b > 0$, $b \neq 1$, and $x > 0$, then

$$\log_a x = \frac{\log_b x}{\log_b a}.$$

Note
As an aid in remembering the change-of-base rule, notice that x is "above" a on both sides of the equation.

Any positive number other than 1 can be used for base b in the change-of-base rule, but usually the only practical bases are e and 10 because calculators give logarithms only for these two bases.

To prove the formula for change of base, let $\log_a x = m$. We use the fact that if $r = s$, $\log_b r = \log_b s$.

$$\log_a x = m$$

$$a^m = x \qquad \text{Change to exponential form.}$$

$$\log_b(a^m) = \log_b x \qquad \text{Take logs on both sides.}$$

$$m \log_b a = \log_b x \qquad \text{Use the power property.}$$

$$(\log_a x)(\log_b a) = \log_b x \qquad \text{Substitute for } m.$$

$$\log_a x = \frac{\log_b x}{\log_b a} \qquad \text{Divide both sides by } \log_b a.$$

(b) $\log_9 121$
(Use natural logarithms.)

E X A M P L E   7   **Using the Change-of-Base Rule**

Find each logarithm.

(a) $\log_5 12$

Use common logarithms and the rule for change of base.

$$\log_5 12 = \frac{\log 12}{\log 5}$$

Now evaluate this quotient.

$$\log_5 12 \approx 1.5440$$

(b) $\log_2 134$

Use natural logarithms and the change-of-base rule.

$$\log_2 134 = \frac{\ln 134}{\ln 2} \approx 7.0661$$

◄◄ **WORK PROBLEM 7 AT THE SIDE.**

12.5 *Exercises*

Fill in the blanks.

1. The base in the expression log x is understood to be _____ .

2. The base in ln x is understood to be _____ .

3. We know that $10^{\quad} = 1$ and $10^{\quad} = 10$. Therefore, log 5 is between _____ and _____ .

4. We know that $e^{\quad} = 2.718$ and $e^{\quad} = 7.390$. Therefore, ln 4 is between _____ and _____ .

5. log $10^{19.2} = $ _____ . (Do not use a calculator.)

6. ln $e^{\sqrt{2}} = $ _____ . (Do not use a calculator.)

You will need a calculator for the remaining exercises in this set.

Find each logarithm. Give an approximation to the nearest ten-thousandth. See Examples 1 and 4.

7. log 43

8. log 98

9. log 328.4

10. log 457.2

11. log .0326

12. log .1741

13. $\log(4.76 \times 10^9)$

14. $\log(2.13 \times 10^4)$

15. ln 7.84

16. ln 8.32

17. ln .0556

18. ln .0217

19. ln 388.1

20. ln 942.6

21. $\ln(8.59 \times e^2)$

22. $\ln(7.46 \times e^3)$ **23.** $\ln 10$ **24.** $\log e$

Use the change-of-base rule (either with common or natural logarithms) to find the following logarithms. Give approximations to the nearest ten-thousandth. See Example 7.

25. $\log_6 13$ **26.** $\log_7 19$ **27.** $\log_{\sqrt{2}} \pi$

28. $\log_\pi \sqrt{2}$ **29.** $\log_{21} .7496$ **30.** $\log_{19} .8325$

31. Let m be the number of letters in your first name, and let n be the number of letters in your last name.
  (a) In your own words, explain what $\log_m n$ means.
  (b) Use your calculator to find $\log_m n$.
  (c) Raise m to the power indicated by the number you found in part (b). What is your result?

32. The equation $5^x = 7$ cannot be solved using the methods described in Section 12.2. However, in solving this equation, we must find the exponent to which 5 must be raised in order to obtain 7: this is $\log_5 7$.
  (a) Use the change-of-base rule and your calculator to find $\log_5 7$.
  (b) Raise 5 to the number you found in part (a). What is your result?
  (c) Using as many decimal places as your calculator gives, write the solution set of $5^x = 7$. (Equations of this type will be studied in more detail in Section 12.6.)

Use the formula $pH = -\log[H_3O^+]$ *to find the pH of the substance with the given hydronium ion concentration. See Example 2.*

33. Ammonia, 2.5×10^{-12} **34.** Sodium bicarbonate, 4.0×10^{-9}

35. Grapes, 5.0×10^{-5} **36.** Tuna, 1.3×10^{-6}

Use the formula for pH to find the hydronium ion concentration of the substance with the given pH. See Example 3.

37. Human blood plasma, 7.4

38. Human gastric contents, 2.0

39. Spinach, 5.4

40. Bananas, 4.6

Solve the following problems. See Examples 5 and 6.

41. Suppose that the amount, in grams, of pluto-nium-241 present in a given sample is deter-mined by the function

$$A(t) = 2.00e^{-.053t},$$

where t is measured in years. Find the amount present in the sample after the given number of years.
(a) 4   **(b)** 10   **(c)** 20
(d) What was the initial amount present?

42. Suppose that the amount, in grams, of radium-226 present in a given sample is determined by the function

$$A(t) = 3.25e^{-.00043t},$$

where t is measured in years. Find the amount present in the sample after the given number of years.
(a) 20   **(b)** 100   **(c)** 500
(d) What was the initial amount present?

43. The number of books, in millions, sold per year in the United States between 1985 and 1990 can be approximated by the function

$$N(t) = 1757e^{.0264t},$$

where $t = 0$ corresponds to the year 1985. Based on this model, how many books were sold in 1994? (*Source:* Book Industry Study Group)

44. Personal consumption expenditures for recre-ation in billions of dollars in the United States during the years 1984 through 1990 can be approximated by the function

$$C(t) = 185.4e^{.0587t},$$

where $t = 0$ corresponds to the year 1984. Based on this model, how much were personal consumption expenditures in 1994? (*Source:* U.S. Bureau of Economic Analysis)

45. The number of Cesarean section deliveries in the United States has increased over the years. According to statistics provided by the U.S. National Center for Health Statistics, between the years 1980 and 1989, the number of such births, in thousands, can be approximated by the function

$$B(t) = 624.6e^{.0516t},$$

where $t = 1$ corresponds to the year 1980. Based on this model, how many such births were there in 1994?

46. According to an article in *The AMATYC Review* (Spring, 1993), the number of students enrolled, in thousands, in intermediate algebra in two-year colleges since 1966 can be approximated by the function

$$E(t) = 39.8e^{.073t},$$

where $t = 1$ corresponds to 1966. Based on this equation, how many students were enrolled in intermediate algebra at the two-year college level in 1995?

INTERPRETING TECHNOLOGY (EXERCISES 47–48)

47. The function $B(x) = 624.6e^{.0156x}$, described in Exercise 45 with $x = t$, is graphed in a graphing calculator-generated window in the accompanying figure. Interpret the meanings of x and y in the display at the bottom in the context of Exercise 45.

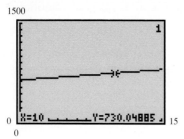

48. The function $E(x) = 39.8e^{.073x}$, described in Exercise 46 with $x = t$, is graphed in a graphing calculator-generated window in the accompanying figure. Interpret the meanings of x and y in the display at the bottom in the context of Exercise 46.

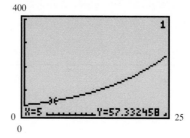

For Exercises 49–52, refer to Example 6 and use the function $N(r) = -5000 \ln r$. Round answers to the nearest hundred.

49. Find $N(.85)$ **50.** Find $N(.35)$ **51.** Find $N(.10)$

52. How many years have elapsed since the split if 75% of the words of the ancestral language are common to both languages today?

12.6 Exponential and Logarithmic Equations and Their Applications

As mentioned at the beginning of this chapter, exponential and logarithmic functions are important in many applications of mathematics. Using these functions in applications requires solving exponential and logarithmic equations. Some simple equations were solved in earlier sections of this chapter. More general methods for solving these equations depend on the following properties.

OBJECTIVES

1 Solve equations involving variable exponents.

2 Solve equations involving logarithms.

3 Solve applied problems involving exponential equations.

FOR EXTRA HELP

Tutorial   Tape 26   SSM, Sec. 12.6

Properties of Exponential and Logarithmic Equations

For all real numbers $b > 0$, $b \neq 1$, and any real numbers x and y:

1. If $x = y$, then $b^x = b^y$.
2. If $b^x = b^y$, then $x = y$.
3. If $x = y$, and $x > 0$, $y > 0$, then $\log_b x = \log_b y$.
4. If $x > 0$, $y > 0$, and $\log_b x = \log_b y$, then $x = y$.

Property 2 was used to solve exponential equations earlier and property 3 was used in Section 12.5.

OBJECTIVE 1 The first examples illustrate a general method for solving exponential equations using property 3.

┌─
E X A M P L E 1   Solving an Exponential Equation

Solve the equation $3^m = 12$.

$$3^m = 12$$
$$\log 3^m = \log 12 \qquad \text{Property 3}$$
$$m \log 3 = \log 12 \qquad \text{Power property}$$
$$m = \frac{\log 12}{\log 3} \qquad \text{Divide by } \log 3.$$

This quotient is the exact solution. To get a decimal approximation for the solution, use a calculator. A calculator gives

$$m \approx 2.262,$$

and the solution set is $\{2.262\}$. Check that $3^{2.262} \approx 12$.
─

┌─
Caution

Be careful: $\dfrac{\log 12}{\log 3}$ is *not* equal to $\log 4$ because $\log 4 \approx .6021$, but $\dfrac{\log 12}{\log 3} \approx 2.262$.
└─

WORK PROBLEM 1 AT THE SIDE. ▶▶

When an exponential equation has e as the base, it is easiest to use base e logarithms.

---

1. Give decimal approximations to the nearest thousandth for the solutions.

(a) $2^p = 9$

(b) $10^k = 4$

---

2. Solve $e^{-.01t} = .38$.

E X A M P L E 2 **Solving an Exponential Equation with Base e**

Solve $e^{.003x} = 40$.

Take base e logarithms on both sides.

$$\ln e^{.003x} = \ln 40$$
$$.003x \ln e = \ln 40 \qquad \text{Power property}$$
$$.003x = \ln 40 \qquad \ln e = \ln e^1 = 1$$
$$x = \frac{\ln 40}{.003} \qquad \text{Divide by .003.}$$
$$x \approx 1230 \qquad \text{Use a calculator.}$$

The solution set is $\{1230\}$. Check that $e^{.003(1230)} \approx 40$.

◄◄ **WORK PROBLEM 2 AT THE SIDE.**

In summary, we can solve exponential equations by one of the following methods. (The method used depends upon the form of the equation.) Examples 1 and 2 illustrate Method 1. We gave examples of Method 2 in Section 12.2.

Methods for Solving Exponential Equations

1. Using property 3, take logarithms to the same base on each side; then use the power property of logarithms on one side or both sides.

2. Using property 2, write both sides as exponentials with the same base; then set the exponents equal.

3. Solve $\log_5 \sqrt{x - 7} = 1$.

Objective 2 The next three examples illustrate some ways to solve logarithmic equations. The properties of logarithms from Section 12.4 are useful here, as is using the definition of a logarithm to change to exponential form.

E X A M P L E 3 **Solving a Logarithmic Equation**

Solve $\log_2(x + 5)^3 = 4$.

$$(x + 5)^3 = 2^4 \qquad \text{Convert to exponential form.}$$
$$(x + 5)^3 = 16$$
$$x + 5 = \sqrt[3]{16} \qquad \text{Take the cube root on both sides.}$$
$$x = -5 + \sqrt[3]{16}$$

Verify that the solution satisfies the equation, so the solution set is $\{-5 + \sqrt[3]{16}\}$.

Caution
Recall that the domain of $y = \log_b x$ is $(0, \infty)$. For this reason, *it is always necessary to check that the solution of a logarithmic equation is in the domain of the logarithmic expression.*

◄◄ **WORK PROBLEM 3 AT THE SIDE.**

Answers
2. $\{96.8\}$
3. $\{32\}$

E X A M P L E   4   Solving a Logarithmic Equation

Solve $\log_2(x + 1) - \log_2 x = \log_2 8$.

$$\log_2(x + 1) - \log_2 x = \log_2 8$$

$$\log_2 \frac{x + 1}{x} = \log_2 8 \qquad \text{Division property}$$

$$\frac{x + 1}{x} = 8 \qquad \text{Property 4}$$

$$8x = x + 1 \qquad \text{Multiply by } x.$$

$$x = \frac{1}{7} \qquad \text{Subtract } x; \text{ divide by 7.}$$

Check this solution by substituting in the given equation. Here, both $x + 1$ and x must be positive. If $x = \frac{1}{7}$, this condition is satisfied, and the solution set is $\{\frac{1}{7}\}$.

WORK PROBLEM 4 AT THE SIDE. ▶▶

E X A M P L E   5   Solving a Logarithmic Equation

Solve $\log x + \log(x - 21) = 2$.

   For this equation, write the left side as a single logarithm. Then write in exponential form and solve the equation.

$$\log x + \log(x - 21) = 2$$

$$\log x(x - 21) = 2 \qquad \text{Product rule}$$

$$x(x - 21) = 10^2 \qquad \begin{array}{l}\text{Log } x = \log_{10} x; \text{ write in}\\ \text{exponential form.}\end{array}$$

$$x^2 - 21x = 100$$

$$x^2 - 21x - 100 = 0 \qquad \text{Standard form}$$

$$(x - 25)(x + 4) = 0 \qquad \text{Factor.}$$

$$x - 25 = 0 \quad \text{or} \quad x + 4 = 0 \qquad \text{Set each factor equal to 0.}$$

$$x = 25 \quad \text{or} \qquad x = -4$$

The value -4 must be rejected as a solution, since it leads to the logarithm of a negative number in the original equation:

$$\log(-4) + \log(-4 - 21) = 2. \qquad \text{The left side is not defined.}$$

The only solution, therefore, is 25, and the solution set is $\{25\}$.

Caution

Do not reject a potential solution just because it is nonpositive. Reject any value which *leads to* the logarithm of a nonpositive number.

WORK PROBLEM 5 AT THE SIDE. ▶▶

4. Solve.
$$\log_8(2x + 5) + \log_8 3 = \log_8 33$$

5. Solve.
$$\log_3 2x - \log_3(3x + 15) = -2$$

ANSWERS

4. $\{3\}$

5. $\{1\}$

6. Find the value of $2000 deposited at 5% compounded annually for 10 years.

In summary, we use the following steps to solve a logarithmic equation.

Solving a Logarithmic Equation

Step 1 Use the multiplication or division properties of logarithms to get a single logarithm on one side.

Step 2 **(a)** Use property 4: If $\log_b x = \log_b y$, then $x = y$. (See Example 4.)

(b) Write the equation in exponential form: if $\log_b x = k$, then $x = b^k$. (See Example 3 and Example 5.)

OBJECTIVE 3 So far in this book, problems involving applications of interest have been limited to simple interest. In most cases, the interest paid or charged is compound (KAHM-pound) interest (interest paid on both principal and interest). The formula for compound interest is an important application of exponential functions.

Compound Interest

If P dollars is deposited in an account paying an annual rate of interest r compounded (paid) n times per year, the account will contain

$$A = P\left(1 + \frac{r}{n}\right)^{nt}$$

dollars after t years.

In the formula, r is usually expressed as a decimal.

EXAMPLE 6 Solving a Compound Interest Problem

How much money will there be in an account at the end of 5 years if $1000 is deposited at 6% compounded quarterly? (Assume no withdrawals are made.)

Because interest is compounded quarterly, $n = 4$. The other values given in the problem are $P = 1000$, $r = .06$ (because $6\% = .06$), and $t = 5$. Substitute into the compound interest formula to get the value of A.

$$A = 1000\left(1 + \frac{.06}{4}\right)^{4 \cdot 5}$$

$$A = 1000(1.015)^{20}$$

Now use the y^x key on a calculator, and round the answer to the nearest cent.

$$A = 1346.86$$

The account will contain $1346.86. (The actual amount of interest earned is $1346.86 − $1000 = $346.86. Do you see why?)

◀◀ **WORK PROBLEM 6 AT THE SIDE.**

─E X A M P L E 7   **Solving an Exponential Decay Problem**

Nuclear energy derived from radioactive isotopes can be used to supply power to space vehicles. The output of the radioactive power supply for a certain satellite is given by the function

$$y = 40e^{-.004t},$$

where y is in watts and t is the time in days.

(a) How much power will be available at the end of 180 days?

Let $t = 180$ in the formula.

$$y = 40e^{-.004(180)}$$

$$y \approx 19.5 \qquad \text{Use a calculator.}$$

About 19.5 watts will be left.

(b) How long will it take for the amount of power to be half of its original strength?

The original amount of power is 40 watts. (Why?) Because half of 40 is 20, replace y with 20 in the formula, and solve for t.

$$20 = 40e^{-.004t}$$

$$.5 = e^{-.004t} \qquad \text{Divide by 40.}$$

$$\ln .5 = \ln e^{-.004t}$$

$$\ln .5 = -.004t \qquad \ln e^k = k$$

$$t = \frac{\ln .5}{-.004}$$

$$t \approx 173 \qquad \text{Use a calculator.}$$

After about 173 days, the amount of available power will be half of its original amount.

■

WORK PROBLEM 7 AT THE SIDE. ▶▶

7. (a) In Example 7, suppose the output is given by $y = 50e^{-.002t}$. How much power will be available in 180 days?

(b) How long will it take for the amount of power to be half its original strength?

Real World

Lesson 8: Inverse Functions

In Section 12.1 we described how inverses of one-to-one functions may be determined algebraically. We also explained how the graph of a one-to-one function f compares to the graph of its inverse f^{-1}: it is a reflection of the graph of f^{-1} across the line $y = x$. In Example 3(a) of that section we showed that the inverse of the one-to-one function $f(x) = 2x + 5$ is given by $f^{-1}(x) = \dfrac{x-5}{2}$. If we use a square viewing window and graph $y_1 = f(x) = 2x + 5$, $y_2 = f^{-1}(x) = \dfrac{x-5}{2}$, and $y_3 = x$, we can see how this reflection appears on the screen. See the first figure at the left.

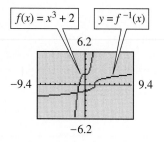

Some graphing calculators have the capability to draw the inverse of a function. The second figure shows the graphs of $f(x) = x^3 + 2$ and its inverse.

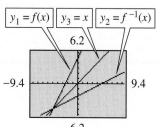

GRAPHING CALCULATOR EXPLORATIONS

1. Refer to the graph of $f(x) = x^3 + 2$ in the second figure.

   (a) Find f^{-1} algebraically.

   (b) Graph $y_1 = f(x) = x^3 + 2$, $y_2 = f^{-1}(x)$, and $y_3 = x$ on the same square viewing window. Describe what you see.

2. Graph $y_1 = 10^x$ and $y_2 = \log x$ on the same square viewing window.

   (a) Trace to the point where $x = .5$ on y_1, and find the approximation for y.

   (b) Find $\sqrt{10}$ using the square root key of your calculator. Compare it to the value you found in part (a) and explain why the approximations are the same.

   (c) Trace to the point where $x = 6$ on y_2, and find the approximation for y.

   (d) Find $\log 6$ using the common logarithm key of your calculator. Compare it to the value you found in part (c) and explain why the approximations are the same.

3. (a) Graph $y_1 = 4^x$ on your calculator.

   (b) The function y_1 is one-to-one and has an inverse: it is $y_2 = \log_4 x$. To graph a logarithmic function that is neither base ten nor base e on a graphing calculator, we must use the change-of-base rule. Graph y_2 as $\log x / \log 4$.

   (c) Trace to the point where $x = .5$ on y_1. What is the value of y?

   (d) Trace to the point on y_2 where x is equal to the value of y in part (c). What is the new value of y?

   (e) How do the results of parts (c) and (d) support the fact that y_1 and y_2 are inverses?

12.6 Exercises

The following equations were solved in Section 12.2 by writing both sides with the same base and then setting the exponents equal. Solve each equation using the method of Example 1 of this section.

1. $5^x = 125$

2. $9^x = 27$

3. By inspection, determine the solution set of $2^x = 64$. Now, without using a calculator, give the exact value of $\dfrac{\log 64}{\log 2}$.

4. Which one of the following is *not* a solution of $7^x = 23$?

    (a) $\dfrac{\log 23}{\log 7}$ **(b)** $\dfrac{\ln 23}{\ln 7}$

    (c) $\log_7 23$ **(d)** $\log_{23} 7$

Solve each of the following equations. Give solutions to the nearest thousandth. See Example 1.

5. $7^x = 5$

6. $4^x = 3$

7. $9^{-x+2} = 13$

8. $6^{-t+1} = 22$

9. $2^{y+3} = 5^y$

10. $6^{m+3} = 4^m$

Use natural logarithms to solve each of the following equations. Give solutions to the nearest thousandth. See Example 2.

11. $e^{.006x} = 30$

12. $e^{.012x} = 23$

13. $e^{-.103x} = 7$

14. $e^{-.205x} = 9$

15. $\ln e^{.04x} = \sqrt{3}$

16. $\ln e^{.45x} = \sqrt{7}$

17. Try solving one of the equations in Exercises 11–16 using common logarithms rather than natural logarithms. (You should get the same solution.) Explain why using natural logarithms is a better choice.

18. If you were asked to solve $10^{.0025x} = 75$, would natural or common logarithms be a better choice? Explain your answer.

Solve each of the following equations. Give the exact solution. See Example 3.

19. $\log_3(6x + 5) = 2$

20. $\log_5(12x - 8) = 3$

21. $\log_7(x + 1)^3 = 2$

22. $\log_4(y - 3)^3 = 4$

23. Suppose that in solving a logarithmic equation having the term $\log_4(x - 3)$ you obtain an apparent solution of 2. All algebraic work is correct. Explain why you must reject 2 as a solution of the equation.

24. Suppose that in solving a logarithmic equation having the term $\log_7(3 - x)$ you obtain an apparent solution of -4. All algebraic work is correct. Should you reject -4 as a solution of the equation? Explain why or why not.

Solve each of the following equations. Give exact solutions. See Examples 4 and 5.

25. $\log(6x + 1) = \log 3$

26. $\log(7 - x) = \log 12$

27. $\log_5(3t + 2) - \log_5 t = \log_5 4$

28. $\log_2(x + 5) - \log_2(x - 1) = \log_2 3$

29. $\log 4x - \log(x - 3) = \log 2$

30. $\log(-x) + \log 3 = \log(2x - 15)$

31. $\log_2 x + \log_2(x - 7) = 3$

32. $\log(2x - 1) + \log 10x = \log 10$

33. $\log 5x - \log(2x - 1) = \log 4$

34. $\log_3 x + \log_3(2x + 5) = 1$

35. $\log_2 x + \log_2(x - 6) = 4$

36. $\log_2 x + \log_2(x + 4) = 5$

Solve each problem. See Examples 6 and 7.

37. How much money will there be in an account at the end of 6 years if \$2000.00 is deposited at 4% compounded quarterly? (Assume no withdrawals are made.)

38. How much money will there be in an account at the end of 7 years if \$3000.00 is deposited at 3.5% compounded quarterly? (Assume no withdrawals are made.)

39. A sample of 400 grams of lead-210 decays to polonium-210 according to the function

$$A(t) = 400e^{-.032t},$$

where t is time in years. How much lead will be left in the sample after 25 years?

40. How long will it take the initial sample of lead in Exercise 39 to decay to half of its original amount? (This is called the *half-life* of the substance.)

*Banks sometimes compute interest based on what is known as **continuous compounding.**
Rather than paying interest a finite number of times per year (as explained in the text just
before Example 6), interest is earned at all times. As a result, the formula in this section
cannot be applied because n is approaching infinity. The formula used to determine the amount
A in an account having initial principal P compounded continuously at an annual rate r for t
years is*

$$A = Pe^{rt}.$$

41. What will be the amount A in an account with
initial principal $4000.00 if interest is com-
pounded continuously at an annual rate of
3.5% for 6 years?

42. Refer to Exercise 38. Does the money grow to
a larger value under those conditions, or when
invested for 7 years at 3% compounded continuously?

43. How long would it take an initial principal P
to double if it is invested at 4.5% compounded
continuously? (This is called the *doubling time*
of the money.)

44. How long would it take $4000 to grow to
$6000 at 3.25% compounded continuously?

INTERPRETING TECHNOLOGY (EXERCISES 45–46)

---

45. The screens show the graphs of $f(x) = \log x^2$ and $g(x) = 2 \log x$. Why does the power
rule from Section 12.4 not apply here?

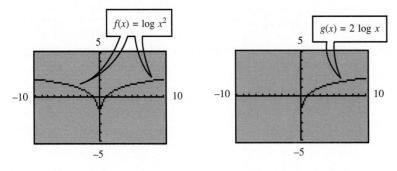

46. One of the graphs shown below is the graph of $f(t) = 2e^{-.125t}$; the other is the graph of
$g(t) = 300e^{.4t}$. Which is the graph of f?

A.   **B.**

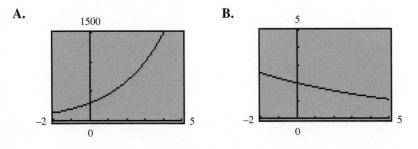

KEY TERMS

12.1 **one-to-one function**
A one-to-one function is a function in which each x-value corresponds to just one y-value and each y-value corresponds to just one x-value.

inverse of a function f
If f is a one-to-one function, the inverse of f is the set of all ordered pairs of the form (y, x), where (x, y) belongs to f.

12.2 **exponential equation**
An equation involving an exponential, where the variable is in the exponent, is an exponential equation.

12.3 **logarithm**
A logarithm is an exponent; $\log_a x$ is the exponent on the base a that gives the number x.

logarithmic equation
A logarithmic equation is an equation with a logarithm in at least one term.

12.5 **common logarithm**
A common logarithm is a logarithm to the base 10.

natural logarithm
A natural logarithm is a logarithm to the base e.

NEW SYMBOLS

f^{-1}	the inverse of f
$\log_a x$	the logarithm of x to the base a
$\log x$	common (base 10) logarithm of x
$\ln x$	natural (base e) logarithm of x
e	a constant, approximately 2.7182818

QUICK REVIEW

Concepts	Examples

12.1 Inverse Functions

Horizontal Line Test
If a horizontal line intersects the graph of a function in no more than one point, then the function is one-to-one.

Find f^{-1} if $f(x) = 2x - 3$. The graph of f is a straight line, so f is one-to-one by the horizontal line test.

Inverse Functions
For a one-to-one function f defined by an equation $y = f(x)$, the defining equation of the inverse function f^{-1} is found by exchanging x and y, solving for y, and replacing y with $f^{-1}(x)$.

Exchange x and y in the equation $y = 2x - 3$.

$$x = 2y - 3$$

Solve for y to get $\quad y = \dfrac{1}{2}x + \dfrac{3}{2}$.

Therefore, $f^{-1}(x) = \dfrac{1}{2}x + \dfrac{3}{2}$.

The graph of f^{-1} is a mirror image of the graph of f with respect to the line $y = x$.

The graphs of a function f and its inverse f^{-1} are given below.

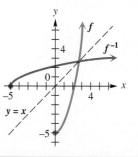

Concepts	Examples
### 12.2 Exponential Functions For $a > 0$, $a \neq 1$, $f(x) = a^x$ is an exponential function with base a. **Graph of $f(x) = a^x$** The graph contains the point $(0, 1)$. When $a > 1$, the graph rises from left to right. When $0 < a < 1$, the graph falls from left to right. The x-axis is an asymptote. The domain is $(-\infty, \infty)$; the range is $(0, \infty)$.	$f(x) = 10^x$ is an exponential function with base 10. $g(x) = \left(\frac{1}{2}\right)^x$ $f(x) = 2^x$
### 12.3 Logarithmic Functions $y = \log_a x$ means $x = a^y$. For $b > 0$, $b \neq 1$, $\log_b b = 1$ and $\log_b 1 = 0$. For $a > 0$, $a \neq 1$, $x > 0$, $f(x) = \log_a x$ is the logarithmic function with base a. **Graph of $g(x) = \log_a x$** The graph contains the point $(1, 0)$. When $a > 1$, the graph rises from left to right. When $0 < a < 1$, the graph falls from left to right. The y-axis is an asymptote. The domain is $(0, \infty)$; the range is $(-\infty, \infty)$.	$y = \log_2 x$ means $x = 2^y$. $\log_3 3 = 1$ $\log_5 1 = 0$ $f(x) = \log_6 x$ is the logarithmic function with base 6. $f(x) = \log_2 x$ $g(x) = \log_{1/2} x$
### 12.4 Properties of Logarithms **Multiplication Property** $$\log_a xy = \log_a x + \log_a y$$ **Division Property** $$\log_a \frac{x}{y} = \log_a x - \log_a y$$ **Power Property** $$\log_a x^r = r \log_a x$$ **Special Properties** $$b^{\log_b x} = x \quad \text{and} \quad \log_b b^x = x$$	$$\log_2 3m = \log_2 3 + \log_2 m$$ $$\log_5 \frac{9}{4} = \log_5 9 - \log_5 4$$ $$\log_{10} 2^3 = 3 \log_{10} 2$$ $$6^{\log_6 10} = 10 \qquad \log_3 3^4 = 4$$

Concepts	Examples
12.5 Evaluating Logarithms **Change-of-Base Rule** If $a > 0$, $a \neq 1$, $b > 0$, $b \neq 1$, $x > 0$, then $$\log_a x = \frac{\log_b x}{\log_b a}.$$	$$\log_3 17 = \frac{\ln 17}{\ln 3} = \frac{\log 17}{\log 3}$$
12.6 Exponential and Logarithmic Equations and Their Applications To solve exponential equations, use these properties $(b > 0, b \neq 1)$. **1.** If $b^x = b^y$, then $x = y$.	Solve $\quad 2^{3x} = 2^5$. $$3x = 5$$ $$x = \frac{5}{3}$$ Solution set: $\left\{ \dfrac{5}{3} \right\}$
2. If $x = y$, $(x > 0, y > 0)$, then $\log_b x = \log_b y$.	Solve $\quad 5^m = 8$. $$\log 5^m = \log 8$$ $$m \log 5 = \log 8$$ $$m = \frac{\log 8}{\log 5}$$ Solution set: $\left\{ \dfrac{\log 8}{\log 5} \right\}$
To solve logarithmic equations, use these properties, where $b > 0$, $b \neq 1$, $x > 0$, $y > 0$. First use the properties of 12.4, if necessary, to get the equation in the proper form. **1.** If $\log_b x = \log_b y$, then $x = y$.	Solve $\quad \log_3 2x = \log_3(x + 1)$. $$2x = x + 1$$ $$x = 1$$ Solution set: $\{1\}$
2. If $\log_b x = y$, then $b^y = x$.	Solve $\quad \log_2(3a - 1) = 4$. $$3a - 1 = 2^4 = 16$$ $$3a = 17$$ $$a = \frac{17}{3}$$ Solution set: $\left\{ \dfrac{17}{3} \right\}$

NUMBERS IN THE

Real World
a graphing calculator
minicourse

Lesson 9: Solution of Exponential and Logarithmic Equations

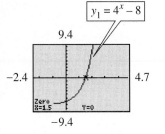

In Lessons 1 and 6 we saw how the x-intercepts of the graph of a function f correspond to the real solutions of the equation $f(x) = 0$. The ideas presented there dealt with linear and quadratic functions. We can extend those ideas to exponential and logarithmic functions. For example, consider the equation $4^x = 8$. Using traditional methods, we can determine that the solution set of this equation is $\{1.5\}$. Now if we write the equation with 0 on one side, we get $4^x - 8 = 0$. Graphing $y_1 = 4^x - 8$ and finding the x-intercept supports this solution, as seen in the first figure at the left. The x-intercept is 1.5, as expected.

Exercise 32 of Section 12.6 requires the solution of the logarithmic equation $\log (2x - 1) + \log 10x = \log 10$. If we graph $y_1 = \log (2x - 1) + \log 10x - \log 10$, we see that the x-intercept is 1, supporting the result obtained when the equation is solved algebraically. See the second figure at the left.

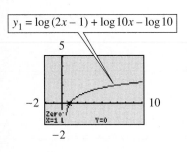

GRAPHING CALCULATOR EXPLORATIONS

1. The equation solved in Example 1 of Section 12.6 is equivalent to $3^x = 12$.

   (a) Because x is the exponent to which 3 must be raised in order to obtain 12, x is the _____ to the base _____ of _____.

   (b) Graph $y_1 = 3^x - 12$, and find an approximation for the x-intercept.

   (c) Use the common logarithm key of your calculator to find an approximation for $\log 12 / \log 3$. How does it compare to the x-intercept in part (b)?

   (d) Clear your screen, and graph $y_1 = \log_3 x = \log x / \log 3$. Then trace to where $x = 12$, and find the corresponding value of y. How does it compare to your values found in parts (b) and (c)?

2. Use a graph to support the fact that the solution set of $\log_3 x + \log_3(2x + 5) = 1$ is $\{\frac{1}{2}\}$. (*Hint:* Graph $y_1 = \log_3 x + \log_3(2x + 5) - 1$, and find the x-intercept.)

3. The formula for continuous compounding of interest is given just prior to Exercises 41–44 in Section 12.6. If we graph the function $y_1 = 1000e^{.04x}$, using a minimum x-value of 0, a maximum x-value of 20, a minimum y-value of 0, and a maximum y-value of 2500, we can observe how \$1000 will grow if compounded continuously at a rate of 4%. Use the graph to find the amount of money that will be in the account after (a) 10 years and (b) 15 years.

[12.1] *Determine whether the graph is the graph of a one-to-one function.*

1.

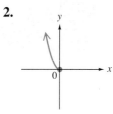

2.

Determine whether the function is one-to-one. If it is, find its inverse.

3. $f(x) = -3x + 7$

4. $f(x) = \sqrt[3]{6x - 4}$

5. $f(x) = -x^2 + 3$

6. $f(x) = x$

Each function graphed below is one-to-one. Graph its inverse on the same set of axes.

7.

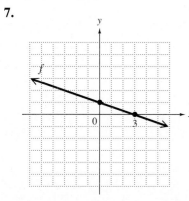

8.

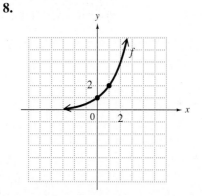

[12.2] *Graph each of the following functions.*

9. $f(x) = 3^x$

10. $f(x) = \left(\dfrac{1}{3}\right)^x$

11. $y = 3^{x+1}$

12. $y = 2^{2x+3}$

Solve each of the following equations.

13. $4^{3x} = 8^{x+4}$

14. $\left(\dfrac{1}{27}\right)^{x-1} = 9^{2x}$

15. $5^x = 1$

16. What is the y-intercept of the graph of $y = a^x$ $(a > 0, a \neq 1)$?

17. How does the answer to Exercise 16 reinforce the definition of 0 as an exponent?

[12.3] *Graph each of the following functions.*

18. $g(x) = \log_3 x$ (*Hint:* See Exercise 9.)

19. $g(x) = \log_{1/3} x$ (*Hint:* See Exercise 10.)

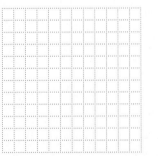

20. (a) Write in exponential form: $\log_5 625 = 4$.
(b) Write in logarithmic form: $5^{-2} = .04$.

21. (a) In your own words, explain the meaning of $\log_b a$.
(b) Based on your explanation above, simplify the expression $b^{\log_b a}$.

[12.4] *Apply the properties of logarithms introduced in Section 12.4 to express each of the following as a sum or difference of logarithms. Assume that all variables represent positive real numbers.*

22. $\log_4 3x^2$

23. $\log_2 \dfrac{p^2 r}{\sqrt{z}}$

Use the properties of logarithms introduced in Section 12.4 to express each of the following as a single logarithm. Assume that all variables represent positive real numbers, $b \neq 1$.

24. $\log_b 3 + \log_b x - 2 \log_b y$

25. $\log_3(x + 7) - \log_3(4x + 6)$

[12.5] *Find each logarithm. Give approximations to four decimal places.*

26. $\log 28.9$

27. $\log .257$

28. $\log 10^{4.8613}$

29. $\ln 28.9$

30. $\ln .257$

31. $\ln e^{4.8613}$

32. Use your calculator to find approximations of the following logarithms.
   (a) $\log 356.8$ **(b)** $\log 35.68$ **(c)** $\log 3.568$
   (d) Observe your answers and make a conjecture concerning the decimal values of the common logarithms of numbers greater than 1 that have the same digits.

Use the change-of-base rule (either with common or natural logarithms) to find each of the following logarithms. Give approximations to four decimal places.

33. $\log_{16} 13$

34. $\log_4 12$

35. $\log_{\sqrt{6}} \sqrt{13}$

36. A population of hares in a specific area is growing according to the function

$$H(t) = 500 \log_3(2t + 3),$$

where t is time in years after the population was introduced into the area. Find the number of hares for the following times.
   (a) $t = 0$ **(b)** $t = 3$ **(c)** $t = 12$
   (d) Is the change-of-base rule needed to work this problem?

Use the formula $\text{pH} = -\log[\text{H}_3\text{O}^+]$ *to find the pH of the substances with the given hydronium ion concentrations.*

37. Milk, 4.0×10^{-7}

38. Crackers, 3.8×10^{-9}

39. The population of Cairo, in millions, is approximated by the function $C(t) = 9e^{.026t}$, where $t = 0$ represents 1990. Find the population in the following years.
   (a) 2000   **(b)** 1970

40. Suppose the quantity, measured in grams, of a radioactive substance present at time t is given by

$$Q(t) = 500e^{-.05t},$$

where t is measured in days. Find the quantity present at the following times.
   (a) $t = 0$ **(b)** $t = 4$

[12.6] *Solve each equation. Give solutions to the nearest thousandth.*

41. $3^x = 9.42$ **42.** $2^{x-1} = 15$ **43.** $e^{.06x} = 3$

Solve each equation. Give exact solutions.

44. $\log_3(9x + 8) = 2$ **45.** $\log_5(y + 6)^3 = 2$

46. $\log_3(p + 2) - \log_3 p = \log_3 2$ **47.** $\log(2x + 3) = \log x + 1$

48. $\log_4 x + \log_4(8 - x) = 2$ **49.** $\log_2 x + \log_2(x + 15) = 4$

Solve each problem.

50. Refer to Exercise 39. In what year did the population reach 10 million?

51. Refer to Exercise 40. What is the half-life of the substance? (That is, how long would it take for the initial amount to become half of what it was?)

52. How much would be in an account after 3 years if $6500.00 was invested at 3% annual interest, compounded daily (use $n = 365$)?

53. Which is a better plan?
Plan A: Invest $1000.00 at 4% compounded quarterly for 3 years
Plan B: Invest $1000.00 at 3.9% compounded monthly for 3 years

---

MIXED REVIEW EXERCISES

Solve.

54. $\log_3(x + 9) = 4$ **55.** $\log_2 32 = x$ **56.** $\log_x \dfrac{1}{81} = 2$ **57.** $27^x = 81$

58. $2^{2x-3} = 8$ **59.** $\log_3(x + 1) - \log_3 x = 2$ **60.** $\log(3x - 1) = \log 10$

CHAPTER 12 TEST

1. Decide whether the function is or is not one-to-one.
   (a) $f(x) = x^2 + 9$ (b)

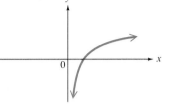

2. Find $f^{-1}(x)$ for the one-to-one function $f(x) = \sqrt[3]{x + 7}$.

3. Graph the inverse of f, given the graph of f below.

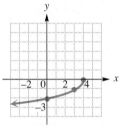

Graph the following functions.

4. $y = 6^x$

5. $y = \log_6 x$

6. How can the graph of the function in Exercise 5 be obtained from the graph of the function in Exercise 4?

1. (a) _____

   (b) _____

2. _____

3.

4.

5.

6. _____

Solve each equation. Give the exact solution.

7. _____

7. $5^x = \dfrac{1}{625}$

8. _____

8. $2^{3x-7} = 8^{2x+2}$

Solve the following problem.

9. **(a)** _____

  (b) _____

  (c) _____

  (d)

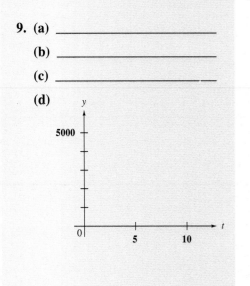

9. A small business estimates that the value y of a copy machine is decreasing according to the function

$$y = 5000(2)^{-.15t},$$

where t is the number of years that have elapsed since the machine was purchased, and y is in dollars.
  (a) What was the original value of the machine? (*Hint:* Let $t = 0$.)
  (b) What is the value of the machine 5 years after purchase? Give your answer to the nearest dollar.
  (c) What is the value of the machine 10 years after purchase? Give your answer to the nearest dollar.
  (d) Graph the function.

10. _____

10. Write in logarithmic form: $4^{-2} = .0625$

11. _____

11. Write in exponential form: $\log_7 49 = 2$

Solve each equation.

12. $\log_{1/2} x = -5$

12. _____

13. $x = \log_9 3$

13. _____

14. $\log_x 16 = 4$

14. _____

15. Fill in the blanks with the correct responses: The value of $\log_2 32$ is _____. This means that if we raise _____ to the _____ power, the result is _____ .

15. _____

Use properties of logarithms to write the expressions in Exercises 16 and 17 as sums or differences of logarithms. Assume variables represent positive numbers.

16. $\log_3 x^2 y$

16. _____

17. $\log_5\left(\dfrac{\sqrt{x}}{yz}\right)$

17. _____

Use properties of logarithms to write the expressions in Exercises 18 and 19 as single logarithms. Assume variables represent positive real numbers, $b \neq 1$.

18. $3 \log_b s - \log_b t$

18. _____

19. $\dfrac{1}{4} \log_b r + 2 \log_b s - \dfrac{2}{3} \log_b t$

19. _____

Use a calculator to find an approximation to the nearest ten-thousandth for each of the following logarithms.

20. _____

20. log 21.3

21. _____

21. ln .43

22. (a) _____

(b) _____

22. (a) Between what two consecutive integers must the value of $\log_6 45$ be?

(b) Use a calculator to find an approximation of $\log_6 45$ to the nearest ten-thousandth.

23. _____

23. Solve for x, and give the solution correct to the nearest ten-thousandth.

$$3^x = 78$$

24. _____

24. Solve: $\log_8(x + 5) + \log_8(x - 2) = \log_8 8$.

25. (a) _____

(b) _____

25. Another way of writing the function of item 9 (describing the value of the office copy machine) is

$$y = 5000e^{-.104t}.$$

(a) What will be the value of the copy machine 15 years after purchase? Give your answer to the nearest dollar.

(b) To the nearest whole number, after how many years will the value of the machine be half of its original value?

Note: This cumulative review exercise set may be considered as a final examination for the course.

Let $S = \{-\frac{9}{4}, -2, -\sqrt{2}, 0, .6, \sqrt{11}, \sqrt{-8}, 6, \frac{30}{3}\}$. List the elements of S that are members of the following sets.

1. Integers

2. Rational numbers

3. Irrational numbers

4. Real numbers

Simplify the following.

5. $|-8| + 6 - |-2| - (-6 + 2)$

6. $-12 - |-3| - 7 - |-5|$

7. $2(-5) + (-8)(4) - (-3)$

Solve the following.

8. $7 - (3 + 4a) + 2a = -5(a - 1) - 3$

9. $2m + 2 \leq 5m - 1$

10. $|2x - 5| = 9$

11. $|3p| - 4 = 12$

12. $|3k - 8| \le 1$

13. $|4m + 2| > 10$

Perform the indicated operations.

14. $(2p + 3)(3p - 1)$

15. $(4k - 3)^2$

16. $(3m^3 + 2m^2 - 5m) - (8m^3 + 2m - 4)$

17. Divide $6t^4 + 17t^3 - 4t^2 + 9t + 4$ by $3t + 1$.

Factor as completely as possible.

18. $8x + x^3$

19. $24y^2 - 7y - 6$

20. $5z^3 - 19z^2 - 4z$

21. $16a^2 - 25b^4$

22. $3x^2 + 3xy + 2xy + 2y^2$

23. $16r^2 + 56rq + 49q^2$

Simplify as much as possible in Exercises 24–27.

24. $\dfrac{(5p^3)^4(-3p^7)}{2p^2(4p^4)}$

25. $\dfrac{x^2 - 9}{x^2 + 7x + 12} \div \dfrac{x - 3}{x + 5}$

26. $\dfrac{2}{k + 3} - \dfrac{5}{k - 2}$

27. $\dfrac{3}{p^2 - 4p} - \dfrac{4}{p^2 + 2p}$

28. Candy worth $1.00 per pound is to be mixed with 10 pounds of candy worth $1.96 per pound to get a mixture that will be sold for $1.60 per pound. How many pounds of the $1.00 candy should be used?

Simplify in Exercises 29–31.

29. $\left(\dfrac{5}{4}\right)^{-2}$

30. $\dfrac{6^{-3}}{6^2}$

31. $2\sqrt{32} - 5\sqrt{98}$

32. Multiply: $(5 + 4i)(5 - 4i)$.

Solve the equations or inequalities in Exercises 33 and 34.

33. $10p^2 + p - 2 = 0$

34. $k^2 + 2k - 8 > 0$

35. Recently the U.S. population has been growing according to the equation

$$y = 1.7x + 230$$

where y gives the population (in millions) in year x, measured from year 1980. For example, in 1980 $x = 0$ and $y = 1.7(0) + 230 = 230$. This means that the population was about 230 million in 1980. To find the population in 1985, let $x = 5$, and so on. Find the population in each of the following years.
(a) 1982   **(b)** 1985   **(c)** 1990
(d) In what year will the population reach 315 million, based on this equation?

36. The long-term debt of the Port of New Orleans dropped from \$70,000,000 in 1986 to \$25,300,000 in 1994. What was the rate of change in debt? (*Source:* Division of Finance and Accounting, Port of New Orleans)

37. Find the standard form of the equation of the line through $(5, -1)$ and parallel to the line with equation $3x - 4y = 12$.

Graph the following.

38. $y = -2.5x + 5$

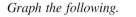
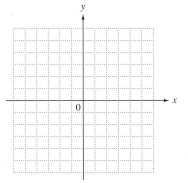

39. $-4x + y \le 5$

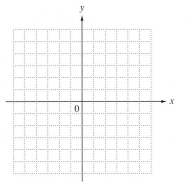

40. $y = \dfrac{1}{3}(x - 1)^2 + 2$

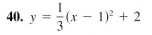
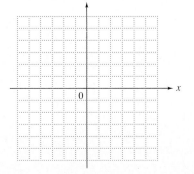

41. $\dfrac{x^2}{9} + \dfrac{y^2}{16} = 1$

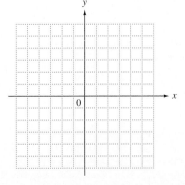

42. $25x^2 - 16y^2 = 400$

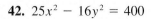
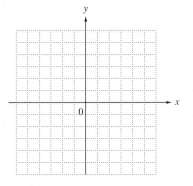

43. (a) Which of the *equations* in Exercises 38–42 define functions?

(b) Of the functions described in part (a), which have inverses?

Solve each system of equations.

44. $5x - 3y = 14$
$2x + 5y = 18$

45. $x + 2y + 3z = 11$
$3x - y + z = 8$
$2x + 2y - 3z = -12$

46. Evaluate $\begin{vmatrix} 2 & 4 & 5 \\ 1 & 3 & 0 \\ 0 & -1 & -2 \end{vmatrix}$.

47. Graph $f(x) = 2^x$.

48. Solve $5^{x+3} = \left(\dfrac{1}{25}\right)^{3x+2}$.

49. Graph $f(x) = \log_3 x$.

50. Solve $\log_5 x + \log_5(x + 4) = 1$.

In this section we provide the answers that we think most students will obtain when they work the exercises using the methods explained in the text. If your answer does not look exactly like the one given here, it is not necessarily wrong. In many cases there are equivalent forms of the answer that are correct. For example, if the answer section shows $\frac{3}{4}$ and your answer is .75, you have obtained the right answer but written it in a different (yet equivalent) form. Unless the directions specify otherwise, .75 is just as valid an answer as $\frac{3}{4}$.

In general, if your answer does not agree with the one given in the text, see whether it can be transformed into the other form. If it can, then it is the correct answer. If you still have doubts, talk with your instructor.

CHAPTER 1

SECTION 1.1 (page 7)

1. False; 3^5 means $3 \cdot 3 \cdot 3 \cdot 3 \cdot 3$. The base is 3 and the exponent is 5. **3.** true **5.** 49 **7.** 144 **9.** 64
11. 1000 **13.** 81 **15.** 1024 **17.** $\frac{16}{81}$
19. .000064 **21.** To evaluate a number to a power, use the number as the base as many times as the power: $6^3 = 6 \cdot 6 \cdot 6 = 216$. **23.** 32 **25.** $\frac{49}{30}$ **27.** 12
29. 23.01 **31.** 95 **33.** 90 **35.** 14 **37.** 9
39. The rule for order of operations says that the multiplication is performed before the addition.
41. false **43.** true **45.** true **47.** false
49. false **51.** true **53.** $15 = 5 + 10$
55. $9 > 5 - 4$ **57.** $16 \neq 19$ **59.** $2 \leq 3$
61. Seven is less than nineteen. True **63.** Three is not equal to six. True **65.** Eight is greater than or equal to eleven. False **67.** Answers will vary. One example is $5 + 3 \geq 2 \cdot 2$. **69.** $30 > 5$ **71.** $3 \leq 12$
73. *The Lion King* and *Forrest Gump* **75.** *The Santa Clause* and *The Flintstones*

SECTION 1.2 (page 13)

1. expression **3.** equation **5.** equation **7.** The first is translated $x - 5$, while the second is translated $5 < x$. **9. (a)** 64 **(b)** 144 **11. (a)** $\frac{7}{8}$ **(b)** $\frac{13}{12}$
13. (a) 9.569 **(b)** 14.353 **15. (a)** 52 **(b)** 114
17. (a) 12 **(b)** 33 **19. (a)** 6 **(b)** $\frac{9}{5}$ **21. (a)** $\frac{4}{3}$
(b) $\frac{13}{6}$ **23. (a)** $\frac{2}{7}$ **(b)** $\frac{16}{27}$ **25. (a)** 12 **(b)** 55

27. (a) 1 **(b)** $\frac{28}{17}$ **29. (a)** 3.684 **(b)** 8.841
31. $12x$ **33.** $x - 2$ **35.** $7 - 4x$ **37.** $2x - 6$
39. $\frac{12}{x + 3}$ **41.** $6(x - 4)$ **43.** The word "and" does not signify addition here. In the phrase "a number and 6," "and" connects two quantities to be multiplied.
45. Two possible pairs are $x = 0$, $y = 6$ and $x = 1$, $y = 4$. **47.** no **49.** yes **51.** yes **53.** no
55. yes **57.** yes **59.** $x + 8 = 18$
61. $2x + 5 = 5$ **63.** $16 - \frac{3}{4}x = 13$
65. $3x = 2x + 8$ **67.** 700 **68.** 3900
69. 7100 **70.** 10,300

SECTION 1.3 (page 23)

1. true **3.** true **5.** false **7.** true **9.** false
11. $-683,226$ **13.** -8
15. [number line from -6 to 3] **17.** [number line from -6 to 4]
19. [number line]
21. -11 **23.** -21 **25.** -100 **27.** $-\frac{2}{3}$
29. false **31.** true **33. (a)** 2 **(b)** 2
35. (a) -6 **(b)** 6 **37. (a)** $\frac{3}{4}$ **(b)** $\frac{3}{4}$ **39.** 7
41. 4 **43.** -12 **45.** -14 **47.** 9
49. No, the statement is false for one number, 0.
51. food in 1992 **53.** true

SECTION 1.4 (page 29)

1. negative **3.** negative **5.** 2 **7.** -3
9. -10 **11.** -13 **13.** -15.9 **15.** 5 **17.** 13
19. 0 **21.** -8 **23.** $\frac{3}{10}$ **25.** $\frac{1}{2}$ **27.** $-\frac{3}{4}$
29. -1.6 **31.** -8.7 **33.** -25 **35.** No
37. true **39.** false **41.** true **43.** false
45. true **47.** false **49.** It must be negative and have the larger absolute value. **50.** The sum of a positive number and 5 cannot be -7. **51.** It must be positive and have the larger absolute value. **52.** The sum of a negative number and -8 cannot be 2.
53. $-5 + 12 + 6$; 13
55. $[-19 + (-4)] + 14$; -9
57. $[-4 + (-10)] + 12$; -2
59. $[8 + (-18)] + 4$; -6 **61.** $-\$80$
63. -184 meters **65.** $112°F$ **67.** 37 yards
69. $\$107$ **71.** $\$286.60$ **73.** -3 **75.** Answers will vary. One example is $x + 9 = 7$.

Section 1.5 (page 37)

1. -8; -6 **3.** $7 - 12$; $12 - 7$ **5.** -4
7. -10 **9.** -16 **11.** 11 **13.** 19 **15.** -4
17. 5 **19.** 0 **21.** $\dfrac{3}{4}$ **23.** $-\dfrac{11}{8}$ **25.** $\dfrac{15}{8}$
27. 13.6 **29.** -11.9 **31.** -2.8 **33.** -6.3
35. To subtract signed numbers, add the opposite of the second number to the first number.   **37.** -14
39. -24 **41.** -16
43. $-\dfrac{17}{8}$ **45.** -48.98
47. For example, $-8 - (-2) = -6$
49. $4 - (-8)$; 12 **51.** $-2 - 8$; -10
53. $[9 + (-4)] - 7$; -2 **55.** $[8 - (-5)] - 12$; 1
57. $-41°$ **59.** $14{,}776$ feet   **61.** $-\$80$
63. $\$105{,}000$ **65.** -10 **67.** -12 **69.** positive
71. positive

Section 1.6 (page 45)

1. positive   **3.** negative   **5.** positive   **7.** -28
9. -30 **11.** -48 **13.** $\dfrac{5}{6}$ **15.** -2.38 **17.** $\dfrac{3}{2}$
19. -11 **21.** -2 **23.** -60 **25.** 35 **27.** 6
29. -18 **31.** 67 **33.** -8 **35.** 64
37. Substitute -3 for x and 4 for y to get $3(-3) + 2(4)$. Next, find the two products: $3(-3) = -9$ and $2(4) = 8$. Finally add the products to get -1.
39. 47 **41.** 72 **43.** $-\dfrac{78}{25}$ **45.** 0 **47.** -23
49. -9 **51.** $9 + (-9)(2)$; -9
53. $-4 - 2(-1)(6)$; 8 **55.** $7(-12) - 9$; -93
57. $12[9 - (-8)]$; 204 **59.** $\dfrac{4}{5}[-8 + (-2)]$; -8
61. 3 **63.** -2 **65.** positive   **66.** negative
67. -15 **68.** -15; The product of a negative number and a positive number must be a negative number.

Section 1.7 (page 55)

1. true   **3.** false   **5.** false
7. $\dfrac{1}{11}$ **9.** $-\dfrac{1}{5}$ **11.** $\dfrac{6}{5}$ **13.** no reciprocal
15. $-\dfrac{7}{8}$ **17.** 2.5 **19.** (c)   **21.** -3 **23.** -2
25. 16 **27.** 0 **29.** 25.63 **31.** $\dfrac{3}{2}$ **33.** 3.4
35. -3 **37.** -5 **39.** 4 **41.** 3 **43.** 7
45. 4 **47.** -3 **49.** $\dfrac{1}{2}$ **51.** 10 **53.** $\dfrac{-36}{-9}$; 4
55. $\dfrac{-12}{-5 + (-1)}$; 2 **57.** $\dfrac{15 + (-3)}{4(-3)}$; -1
59. $\dfrac{-34(7)}{-14}$; 17 **61.** $6x = -42$ **63.** $\dfrac{x}{3} = -3$
65. $6 = x - 2$ **67.** $\dfrac{15}{x} = -5$

69. (a) $3{,}473{,}986$ is divisible by 2 because its last digit is divisible by 2.   **(b)** $4{,}336{,}879$ is not divisible by 2 because its last digit is not divisible by 2.
70. (a) $4{,}799{,}232$ is divisible by 3 because the sum of its digits is 36, which is divisible by 3.   **(b)** $2{,}443{,}871$ is not divisible by 3 because the sum of its digits is 29, which is not divisible by 3.   **71. (a)** The last two digits are 64, which is divisible by 4.   **(b)** The last two digits are 35, which is not divisible by 4.   **72. (a)** The last digit is 5, which is divisible by 5.   **(b)** The last digit is 3, which is not divisible by 5.   **73. (a)** The last digit 2 is divisible by 2, and the sum of the digits is 24, which is divisible by 3.   **(b)** The last digit 0 is divisible by 2, but the sum of the digits is 34, which is not divisible by 3.

Section 1.8 (page 67)

1. 0 **3.** additive   **5.** additive inverse   **7.** $-a$
9. commutative property   **11.** associative property
13. inverse property   **15.** inverse property
17. identity property   **19.** commutative property
21. distributive property   **23.** identity property
25. distributive property   **27.** identity property
29. For example, $(5 - 3) - 1 = 2 - 1 = 1$, while $5 - (3 - 1) = 5 - 2 = 3$. Because $1 \neq 3$, the associative property is not satisfied.   **31.** $7 + r$ **33.** s
35. $-6x + (-6) \cdot 7$; $-6x - 42$
37. $w + [5 + (-3)]$; $w + 2$ **39.** $3x + 16$ **41.** 11
43. 7 **45.** 0 **47.** $-.38$ **49.** 1
51. $(5 + 1)x$; $6x$ **53.** $4t + 12$ **55.** $-8r - 24$
57. $-5y + 20$ **59.** $-16y - 20z$ **61.** $8(z + w)$
63. $7(2v + 5r)$ **65.** $24r + 32s - 40y$
67. $(1 + 1 + 1)q$; $3q$ **69.** $(-5 + 1)x$; $-4x$
71. $-4t - 5m$ **73.** $5c + 4d$ **75.** $3q - 5r + 8s$
77. Answers will vary. For example, "putting on your socks" and "putting on your shoes."   **79.** false
81. (foreign sales) clerk; foreign (sales clerk)

Section 1.9 (page 75)

1. (c)   **3.** (a)   **5.** $4r + 11$ **7.** $32q - 24t$
9. $5 + 2x - 6y$ **11.** $-7 + 3p$ **13.** 14
15. -12 **17.** 5 **19.** 1 **21.** -1 **23.** 74
25. Answers will vary. For example, $-3x$ and $4x$.
27. like   **29.** unlike   **31.** like   **33.** unlike
35. We cannot "add" unlike terms, so we must be able to identify like terms in order to combine them.
37. $9k - 5$ **39.** $-\dfrac{1}{3}t - \dfrac{28}{3}$ **41.** $-4.1r + 5.6$
43. $-2y^2 + 3y^3$ **45.** $-19p + 16$ **47.** $-4y + 22$
49. $-16y + 63$ **51.** $(x + 3) + 5x$; $6x + 3$
53. $(13 + 6x) - (-7x)$; $13 + 13x$
55. $2(3x + 4) - (-4 + 6x)$; 12 **57.** Wording may vary. One example is "the difference between 9 times a number and the sum of the number and 2."
59. $1000 + 5x$ (dollars)   **60.** $750 + 3x$ (dollars)
61. $1000 + 5x + 750 + 3x$ (dollars)
62. $1750 + 8x$ (dollars)

CHAPTER 1 REVIEW EXERCISES (page 81)

1. 625 **2.** .00000081 **3.** .009261 **4.** $\dfrac{125}{8}$

5. 27 **6.** 200 **7.** -7 **8.** $\dfrac{20}{3}$ **9.** $13 < 17$

10. $5 + 2 \neq 10$ **11.** 30 **12.** 60 **13.** 14
14. 13 **15.** $x + 6$ **16.** $8 - x$ **17.** $6x - 9$

18. $12 + \dfrac{3}{5}x$ **19.** yes **20.** no **21.** $2x - 6 = 10$

22. $4x = 8$

23. (number line with points at $-\tfrac{1}{2}$ and 2.5)

24. (number line)

25. (number line with points $-3\tfrac{1}{4}$, $-1\tfrac{1}{8}$, $\tfrac{5}{6}$, $2\tfrac{4}{5}$)

26. (number line)

27. -10 **28.** -9 **29.** $-\dfrac{3}{4}$ **30.** $-|23|$

31. true **32.** true **33.** true **34.** false
35. (a) 9 (b) 9 **36.** (a) 0 (b) 0 **37.** (a) -6

(b) 6 **38.** (a) $\dfrac{5}{7}$ (b) $\dfrac{5}{7}$ **39.** 12 **40.** -3

41. -19 **42.** -7 **43.** -6 **44.** -4 **45.** -17

46. $-\dfrac{29}{36}$ **47.** -10 **48.** -19

49. $(-31 + 12) + 19; 0$ **50.** $[-4 + (-8)] + 13; 1$
51. $-\$8$ **52.** $87°F$ **53.** -11 **54.** -1

55. 7 **56.** $-\dfrac{43}{35}$ **57.** 10.31 **58.** -12

59. 2 **60.** -3 **61.** $-4 - (-6); 2$
62. $[4 + (-8)] - 5; -9$ **63.** $-\$29$
64. $-10°$ **65.** The first step is to change subtracting -6 to adding its opposite, 6. So the problem becomes $-8 + 6$. This sum is -2. **66.** Yes, for example, $-2 - (-6) = -2 + 6 = 4$, a positive number.
67. -40 **68.** -13 **69.** 14 **70.** 12

71. 36 **72.** -105 **73.** $\dfrac{1}{2}$ **74.** 10.08

75. -20 **76.** -10 **77.** -24 **78.** -35
79. -18 **80.** -18 **81.** 125 **82.** -423

83. $-4(5) - 9; -29$ **84.** $\dfrac{5}{6}[12 + (-6)]; 5$

85. 4 **86.** -20 **87.** $-\dfrac{3}{4}$ **88.** 11.3 **89.** -1

90. 2 **91.** 1 **92.** .5 **93.** $\dfrac{12}{8 + (-4)}; 3$

94. $\dfrac{-20(12)}{15 - (-15)}; -8$ **95.** $8x = -24$ **96.** $\dfrac{x}{3} = -2$

97. $x - 3 = -7$ **98.** $x + 5 = -6$
99. identity property **100.** identity property
101. inverse property **102.** inverse property
103. associative property **104.** associative property
105. distributive property **106.** commutative property

107. $(7 + 1)y; 8y$ **108.** $-12 \cdot 4 - (-12)t$; $-48 + 12t$ **109.** $3(2s + 4y); 6s + 12y$
110. $-1(-4r + 5s) = -1(-4r) + (-1)(5s); 4r - 5s$
111. Because $25 - (5 - 2) = 25 - 3 = 22$ and $(25 - 5) - 2 = 20 - 2 = 18, 25 - (5 - 2) \neq (25 - 5) - 2$. The associative property would require these expressions to be equal. **112.** Because $180 \div (15 \div 5) = 180 \div 3 = 60$ and $(180 \div 15) \div 5 = 12 \div 5 = 12/5, 180 \div (15 \div 5) \neq (180 \div 15) \div 5$. The associative property would require these expressions to be equal. **113.** $17p^2$ **114.** $16r^2 + 7r$
115. $-19k + 54$ **116.** $5s - 6$ **117.** $-45t - 23$
118. $-45t^2 - 23.4t$ **119.** $-2(3x) - 7x; -13x$

120. $\dfrac{x + 9}{x - 6}$ **121.** No. The use of *and* there indicates the two quantities to be multiplied. **122.** Answers may vary. For example, "3 times the difference between 4 times a number and 6."

123. 16 **124.** $\dfrac{25}{36}$ **125.** -26 **126.** $\dfrac{8}{3}$

127. $-\dfrac{1}{24}$ **128.** $\dfrac{7}{2}$ **129.** 2 **130.** 77.6

131. $-1\dfrac{1}{2}$ **132.** 11 **133.** $-\dfrac{28}{15}$

134. 24 **135.** -11 **136.** -6

137. $\$13,600 - \$1400; \$12,200$ **138.** $\dfrac{x}{3x - 14}$

CHAPTER 1 TEST (page 87)

1. true **2.** false **3.** (number line with points at $-3, -1, 1, 4$)

4. $-|-8|$ (or -8) **5.** -1.277

6. $\dfrac{-6}{2 + (-8)}; 1$ **7.** negative **8.** 4

9. $-2\dfrac{5}{6}$ **10.** 2 **11.** 6 **12.** 108 **13.** 3

14. $\dfrac{30}{7}$ **15.** -70 **16.** 3 **17.** $178°F$

18. B **19.** D **20.** E **21.** A **22.** C
23. $-9x^2 - 6x - 8$ **24.** Distributive
25. (a) -18 (b) -18 (c) The distributive property tells us that the two methods produce equal results.

CHAPTER 2
SECTION 2.1 (page 93)

1. linear **3.** linear **5.** not linear **7.** If $A = 0$, the equation $Ax + B = C$ becomes $B = C$; with no variable it is not a linear equation. **9.** $\{12\}$
11. $\{-10\}$ **13.** $\{3\}$ **15.** $\{-2\}$ **17.** $\{4\}$
19. $\{0\}$ **21.** $\emptyset$ **23.** $\{\text{all real numbers}\}$
25. $\{4\}$ **27.** $\emptyset$ **29.** $\{\text{all real numbers}\}$

31. $\left\{\dfrac{7}{15}\right\}$ **33.** $\{7\}$ **35.** $\{-4\}$ **37.** $\{13\}$

39. $\{\text{all real numbers}\}$ **41.** $\{18\}$ **43.** $\{12\}$
45. To find the solution to $-x = 5$, either add $x - 5$ to both sides and simplify, or use the rule: if $-x = a$, then $x = -a$. **47.** One example is $x - 6 = -8$.

SECTION 2.2 (page 99)

1. expression **3.** equation **5.** $\dfrac{3}{2}$ **7.** 10

9. $-\dfrac{2}{9}$ **11.** -1 **13.** 6 **15.** -4 **17.** .12

19. -1 **21.** If both sides of an equation are multiplied by 0, the resulting equation is $0 = 0$. This is true, but does not help to solve the equation. **23.** $\left\{\dfrac{15}{2}\right\}$ **25.** $\{-5\}$

27. $\left\{-\dfrac{18}{5}\right\}$ **29.** $\{12\}$ **31.** $\{0\}$ **33.** $\{-12\}$

35. $\{40\}$ **37.** $\{-48\}$ **39.** $\{-35\}$ **41.** $\left\{-\dfrac{27}{35}\right\}$

43. $\{3\}$ **45.** $\{-5\}$ **47.** $\{7\}$ **49.** $\{-2\}$

51. $\left\{-\dfrac{3}{5}\right\}$ **53.** Answers will vary. For example, $\dfrac{3}{2}x = -6$. **55.** $3x = 18 + 5x$; solution set: $\{-9\}$; -9

SECTION 2.3 (page 105)

1. $\{-1\}$ **3.** $\{5\}$ **5.** $\{1\}$ **7.** $\left\{-\dfrac{5}{3}\right\}$ **9.** $\{-1\}$

11. $\emptyset$ **13.** {all real numbers} **15.** No, it is incorrect to divide both sides by a variable. If $-3x$ is added to both sides, the equation becomes $4x = 0$, so $x = 0$ is the correct solution, and $\{0\}$ is the solution set.
17. Clear parentheses. Use the addition property to get all variable terms on one side of the equation and all constants on the other, then collect terms. Use the multiplication property to get the equation in the form $x = $ a number. Check the solution.

19. $\{5\}$ **21.** $\{0\}$ **23.** $\left\{-\dfrac{7}{5}\right\}$

25. $\{120\}$ **27.** $\{6\}$ **29.** $\{15,000\}$
31. 800 **32.** Yes, you will get $(100 \cdot 2) \cdot 4 = 800$. This is a result of the associative property of multiplication. **33.** No, because $(100a)(100b) = 10,000ab \neq 100ab$. **34.** The distributive property involves the operation of addition as well. **35.** Yes, the associative property of multiplication is used. **36.** no

37. $\left\{-\dfrac{13}{8}\right\}$ **39.** $\{0\}$ **41.** $\{4\}$ **43.** $\{20\}$

45. {all real numbers} **47.** $\emptyset$ **49.** $11 - q$

51. $\dfrac{t}{5}$

SECTION 2.4 (page 115)

1. Some examples are *is, are, were,* and *was.* **3.** (b)
5. 3 **7.** -4 **9.** 57 Democrats, 43 Republicans
11. 1037 **13.** shorter piece: 15 inches; longer piece: 24 inches **15.** Murphy: 9; Frank: 3; Corky: 1
17. 36 million miles **19.** Smoltz: $253\frac{2}{3}$; Maddux: 245; Wohlers: $77\frac{1}{3}$ **21.** A and B: 40 degrees; C: 100 degrees
23. sheep: 64,000; kangaroos: 24,000 **25.** exertion: 9443; regulating body temperature: 1757 **27.** $45°$

29. $26°$ **31.** $55°$ **33.** 121 and 122 **35.** 20 and 22 **37.** 1990: 1.45 billion dollars; 1991: 1.95 billion dollars; 1992: 2.20 billion dollars

SECTION 2.5 (page 125)

1. The perimeter of a geometric figure is the distance around the figure. **3.** The area of a geometric figure measures the surface covered or enclosed by the figure.
5. area **7.** perimeter **9.** area **11.** area
13. $P = 20$ **15.** $A = 70$ **17.** $c = 5$
19. $r = 40$ **21.** $I = 875$ **23.** $r = 1.3$
25. $A = 452.16$ **27.** The formula for the area of a rectangle is $A = lw$. For a square, $l = w = s$, so lw becomes $s \cdot s$, or s^2, leading to $A = s^2$. **29.** $V = 384$
31. $V = 48$ **33.** $V = 904.32$ **35.** perimeter: 172 inches; area: 1785 square inches **37.** 1979 feet

39. 23,800.10 square feet **41.** $\dfrac{729}{32}$ or $22\dfrac{25}{32}$ cubic inches **43.** $107°, 73°$ **45.** $75°, 75°$

47. $139°, 139°$ **49.** $r = \dfrac{d}{t}$ **51.** $L = \dfrac{A}{W}$

53. $a = P - b - c$ **55.** $p = \dfrac{I}{rt}$ **57.** $b = \dfrac{2A}{h}$

59. $r = \dfrac{A - p}{pt}$ **61.** $h = \dfrac{V}{\pi r^2}$

63. $C = \dfrac{5}{9}(F - 32)$ or $C = \dfrac{5F - 160}{9}$

64. (a) $P - 2L = 2W$ **(b)** $\dfrac{P - 2L}{2} = W$

65. (a) $\dfrac{P}{2} = L + W$ **(b)** $\dfrac{P}{2} - L = W$

66. (a) Multiplicative identity property **(b)** A number divided by 1 is equal to itself. **(c)** Multiplication of fractions **(d)** Subtraction of fractions

SECTION 2.6 (page 135)

1. (c) **3.** $\dfrac{4}{3}$ **5.** $\dfrac{4}{3}$ **7.** $\dfrac{15}{2}$ **9.** $\dfrac{1}{5}$ **11.** $\dfrac{24}{5}$
13. 8-ounce size **15.** 13-ounce size **17.** 28-ounce size **19.** A ratio is a comparison, while a proportion is a statement that two ratios are equal. **21.** $\{27\}$

23. $\{16\}$ **25.** $\{10\}$ **27.** $\left\{-\dfrac{31}{5}\right\}$ **29.** \$14.58

31. 510 calories **33.** 12,500 fish **35.** $25\frac{2}{3}$ inches
37. 124.8 **39.** 120% **41.** 600 **43.** 1.4%

45. \$533 **47.** 16 **49.** $16\frac{2}{3}\%$ **51.** \$384

53. \$52,307.69 **55.** .564 **57.** .527 **59.** .5 is 50%, not .5%. **61.** 30

62. (a) $5x = 12$ **(b)** $\left\{\dfrac{12}{5}\right\}$ **63.** $\left\{\dfrac{12}{5}\right\}$

64. Both methods give the same solution set.

CHAPTER 2 REVIEW EXERCISES (page 143)

1. $\{9\}$ **2.** $\{4\}$ **3.** $\{-6\}$ **4.** $\left\{\dfrac{3}{2}\right\}$ **5.** $\{20\}$

6. $\left\{-\dfrac{61}{2}\right\}$ **7.** $\{15\}$ **8.** $\{0\}$ **9.** $\emptyset$ **10.** $\{20\}$

11. $-\dfrac{7}{2}$ **12.** 20 **13.** Hawaii: 6425 square miles; Rhode Island: 1212 square miles **14.** Seven Falls: 300 feet; Twin Falls: 120 feet **15.** 80° **16.** Saberhagen: 257 innings; Morris: 266 innings; Langston: 272 innings **17.** $h = 11$ **18.** $A = 28$

19. $r = 4.75$ **20.** $V = 904.32$ **21.** $L = \dfrac{A}{W}$

22. $h = \dfrac{2A}{b + B}$ **23.** 135°, 45° **24.** 100°, 100° **25.** perimeter: 326.5 feet; area: 6538.875 square feet **26.** diameter: approximately 19.9 feet; radius: approximately 9.95 feet; area: approximately 311 square

feet **27.** $\dfrac{3}{2}$ **28.** $\dfrac{5}{14}$ **29.** $\dfrac{3}{4}$ **30.** $\dfrac{1}{12}$ **31.** $\left\{\dfrac{7}{2}\right\}$

32. $\left\{-\dfrac{8}{3}\right\}$ **33.** $\left\{\dfrac{25}{19}\right\}$ **34.** 40% means $\frac{40}{100}$ or $\frac{2}{5}$. It is

the same as the ratio of 2 to 5. **35.** $6\dfrac{2}{3}$ pounds

36. 36 ounces **37.** $3.06 **38.** 375 kilometers

39. 17.48 **40.** 175% **41.** $33\dfrac{1}{3}\%$ **42.** 2500

43. $1118 **44.** 437 miles per tank **45.** $\{7\}$

46. $r = \dfrac{I}{pt}$ **47.** $\{70\}$ **48.** $\left\{\dfrac{13}{4}\right\}$

49. $\emptyset$ **50.** {all real numbers} **51.** 6 **52.** 12 **53.** Buddy: 1200 votes; Bob: 600 votes **54.** Norman: 84 miles; Janice: 28 miles **55.** 12 pounds **56.** 125 miles **57.** 35 geometry tests **58.** 37 small cars **59.** 70 feet **60.** 44 meters **61.** 26 inches

62. $20\dfrac{1}{2}$ inches **63.** 51°, 51° **64.** 92 or more

CHAPTER 2 TEST (page 147)

1. $\{6\}$ **2.** $\{-6\}$ **3.** $\left\{\dfrac{13}{4}\right\}$ **4.** $\{-10.8\}$ **5.** $\emptyset$

6. $\{21\}$ **7.** $\{30\}$ **8.** {all real numbers} **9.** 7 **10.** Golden Gate Bridge: 4200 feet; Brooklyn Bridge: 1595 feet **11.** 4 centimeters; 9 centimeters;

27 centimeters **12.** 18 **13.** $W = \dfrac{P - 2L}{2}$

or $W = \dfrac{P}{2} - L$ **14.** 100°, 80° **15.** 75°, 75°

16. 50° **17.** $\{6\}$ **18.** $\{-29\}$ **19.** 8 slices for $2.19 **20.** 2300 miles

CUMULATIVE REVIEW 1–2 (page 149)

1. $\dfrac{3}{8}$ **2.** $\dfrac{3}{4}$ **3.** $\dfrac{31}{20}$ **4.** $\dfrac{551}{40}$ or $13\dfrac{31}{40}$ **5.** 6

6. $\dfrac{6}{5}$ **7.** 34.03 **8.** 27.31 **9.** 30.51 **10.** 56.3

11. 35 yards **12.** $4\dfrac{1}{6}$ cups **13.** $20\dfrac{23}{24}$ pounds

14. $1187.65 **15.** true **16.** true **17.** 7 **18.** 1 **19.** 13 **20.** -40 **21.** -12 **22.** undefined

23. -6 **24.** 28 **25.** 1 **26.** 0 **27.** $\dfrac{73}{18}$

28. -64 **29.** -134 **30.** $-\dfrac{29}{6}$

31. distributive property **32.** commutative property **33.** inverse property **34.** identity property **35.** $7p - 14$ **36.** $2k - 11$ **37.** $\{7\}$

38. $\{-4\}$ **39.** $\{-1\}$ **40.** $\left\{-\dfrac{3}{5}\right\}$

41. $\{2\}$ **42.** $\{-13\}$ **43.** $\{26\}$ **44.** $\{-12\}$

45. $c = P - a - b$ **46.** $s = \dfrac{P}{4}$ **47.** $1261.88

48. $3750 **49.** $49.50 **50.** $98.45 **51.** 30 centimeters **52.** 16 inches

CHAPTER 3

SECTION 3.1 (page 157)

1. D **3.** B **5.** F **7.** Use a parenthesis when an endpoint is not included; use a bracket when it is included.

9. $(-\infty, -3]$

11. $[5, \infty)$

13. $(7, \infty)$

15. $(-4, \infty)$

17. $(-\infty, -40]$

19. $[3, \infty)$

21. $(-\infty, 4]$

23. $\left(-\infty, -\dfrac{15}{2}\right)$

25. $\left[\dfrac{1}{2}, \infty\right)$

27. $(3, \infty)$

29. $(-\infty, 4)$

31. $\{-9\}$

32. $(-9, \infty)$

33. $(-\infty, -9)$

34. We obtain the set of all real numbers.

35. $(-\infty, -3)$ **36.** On the number line, the point that corresponds to a number separates those points on the line that correspond to numbers less than that number from those that are greater than that number.

37. $(1, 11)$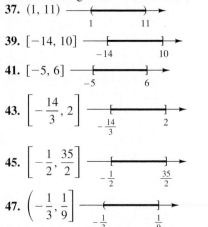

39. $[-14, 10]$

41. $[-5, 6]$

43. $\left[-\dfrac{14}{3}, 2\right]$

45. $\left[-\dfrac{1}{2}, \dfrac{35}{2}\right]$

47. $\left(-\dfrac{1}{3}, \dfrac{1}{9}\right]$

49. April, May, June, July
51. January, February, March, August, September, October, November, December   **53.** at least 80
55. 50 miles   **57.** 26 tapes

SECTION 3.2 (page 165)

1. true   **3.** true   **5.** False; 6 is not included in the union.   **7.** $\{4\}$ or D **9.** $\emptyset$ **11.** $\{1, 2, 3, 4, 5, 6\}$ or A **13.** $\{1, 3, 5, 6\}$ **15.** $\{1, 4, 6\}$ **17.** Each is equal to $\{1\}$. This illustrates the associative property of set intersection.   **19.** The intersection of two streets is the region common to *both* streets.

21. $(-3, 2)$

23. $(-\infty, 2]$

25. $\emptyset$

27. $[4, \infty)$

29. $[-1, 3]$

31. $[5, 9]$

33. $(-\infty, 4]$

35. $(-\infty, 2] \cup [4, \infty)$

37. $(-\infty, 8]$

39. $(-\infty, 1] \cup [10, \infty)$

41. $[-2, \infty)$

43. $(-\infty, \infty)$

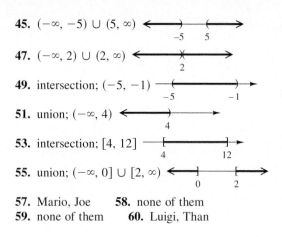

45. $(-\infty, -5) \cup (5, \infty)$

47. $(-\infty, 2) \cup (2, \infty)$

49. intersection; $(-5, -1)$

51. union; $(-\infty, 4)$

53. intersection; $[4, 12]$

55. union; $(-\infty, 0] \cup [2, \infty)$

57. Mario, Joe   **58.** none of them
59. none of them   **60.** Luigi, Than

SECTION 3.3 (page 175)

1. E; C; D; B; A   **3.** Use *or* for the equality statement and the $>$ statement. Use *and* for the $<$ statement.
5. $\{-12, 12\}$ **7.** $\{-5, 5\}$ **9.** $\{-6, 12\}$
11. $\{-4, 3\}$
13. $\left\{-3, \dfrac{11}{2}\right\}$ **15.** $\left\{-\dfrac{19}{2}, \dfrac{9}{2}\right\}$ **17.** $\{-10, -2\}$
19. $\left\{-8, \dfrac{32}{3}\right\}$
21. $(-\infty, -3) \cup (3, \infty)$
23. $(-\infty, -4] \cup [4, \infty)$
25. $(-\infty, -12) \cup (8, \infty)$
27. $\left(-\infty, -\dfrac{7}{3}\right] \cup [3, \infty)$
29. $(-\infty, -2) \cup (8, \infty)$
31. (a)

   (b)

33. $[-3, 3]$

35. $(-4, 4)$

37. $[-12, 8]$

39. $\left(-\dfrac{7}{3}, 3\right)$

41. $[-2, 8]$

43. $(-\infty, -5) \cup (13, \infty)$

45. $\{-6, -1\}$

47. $\left[-\dfrac{10}{3}, 4\right]$

49. $\left[-\dfrac{7}{6}, -\dfrac{5}{6}\right]$

51. $\{-5, 5\}$ **53.** $\{-5, -3\}$ **55.** $(-\infty, -3) \cup (2, \infty)$
57. $[-10, 0]$ **59.** $\{-1, 3\}$

61. $\left\{-3, \dfrac{5}{3}\right\}$ **63.** $\left\{-\dfrac{1}{3}, -\dfrac{1}{15}\right\}$ **65.** $\left\{-\dfrac{5}{4}\right\}$

67. $\varnothing$ **69.** $\left\{-\dfrac{1}{4}\right\}$ **71.** $\varnothing$ **73.** $(-\infty, \infty)$

75. $\left\{-\dfrac{3}{7}\right\}$ **77.** $(-\infty, \infty)$

79. $\left(-\infty, -\dfrac{7}{10}\right) \cup \left(-\dfrac{7}{10}, \infty\right)$

81. 460.2 feet   **82.** Federal Office Building, City Hall, Kansas City Power and Light, Hyatt Regency
83. Southwest Bell Telephone, City Center Square, Commerce Tower, Federal Office Building, City Hall, Kansas City Power and Light, Hyatt Regency
84. **(a)** $|x - 460.2| \geq 75$ **(b)** $x \geq 535.2$ or $x \leq 385.2$ **(c)** Pershing Road Associates, AT&T Town Pavilion, One Kansas City Place   **(d)** It makes sense because it includes all buildings *not* listed earlier.

SUMMARY ON SOLVING LINEAR AND ABSOLUTE VALUE EQUATIONS AND INEQUALITIES (page 181)

1. $\{12\}$ **3.** $\{7\}$ **5.** $\varnothing$

7. $\left[-\dfrac{2}{3}, \infty\right)$ **9.** $\{-3\}$ **11.** $(-\infty, 5]$ **13.** $\{2\}$

15. $\varnothing$ **17.** $(-5.5, 5.5)$ **19.** $\left\{-\dfrac{96}{5}\right\}$

21. $(-\infty, -24)$ **23.** $\left\{\dfrac{7}{2}\right\}$ **25.** $(-\infty, \infty)$

27. $(-\infty, -4) \cup (7, \infty)$ **29.** $\left\{-\dfrac{1}{5}\right\}$ **31.** $\left[-\dfrac{1}{3}, 3\right]$

33. $\left\{-\dfrac{1}{6}, 2\right\}$ **35.** $(-\infty, -1] \cup \left[\dfrac{5}{3}, \infty\right)$

37. $\left\{-\dfrac{5}{2}\right\}$ **39.** $\left[-\dfrac{9}{2}, \dfrac{15}{2}\right]$ **41.** $(-\infty, \infty)$
43. $(-\infty, \infty)$ **45.** $\{-2\}$ **47.** $(-\infty, -1) \cup (2, \infty)$

CHAPTER 3 REVIEW EXERCISES (page 185)

1. $(-9, \infty)$

2. $(-\infty, -3]$

3. $\left(\dfrac{3}{2}, \infty\right)$

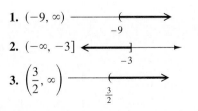

4. $\left(-\infty, -\dfrac{14}{9}\right)$

5. $[-3, \infty)$

6. $[-3, 12]$

7. $[3, 5)$

8. $\left(-3, \dfrac{7}{2}\right)$

9. any grade greater than or equal to 61%
10. Because the statement $-8 < -13$ is *false*, the inequality has no solution.   **11.** $\{3, 9\}$
12. $\{1, 3, 5, 6, 7, 9, 12\}$
13. $(6, 9)$

14. $(8, 14)$

15. $(-\infty, -3] \cup (5, \infty)$

16. $(-\infty, \infty)$

17. $\varnothing$

18. $(-\infty, -2] \cup [7, \infty)$

19. 1988, 1989, 1990, 1991, 1992
20. 1981, 1982, 1983, 1984, 1985, 1988, 1989, 1990, 1991, 1992   **21.** $\{-7, 7\}$ **22.** $\{-11, 7\}$

23. $\left\{-\dfrac{1}{3}, 5\right\}$ **24.** $\varnothing$ **25.** $\{0, 7\}$ **26.** $\left\{-\dfrac{3}{2}, \dfrac{1}{2}\right\}$

27. $\left\{-\dfrac{3}{4}, \dfrac{1}{2}\right\}$ **28.** $\left\{-\dfrac{1}{2}\right\}$

29. $(-14, 14)$

30. $[-1, 13]$

31. $[-3, -2]$

32. $(-\infty, \infty)$

33. $\left(-\infty, -\dfrac{8}{5}\right) \cup (2, \infty)$

34. $(-\infty, \infty)$

35. $\left(-\infty, \dfrac{7}{6}\right]$ **36.** $[-4, 5)$ **37.** $\varnothing$
38. 6 inches, 12 inches, 16 inches   **39.** $(-\infty, 2]$

40. $(-\infty, -1) \cup \left(\dfrac{11}{7}, \infty\right)$ **41.** $\{-5, 15\}$

42. $[-16, 10]$ **43.** 6 inches
44. 683 votes and 532 votes

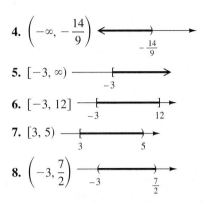

CHAPTER 3 TEST (page 189)

1. We must reverse the direction of the inequality symbol.

2. $[1, \infty)$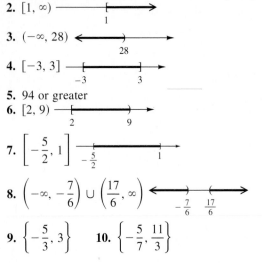

3. $(-\infty, 28)$

4. $[-3, 3]$

5. 94 or greater

6. $[2, 9)$

7. $\left[-\frac{5}{2}, 1\right]$

8. $\left(-\infty, -\frac{7}{6}\right) \cup \left(\frac{17}{6}, \infty\right)$

9. $\left\{-\frac{5}{3}, 3\right\}$ **10.** $\left\{-\frac{5}{7}, \frac{11}{3}\right\}$

CUMULATIVE REVIEW 1–3 (page 191)

1. 9, $\sqrt{36}$ (or 6) **2.** 0, 9, $\sqrt{36}$ (or 6)
3. $-8, 0, 9, \sqrt{36}$ (or 6)

4. $-8, -\frac{2}{3}, 0, \frac{4}{5}, 9, \sqrt{36}$ (or 6) **5.** $-\sqrt{6}$

6. All are real numbers. **7.** $-\frac{22}{21}$ **8.** 8

9. 8 **10.** 0 **11.** -243 **12.** $\frac{216}{343}$ **13.** 1

14. -4096 **15.** $\sqrt{-36}$ is not a real number.

16. $\frac{4 + 4}{4 - 4}$ is undefined. **17.** -16 **18.** -34

19. 184 **20.** $\frac{27}{16}$ **21.** $-20r + 17$ **22.** $13k + 42$

23. commutative property **24.** distributive property

25. inverse property **26.** $-\frac{3}{2}$ **27.** $\{5\}$ **28.** $\{30\}$

29. $\{15\}$ **30.** $b = P - a - c$

31. $[-14, \infty)$

32. $\left[\frac{5}{3}, 3\right)$

33. $(-\infty, 0) \cup (2, \infty)$

34. $\left(-\infty, -\frac{1}{7}\right] \cup [1, \infty)$

35. managerial and professional specialty
36. mathematical and computer scientists **37.** 2 liters
38. 9 cents, 12 nickels, 8 quarters

CHAPTER 4

SECTION 4.1 (page 199)

1. (a) between 1989 and 1990 **(b)** between 1991 and 1992 **(c)** 1991 **3.** Another name is the Cartesian system, named after René Descartes. **5.** origin
7. y; x **9.** two **11. (a)** I **(b)** III **(c)** II
(d) IV **(e)** none **13. (a)** I or III **(b)** II or IV
(c) II or IV **(d)** I or III

15–24. **25.** -3; 3; 2; -1

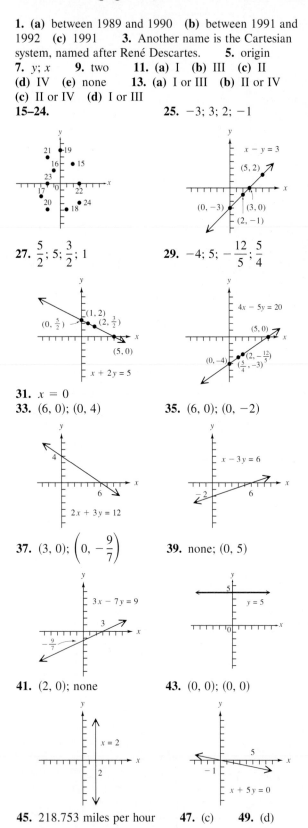

27. $\frac{5}{2}$; 5; $\frac{3}{2}$; 1 **29.** -4; 5; $-\frac{12}{5}$; $\frac{5}{4}$

31. $x = 0$
33. $(6, 0)$; $(0, 4)$ **35.** $(6, 0)$; $(0, -2)$

37. $(3, 0)$; $\left(0, -\frac{9}{7}\right)$ **39.** none; $(0, 5)$

41. $(2, 0)$; none **43.** $(0, 0)$; $(0, 0)$

45. 218.753 miles per hour **47.** (c) **49.** (d)

SECTION 4.2 (page 209)

1. 2   **3.** undefined   **5.** (a), (b), (c), (d), (f)   **7.** 8
9. $\frac{5}{6}$ **11.** 0

13. undefined   **15.** 0

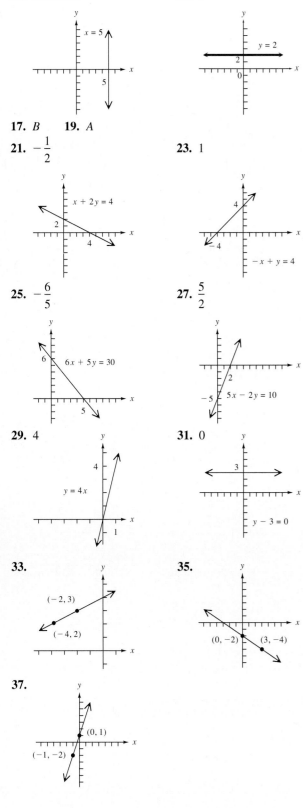

17. *B*   **19.** *A*
21. $-\frac{1}{2}$ **23.** 1

25. $-\frac{6}{5}$ **27.** $\frac{5}{2}$

29. 4   **31.** 0

33.   **35.**

37.

39. perpendicular   **41.** parallel   **43.** neither
45. $\frac{7}{10}$ **47. (a)** $.92   **(b)** It means an *increase* in
price.   **49.** 6.25 billion minutes per year
51. $\frac{1}{3}$ **52.** $\frac{1}{3}$ **53.** $\frac{1}{3}$ **54.** $\frac{1}{3} = \frac{1}{3} = \frac{1}{3}$ is true.
55. They are collinear.   **56.** They are not collinear.

SECTION 4.3 (page 223)

1. A   **3.** C   **5.** H   **7.** B   **9.** $3x + 4y = 10$
11. $2x + y = 18$ **13.** $x - 2y = -13$ **15.** $y = 12$
17. $x = 9$ **19.** $x = .5$ **21.** $2x - y = 2$
23. $x + 2y = 8$ **25.** $2x - 13y = -6$ **27.** $y = 5$
29. $x = 7$ **31.** $y = -\frac{5}{2}x + 10$; $-\frac{5}{2}$; (0, 10)
33. $y = \frac{2}{3}x - \frac{10}{3}$; $\frac{2}{3}$; $\left(0, -\frac{10}{3}\right)$
35. $y = 5x + 15$ **37.** $y = -\frac{2}{3}x + \frac{4}{5}$
39. $y = \frac{2}{5}x + 5$ **41.** $3x - y = 19$
43. $x - 2y = 2$ **45.** $x + 2y = 18$ **47.** $y = 7$
49. $y = 45x$; (0, 0), (5, 225), (10, 450)
51. $y = 1.30x$; (0, 0), (5, 6.50), (10, 13.00)
53. $y = 3x + 15$; (0, 15), (5, 30), (10, 45)
55. $y = .10x + 25.00$; (0, 25.00), (5, 25.50), (10, 26.00)
57. $69 = 3x + 15$; 18 days
59. $42.30 = .10x + 25.00$; 173 miles
61. $y = -\frac{829}{8}x + 29{,}557$ **63.** *A* is y_1 and *B* is y_2.
65. 8   **67.** 32; 212   **68.** (0, 32) and (100, 212)
69. $\frac{9}{5}$ **70.** $F = \frac{9}{5}C + 32$ **71.** $C = \frac{5}{9}(F - 32)$
72. When the Celsius temperature is 50°, the Fahrenheit temperature is 122°.

SECTION 4.4 (page 235)

1. solid; below   **3.** dashed; above
5.   **7.**

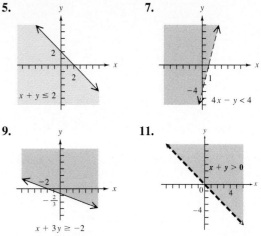

9.   **11.**

13.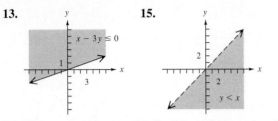

15.

17. If the connecting word is "and," use intersection. If the connecting word is "or," use union.

19.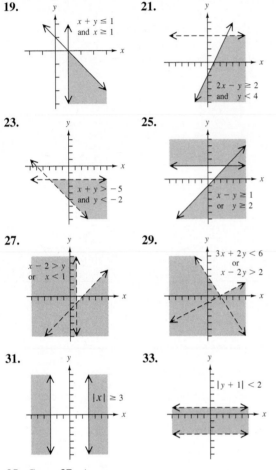

21.

23.

25.

27.

29.

31.

33.

35. C   **37.** A

SECTION 4.5 (page 247)

1. (1990, 746,220), (1992, 661,391), (1993, 590,324)
3. (1990, 285.7), (1992, 318.8), (1993, 339.9)
5. The vertical line test is used to determine whether a graph is that of a function. Any vertical line will intersect the graph of a function in at most one point.
7. function; domain: {5, 3, 4, 7}; range: {1, 2, 9, 3}
9. not a function; domain: {2, 0}; range: {4, 2, 6}
11. not a function; domain: {1, 2, 3, 5};
range: {10, 15, 19, −27}   **13.** function; domain:
$(-\infty, \infty)$; range: $(-\infty, 4]$ **15.** not a function; domain:
[−4, 4]; range: [−3, 3]
17. function; domain: $(-\infty, \infty)$ **19.** not a function;
domain: [0, ∞)   **21.** not a function; domain: $(-\infty, \infty)$

23. function; domain [0, ∞)   **25.** function; domain:
$(-\infty, 0) \cup (0, \infty)$ **27.** function (also a linear function);
domain: $(-\infty, \infty)$

29. function; domain: $\left[-\dfrac{1}{2}, \infty \right)$

31. function; domain: $(-\infty, 9) \cup (9, \infty)$
33. 4   **35.** −11   **37.** $-3p + 4$ **39.** $3x + 4$
41. $-3x - 2$ **43.** −8   **45.** No. In general,
$f(g(x)) \neq g(f(x))$.
47. line; −2; linear; $-2x + 4$; −2; 3; −2
49. domain: $(-\infty, \infty)$; range: $(-\infty, \infty)$

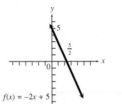

51. domain: $(-\infty, \infty)$; range: $(-\infty, \infty)$

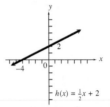

53. domain: $(-\infty, \infty)$; range: {2}

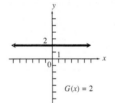

55. (a) 29,562   (b) 29,316   (c) 28,947
57. In 1986 (when $x = 2$), the number of post offices was approximately 29,439.
59. $f(3) = 7$ **61.** $f(x) = -3x + 5$

SECTION 4.6 (page 259)

1. 36   **3.** .625   **5.** $222\dfrac{2}{9}$ **7.** increases; decreases

9. 1.09\dfrac{9}{10}$ **11.** 8 pounds

13. 100 cycles per second   **15.** 3 footcandles
17. $420   **19.** 800 gallons   **21.** 25   **23.** 9
25. (0, 0), (1, 1.25)   **26.** 1.25
27. $y = 1.25x + 0$ or $y = 1.25x$ **28.** $a = 1.25, b = 0$
29. It is the price per gallon and the slope of the line.
30. It can be written in the form $y = kx$ (where $k = a$).
The value of a is called the constant of variation.

Chapter 4 Review Exercises (page 267)

1. $3; 2; 0; \dfrac{10}{3}$ **2.** $-4; 3; -5; 4$

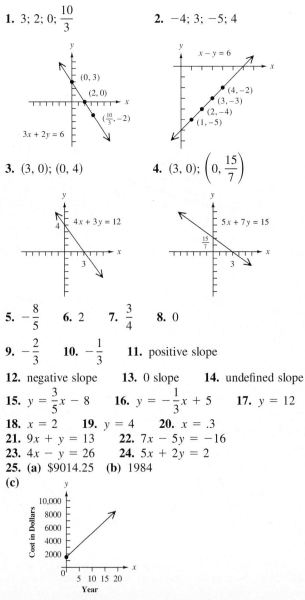

3. $(3, 0); (0, 4)$ **4.** $(3, 0); \left(0, \dfrac{15}{7}\right)$

5. $-\dfrac{8}{5}$ **6.** 2 **7.** $\dfrac{3}{4}$ **8.** 0

9. $-\dfrac{2}{3}$ **10.** $-\dfrac{1}{3}$ **11.** positive slope

12. negative slope   **13.** 0 slope   **14.** undefined slope

15. $y = \dfrac{3}{5}x - 8$ **16.** $y = -\dfrac{1}{3}x + 5$ **17.** $y = 12$

18. $x = 2$ **19.** $y = 4$ **20.** $x = .3$

21. $9x + y = 13$ **22.** $7x - 5y = -16$

23. $4x - y = 26$ **24.** $5x + 2y = 2$

25. (a) $9014.25   **(b)** 1984

(c)

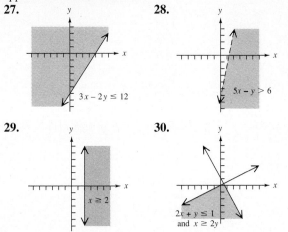

26. (a) 216.25 miles per hour   **(b)** .860 mile per hour; there is a discrepancy because the equation is only an *approximate* linear model.

27.   **28.**

29.   **30.**

31.   **32.**

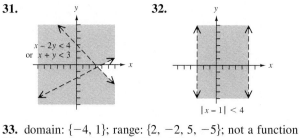

$x - 2y < 4$
or $x + y < 3$

$|x - 1| < 4$

33. domain: $\{-4, 1\}$; range: $\{2, -2, 5, -5\}$; not a function

34. domain: $\{-14, 91, 17, 75, -23\}$; range: $\{9, 12, 18, 70, 56, 5\}$; not a function

35. domain: $[-4, 4]$; range: $[0, 2]$; function

36. -6 **37.** -15 **38.** -96

39. $-8p^2 + 6p - 6$ **40.** function; linear function; domain: $(-\infty, \infty)$ **41.** not a function; domain: $(-\infty, \infty)$

42. function; domain: $(-\infty, \infty)$

43. function; domain: $\left[-\dfrac{7}{4}, \infty\right)$

44. not a function; domain: $[0, \infty)$

45. function; domain: $(-\infty, 36) \cup (36, \infty)$

46.

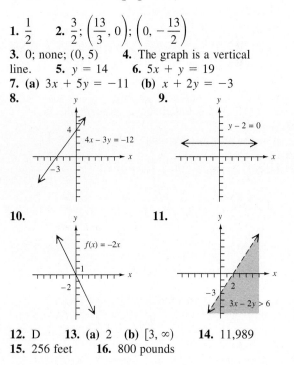

$f(x) = -\dfrac{3}{2}x + \dfrac{7}{2}$

47. 15 **48.** 5 **49.** .850 ohm   **50.** 430 millimeters

51. 5.625 kilometers per second   **52.** $.71\pi$ seconds

Chapter 4 Test (page 271)

1. $\dfrac{1}{2}$ **2.** $\dfrac{3}{2}; \left(\dfrac{13}{3}, 0\right); \left(0, -\dfrac{13}{2}\right)$

3. 0; none; $(0, 5)$ **4.** The graph is a vertical line.   **5.** $y = 14$ **6.** $5x + y = 19$

7. (a) $3x + 5y = -11$ **(b)** $x + 2y = -3$

8.   **9.**

$4x - 3y = -12$

$y - 2 = 0$

10.   **11.**

$f(x) = -2x$

$3x - 2y > 6$

12. D   **13. (a)** 2 **(b)** $[3, \infty)$ **14.** $11{,}989$

15. 256 feet   **16.** 800 pounds

CUMULATIVE REVIEW 1–4 (page 273)

1. true   **2.** true   **3.** false
4. 4   **5.** .64   **6.** −4
7. $\dfrac{8}{5}$ **8.** $-2m + 6$ **9.** $4m − 3$
10. $2x^2 + 5x + 4$ **11.** $(2, \infty)$ **12.** $(-\infty, 1]$
13. $(−3, 5]$ **14.** no   **15.** −24
16. 204   **17.** 56   **18.** undefined   **19.** 10
20. $\left\{\dfrac{7}{6}\right\}$ **21.** $\{−1\}$ **22.** $\left(-\infty, \dfrac{15}{4}\right]$
23. $\left(-\dfrac{1}{2}, \infty\right)$ **24.** $(2, 3)$ **25.** $(-\infty, 2) \cup (3, \infty)$
26. $\left\{-\dfrac{16}{5}, 2\right\}$ **27.** $(−11, 7)$
28. $(-\infty, −2] \cup [7, \infty)$ **29.** 6   **30.** 4 white pills
31. $(0, −3), (4, 0), \left(2, -\dfrac{3}{2}\right)$
32. x-intercept: $(−2, 0)$; y-intercept: $(0, 4)$

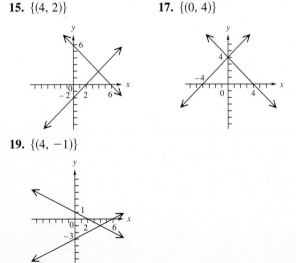

33. $-\dfrac{3}{2}$ **34.** $-\dfrac{3}{4}$ **35.** $3x + 4y = −4$
36. $y = −2$ **37.** $4x + 3y = 7$
38. (a) $(-\infty, \infty)$ **(b)** 22   **39.** $105{,}666\dfrac{2}{3}$
40. the segment for 1988 through 1991

CHAPTER 5
SECTION 5.1 (page 279)

1. system   **3.** consistent   **5.** no   **7.** yes   **9.** yes
11. no   **13.** (b), because the ordered pair must be in Quadrant II
We show the graphs here only for Exercises 15, 17, and 19.
15. $\{(4, 2)\}$ **17.** $\{(0, 4)\}$

19. $\{(4, −1)\}$

SECTION 5.2 (page 289)

1. true   **3.** true   **5.** $\{(−1, 3)\}$ **7.** $\{(−1, −3)\}$
9. $\{(−2, 3)\}$ **11.** $\left\{\left(\dfrac{1}{2}, 4\right)\right\}$ **13.** $\{(3, −6)\}$
15. $\{(7, 4)\}$ **17.** $\{(0, 4)\}$ **19.** $\{(−4, 0)\}$
21. $\{(0, 0)\}$ **23.** $\{(−6, 5)\}$ **25.** $\{(−3, 2)\}$
27. $\left\{\left(\dfrac{1}{8}, -\dfrac{5}{6}\right)\right\}$ **29.** $\{(11, 15)\}$ **31.** $\emptyset$
33. infinite number of solutions   **35.** $\emptyset$ **37.** infinite number of solutions   **39.** 2   **40.** 5
41. $\{(2, 5)\}$

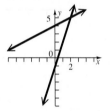

42. The x-coordinate, 2, is equal to the solution of the equation.   **43.** The y-coordinate, 5, is equal to the value we obtained on both sides when checking.   **44.** 5; 3; 5

SECTION 5.3 (page 297)

1. No, it is not correct, because the solution set is $\{(3, 0)\}$. The y-value must also be determined.   **3.** The first student had less work to do, because the coefficient of y in the first equation is −1. The second student had to divide by 2, introducing fractions into the expression for x.
5. $\{(7, 3)\}$ **7.** $\{(−2, 4)\}$ **9.** $\{(−4, 8)\}$
11. $\{(3, −2)\}$ **13.** infinite number of solutions
15. $\left\{\left(\dfrac{1}{3}, -\dfrac{1}{2}\right)\right\}$ **17.** $\emptyset$ **19.** infinite number of solutions   **21. (a)** $\{(1, 4)\}$ **(b)** $\{(1, 4)\}$ **(c)** While it is a matter of preference, most would prefer the addition method.   **23.** $\{(4, −6)\}$ **25.** $\{(7, 0)\}$

21. $\{(1, 3)\}$ **23.** $\{(0, 2)\}$ **25.** $\{(4, −3)\}$
27. Two lines will intersect in only one point (if they are distinct) or infinitely many points (if they are the same). They cannot intersect in exactly two points.
29. $\emptyset$ **31.** infinite number of solutions   **33.** $\emptyset$
35. $(4, −3)$ satisfies both equations.   **37.** $(−4, 8)$ satisfies both equations.   **39.** no solution; infinitely many solutions   **41.** The answer cannot be checked because it is too difficult to read the exact coordinates.

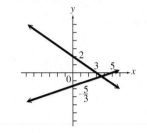

27. infinite number of solutions   **29.** ∅   **31.** yes
33. {(0, 3)}   **35.** {(24, −12)}   **37.** {(3, 2)}
39. To find the total cost, multiply the number of bicycles
(x) by the cost per bicycle (400 dollars), and add the fixed
cost (5000 dollars). Thus, $y_1 = 400x + 5000$ gives this
total cost (in dollars).   **40.** $600x$; $y_2 = 600x$
41. $y_1 = 400x + 5000$; $y_2 = 600x$; Solution:
{(25, 15,000)}   **42.** 25; 15,000; 15,000

SECTION 5.4 (page 307)

1. The statement means that when −1 is substituted for x,
2 is substituted for y, and 3 is substituted for z in the three
equations, the resulting three statements are true.

3. {(1, 4, −3)}   **5.** {(0, 2, −5)}   **7.** $\left\{\left(-\dfrac{7}{3}, \dfrac{22}{3}, 7\right)\right\}$

9. {(4, 5, 3)}   **11.** {(2, 2, 2)}   **13.** $\left\{\left(\dfrac{8}{3}, \dfrac{2}{3}, 3\right)\right\}$

15. Answers will vary. Some possible answers are   **(a)** two
perpendicular walls and the ceiling in a normal room
(b) the floors of three different levels of an office
building   **(c)** three pages of this book (since they intersect
in the spine).   **17.** ∅   **19.** infinite number of
solutions   **21.** infinite number of solutions
23. {(0, 0, 0)}
25. $128 = a + b + c$ **26.** $140 = 2.25a + 1.5b + c$
27. $80 = 9a + 3b + c$ **28.** $a + b + c = 128$,
$2.25a + 1.5b + c = 140$, $9a + 3b + c = 80$;
{(−32, 104, 56)}   **29.** $f(x) = -32x^2 + 104x + 56$
30. 56 feet   **31.** {(2, 1, 5, 3)}

SECTION 5.5 (page 317)

1. Texas: 93; Florida: 41   **3.** $x = 40$, $y = 50$, so the
angles measure 40° and 50°.   **5.** length: 78 feet; width:
36 feet   **7.** CGA monitor: $400; VGA monitor: $500
9. 6 units of yarn; 2 units of thread   **11.** dark clay: $5
per kilogram; light clay: $4 per kilogram
13. (a) 3.2 ounces **(b)** 8 ounces **(c)** 12.8 ounces
(d) 16 ounces   **15.** $.89x$ **17.** 6 gallons of 25%;
14 gallons of 35%   **19.** 6 liters of pure acid; 48 liters of
10% acid   **21.** 14 kilograms of nuts; 16 kilograms of
cereal   **23.** 76 general admission; 108 with student ID
25. 28 dimes; 66 quarters   **27.** $1000 at 2%; $2000 at
4%   **29.** $25y$ miles   **31.** freight train: 50 kilometers
per hour; express train: 80 kilometers per hour   **33.** top
speed: 2100 miles per hour; wind speed: 300 miles per
hour   **35.** $x + y + z = 180$; angle measures: 70°, 30°,
80°   **37.** first: 20°; second: 70°; third: 90°
39. shortest: 12 centimeters; middle: 25 centimeters;
longest: 33 centimeters   **41.** A: 180 cases; B: 60 cases;
C: 80 cases   **43.** $2a + b + c = -5$
44. $-a + c = -1$ **45.** $3a + 3b + c = -18$
46. $a = 1$, $b = -7$, $c = 0$; $x^2 + y^2 + x - 7y = 0$
47. It is not a function because a vertical line intersects its
graph more than once.

SECTION 5.6 (page 327)

1. true   **3.** true   **5.** false   **7.** −3   **9.** 14
11. 0   **13.** 59   **15.** 14   **17.** 1   **19.** −22
21. 0   **23.** 0   **25.** 20   **27.** 0   **29.** −22
31. 0   **33.** $\dfrac{y_2 - y_1}{x_2 - x_1}$ **34.** $y - y_1 = \dfrac{y_2 - y_1}{x_2 - x_1}(x - x_1)$
35. $x_2 y - x_1 y - x_2 y_1 - xy_2 + x_1 y_2 + xy_1 = 0$
36. The result is the same as in Exercise 35.
37. 52   **39.** 9

SECTION 5.7 (page 335)

1. {(4, 6)}   **3.** {(−5, 2)}   **5.** {(1, 0)}
7. $\left\{\left(\dfrac{11}{58}, -\dfrac{5}{29}\right)\right\}$ **9.** {(−2, 3, 5)}   **11.** {(5, 0, 0)}
13. Cramer's rule does not apply.
15. {(20, −13, −12)}   **17.** $\left\{\left(\dfrac{62}{5}, -\dfrac{1}{5}, \dfrac{27}{5}\right)\right\}$
19. {(−1, 2, 5, 1)}   **21.** The systems will vary.
One such system is $\begin{aligned}5x + 6y &= 7\\8x + 9y &= 10.\end{aligned}$
In all cases, the solution set is {(−1, 2)}.
22.

23. $\dfrac{1}{2}\begin{vmatrix} 0 & 0 & 1 \\ -3 & -4 & 1 \\ 2 & -2 & 1 \end{vmatrix}$ **24.** 7

CHAPTER 5 REVIEW EXERCISES (page 341)

1. yes   **2.** yes   **3.** no   **4.** no   **5.** {(3, 1)}
6. {(0, −2)}   **7.** infinite number of solutions   **8.** ∅
9. It would be difficult to read the exact coordinates from
the graph.   **10.** It is not a solution of the system
because it is not a solution of the second equation,
$2x + y = 4$.   **11.** {(7, 1)}   **12.** {(−5, −2)}
13. {(−3, 2)}   **14.** {(−4, 3)}   **15.** infinite number of
solutions   **16.** ∅   **17.** (c)   **18. (a)** 2   **(b)** 9
19. {(2, 1)}   **20.** {(3, 5)}   **21.** {(6, 4)}   **22.** ∅
23. {(−4, 1)}   **24.** infinite number of solutions
25. {(9, 2)}   **26.** $\left\{\left(\dfrac{10}{7}, -\dfrac{9}{7}\right)\right\}$ **27.** {(8, 9)}
28. {(2, 1)}   **29.** {(1, −5, 3)}   **30.** ∅
31. {(1, 2, 3)}   **32.** {(0, 0, 0)}   **33.** length: 6 feet;
width: 4 feet   **34.** 3 weekend days; 5 weekdays
35. plane: 300 miles per hour; wind: 20 miles per hour
36. 30 pounds of $2-a-pound candy; 70 pounds of
$1-a-pound candy   **37.** 85°; 60°; 35°   **38.** $40,000
at 10%; $100,000 at 6%; $140,000 at 5%
39. 10 pounds of jelly beans; 2 pounds of chocolate eggs;
3 pounds of marshmallow chicks   **40.** 8 fish at $20;
15 fish at $40; 6 fish at $65   **41.** 80   **42.** −21

43. -38 **44.** 0 **45.** 2 **46.** 0 **47.** Cramer's rule does not apply if $D = 0$.

48. $\left\{\left(\dfrac{37}{11}, \dfrac{14}{11}\right)\right\}$ **49.** $\left\{\left(\dfrac{39}{10}, \dfrac{6}{5}\right)\right\}$ **50.** $\{(3, -2, 1)\}$

51. $\left\{\left(\dfrac{172}{67}, -\dfrac{14}{67}, -\dfrac{87}{67}\right)\right\}$ **52.** $\{(12, 9)\}$

53. $\emptyset$ **54.** $\{(3, -1)\}$ **55.** $\{(5, 3)\}$ **56.** $\{(0, 4)\}$

57. $\left\{\left(\dfrac{82}{23}, -\dfrac{4}{23}\right)\right\}$ **58.** 20 liters   **59.** $38°, 66°, 76°$

CHAPTER 5 TEST (page 347)

1. $\{(6, 1)\}$

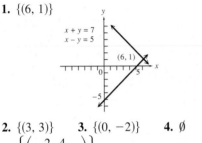

2. $\{(3, 3)\}$ **3.** $\{(0, -2)\}$ **4.** $\emptyset$

5. $\left\{\left(-\dfrac{2}{3}, \dfrac{4}{5}, 0\right)\right\}$ **6.** $\{(6, -4)\}$ **7.** infinite number of solutions   **8.** 180 votes; 225 votes   **9.** 4 liters of 20%; 8 liters of 50%   **10.** AC adaptor: \$8; rechargeable flashlight: \$15   **11.** 60 ounces of Orange Pekoe; 30 ounces of Irish Breakfast; 10 ounces of Earl Grey

12. 3 **13.** 0 **14.** $\left\{\left(-\dfrac{9}{4}, \dfrac{5}{4}\right)\right\}$ **15.** $\{(-1, 2, 3)\}$

CUMULATIVE REVIEW 1–5 (page 349)

1. 81 **2.** -81 **3.** -81 **4.** $.7$ **5.** $-.7$
6. not a real number   **7.** 4 **8.** -4 **9.** -199
10. 455 **11.** 14 **12.** $\left\{-\dfrac{15}{4}\right\}$ **13.** $\left\{\dfrac{2}{3}, 2\right\}$

14. $x = \dfrac{d - by}{a - c}$ or $x = \dfrac{by - d}{c - a}$ **15.** $\{11\}$

16. $\left(-\infty, \dfrac{240}{13}\right]$ **17.** $\left[-2, \dfrac{2}{3}\right]$ **18.** $(-\infty, \infty)$

19. 45 miles per hour, 75 miles per hour   **20.** 6 meters
21. 35 pennies, 29 nickels, 30 dimes   **22.** $46°, 46°, 88°$

23. $y = 6$ **24.** $x = 4$ **25.** $-\dfrac{4}{3}$ **26.** $\dfrac{3}{4}$

27. $4x + 3y = 10$ **28.** $f(x) = -\dfrac{4}{3}x + \dfrac{10}{3}$

29. $5x - 3y = -15$
30.
31.

32. $\{(3, -3)\}$ **33.** $\{(5, 3, 2)\}$ **34.** 33
35. 5 pounds of oranges; 1 pound of apples   **36.** video rental firm: \$60,000; bond and money market fund: \$10,000 each   **37.** $x = 8$ or 800 parts; \$3000
38. about \$500

CHAPTER 6

SECTION 6.1 (page 357)

1. 3^7 **3.** $(-2)^5$ **5.** w^6 **7.** $\left(-\dfrac{1}{4}\right)^5$ **9.** $(-7x)^4$

11. The expression $(5x)^3 = (5x)(5x)(5x) = 125x^3$, while $5x^3 = 5x \cdot x \cdot x \neq 125x^3$.
13. base: 2; exponent: 7; 128
15. base: -2; exponent: 7; -128
17. base: $-8x$; exponent: 4
19. base: x; exponent: 4
21. The notation $a\text{^}b$ means a^b.
23. 1 **25.** 5^8 **27.** 4^{12} **29.** $(-7)^9$
31. t^{24} **33.** $-56r^7$ **35.** $42p^{10}$
37. The product rule applies only to exponentials with the same base. $3^2 \cdot 4^3 = 9 \cdot 64 = 576$
39. $14x^4; 45x^8$ **41.** $5a^2; -140a^6$ **43.** 4^6
45. t^{20} **47.** 7^3r^3 **49.** $5^5x^5y^5$
51. 5^{12} **53.** -8^{15} **55.** $8q^3r^3$ **57.** $\dfrac{1}{2^3}$ **59.** $\dfrac{a^3}{b^3}$

61. $\dfrac{9^8}{5^8}$ **63.** $\dfrac{5^5}{2^5}$ **65.** $\dfrac{9^5}{8^3}$ **67.** $2^{12}x^{12}$
69. $(-6)^5p^5$ **71.** $6^5x^{10}y^{15}$ **73.** x^{21}
75. $2^2w^4x^{26}y^7$ **77.** $-r^{18}s^{17}$ **79.** $\dfrac{5^3a^6b^{15}}{c^{18}}$
81. No, $(10^2)^3 = 10^{(2)(3)} = 10^6 = 1,000,000$.

SECTION 6.2 (page 365)

1. 1 **3.** -1 **5.** 0 **7.** 1 **9.** 2 **11.** $\dfrac{1}{64}$

13. 16 **15.** $\dfrac{49}{36}$ **17.** $\dfrac{1}{81}$ **19.** $\dfrac{8}{15}$ **21.** negative

23. negative   **25.** positive   **27.** zero   **29.** 1

30. $\dfrac{5^2}{5^2}$ **31.** 5^0 **32.** $5^0 = 1$; this supports the definition of a zero exponent.

33. $\dfrac{1}{9}$ **35.** $\dfrac{1}{6^5}$ **37.** 6^3 **39.** $2r^4$ **41.** $\dfrac{5^2}{4^3}$

43. $\dfrac{p^5}{q^8}$ **45.** r^9 **47.** $\dfrac{x^5}{6}$ **49.** 7^3 **51.** $\dfrac{1}{x^2}$

53. $\dfrac{4^3x}{3^2}$ **55.** $\dfrac{x^2z^4}{y^2}$ **57.** $6x$ **59.** $\dfrac{1}{m^{10}n^5}$

61. $\dfrac{5}{16x^5}$ **63.** $\dfrac{36q^2}{m^4p^2}$

SECTION 6.3 (page 369)

1. \$3 $\times 10^5$; \$2.8 $\times 10^6$ **3.** 7.52×10^8; 3.94×10^8
5. in scientific notation   **7.** not in scientific notation; 5.6×10^6 **9.** not in scientific notation; 8×10^1

11. not in scientific notation; 4×10^{-3} **13.** A number is written in scientific notation if it is the product of a number whose absolute value is between 1 and 10 (inclusive of 1) and a power of 10.
15. 5.876×10^9 **17.** 8.235×10^4
19. 7×10^{-6} **21.** -2.03×10^{-3} **23.** $750{,}000$
25. $5{,}677{,}000{,}000{,}000$ **27.** -6.21 **29.** $.00078$
31. $.000000005134$ **33.** 6×10^{11}; $600{,}000{,}000{,}000$
35. 1.5×10^7; $15{,}000{,}000$ **37.** 6.426×10^4; $64{,}260$
39. 3×10^{-4} **41.** 4×10^1 **43.** 2.6×10^{-3}
45. $635{,}000{,}000{,}000$ **47.** $\$4.55 \times 10^7$
49. $.00023$; $.006$

Section 6.4 (page 377)

1. sometimes **3.** always **5.** never **7.** always
9. $1; 6$ **11.** $1; 1$ **13.** $2; -19, -1$ **15.** $2; 1, 8$
17. $2m^5$ **19.** $-r^5$ **21.** cannot be simplified
23. $-5x^5$ **25.** $5p^9 + 4p^7$ **27.** $-2y^2$
29. already simplified; 4; binomial **31.** already simplified; $6m^5 + 5m^4 - 7m^3 - 3m^2$; 5; none of these
33. $x^4 + \frac{1}{3}x^2 - 4$; 4; trinomial **35.** $7; 0$; monomial
37. $4; \$5.00$ **38.** $6; \$27$ **39.** $2.5; 130$
40. $3; 104$ **41.** $5m^2 + 3m$ **43.** $4x^4 - 4x^2$
45. $\frac{7}{6}x^2 - \frac{2}{15}x + \frac{5}{6}$ **47.** $15m^3 - 13m^2 + 8m + 11$
49. $8r^2 + 5r - 12$ **51.** $5m^2 - 14m + 6$
53. $4x^3 + 2x^2 + 5x$
55. $-18y^5 + 7y^4 + 5y^3 + 3y^2 + y$
57. $-2m^3 + 7m^2 + 8m - 10$
59. $8x^2 + 8x + 6$ **61.** $8t^2 + 8t + 13$
63. $-11x^2 - 3x - 3$ **65.** The degree of a term is determined by the exponent on the *variable,* but 3^4 is not a variable. The degree of $3^4 = 3^4x^0$ is 0.
67. $13a^2b - 7a^2 - b$; degree 3
69. $c^4d - 5c^2d^2 + d^2$; degree 5
71. $12m^3n - 11m^2n^2 - 4mn^2$; degree 4

Section 6.5 (page 385)

1. distributive **3.** $-6m^2 - 4m$
5. $6p - \frac{9}{2}p^2 + 9p^4$ **7.** $6y^5 + 4y^6 + 10y^9$
9. $n^2 + n - 6$ **11.** $8r^2 - 10r - 3$
13. $9x^2 - 4$ **15.** $9q^2 + 6q + 1$
17. $6t^2 + 23st + 20s^2$ **19.** $-.3t^2 + .22t + .24$
21. The answers are $x^2 - 16$, $y^2 - 4$, and $r^2 - 49$. All three products are the difference of the square of the first term and the square of the last term of the binomials.
23. $12x^3 + 26x^2 + 10x + 1$
25. $20m^4 - m^3 - 8m^2 - 17m - 15$
27. $5x^4 - 13x^3 + 20x^2 + 7x + 5$
29. $x^2 + 24x + 144$ **31.** $9t^2 + 6t + 1$
33. $25x^2 - 20xy + 4y^2$ **35.** $h^3 - 15h^2 + 75h - 125$
37. $r^4 + 4r^3 + 6r^2 + 4r + 1$
39. $9x^4 + 6x^3 - 23x^2 - 8x + 16$
41. $30x + 60$ **42.** $30x + 60 = 600$; 18
43. 10 yards by 60 yards **44.** $\$2100$
45. 140 yards **46.** $\$1260$

Section 6.6 (page 391)

1. **(a)** $2x^2$ **(b)** $-10x$ **(c)** $3x$ **(d)** -15 **(e)** $-7x$
(f) $2x^2 - 7x - 15$ **3.** $r^2 + 4r + 3$
5. $w^2 - 10w + 24$ **7.** $x^2 + .5x - .24$
9. $x^2 - \frac{5}{12}x - \frac{1}{6}$ **11.** $27x^2 + 69x + 14$
13. $9m^2 + 36m + 35$ **15.** $15 - 19x + 6x^2$
17. $-15 + 23z - 6z^2$ **19.** $\frac{16}{9} + \frac{2}{3}k - 3k^2$
21. $27w^2 + 15wz - 2z^2$ **23.** $30y^2 + 7yz - 15z^2$
25. $-.16p^2 - .74ps + .3s^2$ **27.** **(a)** $4x^2$ **(b)** $12x$
(c) 9 **(d)** $4x^2 + 12x + 9$ **29.** $a^2 - 2ac + c^2$
31. $p^2 + 4p + 4$ **33.** $16x^2 - 24x + 9$
35. $.64t^2 + 1.12ts + .49s^2$ **37.** $25x^2 + 4xy + \frac{4}{25}y^2$
39. **(a)** $49x^2$ **(b)** 0 **(c)** $-9y^2$ **(d)** $49x^2 - 9y^2$; Because 0 is the identity element for addition, it is not necessary to write "+ 0." **41.** $q^2 - 4$
43. $4w^2 - 25$ **45.** $100x^2 - 9y^2$ **47.** $4x^4 - 25$
49. $49x^2 - \frac{9}{49}$ **51.** 1225 **52.** $30^2 + 2(30)(5) + 5^2$
53. 1225 **54.** They are equal.

Section 6.7 (page 395)

1. To use the method of this section, the denominator must be just one term. This is true of the first problem, but not the second. **3.** $8; 13$; no **5.** $5m^4 - 8m^3 + 4m^2$
7. $4m^4 - 2m^2 + 2m$ **9.** $\frac{m^4}{2} - 2m + \frac{4}{m}$
11. $4x^3 - 3x^2 + 2x$ **13.** $1 + 5x - 9x^2$
15. $\frac{12}{x} + 8 + 2x$ **17.** $\frac{4x^2}{3} + x + \frac{2}{3x}$
19. $9r^3 - 12r^2 - 2r + 1 - \frac{2}{3r}$ **21.** $-m^2 + 3m - \frac{4}{m}$
23. $4 - 3a + \frac{5}{a}$ **25.** $\frac{12}{x} - \frac{6}{x^2} + \frac{14}{x^3} - \frac{10}{x^4}$
27. No, $\frac{2}{3}x$ means $\frac{2x}{3}$ which is not the same as $\frac{2}{3x}$. In the first case we multiply by x; in the second case we divide by x. Yes, $\frac{4}{3}x^2 = \frac{4x^2}{3}$. In both cases we are multiplying by x^2. **29.** $15x^5 - 35x^4 + 35x^3$ **31.** 1423
32. $(1 \times 10^3) + (4 \times 10^2) + (2 \times 10^1) + (3 \times 10^0)$
33. $x^3 + 4x^2 + 2x + 3$ **34.** They are similar in that the coefficients of the powers of ten are equal to the coefficients of the powers of x. They are different in that one is a number while the other is a polynomial. They are equal if $x = 10$.

Section 6.8 (page 401)

1. The divisor is $2x + 5$; the quotient is $2x^3 - 4x^2 + 3x + 2$.
3. Divide $12m^2$ by $2m$ to get $6m$. **5.** $x + 2$
7. $2y - 5$ **9.** $p - 4 + \frac{44}{p + 6}$ **11.** $r - 5$
13. $2a - 14 + \frac{74}{2a + 3}$ **15.** $4x^2 - 7x + 3$

17. 33   **18.** 33   **19.** They are the same.
20. The answers should agree.   **21.** $3y^2 - 2y + 2$
23. $3k - 4 + \dfrac{2}{k^2 - 2}$ **25.** $x^2 + 1$
27. $2p^2 - 5p + 4 + \dfrac{6}{3p^2 + 1}$ **29.** $x^3 + 3x^2 - x + 5$
31. $x^2 + 1$

CHAPTER 6 REVIEW EXERCISES (page 405)

1. 4^{11} **2.** $(-5)^{11}$ **3.** $-72x^7$ **4.** $10x^{14}$
5. $19^5 x^5$ **6.** $(-4)^7 y^7$ **7.** $5p^4 t^4$ **8.** $\dfrac{7^6}{5^6}$
9. $3^3 x^6 y^9$ **10.** t^{42} **11.** $6^2 x^{16} y^4 z^{16}$ **12.** The
product rule for exponents does not apply here because we
want the sum of 7^2 and 7^4, not their product.
13. 2   **14.** $\dfrac{1}{32}$ **15.** $\dfrac{5^2}{6^2}$ or $\dfrac{25}{36}$ **16.** $-\dfrac{3}{16}$
17. 6^2 **18.** x^2 **19.** $\dfrac{1}{p^{12}}$ **20.** r^4 **21.** 2^8
22. $\dfrac{1}{9^6}$ **23.** 5^8 **24.** $\dfrac{1}{8^{12}}$ **25.** $\dfrac{1}{m^2}$ **26.** y^7
27. r^{13} **28.** $(-5)^2 m^6$ **29.** $\dfrac{y^{12}}{2^3}$ **30.** $\dfrac{1}{a^3 b^5}$
31. $2 \cdot 6^2 \cdot r^5$ **32.** $\dfrac{2^3 n^{10}}{3m^{13}}$ **33.** 4.8×10^7
34. 2.8988×10^{10} **35.** 6.5×10^{-5}
36. 8.24×10^{-8} **37.** 24,000   **38.** 78,300,000
39. .000000897   **40.** .00000000000995   **41.** 800
42. 4,000,000   **43.** .025   **44.** .01   **45.** $22m^2$; 2;
monomial   **46.** $p^3 - p^2 + 4p + 2$; 3; none of these
47. $8r^5 + 4r^4$; 5; binomial   **48.** $15a^5 + 10a^4 + 10a^3$;
5; trinomial   **49.** $-5a^3 + 4a^2$ **50.** $2r^3 - 3r^2 + 9r$
51. $11y^2 - 10y + 9$ **52.** $-13k^4 - 15k^2 - 4k - 6$
53. $10m^3 - 6m^2 - 3$ **54.** $-y^2 - 4y + 26$
55. $10p^2 - 3p - 11$ **56.** $7r^4 - 4r^3 - 1$
57. $10x^2 + 70x$ **58.** $-6p^5 + 15p^4$
59. $m^2 - 7m - 18$ **60.** $6k^2 - 9k - 6$
61. $6r^3 + 8r^2 - 17r + 6$ **62.** $8y^3 + 27$
63. $r^3 + 6r^2 + 12r + 8$
64. $(a + b)^2 = a^2 + 2ab + b^2$. The term $2ab$ is not in
$a^2 + b^2$.
65. $6k^2 + 11k + 3$ **66.** $2a^2 + 5ab - 3b^2$
67. $12k^2 - 48kq + 21q^2$ **68.** $a^2 + 8a + 16$
69. $9p^2 - 12p + 4$ **70.** $4r^2 + 20rs + 25s^2$
71. $36m^2 - 25$ **72.** $4z^2 - 49$ **73.** $25a^2 - 36b^2$
74. $4x^4 - 25$ **75.** $\dfrac{5y^2}{3}$ **76.** $-2x^2 y$
77. $-y^3 + 2y - 3$ **78.** $p - 3 + \dfrac{5}{2p}$
79. $-x^9 + 2x^8 - 4x^3 + 7x$
80. $-2m^2 n + mn^2 + \dfrac{6n^3}{5}$ **81.** $2r + 7$
82. $4m + 3 + \dfrac{5}{3m - 5}$ **83.** $2a + 1 + \dfrac{-8a + 12}{5a^2 - 3}$
84. $k^2 + 2k + 4 + \dfrac{-2k - 12}{2k^2 + 1}$
85. 0   **86.** $\dfrac{3^5}{p^3}$ **87.** $\dfrac{1}{7^2}$ **88.** $49 - 28k + 4k^2$

89. $y^2 + 5y + 1$ **90.** $\dfrac{6^4 r^8 s^4}{5^4}$
91. $-8m^7 - 10m^6 - 6m^5$ **92.** 2^5
93. $5xy^3 - \dfrac{8y^2}{5} + 3x^2 y$ **94.** $\dfrac{r^2}{6}$
95. $8x^3 + 12x^2 y + 6xy^2 + y^3$ **96.** $\dfrac{3}{4}$
97. $a^3 - 2a^2 - 7a + 2$ **98.** $8y^3 - 9y^2 + 5$
99. $10r^2 + 21r - 10$ **100.** $144a^2 - 1$

CHAPTER 6 TEST   (page 409)

1. $\dfrac{1}{625}$ **2.** 2   **3.** $\dfrac{7}{12}$ **4.** 8^5 **5.** $x^2 y^6$
6. (a) 3.44×10^{11} **(b)** 5.57×10^{-6}
7. (a) 29,600,000   **(b)** .0000000607   **8.** $-7x^2 + 8x$;
2; binomial   **9.** $4n^4 + 13n^3 - 10n^2$; 4; trinomial
10. $4t^4 + t^3 - 6t^2 - t$ **11.** $-2y^2 - 9y + 17$
12. $-12t^2 + 5t + 8$ **13.** $-27x^5 + 18x^4 - 6x^3 + 3x^2$
14. $t^2 - 5t - 24$ **15.** $8x^2 + 2xy - 3y^2$
16. $25x^2 - 20xy + 4y^2$ **17.** $100v^2 - 9w^2$
18. $2r^3 + r^2 - 16r + 15$ **19.** $x^3 + 3x^2 + 3x + 1$
20. $4y^2 - 3y + 2 + \dfrac{5}{y}$ **21.** $-3xy^2 + 2x^3 y^2 + 4y^2$
22. $2x + 9$ **23.** $3x^2 + 6x + 11 + \dfrac{26}{x - 2}$
24. $9x^2 + 54x + 81$ **25.** Answers will vary. One
example is $(-4x^4 + 3x^3 + 2x + 1) +$
$(4x^4 - 8x^3 + 2x + 7) = -5x^3 + 4x + 8$.

CUMULATIVE REVIEW 1–6 (page 411)

1. false   **2.** -25 **3.** 12   **4.** $-57\dfrac{5}{12}$
5. 16   **6.** -300 **7.** $\dfrac{1}{3}$ **8.** commutative
9. $-5t^2 + 21t - 52$ **10.** $\{0\}$ **11.** $2(5x) + 3$
12. $\{-48\}$ **13.** pigs: 66; goats: 32   **14.** $W = \dfrac{A}{L}$
15. 62 pounds   **16.** 300%
17. $\left[-\dfrac{4}{5}, \infty\right)$ **18.** $\{6, 9\}$ **19.** $(2, 10]$
20. $\left\{1, \dfrac{5}{2}\right\}$ **21.** $(-\infty, -4) \cup (-1, \infty)$
22. III   **23.** $\dfrac{3}{11}$ **24.** $\dfrac{3}{5}$
25.

26.

27. -52 **28.** 22.5   **29.** $\{(2, 1)\}$ **30.** $\{(1, 4)\}$
31. $\{(2, 1, 4)\}$ **32.** peanuts: $2 per pound;
cashews: $4 per pound   **33.** -18 **34.** 82   **35.** $\dfrac{17}{72}$

36. -5.88×10^{-8} **37.** $5x^3 - 22x^2 + 21x + 4$

38. $15r^3 - 8r^2 + 23r - 42$ **39.** $5w + 2 - \dfrac{3}{w}$

40. $4x^2 + 2x + 8$

CHAPTER 7

SECTION 7.1 (page 419)

1. no; 6   **3.** One example is 7, 14, 21.   **5.** 4
7. 4   **9.** 6   **11.** 1   **13.** 8   **15.** $10x^3$ **17.** xy^2
19. 6   **21.** yes; x^3y^2 **23.** $3m^2$ **25.** $2z^4$
27. $2mn^4$ **29.** $-7x^3y^2$ **31.** $5y^6(13y^4 + 7)$
33. no common factor (except 1)   **35.** $8m^2n^2(n + 3)$
37. $13y^2(y^6 + 2y^2 - 3)$ **39.** $9qp^3(5q^3p^2 + 4p^3 + 9q)$
41. $(2a + b)(a^2 + b^2)$ **43.** $(p + 4)(p + 3)$
45. $(a - 2)(a + 5)$ **47.** $(z + 2)(7z - a)$
49. $(3r + 2y)(6r - x)$ **51.** $(a^2 + b^2)(3a + 2b)$
53. $(1 - a)(1 - b)$ **55.** $(4m - p^2)(4m^2 - p)$
57. the commutative and associative properties
58. $2x(y - 4) - 3(y - 4)$ **59.** No, because it is not a
product. It is the difference between two terms, $2x(y - 4)$
and $3(y - 4)$.   **60.** $(2x - 3)(y - 4)$; yes
61. (a) yes (b) Both answers are acceptable because
both factorizations have the same product.

SECTION 7.2 (page 425)

1. a and b must have different signs.   **3.** A prime
polynomial is one that cannot be factored with integer
coefficients.   **5.** 1 and 12, -1 and -12, 2 and 6, -2
and -6, 3 and 4, -3 and -4; the pair with a sum of 7 is 3
and 4.   **7.** 1 and -24, -1 and 24, 2 and -12, -2 and
12, 3 and -8, -3 and 8, 4 and -6, -4 and 6; the pair
with a sum of -5 is 3 and -8.   **9.** 1 and 27, -1 and
-27, 3 and 9, -3 and -9; the pair with a sum of 28 is 1
and 27.   **11.** (c)   **13.** $x + 11$ **15.** $x - 8$
17. $y - 5$ **19.** $x + 11$ **21.** $y - 9$
23. $(y + 8)(y + 1)$ **25.** $(b + 3)(b + 5)$
27. $(m + 5)(m - 4)$ **29.** $(y - 5)(y - 3)$
31. $(r - 6)(r + 5)$ **33.** $(t - 4)^2$ or $(t - 4)(t - 4)$
35. $(r + 2a)(r + a)$ **37.** $(t + 2z)(t - 3z)$
39. $(x + y)(x + 3y)$ **41.** $(v - 5w)(v - 6w)$
43. $4(x + 5)(x - 2)$ **45.** $2t(t + 1)(t + 3)$
47. $2x^4(x - 3)(x + 7)$ **49.** $mn(m - 6n)(m - 4n)$
51. The factorization $(2x + 4)(x - 3)$ is incorrect because
$2x + 4$ has a common factor which must be factored out
for the trinomial to be completely factored.
53. $a^2 + 13a + 36$

SECTION 7.3 (page 433)

1. 2, -21 **3.** $-6x$, $7x$ or $7x$, $-6x$
5. (b)   **7.** (a)   **9.** (a)   **11.** $2a + 5b$
13. $x^2 + 3x - 4$; $x + 4$, $x - 1$, or $x - 1$, $x + 4$
15. $2z^2 - 5z - 3$; $2z + 1$, $z - 3$, or $z - 3$, $2z + 1$
17. The binomial $2x - 6$ cannot be a factor because it has
a common factor of 2, but the polynomial does not.
19. $(2x + 1)(x + 3)$ **21.** $(3a + 7)(a + 1)$
23. $(4r - 3)(r + 1)$ **25.** $(3m - 1)(5m + 2)$

27. $(4m + 1)(2m - 3)$ **29.** $(4x + 3)(5x - 1)$
31. $(3m + 1)(7m + 2)$ **33.** $(4y - 1)(5y + 11)$
35. $(2b + 1)(3b + 2)$ **37.** $3(4x - 1)(2x - 3)$
39. $q(5m + 2)(8m - 3)$ **41.** $2m(m - 4)(m + 5)$
43. $3n^2(5n - 3)(n - 2)$ **45.** $3x^3(2x + 5)(3x - 5)$
47. $y^2(5x - 4)(3x + 1)$ **49.** $(3p + 4q)(4p - 3q)$
51. $(5a + 2b)(5a + 3b)$ **53.** $(3a - 5b)(2a + b)$
55. $m^4n(3m + 2n)(2m + n)$ **57.** $(5 - x)(1 - x)$
59. $(4 + 3x)(4 + x)$ **61.** $-1(x + 7)(x - 3)$
63. $-1(3x + 4)(x - 1)$ **65.** $-1(a + 2b)(2a + b)$
67. $5 \cdot 7$ **68.** $(-5)(-7)$ **69.** The product of
$3x - 4$ and $2x - 1$ is $6x^2 - 11x + 4$.   **70.** The
product of $4 - 3x$ and $1 - 2x$ is $6x^2 - 11x + 4$.
71. The factors in Exercise 69 are the opposites of the
factors in Exercise 70.   **72.** $(3 - 7t)(5 - 2t)$

SECTION 7.4 (page 439)

1. 1; 4; 9; 16; 25; 36; 49; 64; 81; 100; 121; 144; 169; 196;
225; 256; 289; 324; 361; 400   **3.** 2
5. $(y + 5)(y - 5)$ **7.** $(3r + 2)(3r - 2)$
9. $\left(6m + \dfrac{4}{5}\right)\left(6m - \dfrac{4}{5}\right)$ **11.** $4(3x + 2)(3x - 2)$
13. $(14p + 15)(14p - 15)$ **15.** $(4r + 5a)(4r - 5a)$
17. prime   **19.** $(p^2 + 7)(p^2 - 7)$
21. $(x^2 + 1)(x + 1)(x - 1)$
23. $(p^2 + 16)(p + 4)(p - 4)$
25. The teacher was justified, because it was not factored
completely; $x^2 - 9$ can be factored as $(x + 3)(x - 3)$. The
complete factored form is $(x^2 + 9)(x + 3)(x - 3)$.
27. $(w + 1)^2$ **29.** $(x - 4)^2$ **31.** $\left(t + \dfrac{1}{2}\right)^2$
33. $(x - .5)^2$ **35.** $2(x + 6)^2$ **37.** $(4x - 5)^2$
39. $(7x - 2y)^2$ **41.** $(8x + 3y)^2$ **43.** $-2(5h - 2y)^2$
45. No, it is not a perfect square because the middle term
should be $30y$, not $14y$.
47. $(2x + 3)(5x - 2)$ **48.** $5x - 2$ **49.** Yes,
because we saw in Exercise 47 that $(2x + 3)(5x - 2) =$
$10x^2 + 11x - 6$.   **50.** The quotient is $x^2 + x + 1$, so
$x^3 - 1 = (x - 1)(x^2 + x + 1)$.

SUMMARY ON FACTORING (page 441)

1. $8m^3(4m^6 + 2m^2 + 3)$ **3.** $7k(2k + 5)(k - 2)$
5. $(6z + 1)(z + 5)$ **7.** $(7z + 4y)(7z - 4y)$
9. $4(4x + 5)$ **11.** $(5y - 6z)(2y + z)$
13. $(m - 3)(m + 5)$ **15.** $8z(4z - 1)(z + 2)$
17. $(z - 6)^2$ **19.** $(y - 6k)(y + 2k)$
21. $6(y - 2)(y + 1)$ **23.** $(p - 6)(p - 11)$
25. prime   **27.** $(z + 2a)(z - 5a)$
29. $(2k - 3)^2$ **31.** $(4r + 3m)^2$
33. $3k(k + 1)(k - 5)$ **35.** $4(2k - 3)^2$
37. $6y^4(3y + 4)(2y - 5)$ **39.** $5z(z - 2)(z - 7)$
41. $6(3m + 2z)(3m - 2z)$ **43.** $2(3a - 1)(a + 2)$
45. $(m + 9)(m - 9)$ **47.** $5m^2(5m - 13n)(5m - 3n)$
49. $(m - 2)^2$ **51.** $6k^2p(4k^2 + 10kp + 25p^2)$
53. $(3p - 2q)(4p + 3q)$ **55.** $4(4p + 5m)(4p - 5m)$
57. $(10a + 9y)(10a - 9y)$ **59.** $(a + 4)^2$

SECTION 7.5 (page 449)

1. $ax^2 + bx + c$ **3.** $\{-5, 2\}$ **5.** $\left\{3, \dfrac{7}{2}\right\}$ **7.** $\left\{0, \dfrac{4}{3}\right\}$

9. To solve the equation $2x(3x - 4) = 0$, set each *variable* factor equal to 0, to get $x = 0$ or $3x - 4 = 0$. Solve $3x - 4 = 0$, getting $\frac{4}{3}$. The solution set is $\{0, \frac{4}{3}\}$.
11. Because $(x - 9)^2 = (x - 9)(x - 9) = 0$ leads to two solutions of 9, we call 9 a double solution.   **13.** $\{-2, -1\}$
15. $\{1, 2\}$ **17.** $\{-8, 3\}$ **19.** $\{-1, 3\}$ **21.** $\{-2, -1\}$
23. $\{-4\}$ **25.** $\left\{-2, \dfrac{1}{3}\right\}$ **27.** $\left\{-\dfrac{4}{3}, \dfrac{1}{2}\right\}$
29. $\left\{-\dfrac{2}{3}\right\}$ **31.** $\{-3, 3\}$ **33.** $\left\{-\dfrac{7}{4}, \dfrac{7}{4}\right\}$
35. $\{-11, 11\}$ **37.** -11 is another solution.   **39.** $\{0, 7\}$
41. $\left\{0, \dfrac{1}{2}\right\}$ **43.** $\{2, 5\}$ **45.** $\left\{-4, \dfrac{1}{2}\right\}$
47. $\left\{-12, \dfrac{11}{2}\right\}$ **49.** $\{-1, 3\}$ **51.** $\left\{-\dfrac{5}{2}, \dfrac{1}{3}, 5\right\}$
53. $\left\{-\dfrac{7}{2}, -3, 1\right\}$ **55.** $\left\{-\dfrac{7}{3}, 0, \dfrac{7}{3}\right\}$

SECTION 7.6 (page 455)

1. a variable; the unknown   **3.** an equation
5. (a) $60 = \dfrac{1}{2}(3x + 6)(x + 5)$ (b) 3   (c) base: 15 units; height: 8 units   **7.** length: 7 inches; width: 4 inches   **9.** length: 11 inches; width: 8 inches
11. mirror: 7 feet; painting: 9 feet   **13.** (a) 1 second
(b) $\dfrac{1}{2}$ second and $1\dfrac{1}{2}$ seconds   (c) 3 seconds   (d) The negative solution, -1, does not make sense since t represents time, and t cannot be negative.
15. (a) 256 feet   (b) 1024 feet   **17.** $-3, -2$ or 4, 5
19. 7, 9, 11   **21.** $-2, 0, 2$ or 6, 8, 10   **23.** 12 miles
25. 8 feet   **27.** height: 6 meters; area of the base: 16 square meters   **29.** $y^2 - 4xy + 4x^2$
31. c^2 **32.** b^2 **33.** a^2 **34.** $a^2 + b^2 = c^2$; This is the Pythagorean formula.

CHAPTER 7 REVIEW EXERCISES (page 461)

1. $7(t + 2)$ **2.** $30z(2x^2 + 1)$ **3.** $35x^2(x + 2)$
4. $(x - 4)(2y + 3)$ **5.** $50m^2n^2(2n - mn^2 + 3)$
6. $(3y + 2)(2y + 3)$ **7.** $(x + 3)(x + 2)$
8. $(y - 5)(y - 8)$ **9.** $(q + 9)(q - 3)$
10. $(r - 8)(r + 7)$ **11.** $(r + 8s)(r - 12s)$
12. $(p + 12q)(p - 10q)$ **13.** $8p(p + 2)(p - 5)$
14. $3x^2(x + 2)(x + 8)$ **15.** $(m + 3n)(m - 6n)$
16. $(y - 3z)(y - 5z)$ **17.** $p^5(p - 2q)(p + q)$
18. $3r^3(r + 3s)(r - 5s)$ **19.** prime
20. $3(x^2 + 2x + 2)$ **21.** r and $6r$, $2r$ and $3r$
22. Factor out z.   **23.** $(2k - 1)(k - 2)$
24. $(3r - 1)(r + 4)$ **25.** $(3r + 2)(2r - 3)$
26. $(5z + 1)(2z - 1)$ **27.** $(v + 3)(8v - 7)$
28. $4x^3(3x - 1)(2x - 1)$ **29.** $-3(x + 2)(2x - 5)$
30. $rs(5r + 6s)(2r + s)$ **31.** (b)   **32.** (d)
33. $(n + 7)(n - 7)$ **34.** $(5b + 11)(5b - 11)$
35. $(7y + 5w)(7y - 5w)$ **36.** $36(2p + q)(2p - q)$

37. prime   **38.** $(z + 5)^2$ **39.** $(r - 6)^2$
40. $(3t - 7)^2$ **41.** $(4m + 5n)^2$ **42.** $6x(3x - 2)^2$
43. $\left\{-\dfrac{3}{4}, 1\right\}$ **44.** $\{-7, 4, -3\}$ **45.** $\left\{0, \dfrac{5}{2}\right\}$
46. $\{-3, -1\}$ **47.** $\{1, 4\}$ **48.** $\{3, 5\}$
49. $\left\{-\dfrac{4}{3}, 5\right\}$ **50.** $\left\{-\dfrac{8}{9}, \dfrac{8}{9}\right\}$ **51.** $\{0, 8\}$
52. $\{-1, 6\}$ **53.** $\{7\}$ **54.** $\{6\}$
55. $\left\{-\dfrac{2}{5}, -2, -1\right\}$ **56.** $\{-3, 3\}$
57. length: 10 feet; width: 4 feet   **58.** length: 6 meters; width: 2 meters   **59.** length: 10 centimeters; width: 8 centimeters   **60.** length: 6 meters; height: 5 meters
61. 26 miles   **62.** 5 feet   **63.** 6, 7 or $-5, -4$
64. 6 inches, 8 inches, 10 inches   **65.** 112 feet
66. 192 feet   **67.** 256 feet   **68.** after 8 seconds
69. 2 inches   **70.** 2 centimeters
71. $(z - x)(z - 10x)$ **72.** $(3k + 5)(k + 2)$
73. $(3m + 4p)(5m - 4p)$
74. $(y^2 + 25)(y + 5)(y - 5)$
75. $3m(2m + 3)(m - 5)$
76. $8abc(3b^2c - 7ac^2 + 9ab)$
77. prime   **78.** $6xyz(2xz^2 + 2y - 5x^2yz^3)$
79. $2a^3(a + 2)(a - 6)$ **80.** $(2r + 3q)(6r - 5q)$
81. $(10a + 3)(10a - 3)$ **82.** $(7t + 4)^2$
83. $\{0, 7\}$ **84.** $\{-5, 2\}$ **85.** $\left\{-\dfrac{2}{5}\right\}$
86. 15 meters, 36 meters, 39 meters
87. length: 6 meters; width: 4 meters
88. $-5, -4, -3$ or 5, 6, 7   **89.** $-6, -4$ or 6, 8
90. width: 10 meters; length: 17 meters   **91.** 6 meters

CHAPTER 7 TEST (page 467)

1. (d)   **2.** $6x(2x - 5)$ **3.** $m^2n(2mn + 3m - 5n)$
4. $(x + 3)(x - 8)$ **5.** $(x - 7)(x - 2)$
6. $(2x + 3)(x - 1)$ **7.** $(3x + 1)(2x - 7)$
8. $3(x + 1)(x - 5)$ **9.** $(5z - 1)(2z - 3)$
10. prime   **11.** prime   **12.** $(y + 7)(y - 7)$
13. $(3y + 8)(3y - 8)$ **14.** $(x + 8)^2$
15. $(2x - 7y)^2$ **16.** $-2(x + 1)^2$
17. $3t^2(2t + 9)(t - 4)$ **18.** prime   **19.** $4t(t + 4)^2$
20. $(x^2 + 9)(x + 3)(x - 3)$
21. $(p + 3)(p + 3) = p^2 + 6p + 9 \neq p^2 + 9$
22. $\{-3, 9\}$ **23.** $\{\frac{1}{2}, 6\}$ **24.** $\{-\frac{2}{5}, \frac{2}{5}\}$ **25.** $\{10\}$
26. $\{0, 3\}$ **27.** $-\frac{2}{3}$ is also a solution.
28. 6 feet by 9 feet   **29.** 17 feet   **30.** -2 and -1

CUMULATIVE REVIEW 1–7 (page 469)

1. $\dfrac{5}{36}$ **2.** 22   **3.** false   **4.** -1
5. distributive   **6.** $-38r + 138$ **7.** $\{0\}$
8. 7, 8, 9   **9.** 5000 cubic feet   **10.** 27 heads
11. $\left[\dfrac{1}{2}, \infty\right)$ **12.** $(-\infty, 1) \cup [3, \infty)$ **13.** $[-1, 4]$
14. $\left\{\dfrac{6}{7}, \dfrac{4}{5}\right\}$ **15.** vertical   **16.** negative

17. **18.**

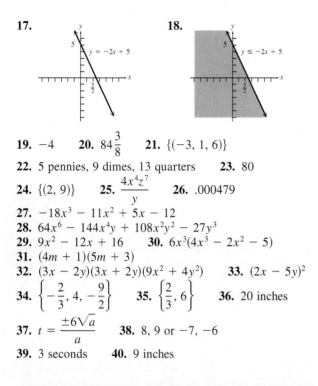

19. -4 **20.** $84\frac{3}{8}$ **21.** $\{(-3, 1, 6)\}$

22. 5 pennies, 9 dimes, 13 quarters   **23.** 80

24. $\{(2, 9)\}$ **25.** $\dfrac{4x^4z^7}{y}$ **26.** .000479

27. $-18x^3 - 11x^2 + 5x - 12$
28. $64x^6 - 144x^4y + 108x^2y^2 - 27y^3$
29. $9x^2 - 12x + 16$ **30.** $6x^3(4x^3 - 2x^2 - 5)$
31. $(4m + 1)(5m + 3)$
32. $(3x - 2y)(3x + 2y)(9x^2 + 4y^2)$ **33.** $(2x - 5y)^2$
34. $\left\{-\dfrac{2}{3}, 4, -\dfrac{9}{2}\right\}$ **35.** $\left\{\dfrac{2}{3}, 6\right\}$ **36.** 20 inches

37. $t = \dfrac{\pm 6\sqrt{a}}{a}$ **38.** 8, 9 or $-7, -6$

39. 3 seconds   **40.** 9 inches

CHAPTER 8
SECTION 8.1 (page 477)

1. division by zero is undefined   **3.** $1; 1; \dfrac{P}{Q}$ **5.** 0

7. $\dfrac{5}{3}$ **9.** $-3, 2$ **11.** never undefined   **13. (a)** 1

(b) $\dfrac{17}{12}$ **15. (a)** 0   **(b)** $-\dfrac{10}{3}$ **17. (a)** $\dfrac{9}{5}$

(b) undefined   **19. (a)** $\dfrac{2}{7}$ **(b)** $\dfrac{13}{3}$ **21.** No, not if

the number is 0. Division by 0 is undefined.   **23.** $3r^2$

25. $\dfrac{2}{5}$ **27.** $\dfrac{x-1}{x+1}$ **29.** $\dfrac{7}{5}$ **31.** $m - n$

33. $\dfrac{3(2m+1)}{4}$ **35.** $\dfrac{3m}{5}$ **37.** $\dfrac{3r-2s}{3}$ **39.** $\dfrac{z-3}{z+5}$

41. $\dfrac{x+1}{x-1}$ **43.** -1 **45.** $-(m+1)$ **47.** -1

49. $-\dfrac{7}{5}$ **50.** They are equal.   **51.** $-\dfrac{4-3x}{7}$

52. $\dfrac{-4+3x}{7}, \dfrac{4-3x}{-7}$

SECTION 8.2 (page 483)

1. $\dfrac{4m}{3}$ **3.** $\dfrac{40y^2}{3}$ **5.** $\dfrac{2}{c+d}$ **7.** $\dfrac{16q}{3p^3}$

9. $\dfrac{7}{r^2+rp}$ **11.** $\dfrac{z^2-9}{z^2+7z+12}$ **13.** 5   **15.** $-\dfrac{3}{2t^4}$

17. $\dfrac{1}{4}$ **19.** To multiply two rational expressions,

multiply the numerators and multiply the denominators.
Simplify the answer if possible.

21. $\dfrac{10}{9}$ **23.** $-\dfrac{3}{4}$ **25.** $-\dfrac{9}{2}$ **27.** $\dfrac{-9(m-2)}{m+4}$

29. $\dfrac{p+4}{p+2}$ **31.** $\dfrac{(k-1)^2}{(k+1)(2k-1)}$ **33.** $\dfrac{4k-1}{3k-2}$

35. $\dfrac{m+4p}{m+p}$ **37.** $\dfrac{10}{x+10}$ **39.** $x \neq -7$, because

division requires multiplying by the reciprocal of the
second fraction. In the reciprocal, $x + 7$ is in the
denominator.

SECTION 8.3 (page 489)

1. true   **3.** false   **5.** 30   **7.** x^7 **9.** $72q$
11. $84r^5$ **13.** $2^3 \cdot 3 \cdot 5$ **15.** The least common
denominator is their product.
17. $28m^2(3m - 5)$ **19.** $30(b - 2)$
21. $c - d$ or $d - c$ **23.** $k(k + 5)(k - 2)$
25. $(p + 3)(p + 5)(p - 6)$ **27.** $\dfrac{20}{55}$ **29.** $\dfrac{-45}{9k}$

31. $\dfrac{26y^2}{80y^3}$ **33.** $\dfrac{35t^2r^3}{42r^4}$ **35.** $\dfrac{20}{8(m+3)}$

37. $\dfrac{-8t}{6t-12}$ **39.** $\dfrac{14(z-2)}{z(z-3)(z-2)}$

41. $\dfrac{2(b-1)(b+2)}{b^3+3b^2+2b}$

SECTION 8.4 (page 497)

1. numerators; same denominator   **3.** $\dfrac{11}{m}$ **5.** b

7. x **9.** $y - 6$ **11.** To add rational expressions
with different denominators, first find the LCD. Then
convert each rational expression to an equivalent
expression with the LCD as denominator. Add numerators
and simplify if possible. The LCD is the denominator of
the sum. Finally, write the result in lowest terms.

13. $\dfrac{3z+5}{15}$ **15.** $\dfrac{10-7r}{14}$ **17.** $\dfrac{-3x-2}{4x}$

19. $\dfrac{11(1+k)}{8}$ **21.** $\dfrac{4(b+4)}{3b}$ **23.** $\dfrac{7-6p}{3p^2}$

25. $\dfrac{-2(p-3)}{p(p-2)}$ **27.** $\dfrac{-2}{x-5}$ or $\dfrac{2}{5-x}$

29. $\dfrac{-2}{y-1}$ or $\dfrac{2}{1-y}$ **31.** $\dfrac{-2(c-14)}{(c-2)(c+2)}$

33. $\dfrac{-(m+n)}{2(m-n)}$ **35.** $\dfrac{7x-22}{4(x+2)(x-2)}$

37. $m - 2$ or $2 - m$
39. $\dfrac{-a^2+a+5}{(a+1)(a-1)(a+4)}$ **41.** $\dfrac{43m+1}{5m(m-2)}$

43. $\dfrac{11y^2-y-11}{(2y-1)(y+3)(3y+2)}$

SECTION 8.5 (page 503)

1. division   **3.** -6 **5.** $\dfrac{1}{pq}$ **7.** $\dfrac{1}{xy}$ **9.** $\dfrac{2a^2b}{3}$

11. $\dfrac{m(m+2)}{3(m-4)}$ **13.** $\dfrac{2}{x}$ **15.** $\dfrac{8}{x}$ **17.** $\dfrac{a^2-5}{a^2+1}$

19. $\dfrac{3(p+2)}{2(2p+3)}$ **21.** $\dfrac{t(t-2)}{4}$ **23.** $\dfrac{-k}{2+k}$

25. $\dfrac{3m(m-3)}{(m-1)(m-8)}$ **27.** $\dfrac{6}{5}$

29. $\dfrac{\frac{3}{8}+\frac{5}{6}}{2}$ **30.** $\dfrac{29}{48}$ **31.** $\dfrac{29}{48}$

32. The choice of method is a personal preference.

Section 8.6 (page 511)

1. expression; $\dfrac{43}{40}x$ **3.** equation; $\left\{\dfrac{40}{43}\right\}$

5. expression; $-\dfrac{1}{10}y$

7. When solving an equation, we multiply both sides by the LCD, which eliminates all denominators. When adding or subtracting fractions, we multiply by 1 in the form LCD/LCD. The denominators are not eliminated.

9. $\left\{-\dfrac{3}{4}\right\}$ **11.** $\{24\}$ **13.** $\{-15\}$ **15.** $\{7\}$

17. $\{-15\}$ **19.** $\{-5\}$ **21.** $\{-6\}$ **23.** $\emptyset$

25. $\{5\}$ **27.** $\{12\}$ **29.** $\{2\}$ **31.** 0 and 4

33. $\{3\}$ **35.** $\{3\}$ **37.** $\{-2, 12\}$ **39.** $\emptyset$

41. $\left\{-6, \dfrac{1}{2}\right\}$ **43.** $\left\{-\dfrac{1}{5}, 3\right\}$ **45.** $\left\{-\dfrac{3}{5}, 3\right\}$

47. $F = \dfrac{ma}{k}$ **49.** $a = \dfrac{kF}{m}$ **51.** $R = \dfrac{E-Ir}{I}$

53. $A = \dfrac{h(B+b)}{2}$ **55.** $a = \dfrac{2S-ndL}{nd}$

Summary on Rational Expressions (page 513)

1. $\dfrac{10}{p}$ **3.** $\dfrac{1}{2x^2(x+2)}$ **5.** $\dfrac{y+2}{y-1}$ **7.** $\{39\}$

9. $\dfrac{13}{3(p+2)}$ **11.** $\left\{\dfrac{1}{7}, 2\right\}$ **13.** $\dfrac{7}{12z}$

15. $\dfrac{3m+5}{(m+2)(m+3)(m+1)}$

Section 8.7 (page 523)

1. (a) $\dfrac{1}{3}x$ (b) $\dfrac{1}{6}x + 2$ (c) $\dfrac{1}{3}x = \dfrac{1}{6}x + 2$ **3.** $\dfrac{9}{5}$

5. \$99.7 million **7.** Germany: \$.18 billion; United States: \$1.82 billion **9.** 24.15 kilometers per hour

11. $\dfrac{D}{R} = \dfrac{d}{r}$ **13.** 900 miles

15. N'Deti: 12.02 miles per hour; McDermott: 8.77 miles per hour **17.** 8 miles per hour

19. $\dfrac{1}{10}$ job per hour **21.** $\dfrac{72}{17}$ or 4.24 hours

23. $\dfrac{60}{11}$ or 5.45 hours **25.** 3 hours **27.** increases

29. \$40.32 **31.** $106\dfrac{2}{3}$ miles per hour

33. 8.68 pounds per square inch **35.** 20 pounds per square foot

Chapter 8 Review Exercises (page 531)

1. 3 **2.** 0 **3.** $-1, 3$ **4.** $-5, -\dfrac{2}{3}$

5. (a) $-\dfrac{4}{7}$ (b) -16 **6.** (a) $\dfrac{11}{8}$ (b) $\dfrac{13}{22}$

7. (a) undefined (b) 1 **8.** (a) undefined (b) $\dfrac{1}{2}$

9. $\dfrac{b}{3a}$ **10.** -1 **11.** $\dfrac{-(2x+3)}{2}$ **12.** $\dfrac{2p+5q}{5p+q}$

13. $\dfrac{72}{p}$ **14.** 2 **15.** $\dfrac{2}{3m^6}$ **16.** $\dfrac{5}{8}$ **17.** $\dfrac{3}{2}$

18. $\dfrac{r+4}{3}$ **19.** $\dfrac{3a-1}{a+5}$ **20.** $\dfrac{y-2}{y-3}$ **21.** $\dfrac{p+5}{p+1}$

22. $\dfrac{3z+1}{z+3}$ **23.** 96 **24.** $108y^4$

25. $m(m+2)(m+5)$ **26.** $(x+3)(x+1)(x+4)$

27. $\dfrac{35}{56}$ **28.** $\dfrac{40}{4k}$ **29.** $\dfrac{15a}{10a^4}$ **30.** $\dfrac{-54}{18-6x}$

31. $\dfrac{15y}{50-10y}$ **32.** $\dfrac{4b(b+2)}{(b+3)(b-1)(b+2)}$ **33.** $\dfrac{15}{x}$

34. $-\dfrac{2}{p}$ **35.** $\dfrac{4k-45}{k(k-5)}$ **36.** $\dfrac{28+11y}{y(7+y)}$

37. $\dfrac{-2-3m}{6}$ **38.** $\dfrac{3(16-x)}{4x^2}$

39. $\dfrac{7a+6b}{(a-2b)(a+2b)}$ **40.** $\dfrac{-k^2-6k+3}{3(k+3)(k-3)}$

41. $\dfrac{5z-16}{z(z+6)(z-2)}$ **42.** $\dfrac{-13p+33}{p(p-2)(p-3)}$ **43.** $\dfrac{a}{b}$

44. $\dfrac{4(y-3)}{y+3}$ **45.** $\dfrac{6(3m+2)}{2m-5}$ **46.** $\dfrac{(q-p)^2}{pq}$

47. $\dfrac{xw+1}{xw-1}$ **48.** $\dfrac{1-r-t}{1+r+t}$ **49.** $\left\{\dfrac{35}{6}\right\}$

50. $\{-16\}$ **51.** $\{-4\}$ **52.** $\emptyset$ **53.** $\{3\}$

54. $t = \dfrac{Ry}{m}$ **55.** $y = \dfrac{4x+5}{3}$ **56.** $m = \dfrac{4+p^2q}{3p^2}$

57. 12 **58.** small car: \$120; large car: \$180

59. about 6.7 hours **60.** $\dfrac{40}{13}$ or about 3.1 hours

61. 2 hours **62.** 10 feet **63.** 4 centimeters

64. $\dfrac{(5+2x-2y)(x+y)}{(3x+3y-2)(x-y)}$ **65.** $\dfrac{m+7}{(m-1)(m+1)}$

66. $8p^2$ **67.** $\dfrac{1}{6}$ **68.** 3 **69.** $\dfrac{z+7}{(z+1)(z-1)^2}$

70. $d = \dfrac{k+FD}{F}$ or $d = \dfrac{k}{F} + D$ **71.** $\{-2, 3\}$

72. about 17.4 million **73.** $\dfrac{27}{10}$ or $2\dfrac{7}{10}$ hours

74. 150 kilometers per hour **75.** about \$101 billion

CHAPTER 8 TEST (page 535)

1. $-2, 4$ 2. $\dfrac{2m}{3}$ 3. $\dfrac{3a+2}{a-1}$ 4. x 5. $\dfrac{25}{27}$

6. $\dfrac{3k-2}{3k+2}$ 7. $\dfrac{a-1}{a+4}$ 8. $\dfrac{240p^2}{64p^3}$ 9. $\dfrac{21}{42m-84}$

10. $\dfrac{3}{c}$ 11. $\dfrac{-14}{5(y+2)}$ 12. $\dfrac{-m^2+7m+2}{(2m+1)(m-5)(m-1)}$

13. $\dfrac{2k}{3p}$ 14. $\dfrac{-2-x}{4+x}$ 15. $-1, 4$ 16. $\left\{-\dfrac{58}{11}\right\}$

17. $\{1\}$ 18. $\dfrac{1}{4}$ or $\dfrac{1}{2}$ 19. 3 miles per hour

20. 100 amperes

CUMULATIVE REVIEW 1–8 (page 537)

1. $(-\infty, \infty)$

2. $[-3, 3]$ ⊢━━━━━┥➤
          -3 3

3. \$4000 at 4%; \$8000 at 3%
4. 78 or greater   5. $\varnothing$ 6. 0
7. $(-\infty, 1] \cup [2, \infty)$ 8. $\{11\}$
9. $(-6, \infty)$ ━━━━━(━━➤
                    -6

10. $(-5, 4)$ ━━(━━)━━➤
              -5 4

11. x-intercept: $(-5, 0)$; y-intercept: $(0, 4)$
12. no x-intercept; y-intercept: $(0, -4)$
13. x-intercept and y-intercept: $(0, 0)$ 14. $\{(4, -2)\}$
15. 3   16. 7.6×10^{-5} 17. 5,600,000,000

18. $\dfrac{21y^7}{x^9}$ 19. $\dfrac{a^{10}}{b^{10}}$ 20. $\dfrac{m}{n}$

21. $-25x^3 - 2x^2 - 36x + 114$ 22. $12f^2 + 5f - 3$

23. $x^3 + y^3$ 24. $49t^6 - 64$ 25. $\dfrac{1}{16}x^2 + \dfrac{5}{2}x + 25$

26. 20   27. $2x^3 + 5x^2 - 3x - 2$
28. $(2x + 5)(x - 9)$ 29. $25(2t^2 + 1)(2t^2 - 1)$

30. $\left\{-\dfrac{7}{3}, 1\right\}$ 31. $-1, 4$ 32. $\dfrac{2x-3}{2(x-1)}$

33. 3   34. $\dfrac{-a-5b}{(a+b)(a-b)}$ 35. $\dfrac{2(x+2)}{2x-1}$

36. $\dfrac{3-2x-2y}{(x+y)(x-y)}$ 37. $\{-4\}$

38. $q = \dfrac{fp}{p-f}$ or $q = \dfrac{-fp}{f-p}$

39. 150 miles per hour   40. $\dfrac{6}{5}$ or $1\dfrac{1}{5}$ hours

CHAPTER 9

SECTION 9.1 (page 547)

1. False. Zero has only one square root.
3. true   5. true   7. $-4, 4$ 9. $-12, 12$

11. $-\dfrac{5}{14}, \dfrac{5}{14}$ 13. $-30, 30$ 15. 100   17. 19

19. $3x^2 + 4$ 21. 7   23. -11 25. $-\dfrac{12}{11}$

27. not a real number   29. a must be positive.

31. a must be negative.   33. rational; 5
35. irrational; 5.385   37. rational; -8
39. irrational; -17.321 41. not a real number
43. The answer to Exercise 23 is the negative square root of a positive number. However, in Exercise 27, the square root of a negative number is not a real number.
45. $c = 17$ 47. $b = 8$ 49. $c = 11.705$
51. 24 centimeters   53. 80 feet   55. 195 miles
57. 9.434   59. Answers will vary. For example, if $a = 2$ and $b = 7$, $\sqrt{a^2 + b^2} = \sqrt{2^2 + 7^2} = \sqrt{53}$, while $\sqrt{a^2 + b^2} \neq a + b$, because $2 + 7 = 9$ and $\sqrt{53} \neq 9$.
61. 10   63. 5   65. 5   67. not a real number
69. -3 71. 3

SECTION 9.2 (page 555)

1. false   3. 9   5. $3\sqrt{10}$ 7. 13   9. $\sqrt{13r}$
11. (a)   13. $3\sqrt{5}$ 15. $3\sqrt{10}$ 17. $5\sqrt{3}$
19. $5\sqrt{5}$ 21. $-10\sqrt{7}$ 23. $9\sqrt{3}$ 25. $3\sqrt{6}$
27. 24   29. $6\sqrt{10}$ 31. $\sqrt{8} \cdot \sqrt{32} = \sqrt{8 \cdot 32} = \sqrt{256} = 16$. Also, $\sqrt{8} = 2\sqrt{2}$ and $\sqrt{32} = 4\sqrt{2}$, so $\sqrt{8} \cdot \sqrt{32} = 2\sqrt{2} \cdot 4\sqrt{2} = 8 \cdot 2 = 16$. Both methods give the same answer, and the correct answer can always be obtained using either method.   33. $\dfrac{4}{15}$ 35. $\dfrac{\sqrt{7}}{4}$

37. 5   39. $\dfrac{25}{4}$ 41. $6\sqrt{5}$ 43. m 45. y^2

47. $6z$ 49. $20x^3$ 51. $z^2\sqrt{z}$ 53. x^3y^6
55. $2\sqrt[3]{5}$ 57. $3\sqrt[3]{2}$ 59. $4\sqrt[3]{2}$ 61. $2\sqrt[4]{5}$

63. $\dfrac{2}{3}$ 65. $-\dfrac{6}{5}$

In Exercise 67, the number of displayed digits will vary among calculator models. Also, less sophisticated models may exhibit round-off error in the final decimal place.
67. (a) 4.472135955   (b) 4.472135955
(c) The numerical results are not a proof because both answers are approximations, and they might differ if calculated to more decimal places.
69. 6 centimeters   71. 6 inches
73. The product rule for radicals requires that both a and b must be nonnegative. Otherwise $\sqrt{a}$ and $\sqrt{b}$ would not be real numbers.

SECTION 9.3 (page 561)

1. $-5\sqrt{7}$ 3. $5\sqrt{17}$ 5. $5\sqrt{7}$ 7. $11\sqrt{5}$
9. $15\sqrt{2}$ 11. $-6\sqrt{2}$ 13. $17\sqrt{7}$
15. $-16\sqrt{2} - 8\sqrt{3}$ 17. $20\sqrt{2} + 6\sqrt{3} - 15\sqrt{5}$
19. $4\sqrt{2}$ 21. $2\sqrt{3} + 4\sqrt{3} = (2 + 4)\sqrt{3} = 6\sqrt{3}$
23. $11\sqrt{3}$ 25. $5\sqrt{x}$ 27. $3x\sqrt{6}$ 29. 0
31. $-20\sqrt{2k}$ 33. $42x\sqrt{5z}$ 35. $-\sqrt[3]{2}$

37. $6\sqrt[3]{p^2}$ 39. $21\sqrt[4]{m^3}$ 41. Answers will vary. One example is $\sqrt{36} + \sqrt[3]{64} = 6 + 4 = 10$.   43. $-6x^2y$
44. $-6(p - 2q)^2(a + b)$ 45. $-6a^2\sqrt{xy}$
46. The answers are alike because the numerical coefficient of the three answers is the same: -6. Also, the first variable factor is raised to the second power, and the second variable factor is to the first power. They are different because the variable factors are different: x and y, then $p - 2q$ and $a + b$, and then a and $\sqrt{xy}$.

SECTION 9.4 (page 567)

1. $4\sqrt{2}$ **3.** $\dfrac{-\sqrt{33}}{3}$ **5.** $\dfrac{7\sqrt{15}}{5}$ **7.** $\dfrac{\sqrt{30}}{2}$

9. $\dfrac{16\sqrt{3}}{9}$ **11.** $\dfrac{-3\sqrt{2}}{10}$ **13.** $\dfrac{21\sqrt{5}}{5}$ **15.** $\sqrt{3}$

17. $\dfrac{\sqrt{2}}{2}$ **19.** $\dfrac{\sqrt{65}}{5}$ **21.** We are actually multiplying by 1. The identity property of multiplication justifies our result. **23.** $\dfrac{\sqrt{21}}{3}$ **25.** $\dfrac{3\sqrt{14}}{4}$ **27.** $\dfrac{1}{6}$ **29.** 1

31. $\dfrac{\sqrt{7x}}{x}$ **33.** $\dfrac{2x\sqrt{xy}}{y}$ **35.** $\dfrac{x\sqrt{3xy}}{y}$

37. $\dfrac{3ar^2\sqrt{7rt}}{7t}$ **39. (a)** $\dfrac{9\sqrt{2}}{4}$ seconds

(b) 3.182 seconds **41. (b)**

43. $\dfrac{\sqrt[3]{12}}{2}$ **45.** $\dfrac{\sqrt[3]{196}}{7}$ **47.** $\dfrac{\sqrt[3]{6y}}{2y}$ **49.** $\dfrac{\sqrt[3]{42mn^2}}{6n}$

SECTION 9.5 (page 573)

1. 13 **3.** 4 **5.** $9\sqrt{5}$ **7.** $16\sqrt{2}$
9. $\sqrt{15} - \sqrt{35}$ **11.** $2\sqrt{10} + 30$ **13.** $4\sqrt{7}$
15. $57 + 23\sqrt{6}$ **17.** $81 + 14\sqrt{21}$
19. $37 + 12\sqrt{7}$ **21.** 23 **23.** 1
25. $2\sqrt{3} - 2 + 3\sqrt{2} - \sqrt{6}$ **27.** $15\sqrt{2} - 15$
29. $\sqrt{30} + \sqrt{15} + 6\sqrt{5} + 3\sqrt{10}$ **31.** Because multiplication must be performed before addition, it is incorrect to add -37 and -2. Since $-2\sqrt{15}$ cannot be simplified, the expression cannot be written in a simpler form, and the final answer is $-37 - 2\sqrt{15}$.

33. $\dfrac{3 - \sqrt{2}}{7}$ **35.** $-4 - 2\sqrt{11}$

37. $1 + \sqrt{2}$ **39.** $-\sqrt{10} + \sqrt{15}$ **41.** $3 - \sqrt{3}$
43. $2\sqrt{5} + \sqrt{15} + 4 + 2\sqrt{3}$ **45.** $\sqrt{11} - 2$

47. $\dfrac{\sqrt{3} + 5}{8}$ **49.** $\dfrac{6 - \sqrt{10}}{2}$ **51.** $30 + 18x$

52. They are not like terms. **53.** $30 + 18\sqrt{5}$
54. They are not like radicals. **55.** Make the first term $30x$, so that $30x + 18x = 48x$; make the first term $30\sqrt{5}$, so that $30\sqrt{5} + 18\sqrt{5} = 48\sqrt{5}$. **56.** Both like terms and like radicals are combined by adding their numerical coefficients. The variables in like terms are replaced by radicals in like radicals. **57.** 4 inches

SECTION 9.6 (page 583)

1. Because $\sqrt{x}$ must be greater than or equal to 0 for any real number x, it cannot equal -8. **3.** $\{49\}$ **5.** $\{7\}$

7. $\{85\}$ **9.** $\{-45\}$ **11.** $\left\{-\dfrac{3}{2}\right\}$ **13.** $\emptyset$

15. $\{121\}$ **17.** $\{8\}$ **19.** $\{1\}$ **21.** $\{6\}$ **23.** $\emptyset$
25. $\{5\}$ **27.** The square of the right side is actually $x^2 - 14x + 49$. **29.** $\{12\}$ **31.** $\{5\}$ **33.** $\{0, 3\}$
35. $\{-1, 3\}$ **37.** $\{8\}$ **39.** $\{4\}$ **41. (a)** 70.5 miles per hour **(b)** 59.8 miles per hour **(c)** 53.9 miles per hour **43.** 158.6 feet

SECTION 9.7 (page 593)

1. i **3.** $\sqrt{b}$ **5.** $a + bi$ is a complex number if a and b are real numbers and i is the imaginary unit. Therefore, for every real number a, if $b = 0$, $a = a + 0i$ is a complex number.
7. $13i$ **9.** $-12i$ **11.** $i\sqrt{5}$ **13.** $4i\sqrt{3}$
15. -15 **17.** -10 **19.** $\sqrt{3}$ **21.** $5i$
23. $-1 + 7i$ **25.** $0 + 0i$ **27.** $7 + 3i$
29. $-2 + 0i$ **31.** $1 + 13i$ **33.** $6 + 6i$
35. $4 + 2i$ **37.** -81 **39.** -16 **41.** $-10 - 30i$
43. $10 - 5i$ **45.** $-9 + 40i$ **47.** 153
49. (a) $a - bi$ **(b)** a^2; b^2 **51.** $1 + i$ **53.** $-1 + 2i$

55. $2 + 2i$ **57.** $-\dfrac{5}{13} - \dfrac{12}{13}i$

59. (a) $4x + 1$ **(b)** $4 + i$ **60. (a)** $-2x + 3$
(b) $-2 + 3i$ **61. (a)** $3x^2 + 5x - 2$ **(b)** $5 + 5i$

62. (a) $-\sqrt{3} + \sqrt{6} + 1 - \sqrt{2}$ **(b)** $\dfrac{1}{5} - \dfrac{7}{5}i$

63. Because $i^2 = -1$, two pairs of like terms can be combined in Exercise 61(b). **64.** Because $i^2 = -1$, terms can be combined in the numerator, and the

denominator is changed. **65.** $\dfrac{5}{41} + \dfrac{4}{41}i$ **67.** -1

69. i **71.** 1 **73.** $-i$
75. Since $i^{20} = (i^4)^5 = 1^5 = 1$, the student multiplied by 1, which is justified by the identity property of multiplication.

77. $\dfrac{1}{2} + \dfrac{1}{2}i$ **79.** $(1 + 5i)^2 - 2(1 + 5i) + 26$ will

simplify to 0 when the operations are applied.

CHAPTER 9 REVIEW EXERCISES (page 601)

1. $-7, 7$ **2.** $-9, 9$ **3.** $-14, 14$ **4.** $-11, 11$
5. $-15, 15$ **6.** $-27, 27$ **7.** 4 **8.** -6 **9.** 10

10. 3 **11.** not a real number **12.** -65 **13.** $\dfrac{7}{6}$

14. $\dfrac{10}{9}$ **15.** a must be negative. **16.** 8

17. irrational; 4.796 **18.** rational; 13 **19.** rational; -5 **20.** not a real number **21.** $\sqrt{14}$ **22.** 6
23. $5\sqrt{3}$ **24.** 12 **25.** $-3\sqrt{3}$ **26.** $4\sqrt{3}$
27. $4\sqrt{10}$ **28.** -5 **29.** 12 **30.** 18

31. $16\sqrt{6}$ **32.** $25\sqrt{10}$ **33.** $\dfrac{3}{2}$ **34.** $-\dfrac{11}{20}$

35. $\dfrac{\sqrt{3}}{7}$ **36.** $\dfrac{\sqrt{7}}{13}$ **37.** $\dfrac{\sqrt{5}}{6}$ **38.** $\dfrac{2}{15}$ **39.** $3\sqrt{2}$

40. 8 **41.** $2\sqrt{2}$ **42.** p **43.** $\sqrt{km}$ **44.** r^9

45. x^5y^8 **46.** $x^4\sqrt{x}$ **47.** $\dfrac{6}{p}$ **48.** $a^7b^{10}\sqrt{ab}$

49. $11x^3y^5$ **50.** Yes, because both approximations are .7071067812. **51.** $2\sqrt{11}$ **52.** $9\sqrt{2}$ **53.** $21\sqrt{3}$
54. $12\sqrt{3}$ **55.** 0 **56.** $3\sqrt{7}$ **57.** $2\sqrt{3} + 3\sqrt{10}$
58. $2\sqrt{2}$ **59.** $6\sqrt{30}$ **60.** $5\sqrt{x}$ **61.** 0

62. $-m\sqrt{5}$ **63.** $11k^2\sqrt{2n}$ **64.** $\dfrac{10\sqrt{3}}{3}$

65. $\dfrac{15\sqrt{2}}{2}$ **66.** $\dfrac{8\sqrt{10}}{5}$ **67.** $\sqrt{5}$ **68.** $\sqrt{6}$

69. $\dfrac{\sqrt{30}}{15}$ **70.** $\dfrac{\sqrt{10}}{5}$ **71.** $\sqrt{10}$ **72.** $\dfrac{\sqrt{42}}{21}$

73. $\dfrac{r\sqrt{x}}{4x}$ **74.** $\dfrac{\sqrt[3]{9}}{3}$ **75.** $\dfrac{\sqrt[3]{98}}{7}$ **76.** To show this, simplify the second expression using the guidelines given in Section 9.5.

$$\sqrt{\dfrac{48}{128}} = \dfrac{\sqrt{48}}{\sqrt{128}} = \dfrac{\sqrt{16\cdot 3}}{\sqrt{64\cdot 2}} = \dfrac{4\sqrt{3}}{8\sqrt{2}} = \dfrac{4\sqrt{3}\cdot\sqrt{2}}{8\sqrt{2}\cdot\sqrt{2}}$$

$$= \dfrac{4\sqrt{6}}{8\cdot 2} = \dfrac{\sqrt{6}}{4}$$

77. $-\sqrt{15} - 9$ **78.** $3\sqrt{6} + 12$ **79.** $22 - 16\sqrt{3}$
80. $179 + 20\sqrt{7}$ **81.** -2 **82.** -13
83. $2\sqrt{21} - \sqrt{14} + 12\sqrt{2} - 4\sqrt{3}$ **84.** $-2 + \sqrt{5}$
85. $\dfrac{-2\sqrt{2} - 6}{7}$ **86.** $\dfrac{-2 + 6\sqrt{2}}{17}$ **87.** $\dfrac{-\sqrt{3} + 3}{2}$
88. $\dfrac{-\sqrt{10} + 3\sqrt{5} + \sqrt{2} - 3}{7}$
89. $\dfrac{2\sqrt{3} + 2 + 3\sqrt{2} + \sqrt{6}}{2}$ **90.** $\dfrac{3 + 2\sqrt{6}}{3}$
91. $\dfrac{1 + 3\sqrt{7}}{4}$ **92.** $3 + 4\sqrt{3}$ **93.** $\emptyset$ **94.** $\emptyset$
95. $\{48\}$ **96.** $\{1\}$ **97.** $\{2\}$ **98.** $\{6\}$ **99.** $\{-2\}$
100. $\{-2\}$ **101.** $\{4\}$ **102.** $\{-2\}$ **103.** $\{3\}$
104. $\{-1\}$ **105.** $5i$ **106.** $10i\sqrt{2}$ **107.** no
108. $-10 - 2i$ **109.** $14 + 7i$ **110.** $-\sqrt{35}$
111. -45 **112.** 3 **113.** $5 + i$ **114.** $32 - 24i$
115. $1 - i$ **116.** $4 + i$ **117.** $-i$ **118.** 1
119. $-i$ **120.** 9 **121.** $11\sqrt{3}$ **122.** $\dfrac{11}{t}$
123. $\dfrac{5 - \sqrt{2}}{23}$ **124.** $\dfrac{2\sqrt{10}}{5}$ **125.** $5y\sqrt{2}$
126. -5 **127.** $-\sqrt{10} - 5\sqrt{15}$ **128.** $\dfrac{4r\sqrt{3rs}}{3s}$
129. $\dfrac{2 + \sqrt{13}}{2}$ **130.** $-7\sqrt{2}$ **131.** $7 - 2\sqrt{10}$
132. 1 **133.** 26 **134.** $7i\sqrt{2}$ **135.** $\{7\}$ **136.** $\emptyset$
137. $\{8\}$

CHAPTER 9 TEST (page 607)

1. $-14, 14$ **2. (a)** irrational **(b)** 11.916 **3.** 6
4. $-3\sqrt{3}$ **5.** $\dfrac{8\sqrt{2}}{5}$ **6.** $2\sqrt[3]{4}$ **7.** $4\sqrt{6}$
8. $9\sqrt{7}$ **9.** $-5\sqrt{3x}$ **10.** $2y\sqrt[3]{4x^2}$ **11.** 31
12. $6\sqrt{2} + 2 - 3\sqrt{14} - \sqrt{7}$ **13.** $11 + 2\sqrt{30}$
14. (a) $6\sqrt{2}$ inches **(b)** 8.485 inches **15.** $\dfrac{5\sqrt{14}}{7}$
16. $\dfrac{\sqrt{6x}}{3x}$ **17.** $-\sqrt[3]{2}$ **18.** $\dfrac{-12 - 3\sqrt{3}}{13}$ **19.** $\{3\}$
20. $\left\{\dfrac{1}{4}, 1\right\}$ **21. (a)** $-5 - 8i$ **(b)** $3 + 4i$ **22.** $-i$

CUMULATIVE REVIEW 1–9 (page 609)

1. $\left\{\dfrac{4}{5}\right\}$ **2.** $\left\{\dfrac{11}{10}, \dfrac{7}{2}\right\}$ **3.** $(-6, \infty)$ **4.** $(1, 3)$
5. $(-2, 1)$ **6.** $12x + 11y = 18$

7. (c) **8. (a)** $(0, 6)$ **(b)** $(2, 0)$ **9.** $\$120$
10.

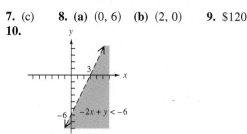

11. Both angles measure $80°$. **12.** $\emptyset$ **13.** infinite number of solutions **14.** 0 **15.** -27
16. $-k^3 - 3k^2 - 8k - 9$ **17.** $8x^2 + 17x - 21$
18. $z - 2 + \dfrac{3}{z}$ **19.** $3y^3 - 3y^2 + 4y + 1 + \dfrac{-10}{2y + 1}$
20. $(2p - 3q)(p - q)$ **21.** $(3k^2 + 4)(6k^2 - 5)$
22. $(5x + 7y)(5x - 7y)$ **23.** $\dfrac{y}{y + 5}$
24. $\dfrac{4x + 2y}{(x + y)(x - y)}$ **25.** $-\dfrac{9}{4}$ **26.** $\dfrac{-1}{a + b}$
27. $\left\{-3, -\dfrac{5}{2}\right\}$ **28.** $\left\{-\dfrac{2}{5}, 1\right\}$ **29.** -12
30. -4 **31.** $8\sqrt{5}$ **32.** $\dfrac{-9\sqrt{5}}{20}$
33. $4(\sqrt{6} + \sqrt{5})$ **34.** $6\sqrt[3]{4}$ **35.** $i\sqrt{29}$ **36.** $\{6\}$
37. 15 miles per hour **38.** Brenda: 8 miles per hour; Chuck: 4 miles per hour **39.** 17 dimes and 12 quarters
40. $\dfrac{80}{39}$ or $2\dfrac{2}{39}$ liters

CHAPTER 10
SECTION 10.1 (page 615)

1. rational **3. (d)** **5.** Because the product of either two positive numbers or two negative numbers is positive and zero squared is zero, the square of a real number cannot be negative. **7.** No, the equation is equivalent to $x^2 = -16$, which has no real solutions. **9.** $\{-9, 9\}$
11. $\{-\sqrt{14}, \sqrt{14}\}$ **13.** $\{-4\sqrt{3}, 4\sqrt{3}\}$
15. $\left\{-\dfrac{5}{2}, \dfrac{5}{2}\right\}$ **17.** $\{-1.5, 1.5\}$ **19.** $\{-\sqrt{3}, \sqrt{3}\}$
21. $\{-2, 8\}$ **23.** $\{-5 + 4i, -5 - 4i\}$
25. $\{8 + 3\sqrt{3}, 8 - 3\sqrt{3}\}$ **27.** $\left\{-3, \dfrac{5}{3}\right\}$
29. $\left\{0, \dfrac{3}{2}\right\}$ **31.** $\left\{\dfrac{5 + \sqrt{30}}{2}, \dfrac{5 - \sqrt{30}}{2}\right\}$
33. $\left\{\dfrac{-1 + 3\sqrt{2}}{3}, \dfrac{-1 - 3\sqrt{2}}{3}\right\}$
35. $\{-10 + 4\sqrt{3}, -10 - 4\sqrt{3}\}$
37. $\left\{\dfrac{1 + 4\sqrt{3}}{4}, \dfrac{1 - 4\sqrt{3}}{4}\right\}$
39. The answers are equivalent. If the answers of either student are multiplied by $\dfrac{-1}{-1}$ they will equal those of the other student. **41.** $\{-7, 3\}$ **43.** about $\dfrac{1}{2}$ second
45. 424 cards

SECTION 10.2 (page 623)

1. Divide both sides by 4. **3.** $\{1, 3\}$
5. $\{-1 + \sqrt{6}, -1 - \sqrt{6}\}$ **7.** $\{-3\}$ **9.** 49
11. $\dfrac{25}{4}$ **13.** $\dfrac{1}{16}$ **15.** (d)
17. $\left\{\dfrac{2 + \sqrt{14}}{2}, \dfrac{2 - \sqrt{14}}{2}\right\}$ **19.** $\left\{-\dfrac{3}{2}, \dfrac{1}{2}\right\}$
21. $\left\{-1 + \dfrac{1}{4}i, -1 - \dfrac{1}{4}i\right\}$
23. $\left\{\dfrac{-7 + \sqrt{97}}{6}, \dfrac{-7 - \sqrt{97}}{6}\right\}$ **25.** $\{-4, 2\}$
27. $\{1 + \sqrt{6}, 1 - \sqrt{6}\}$ **29.** 1 and 5 seconds
31. $\{-1, 6\}$

SECTION 10.3 (page 631)

1. true **3.** true **5.** true **7.** $\{3, 5\}$
9. $\left\{\dfrac{-2 + \sqrt{2}}{2}, \dfrac{-2 - \sqrt{2}}{2}\right\}$ **11.** $\left\{\dfrac{1 + \sqrt{3}}{2}, \dfrac{1 - \sqrt{3}}{2}\right\}$
13. $\{5 + \sqrt{7}, 5 - \sqrt{7}\}$
15. $\left\{\dfrac{-2 + \sqrt{10}}{2}, \dfrac{-2 - \sqrt{10}}{2}\right\}$
17. $\{-1 + 3\sqrt{2}, -1 - 3\sqrt{2}\}$ **19.** $\{-i\sqrt{47}, i\sqrt{47}\}$
21. $\{3 + i\sqrt{5}, 3 - i\sqrt{5}\}$ **23.** $\left\{\dfrac{1 + i\sqrt{6}}{2}, \dfrac{1 - i\sqrt{6}}{2}\right\}$
25. $\left\{\dfrac{-2 + i\sqrt{2}}{3}, \dfrac{-2 - i\sqrt{2}}{3}\right\}$
27. $\{4 + 3i\sqrt{2}, 4 - 3i\sqrt{2}\}$
29. $\left\{\dfrac{-1 + i\sqrt{3}}{2}, \dfrac{-1 - i\sqrt{3}}{2}\right\}$
31. at 3 seconds and 5 seconds **33.** It reaches its
maximum height at 5 seconds because this is the only time
it reaches 400 feet. **35.** 2117 **36.** 2243
37. 2356 **38.** Find $f(3)$. **39.** 1992 ($x = 4$
corresponds to 1992) **40.** 1999 **41.** (b) **43.** (c)
45. (a) **47.** (d) **49.** $(6x + 5)(4x - 9)$
51. cannot be factored **53.** $(12x + 1)(x - 7)$
55. -4 and 4

SUMMARY ON QUADRATIC EQUATIONS (page 635)

1. $\{-6, 6\}$ **3.** $\left\{-\dfrac{10}{9}, \dfrac{10}{9}\right\}$ **5.** $\{1, 3\}$ **7.** $\{4, 5\}$
9. $\left\{-\dfrac{1}{3}, \dfrac{5}{3}\right\}$ **11.** $\{-17, 5\}$
13. $\left\{\dfrac{7 + 2\sqrt{6}}{3}, \dfrac{7 - 2\sqrt{6}}{3}\right\}$
15. $\left\{\dfrac{-1 + i\sqrt{3}}{2}, \dfrac{-1 - i\sqrt{3}}{2}\right\}$ **17.** $\left\{-\dfrac{1}{2}, 2\right\}$
19. $\left\{-\dfrac{5}{4}, \dfrac{3}{2}\right\}$ **21.** $\{1 + \sqrt{2}, 1 - \sqrt{2}\}$ **23.** $\left\{\dfrac{2}{5}, 4\right\}$
25. $\left\{\dfrac{-3 + \sqrt{41}}{2}, \dfrac{-3 - \sqrt{41}}{2}\right\}$ **27.** $\left\{\dfrac{1}{4}, 1\right\}$
29. $\left\{\dfrac{-2 + \sqrt{11}}{3}, \dfrac{-2 - \sqrt{11}}{3}\right\}$

31. $\left\{\dfrac{-7 + \sqrt{5}}{4}, \dfrac{-7 - \sqrt{5}}{4}\right\}$
33. $\left\{\dfrac{8 + 8\sqrt{2}}{3}, \dfrac{8 - 8\sqrt{2}}{3}\right\}$
35. $\left\{\dfrac{-1 + i\sqrt{11}}{2}, \dfrac{-1 - i\sqrt{11}}{2}\right\}$ **37.** $\left\{-\dfrac{2}{3}, 2\right\}$
39. $\left\{-4, \dfrac{3}{5}\right\}$ **41.** $\left\{-\dfrac{2}{3}, \dfrac{2}{5}\right\}$

SECTION 10.4 (page 643)

1. Multiply by the LCD, x. **3.** Make a substitution for
$r^2 + r$. **5.** $\{-4, 7\}$
7. $\left\{-\dfrac{2}{3}, 1\right\}$ **9.** $\left\{-\dfrac{14}{17}, 5\right\}$ **11.** $\left\{-\dfrac{11}{7}, 0\right\}$
13. $\left\{\dfrac{-1 + \sqrt{13}}{2}, \dfrac{-1 - \sqrt{13}}{2}\right\}$ **15.** $\dfrac{1}{m}$ job per hour
17. 25 miles per hour **19.** 80 miles per hour
21. 9 minutes **23.** $\{3\}$
25. $\left\{\dfrac{8}{9}\right\}$ **27.** $\{16\}$ **29.** $\left\{\dfrac{2}{5}\right\}$ **31.** $\{-3, 3\}$
33. $\left\{-\dfrac{3}{2}, -1, 1, \dfrac{3}{2}\right\}$ **35.** $\{-6, -5\}$ **37.** $\{-4, 1\}$
39. $\left\{-\dfrac{1}{3}, \dfrac{1}{6}\right\}$ **41.** $\left\{-\dfrac{1}{2}, 3\right\}$ **43.** $\{-8, 1\}$
45. $\{25\}$ **47.** It would cause both denominators to be 0,
and division by 0 is undefined.
48. The solution is $\dfrac{12}{5}$.
49. $\left(\dfrac{x}{x - 3}\right)^2 + 3\left(\dfrac{x}{x - 3}\right) - 4 = 0$
50. The numerator can never equal the denominator, since
the denominator is 3 less than the numerator.
51. $\left\{\dfrac{12}{5}\right\}$; The values for t are -4 and 1. The value 1 is
impossible because it leads to a contradiction
$\left(\text{since } \dfrac{x}{x - 3} \text{ is never equal to 1}\right)$.
52. $\left\{\dfrac{12}{5}\right\}$; The values for s are $\dfrac{1}{x}$ and $\dfrac{-4}{x}$. The value $\dfrac{1}{x}$ is
impossible, since $\dfrac{1}{x} \neq \dfrac{1}{x - 3}$ for all x.

SECTION 10.5 (page 651)

1. $m = \sqrt{p^2 - n^2}$
3. $t = \dfrac{\pm\sqrt{dk}}{k}$ **5.** $d = \dfrac{\pm\sqrt{skI}}{I}$
7. $v = \dfrac{\pm\sqrt{kAF}}{F}$ **9.** $r = \dfrac{\pm\sqrt{3\pi Vh}}{\pi h}$
11. $t = \dfrac{-B \pm \sqrt{B^2 - 4AC}}{2A}$
13. $h = \dfrac{D^2}{k}$ **15.** $\ell = \dfrac{p^2 g}{k}$

17. If g is positive, the only way to have a real value for p is to have $k\ell$ positive, since the quotient of two positive numbers is positive. If k and ℓ have different signs, their product is negative, leading to a negative radicand.
19. eastbound ship: 80 miles; southbound ship: 150 miles
21. 20 meters, 21 meters, 29 meters
23. 12 centimeters by 20 centimeters   **25.** 1 foot
27. 4.3 hours   **29.** 7.8 hours and 8.3 hours
31. $4814.10   **33.** 1987   **35.** 9.2 seconds
37. 1 second and 8 seconds   **39.** .035 or 3.5%
41. 5.5 meters per second   **43.** 5 or 14

SECTION 10.6 (page 661)

1. Include the endpoints if the symbol is $\geq$ or $\leq$. Exclude the endpoints if the symbol is $>$ or $<$.
3. $(-\infty, -4] \cup [3, \infty)$
5. $(-\infty, -1) \cup (5, \infty)$
7. $(-4, 6)$
9. $(-\infty, 1] \cup [3, \infty)$
11. $\left(-\infty, -\dfrac{3}{2}\right] \cup \left[\dfrac{3}{5}, \infty\right)$
13. $\left(-\dfrac{2}{3}, \dfrac{1}{3}\right)$
15. $\left(-\infty, -\dfrac{1}{2}\right] \cup \left[\dfrac{1}{3}, \infty\right)$
17. $(-\infty, 3 - \sqrt{3}] \cup [3 + \sqrt{3}, \infty)$
19. $(-\infty, 1) \cup (2, 4)$
21. $\left[-\dfrac{3}{2}, \dfrac{1}{3}\right] \cup [4, \infty)$
23. $(-\infty, 1) \cup (4, \infty)$
25. $\left[-\dfrac{3}{2}, 5\right)$
27. $(2, 6]$
29. $\left(-\infty, \dfrac{1}{2}\right) \cup \left(\dfrac{5}{4}, \infty\right)$
31. $[-4, -2)$
33. $\left(0, \dfrac{1}{2}\right) \cup \left(\dfrac{5}{2}, \infty\right)$

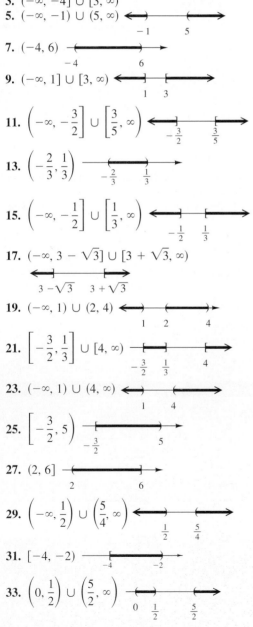

35. $(-\infty, \infty)$ **37.** $\emptyset$
39. 3 seconds and 13 seconds   **40.** between 3 seconds and 13 seconds   **41.** at 0 seconds (the time when it is initially projected) and at 16 seconds (the time when it hits the ground)   **42.** between 0 and 3 seconds and also between 13 and 16 seconds

CHAPTER 10 REVIEW EXERCISES (page 669)

1. $\{-12, 12\}$ **2.** $\{-\sqrt{37}, \sqrt{37}\}$ **3.** $\{-8\sqrt{2}, 8\sqrt{2}\}$
4. $\{-7, 3\}$ **5.** $\{3 + \sqrt{10}, 3 - \sqrt{10}\}$
6. $\left\{\dfrac{-1 + \sqrt{14}}{2}, \dfrac{-1 - \sqrt{14}}{2}\right\}$ **7.** $\left\{-\dfrac{2}{3} + i, -\dfrac{2}{3} - i\right\}$
8. $(ax + b)^2 =$ a number   **9.** $\{-5, -1\}$
10. $\{-2 + \sqrt{11}, -2 - \sqrt{11}\}$
11. $\{-1 + \sqrt{6}, -1 - \sqrt{6}\}$
12. $\left\{\dfrac{-4 + \sqrt{22}}{2}, \dfrac{-4 - \sqrt{22}}{2}\right\}$ **13.** $\left\{-\dfrac{2}{5}, 1\right\}$
14. $\left\{\dfrac{3 + i\sqrt{87}}{8}, \dfrac{3 - i\sqrt{87}}{8}\right\}$ **15.** $2\dfrac{1}{2}$ seconds
16. 4 seconds
17. $\left\{-\dfrac{7}{2}, 3\right\}$ **18.** $\left\{\dfrac{-5 + \sqrt{53}}{2}, \dfrac{-5 - \sqrt{53}}{2}\right\}$
19. $\left\{\dfrac{1 + \sqrt{41}}{2}, \dfrac{1 - \sqrt{41}}{2}\right\}$ **20.** $\left\{\dfrac{7}{3}\right\}$
21. $\left\{\dfrac{2 + i\sqrt{2}}{3}, \dfrac{2 - i\sqrt{2}}{3}\right\}$
22. $\left\{\dfrac{-7 + \sqrt{37}}{2}, \dfrac{-7 - \sqrt{37}}{2}\right\}$
23. $\left\{\dfrac{-3 + i\sqrt{23}}{4}, \dfrac{-3 - i\sqrt{23}}{4}\right\}$
24. The student was incorrect, since the fraction bar should extend under the term $-b$.   **25.** 4 seconds and 12 seconds   **26.** The rock reaches the height 768 feet twice: once on its way up and again on its way down.
27. $k = 1$ **28.** all k such that $k < 1$
29. (c)   **30.** (a)   **31.** (d)   **32.** (b)
33. It factors as $(6x - 5)(4x - 9)$.
34. It cannot be factored.
35. $\left\{-\dfrac{5}{2}, 3\right\}$ **36.** $\left\{-\dfrac{1}{2}, 1\right\}$ **37.** $\left\{-\dfrac{11}{6}, -\dfrac{19}{12}\right\}$
38. $\{-4\}$ **39.** $\left\{-\dfrac{343}{8}, 64\right\}$ **40.** $\{-3, -2, 1, 2\}$
41. 40 miles per hour   **42.** Linda: 7.2 hours; Ed: 5.2 hours   **43.** Because x appears on the left side alone, and because it is equal to the nonnegative square root of $2x + 4$, it cannot be negative.   **44.** Take the positive and negative square roots of each value obtained from the formula.
45. $d = \dfrac{\pm\sqrt{SkI}}{I}$ **46.** $v = \dfrac{\pm\sqrt{rFkw}}{kw}$

47. $r = \dfrac{-\pi h \pm \sqrt{\pi^2 h^2 + 2\pi S}}{2\pi}$

48. $t = \dfrac{3m \pm \sqrt{9m^2 + 24m}}{2m}$

49. 5.2 seconds   **50.** .7 second and 4.0 seconds
51. 10 meters by 12 meters   **52.** 9 feet, 12 feet, 15 feet
53. 68 and 69   **54.** 1 inch   **55.** 3 minutes
56. \$.50   **57.** 4.5%   **58.** Jim: 7.5 hours;
Jake: 8.5 hours

59. $\left(-\infty, -\dfrac{3}{2}\right) \cup (4, \infty)$ ⟵———)———(———⟶
                                   $-\dfrac{3}{2}$ 4

60. $[-4, 3]$ [———————]——⟶
                  -4 3

61. $\left(-\infty, -\dfrac{1}{2}\right) \cup (3, \infty)$ ⟵———)———(———⟶
                                 $-\dfrac{1}{2}$ 3

62. $[-3, 2)$ [———————)——⟶
                -3 2

63. $(-\infty, -5] \cup [-2, 3]$ ⟵]——[———]——⟶
                        -5 -2 3

64. $\emptyset$ ———————⟶

65. $R = \dfrac{\pm\sqrt{Vh - r^2 h}}{h}$ **66.** $\left\{\dfrac{3 + i\sqrt{3}}{3}, \dfrac{3 - i\sqrt{3}}{3}\right\}$

67. $\{-2, -1, 3, 4\}$ **68.** $(-\infty, -6) \cup \left(-\dfrac{3}{2}, 1\right)$

69. $\left\{\dfrac{-11 + \sqrt{7}}{3}, \dfrac{-11 - \sqrt{7}}{3}\right\}$ **70.** $y = \dfrac{6p^2}{z}$

71. $\left\{\dfrac{3}{5}, 1\right\}$ **72.** $\left\{-\dfrac{5}{3}, -\dfrac{3}{2}\right\}$ **73.** $\left(-5, -\dfrac{23}{5}\right]$

74. 3 hours and 6 hours   **75.** 7 miles per hour

76. $\left\{\dfrac{4}{11}, 5\right\}$ **77.** $\{-11i\sqrt{2}, 11i\sqrt{2}\}$ **78.** $(-\infty, \infty)$

CHAPTER 10 TEST (page 675)

1. $\{-3\sqrt{6}, 3\sqrt{6}\}$ **2.** $\left\{-\dfrac{8}{7}, \dfrac{2}{7}\right\}$

3. $\{-1 + \sqrt{2}, -1 - \sqrt{2}\}$ **4.** $\left\{\dfrac{3 + \sqrt{17}}{4}, \dfrac{3 - \sqrt{17}}{4}\right\}$

5. $\left\{\dfrac{2 + i\sqrt{11}}{3}, \dfrac{2 - i\sqrt{11}}{3}\right\}$ **6.** $\left\{\dfrac{2}{3}\right\}$

7. Maretha: 11.1 hours; Lillaana: 9.1 hours   **8.** (a)
9. discriminant: 88; There are two irrational solutions.

10. $\left\{-\dfrac{2}{3}, 6\right\}$ **11.** $\left\{\dfrac{-7 + \sqrt{97}}{8}, \dfrac{-7 - \sqrt{97}}{8}\right\}$

12. $\left\{-2, -\dfrac{1}{3}, \dfrac{1}{3}, 2\right\}$ **13.** $\left\{-\dfrac{5}{2}, 1\right\}$

14. $r = \dfrac{\pm\sqrt{\pi S}}{2\pi}$ **15.** 1985   **16.** 7 miles per hour
17. 2 feet   **18.** 16 meters

19. $(-\infty, -5) \cup \left(\dfrac{3}{2}, \infty\right)$ ⟵———)———(———⟶
                              -5 $\dfrac{3}{2}$

20. $(-\infty, 4) \cup [9, \infty)$ ⟵———)———[———⟶
                               4   9

CUMULATIVE REVIEW 1–10 (page 677)

1. $-2, 0, 7$ **2.** $-\dfrac{7}{3}, -2, 0, .7, 7, \dfrac{32}{3}$
3. all except $\sqrt{-8}$ **4.** All are complex numbers.
5. 6   **6.** 41   **7.** $\{1\}$

8. $\left[2, \dfrac{8}{3}\right]$ **9.** $\left(-\infty, -\dfrac{9}{4}\right) \cup \left(\dfrac{5}{4}, \infty\right)$

10. slope: $\dfrac{1}{2}$; y-intercept: $\left(0, -\dfrac{7}{4}\right)$ **11.** $x + 3y = -1$

12. $\dfrac{x^8}{y^4}$ **13.** $\dfrac{4}{xy^2}$ **14.** $\sqrt{15}$; $\left[\dfrac{5}{2}, \infty\right)$

15. $\{(1, -4)\}$ **16.** $\left\{\left(\dfrac{1}{3}, \dfrac{1}{4}, -\dfrac{1}{2}\right)\right\}$

17. $\dfrac{4}{9}t^2 + 12t + 81$ **18.** $-3t^3 + 5t^2 - 12t + 15$

19. $4x^2 - 6x + 11 + \dfrac{4}{x + 2}$ **20.** $x(4 + x)(4 - x)$
21. $(4m - 3)(6m + 5)$ **22.** $(3x - 5y)^2$

23. $(2a + b)(x - 5)$ **24.** $\dfrac{x - 5}{x + 5}$ **25.** $-\dfrac{5}{18}$

26. $-\dfrac{8}{k}$ **27.** $\dfrac{r - s}{r}$ **28.** $\dfrac{3\sqrt[3]{4}}{4}$ **29.** $\sqrt{7} + \sqrt{5}$

30. $\left\{\dfrac{2}{3}\right\}$ **31.** $\emptyset$ **32.** $\left\{\dfrac{7 + \sqrt{177}}{4}, \dfrac{7 - \sqrt{177}}{4}\right\}$

33. $\{-2, -1, 1, 2\}$ **34.** 7 inches by 3 inches
35. southbound car: 57 miles; eastbound car: 76 miles
36. suppliers I and III: 22 units; supplier II: 56 units
37. .3 second and 4.7 seconds   **38.** $\{-3, 5\}$

39. $\left\{\dfrac{2 + \sqrt{10}}{2}, \dfrac{2 - \sqrt{10}}{2}\right\}$ **40.** $(5, 13)$

CHAPTER 11

SECTION 11.1 (page 687)

1. (a) B   (b) C   (c) A   (d) D   **3.** $(0, 0)$
5. $(0, 4)$ **7.** $(1, 0)$ **9.** $(-3, -4)$
11. downward; wider   **13.** upward; narrower
15. In Exercise 9, the parabola is shifted 3 units to the left
and 4 units down. The parabola in Exercise 10 is shifted
5 units to the right and 8 units down.   **17.** (a) I
(b) IV   (c) II   (d) III
19.   **21.**

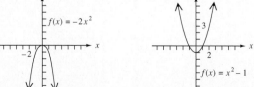

23.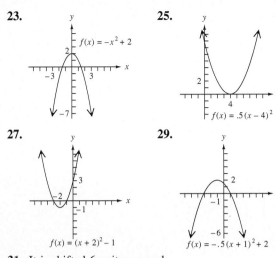

25.

27.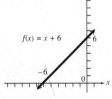

29.

31. It is shifted 6 units upward.

32.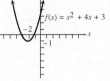

33. It is shifted 6 units upward.

34. It is shifted 6 units to the right.

35.

36. It is shifted 6 units to the right.

SECTION 11.2 (page 697)

1. If x is squared, it has a vertical axis; if y is squared, it has a horizontal axis.   **3.** Use the discriminant of the corresponding quadratic equation. If it is positive, there are two x-intercepts. If it is zero, there is just one intercept (the vertex), and if it is negative, there are no x-intercepts.

5. $(-1, 3)$; upward; narrower; no x-intercepts

7. $\left(\frac{5}{2}, \frac{37}{4}\right)$; downward; same; two x-intercepts

9. $(-3, -9)$; to the right; wider

11. domain: $(-\infty, \infty)$; range: $[-1, \infty)$

13. domain: $(-\infty, \infty)$; range: $(-\infty, -3]$

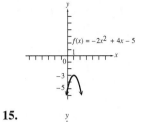

15.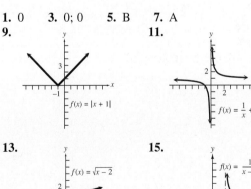

17.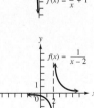

19. B   **21.** A   **23.** 160 feet by 320 feet
25. 16 feet; 2 seconds   **27.** 30 and 30
29. The parabola opens downward, indicating $a < 0$.
31. Since $x = 2.1 \approx 2$, the year corresponding to 2, which is 1990, is the year of maximum number of jobs.
33. (a) $\{1, 3\}$ (b) $(-\infty, 1) \cup (3, \infty)$ (c) $(1, 3)$

34. (a) $\left\{-4, \frac{2}{3}\right\}$ (b) $\left(-\infty, -4\right] \cup \left[\frac{2}{3}, \infty\right)$

(c) $\left(-4, \frac{2}{3}\right)$ **35.** (a) $\{-2, 5\}$ (b) $[-2, 5]$

(c) $(-\infty, -2] \cup [5, \infty)$ **36.** (a) $\left\{-3, \frac{5}{2}\right\}$

(b) $\left[-3, \frac{5}{2}\right]$ (c) $(-\infty, -3] \cup \left[\frac{5}{2}, \infty\right)$

SECTION 11.3 (page 709)

1. 0   **3.** 0; 0   **5.** B   **7.** A
9.

11.

13.

15.

17.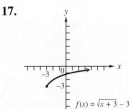

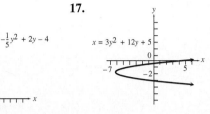

19. B   **21.** D

23. (a) 5   (b) $2\sqrt{10}$ (c) $5\sqrt{2}$ (d) $8\sqrt{2}$

25. $(4, 6)$, $r = 7$ **27.** $(1, -2)$, $r = 3$

29. $(-5, -4)$, $r = 6$

31.

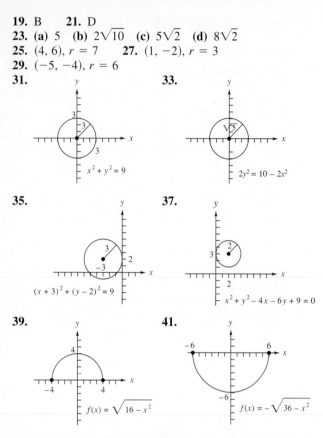

$x^2 + y^2 = 9$

33.

$2y^2 = 10 - 2x^2$

35.

$(x + 3)^2 + (y - 2)^2 = 9$

37.

$x^2 + y^2 - 4x - 6y + 9 = 0$

39.

$f(x) = \sqrt{16 - x^2}$

41.

$f(x) = -\sqrt{36 - x^2}$

43. (a) $|x + p|$ (b) The distance from the point to the origin should equal the distance from the line to the origin. (c) $\sqrt{(x - p)^2 + y^2}$ (d) $y^2 = 4px$

SECTION 11.4 (page 721)

1. C   **3.** D   **5.** When written in one of the forms given in the box "Equations of Hyperbolas" in this section, it will open up and down if the $-$ sign precedes the x^2 term; it will open left and right if the $-$ sign precedes the y^2 term.

7.

$\dfrac{x^2}{9} + \dfrac{y^2}{25} = 1$

9.

$\dfrac{x^2}{36} + \dfrac{y^2}{16} = 1$

11.

$\dfrac{x^2}{49} + \dfrac{y^2}{25} = 1$

13. The points (a, b), $(a, -b)$, $(-a, -b)$, $(-a, b)$ or the points (b, a), $(-b, a)$, $(-b, -a)$, $(b, -a)$ are used as corners of a rectangle. The diagonals of the rectangle are drawn, and they are used as asymptotes for the hyperbola.

15.

$\dfrac{x^2}{16} - \dfrac{y^2}{9} = 1$

17.

$\dfrac{y^2}{9} - \dfrac{x^2}{9} = 1$

19.

$\dfrac{x^2}{25} - \dfrac{y^2}{36} = 1$

21. hyperbola

$x^2 - y^2 = 16$

23. ellipse

$4x^2 + y^2 = 16$

25. circle

$y^2 = 36 - x^2$

27. hyperbola

$9x^2 = 144 + 16y^2$

29. hyperbola

$y^2 = 4 + x^2$

31.

$y = \sqrt{\dfrac{x + 4}{2}}$

33. The two thumbtacks act as foci, and the length of the string is constant, satisfying the requirements of the definition of an ellipse.

35. (a) $\sqrt{43} + \sqrt{5013} \approx 77.4$ million miles
(b) $\sqrt{5013} - \sqrt{43} \approx 64.2$ million miles

SECTION 11.5 (page 729)

1. Substitute $x - 1$ for y in the first equation. Then solve for x. Find the corresponding y-values by substituting back into $y = x - 1$.   **3.** No, it is not possible. The maximum number of points of intersection is two.

5. $\left\{(0, 0), \left(\dfrac{1}{2}, \dfrac{1}{2}\right)\right\}$ **7.** $\{(-6, 9), (-1, 4)\}$

9. $\left\{\left(-\frac{1}{5}, \frac{7}{5}\right), (1, -1)\right\}$ **11.** $\left\{(-2, -2), \left(-\frac{4}{3}, -3\right)\right\}$

13. $\{(-3, 1), (1, -3)\}$ **15.** $\left\{\left(-\frac{3}{2}, -\frac{9}{4}\right), (-2, 0)\right\}$

17. $\{(-\sqrt{3}, 0), (\sqrt{3}, 0), (-\sqrt{5}, 2), (\sqrt{5}, 2)\}$
19. $\{(-2, 0), (2, 0)\}$
21. $\{(i\sqrt{2}, -3i\sqrt{2}), (-i\sqrt{2}, 3i\sqrt{2}), (-\sqrt{6}, -\sqrt{6}),$ $(\sqrt{6}, \sqrt{6})\}$
23. $\{(-2i\sqrt{2}, -2\sqrt{3}), (-2i\sqrt{2}, 2\sqrt{3}), (2i\sqrt{2}, -2\sqrt{3}),$ $(2i\sqrt{2}, 2\sqrt{3})\}$
25. $\{(-\sqrt{5}, -\sqrt{5}), (\sqrt{5}, \sqrt{5})\}$
27. $\{(i, 2i), (-i, -2i), (2, -1), (-2, 1)\}$ **29.** $(-1, 2)$
31. $(-4, 8)$ **33.** length: 12 feet; width: 7 feet
35. $20; 800 calculators

Section 11.6 (page 737)

1. (a) B **(b)** D **(c)** A **(d)** C
3.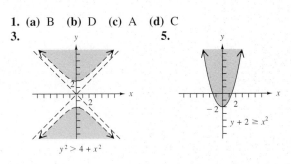

$y^2 > 4 + x^2$

5.

$y + 2 \geq x^2$

7.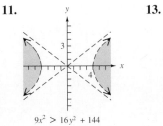

$2y^2 \geq 8 - x^2$

9.

$y \leq x^2 + 4x + 2$

11.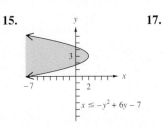

$9x^2 > 16y^2 + 144$

13.

$x^2 - 4 \geq -4y^2$

15.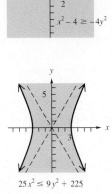

$x \leq -y^2 + 6y - 7$

17.

$25x^2 \leq 9y^2 + 225$

19. (c)

21.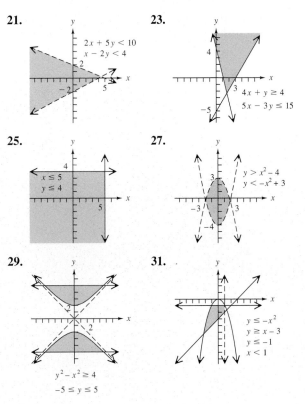

$2x + 5y < 10$
$x - 2y < 4$

23.

$4x + y \geq 4$
$5x - 3y \leq 15$

25.

$x \leq 5$
$y \leq 4$

27.

$y > x^2 - 4$
$y < -x^2 + 3$

29.

$y^2 - x^2 \geq 4$
$-5 \leq y \leq 5$

31.

$y \leq -x^2$
$y \geq x - 3$
$y \leq -1$
$x < 1$

33. A   **35.** B

Chapter 11 Review Exercises (page 747)

1. $(0, -2)$ **2.** $(0, 6)$ **3.** $(-2, 0)$ **4.** $(1, 0)$
5. $(3, 7)$ **6.** $(2, 1)$ **7.** $(3, -2)$ **8.** $(-4, 3)$
9. domain: $(-\infty, \infty)$; range: $[-3, \infty)$

$f(x) = 4x^2 + 4x - 2$

10. domain: $(-\infty, \infty)$; range: $(-\infty, 3]$

$f(x) = -2x^2 + 8x - 5$

11.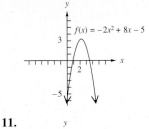

$x = -\frac{1}{2}y^2 + 6y - 14$

12.

$x = 2y^2 + 8y + 3$

13. If the discriminant $b^2 - 4ac$ is positive, there are two x-intercepts. If it is zero, there is only one x-intercept (the vertex). If it is negative, there are no x-intercepts.
14. (a)   **15.** 1986; $.1 billion   **16.** 5 seconds; 400 feet   **17.** length: 50 meters; width: 50 meters
18. 5 and 5

19.   **20.**

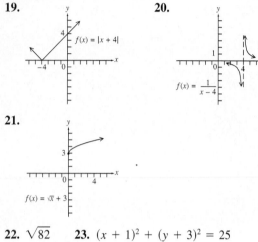

$f(x) = |x + 4|$

$f(x) = \dfrac{1}{x-4}$

21.

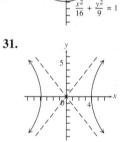

$f(x) = \sqrt{x} + 3$

22. $\sqrt{82}$ **23.** $(x + 1)^2 + (y + 3)^2 = 25$
24. $(x - 4)^2 + (y - 2)^2 = 36$
25. $(-3, 2), r = 4$ **26.** $(4, 1), r = 2$
27. $(-1, -5), r = 3$ **28.** $(3, -2), r = 5$
29.   **30.**

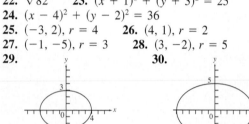

$\dfrac{x^2}{16} + \dfrac{y^2}{9} = 1$

$\dfrac{x^2}{49} + \dfrac{y^2}{25} = 1$

31.   **32.**

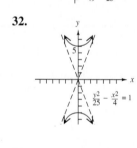

$\dfrac{x^2}{16} - \dfrac{y^2}{25} = 1$

$\dfrac{y^2}{25} - \dfrac{x^2}{4} = 1$

33.   **34.**

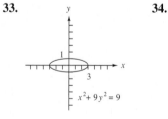

$x^2 + 9y^2 = 9$

$f(x) = -\sqrt{16 - x^2}$

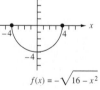

35. circle   **36.** parabola   **37.** hyperbola
38. ellipse   **39.** parabola   **40.** hyperbola
41. $\{(6, -9), (-2, -5)\}$ **42.** $\{(1, 2), (-5, 14)\}$
43. $\{(4, 2), (-1, -3)\}$ **44.** $\{(-2, -4), (8, 1)\}$
45. $\{(-\sqrt{2}, 2), (-\sqrt{2}, -2), (\sqrt{2}, -2), (\sqrt{2}, 2)\}$
46. $\{(-\sqrt{6}, -\sqrt{3}), (-\sqrt{6}, \sqrt{3}), (\sqrt{6}, -\sqrt{3}), (\sqrt{6}, \sqrt{3})\}$
47. 0, 1, or 2   **48.** 0, 1, 2, 3, or 4

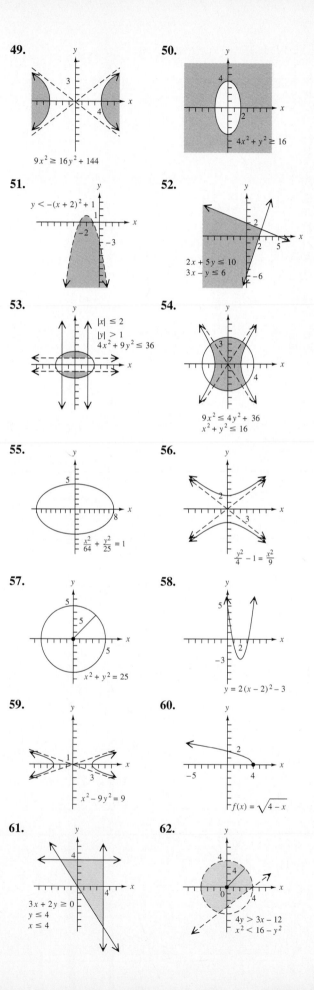

49.

$9x^2 \geq 16y^2 + 144$

50.

$4x^2 + y^2 \geq 16$

51.

$y < -(x + 2)^2 + 1$

52.

$2x + 5y \leq 10$
$3x - y \leq 6$

53.

$|x| \leq 2$
$|y| > 1$
$4x^2 + 9y^2 \leq 36$

54.

$9x^2 \leq 4y^2 + 36$
$x^2 + y^2 \leq 16$

55.

$\dfrac{x^2}{64} + \dfrac{y^2}{25} = 1$

56.

$\dfrac{y^2}{4} - 1 = \dfrac{x^2}{9}$

57.

$x^2 + y^2 = 25$

58.

$y = 2(x - 2)^2 - 3$

59.

$x^2 - 9y^2 = 9$

60.

$f(x) = \sqrt{4 - x}$

61.

$3x + 2y \geq 0$
$y \leq 4$
$x \leq 4$

62.

$4y > 3x - 12$
$x^2 < 16 - y^2$

CHAPTER 11 TEST (page 753)

1. $(0, -2)$

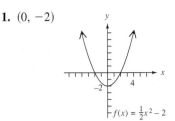

$f(x) = \frac{1}{2}x^2 - 2$

2. vertex: $(2, 3)$; domain: $(-\infty, \infty)$; range: $(-\infty, 3]$

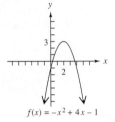

$f(x) = -x^2 + 4x - 1$

3. (a) 31,600 (b) 1984; 36,000   **4.** B

5.

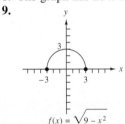

$f(x) = |x - 3| + 4$

6. center: $(2, -3)$; radius: 4

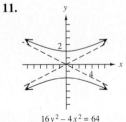

$(x - 2)^2 + (y + 3)^2 = 16$

7. center: $(-4, 1)$; radius: 5
8. The graph has no x-intercepts.

9.

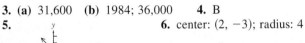

$f(x) = \sqrt{9 - x^2}$

10.

$4x^2 + 9y^2 = 36$

11.

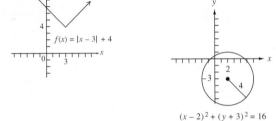

$16y^2 - 4x^2 = 64$

12.

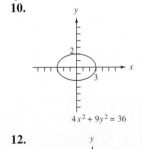

$x = -(y - 2)^2 + 2$

13. ellipse   **14.** hyperbola   **15.** parabola

16. $\left\{ \left(-\frac{1}{2}, -10 \right), (5, 1) \right\}$

17. $\left\{ (-2, -2), \left(\frac{14}{5}, -\frac{2}{5} \right) \right\}$

18. $\{ (-\sqrt{22}, -\sqrt{3}), (-\sqrt{22}, \sqrt{3}), (\sqrt{22}, -\sqrt{3}),$
$(\sqrt{22}, \sqrt{3}) \}$

19.

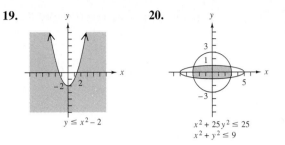

$y \le x^2 - 2$

20.

$x^2 + 25y^2 \le 25$
$x^2 + y^2 \le 9$

CUMULATIVE REVIEW 1–11 (page 757)

1. -4 **2.** $\left\{ \frac{2}{3} \right\}$ **3.** $\left(-\infty, \frac{3}{5} \right]$
4. $\{-4, 4\}$ **5.** $(-\infty, -5) \cup (10, \infty)$
6. $\frac{2}{3}$ **7.** $3x + 2y = -13$ **8.** $25y^2 - 30y + 9$
9. $12r^2 + 40r - 7$
10. $4x^3 - 4x^2 + 3x + 5 + \dfrac{3}{2x + 1}$
11. $(3x + 2)(4x - 5)$ **12.** $(2y^2 - 1)(y^2 + 3)$
13. $(z^2 + 1)(z + 1)(z - 1)$
14. $(10x^2 + 9z)(10x^2 - 9z)$ **15.** $\dfrac{40}{9}$
16. $\dfrac{y - 1}{y(y - 3)}$ **17.** $\dfrac{3c + 5}{(c + 5)(c + 3)}$ **18.** $\dfrac{1}{p}$
19. $\frac{6}{5}$ or $1\frac{1}{5}$ hours   **20.** $\dfrac{3}{4}$ **21.** $\dfrac{a^5}{4}$ **22.** $2\sqrt[3]{2}$
23. $\dfrac{3\sqrt{10}}{2}$ **24.** $\dfrac{7}{5} + \dfrac{11}{5}i$ **25.** $\emptyset$ **26.** $\left\{ \dfrac{1}{5}, -\dfrac{3}{2} \right\}$
27. $\left\{ \dfrac{1 + 2\sqrt{2}}{4}, \dfrac{1 - 2\sqrt{2}}{4} \right\}$
28. $\left\{ \dfrac{3 + \sqrt{33}}{6}, \dfrac{3 - \sqrt{33}}{6} \right\}$
29. $\left\{ -\dfrac{\sqrt{6}}{2}, \dfrac{\sqrt{6}}{2}, -\sqrt{7}, \sqrt{7} \right\}$ **30.** $v = \dfrac{\pm\sqrt{rFkw}}{kw}$
31.

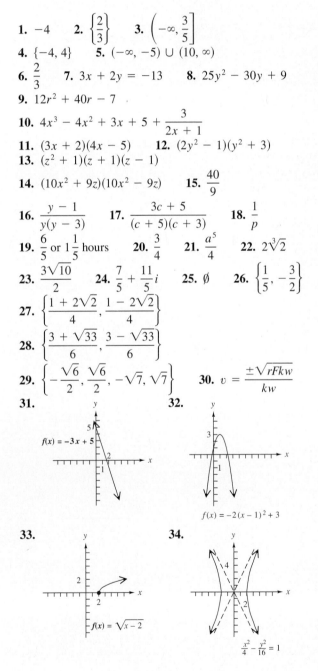

$f(x) = -3x + 5$

32.

$f(x) = -2(x - 1)^2 + 3$

33.

$f(x) = \sqrt{x - 2}$

34.

$\dfrac{x^2}{4} - \dfrac{y^2}{16} = 1$

35.

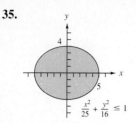

$$\frac{x^2}{25} + \frac{y^2}{16} \le 1$$

36. $\{(3, -3)\}$ **37.** $\{(4, 1, -2)\}$

38. $\left\{(-1, 5), \left(\frac{5}{2}, -2\right)\right\}$

39. 40 miles per hour   **40.** 15 fives and 7 twenties

CHAPTER 12
SECTION 12.1 (page 767)

1. no   **3.** (a)   **5.** Two or more siblings might be in the class.   **7.** $\{(6, 3), (10, 2), (12, 5)\}$ **9.** not a one-to-one function   **11.** $f^{-1}(x) = \dfrac{x - 4}{2}$

13. $g^{-1}(x) = x^2 + 3, x \ge 0$ **15.** not one-to-one
17. $f^{-1}(x) = \sqrt[3]{x + 4}$ **19.** Yes, by the horizontal line test, $f(x) = mx + b$ is one-to-one and so it has an inverse.
21. (a) 8 (b) 3   **23.** (a) 1 (b) 0
25. (a) one-to-one (b)

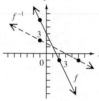

27. (a) not one-to-one   **29.** (a) one-to-one
(b)

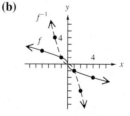

31.

33.

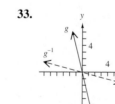

35. 0, 1, 2   **37.** $-3, -2, -1, 6$

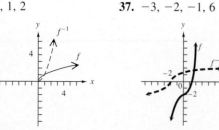

39. inverses; $y_2 = \sqrt[3]{x - 5}$
41. It is not a function.

SECTION 12.2 (page 777)

1. rises; falls   **3.** does not
5.   **7.**

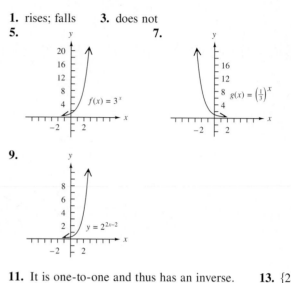

9.

11. It is one-to-one and thus has an inverse.   **13.** $\{2\}$

15. $\left\{\dfrac{3}{2}\right\}$ **17.** $\{7\}$ **19.** $\{-3\}$ **21.** $\{-1\}$

23. $\{-3\}$ **25.** (a) 100 grams (b) 31.25 grams
(c) .30 gram (to the nearest hundredth)   **27.** about $360,000   **29.** In $1976 + 15 = 1991$, the average salary was about $800,000.

SECTION 12.3 (page 783)

1. (a) 2 (b) 4 (c) -1 (d) -2 (e) $\frac{1}{2}$ (f) 0
3. $\log_4 1024 = 5$ **5.** $\log_{1/2} 8 = -3$
7. $\log_{10} .001 = -3$ **9.** $4^3 = 64$
11. $10^{-4} = \dfrac{1}{10,000}$ **13.** $6^0 = 1$
15. Since the radical $\sqrt{9} = 9^{1/2} = 3$, the exponent to which 9 must be raised is $1/2$.

17. $\left\{\dfrac{1}{3}\right\}$ **19.** $\{81\}$ **21.** $\left\{\dfrac{1}{5}\right\}$ **23.** $\{1\}$

25. $\{x \mid x > 0, x \ne 1\}$ **27.** $\{5\}$ **29.** $\left\{\dfrac{5}{3}\right\}$ **31.** $\{4\}$

33. $\left\{\dfrac{3}{2}\right\}$ **35.**

$y = \log_3 x$

37.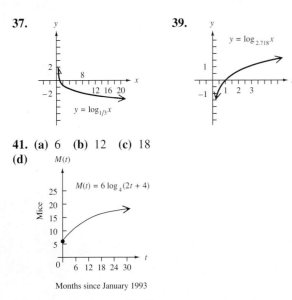

39.

$y = \log_{2.718} x$

41. (a) 6   **(b)** 12   **(c)** 18
(d)

$M(t) = 6\log_4(2t + 4)$

Months since January 1993

43. 8   **45.** 24   **47.** Since every real number power of 1 equals 1, $f(x) = \log_1 x$ is not a logarithmic function.
49. $(0, \infty); (-\infty, \infty)$ **51.** about 4 times more powerful

SECTION 12.4 (page 791)

1. sum   **3.** k **5.** $\log_7 4 - \log_7 5$
7. $\dfrac{1}{4}\log_2 8$ or $\dfrac{3}{4}$ **9.** $\log_4 3 + \dfrac{1}{2}\log_4 x - \log_4 y$
11. $\dfrac{1}{3}\log_3 4 - 2\log_3 x - \log_3 y$
13. $\dfrac{1}{2}\log_3 x + \dfrac{1}{2}\log_3 y - \dfrac{1}{2}\log_3 5$
15. $\dfrac{1}{3}\log_2 x + \dfrac{1}{5}\log_2 y - 2\log_2 r$
17. The distributive property tells us that the *product* $a(x + y)$ equals the sum $ax + ay$. In the notation, $\log_a(x + y)$, the parentheses do not indicate multiplication. They indicate that $x + y$ is the result of raising a to some power.   **19.** $\log_b xy$ **21.** $\log_a \dfrac{m^3}{n}$ **23.** $\log_a \dfrac{rt^3}{s}$
25. $\log_a \dfrac{125}{81}$ **27.** $\log_{10}(x^2 - 9)$ **29.** $\log_p \dfrac{x^3 y^{1/2}}{z^{3/2} a^3}$
31. For the power rule $\log_b x^r = r\log_b x$ to be true, x must be in the domain of $g(x) = \log_b x$.
33. 4   **34.** It is the exponent to which 3 must be raised in order to obtain 81.   **35.** 81   **36.** It is the exponent to which 2 must be raised in order to obtain 19.
37. 19   **38.** m

SECTION 12.5 (page 797)

1. 10   **3.** 0; 1; 0; 1   **5.** 19.2   **7.** 1.6335
9. 2.5164   **11.** -1.4868 **13.** 9.6776
15. 2.0592   **17.** -2.8896 **19.** 5.9613
21. 4.1506   **23.** 2.3026   **25.** 1.4315   **27.** 3.3030
29. $-.0947$ **31.** Answers will vary. Suppose the name is Paul Bunyan, with $m = 4$ and $n = 6$.   **(a)** $\log_4 6$ is the exponent to which 4 must be raised in order to obtain 6.
(b) 1.29248125   **(c)** 6 (the value of n)   **33.** 11.6

35. 4.3   **37.** 4.0×10^{-8} **39.** 4.0×10^{-6}
41. (a) 1.62 grams   **(b)** 1.18 grams   **(c)** .69 gram
(d) 2.00 grams   **43.** 2228 million (or approximately 2.2 billion)   **45.** 1354.4 thousand (or 1,354,400)
47. In 1989 (when $x = 10$), the number of births was about 730,000.   **49.** about 800 years
51. about 11,500 years

SECTION 12.6 (page 807)

1. $\{3\}$ **3.** $\{6\}$; $\dfrac{\log 64}{\log 2} = 6$ **5.** $\{.827\}$ **7.** $\{.833\}$
9. $\{2.269\}$ **11.** $\{566.866\}$ **13.** $\{-18.892\}$
15. $\{43.301\}$ **17.** Natural logarithms are a better choice because $\ln e^x = x$.   **19.** $\left\{\dfrac{2}{3}\right\}$
21. $\{-1 + \sqrt[3]{49}\}$ **23.** 2 cannot be a solution because $\log(2 - 3) = \log(-1)$, which is not acceptable.
25. $\left\{\dfrac{1}{3}\right\}$ **27.** $\{2\}$ **29.** $\varnothing$ **31.** $\{8\}$
33. $\left\{\dfrac{4}{3}\right\}$ **35.** $\{8\}$ **37.** $\$2539.47$ **39.** about 180 grams   **41.** $\$4934.71$ **43.** 15.4 years
45. The domain of f is $(-\infty, 0) \cup (0, \infty)$, while the domain of g is $(0, \infty)$.

CHAPTER 12 REVIEW EXERCISES (page 815)

1. not one-to-one   **2.** one-to-one
3. $f^{-1}(x) = \dfrac{x - 7}{-3}$ or $\dfrac{7 - x}{3}$ **4.** $f^{-1}(x) = \dfrac{x^3 + 4}{6}$
5. not one-to-one   **6.** $f^{-1}(x) = x$
7.   **8.**

9.   **10.**

$f(x) = 3^x$ $f(x) = \left(\dfrac{1}{3}\right)^x$

11.   **12.**

$y = 3^{x+1}$ $y = 2^{2x+3}$

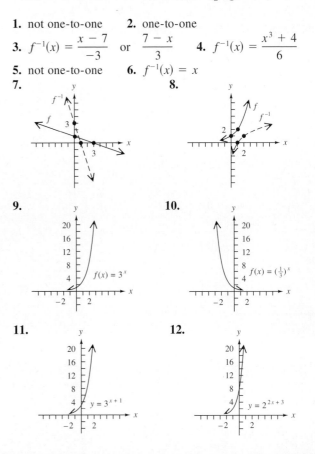

13. $\{4\}$ **14.** $\left\{\dfrac{3}{7}\right\}$ **15.** $\{0\}$ **16.** $(0, 1)$

17. It says that $a^0 = 1$, which agrees with the definition.

18.

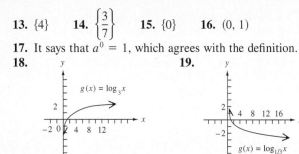

19.

20. (a) $5^4 = 625$ (b) $\log_5 .04 = -2$ **21.** (a) $\log_b a$ represents the exponent on b that yields a.   (b) a

22. $\log_4 3 + 2\log_4 x$ **23.** $2\log_2 p + \log_2 r - \dfrac{1}{2}\log_2 z$

24. $\log_b \dfrac{3x}{y^2}$ **25.** $\log_3 \dfrac{x+7}{4x+6}$ **26.** 1.4609

27. $-.5901$ **28.** 4.8613 **29.** 3.3638
30. -1.3587 **31.** 4.8613 **32.** (a) 2.552424846
(b) 1.552424846 (c) 0.552424846 (d) The whole
number parts will vary, but the decimal parts are the same.
33. $.9251$ **34.** 1.7925 **35.** 1.4315 **36.** (a) 500
(b) 1000 (c) 1500 (d) no   **37.** 6.4 **38.** 8.4
39. (a) 11.7 (million)   (b) 5.4 (million)
40. (a) 500 grams   (b) about 409 grams
41. $\{2.042\}$ **42.** $\{4.907\}$ **43.** $\{18.310\}$
44. $\left\{\dfrac{1}{9}\right\}$ **45.** $\{-6 + \sqrt[3]{25}\}$ **46.** $\{2\}$ **47.** $\left\{\dfrac{3}{8}\right\}$
48. $\{4\}$ **49.** $\{1\}$ **50.** about 1994 **51.** almost
14 years   **52.** $\$7112.11$ **53.** Plan A; it would pay
$\$2.92$ more.   **54.** $\{72\}$ **55.** $\{5\}$
56. $\left\{\dfrac{1}{9}\right\}$ **57.** $\left\{\dfrac{4}{3}\right\}$ **58.** $\{3\}$ **59.** $\left\{\dfrac{1}{8}\right\}$
60. $\left\{\dfrac{11}{3}\right\}$

CHAPTER 12 TEST (page 819)

1. (a) not one-to-one   (b) one-to-one
2. $f^{-1}(x) = x^3 - 7$
3.

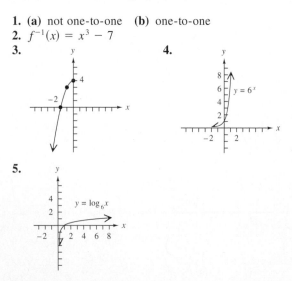

4.

5.

6. Interchange the x- and y-values of the ordered pairs,
because they are inverses of each other.
7. $\{-4\}$ **8.** $\left\{-\dfrac{13}{3}\right\}$

9. (a) $\$5000$ (b) $\$2973$ (c) $\$1768$
(d)

10. $\log_4 .0625 = -2$ **11.** $7^2 = 49$ **12.** $\{32\}$
13. $\left\{\dfrac{1}{2}\right\}$ **14.** $\{2\}$ **15.** 5; 2; 5th; 32
16. $2\log_3 x + \log_3 y$ **17.** $\dfrac{1}{2}\log_5 x - \log_5 y - \log_5 z$
18. $\log_b \dfrac{s^3}{t}$ **19.** $\log_b \dfrac{r^{1/4}s^2}{t^{2/3}}$ **20.** 1.3284
21. $-.8440$ **22.** (a) 2 and 3 (b) 2.1245
23. $\{3.9656\}$ **24.** $\{3\}$ **25.** (a) $\$1051$ (b) after
7 years (rounded up)

CUMULATIVE REVIEW 1–12 (page 823)

1. $-2, 0, 6, \dfrac{30}{3}$ (or 10)   **2.** $-\dfrac{9}{4}, -2, 0, .6, 6, \dfrac{30}{3}$ (or 10)
3. $-\sqrt{2}, \sqrt{11}$ **4.** all except $\sqrt{-8}$ **5.** 16
6. -27 **7.** -39 **8.** $\left\{-\dfrac{2}{3}\right\}$ **9.** $[1, \infty)$
10. $\{-2, 7\}$ **11.** $\left\{-\dfrac{16}{3}, \dfrac{16}{3}\right\}$ **12.** $\left[\dfrac{7}{3}, 3\right]$
13. $(-\infty, -3) \cup (2, \infty)$ **14.** $6p^2 + 7p - 3$
15. $16k^2 - 24k + 9$ **16.** $-5m^3 + 2m^2 - 7m + 4$
17. $2t^3 + 5t^2 - 3t + 4$
18. $x(8 + x^2)$ **19.** $(3y - 2)(8y + 3)$
20. $z(5z + 1)(z - 4)$ **21.** $(4a + 5b^2)(4a - 5b^2)$
22. $(x + y)(3x + 2y)$ **23.** $(4r + 7q)^2$
24. $-\dfrac{1875p^{13}}{8}$ **25.** $\dfrac{x+5}{x+4}$
26. $\dfrac{-3k - 19}{(k+3)(k-2)}$ **27.** $\dfrac{22 - p}{p(p-4)(p+2)}$
28. 6 pounds
29. $\dfrac{16}{25}$ **30.** $\dfrac{1}{6^5}$ **31.** $-27\sqrt{2}$
32. 41 **33.** $\left\{-\dfrac{1}{2}, \dfrac{2}{5}\right\}$ **34.** $(-\infty, -4) \cup (2, \infty)$
35. (a) 233.4 million   (b) 238.5 million
(c) 247 million   (d) 2030
36. $-5{,}587{,}500$ dollars per year   **37.** $3x - 4y = 19$
38.

39.

40.

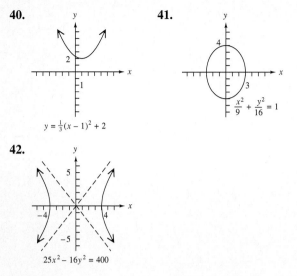

$y = \frac{1}{3}(x-1)^2 + 2$

41.

$\frac{x^2}{9} + \frac{y^2}{16} = 1$

46. -9

47.

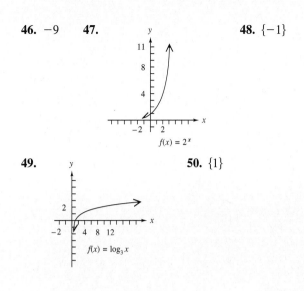

$f(x) = 2^x$

48. $\{-1\}$

42.

$25x^2 - 16y^2 = 400$

49.

$f(x) = \log_3 x$

50. $\{1\}$

43. (a) those in Exercises 38 and 40   **(b)** only the one in Exercise 38   **44.** $\{(4, 2)\}$ **45.** $\{(1, -1, 4)\}$

Index

A

Absolute value, 21
  evaluating, 21
Absolute value equations, 169
  solving, 170
Absolute value function, 703
Absolute value inequalities, 169, 232
  solving, 171
Addition
  associative property of, 59
  commutative property of, 59
  of complex numbers, 589
  identity element for, 61
  identity property of, 61
  inverse for, 62
  inverse property of, 62
  of like terms, 371
  of multivariable polynomials, 376
  with the number line, 25
  of polynomials, 373
  properties of, 59
  of radicals, 559
  of rational expressions, 491
  of real numbers, 25
  of signed numbers, 25
Addition method for linear systems, 283
  steps to solve by, 286
Addition property
  of equality, 90
  of inequality, 151
Additive identity element, 61
Additive inverse, 62
Algebraic expressions, 9
  distinguishing from equations, 12
  evaluating, 9
  from word phrases, 10
  simplifying, 71
Angles
  complementary, 112
  measure of, 112
  right, 112
  straight, 112
  supplementary, 112
Area, 119, 451
Array of signs for a determinant, 325
Associative property
  of addition, 59
  of multiplication, 59
Asymptotes, 703
  of a hyperbola, 716
Average rate of change, 208

Axis
  of a parabola, 682, 686, 696
  x-, 194
  y-, 194

B

Bar graph, 193
Base of an exponential expression, 1, 351
Binomials, 372
  cube of, 384
  multiplication of, 387
  square of, 389
Boundary line, 229
Braces, 90
Brackets, 3

C

Cartesian coordinate system, 194
Celsius to Fahrenheit equation, 228
Center of a circle, 706
Change-of-base rule, 796
Circle, 681, 706
  center of, 706
  equation of, 707
  graph of, 706
  radius of, 706
Coefficient, 71, 371
Collinear, 214
Combined variation, 257
Combining terms, 72, 371
Common factor, 413
Common logarithms, 793
  evaluating, 793
Commutative property
  of addition, 59
  of multiplication, 59
Complementary angles, 112
Completing the square, 617, 691
Complex form of a rational expression, 499
Complex fractions, 499
Complex numbers, 588
  addition of, 589
  conjugate of, 590
  division of, 590
  imaginary part of, 588
  multiplication of, 590
  real part of, 588
  subtraction of, 589

Components of an ordered pair, 194
Compound inequalities, 155, 161
Compound interest, 804
Conic sections, 681
    summary of, 718
Conjugates
    of a complex number, 590
    of an irrational number, 571
Consecutive integers, 118, 452
    even, 118, 452
    odd, 452
Consistent system, 277
Constant of variation, 253
Continuous compounding, 810
Coordinate system
    Cartesian, 194
    rectangular, 194
Coordinates of a point in a plane, 194
Cramer's rule, 331
Cross products, 130
Cube
    of a binomial, 384
    perfect, 546, 554
    volume of, 558
Cube root, 545

D

Decimal numbers, linear equations with, 97, 104
Degree, 112
    of a multivariable polynomial, 375
    of a polynomial, 372
    of a term, 372
Denominator
    least common, 485
    rationalizing, 563
Dependent equations, 277
Dependent variable, 240
Descartes, René, 194
Descending powers, 372
Determinant method for systems, 331
Determinants, 323
    array of signs for, 325
    Cramer's rule for, 331
    evaluating, 323, 324
    expansion of, 324
    minor of, 324
Difference, 11, 33
Difference of two squares, factoring, 435
Direct variation, 253, 520
    as a power, 254
Discriminant, 628, 694
Distance formula, 707
Distance, rate, and time problems, 516
Distance, rate, and time relationship, 313, 515,
    638
Distributive property, 62, 91

Divisibility tests for numbers, 58
Division
    of complex numbers, 590
    definition of, 49
    of polynomials, 393, 397
    of radicals, 553
    of rational expressions, 480
    of real numbers, 49
    of signed numbers, 49
    word phrases for, 52
    by zero, 49
Division property for logarithms, 788
Domain of a function, 241
Double negative rule, 20

E

e, 794
Elimination method for systems, 302
Ellipse, 681, 713
    equation of, 713
    foci of, 713
    graph of, 713
    intercepts of, 713
Empty set, 92
Equality
    addition property of, 90
    multiplication property of, 95
    squaring property of, 577
Equation(s), 11. See also System of equations
    absolute value, 169
    Celsius to Fahrenheit, 228
    of a circle, 707
    dependent, 277
    distinguishing from expressions, 12
    of an ellipse, 713
    equivalent, 89
    exponential, 773, 801
    extraneous solution of, 578
    Fahrenheit to Celsius, 228
    from word statements, 12
    of a horizontal line, 197, 216
    of a hyperbola, 716
    identity, 92
    independent, 277
    of an inverse function, 764
    linear in one variable, 89
    linear in two variables, 194, 215
    logarithmic, 780, 802
    with no solution, 92
    nonlinear, 725
    of a parabola, 682, 712
    quadratic, 443, 611
    quadratic in form, 637
    with radicals, 577
    with rational expressions, 505
    solution of, 11

square root property of, 611
of a vertical line, 198, 216
Equivalent equations, 89
Equivalent inequalities, 151
Evaluating
algebraic expressions, 9
determinants, 323, 324
logarithms, 793
Even consecutive integers, 118, 452
Expansion of a determinant, 324
Exponential equation, 773, 801
properties of, 801
solving, 774, 801
Exponential expressions, 1, 351
base of, 1, 351
division of, 362
evaluating, 1, 351
multiplication of, 352
Exponential function, 771
graph of, 771
Exponents, 1, 351
applications of, 367
integer, 359
negative, 360
power rule for, 353
product rule for, 352
quotient rule for, 362
rules for, 355, 363
and scientific notation, 367
zero, 359
Expressions
algebraic, 9
exponential, 1, 351
mathematical, 6
radical, 542
rational, 471
simplifying, 71
Extraneous solution, 578

First-degree equation. *See* Linear Equation
Foci of an ellipse, 713
FOIL, 387, 421
inner product of, 387
outer product of, 387
Formulas, 119
distance, 707
Pythagorean, 453, 544, 647
quadratic, 625
Fourth root, 545
Fractional inequalities, 657
solving, 657
Fractions
complex, 499
inequalities with, 657
linear equations with, 97, 103
reciprocals of, 49, 480
Functions, 239
absolute value, 703
domain of, 241
exponential, 771
graph of, 244
inverse of, 762, 764
linear, 245
logarithmic, 781
notation for, 244
one-to-one, 761
quadratic, 682
range of, 241
rational, 255
reciprocal, 703
square root, 704
squaring, 703
translation of, 704
vertical line test for, 243
Fundamental property of rational expressions, 473
Fundamental rectangle of a hyperbola, 717
$f(x)$ notation, 244

F

Factor
common, 413
greatest common, 413, 428
Factored form of a number, 413
Factoring, 64, 413
difference of two squares, 435
by grouping, 416
perfect square trinomials, 436
quadratic equations, 443, 611
trinomials, 421, 427
Factoring method of solving quadratic equations,
443, 611
Factoring trinomials, 421
by grouping, 427
by trial and error, 429
Fahrenheit to Celsius equation, 228

G

Gains or loses, interpreting, 28
Geometry applications, 119
Grade, 203
Graph
of a circle, 706
of an ellipse, 713
of an equation, 195
of an exponential function, 771
of a function, 244
of a horizontal line, 197
of a hyperbola, 715
of inverses, 766
of a linear equation, 195
of a linear function, 245
of a linear inequality, 229
of a logarithmic function, 781

Graph (*continued*)
  of a number, 18
  of a parabola, 682, 693
  of rational numbers, 18
  of a system, 735
  of a vertical line, 198
Graphical solution of a system of linear equations,
    276
Greater than, 4
Greater than or equal to, 4
Greatest common factor, 413, 428
  factoring out, 415
  of numbers, 413
  steps to find, 415
  of variable terms, 414
Grouping, factoring by, 416
  factoring trinomials by, 427
Grouping symbols, 3
Growth and decay, 775

H

Half-life, 809
Horizontal line, 197
  equation of, 197, 216
  graph of, 197
  slope of, 206
Horizontal line test, 763
Horizontal parabola, 695
  axis of, 696
  vertex of, 696
Horizontal shift of a parabola, 684
Hyperbola, 681, 715
  asymptotes of, 716
  equations of, 716
  fundamental rectangle of, 717
  graph of, 715
  intercepts of, 715
Hypotenuse of a right triangle, 453, 647

I

i, 587
  powers of, 591
Identity element
  for addition, 61
  for multiplication, 61
Identity equation, 92
Identity properties, 61
Imaginary part of a complex number, 588
Imaginary unit, 587
Inconsistent system, 277
Independent equations, 277
Independent variable, 240
Index of a radical, 545
Inequalities, 4, 151
  absolute value, 169, 232
  addition property of, 151

compound, 155, 161
  equivalent, 151
  with fractions, 657
  linear in one variable, 151
  linear in two variables, 229
  multiplication property of, 152
  nonlinear, 655, 733
  nonstrict, 155
  polynomial, 656
  quadratic, 655
  second-degree, 733
  strict, 155
  symbols of, 4, 6
  system of, 734
  three-part, 155
Integers, 17
  consecutive, 118, 452
  consecutive even, 118, 452
  consecutive odd, 452
  as exponents, 359
Intercepts
  of an ellipse, 713
  of a hyperbola, 715
  x-, 196
  y-, 196
Interest
  compound, 804
  simple, 318
Intersection
  of linear inequalities, 231
  of sets, 161
Inverse of a function, 762, 764
  equation of, 764
  graph of, 766
  notation for, 762
Inverse properties, 62
Inverse variation, 255, 521
  as a power, 255
Irrational numbers, 19, 543
  conjugate of, 571
  decimal approximation of, 543

J

Joint variation, 256

L

Least common denominator (LCD), 485
Legs of a right triangle, 453, 647
Less than, 4
  definition of, 20
Less than or equal to, 4
Like radicals, 559
Like terms, 72, 371
  combining, 72, 371
Line graph, 193

Linear equations in one variable, 89
  applications of, 109, 119
  with decimals, 97, 104
  with fractions, 97, 103
  geometric applications of, 119
  solving, 90, 101
Linear equations in two variables, 194, 215
  graph of, 195
  standard form of, 196, 218
  summary of forms, 220
  systems of, 275
Linear function, 245
  graph of, 245
Linear inequalities in one variable, 151
  solving, 151
  steps for solving, 153
Linear inequalities in two variables, 229
  graph of, 229
  intersection of, 231
  union of, 232
Linear system of equations. *See* System of linear
  equations
Logarithm, 779
  change-of-base rule for, 796
  common, 793
  definition of, 779
  division property for, 788
  evaluating, 793
  exponential form of, 779
  multiplication property for, 787
  natural, 794
  power property for, 788
  properties of, 780
Logarithmic equation, 780, 802
  properties of, 801
  solving, 780, 802
Logarithmic function, 781
  graph of, 781
Losses or gains, interpreting, 28
Lowest terms of a rational expression, 473

M

Mapping, 240
Mathematical expressions, 6
Mathematical sentences, 6
Matrices, 323
  determinant of, 323
  square, 323
Measure of angles, 112
Minor of a determinant, 324
Mixture problems, 312
Monomial, 372
Motion problems, 313, 516, 638
Multiplication
  associative property of, 59
  of binomials, 387

  commutative property of, 59
  of complex numbers, 590
  of exponential expressions, 352
  FOIL method of, 387
  identity element for, 61
  identity property of, 61
  inverse for, 62
  inverse property of, 62
  of polynomials, 381
  properties of, 59
  of radicals, 551
  of rational expressions, 479
  of real numbers, 41
  of signed numbers, 41
  of sum and difference of two terms, 390
  of two binomials, 381
  word phrases for, 43
Multiplication property
  of equality, 95
  of inequality, 152
  for logarithms, 787
Multiplicative identity element, 61
Multiplicative inverse, 62
Multivariable polynomial, 375
  addition of, 376
  degree of, 375
  subtraction of, 376

N

Natural logarithms, 794
  applications of, 795
Natural numbers, 17
Negative exponents, 360
Negative numbers, 17
  applications of, 18
Negative of a number, 17
Negative slope, 206
Nonlinear equation, 725
Nonlinear inequality, 655, 733
Nonlinear system of equations, 725
  solving, 725
Nonstrict inequalities, 155
Not equal, 4
Notation, functional, 244
Null set, 92
Number(s). *See also* Real numbers
  absolute value of, 21
  complex, 588
  divisibility tests for, 58
  factors of, 413
  graph of, 18
  greatest common factor of, 413
  imaginary, 588
  integer, 17
  irrational, 19, 543
  natural, 17

Number(s) (*continued*)
negative, 17
negative of, 17
opposite of, 17, 21
ordering of, 20
perfect square, 541
positive, 17
rational, 18, 543
real, 17, 19
square of, 541
square roots of, 541
whole, 17
Number line, 17
addition with, 25
subtraction with, 33
Numerical coefficients, 71, 371

O

Odd consecutive integers, 452
Ohm's law, 596
One-to-one function, 761
horizontal line test for, 763
Operations
order of, 2
set, 161
Operations with signed numbers, summary of, 51
Opposite of a number, 17, 21
Order
of operations, 2
of a radical, 545
Ordered pairs, 194
applications of, 221
components of, 194
Ordered triple, 301
Ordering of real numbers, 20
Origin, 194

P

Pairs, ordered, 194
Parabola, 681, 682
application of, 695
axis of, 682, 686, 696
downward opening, 685
equation of, 682, 712
graph of, 682, 693
horizontal, 695
horizontal shift of, 684
vertex formula for, 692
vertex of, 682, 686, 691, 696
vertical shift of, 683
x-intercepts of, 694
Parallel lines, slope of, 206
Parentheses, 3
Pendulum, period of, 568

Percent, 132
applications of, 132
Perfect cube, 546, 554
Perfect square numbers, 541
Perfect square trinomial, 436
factoring of, 436
Perimeter, 120
Period of a pendulum, 568
Perpendicular lines, slope of, 207
pH, 793
Pie chart, 193
Plane, coordinates of points in, 194
Plotting points, 194
Plus or minus symbol, 612
Point-slope form, 215
Polynomial inequality, 656
Polynomials, 372
addition of, 373
binomial, 372
classifying, 372
degree of, 372
descending powers of, 372
division by a monomial, 393
division of, 393, 397
monomial, 372
multiplication by a monomial, 381
multiplication of, 381
multivariable, 375
numerical coefficients of, 371
prime, 422
square of, 383
steps to factor, 441
subtraction of, 374
terms of, 371
trinomial, 372
Positive numbers, 17
Positive slope, 206
Power property for logarithms, 788
Power rule for exponents, 353
Powers, 1, 351
descending, 372
Powers of *i*, 591
Price per unit, 129
Prime polynomials, 422
Principal roots, 546
Product, 10, 41
of the sum and difference of two terms, 390
Product rule
for exponents, 352
for radicals, 551
Proportions, 130
applications of, 131
cross products of, 130
solving, 131
terms of, 130
Pyramid, volume of, 458
Pythagorean formula, 453, 544, 647
application of, 545

Q

Quadrants, 194
Quadratic equations, 443, 611
  applications of, 451, 647
  completing the square method of solving, 617
  discriminant of, 628, 694
  factoring method of solving, 443, 611
  with imaginary solutions, 613
  quadratic formula method of solving, 625
  square root method of solving, 612
  standard form of, 443
Quadratic formula, 625
Quadratic function, 682
Quadratic inequality, 655, 702
  solving, 655
Quadratic in form equations, 637
Quotient, 49
Quotient rule
  for exponents, 362
  for radicals, 553

R

Radical expression, 542
Radical sign, 542
Radicals, 542
  addition of, 559
  division of, 553
  equations with, 577
  index of, 545
  like, 559
  multiplication of, 551
  order of, 545
  product rule for, 551
  properties of, 554
  quotient rule for, 553
  simplifying, 551, 553, 559, 564, 569
  subtraction of, 559
  unlike, 559
Radicand, 542
Radius of a circle, 706
Range of a function, 241
Rate of change, average, 208
Rate of work, 519
Ratio, 129
Rational expressions, 471
  addition of, 491
  applications of, 515
  complex form of, 499
  division of, 480
  equations with, 505
  evaluating, 472
  fundamental property of, 473
  lowest terms of, 473
  multiplication of, 479
  subtraction of, 493
  with zero denominator, 471

Rational functions, 255
Rational numbers, 18, 543
  graph of, 18
Rationalizing the denominator, 563
Real number(s), 17, 19. *See also* Numbers
  addition of, 25
  chart of, 19
  division of, 49
  multiplication of, 41
  ordering of, 20
  subtraction of, 33
Real part of a complex number, 588
Reciprocal function, 703
Reciprocals of fractions, 49, 480
Rectangular coordinate system, 194
  plotting points in, 194
  quadrants of, 194
Relation, 240
  domain of, 241
  range of, 241
Richter scale, 786
Right angle, 112
Right triangle, 453, 647
  hypotenuse of, 453, 647
  legs of, 453, 647
Root(s)
  cube, 545
  fourth, 545
  principal, 546
  square, 541
Rules for exponents, 355, 363

S

Scientific notation, 367
Second-degree equation. *See* Quadratic equation
Second-degree inequalities, 733
Semicircle, 708
Sentence, mathematical, 6
Set(s)
  empty, 92
  intersection of, 161
  null, 92
  union of, 163
Set operations, 161
Signed numbers
  addition of, 25
  division of, 49
  multiplication of, 41
  subtraction of, 33
  summary of operations with, 51
Simple interest, 318
Simplifying expressions, 71
Simplifying radicals, 551, 553, 559, 564, 569
Slope
  definition of, 203
  of a horizontal line, 206
  of a line, 203

Slope (*continued*)
  negative, 206
  of parallel lines, 206
  of perpendicular lines, 207
  positive, 206
  of a vertical line, 206
Slope-intercept form, 217
Solution of an equation, 11
Solving for a specified variable, 123, 509
Sphere, volume of, 558
Square
  of a binomial, 389
  completing the, 617, 691
  of a number, 541
  of a polynomial, 383
Square brackets, 3
Square matrix, 323
Square root function, 704
Square root method of solving quadratic equations, 612
Square root property of equations, 611
Square roots, 541
  approximating, 543
  symbol for, 541
Squaring function, 703
Squaring property of equality, 577
Standard form
  of a linear equation, 196, 218
  of a quadratic equation, 443
Straight angle, 112
Strict inequalities, 155
Substitution method for linear systems, 293
Subtraction
  of complex numbers, 589
  definition of, 33
  with grouping symbols, 34
  of a multivariable polynomial, 376
  with the number line, 33
  of polynomials, 374
  of radicals, 559
  of rational expressions, 493
  of real numbers, 33
  of signed numbers, 33
  word phrases for, 35
Sum, 10
Supplementary angles, 112
Symbols of inequality, 4, 6
  converting between, 5
System of inequalities, 734
  graph of, 735
System of linear equations, 275
  addition method for solving, 283
  applications of, 311
  consistent, 277
  Cramer's rule method for solving, 331
  determinant method for solving, 331
  elimination method for solving, 302
  graphing method for solving, 276

  inconsistent, 277
  solution of, 275
  substitution method for solving, 293
  with three variables, 301
System of nonlinear equations, 725

T

Table of values, 197
Terms, 71
  combining, 72, 371
  degree of, 372
  like, 72, 371
  numerical coefficient of, 71, 371
  of a polynomial, 371
  of a proportion, 130
  unlike, 72, 372
Test for divisibility, 58
Three-part inequalities, 155
  solving, 155
Translation of a graph of a function, 704
Triangles, right, 453, 647
Trinomials, 372
  factoring of, 421, 427
  perfect square, 436
  prime, 422
Triple, ordered, 301

U

Union
  of linear inequalities, 232
  of sets, 163
Unit cost, 129
Unlike radicals, 559
Unlike terms, 72, 372

V

Variable(s), 9
  dependent, 240
  independent, 240
  solving for specified, 123, 509
Variation, 253, 520
  combined, 257
  constant of, 253
  direct, 253, 520
  inverse, 255, 521
  joint, 256
Vertex formula for a parabola, 692
Vertex of a parabola, 682, 686, 691, 696
Vertical line, 198
  equation of, 198, 216
  graph of, 198
  slope of, 206

Vertical line test for a function, 243
Vertical shift of a parabola, 683
Volume, 126
    of a cube, 558
    of a pyramid, 458
    of a sphere, 558

W

Whole numbers, 17
Word phrases to algebraic expressions, 10
Word statements to equations, 12
Words to symbols conversions, 5
Work problems, 639, 649
Work rate problems, 519
Written expressions to algebraic equivalent, 156

X

x-axis, 194
x-intercept, 196
    of a parabola, 694

Y

y-axis, 194
y-intercept, 196

Z

Zero, division by, 49
Zero denominator in a rational expression, 471
Zero exponents, 359
Zero-factor property, 443

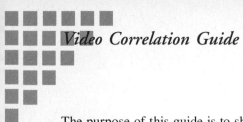

The purpose of this guide is to show those exercises from the text that are used in the Real to Reel videotape series that accompanies *Introductory and Intermediate Algebra.*

Section	Exercises	Section	Exercises
R.1	19, 35, 43, 51, 57, 69	6.4	19, 25, 49
R.2	11, 13, 19, 25, 31, 33	6.5	7, 17, 19, 35
		6.6	7, 21, 37, 45
1.1	41	6.7	15, 21
1.2	63	6.8	7, 13, 27, 31
1.3	13		
1.4	57	7.1	39, 49
1.5	23	7.2	47, 31
1.6	53	7.3	53, 55, 65
1.7	55	7.4	23, 31, 43
1.8	75	7.5	25, 29
1.9	55	7.6	25
2.1	15	8.1	41
2.2	39, 43	8.2	13
2.3	43	8.3	25
2.4	25, 27	8.4	29
2.5	47	8.5	
2.6	27, 31	8.6	23, 37
2.7	17, 43	8.7	
3.1	19, 45	9.1	45
3.2	7	9.2	5, 35
3.3	29, 55	9.3	7, 23
3.4	19, 25, 49	9.4	15, 25
3.5	7, 17, 19, 35	9.5	
3.6	7, 21, 37, 45	9.6	37
3.7	15, 21		
3.8	7, 13, 27, 31	10.1	33
		10.2	19
4.1	35	10.3	31
4.2	35	10.4	13, 27, 43
4.3	59	10.5	21, 35
4.4	19, 33	10.6	31
4.5	1, 11, 19		
4.6	11, 15	11.1	29
		11.2	11
5.1	21	11.3	9
5.2	19	11.4	17
5.3	15	11.5	27, 33
5.4	13	11.6	27
5.5	19		
5.6	19, 29	12.1	17, 21
5.7	3, 13	12.2	25
		12.3	27
6.1	11, 69, 73	12.4	9, 23
6.2	19, 45, 51	12.5	33, 43
6.3	45	12.6	7, 19, 37